# THE STRIATED MUSCLE

INTERNATIONAL ACADEMY OF PATHOLOGY

# MONOGRAPHS IN PATHOLOGY

*SERIES EDITOR*, **F. K. Mostofi, M.D.**

*Chief, Genitourinary Pathology Branch, Armed Forces Institute of Pathology, and Veterans Administration Special Laboratory for Pathology at the Armed Forces Institute of Pathology, Washington, D.C.*

No. 1. **The Lymphocyte and Lymphocytic Tissue**
John W. Rebuck *and* Robert E. Stowell, *Editors*

No. 2. **The Adrenal Cortex**
Henry D. Moon *and* Robert E. Stowell, *Editors*

No. 3. **The Ovary**
Hugh G. Grady *and* David E. Smith, *Editors*

No. 4. **The Peripheral Blood Vessels**
J. Lowell Orbison *and* David E. Smith, *Editors*

No. 5. **The Thyroid**
J. Beach Hazard *and* David E. Smith, *Editors*

No. 6. **The Kidney**
F. K. Mostofi *and* David E. Smith, *Editors*

No. 7. **The Connective Tissue**
Bernard M. Wagner *and* David E. Smith, *Editors*

No. 8. **The Lung**
Averill A. Liebow *and* David E. Smith, *Editors*

No. 9. **The Brain**
Orville T. Bailey *and* David E. Smith, *Editors*

No. 10. **The Skin**
E. B. Helwig *and* F. K. Mostofi, *Editors*

No. 11. **The Platelet**
K. M. Brinkhous, R. W. Shermer, *and* F. K. Mostofi, *Editors*

No. 12. **The Striated Muscle**
Carl M. Pearson *and* F. K. Mostofi, *Editors*

No. 13. **The Liver**
E. A. Gall *and* F. K. Mostofi, *Editors*

**Succeeding Volumes to be Announced**

International Academy of Pathology Monograph

# THE STRIATED MUSCLE

## by 28 authors

EDITED BY **CARL M. PEARSON, M.D.**

*Professor of Medicine, The University of California School of Medicine, The Center for the Health Sciences, Los Angeles, California*

and **F. K. MOSTOFI, M.D.**

*Chief, Genitourinary Pathology Branch, Armed Forces Institute of Pathology, and Veterans Administration Special Laboratory for Pathology at the Armed Forces Institute of Pathology, Washington, D.C.*

**THE WILLIAMS & WILKINS COMPANY**

Baltimore, 1973

THE WILLIAMS & WILKINS COMPANY
428 E. Preston Street
Baltimore, Md. 21202, U.S.A.



Made in the United States of America

Library of Congress Catalog Card Number 73-6903
SBN 0-683-06811-3

Composed and Printed at the
Waverly Press, Inc.
Mt. Royal and Guilford Aves.
Baltimore, Md. 21202, U.S.A.

# TABLE OF CONTENTS

## THE STRIATED MUSCLE Dr. Carl M. Pearson

*Chapter*

1. Skeletal Muscle: Diversity and Mutability of its Histochemical, Electron Microscopic, Biochemical, and Physiologic Properties. . . . . . . . . . . . . . . . . . . . . . . . . . . . . . . . . . . . . . *J. B. Peter, M.D.* 1
2. Techniques of Muscle Biopsy and Staining Methods with Particular Emphasis on Commonly Encountered Artifacts. *Pasquale A. Cancilla, M.D.* 19
3. Histoenzymatic Methods Applied to Mammalian Striated Muscle. *M. A. Verity, M.D., and Ralph F. Coleman, M.D.* 28
4. Immunochemical and Ultrastructural Studies of Myogenesis. *E. Raworth Allen, M.D.* 40
5. The Ultrastructure of Human Skeletal Muscle: Variation from Archetypal Morphology. *D. A. Fischmann, M.D., H. Y. Meltzer, M.D., and R. W. Poppei, M.D.* 58
6. Histochemical Fiber Types in Clinical Disorders of Muscle *Pasquale A. Cancilla, M.D., and M. Anthony Verity, M.D.* 77
7. The Pathologic Interpretation of Muscle Histochemistry. *Michael H. Brooke, M.D.* 86
8. Cytochemical Properties of the Basement Lamina of Myofibers and Neuromuscular Junctions. *Sumner I. Zacks, M.D., Atsushi Sainto, and M. F. Scheff.* 123
9. Ultrastructural Pathologic Characteristics of the Skeletal Muscle Fiber: An Introductory Survey . . . . *Harold M. Price, M.D.* 144
10. Current Concepts of Skeletal Muscle Regeneration. *M. Reznik, M.D.* 185
11. Ultrastructural Aspects of the Normal Mammalian Muscle Spindle. . . . *Howard A. Scalzi, M.D., and Harold M. Price, M.D.* 226
12. The Influence of the Nervous System on the Muscle Structure on Function. . . . . . . . . . . . . . . . . . . . . . . . . . . . . . . . . . . . *V. Dubowitz, M.D.* 236
13. The Histopathology of Neurogenic Muscular Atrophy. *Jon D. Dorman, M.D.* 249
14. Progressive Muscular Dystrophy: Structural Alterations in Various Stages and in Carriers of Duchenne Dystrophy. *John N. Walton, M.D.* 263
15. The Pathologic Substratum of Polymyositis. *Raymond D. Adams, M.D.* 292
16. Vacuolar Myopathies: Multiple Etiologies and Sequential Structural Studies. . . . . . . . . . . . . . . . . . . . . . . . . *Andrew G. Engel, M.D.* 301

17. Normal Ultrastructural Aspect of Human Motor End-Plate and its Pathologic Modifications. . . . . . . . . . . . . *Michel Fardeau, M.D.* 342
18. Myopathies of Animals. . . . . . . . . . . . . . . . . . . . . . . . *W. J. Hadlow, M.D.* 364
19. Ultrastructure of Muscle in Glycogen Storage Diseases. *Donald L. Schotland, M.D.* 410
20. The Pathologic Features of the Periodic Paralyses. *Carl M. Pearson, M.D.* 427
21. Congenital Myopathies. . . . . . . . . . . . . . *Theodore L. Munsat, M.D.* 442
22. Duchenne Muscular Dystrophy: A Histologically Based Ischemia Hypothesis and Comparison with Experimental Ischemia Myopathy. . . . . . . . . . . . . . . . . . . . . . . . . . . . . . . . . *W. King Engel, M.D.* 453
23. Some Observations on Normal and Diseased Human Muscle in Tissue Culture. . . . . . . . . . . . . . . . . . . . . . . . . . . . *Victor Dubowitz, M.D.* 473
24. Observations on Etiology of Polymyositis. *Byron A. Kakulas, M.D.* 485
25. Myasthenia Gravis: Evidence for Thymus Regulation of Neuromuscular Transmission. . . . . . . . . . . . . . . . . *Gideon Goldstein, M.D.* 498

# STRIATED MUSCLE Contributors

RAYMOND D. ADAMS, M.D.
*Department of Nephropathology, Harvard Medical School of Medicine, Massachusetts General Hospital, Boston, Massachusetts*

E. RAWORTH ALLEN, PH.D.
*Department of Anatomy, Louisiana State University Medical Center, 1190 Florida Avenue, New Orleans, Louisiana*

MICHAEL H. BROOKE, M.D.
*Division of Neurology, University of Colorado School of Medicine, Denver, Colorado*

PASQUALE A. CANCILLA, M.D.
*Department of Pathology (Neuropathology), UCLA School of Medicine, Los Angeles, California*

RALPH F. COLEMAN, M.D.
*Camarillo State Hospital, Camarillo, California*

JON D. DORMAN, M.D.
*704 Shaws Flat Road, Sonora, California*

VICTOR DUBOWITZ, B.SC., M.D., PH.D., M.R.C.P., D.C.H.
*University of Sheffield, Sheffield Children's Hospital, Department of Child Health, Children's Hospital, West Bank, Sheffield S10 2th, England*

ANDREW G. ENGEL, M.D.
*Mayo Graduate School of Medicine, University of Minnesota. Consultant in Neurology, Mayo Clinic, Rochester, Minnesota*

W. KING ENGEL, M.D.
*Medical Neurology Branch, National Institute of Neurological Diseases and Stroke, National Institutes of Health, Bethesda, Maryland*

MICHEL FARDEAU, M.D.
*C.N.R.S., Service de Microscopie Electronique, Hôpital de la Salpétriere, 75 Paris XIIIe France*

DONALD A. FISCHMANN, M.D.
*The University of Chicago, The Division of the Biological Sciences and The Pritzker School of Medicine, 950 East 59th Street, Chicago, Illinois*

GIDEON GOLDSTEIN, M.D., PH.D.
*Irvington House Institute, Research Associate Professor of Pathology, New York University School of Medicine, New York*

W. J. HADLOW, D.V.M.
*U.S. Department of Health, Education and Welfare, Public Health Service, National Institutes of Health, National Institute of Allergy and Infectious Diseases, Rocky Mountain Laboratory, Hamilton, Montana*

BYRON A. KAKULAS, M.D.
*Department of Pathology, University of Western Australia, School of Medicine, Royal Perth Hospital, Perth, Australia*

Herbert Y. Meltzer, M.D.
*The University of Chicago, The Division of the Biological Sciences and The Pritzker School of Medicine, 950 East 59th Street, Chicago, Illinois*

Theodore L. Munsat, M.D.
*Division of Neurology, UCLA School of Medicine, Los Angeles, California*

Carl M. Pearson, M.D.
*Department of Medicine, UCLA School of Medicine, Los Angeles, California*

James B. Peter, M.D., Ph.D.
*Department of Medicine, UCLA School of Medicine, Los Angeles, California*

Robert W. Poppei,
*The University of Chicago, The Division of the Biological Sciences and The Pritzker School of Medicine, 950 East 59th Street, Chicago, Illinois*

Harold M. Price, M.D.
*University of Pittsburgh, School of Medicine, Magee-Womens Hospital, Pittsburgh, Pennsylvania*

Michel Reznik, M.D.
*Université de Liège, Service d'Anatomie Pathologique, 1 rue des Bonnes Villes, Liège, Belgium*

Atsushi Sainto, M.D.
*Department of Pathology, University of Pennsylvania, School of Medicine, Philadelphia, Pennsylvania*

Howard A. Scalzi, M.D.
*Department of Anatomy, Kansas City College of Osteopathy and Surgery, Kansas City, Missouri*

M. F. Scheff, M.D.
*Department of Pathology, University of Pennsylvania, School of Medicine, Philadelphia, Pennsylvania*

Donald L. Schotland, M.D.
*University of Pennsylvania, School of Medicine, Neurological Clinical Residence Center and The Spiller Neurological Unit, University of Pennsylvania Hospital, Philadelphia, Pennsylvania*

M. Anthony Verity, M.D.
*Department of Pathology (Neuropathology), UCLA School of Medicine, Los Angeles, California*

John N. Walton, T.D., M.D., F.R.C.P.
*Professor of Neurology, University of Newcastle-Upon-Tyne, Neurologist, Regional Neurological Centre, General Hospital, Westgate Road, Newcastle-Upon-Tyne, NE4 6BE, England*

Sumner I. Zacks, M.D.
*Department of Pathology, University of Pennsylvania, School of Medicine, Philadelphia, Pennsylvania*

# FOREWORD

The Striated Muscle is the 12th of a series of *Monographs in Pathology* initiated by the International Academy of Pathology in 1959. Its various chapters are expansions of the lectures given as a course entitled "Pathologic Physiology and Anatomy of the Striated Muscle" which was conducted at the 59th Annual Meeting of the United States-Canadian Division of the International Academy of Pathology by Dr. Carl M. Pearson, Professor of Medicine and Director, Division of Rheumatology, University of California, Los Angeles.

Despite a number of recent books dealing with diseases of the muscle, Dr. Pearson's careful selection of subjects and faculty and our joint efforts at editing the monograph have resulted in a valuable reference book not only for pathologists but for all other students of diseases of the muscle.

The Academy is grateful to Dr. Pearson and his distinguished faculty who have contributed to this monograph, which contains a remarkable variety of subjects comprising research and clinical aspects of normal and abnormal skeletal muscle. Although the publication of this monograph was delayed, the authors have endeavored to bring the contents up-to-date.

The Council of the International Academy of Pathology has sponsored the publication of this and previous monographs to provide a permanent reference source for the subjects especially presented at the meetings of the Academy. The goals are to assemble material of the finest scientific standards and educational value relative to the normal and abnormal structure and function of the muscle and to present the information in a manner that will be of continuing benefit to the entire scientific community, as well as to those who attended the meeting.

F. K. Mostofi, M.D.
*Series Editor*

Washington, D. C.

# PREFACE

As has been happening in all of medical research, investigation into the structure and function of striated skeletal muscles, has accelerated at a remarkable rate within the past two decades.

This Symposium has been designed for use by investigators as well as for those interested in diseases of the muscle. It covers a wide variety of topics which are the prime subjects of today's investigative thrusts.

Much of what is known today about the biochemistry of *all* cell function was begun more than a century ago on muscle tissue since muscle has the remarkable capacity of transferring chemical and electrochemical energy into mechanical energy, thereby producing movement and work.

In the 19th century, physiologists first came to realize this and utilized it as a way of assessing the sequence of chemical events that occur when a cell (muscle fiber) is activated. They reasoned, and rightly so, that there was close temporal proximity between the chemical events that were occurring in the muscle fiber and its contraction. One of the first products of muscle contraction, which was isolated in the 1850s, was lactic acid which we now know is a significant by-product of carbohydrate metabolism.

Very little else occurred in muscle research, aside from defining clinical syndromes, until the 1940s when the Nobel Laureate Dr. Albert Szent-Gyorgyi discovered a contractile protein of muscle that he called actomyosin.

Within the past two decades literally a flood of new observations has occurred in regards to mammalian striated muscle. For instance, the serial events that occur in order to produce energy in muscle in both the oxidative and glycolytic circuits have been described (some diseases have now been associated with defects in certain enzymes or chemical constituents in these systems). Moreover the contractile proteins of muscle, it is now realized, are not comprised only of the "single protein actomyosin" but rather of a series of interacting proteins which are necessary for muscle contraction. These contractile proteins number now between 8 and 11, depending upon which information one is willing to accept in 1972.

The electron microscope has also, during this era, opened up our understanding of the intricate structural composition of the striated muscle fiber. Not only has the overall structure of the motor end-plates been reasonably well delineated, as described in this Symposium, but also the sarcolemmal membrane has been shown to be directly continuous with the transverse tubules leading to the interior muscle fiber, and the longitudinal sarcotubular system has been shown to be intimately related to the muscle contraction-relaxation sequence which involves a complex series of biochemical reactions.

Electron microscopy has also revealed that what was thought initially to be only one type of muscle fiber nucleus has now shown that each multinucleated muscle fiber really contains two different kinds of nuclei. One of these, the more frequent one, is the sarcolemmal nucleus that does the everyday housekeeping of the muscle fiber. The other is called the satellite cell. This cell is apparently in the resting stage during normal muscle function but appears to be rapidly called into action when muscle is in some fashion or another injured. Therefore, it has been recently realized that the satellite cell plays an extremely crucial role in the repair of muscle.

Tissue culture techniques have advanced rapidly during the past decade so that it is now possible to display the entire sequence of events *in vitro*, from a single myoblast or satellite cell to the formation of a mature, cross-striated, multinucleate muscle fiber.

The role of the nervous system in maintaining muscle in an intact and healthy fashion has only been shown within the past 5 years, although it was suspected for many years before that. The cross-innervation experiments of Eccles and Buller demonstrated that the nerve leading to muscle dictates the ultimate biochemical makeup of that muscle fiber.

Muscle histochemistry has also come into its own during this time, and this discipline has added considerably to our understanding of patterns of chemically distinct muscle fiber distribution within individual muscles as well as showing us different histochemical and chemical patterns in certain disease processes.

In regards to muscle diseases, much is yet to be learned. We have an understanding of some of the most common muscle diseases but we do not have cures, unfortunately, for any except one or two of them. Polymyositis may be treated satisfactorily with the use of corticosteroids and/or immunosuppressive agents. However, the etiology of the disease processes that we most wish to understand, namely the muscular dystrophies, has completely escaped our understanding so far. Some of the neurogenic atrophies fall midway between these extremes.

This Symposium covers not only current knowledge from experts in the field about the basic aspects of muscle structure and function but also what is currently understood about certain muscle diseases. Generally speaking there is no significant geographical preference, or lack thereof, for any of the diseases. All muscle diseases appear to be almost universally distributed with a relatively similar preference. There are one or two exceptions to this rule which are probably dictated by genetic preponderance in certain races. For instance, the simultaneous occurrence of hyperthyroidism, thyrotoxic myopathy, and hypokalemic periodic paralysis is far more prevalent in Japan, and probably in China, than anywhere else in the world. Certain forms of muscle necrosis associated with exogenous agents, such as ingestion of alkaloids or from the venom of certain poisonous vertebrates (such as the sea snake in the waters off of Malaysia) do induce muscle disease at a relatively high rate in those areas. The vertebrate animal kingdom also suffers from diseases similar to those which afflict human beings, and, therefore, much can be learned from further studies on these models or defects of nature as have been described in Chapter 18.

In summary, this Symposium has brought together many of the world's leading experts on muscle structure, function, and disease, and it serves as a testimonial to the accelerated rate at which research has progressed so far, as well as defining the magnitude of the job that lies ahead.

It can be predicted with a good deal of confidence that many of the diseases that we now consider to be so mysterious will be understood, and their cures will possibly be available, within the next two decades.

Carl M. Pearson, M.D.

*Los Angeles, California*

Chapter 1

# Skeletal Muscle: Diversity and Mutability of Its Histochemical, Electron Microscopic, Biochemical, and Physiologic Properties*

J. B. PETER

Normal muscle works as an integrated unit. Investigators of muscle likewise function as units, but the integration of knowledge derived by investigative units devoted to diverse scientific disciplines is sometimes erratic. This failure of integration is understandable in view of the asynchronous advances in various disciplines of muscle biology. This review will incorporate some of this new knowledge into the well known, well established framework of muscle biology.

The purpose of the review is to provide a broad, but necessarily simplified, overview of our current concepts of the biology of muscle with special reference to recent advances in muscle biochemistry. The review is directed to the pathologists who must assess biopsies of skeletal muscle. Hopefully it will promote correct (or even "new") insights and will at least facilitate the intellectual gratification which results from some understanding of how the biopsied muscle functioned or malfunctioned *in vivo*.

## HISTOCHEMISTRY

Human skeletal muscle is generally considered to be composed of at least two histochemically distinct types of fibers.[13, 16, 17, 49] Human skeletal muscle fibers classified as type I are generally smaller, contain greater activities of oxidative enzymes (indicating the presence of more mitochondria), and are less active with techniques to demonstrate phosphorylase. Type II fibers exhibit what Dubowitz and Pearse[14] termed a reciprocal histochemical relationship to type I fibers, *i.e.*, low oxidative enzymes and high phosphorylase activity. Likewise, type I fibers stain light with myofibrillar ATPase at pH 9.4 and type II fibers stain dark, probably indicating differences in the alkali sensitivity of this ATPase in different fiber types as discussed below.[2, 24]

* This investigation was supported by National Institutes of Health Grants NS 07587 and NS 10403.

It is generally considered that type I and type II fibers correspond, respectively, to the red and white fibers of lower animals and birds, some species of which contain muscles composed almost wholly of grossly red or white fibers. On the other hand, in addition to fibers resembling type I and II fibers a third type has been recognized in animal muscle for several years.[15, 16, 22, 24, 30, 49, 52] This has been variously called the red fiber, C fiber, and "group" II fibers. Recent studies have clearly established the histochemical,[2, 15, 24] physiological,[2] and biochemical[36] characteristics of all three types of fibers in the rat and guinea pig (*cf.* Table 2).

Figure 1 shows serial sections through the gastrocnemius of a guinea pig. Note how the fibers which stain dark for myofibrillar ATPase at pH 9.4 (fast-twitch fibers, *cf.* Tables 2, 3, and below) may stain either dark ("red" fibers) or light ("white" fibers) with oxidative enzymes such as malate dehydrogenase. The fiber which stains light for myofibrillar ATPase at pH 9.4 (defined sometimes as the "intermediate" fibers) also stains moderately dark with oxidative enzymes and is more precisely termed a slow-twitch-oxidative fiber. Thus, in guinea pigs, at least, fibers which stain dark with oxidative enzymes may stain either dark (fast-twitch-oxidative-glycolytic) or light (slow-twitch-oxidative) with myofibrillar ATPase at pH 9.4[2, 36] (*cf.* Table 2).

The question is whether these three distinctive types of fibers (fast-twitch-oxidative-glycolytic; fast-twitch-glycolytic; and slow-twitch-oxidative) are also represented in human muscle or whether it contains only type I and type II fibers, which presumably correspond to slow-twitch-oxidative and fast-twitch-glycolytic fibers, respectively. Recent evidence[6] suggests that in human muscle the fibers which stain darkest with Sudan black B (presumably reflecting the greatest number of mitochondria) are the slow-twitch fibers, whereas fibers staining light or moderate ("intermediate") with sudan black B contract rapidly. This excellent work, however, tends to confuse the terminology problem further, because the fibers staining intermediate with Sudan and assumed to contract rapidly were equated with type B fibers, which are known to stain light with myofibrillar ATPase at pH 9.4[15, 52] and to contract slowly.[2] One must conclude, therefore, that the human fibers which have been called intermediate with oxidative enzyme stains or Sudan black B do not correspond to the slow-twitch-oxidative[36] (type B[52]) fibers of rats and guinea pigs. It may be that in human muscle there is less variation in mitochondrial content of fast-twitch fibers compared with guinea pigs and that, in contrast to the guinea pig, it is the fiber with the most mitochondria which contracts most slowly in human muscle. It is crucial that this problem be resolved with serial sections of human muscle to see whether the "intermediate" fiber (intermediate staining intensity with oxidative enzymes) actually gives a dark reaction for myofibrillar ATPase at pH 9.4, as would be predicted from the study described above.[6] If this is the case (and if fibers of human limb muscles staining dark with myofibrillar ATPase at pH 9.4 are fast-twitch relative to those staining light at pH 9.4), then the slow-twitch-oxidative fiber of human muscle will be the counterpart of the slow-twitch-oxidative fiber of guinea pigs. In this situation human fibers, like those of rats and guinea pigs, could be described as fast-twitch-glycolytic, fast-twitch-oxidative-glycolytic, and slow-twitch-oxida-

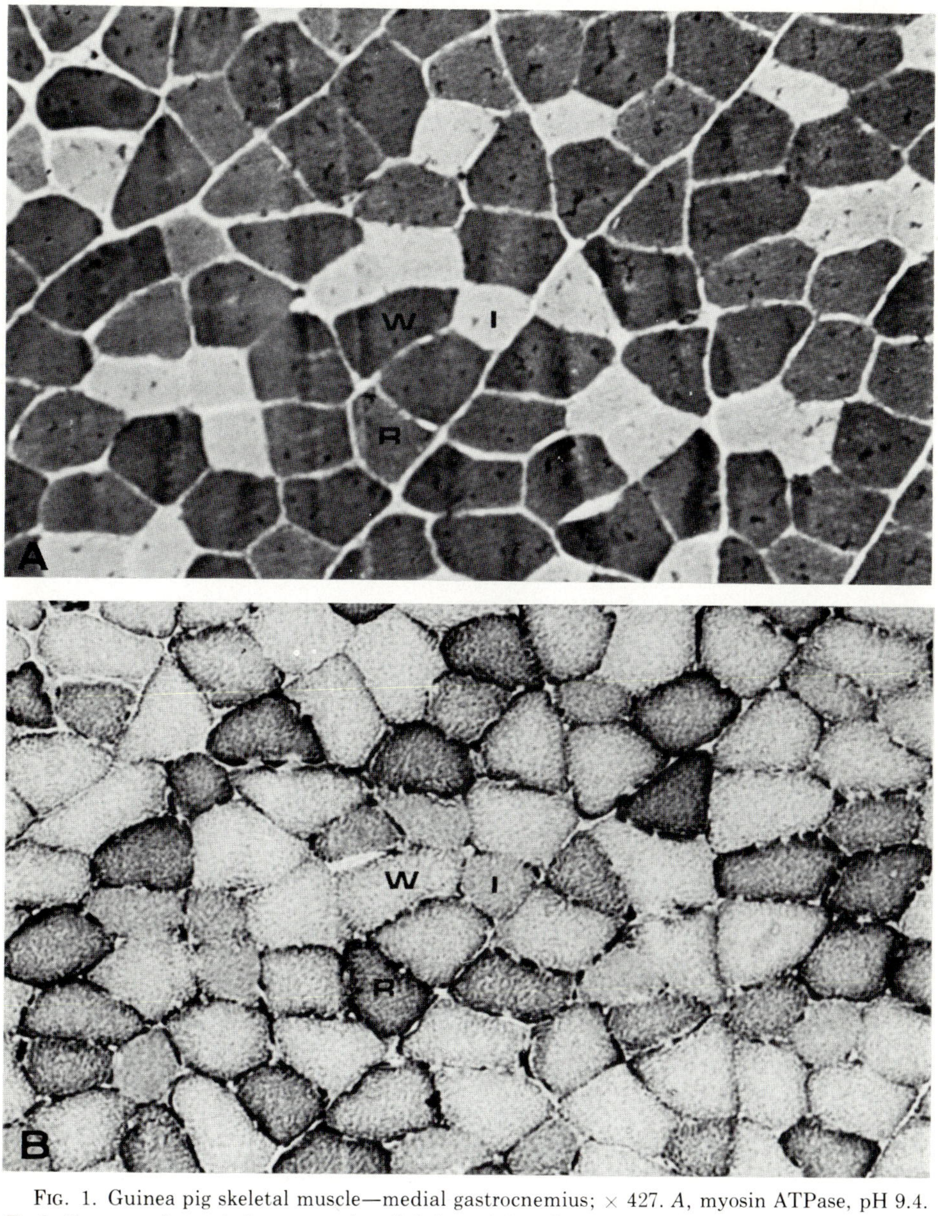


FIG. 1. Guinea pig skeletal muscle—medial gastrocnemius; × 427. *A*, myosin ATPase, pH 9.4. Dark fibers are fast-twitch compared to lightly staining slow-twitch fibers. *B*, malate dehydrogenase. This consecutive section, when compared with *A*, shows that fast-twitch fibers (pH 9.4 myosin ATPase = dark) may stain either light (*W*) or dark (*R*) for malate dehydrogenase and other mitochondrial enzymes. The slow-twitch fiber (light myosin reaction) invariably stains moderately dark (*I*) with oxidative enzyme reactions.

tive, and the major difference between fiber types of guinea pig and human muscle would reflect a greater range of mitochondrial content in fast-twitch fibers of the guinea pig (very low to high) than in fast-twitch fibers of humans (very low to moderate).

The investigator must beware of "special cases." For example, histochemical studies of certain specialized muscles with dual innervation, *e.g.*, extraocular, suggest that two types of myosin ATPase may be detected in a single fiber,[58] as might be predicted from tonic, graded contractions of such muscles. Finally, it should be reemphasized that a necessary correlation between histochemical reactions for "myofibrillar ATPase" and speed of muscle contraction cannot be assumed but must be proved.

## ELECTRON MICROSCOPY

The major subcellular components of skeletal muscle fibers are listed in Table 1. Each fiber is composed of hundreds of myofibrils ($\sim$1 $\mu$ diameter) which are surrounded by a lacelike network of tubules. A myofibril is illustrated schematically in Figure 2 which is designed to emphasize that the intermyofibrillar membranous tubular complex is composed of two embryologically distinct systems of tubules known as the sarcoplasmic reticulum and the T (transverse) system. The T system is an invagination of the plasma membrane of the cell; its function is to conduct the electrical impulse from the cell membrane into the fiber to trigger contraction by the release of calcium which is stored in the sarcoplasmic reticulum in relaxed muscle. As illustrated in Figure 2, the T system surrounds the myofibril midway between the Z lines and the centers of the A bands; the T system is approximated to, but apparently not continuous with, dilated elements (terminal cisternae) of the sarcoplasmic reticulum on either side. Thus in human muscle, each sarcomere (the repeating Z-line-to-Z-line unit of the fibril) contains two triads each composed of a pair of terminal cisternae on each side of the T tubule.

The number and distribution of mitochondria differs greatly in different types of skeletal muscle fibers of man and lower animals.[22, 36, 46] Some fibers, variously called type II in human muscle or fast-twitch-glycolytic in lower animals,[36] have relatively small numbers of subsarcolemmal mitochondria and few or no long rows of intermyofibrillar mitochondria. Type II fibers do exhibit

TABLE 1. SUBCELLULAR COMPONENTS OF MUSCLE

- Sarcolemma
- Nuclei; ribosomes
- Myofibrils
  - Intermyofibrillary tubular complex
    - T (Transverse) system
    - Sarcoplasmic reticulum
  - Contractile proteins of the myofilaments: myosin, actin, tropomyosin, troponin, actinin, M protein
- Mitochondria
- Lysosomes
- Golgi complex
- Glycogen particles; lipid bodies
- Soluble proteins, *e.g.*, enzymes of carbohydrate metabolism

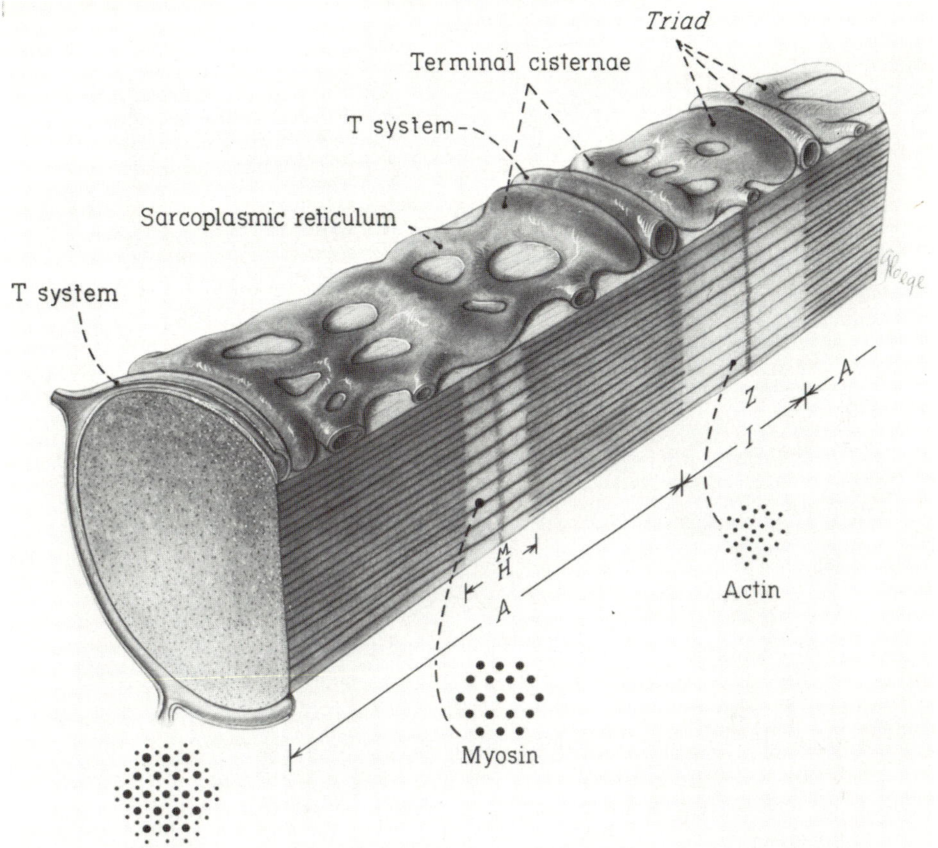


FIG. 2. Schematic illustration of a myofibril.

the bracelet-like mitochondria which partially surround the fibril at the level of the I band. Type I fibers (presumably slow-twitch-oxidative fibers in humans[32]) contain many more mitochondria in all three locations, *i.e.*, in dense subsarcolemmal accumulations (especially in the regions of motor end-plates and at the poles of nuclei), in longitudinal rows between myofibrils, and in pairs in a bracelet-like arrangement around myofibrils at the I bands.

The distribution of the contractile proteins, myosin and actin, in the A and I bands are well known and are illustrated in Figure 3. Each double-stranded superhelix of globular actin monomers forming a thin filament contains about 1600 molecules of actin. The thick filaments (myosin polymers) are probably composed of about 360 tadpole-shaped molecules of myosin which are arranged in tail-to-tail manner at the M band with consequent reversal of the polarity of the myosin filaments. Elsewhere than at the M band the thick filament is composed of myosin molecules arranged in a head-to-tail fashion. Less generally known are the distributions of tropomyosin, troponin, $\alpha$-actin, and M protein (Fig. 3). Tropomyosin is a "structural" protein which runs along the grooves of the actin filaments which are composed of globular actin monomers arranged in a double-stranded helical chain. Troponin, another protein complex which can be

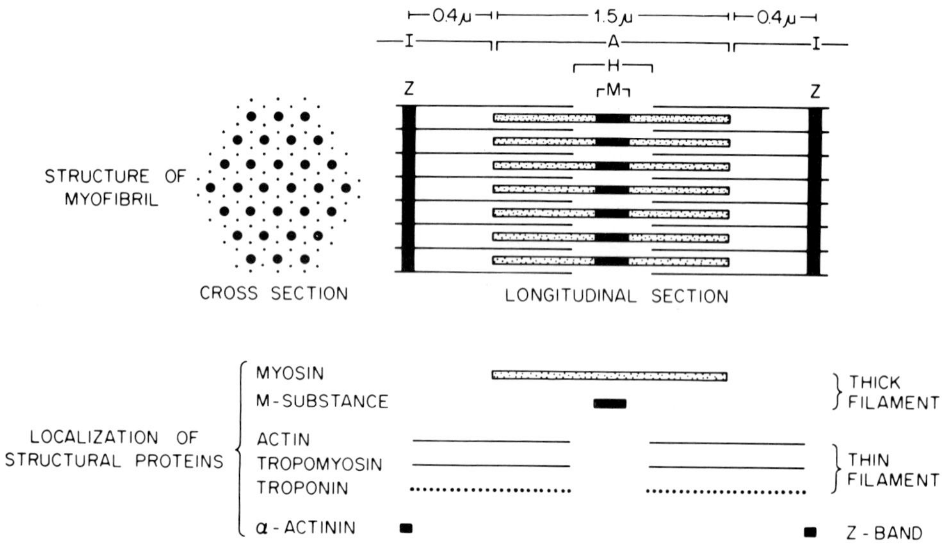


FIG. 3. Localization of tropomyosin, troponin, and other structural proteins relative to actin and myosin.

isolated from actin, is distributed along the actin filament at regular intervals and is considered to be the calcium receptor of muscle[14] which, when combined with calcium, allows actin and myosin to slide together thus producing shortening of the sarcomere and hence of the muscle. Apparently each actin filament (I band filament) contains about 48 pairs of molecules of troponin.

Electron microscopy of muscle has also given us important insights into the embryogenesis of muscle. For example, we now know that the membranes of the sarcoplasmic reticulum are assembled within the fiber, whereas the T system arises as an invagination from the sarcolemma.[31] As expected such information provides valuable background for the study of human disease. For example, in myotonic dystrophy in which a defect of the sarcolemma is suspected, one might not be surprised to find evidence of a biochemically normal sarcoplasmic reticulum[44] since this apparently develops independently of the sarcolemma. For detailed discussion of other electron microscopic features of human muscle the reader is referred elsewhere (*c.f.* Chapter 9).[46]

## BIOCHEMISTRY

Much of the success of modern biochemistry has resulted from techniques which allow fractionation of complicated tissues and cells thereof into their component subcellular parts. It is reasonable to expect that this principle will also be most helpful in the study of human skeletal muscle and its diseases. However, the diversity of skeletal muscle fibers points up the need for simultaneous application of both histochemical and biochemical techniques if definitive information is to be obtained from biochemical analysis of skeletal muscle. To illustrate this point I have listed in Table 2 the activities of certain enzymes in guinea pig muscles composed predominantly or exclusively of one type of fiber. Both the fast-twitch-oxidative-glycolytic and slow-twitch-oxidative fibers are grossly red, but their enzyme activities differ greatly. Indeed, in some cases, *e.g.*, phosphorylase and mitochondrial α-glycerophosphate dehydrogenase, the enzyme activity of one grossly red fiber (fast-twitch-oxidative-glycolytic) more closely resembles that of a grossly white fiber (fast-twitch-glycolytic) than that of another grossly red fiber (the slow-twitch-oxidative). These data,[36] plus the known predilection of certain diseases to affect one fiber type more than another (*cf.* Chapters 6, 7, 16, and others), necessitate concomitant histochemical assessment of the muscle being studied if biochemical data are to be interpreted critically.

One of the most formidable problems in biochemical studies of normal and diseased muscle was the need for efficient, reliable techniques for isolation of subcellular fractions from small samples of muscle. Fortunately for both patient and investigator these techniques are now available.[35, 38, 39, 41, 43] Mitochondria isolated by these techniques from human skeletal muscle exhibit tightly coupled oxidative phosphorylation and most of the other qualitative properties characteristic of mitochondria from liver or skeletal muscle of lower animals. All of the major varieties of muscular dystrophy have been studied and, contrary to other conclusions based on less reliable techniques, mitochondrial biochemistry, including fatty acid oxidation (palmitoyl-L-carnitine as substrate),[42] is entirely normal in human dystrophy,[57] as is electron microscopic morphology of the isolated mitochondria. These techniques have not yet been systematically applied to the diseases previously termed "mitochondrial myopathies," [47] but such efforts are surely indicated. However, our studies with isolated human skeletal muscle mitochondria,[37, 55] like those of others with rat liver and bovine

TABLE 2. CLASSIFICATION OF MUSCLE FIBERS; GUINEA PIG MUSCLE AS A MODEL FOR CHARACTERIZING FIBER TYPES OF HUMAN MUSCLE[a]

| Characteristics | Muscle | | |
|---|---|---|---|
| | Red Vastus | White Vastus | Soleus |
| Suggested nomenclature | Fast-twitch-oxidative-glycolytic | Fast-twitch-glycolytic | Slow-twitch-oxidative |
| Mitochondria | | | |
| Content | High | Low | Medium |
| Distribution | Diffuse-subsarcolemmal | Subsarcolemmal | Diffuse |
| Succinate dehydrogenase | High | Low | Medium |
| α-Glycerophosphate dehydrogenase | Medium | High | Low |
| Myosin ATPase | | | |
| pH 9.4 | High | High | Low |
| Carbohydrate metabolism | | | |
| Phosphorylase | High | High | Low |
| Hexokinase | Medium | Low | High |
| Lactate dehydrogenase | Medium | High | Low |
| Acid hydrolases | Medium | Low | High |
| Twitch time | Fast | Fast | Slow |

[a] See section on "Physiology and Terminology" for application of this system of nomenclature to fiber types of human skeletal muscle.

heart mitochondria, emphasize the hazard of defining a myopathy as "mitochondrial" in the absence of data other than scattered, although interesting, morphologic aberrations of mitochondrial structure. For example, Figure 4, *A* to *C*, shows morphologic changes induced by addition of substrate to mitochondria isolated from skeletal muscle. Similar morphologic alterations in a small proportion of muscle mitochondria *in situ* might not be unexpected and could be erroneously interpreted as being relevant to the pathogenesis of the disease under study.

A number of investigators have studied the biochemistry of fragmented sarcoplasmic reticulum (SR) from normal and diseased human muscle.[39, 43, 44, 51, 53, 55, 56, 57] When muscle is homogenized the SR is fragmented and the fragments reseal into closed vesicles. Fragmented SR (or sarcotubular vesicles) can be isolated by differential centrifugation and used to study the energy-dependent accumulation of calcium, the same function which the SR fulfills *in vivo* by lowering the calcium concentration to levels at which muscle relaxes. Fragmented sarcoplasmic reticulum (FSR) is most critically studied *in vitro* by assaying its calcium-concentrating ability,[18, 56, 57] which refers to the ability of FSR to lower the calcium concentration of the assay medium to less than 1 $\mu$M, the level of calcium at or below which relaxation occurs *in vivo*. FSR from various sources can be compared with this assay because the total calcium-stor-

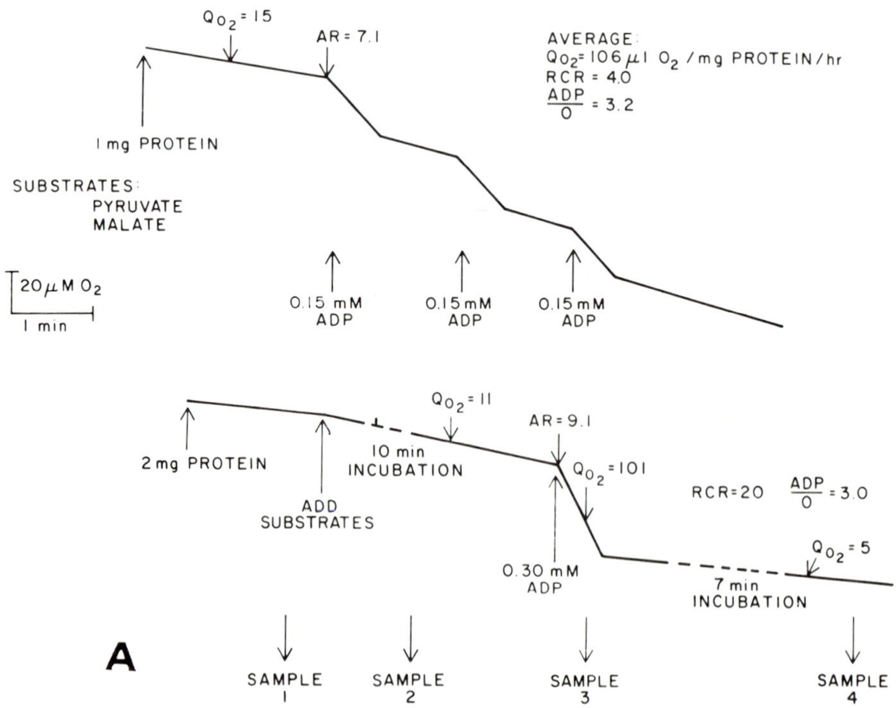


FIG. 4. Conformation changes in isolated mitochondria. *A*, polarigraphic tracing illustrating time of sampling for electron microscopic study. *B*, sample 1. Mitochondria incubated in state I (no substrate); × 9870. *C*, sample 2. The same preparation of mitochondria after incubation with substrate (pyruvate + malate) for 10 min; × 9870.

ing capacity (a measurement without obvious relevance to the function of FSR *in vivo*) is first determined for each preparation and then the assay for calcium-concentrating ability is performed at a total calcium concentration which is about one-third total capacity.[18, 56] Figure 5 shows that FSR prepared from fast-twitch-oxidative-glycolytic and fast-twitch-glycolytic muscle of the guinea pig concentrates calcium at similar rates which are both faster than the rate of slow-twitch-oxidative muscle. This plus the lower yield of FSR from

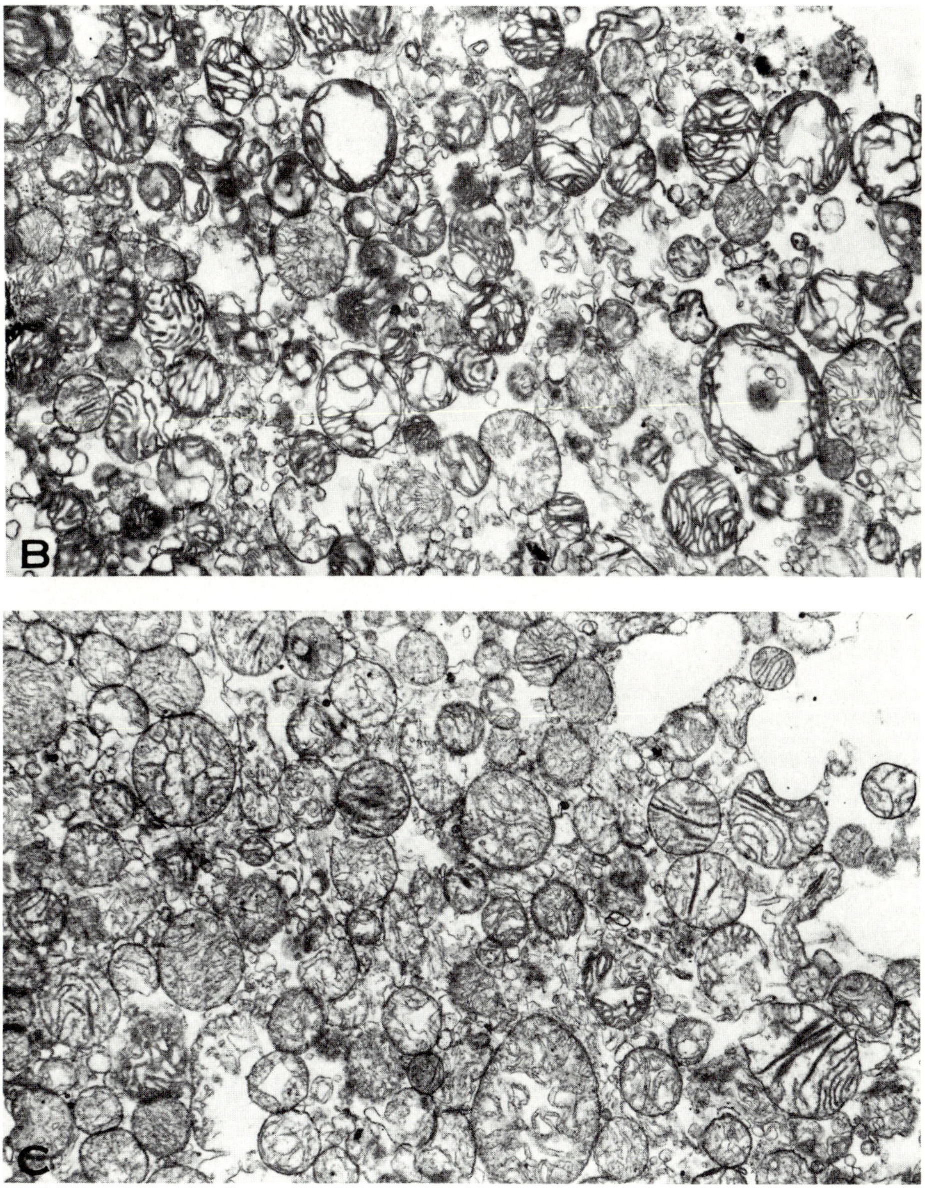


FIG. 4 *B–C*.

slow-twitch-oxidative muscle explains why it relaxes more slowly, but it also shows that FSR from slow-twitch-oxidative muscle can create a powerful calcium vacuum which is undoubtedly important in the relaxation of this type of muscle.[18, 32] The electron microscopic appearance of FSR from all three fiber types is similar (Fig. 6).[38]

Using the assay for calcium-concentrating ability we have now studied FSR

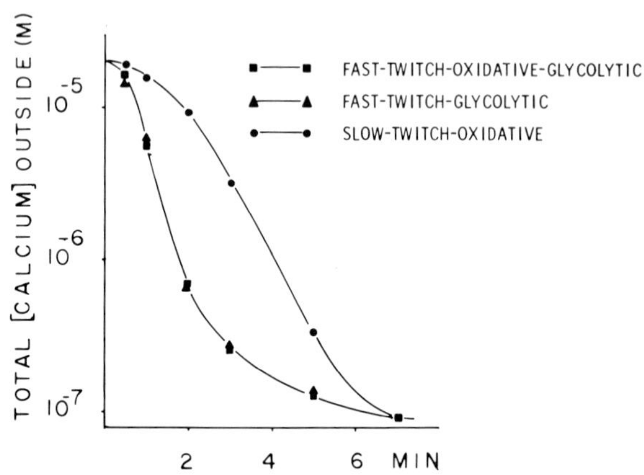


FIG. 5. Assay of the calcium-concentrating ability of FSR prepared from fast-twitch-oxidative-glycolytic, fast-twitch-glycolytic, and slow-twitch-oxidative muscles. Note that the major difference lies in the slower rate at which the FSR from slow-twitch muscle concentrates the calcium. The level of calcium left in the medium is very similar with all three preparations of FSR. Experimental details are reported in Reference 19.

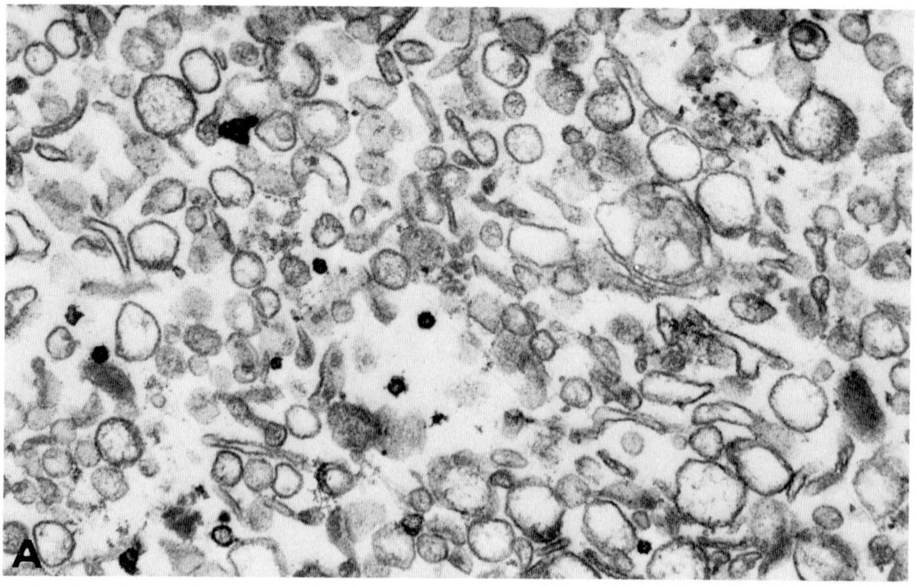


FIG. 6. Electron microscopic appearance of FSR; × 38,250. *A*, FSR from fast-twitch-oxidative-glycolytic muscle. *B*, FSR from fast-twitch-glycolytic muscle. *C*, FSR from slow-twitch-oxidative muscle.

from 12 patients with Duchenne muscular dystrophy as well as from patients with a variety of muscle and neuromuscular diseases. The data,[39] which will be published in detail elsewhere, clearly show that decreased calcium-concentrating ability is not an invariable feature of dystrophic muscle. The erroneous conclusions of other authors that a defect of FSR may be important in the genesis of Duchenne dystrophy undoubtedly reflect misleading results from measurements of total calcium-accumulating capacity, a parameter which is

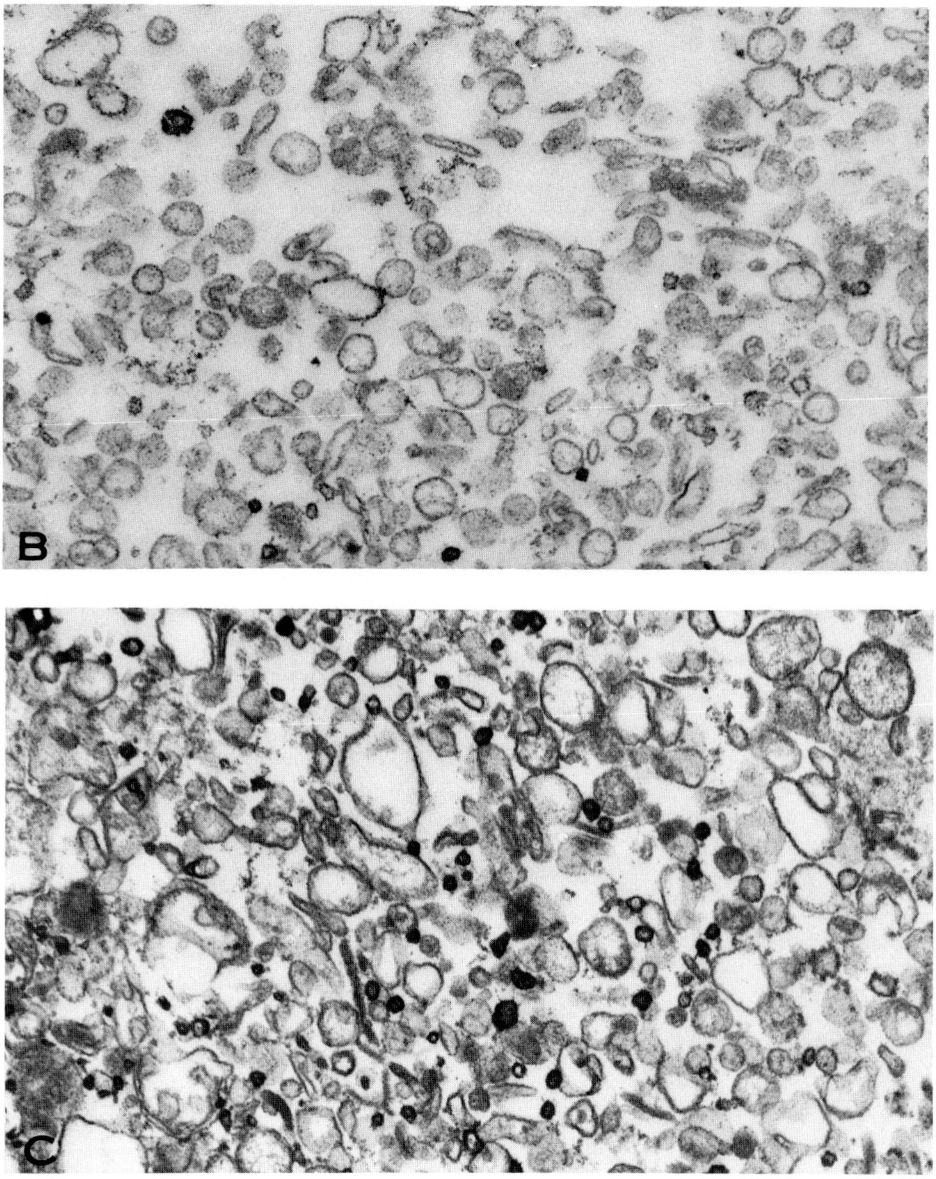


FIG. 6 *B–C*.

obviously influenced by the purity of the preparation being studied. This example emphasizes the importance of a critical choice of assay techniques in the study of diseased muscle, as has been discussed in detail elsewhere.[33]

The proteins which compose the muscle filaments can also be isolated from small samples of human muscle. Studies[2, 19] of actomyosin and myosin from fast-twitch-oxidative-glycolytic, fast-twitch-glycolytic, and slow-twitch-oxidative muscle confirm and extend earlier data[1] correlating myosin ATPase activity with contraction time of muscle. Table 3 shows that the specific activity of myosin ATPase isolated from muscles composed predominantly of fast-twitch-oxidative-glycolytic, or fast-twitch-glycolytic, fibers is high compared with that of slow-twitch-oxidative fibers.[2] In human muscle the fiber types are intermingled, and microtechniques must be developed before the specific activity of myosin or actomyosin ATPase from different fibers can be compared with contraction times. Further studies of myosin and actomyosin ATPase in human muscle disease are surely indicated but must be cautiously interpreted, especially if not combined with histochemical studies. For example, one should not conclude that an abnormal value for the specific activity of myosin ATPase necessarily reflects an abnormal or altered myosin; rather it may reflect differential atrophy or other disease of one fiber type with resultant alteration in the arithmetic mean specific activity of myosin ATPase isolated from the muscle.

Recent studies from this laboratory[20, 21] have revealed striking decreases in the trypsin-sensitive, calcium-binding capacity of natural actomyosin in Duchenne muscular dystrophy. This observation is probably explained by the fact that troponin, the calcium receptor of the contractile protein system,[14] is rapidly digested by trypsin with resultant abolition of the calcium sensitivity of actomyosin ATPase.[20, 21] Thus, the observations with natural actomyosin in Duchenne dystrophy suggest an alteration in the amount or character of its component troponin.[21] This abnormality is present in Duchenne dystrophic muscle even when a number of other labile functions of muscle, *e.g.*, calcium-concentrating ability of FSR, are normal. The troponin abnormality, therefore, is an early change in Duchenne dystrophic muscle, which is not mimicked by other diseases such as polymyositis. The problem now is to evaluate critically its relevance to the pathogenesis of Duchenne dystrophy.

Other studies from this laboratory[34] have resulted in the isolation of purified,

TABLE 3. CORRELATION OF FIBER TYPE AND MYOSIN ATPASE IN GUINEA PIG MUSCLE

| | Percentage of Fiber Types in | | |
|---|---|---|---|
| | Soleus | Flexor Hallucis Longus | Red Vastus |
| Fast-twitch-oxidative-glycolytic | 0 | 10 | 78 |
| Fast-twitch-glycolytic | 0 | 73 | 18 |
| Slow-twitch-oxidative | 100 | 17 | 4 |
| Time-to-peak tension (msec) | 82 | 21 | 19 |
| Half-relaxation time (msec) | 114 | 20 | 19 |
| Myosin ATPase, pH 6.8 ($\mu$moles $P_i$/min/mg) | 0.12 | 0.22 | 0.23 |

recognizable sarcolemma from human skeletal muscle (Fig. 7). These membranes, as expected, contain a ($Na^+$ + $K^+$)-stimulated ATPase which is inhibited by ouabain and diphenylhydantoin. It should soon be possible to study sarcolemma of patients with myotonic dystrophy, a disease which might be predicted to show abnormalities in this membrane and/or its associated enzymes.

Very recently slow-twitch-oxidative fibers in the guinea pig (*e.g.*, soleus) were shown to contain much higher activities of acid hydrolases than fast-twitch-oxidative-glycolytic fibers (red vastus lateralis) or fast-twitch-glycolytic fibers (white vastus lateralis).[40] The higher activities of these enzymes, including aryl-sulfatase A, $\beta$-*N*-acetylglucosaminidase, $\beta$-galactosidase, cathepsin, and acid phosphatase, may provide a new and independent means of classifying muscle fibers. The increased activities may reflect an increased turnover rate of the constituents of these fibers.

The focus of this section has purposely been limited to the areas of muscle biochemistry in which important technical advances have recently been made. The better known muscle diseases due to absence of certain enzymes of carbohydrate metabolism have recently been reviewed elsewhwere.[28] It is likely that similar diseases involving other glycolytic enzymes await the thorough investigator, but it is very unlikely that a defect in any of these accounts for the common types of muscular dystrophy. Other relatively unexplored areas involving structural proteins, nucleic acid, protein, and membrane biosynthesis are in great need of careful study.

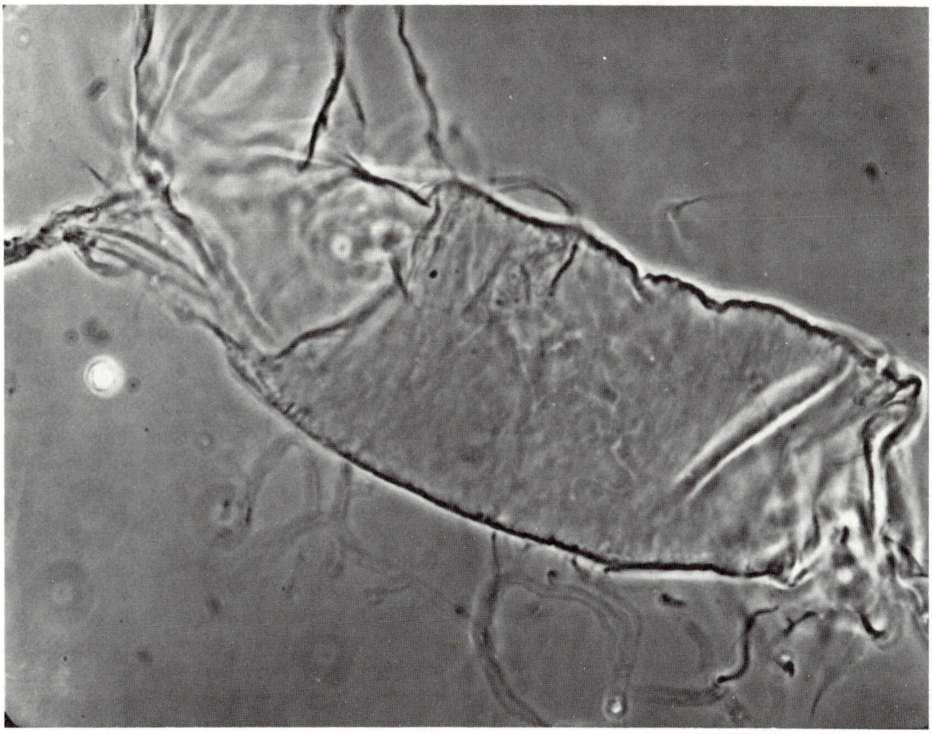

FIG. 7. Sarcolemma prepared from human skeletal muscle; × 666.

## PHYSIOLOGY AND TERMINOLOGY

This section will be devoted to a brief discussion of the relationship of muscle fiber type to the contraction-relaxation kinetics of muscle. It is now entirely clear that grossly red muscles may be relatively fast or slow in contraction time or time-to-peak tension.[2] Although still commonplace there is no longer any excuse for assuming that a grossly red skeletal muscle contracts slowly (*cf.* Table 2). We have proposed elsewhere that muscle fibers be classified in a way meaningful to physiologists and biochemists as well as histochemists.[36] This terminology should also be meaningful to clinicians, pathologists, and students. I have tried to use such terminology throughout this paper and will now try to justify its use.

It seems most reasonable to use the manner in which a muscle performs its primary function, *i.e.*, to do work, as the fundamental basis for classification. Thus, with few exceptions, mammalian extrafusal muscle fibers are either slow-twitch or fast-twitch in contraction-relaxation kinetics. The absolute speeds will, of course, differ among animals and even within the same animal, depending on several variables including the inertia of the object being moved. The pathways by which such fibers derive their energy are considered tangential to this prime function of muscle and should be used to subclassify particular fiber types as oxidative, glycolytic, or oxidative-glycolytic.

The terms "red" and "white" have long been used to connote not only the gross color of a muscle, *i.e.*, red or white, but also, by inference, its dependence on oxidative (red) or glycolytic (white) metabolism for its energy. The inaccuracy and inadequacy of these terms prohibits their continued use in any context other than to describe the gross color of muscle.

Studies in progress in several laboratories should soon allow us to quantitate the terms "fast," "slow," "oxidative," and "glycolytic." It is reasonable to expect that the energy-generating systems available in a given fiber will correlate with the type, intensity, and duration of work required of the motor unit to which it belongs.

It is possible, of course, to assign numbers or letters (*cf.* Reference 36 for review) to designate each of these (and ? other) types of fibers, but unless universally accepted this would not greatly facilitate communication between histochemists, biochemists, and physiologists, and surely it would be no favor to students. For now, I submit that a muscle fiber is best described by how it does its work and from what sources it derives its energy. Speed of contraction and energy-generating systems, since already in common use, are reasonable bases for fiber "typing." Even this classification must be referred to human muscle with caution, however, until the contraction-relaxation times of its different fibers have been carefully correlated with a battery of histochemical and biochemical studies. Studies in lower animals suggest that in mature fast-twitch fibers of human limb muscles histochemical stains for myofibrillar ATPase will be dark at pH 9.4, compared with the light staining of slow-twitch fibers.

## MUTABILITY OF SKELETAL MUSCLE

Slow-twitch and fast-twitch muscle in cats, rats, and guinea pigs can be converted to fast- and slow-twitch muscles, respectively, by transecting and switching their nerve supplies.[8, 10] Histochemical and biochemical studies of such cross-innervated muscles in several species show that not only the contraction kinetics but also metabolic properties are interconverted to a remarkable degree.[9, 12, 23, 29, 45, 48, 50]

For example, the specific activity of myosin ATPase of the fast-twitch flexor hallucis longus muscle decreases greatly after cross-innervation with the nerve to the slow-twitch soleus.[9, 29] Other studies have shown increased activities of mitochondrial enzymes in fast-twitch muscle which are cross-innervated with the nerve to slow-twitch muscle.[45] It is, however, not clear what dictates these changes, which could be mediated by trophic substances from the nerve unit, by hormones whose action on muscle is controlled or mediated by the nervous system, by the pattern of electrical activity from the nerve which might in turn repress and depress synthesis of certain species of nucleic acids and proteins, or by signals from the cell membrane reflecting these influences—among other possibilities (*cf.* 23).

During the past decade we have also learned that the biochemical and kinetic properties of muscle change greatly even after the muscle appears mature by routine light microscopy either in embryo or newborn. In general the contraction times become shorter (*i.e.*, the muscle contracts faster) as muscle matures.[7, 11] Methylation of histidine and lysine residues in the myosin molecule apparently accompanies its maturation.[54] Recent evidence shows that myosin of fast-twitch fibers, but not of slow-twitch fibers, contains 3-methylhistidine,[27] as might be expected if increased methylation correlates with the specific activity of myosin ATPase and hence the contraction speed of the fiber.

Other evidence reviewed in detail elsewhere[5, 25, 32] shows that many of the metabolic characteristics of muscle fibers can be altered by more physiologic means, such as chronic treadmill exercise. In the guinea pig, for example, the amount of mitochondria in skeletal muscle can be increased as determined by histochemical appearance with stains for oxidative enzymes or by actual determination of the concentration of mitochondrial protein[3] or cytochromes.[5] As yet no *physiologic* stimulus has been conclusively shown to convert mature slow-twitch fibers to fast-twitch fibers or vice versa.[4, 32] On the other hand, the process of maturation during which the speed of fiber contraction increases can be accelerated by removing certain animals from their nests in the early newborn period,[26] but one cannot be sure that the extra exercise so produced was in itself the direct stimulus for accelerated maturation as opposed, for example, to systemic hormonal mediators.

Study of biochemical, histochemical, and physiologic responses of muscle to exercise and study of the adaptation of muscle after exercise are obviously relevant to the clinical problems dealt with in this volume. In addition, though perhaps less apparent, such studies are relevant to the proper interpretation of routine histochemical stains of normal and diseased tissue. It is important to recognize that a given variation from normal, *e.g.*, hyperactivity with stains for mitochondrial enzymes, may reflect a reaction, compensatory or otherwise, to disease—rather than the cause of a disease.

## ACKNOWLEDGMENTS

The collaboration of Drs. R. James Barnard, R. F. Dunn, V. Reggie Edgerton, Walter Fiehn, Tetsuo Furukawa, and M. Worsfold in the original work cited in this manuscript is gratefully acknowledged.

## REFERENCES

1. Barany, M.: ATPase activity of myosin correlated with speed of muscle shortening. *J. Gen. Physiol. (Suppl.) 50:* 197–218, 1967.
2. Barnard, R. J., Edgerton, V. R., Furukawa, T., and Peter, J. B.: Histochemical, biochemical and contractile properties of red, white and intermediate fibers. *Am. J. Physiol. 220:* 410–414, 1971.
3. Barnard, R. J., Edgerton, V. R., and Peter, J. B.: Effect of exercise on skeletal muscle I. Biochemical and histochemical properties. *J. Appl. Physiol. 28:* 762–766, 1970.
4. Barnard, R. J., Edgerton, V. R., and Peter, J. B.: Effect of exercise on skeletal muscle II. Contractile properties. *J. Appl. Physiol. 28:* 767–770, 1970.
5. Barnard, R. J., and Peter, J. B.: Effect of exercise on skeletal muscle III. Cytochrome changes. *J. Appl. Physiol. 31:* 904–908, 1971.
6. Buchthal, F., and Schmalbruch, H.: Contraction times and fibre types in intact human muscle. *Acta Physiol. Scand. 79:* 435–452, 1970.
7. Buller, A. J., Eccles, J. C., and Eccles, R. M.: Differentiation of fast and slow muscles in the cat hind limb. *J. Physiol. 150:* 399–416, 1960.
8. Buller, A. J., Eccles, J. C., and Eccles, R. M.: Interactions between motoneurones and muscles in respect of the characteristic speeds of their responses. *J. Physiol. 150:* 417–439, 1960.
9. Buller, A. J., Mommaerts, W. F. H. M., and Seraydarian, K.: Enzymic properties of myosin in fast and slow twitch muscles of the cat following cross-innervation. *J. Physiol. 205:* 581–597, 1969.
10. Close, R.: Dynamic properties of fast and slow skeletal mucles of the rat after nerve cross-union. *J. Physiol. 204:* 331–346, 1969.
11. Close, R.: Dynamic properties of fast and slow skeletal muscles of the rat during development. *J. Physiol. 173:* 74–95, 1964.
12. Dubowitz, V., and Newman, D. L.: Change in enzyme pattern after cross-innervation of fast and slow skeletal muscle. *Nature 214:* 840–841, 1967.
13. Dubowitz, V., and Pearse, A. G. E.: A comparative histochemical study of oxidative enzyme and phosphorylase in skeletal muscle. *Histochemie 2:* 105–117, 1960.
14. Ebashi, S., and Endo, M.: Calcium ion and muscle contraction. *Progr. Biophys. 18:* 123–183, 1968.
15. Edgerton, V. R., and Simpson, D. R.: The intermediate fiber of rats and guinea pigs. *J. Histochem. Cytochem. 17:* 828–838, 1969.
16. Engel, W. K.: Diseases of the neuromuscular junction and muscle. In *Neurohistochemistry*, ed. by Adams, C. W. M., pp. 622–672. Amsterdam, Elsevier Publishing Company, 1965.
17. Engel, W. K.: The essentiality of histo- and cytochemical studies of skeletal muscle in the investigation of neuromuscular disease. *Neurology (Minneap) 12:* 778–794, 1962.
18. Fiehn, W., and Peter, J. B.: Properties of the fragmented sarcoplasmic reticulum from fast-twitch and slow-twitch muscles. *J. Clin. Invest. 50:* 570–573, 1971.
19. Furukawa, T., and Peter, J. B.: Studies of myosin and actomyosin ATPase of human skeletal muscle. Unpublished data.
20. Furukawa, T., and Peter, J. B.: Troponin activity of different types of muscle fibers. *Exp. Neurol. 31:* 214–222, 1971.
21. Furukawa, T., and Peter, J. B.: Muscular dystrophy and other myopathies. Troponin activity of natural actomyosin from skeletal muscle. *Arch. Neurol. 26:* 385–390, 1972.

22. Gauthier, G. F.: On the relationship of ultrastructural and cytochemical features to color in mammalian skeletal muscle. *Z. Zellforsch. Mikrosk. Anat. 95:* 462–482, 1969.
23. Guth, L.: "Trophic" influences of nerve on muscle. *Physiol. Rev. 48:* 645–687, 1968.
24. Guth, L., and Samaha, F. J.: Qualitative differences between actomyosin ATPase of slow and fast mammalian muscle. *Exp. Neurol. 25:* 138–152, 1969.
25. Jeffress, R. N., and Peter, J. B.: Adaptation of muscle to overloading—A review. *Bull. Los Angeles Neurol. Soc. 35:* 134–144, 1970.
26. Kendrick-Jones, J., and Perry, S. V.: Biochemical adaptation in muscle. In *Exploratory Concepts in Muscular Dystrophy and Related Disorders*, ed. by Milhorat, A. T., pp. 64–72. Amsterdam, Excerpta Medica Foundation, 1967.
27. Kuehl, W. M., and Adelstein, R. S.: The absence of 3-methylhistidine in red, cardiac and fetal myosins. *Biochem. Biophys. Res. Commun. 39:* 956–964, 1970.
28. McArdle, B.: Metabolic and endocrine myopathies. In *Disorders of Voluntary Muscle*, ed. by Walton, J. N., 2nd ed., pp. 607–638, London, J. & A. Churchhill, Ltd., 1969.
29. Mommaerts, W. F. H. M., Buller, A. J., and Seraydarian, K.: The modification of some biochemical properties of muscle by cross-innervation. *Proc. Natl. Acad. Sci. U. S. A. 64:* 128–133, 1969.
30. Ogata, T., and Masahiko, M.: Histochemical study of oxidative enzymes in vertebrate muscles. *J. Histochem. Cytochem. 12:* 171–182, 1964.
31. Peachey, L. D.: Muscle. *Annu. Rev. Physiol. 30:* 401–440, 1970.
32. Peter, J. B.: Histochemical, biochemical and physiological studies of skeletal muscle and its adaptation to exercise. In *Contractility of Muscle Cells and Related Processes*, ed. by Podolsky, R. J., pp. 151–173, Englewood Cliffs, N. J., Prentice-Hall, 1972.
33. Peter, J. B.: Biochemical approaches to the study of muscle diseases. In *Muscle Birth Defects, Orig. Art. Ser.* VII, ed. by McKusick, V., pp. 38–42, 1971.
34. Peter, J. B.: A ($Na^+ + K^+$) ATPase of sarcolemma from skeletal muscle. *Biochem. Biophys. Res. Commun. 40:* 1362–1367, 1970.
35. Peter, J. B.: Studies of human skeletal muscle mitochondria. *Biochem. Med. 2:* 179–189, 1968.
36. Peter, J. B., Barnard, R. J., Edgerton, V. R., Gillespie, C. A., and Stempel, K. E.: Metabolic profiles of three fiber types of skeletal muscle in guinea pigs and rabbits. *Biochemistry 11:* 2627–2633, 1972.
37. Peter, J. B., and Dunn, R. F.: Unpublished observations.
38. Peter, J. B., Fiehn, W., and Dunn, R. F.: Biochemistry and morphology of fragmented sarcoplasmic reticulum. In *Electron Microscopy 1970*, ed. by Arceneaux, C. L., page 90, Baton Rouge, Claitor's Publishing Division, 1970.
39. Peter, J. B., Fiehn, W., and Worsfold, M.: Studies of fragmented sarcoplasmic reticulum in muscular dystrophy and related disorders. Unpublished data.
40. Peter, J. B., Kar, N., Barnard, R. J., Pearson, C. M., and Edgerton, V. R.: Distribution of acid hydrolases in guinea pig skeletal muscle. *Biochem. Med. 6:* 257–261, 1972.
41. Peter, J. B., Stempel, K., and Armstrong, J.: Biochemistry and electron microscopy of mitochondria in muscular and neuromuscular diseases. In *Muscle Diseases*, ed. by Walton, J. N., Canal, N., and Scarlato, G., pp. 228–235. Amsterdam, Excerpta Medica Foundation, 1970.
42. Peter, J. B., Stempel, K., and Verhaag, D.: Fatty acid oxidation by mitochondria from normal and dystrophic skeletal muscle. Unpublished data.
43. Peter, J. B., and Worsfold, M.: Muscular dystrophy and other myopathies; sarcotubular vesicles in early disease. *Biochem. Med. 2:* 364–371, 1969.
44. Peter, J. B., and Worsfold, M.: Oxidative phosphorylation and calcium transport by sarcotubular vesicles in myotonic dystrophy. *Biochem. Med. 2:* 457–460, 1969.
45. Prewitt, M. A., and Salafsky, B.: Effect of cross-innervation on biochemical characteristics of skeletal muscles. *Am. J. Physiol. 213:* 295–300, 1967.
46. Price, H. M.: Ultrastructure of the skeletal muscle fiber. In *Disorders of Voluntary Muscle*, ed. by Walton, J. N., pp. 29–56. Ltd., London, J. & A. Churchill, Ltd., 1969.
47. Price, H. M.: Mitochondrial myopathies in man? A review of evidence. In *Exploratory Concepts in Muscular Dystrophy and Related Disorders*, ed. by Milhorat, A. T., pp. 341–350. Amsterdam, Excerpta Medica Foundation, 1967.

48. Robbins, N., Karpati, G., and Engel, W. K.: Histochemical and contractile properties in the cross-innervated guinea pig soleus muscle. *Arch. Neurol. 20:* 318–329, 1969.
49. Romanul, F. C. A.: Enzymes in muscle. I. Histochemical studies of enzymes in individual muscle fibers. *Arch. Neurol. 11:* 355–368, 1964.
50. Romanul, F. C. A., and Van Der Meulen, J. P.: Slow and fast muscles after cross-innervation. Enzymatic and physiological changes. *Arch. Neurol. 17:* 387–402, 1967.
51. Samaha, F. J., and Gergely, J.: Biochemical abnormalities of the sarcoplasmic reticulum in muscular dystrophy. *New Engl. J. Med. 280:* 184–187, 1969.
52. Stein, J. M., and Padykula, H. A.: Histochemical classification of individual skeletal muscle fibers of the rat. *Am. J. Anat. 110:* 103–115, 1962.
53. Sugita, H., Okimoto, K., Ebashi, S., and Okinaka, S.: Biochemical alterations in progressive muscular dystrophy with special reference to the sarcoplasmic reticulum. In *Exploratory Concepts in Muscular Dystrophy and Related Disorders*, ed. by Milhorat, A. T., pp. 321–326. Amsterdam, Excerpta Medica Foundation, 1967.
54. Trayer, I. P., Harris, C. I., and Perry, S. V.: 3-Methyl histidine and adult foetal forms of skeletal muscle myosin. *Nature 217:* 452–453, 1968.
55. Worsfold, M., Dunn, R. F., and Peter, J. B.: A simplified technique for preparing isolated subcellular particles for electron microscopy. *J. Lab. Clin. Med. 74:* 160–165, 1969.
56. Worsfold, M., and Peter, J. B.: Kinetics of calcium transport by fragmented sarcoplasmic reticulum. *J. Biol. Chem. 245:* 5545–5552, 1970.
57. Worsfold, M., Peter, J. B., and Dunn, R. F.: Duchenne muscular dystrophy; distinctive biochemical and electron microscopic abnormalities of the sarcotubular vesicle fraction. In *Muscle Diseases*, ed. by Walton, J. N., pp. 303–308. Amsterdam, Excerpta Medica Foundation, 1970.
58. Yellin, H.: Unique intrafusal and extraocular muscle fibers exhibiting dual actomyosin ATPase activity. *Exp. Neurol. 25:* 153–163, 1969.

Chapter 2

# Techniques of Muscle Biopsy and Staining Methods with Particular Emphasis on Commonly Encountered Artifacts*

PASQUALE A. CANCILLA

When a pathologist examines muscle, his aim is to arrive at a quick and accurate diagnosis. He is often frustrated in this aim by problems with fixation, variability in the staining reactions of the tissue, and a variety of artifacts. In addition, he may not have the leisure to interpret or the trained personnel to perform the complex array of special procedures and stains that may be required in some cases. In this presentation, sampling, fixation, and staining of skeletal muscle will be considered in detail, and methods to avoid some of the more common difficulties will be indicated. Artifacts will still be present in spite of most efforts to overcome them, and we will therefore give special emphasis to those more commonly seen. Stains for enzyme histochemistry will receive brief attention, since they will be considered in other presentations.

## SAMPLING

Since it may be impossible to make an accurate diagnosis from muscle with advanced disease or with minimal involvement, the site to be biopsied must be selected with care. In problem cases, we generally biopsy two sites, selecting muscles that have active, but not advanced, disease by clinical criteria. This policy has increased our awareness of the histologic variations that may be present in muscle disease and the care that must be taken to avoid overinterpretation or prognostication from a single specimen.

The biopsy site should avoid pressure points, should not be the site of previous biopsies or electromyography, and should be of sufficient bulk to provide an adequate sample. Care should be taken to avoid injection of local anesthetic into the muscle, and the tissue should be handled gently during removal. The use of an isometric clamp is recommended to avoid contraction of the specimen[8, 13, 14] (Fig. 1). If a clamp is not available, the specimen may be placed

* This investigation was supported in part by United States Public Health Service Grants GM-15759 and RRW0238 and by a grant from the Muscular Dystrophy Associations of America.

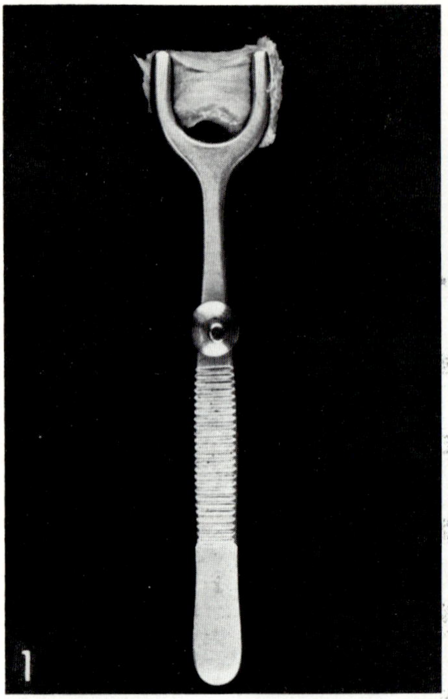

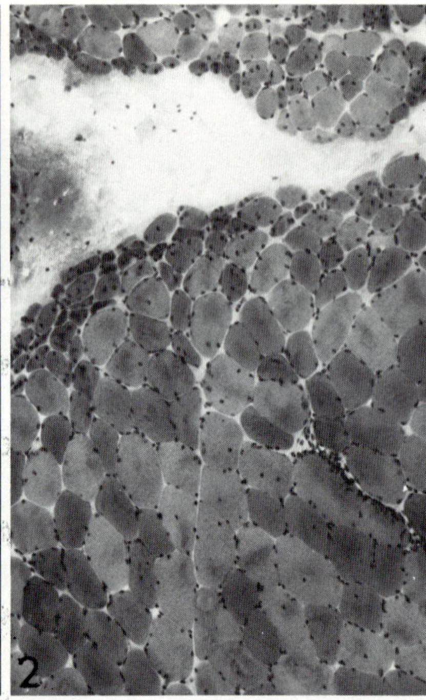

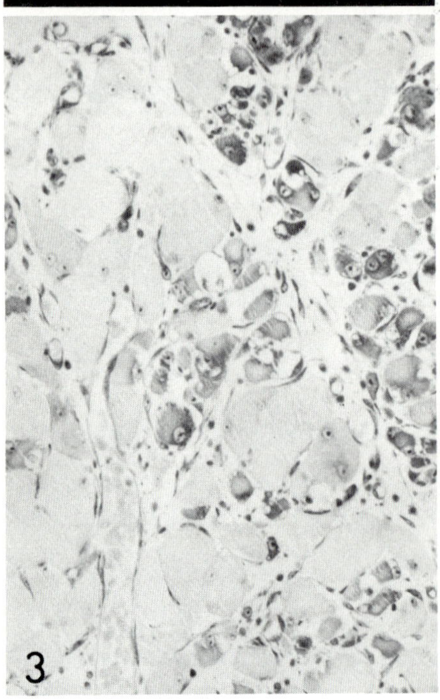

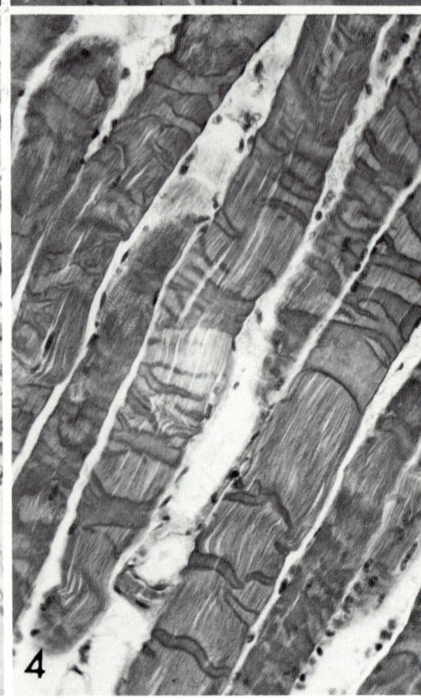

FIG. 1. Isometric muscle clamp.

FIG. 2. Rapidly frozen muscle cut in a freezing microtome and stained with hematoxylin and eosin. The variation in size of the fibers is evident. Dark staining fibers correspond to histochemical type I fibers, and light staining fibers correspond to histochemical type II fibers. × 85.

FIG. 3. Regenerating muscle fibers are dark due to basophilia of the sarcoplasm. These fibers have vesicular nuclei with prominent nucleoli. Azure B bromide stain; × 210.

FIG. 4. Muscle fibers with irregular and hyalinized contraction bands. Hematoxylin and eosin stain; × 210.

in a saline-soaked sponge at room temperature for 15 to 30 minutes before fixation.[3, 8] When enzyme histochemistry is planned, the tissue should be immediately frozen to −160°C with isopentane in liquid nitrogen to prevent loss of enzyme activity or drying of the edges from exposure.[5, 13]

## FIXATION

The methods of preservation of the tissue must be decided before the biopsy is performed because the types of stains and special procedures generally are limited by the methods of preservation selected. Table 1 indicates some of the more commonly used fixatives and their application.

The most useful preservative is a solution of 10% neutral buffered formaldehyde. The biopsy specimen, held in the clamp, is fixed for 24 hours. Then the clamp can be removed and the specimen processed or stored in the fixative. The tissue is cut so that longitudinal and transverse sections are obtained. This is very important in muscle biopsies since each of these pieces of tissue, when properly oriented, yields different but critical information to the examiner.[2] With proper staining of well oriented, formaldehyde-fixed material, the correct diagnosis can usually be made and it may be the only method of fixation required.

In special situations or in problem cases additional information about the muscle is desirable. In our laboratory, we are alerted to these situations by the special request of the clinician for additional examinations. A complete history and physical findings, together with pertinent laboratory data, accompany his request for the examination. Based on this information, a decision is reached as to whether enzyme histochemistry, electron microscopy, or other special procedures will be done. If enzyme histochemistry is required, a second specimen is removed in a clamp and immediately frozen.[3]

For electron microscopy, a small piece of muscle is immersed in 3% buffered glutaraldehyde.[15] After overnight fixation, the tissue is divided into small blocks, postfixed in osmium tetroxide, and embedded in Epon 812. Generally, the blocks of tissue will be stored and no further studies will be done. However, if a lesion is found by light microscopy or enzyme histochemistry that is unusual and might be clarified by the greater magnification and resolution of electron microscopy, then these studies are undertaken. One-micron-thick sections are cut on a special microtome and stained with alkaline toluidine blue or cresyl violet. These are examined with the light microscope to select representative areas for final ultrathin sectioning and examination with the electron micro-

TABLE 1. PRESERVATIVES AND THEIR USE IN THE STUDY OF MUSCLE

| Preservative | Embedding Media | Application |
|---|---|---|
| 10% neutral formalin, Bouin's, Zenker's | Paraffin or paraplast | Light microscopy, standard stains |
| Rapid freezing in isopentane and liquid nitrogen | None | Freezing microtome sections for standard stains and enzyme histochemistry |
| 3% buffered glutaraldehyde or 1% $OsO_4$ | Epon 812 | 1-$\mu$ sections for light microscopy; electron microscopy |

scope. Random sectioning for electron microscopy leads to gross sampling errors and should be avoided if meaningful data are to be obtained from this costly and time-consuming examination.

## STAINING METHODS

Three of the more commonly used stains in the study of muscle disease are hematoxylin and eosin, Gomori's trichrome, and Mallory's phosphotungstic acid hematoxylin.[11] Hematoxylin and eosin is the most familiar and it is often the only stain necessary for a diagnosis. The trichrome and phosphotungstic acid hematoxylin stains are useful for the examination of striations, to indicate degeneration or necrosis of fibers or to demonstrate cytoplasmic alterations such as nemaline rods or target fibers.

Sections of the quick-frozen material stained with hematoxylin and eosin or a modified trichrome stain[9] are invaluable for rapid and reliable examinations of muscle (Fig. 2). Although the sections of the frozen tissue are thicker than those obtained from paraffin-embedded material, they may show alterations that otherwise are not apparent. Such preparations are generally free of shrinkage artifact. Muscle fiber relationships and fiber size are readily apparent, allowing for an accurate appraisal of variations in individual fiber size. It is also possible to anticipate the results of the enzyme histochemistry since the tinctorial qualities of the fibers are such that two fiber types, corresponding to the two histochemical types, may be appreciated.[9] With both staining methods, the darker staining fibers are type I and the lighter staining fibers are type II.

The azure B bromide stain is used to demonstrate regenerating muscle fibers.[16] The basophilia of regenerating fibers is intensified by this stain, and, in instances in which fiber basophilia is equivocal in hematoxylin and eosin stains, the azure B bromide stain may be definitive (Fig. 3). Other stains are used whenever indicated to investigate or demonstrate special features of biopsy material.

Histochemical stains for enzymes are discussed in detail in other presentations in this course. In our laboratory, we routinely use succinic dehydrogenase and DPNH diaphorase, which stain type I fibers dark and type II fibers light, and myosin ATPase and myophosphorylase stains, which stain type I fibers light and type II fibers dark. These histochemical procedures are similar to those recommended by others.[3, 7, 13]

## ARTIFACTS

The problems that are created by artifacts in the preparation and interpretation of muscle have been considered in several recent publications.[1, 4, 6, 8, 10, 12] The number of articles gives some indication of the importance and frequency of the problem. Many of the artifacts are produced at the time of biopsy or shortly thereafter, before the specimen reaches the pathologist. The contractile nature of the tissue is responsible for many of the problems. Gentle dissection of the tissue and the use of an isometric clamp will minimize these technical problems. If a large specimen is obtained, the traumatized edges of the tissue may be discarded and further problems will be avoided.

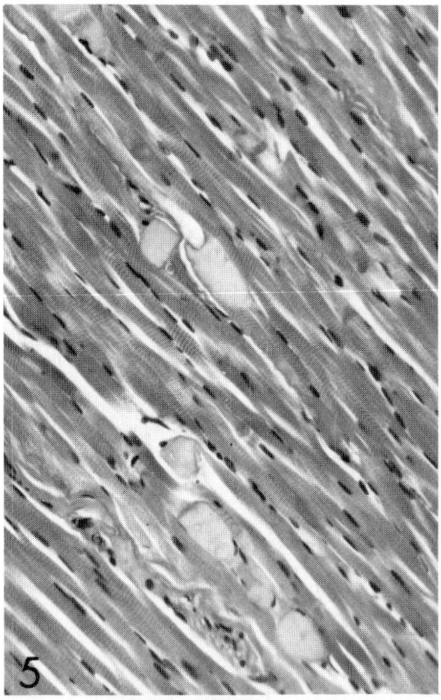


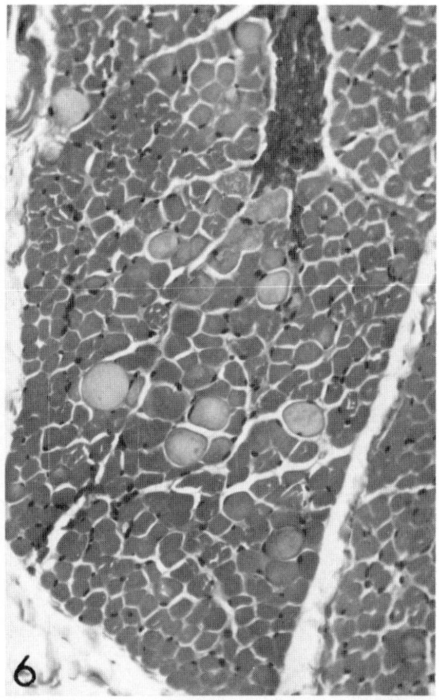


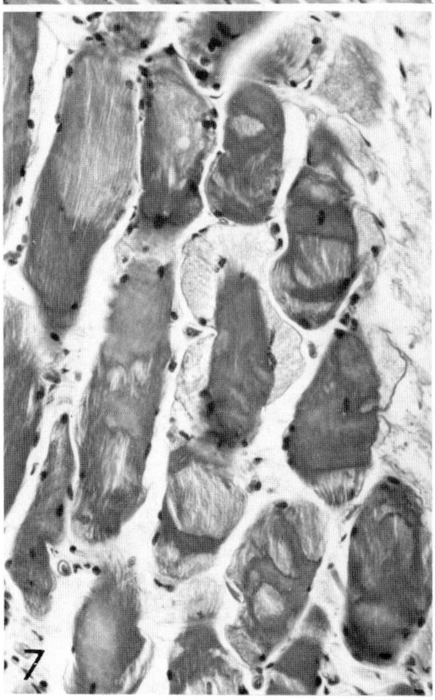


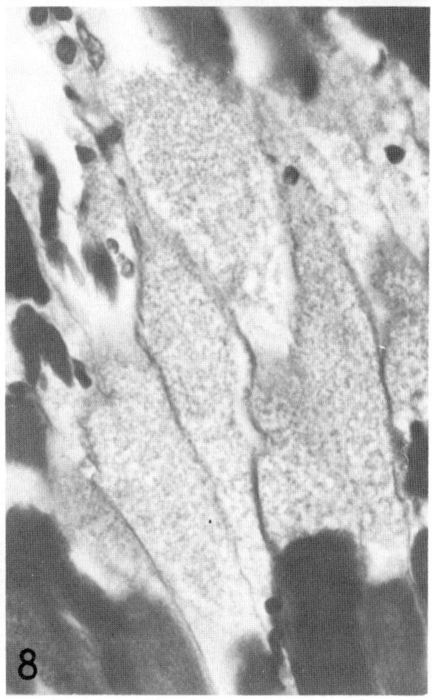


FIG. 5. Muscle fibers with segmental expansion and loss of striations. Hematoxylin and eosin stain; × 210.

FIG. 6. Transverse section of muscle illustrated in Figure 5. The altered fibers are rounded and pale staining. Such fibers may be interpreted as degenerated or necrotic, but in this instance they are an artifact of contraction. Hematoxylin and eosin stain; × 210.

FIG. 7. Oblique section of muscle with irregular contraction bands. The fiber in the center is expanded and there is dissolution of the myofibrils. Hematoxylin and eosin stain; × 210.

FIG. 8. Granular dissolution of muscle fibers associated with excessive contraction. Hematoxylin and eosin stain; × 210.

Although contraction produces an irregular banded pattern which is more prevalent at the edge of the specimen, this artifact may be found anywhere in the specimen (Fig. 4). The more contracted portions of the fiber lose their striations and appear hyalinized in longitudinal section (Fig. 5). When such fibers are viewed in transverse section, they may be pale or eosinophilic and, thus, may be misinterpreted as necrotic (Fig. 6). This change may be accompanied by pyknosis of sarcolemmal nuclei or displacement of sarcolemmal nuclei toward the center of the fiber, increasing the hazard of misinterpretation. Excessive contraction or swelling of the fiber produces a granular change in the cytoplasm or large vacuoles filled with pale eosinophilic granular material (Fig. 7). These alterations are usually accompanied by other artifacts, but, when the biopsy is small and the change is widespread, granular-vacuolar alterations may be difficult to differentiate from those found in the vacuolar myopathies (Fig. 8).

Atrophic fibers, particularly those that occur in the neurogenic atrophies such as Werdnig-Hoffmann disease, are especially vulnerable to a somewhat different artifact. The individual muscle fibers have a peripheral clear zone that is produced by shrinkage of the myofibrils from the sarcolemmal and basement membranes (Fig. 9). Sarcolemmal nuclei are often located in this expanded space. The artifact occurs sometime during the routine processing and embedding of the tissue since rapidly frozen tissue does not exhibit this change.

Incomplete fixation, inadequate dehydration during the processing, or irregular clearing of the tissue, alone or in combination, lead to a variety of artifacts in the routine examination of muscle biopsies. The alterations include a loss of striations, hyalinization and rounding of fibers, and variation in fiber size (Fig. 10). They are more common in the center of the specimen or in compressed segments of the muscle because the problem is related to penetration of the various solutions used in processing the tissue. Identical alterations, however, may be found in diseased muscle where they may be a part of the pathologic process. The differentiation between a real lesion and an artifact may be difficult. As a result, some authors have recommended that myofiber alterations alone should not be used as an indicator of disease.[2]

Many other artifacts may be attributed to the process of dehydration, clearing, and embedding of muscle. Brittle tissue will lead to cracking and fragmentation of muscle fibers during sectioning. Soft embedding media will allow compression of the tissue during sectioning and result in pressure ridges. Hard media will result in shattering of muscle fibers.

## OTHER PROBLEMS IN INTERPRETATION

Familiarity with the range of normal histology and experience with common artifacts of muscle are essential for a proper interpretation of diagnostic biopsies. There are several normal structures in muscle which may be confused with pathologic alterations and which should be noted.

Neurovascular bundles appear as cellular areas intimately associated with muscle fibers when viewed in longitudinal section (Fig. 11). Nuclei in these structures are irregularly distributed and vary in their staining properties and

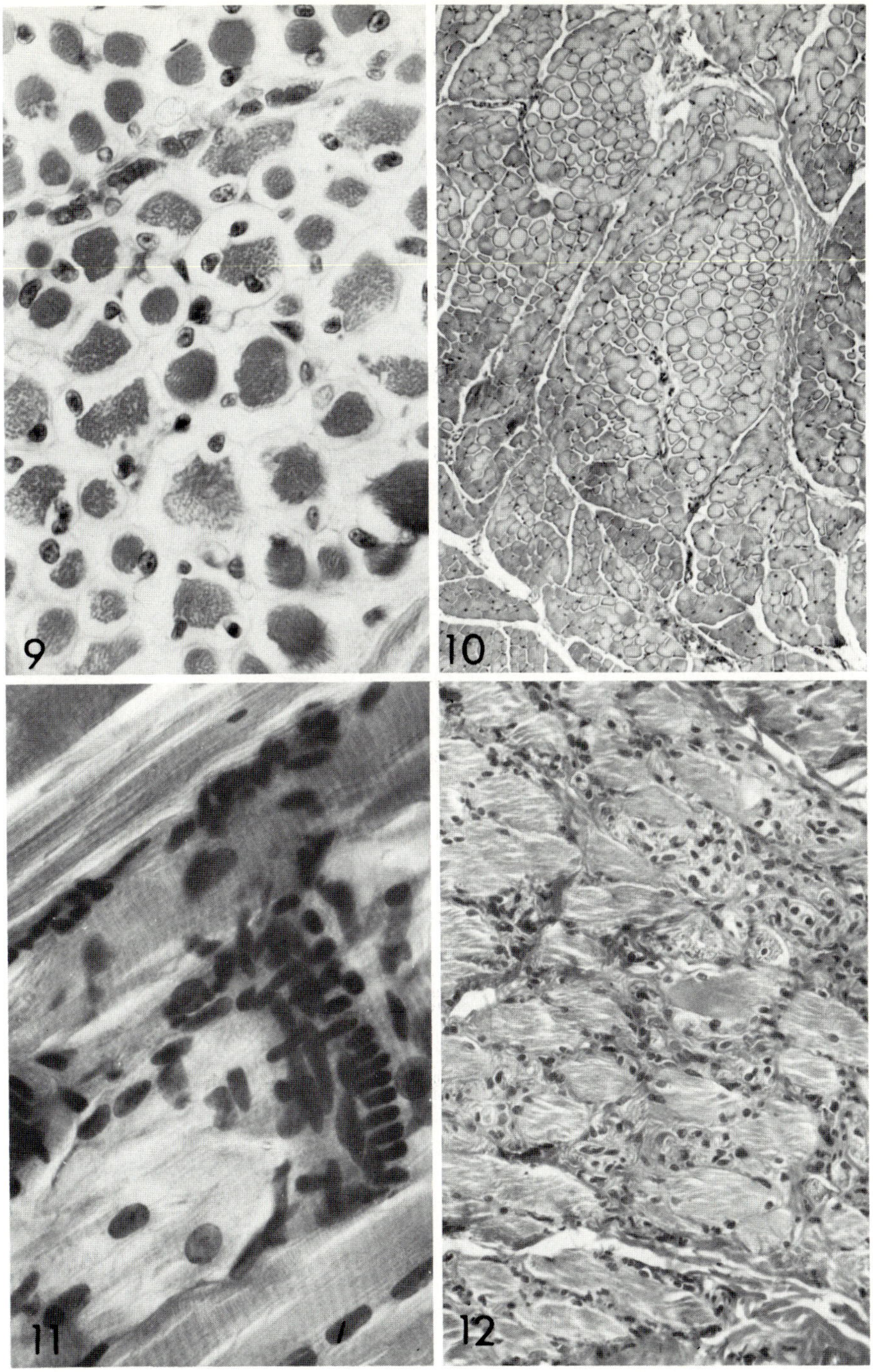


FIG. 9. Fascicles of atrophic fibers. The myofibrils in the center of the fiber are separated from the periphery by a clear area. Sarcolemmal nuclei are present in the clear space. Hematoxylin and eosin stain; × 525.

FIG. 10. Muscle fibers vary in size and tinctorial qualities in the fibers from the center of block of tissue that was not completely fixed before processing. Hematoxylin and eosin stain; × 85.

FIG. 11. Neurovascular bundle appearing as a cellular area in the muscle. Hematoxylin and eosin stain; × 525.

FIG. 12. The cellular areas between these muscle fibers are due to the proliferation of Schwann cells and axons after injury of a nerve. Hematoxylin and eosin. × 210.

appearance. Such cellular areas may be misinterpreted as atrophic fibers with aggregations of sarcolemmal nuclei or as regenerating fibers because of the vesicular nuclei that are present. Injury to intrafascicular nerve fibers from trauma or injection may lead to the formation of traumatic neuromas. Such areas are cellular and may be interpreted as intrafascicular fibrosis or inflammation (Fig. 12). If there is any question, axons can generally be demonstrated with appropriate stains.

Myotendinous junctions are often present in the section. The terminal portion of the muscle fiber may appear enlarged and pale staining. It often has a vesicular nucleus in the center of the fiber. In cross-section, these fibers may be interpreted as abnormal because of their tinctorial qualities and location of nuclei. Normal muscle spindles may vary in their microscopic appearance and should not be confused with diseased fibers.

## SUMMARY

Problems that arise in the sampling, fixation, preparation, and staining of tissue have been described. Gentle handling of the tissue and the use of an isometric clamp at the time of biopsy will help to prevent many of the artifacts. Well oriented longitudinal and transverse sections of muscle each yield different information to the examiner and should be prepared in every biopsy. Unfixed frozen sections stained with hematoxylin and eosin or the modified trichrome method are useful for the assessment of fiber size and fiber relationships. In addition they may even indicate the results of enzyme histochemistry.

Attention has been given to some of the more common artifacts that occur in muscle. Familiarity with these changes will facilitate the interpretation of muscle biopsies and lead to more accurate diagnosis of diseases of muscle.

## ACKNOWLEDGMENTS

The author would like to thank Margaret Hall, Laurel Reed, Stephen Frommes, Joyce Brooks, and Ethel Doucette for very valuable assistance.

### REFERENCES

1. Adams, R. D.: Pathological reactions of the skeletal muscle fibre in man. In *Disorders of Voluntary Muscle*, ed. by Walton, J. N., pp. 143–202.
2. Adams, R. D.: The histopathology of human muscle disease. In *Neuromuscular Disorders*, ed. by Adams, R. D., Eaton, L. M., and Shy, G. M., pp. 318–356. Baltimore, The Williams and Wilkins Co., 1960.
3. Adams, R. D., Coërs, C., and Walton, J. N.: Report of a sub-committee on quantitation of muscle biopsy findings. *J. Neurol. Sci. 6:* 179–188, 1968.
4. Adams, R. D., Denny-Brown, D., and Pearson, C. M.: Common artifacts in microscopic sections of muscle. In *Diseases of Muscle*, pp. 714–717. New York, Harper and Row, 1967.
5. Engel, W. K.: Diseases of the neuromuscular junction and muscle. In *Neurohistochemistry*, ed. by Adams, C. W. M., pp. 622–672. Amsterdam, Elsevier Publishing Co., 1965.
6. Engel, W. K.: Muscle Biopsy. *Clin. Orthop. 39:* 80–105, 1965.
7. Engel, W. K.: Selective and nonselective susceptibility of muscle fiber types. A new approach to human neuromuscular diseases. *Arch. Neurol. 22:* 97–117, 1970.
8. Engel, W. K., and Brooke, M. H.: Muscle biopsy as a clinical diagnostic aid. In *Neurological Diagnostic Techniques*, ed. by Fields, W. S., pp. 90–146. Springfield, Charles C Thomas, Ltd., 1966.

9. Engel, W. K., and Cunningham, G. G.: Rapid examination of muscle tissue. An improved trichrome method for fresh frozen biopsy sections. *Neurology 13:* 919–923, 1963.
10. Kakulas, B. A., Adams, R. D., and McGee, D. A.: The control of artifact in biopsied muscle. In *Proceedings of the Vth International Congress of Neuropathology*, ed. by Lüthy, F., and Bischoff, A., pp. 632–640. Amsterdam, Excerpta Medica Foundation, 1965.
11. Luna, L. G.: *Manual of Histologic Staining Methods of the Armed Forces Institute of Pathology*, 3rd. ed. New York, McGraw Hill Book Co., 1968.
12. Pearson, C. M.: The histopathology of some human myopathies. In *Muscle*, ed. by Paul, W. M., pp. 423–452. Oxford, Pergamon Press, 1965.
13. Pearson, C. M., Coleman, R. F., Fowler, W. M., Mommaerts, W. F. H. M., Munsat, T. L., and Peters, J. B.: Skeletal muscle. Basic and clinical aspects and illustrative new diseases. *Ann. Intern Med. 67:* 614–650, 1967.
14. Price, H. M., Howes, E. L., Sheldon, D. B., Hutson, O. D., Fitzgerald, R. T., Blumberg, J. M., and Pearson, C. M.: An improved biopsy technique for light and electron microscopic studies of human skeletal muscle. *Lab. Invest. 14:* 194–199, 1965.
15. Sabatini, D. D., Bensch, K., and Barrnett, R.: Cytochemistry and electron microscopy and preservation of cellular ultrastructure of enzymes by aldehyde fixation. *J. Cell Biol. 17:* 19–58, 1963.
16. Wood, F. D., and Pearson, C. M.: Azure B bromide staining of regenerating muscle. *Am. J. Clin. Pathol. 45:* 42–45, 1966.

Chapter 3

# Histoenzymatic Methods Applied to Human Striated Muscle Disease*

M. ANTHONY VERITY AND R. F. COLEMAN

---

The basic role of histochemistry in the extension of our knowledge of general pathologic processes is appreciated. More specifically, within the last decade or so, a deliberate attempt has been fostered to transform the purely descriptive aspects of routine muscle biopsies into more meaningful dynamic and functional interpretations. The general science of histochemistry may be loosely divided into the two integrated fields of substrate histochemistry and histoenzymology. In this presentation we wish to draw upon our experiences and those of others in documenting the value, basic principles, techniques, and pitfalls of a coordinated program of histoenzymatic methodology applied to selected human muscle disorders. The techniques of biopsy and an evaluation of artifactual change have been discussed previously and will not be considered further.[4] The aims of our coordinated program are (1) to utilize the skills, techniques, and varied interests of physicians to elucidate the nature and basic pathogenesis of disorders affecting the neuromuscular apparatus; (2) to provide a continuing forum for interchange of results pertaining to human neuromuscular disease and to suggest appropriate therapy. The profile and logistics of the program are summarized in Fig. 1. The key to the program resides in direct communication between physicians and pathologist with the presentation of a clinical summary. Optimal biopsies are usually obtained through the skill and interest of a selected surgeon. The histochemical procedures are performed on a particular day depending upon volume and urgency. This necessitates a provisional report to the physician on information obtained from routine histology.

What may we hope to gain from a histoenzymatic approach to the investigation of muscle disease? The value of such studies is summarized in Table 1. The definition of two fiber types in human striated muscle with a possible intermediate subtype has been well established.[11, 15] Histoenzymatic techniques have revealed a reciprocity in staining of individual fibers through the use of glycolytic and mitochondrial enzyme reactions (Fig. 2). The fibers of smaller

---

* This investigation was supported by a grant from the Muscular Dystrophy Associations of America and by United States Public Health Service Grants CA-05321 and GM-15759.

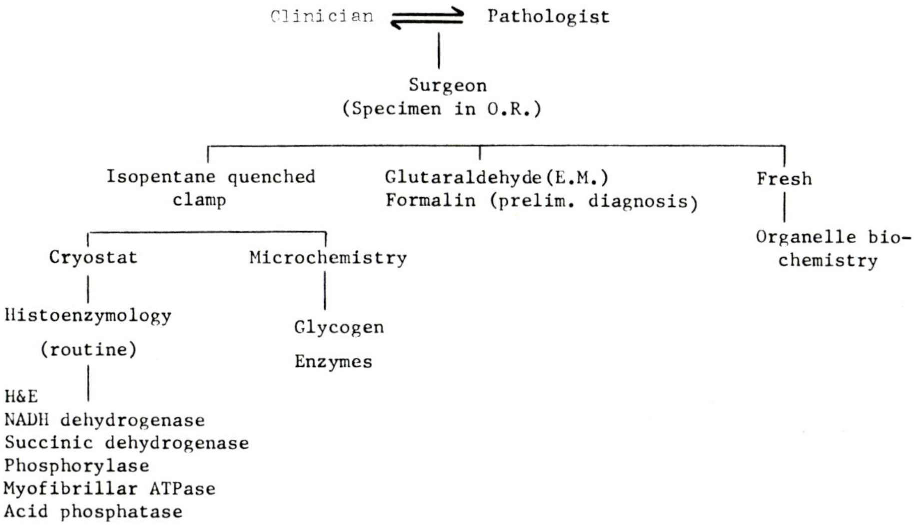


FIG. 1. Logistic flow diagram for the coordinated investigation of neuromuscular disorders in a large teaching hospital.

TABLE 1. VALUE OF HISTOCHEMISTRY IN EVALUATION OF NEUROMUSCULAR DISORDERS

1. Definition of fiber types.
2. Topographic localization of biochemically defined organelles through use of enzyme "markers."
   A. Mitochondria
   B. Lysosome—vacuolar system
   C. Myofibrils
   D. Sarcoplasm
   E. Extramuscular components
3. Semiquantitative estimation of enzyme activity.
4. Evaluation of regenerative activity in different muscle groups, *e.g.*, therapy.
5. Identification of specific inflammatory cells (azure B, acid phosphatase).
6. Determination of collateral reinnervation from fiber type grouping.
7. Morphometric analysis of type-specific fibers.

diameter are usually type I. Type I fibers are those with high activities of the mitochondrial oxidative enzymes and relatively low activities of myosin adenosine triphosphatase and phosphorylase. Proportionately more mitochondria are present in type I than in type II fibers.[39] Type I fibers have a predominantly oxidative metabolism utilizing Krebs' cycle intermediates and fatty acids as primary energy substrates. Type II fibers are thought to rely to a large extent on glycogen as an energy substrate. Functional and histoenzymatic relationships are summarized in Table 2. The acid phosphatase reaction has proved of value in identifying the presence of histiocytic cells in the endomysium in low grade or chronic inflammatory conditions (Fig. 3) and more dramatically in those myopathies in which lysosome activity or autophagic vacuole formation is significant (Fig. 4). The recognition, enumeration, and topographic relation of particular fiber types has been shown to be of value in the elucidation of neurogenic injury during early life,[10, 18, 19] in documenting

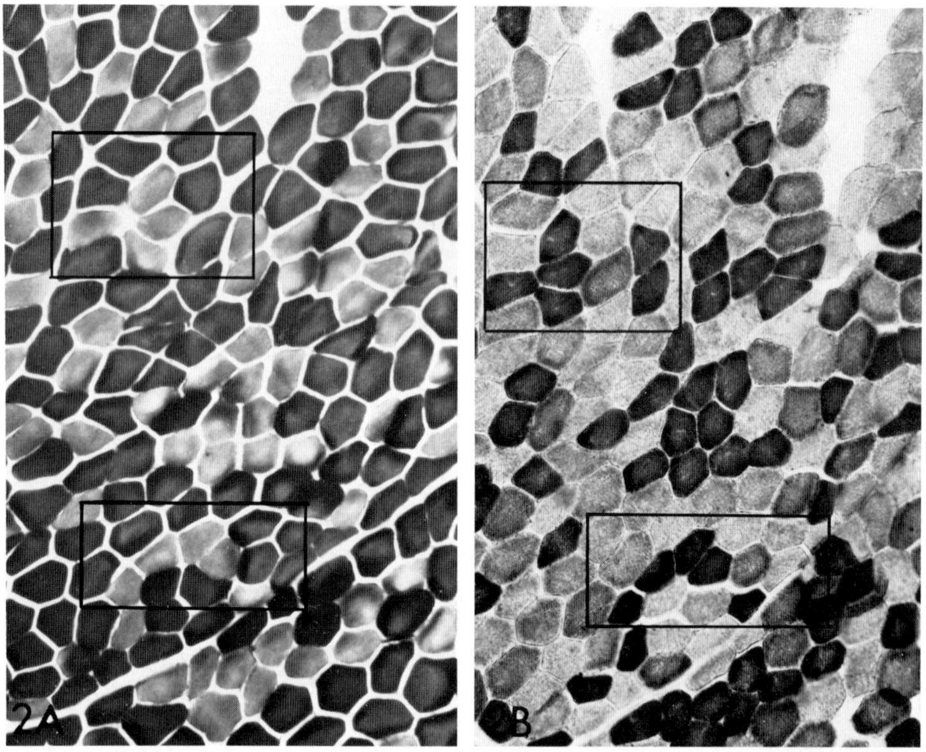


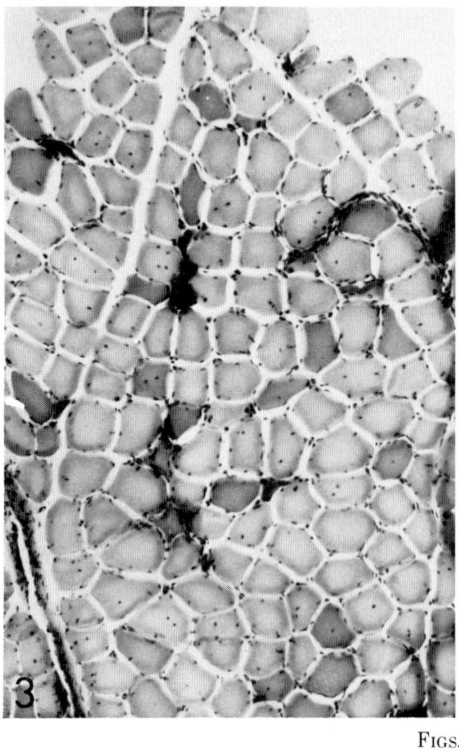


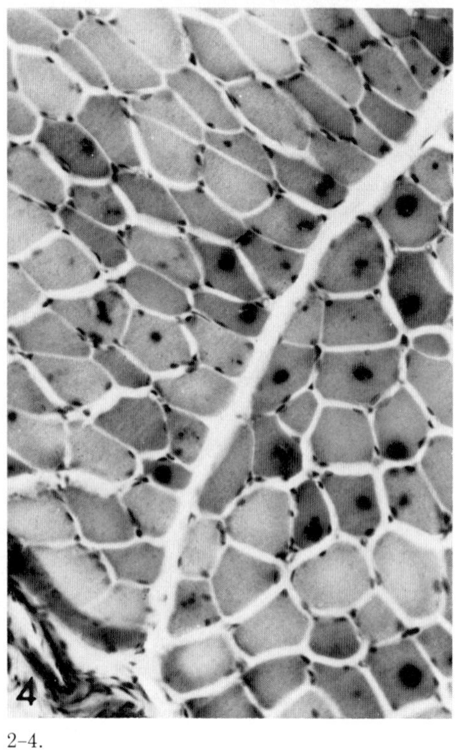


Figs. 2–4.

selective organelle dysfunction changes which predominantly affect one fiber type,[7, 33, 46] and in indicating the presence of collateral reinnervation in fiber grouping.[27] Detailed quantitative estimates of fiber size have been obtained[3] and when plotted as histograms are of value in showing preferential atrophy in a fiber type and the results of therapy.[32]

Should muscle histoenzymology represent a defined specialty existing in isolation or should the techniques and their interpretation be fully dependent upon strict biochemical alignment? We belong to the second school of thought in which a direct relationship between enzyme histochemistry and correlative enzyme microchemistry is ever present. For instance, we must ask the question whether the activity demonstrated histochemically does in fact indicate the site or activity of a known biochemical reaction? As an example of a possible pitfall in this regard, Engel[16] recorded mitochondrial aggregation, due to high reactivity with the NADH-tetrazolium reduction reaction, limited to type II fibers. The aggregates were unstained with succinic dehydrogenase, suggesting mitochondria with an abnormal histoenzymatic complement. Subsequent electron microscopic examination revealed masses of tubules probably arising from the sarcoplasmic reticulum[17] and conferring NADH reduction positivity as a manifestation of microsomal NADH cytochrome *c* reduction activity. Although a selective mitochondrial histoenzymatic "deletion" was not revealed in this example, it is perhaps only a question of time before such a mitochondrial abnormality is uncovered.

TABLE 2. SIGNIFICANT MUSCLE FIBER CHARACTERISTICS

| Features | Type I Fibers | Type II Fibers |
|---|---|---|
| A. Contraction characteristics | Slow contraction, tension sustained | Rapid strong contraction, short duration |
| B. Metabolic characteristics | | |
| Energy source | Oxidative phosphorylation | Glycolysis |
| Metabolites | Fatty acids, glucose | Glycogen |
| C. Enzymatic characteristics | | |
| Mitochondrial enzymes | Rich | Low |
| Cytochrome oxidase | | |
| NADH diaphorase | | |
| β-hydroxybutyric dehydrogenase | | |
| Succinate dehydrogenase | | |
| Krebs cycle enzymes | | |
| Glycolytic enzymes | Low | Rich |
| Phosphorylase | | |
| Lactic dehydrogenase | | |
| Myosin ATPase | Low | Rich |
| D. Isozymes of lactic dehydrogenase | $LD_1$, $LD_2$, $LD_3$ | $LD_4$, $LD_5$ |

FIG. 2. Reciprocal checkerboard staining of myofibrillar ATPase (*A*) and succinic dehydrogenase (*B*) reactions in normal human deltoid muscle; × 75.

FIG. 3. Acid phosphatase reaction of inflammatory cells in the endomysium and necrotic fibers representing lysosome-vacuolar activity. Acute myositis. Gomori; × 75.

FIG. 4. Prominent acid phosphatase reaction product in autophagic vacuoles. α-1,4-Glucosidase deficiency (Pompe's glycogenosis). Gomori acid phosphatase; × 200.

If biochemical existence can be documented, are we demonstrating a single enzyme or a multienzyme complex working as an integrated whole? While very few of these questions can be accurately answered in our studies of muscle histoenzymatic mechanisms, some kinetic studies have revealed that our initial and early attempts are not unwarranted.[23] Quantitative aspects of the relationship between enzyme histochemistry and biochemistry have been approached by making determinations of the $K_m$ of given enzyme systems and their histochemical conditions. The $K_m$ is the concentration of substrate at which the rate of reaction is half that of the maximum rate and, as such, is characteristic for any given enzyme. A second approach is the direct quantitative correlation of enzyme activity in cryostat sliced material by the application of microchemical techniques. Such methods have been developed and refined by Lowry, Linderström-Lang and co-workers, and Pope. In a further extension of histoenzymatic techniques it is to be expected that valuable diagnostic information will be obtained from the zymogram technique of Hunter and Markert.[25] Using gel electrophoresis we have been able to separate particular enzyme proteins of muscle biopsies into a number of different bands. When histochemical enzyme reactions are applied to the bands some bands react strongly with most of the substrates while others only respond to a single substrate. Hence, the isoenzyme pattern of the muscle may be discerned. An alternative to the latter method is to use, under selected histochemical conditions, specific activators and inhibitors of the individual isoenzyme subfractions.[2] In this way we may possibly approach the ideal conditions where enzymes may be identified biochemically and histochemically. Apart from these considerations, a comparison of techniques in enzyme localization in muscle biopsies (Table 3) reveals that microscopic histochemistry may offer good tissue localization with varying degrees of false localization and some degree of detection of cytochondrial heterogeneity. Unfortunately, the degree of quantitation and the application of kinetic studies is limited.

We have discussed briefly the value of histoenzymatic methods to the investigation of muscle disorders. What are the possible mechanisms whereby

TABLE 3. COMPARISON OF TECHNIQUES IN ENZYME LOCALIZATION

| Characteristic | Microscopical Histochemistry | Quantitative Microchemistry | Differential Centrifugation |
|---|---|---|---|
| Ease of performance | Simple | Highly specialized | Specialized |
| Degree of quantitation | Poor | Good | Good |
| Biochemical techniques available | Limited | High | High |
| Fixation | Fixation and freeze-dry | Freeze-dry techniques | None |
| Spatial distribution | | | |
| Tissue | Excellent | Excellent | Poor |
| Cell | Good | | Cytochondrial separation possible |
| False localization with enzyme loss | Moderate | Minimal | Moderate |
| Application of kinetic studies | Very limited | Limited | Good |
| Detection of cytochondrial heterogeneity | Yes | No | No |

this may be performed? The vast majority of histoenzymatic procedures applied to fresh cryostat sections utilize a hydrolytic reaction system or some form of solubility change in the reagent following electron capture. Figure 5 summarizes the possible methodologies. In a hydrolytic reaction sequence whereby substrate AB is transformed into products A + B, a histoenzymatic demonstration demands the capture and localization of one of the products at the site of enzyme reaction by substantive and insolubility changes. In the most common form of reaction involving simultaneous capture of a product, three possible mechanisms may be invoked (Fig. 5). In investigations of the esterases and carboxylases, the cleavage of an indoxyl-linked substrate liberates leucoindigo, which on oxidation forms the brightly colored indigo.[22] In the early Gomori metal-salt precipitation techniques[21] for the demonstration of pyrophosphatases and phosphomonoesterases, the liberated phosphate is captured by lead, precipitated at the site of capture, and transformed into the black, easily visualized lead sulfide. Thirdly, naphthyl substrates liberate free naphthol on hydrolysis, which is capable of reacting with any number of chemically modified diazonium salts to form the highly colored diazo dyes.[28] In a modification of the simultaneous capture reaction the diazonium salt may be added *after* substrate incubation. Such a procedure, referred to as postincubation diazonium coupling,[37] allows for the reaction sequence to occur in the absence of possible inhibition by the diazonium salt.

Dehydrogenase histochemistry received great impetus from the finding that tetrazolium salts, on reduction, formed highly colored insoluble, substantive formazan compounds (see Pearse[29] for review). Moreover, tetrazolium salts can be prepared with high redox potentials, allowing the direct transfer of electrons to the salt. With the mitochondrial dehydrogenases such a reaction sequence allows for the short circuiting of the cytochromes (Fig. 5) and demonstration of individual dehydrogenases in the presence of their specific substrates. A note of

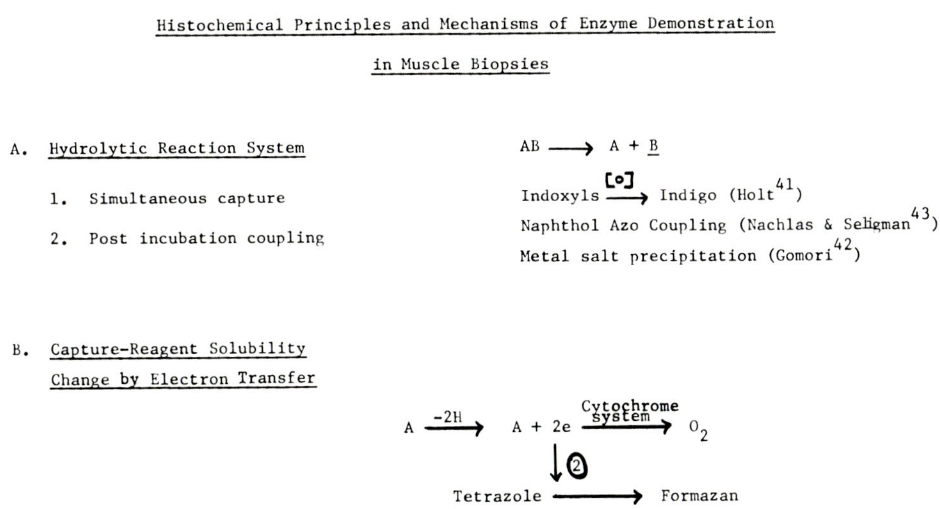


FIG. 5. Basic principles used in the histoenzymatic demonstration of selected enzyme reactions in muscle biopsies.

caution must be interjected. While the mitochondrial NADH diaphorase reaction may be used during the preliminary screening of muscle disorders, other mitochondrial-specific histoenzymatic procedures must be utilized to evaluate any further abnormality in mitochondrial topography or reactivity (Figs. 6 and 7). "Apparent" mitochondrial hyperreactivity has been described on the basis of changes in NADH dehydrogenase, which was more likely due to NADH cytochrome *c* reductase of microsomal (sarcotubular) origin.[16, 17] Succinic dehydrogenase, cytochrome oxidase, or menadione-linked $\alpha$-glycerophosphate dehydrogenase would help clarify the mitochondrial reactivity profile.

## EVALUATION OF MITOCHONDRIAL HISTOENZYMATIC ACTIVITY IN NEUROMUSCULAR DISEASE

Knowledge of the pathologic reactions of subcellular organelles is vital to a further understanding of the primary and secondary disease processes occurring in muscle. The recognized association of oxidative enzyme reaction with mitochondria allows for preliminary data to be obtained on the reactivity of these organelles. Disturbances in the histoenzymatic architecture will highlight specific neuromuscular disorders, allow for some identification of the disturbance in mitochondrial reactivity, and suggest the application of more specific biochemical investigations to elucidate the particular myopathy.[31] Table 4 attempts to classify the changes in mitochondrial reaction product evidenced topographically in a variety of neuromuscular disorders. Such a classification is not all inclusive, but defines major subdivisions of topographic reactivity with examples of described disorders in each category. There is most certainly overlap between divisions. For example, mitochondrial aggregation (Table 4, profile 1b) and intrafiber loss (profile 2) occur in central core disease.[34] To date, no unequivocal example of an abnormal cytochemical enzyme profile (profile 5) has been documented, but it is to be expected that experimental alteration in

TABLE 4. TOPOGRAPHIC EVALUATION OF MITOCHONDRIAL OXIDATIVE ENZYME REACTIVITY IN NEUROMUSCULAR DISORDERS

| Profile | Disorder |
|---|---|
| 1. Abnormal intrafiber location | |
| (a) Dislocation | Glycogenoses types II[1, 6] and V,[30, 38] Vitamin E deficiency,[24] nemaline myopathy,[41, 44] vacuolar myopathies,[36] chloroquin,[20, 36] polysaccharide myopathy[26] |
| (b) Aggregation with or without hyperactivity | Type I myopathy,[7, 33, 46] familial myopathy,[8] target fibers,[35] pseudo-Hurler's,[43] central core disease,[26] type II myopathy |
| 2. Intrafiber loss | Emetine,[13] plasmocid,[34] central core disease,[26] target fibers,[35] familial myofibril lysis,[5] congenital myopathy with multifocal degeneration,[14] acute necrotizing myopathy, *e.g.*, polymyositis,[9] alcoholism[12] |
| 3. Abnormal size and shape (with inclusions) | Megaconial myopathy,[40, 42] pleoconial myopathy[42] |
| 4. Disturbance in fiber differentiation | Chronic peripheral neuropathy with denervation, collateral reinnervation, myotubular,[45] delay in postnatal maturation, Duchenne dystrophy, scapuloperoneal myopathy, myopathy with multifocal degeneration[14] |
| 5. Abnormal mitochondrial enzyme profile | ? vitamin or cofactor deficiency |

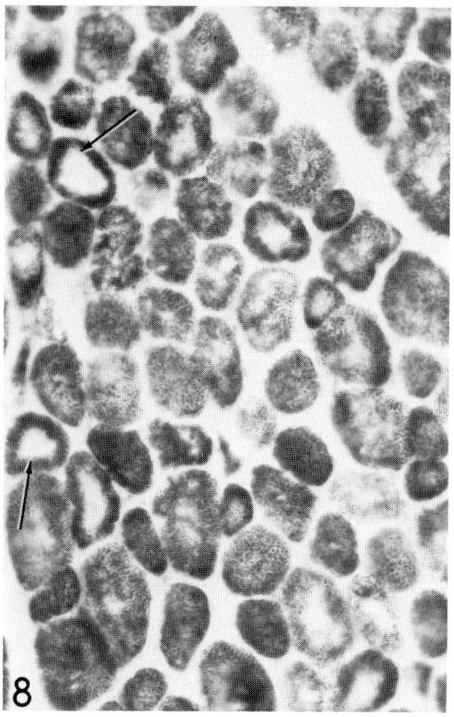

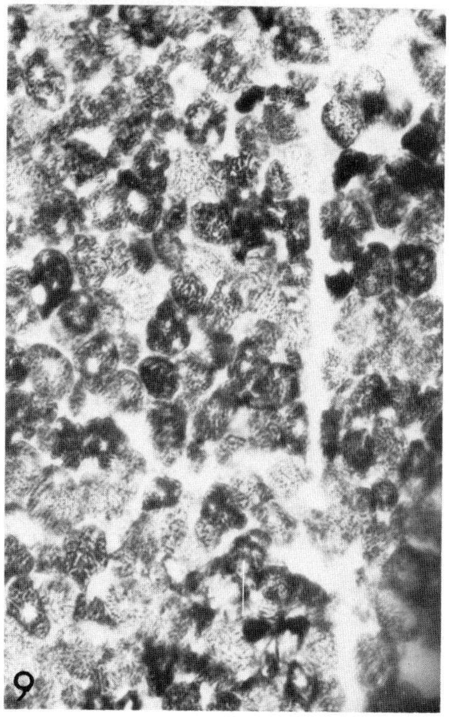

FIG. 6. Normal gastrocnemius muscle. Type I (dark) and type II (light) fibers are clearly seen. Three intermediate fibers may be recognized. NADH dehydrogenase; × 194.

FIG. 7. Comparative section with Figure 6. The punctate (mitochondrial) nature of the reaction product is more easily seen and appears linear in many areas, especially of the intermediate fibers. Succinic dehydrogenase; × 194.

FIG. 8. Small, variably atrophic muscle fibers containing large, membrane lined (*arrows*) autophagic vacuoles. Fiber differentiation is poor. $\alpha$-1,4-Glucosidase deficiency. Succinic dehydrogenase; × 194.

FIG. 9. A 15-month-old girl with cardiomegaly and generalized hypotonia. Excessive glycogen was deposited in muscle fibers. Immature types I and II fibers show vacuolization and dislocation of mitochondrial reaction product to periphery. NADH dehydrogenase; × 194.

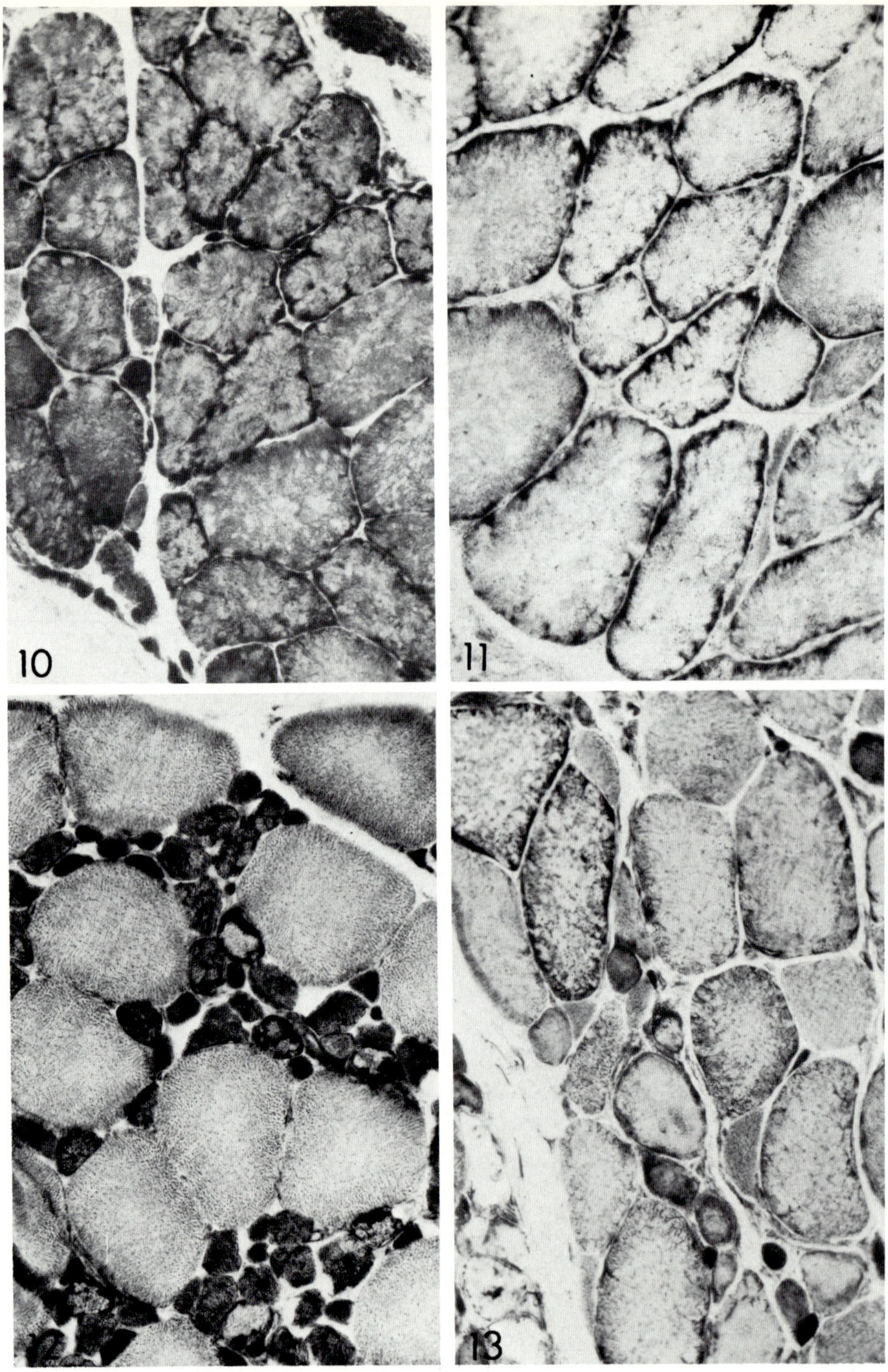


FIG. 10. Nonfamilial, progressive, degenerative myopathy in a 55-year-old woman. Serum enzymes were not elevated. Variable fiber atrophy, especially in type I population, with subsarcolemmal accumulation of reaction product. NADH dehydrogenase; × 194.

FIG. 11. Same as Figure 10, but revealing the "scalloped" nature of the subsarcolemmal reaction especially in the smaller fibers. This is an example of a mitochondrial myopathy with aggregation. Succinic dehydrogenase; × 194.

FIG. 12. Familial myopathy[5] with selective myofibril dissolution in type I fibers. There is marked atrophy of type I fibers, aggregation, and focal sarcolemmal loss of reaction product. NADH dehydrogenase; × 194.

FIG. 13. Duchenne muscular dystrophy. Types I and II fibers are quite variable in size and shape with prominent subsarcolemmal or reticular reaction patterns. NADH dehydrogenase; × 194.

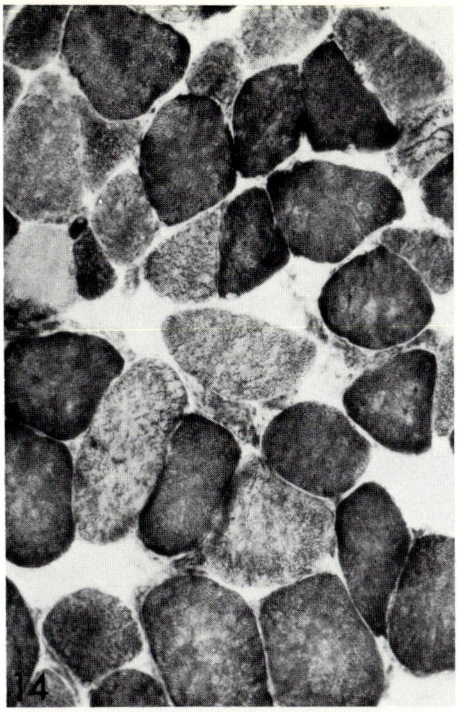

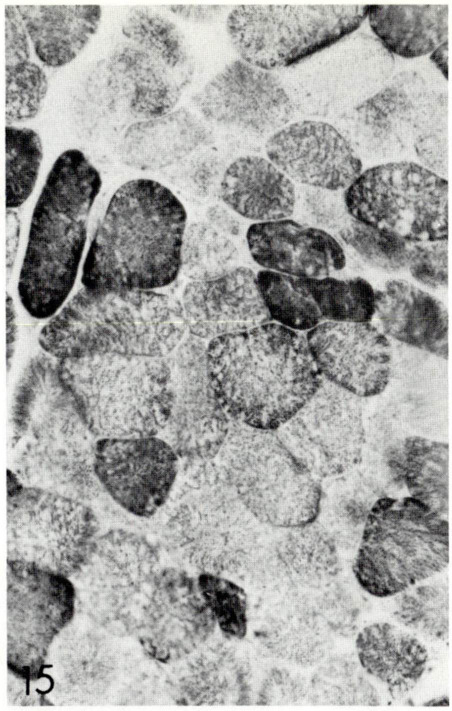

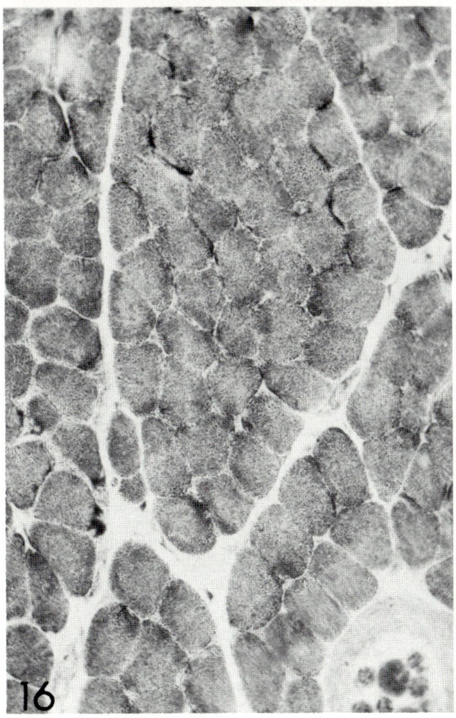

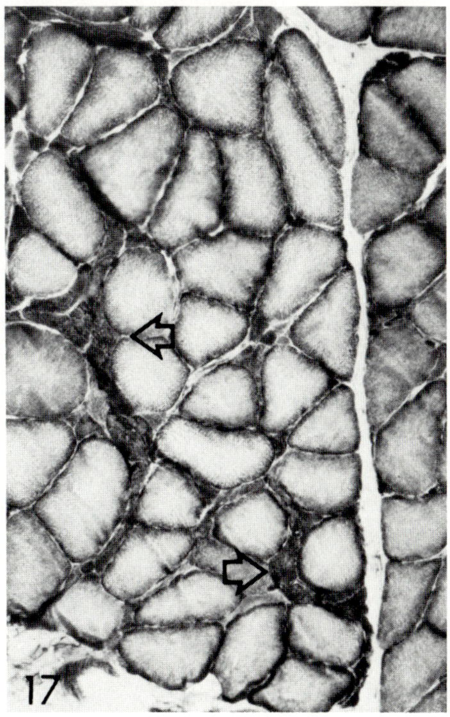

FIG. 14. Chronic polymyositis. Long standing corticosteroid therapy. There is fiber atrophy especially noted in the type II fiber population. Individual fibers are separated but retain a polyhedral configuration. NADH dehydrogenase; × 194.

FIG. 15. Acute polymyositis with necrosis and focal regeneration. Fibers are smaller, rounded, and present a "moth eaten" appearance characteristic of the acute necrotizing myopathies. Succinic dehydrogenase; × 194.

FIG. 16. Scapuloperoneal myopathy. Portions of five fascicles revealing "sameness" of fibers (probably type I). Note the intrafusal fibers at *bottom right*. Succinic dehydrogenase; × 194.

FIG. 17. Grouped fiber atrophy, selectively of type I fibers. Neurogenic atrophy. NADH dehydrogenase; × 194.

mitochondrial membrane biogenesis induced secondarily to vitamin deficiency may be accompanied by deficiency in a selected dehydrogenase activity. Figures 8 to 17 demonstrate some topographic features of mitochondrial reactivity seen in some common neuromuscular disorders.

## ACKNOWLEDGMENTS

We gratefully acknowledge the technical assistance of Margaret Hall and Laurel Reed. The muscle biopsies were obtained by Dr. Fred Marx.

### REFERENCES

1. Baudhuin, P., Hers, H. G., and Loeb, H.: An electron microscopic and biochemical study of type II glycogenosis. *Lab. Invest. 13:* 1139–1152, 1964.
2. Brody, I. A., and Engel, W. K.: Isozyme histochemistry; the display of selective lactate dehydrogenase isozymes in sections of skeletal muscle. *J. Histochem. Cytochem. 12:* 687–695, 1964.
3. Brooke, M. H., and Engel, W. K.: Histographic analysis of human muscle biopsies with regard to fiber type. *Neurology 19:* 221–233, 1969.
4. Cancilla, P. A.: Techniques of muscle biopsy and staining methods with particular emphasis on commonly encountered artifacts. In *The Striated Muscle*, ed. by Pearson, C. M. and Mostofi, F. K. Baltimore, Williams and Wilkins, 1973.
5. Cancilla, P. A., Kalyanaraman, K., Verity, M. A., Munsat, T., and Pearson, C. M.: Familial myopathy with disruption of myofibrils in type I fibers. *Neurology 21:* 580–585, 1971.
6. Cardiff, R. D.: Histochemical and electron microscopic study of skeletal muscle in a case of Pompe's disease. *Pediatrics 37:* 249–259, 1966.
7. Coleman, R. F., Nienhuis, A. W., Brown, W. J., Munsat, T. L., and Pearson, C. M.: New myopathy with mitochondrial enzyme hyperactivity. *J.A.M.A. 199:* 118–124, 1967.
8. D'Agostino, A. N., Ziter, F. A., Rallison, M. L., and Bray, P. F.: Familial myopathy with abnormal muscle mitochondria. *Arch. Neurol. 18:* 388–401, 1968.
9. Denny-Brown, D.: The nature of polymyositis and related muscular diseases. *Trans. Coll. Physicians Philadelphia 28:* 14–29, 1960.
10. Dubowitz, V.: Enzyme histochemistry in the study of normal and diseased muscle. In *Methods and Achievements in Experimental Pathology*, ed. by Bajusz, E., vol. 2, pp. 133–137. Basel and New York, S. Karger, 1967.
11. Dubowitz, V., and Pearse, A. G. E.: A comparative histochemical study of oxidative enzyme and phosphorylase activity in skeletal muscle. *Histochemie 2:* 105–117, 1960.
12. Ekbom, K., Hed, R., Kirstein, L., and Aström, K. E.: Muscular afflictions in chronic alcoholism. *Arch. Neurol. 10:* 449–461, 1964.
13. Engel, A. G.: Ultrastructural reactions in muscle disease. *Med. Clin. North Am. 52:* 909–931, 1968.
14. Engel, A. G., and Gomez, M. R.: Congenital myopathy associated with multifocal degeneration of muscle fibers. *Trans. Am. Neurol. Assoc. 91:* 222–223, 1966.
15. Engel, W. K.: The essentiality of histo- and cytochemical studies of skeletal muscle in the investigation of neuromuscular disease. *Neurology 12:* 778–784, 1962.
16. Engel, W. K.: Mitochondrial aggregates in muscle disease. *J. Histochem. Cytochem. 12:* 46–48, 1964.
17. Engel, W. K.: Selective and nonselective susceptibility of muscle fiber types. *Arch. Neurol. 22:* 97–117, 1970.
18. Engel, W. K., and Karpati, G.: Impaired skeletal muscle maturation following neonatal neurectomy. *Dev. Biol. 17:* 713–723, 1968.
19. Fenichel, G. M., and Engel, W. K.: Histochemistry of muscle in infantile muscular atrophy. *Neurology. 13:* 1059–1067, 1963.
20. Garcia, R., Rondot, P., and Fardeau, M.: Sur les accidents neuromusculaires et en particulier sur une myopathie vasculaire observés au cours d'une traitement prolongé par la chloroquine. *Rev. Neurol. 111:* 177–195, 1964.
21. Gomori, G.: A new histochemical test for glycogen and mucin. *Am. J. Clin. Pathol. 16:* 177–179, 1946.

22. Holt, S. J.: A new approach to the cytochemical localization of enzymes. *Proc. R. Soc. Lond. (Biol.) 142:* 160, 1954.
23. Hopsu, V. K., and McMillan, P. J.: Quantitative characterization of a histochemical enzyme system. *J. Histochem. Cytochem. 12:* 315–324, 1964.
24. Howes, E. L., Price, H. M., and Blumberg, J. M.: Effect of diet producing lipochrome pigment on ultrastructure of skeletal muscle in rat. *Am. J. Pathol. 45:* 599–631, 1964.
25. Hunter, R. L., and Markert, C. L.: Histochemical demonstration of enzymes separated by zone electrophoresis in starch gel. *Science 125:* 1294, 1957.
26. Karpati, G., Carpenter, S., and Wolfe, E.: A peculiar type of polysaccharide accumulation in muscle. *J. Neuropathol. Exp. Neurol. 28:* 157, 1968.
27. Karpati, G., and Engel, W. K.: Type grouping in skeletal muscles after experimental reinnervation. *Neurology 18:* 447–455, 1968.
28. Nachlas, M. M., and Seligman, A. M.: The comparative distribution of esterase in the tissues of five mammals by a histochemical technique. *Anat. Rec. 105:* 677–687, 1949.
29. Pearse, A. G. E.: *Histochemistry. Theoretical and Applied*, 2nd Ed. London, J. & A. Churchill, Ltd., 1960.
30. Pearson, C. M., Rimer, D. G., and Mommaerts, W. F. H.: A metabolic myopathy due to absence of muscle phosphorylase. *Am. J. Med. 30:* 502–517, 1961.
31. Peter, J. B.: Skeletal muscle; diversity and mutability of its histochemical, electron microscopic, biochemical, and physiologic properties. In *The Striated Muscle*, ed. by Pearson, C. M., and Mostofi, F. K. Baltimore, The Williams and Wilkins Co., 1973.
32. Pleasure, D. E., Walsh, G. O., and Engel, W. K.: Atrophy of skeletal muscle in patients with Cushing's syndrome. *Arch. Neurol. 22:* 118–125, 1970.
33. Price, H. M., Gordon, G. B., Munsat, T. L., and Pearson, C. M.: Myopathy with atypical mitochondria in human type I skeletal muscle fibers. *J. Neuropathol. Exp. Neurol. 26:* 475–497, 1967.
34. Price, H. M., Pease, D. C., and Pearson, C. M.: Selective actin filament and Z-band degeneration induced by plasmocid. An electron microscopic study. *Lab. Invest. 11:* 549–592, 1962.
35. Resnick, J. S., and Engel, W. K.: Target fibers; structural and cytochemical aspects and their relationship to neurogenic muscular disease. In *Exploratory Concepts in Muscular Dystrophy and Related Disorders*, ed. by Milhorat, A. T., pp. 255–266. New York, Excerpta Medica Foundation, 1967.
36. Rewcastle, N. B., and Humphrey, J. G.: Vacuolar myopathy. Clinical histochemical and microscopic study. *Arch. Neurol. 12:* 570–582, 1965.
37. Rutenberg, A. M., and Seligman, A. M.: The histochemical demonstration of acid phosphatase by a post-incubation coupling technique. *J. Histochem. Cytochem. 3:* 455–469, 1955.
38. Schotland, D. L., Spiro, D., Rowland, L. P., and Cormel, P.: Ultrastructural studies of muscle in McArdles' disease (deficiency of muscle phosphorylase). *J. Neuropathol. Exp. Neurol. 24:* 629–644, 1965.
39. Shafiq, S. A.: Fine structure of fiber types in normal human muscle. *Anat. Rec. 156:* 283–302, 1966.
40. Shafiq, S. A., Milhorat, A. T., and Gorycki, M. A.: Giant mitochondria in human muscle with inclusions. *Arch. Neurol. 17:* 666–671, 1967.
41. Shy, M. G., Engel, W. K., Somers, J. E., and Wanko, T.: Nemaline myopathy. A new congenital myopathy. *Brain 86:* 793–810, 1963.
42. Shy, G. M., Gonatas, N. K., and Perez, M.: Two childhood myopathies with abnormal mitochondria. *Brain 89:* 133–158, 1966.
43. Shy, G. M., Silberberg, D. H., Appel, S. H., Mishkin, M. M., and Godfrey, E. H.: Generalized disorder of nervous system, muscle and heart resembling Refsum's disease and Hunter's syndrome. *Am. J. Med. 42:* 163–178, 1967.
44. Spiro, A. J., Prineas, J. W., and Moore, C. L.: New mitochondrial myopathy with salt craving. *Arch. Neurol. 22:* 259–269, 1970.
45. Spiro, A. J., Shy, G. M., and Gonatas, N. K.: Myotubular myopathy. *Arch. Neurol. 14:* 1–14, 1966.
46. Wijngaarden, G. K.: Skeletal muscle disease with abnormal mitochondria. *Brain 90:* 577–592, 1967.

Chapter 4

# Immunochemical and Ultrastructural Studies of Myogenesis*

E. RAWORTH ALLEN

The development of striated muscle, the very early stages in particular, has been an extremely controversial topic. Many unusual and often conflicting descriptions of myogenesis appeared in the literature over the first half of the 20th century. There was, however, a great technical limitation because of the available techniques and/or microscopic resolving power. Many investigators in the field of muscle differentiation waited anxiously for the first electron microscopic study of myogenesis. They shared the opinion that this new technique would solve all of the unknown facets. Its introduction[62] did shed light on many areas, but it far from clarified the confused state.

The general opinion from the first ultrastructural studies[27, 62] of muscle development to the early 1960's was that myofibrils formed by some mysterious condensation and polymerization[18, 62] in the sarcoplasm. Almost every cellular structure or organelle found in muscle has been thought responsible for myofibrillogenesis by one early investigator or another.[9, 17, 41, 65]

Myofibrils were shown to be composed of filaments (myofilaments) in early studies of fragmented muscle.[8, 20] Two distinct classes of filaments (thick and thin) were later isolated and characterized both from adult striated muscle and from purified preparations of the two major muscle proteins.[22, 34] These thick and thin filaments are widely accepted as the proteins myosin (from A bands) and actin (from I bands), respectively. In 1949,[37] antibodies to the two major muscle proteins were prepared. A few years later, immunochemical studies were performed using the antibody tagging method[7] to localize muscle proteins within embryonic striated muscle.[12] A thorough investigation of myogenesis in the chick[32] and then in humans[10] using fluorescent antimyosin followed. After the technique was proven valid for myosin, several investigations followed in human pathologic muscle.[28, 36, 39, 58] Ultrastructural visualization of the muscle antibodies was then shown on adult myofibrils and on isolated, individual thick and thin myofilaments.[52, 53] The antibody staining of myofilaments, as seen in

* This work was supported in part by the Muscular Dystrophy Association and by Grant J-1 from the Health Research Services Foundation.

electron microscopy, was compared and correlated with results obtained with myofibrils in fluorescence microscopy.[50, 51] Of the muscle proteins and muscle protein antibodies studied with immunochemical techniques thus far, only myosin and antimyosin results have been widely accepted and will be emphasized in this chapter. Purified preparations of the remaining muscle proteins and antibodies are presently undergoing considerable criticism, speculation, and/or refinement.

A closer examination of the sequence of events within early myogenesis becomes necessary. It is especially apparent that the true origin of myofilaments, their individual appearances, and their organization into functional sarcomeres are necessary core ingredients to understanding the complete differentiation of striated muscle.

## APPEARANCE OF MYOFILAMENTS

The embryonic chick contains myotomes (muscle masses) which make up the trunk musculature. Myotomes are located within somites, which are arranged in serial pairs from the head to the tip of the tail. There is a definite cephalocaudad progression of development, so that the proximal myotomes are larger and more mature than the distal ones. The controlled use of this somitic muscle has offered a convenient and concise system for the study of temporal aspects in myogenesis.[2, 32, 47, 48, 54] By applying the fluorescence antibody technique[7] with antimyosin as a stain, cross-striated myofibrils (A band fluorescence) were detected in myoblasts within somites of the anterior cervical region of stages 14 and 15[21] chick embryos.[32] Embryos of these stages are found after approximately 50 to 55 hours of incubation. The brachial somites did not contain myofibrils with the same banding patterns until 2 to 8 hours later in stages 16 and 17. Embryos were of stage 18 (66 to 68 hours incubation) before cross-striations were observed in the thoracic region. These same studies,[32] indicated a simultaneous appearance of actin and myosin, as did some ultrastructural studies of developing skeletal muscle from a source other than chick somites (amphibian,[23] insect,[57, 61] chick leg muscle[13]). The possibilities exist that (1) the fluorescence antibody staining technique, although superior to other histologic staining methods for light microscopy,[32] is not sensitive enough to detect a small amount of filamentous protein; (2) somitic muscle and muscle from another source may utilize a different mechanism. This latter possibility is also supported by immunochemical data[49] which showed the protein actin to appear before myosin in chick skeletal muscle and myosin first in cardiac muscle. A later immunochemical investigation indicated an inherent difference in skeletal and cardiac muscle from the dog.[15] Biochemical data from the chick embryo obtained *in vivo*[48] (somitic) and *in vitro*[25] (fore limb) have indicated a previous synthesis of actin to myosin in early muscle development.

Ultrastructural investigations[2, 47, 48] have shown thin, 60 to 70 Å filaments (actin) in the sarcoplasm of very young, somitic muscle cells. These filaments appeared wavy and randomly dispersed. The number of these filaments increased with development, until which time the sarcoplasm was virtually saturated (Fig. 1). Only after there was a large concentration of these thin

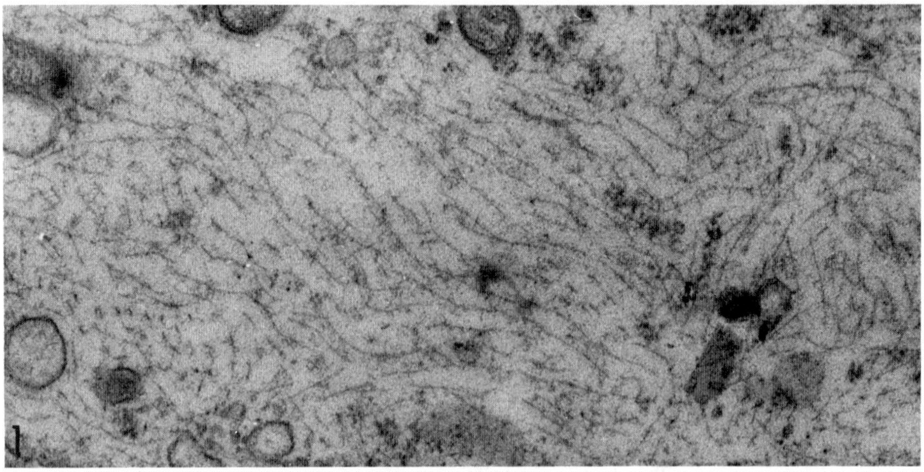

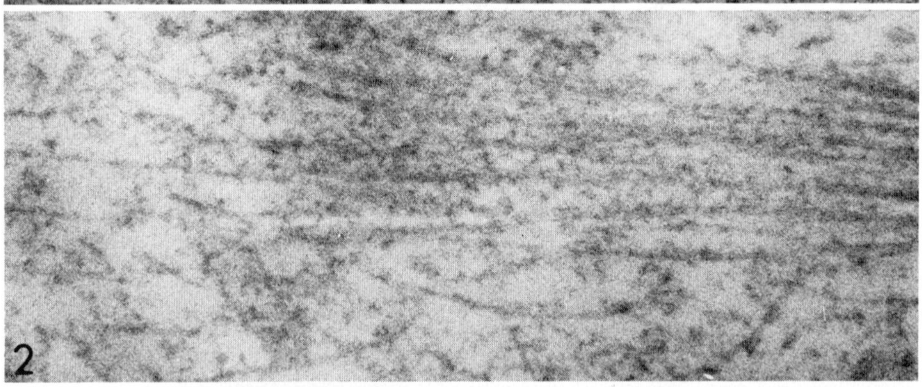

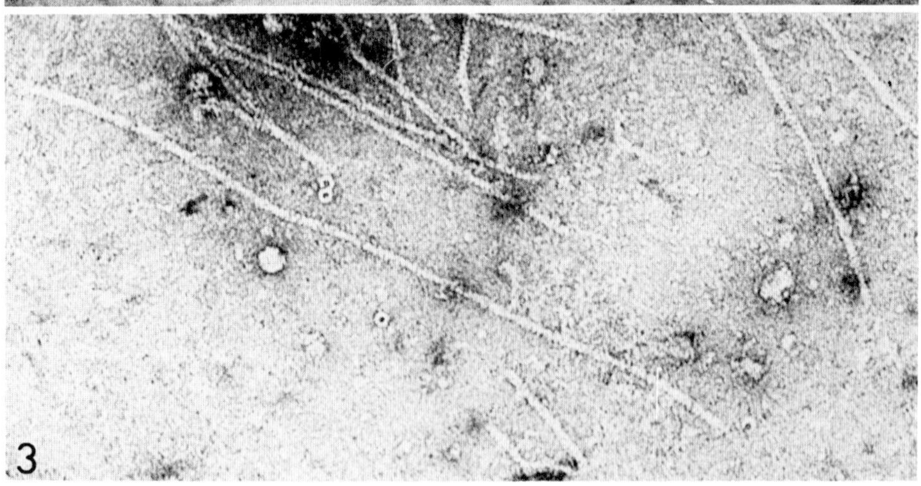

FIG. 1. Electron micrograph of the sarcoplasm of a differentiating myoblast from the upper thoracic myotomes (somites 20 to 25) of a 2-day (stage 16) chick embryo. The early muscle cell is elongate with a sarcoplasm literally filled with randomly oriented and wavy thin myofilaments. No thick (myosin) myofilaments can be detected. Phosphotungstic acid stain; × 30,500.

FIG. 2. Electron micrograph of a differentiating myoblast from the upper thoracic myotomes of a 3-day (stage 18) chick embryo. Thick (myosin) myofilaments are seen, not singly as with the thin myofilaments, but closely associated with other thick and thin myofilaments. Phosphotungstic acid stain; × 144,000.

FIG. 3. Electron micrograph of a negatively stained homogenate of stage 16 embryonic myotomic muscle. The myogenic tissue was obtained from the same numbered contralateral somite as that illustrated in Figure 1. Again, no thick and many thin myofilaments can be seen. Ultrastructural visualization is superior with this technique. Single ribosomes are also visible. 0.1% aqueous uranyl acetate negative stain; × 144,000.

filaments could an occasional thick filament be detected. The thick ones (110 to 160 A) were not seen singly, as in the case of the thin ones, but in close apposition to other thick and thin ones (Fig. 2). They appeared very rigid or stiff, and at high magnification small lateral projections could be seen on their surfaces. Both the thin and the thick filaments appeared identical to those described in sectioned adult striated muscle.[33]

Besides routinely embedded and sectioned techniques for electron microscopy, another technique was developed for the isolation and visualization of myofilaments in adult striated muscle.[34] This method entailed homogenizing the muscle in a relaxing medium and negatively staining[8] the homogenate directly on carbon-coated grids. After a slight solution modification for the use of embryonic tissue, the technique was applied to the same aged embryonic muscle (contralateral somites) as those embedded, sectioned,[2] and illustrated in Figures 1 and 2. As seen in Figure 3, only the thinner filaments were found in this early stage of development. The thin filaments were 55 to 65 Å in diameter and consisted of two tightly wound strands of 50-Å subunits. The strands crossed over one another at approximately 350-Å intervals.[22] In Figure 4 is seen a thick filament isolated and negatively stained from a more advanced myogenic muscle mass (stage 18). These filaments were typically 1.5 $\mu$ in length (A band length), 110 to 120 Å in diameter, and possessed tapered ends. Their surfaces were rough with the exception of a smooth central portion of 0.15 to 0.20 $\mu$ in length. The rough surfaces may correspond to the lateral projections seen on thick filaments in sectioned material. The individual embryonic myofilaments were ultrastructurally identical to those isolated and described by the same technique from adult muscle.[22, 34, 35] Utilizing these two entirely different methods, the thin filaments were seen to appear from 12 to 14 developmental hours (stage 16) prior to the thick filaments (stage 18).[2]

## APPEARANCE OF LARGE POLYRIBOSOMES

Bergman, when studying rat myogenesis,[5] first described in 1962 an association between myofilament differentiation and "free ribosomes." In the same year,[64] "polyribosomes," containing many ribosomes in a chain, were described as being the fabricators of hemoglobin on reticulocytes. Although no significance was given to their presence, "helically arranged ribosomes" were reported in a developing frog striated muscle cell.[63] A year later, long polyribosomes were again associated with myofibrillogenesis in developing rat cardiac muscle.[6] A large polyribosome seen in sectioned embryonic chick muscle is depicted in Figure 5. Approximately 60 to 65 ribosomes can be counted in this polyribosome. Figure 6 shows a large polyribosome in the midst of many thin filaments. This preparation (Fig. 6) was obtained from a negatively stained homogenate of stage 18 embryonic chick muscle. Since these large polyribosomes were necessarily extremely fragile, very gentle homogenization was required in order to maintain their structural integrity. Only a few strokes of a 1-cc syringe (no needle) sufficed. In addition, a hypotonic solution[3] with a salt ratio similar to Huxley's[34] served as the relaxing medium. This solution helped to lyse cells and empty their contents without excessive trauma. The literature soon contained many descriptions of polyribosomes, probably due both to an increase in

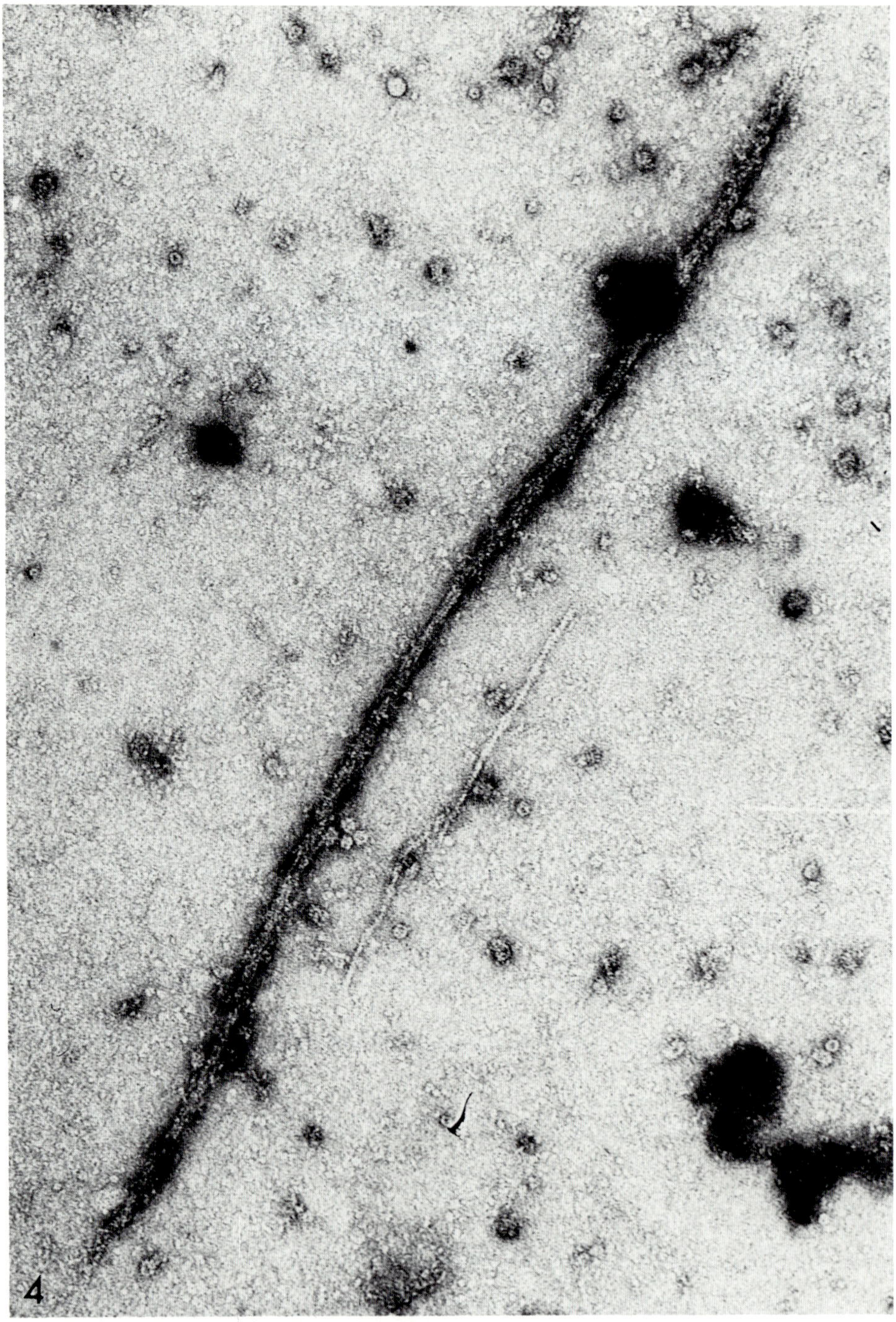


FIG. 4. Electron micrograph of a negatively stained homogenate of stage 18 embryonic somites. This myogenic tissue was obtained from the same numbered contralateral somite as that illustrated in Figure 2. A thick myofilament is depicted beside a broken thin one. Single ribosomes and other small granules are also visible. 0.2% aqueous uranyl acetate negative stain; × 144,000.

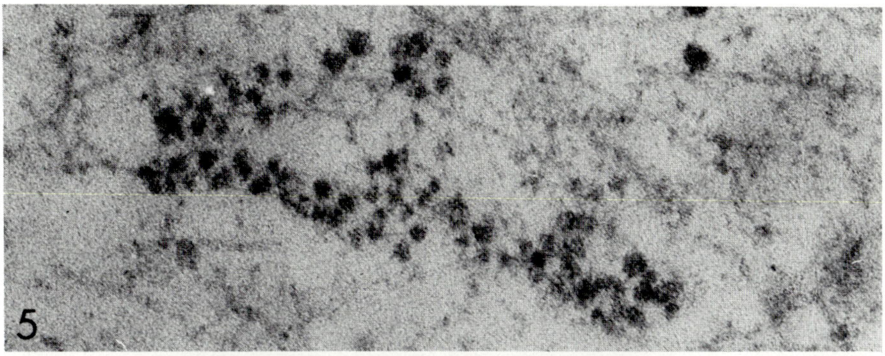

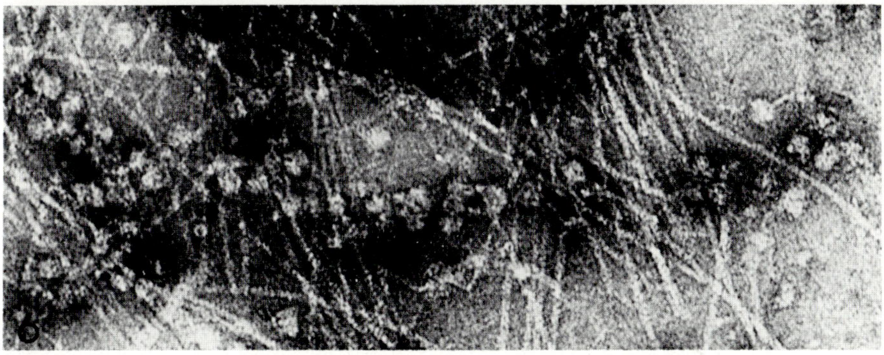

FIG. 5. Electron micrograph of a stage 18 chick embryo somite (upper thoracic) after routine embedding and sectioning. Approximately 60 to 65 ribosomes are seen in the polyribosome. Myofilaments can be seen in the background. Lead tartrate stain; × 144,000.

FIG. 6. Electron micrograph of a negatively stained homogenate from the same aged material as illustrated in Figure 5. A long polyribosome is shown intermingled with a mass of thin myofilaments. 0.1% uranyl acetate negative stain; × 144,000.

awareness of them by investigators and better resolution of the newer electron microscopes. A model for polypeptide synthesis on the polyribosome was proposed.[55] According to this model, the larger the protein synthesized, the larger would be the polyribosome. This was further indicated when long nascent collagen polypeptide chains were isolated in conjunction with long polyribosomes.[44]

## CORRELATION OF LARGE POLYRIBOSOMES AND MYOSIN SYNTHESIS

### MORPHOLOGICAL CORRELATION

A stage in the development of embryonic chick somitic muscle was found (stage 16) when there were thin (actin) and no thick (myosin) filaments present in early myoblasts. Twelve to 14 developmental hours later (stage 18) there occurred both a large concentration of thin filaments and the first appearance of thick ones. In addition, at this same stage, large polyribosomes containing 60 to 65 ribosomes were first detected. The ratio is illustrated in Figure 7, where many randomly dispersed thin filaments, a few linearly arranged thick ones, and a

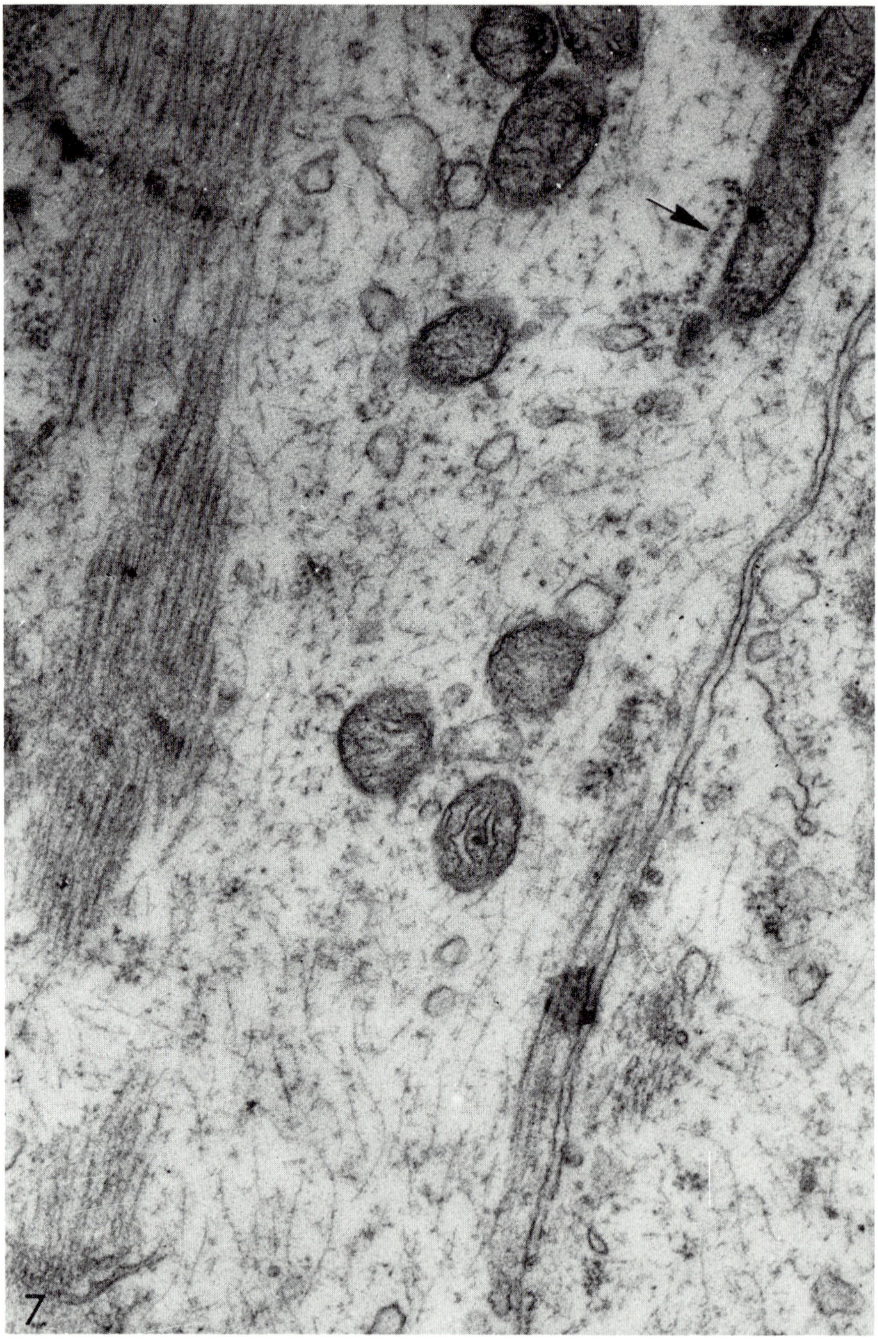


FIG. 7. Electron micrograph of two sectioned myoblasts from upper thoracic somites of a stage 24, 4-day chick embryo. Many thin myofilaments are seen dispersed throughout the sarcoplasm. Groups of thin and thick myofilaments form an early myofibril. One long polyribosome is illustrated (*arrow*). Phosphotungstic acid stain; × 40,000.

large polyribosome are clearly visible. When applying the homogenization negative staining technique to the same numbered somites,[2] again no thick filaments nor long polyribosomes could be found until 12 to 14 hours after thin filaments were seen, then they appeared almost simultaneously.

A class of large polyribosomes, containing 50 to 60 ribosomes, was later isolated from embryonic chick leg muscle.[26] Under ideal synthetic conditions, this same class of large polyribosomes fabricated a protein *in vitro* which acrylamide gel analysis identified as myosin. Estimates for the myosin polypeptide chain size were in good agreement with the physical and chemical measurements of the size of myosin subunits.

### Precipitation experiments

Homogenates were prepared of chick embryo somites located in the same upper thoracic region that were of an age (5 days) known to contain actively synthesizing myosin.[3] The homogenate was briefly centrifuged at 11,000 $\times$ $g$ in order to rid the preparation of heavy and/or unhomogenized material.

A negatively stained preparation of the homogenate was made at this point in order to verify the presence of the large polyribosomes (Fig. 8). This polyribosome had a stretched out appearance rather than the helically arranged one, probably because of a drying down effect on the grid during preparation. After the addition of antimyosin (prepared in rabbits against chicken myosin) to the homogenate, a similar negatively stained preparation was made and the polyribosome illustrated in Figure 9 was seen. The polyribosomes contained, predominantly on one end in many instances, an electron dense material. Assuming the dense material to be antimyosin complexed with myosin, the observed binding sites agreed with the expected location of large nascent polypeptide chains as suggested by Rich's model.[55] The long polypeptide chains, having many available antigenic sites, would be located at one end of the polyribosome. The lengths would gradually taper to shorter polypeptide chains which would show little or no precipitation at the other end of the polyribosome. This was the observed relationship.

*Spectrophotometric analysis.* Small aliquots of the above mentioned homogenate were placed into four separate centrifuge tubes. Antimyosin was added to one tube and equal amounts of antitropomyosin, normal rabbit $\gamma$-globulin, and the original hypotonic standard salt solution were added to the other three. The tubes were centrifuged at 1,700 $\times$ $g$ for 5 minutes. The supernatants of all four samples were read at a wavelength of 260 mu (absorption peak for ribonucleic acid) before and after the addition. Results of the spectrophotometric readings are given in Table 1. The only significant difference in ribonucleic acid (RNA) content was seen in the tube treated with antimyosin. It showed a 30% difference. A pellet was also visible with the naked eye only in this tube. The RNA, which had thus precipitated with the antimyosin, was presumed to be polyribosomal. After negatively staining a sample of the precipitate, a polyribosomal complex similar to the one illustrated in Figure 9 was seen. No precipitations were observed in the other experimental tubes.

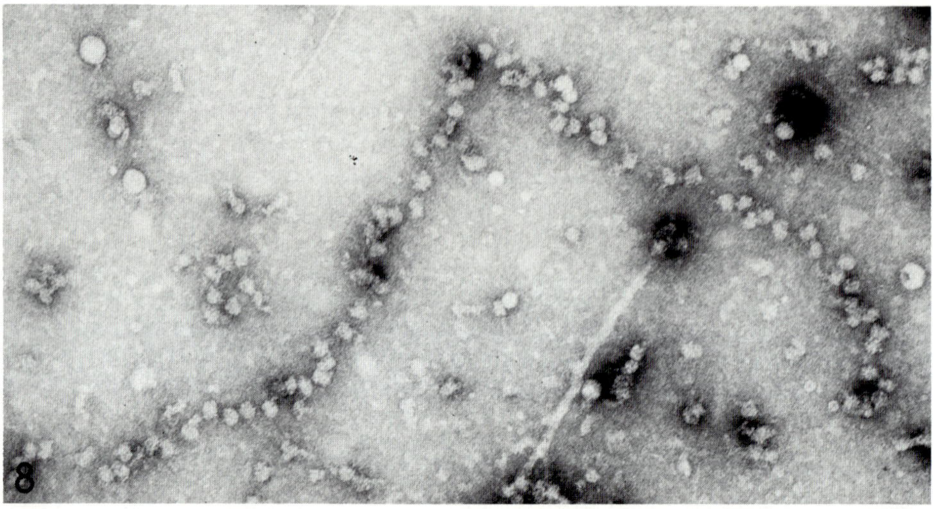

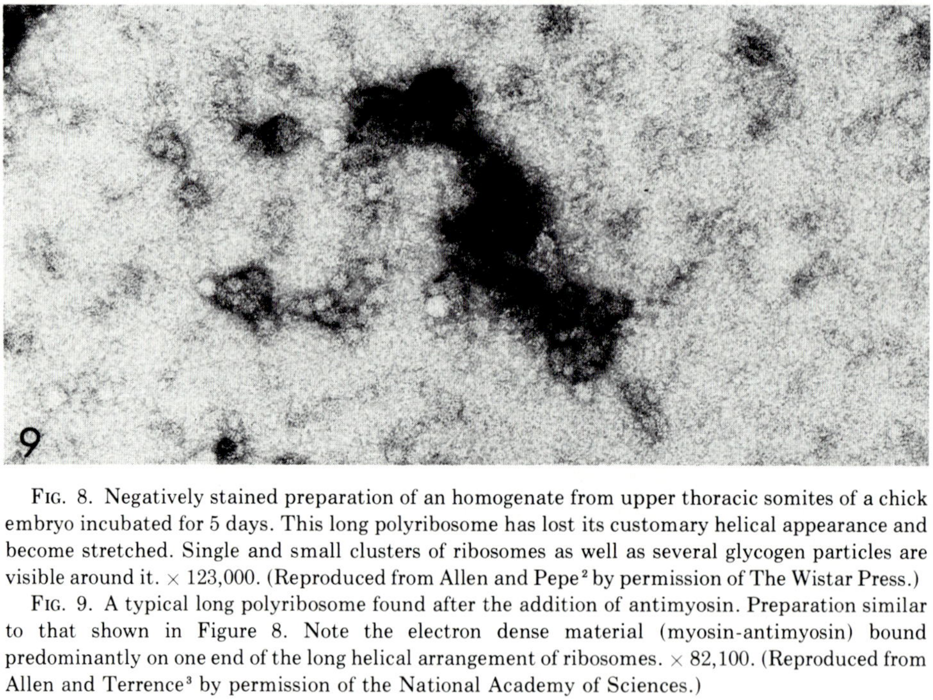

FIG. 8. Negatively stained preparation of an homogenate from upper thoracic somites of a chick embryo incubated for 5 days. This long polyribosome has lost its customary helical appearance and become stretched. Single and small clusters of ribosomes as well as several glycogen particles are visible around it. × 123,000. (Reproduced from Allen and Pepe[2] by permission of The Wistar Press.)

FIG. 9. A typical long polyribosome found after the addition of antimyosin. Preparation similar to that shown in Figure 8. Note the electron dense material (myosin-antimyosin) bound predominantly on one end of the long helical arrangement of ribosomes. × 82,100. (Reproduced from Allen and Terrence[3] by permission of the National Academy of Sciences.)

TABLE 1. SPECTROPHOTOMETRIC ANALYSIS OF THE SUPERNATANT FLUIDS FROM MUSCLE HOMOGENATES

| Substance Added to Homogenate | Optical Density[a] | Optical Density[b] | Difference |
|---|---|---|---|
| Antimyosin | 0.245 | 0.170 | −75 (−30%) |
| Normal γ-globulin | 0.245 | 0.250 | +5 (+2%) |
| Standard salt | 0.245 | 0.230 | −15 (−6%) |
| Antitropomyosin | 0.245 | 0.250 | +5 (+2%) |

[a] Before specified additions and centrifugations.

[b] After specified additions and centrifugations.

*Separation and identification of precipitates.* Antimyosin was tagged with fluorescein isothiocyanate[12] prior to the addition of the antibody to a similar somitic muscle homogenate. The precipitate was then isolated by virtue of its ability to pass through a 40% sucrose solution and form a pellet when ultracentrifuged at 90,000 × *g* for 2 hours.[3] No other combination attempted would precipitate and pass through the sucrose. Homogenates from adult muscle, liver, and yeast would not combine with the antibody and be isolated. Likewise, attempts to combine the embryonic somitic muscle homogenates with other antibodies and solutions (antiactin, antitropomyosin, salt solutions, normal $\gamma$-globulin) were unsuccessful. Figure 10 is an electron micrograph of a diluted and negatively stained preparation from this pellet. The large polyribosomes were completely covered by and associated with a homogenous darkly staining material rather than associated predominantly on one end as seen in the complex prior to ultracentrifugation through a sucrose solution.

Two procedures were designed[3] so as to separate the complex from the polyribosomes.

1. Pancreatic ribonuclease (RNase) was added to one precipitated sample. After a 5-minute incubation at a temperature of 37.5°C, the reaction was stopped by exposure to 2–3°C and was centrifuged for 5 minutes at 1,700 × *g*.

2. Another sample was incubated at 25°C for 5 minutes with the antibiotic puromycin. It has been shown that puromycin, if added during active synthesis, will release nascent polypeptide chains from polyribosomes.[4, 46] Additional adenosine 5-triphosphate and some of the original 90,000 × *g* supernatant fluid was added in order to ensure peptide bond synthesis. This incubation mixture was also centrifuged for 5 minutes at 1,700 × *g*.

Samples from the supernatant fluids and pellets of the two experimental tubes were negatively stained for electron microscopy. A similar unidentifiable mass of electron dense material was seen in the pellets of both tubes (Fig. 11). These pellets were very fluorescent when smeared on glass slides and examined under the fluorescence microscope. Since the antibody was tagged with fluorescein, this indicated the presence of antimyosin. These same pellets also showed a high ATPase activity,[14] a good marker for myosin. This indicated the presence of nascent myosin, since ATPase activity is not inhibited on combination and precipitation with specific antibodies.[56] A negatively stained preparation of the supernatant fluid from the puromycin experiment is illustrated in Figure 12, and after RNase in Figure 13. Polyribosomes, relatively free of the antimyosin-myosin complex were recovered following puromycin exposure. Previous work[26, 38, 44] suggested the observed release of single ribosomes rather than polyribosomes following RNase (Fig. 13). Neither of the supernatant fluids contained significant fluorescence or ATPase activity. Cycloheximide, another antibiotic which inhibits protein synthesis, has been shown not to effect the detachment of nascent polypeptide chains from polyribosomes.[16, 60] The substitution of this antibiotic was a successful control for the separation of the antimyosin-myosin complex from long polyribosomes with puromycin. The precipitates (myosin, antimyosin—long polyribosomes) could neither be separated from one another nor characterized with this antibiotic.

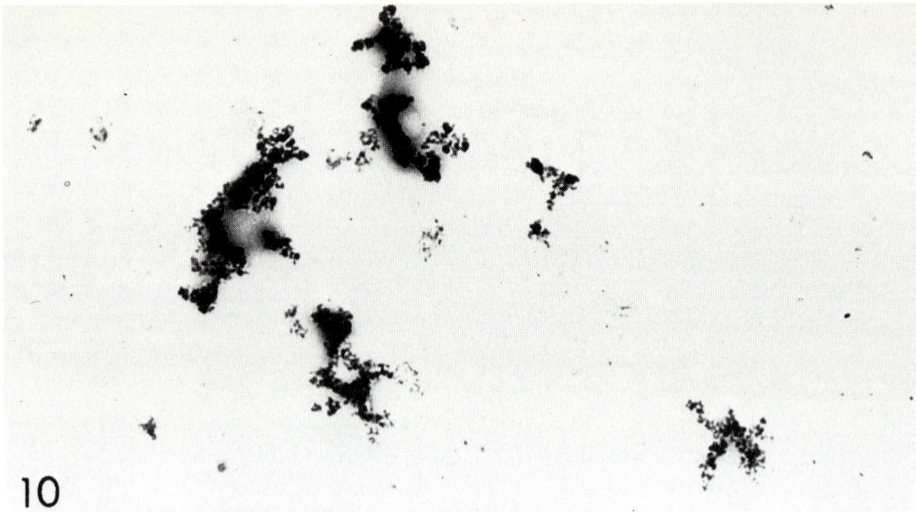

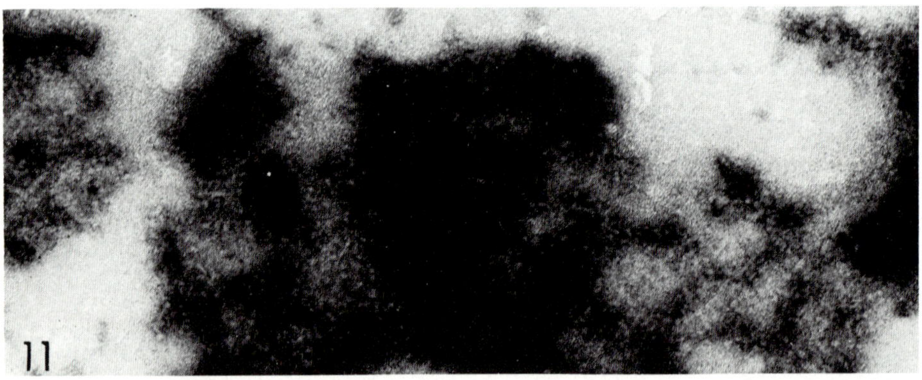

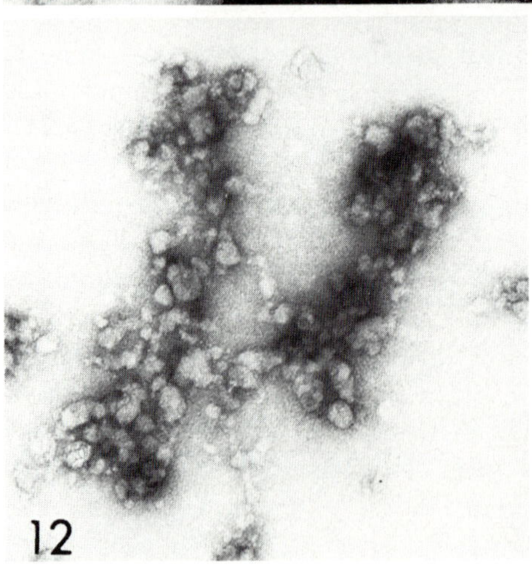

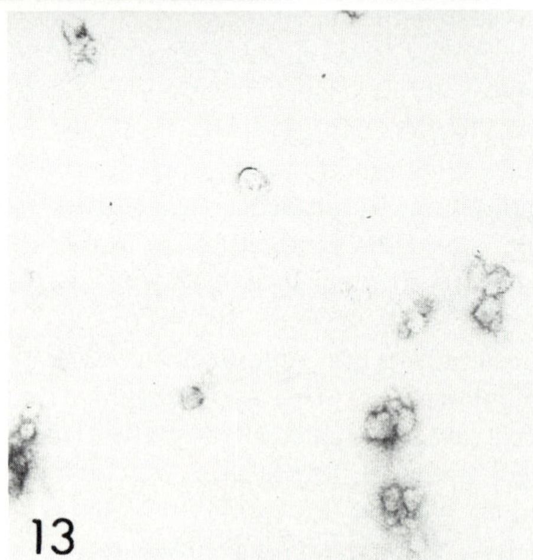

FIG. 10. Low power electron micrograph of a negatively stained homogenate of the pellet after isolating polyribosome-myosin-antimyosin complexes with precipitation and ultracentrifugation through 40% sucrose. Long polyribosomes have a uniform darkly staining appearance. 0.2% aqueous uranyl acetate negative stain; × 12,460.

FIG. 11. Negatively stained sample of pellet separated from polyribosomes following puromycin. An identical electron dense amorphous material resulted after RNase treatment. 0.2% aqueous uranyl acetate negative stain; × 82,100.

FIG. 12. Negatively stained sample of the supernatant fluid following incubation (separation) with puromycin. Two or possibly three polyribosomes are clumped together. × 82,100.

FIG. 13. Negatively stained sample of the supernatant fluid following incubation with RNase. Only single ribosomes are visible after the enzymatic treatment. × 82,100.

## SIMILARITIES OF EMBRYONIC AND ADULT MYOSIN

Thick myofilaments are composed of hundreds of myosin molecules.[35] These filaments have been shown to be ultrastructurally identical, whether artificially synthesized from purified preparations of myosin[35] or isolated from either embryonic[2] or adult skeletal muscle.[34]

After applying the fluorescent antimyosin staining technique,[7] the antibody was found to localize identically within sarcomeres of myofibrils (containing many myofilaments) from both embryonic and adult cardiac and skeletal muscle grown *in vitro*.[31] Both types of muscle contracted when treated with ATP,[30] but neither would contract when previously treated with antimyosin.[12, 29] In addition, neither embryonic nor adult myosin could be salt extracted when pretreated with antimyosin.[29] Nascent myosin, while still attached to the synthesizing polyribosomes in embryonic somitic muscle, also reacted strongly with the same antimyosin as does myosin of mature striated muscle.[3]

Trypsin and papain are two proteolytic enzymes that were found to attack and degrade myosin.[45, 59] Nascent myosin still attached to the polyribosomes was precipitated with antimyosin and then exposed to trypsin, papain, collagenase, and elastase at optimal concentrations, temperatures, and pH levels for proteolytic digestion at various lengths of time.[42, 45] After centrifugation and spectrophotometric examination, only trypsin and papain were observed to have a digestive effect.[1] Both of these proteolytic enzymes attacked myosin so as to inhibit any precipitation with the remaining polyribosomes. The polyribosomes, as shown previously, were left intact after proteolytic digestion.[43] Results are given in Table 2. In the case of trypsin, precipitation and, therefore, no inhibition was noted after 2 and 5 minutes of incubation. However, no precipitation, an indication of polypeptide digestion, was observed after 8 and 10 or more minutes.[1] This agrees with results obtained with trypsinized adult skeletal muscle.[45] Neither elastase nor collagenase had any effect, and precipitation occurred. Subsequent negatively stained preparations served to monitor the experiments in electron microscopy.

## ORGANIZATION OF MYOFILAMENTS INTO MYOFIBRILS

Developing somitic myoblasts contain a large concentration of thin actin filaments before the first thick ones are detected.[2, 47, 48] As seen in Figure 14, when only one thick filament is present, it is immediately surrounded by many thin ones. An isolated thick filament has never been reported in ultrastructural

TABLE 2. OPTICAL DENSITY (260 MU) OF THE SUPERNATANT FLUIDS AFTER PROTEOLYTIC DIGESTION

| Enzyme | Before Centrifugation | After 10 min Centrifugation | Difference |
|---|---|---|---|
| Trypsin | 1.65 | 1.65 | 0% |
| Papain | 0.96 | 0.96 | 0% |
| Collagenase | 0.92 | 0.74 | +20% |
| Elastase | 0.80 | 0.60 | +25% |
| Control[a] | 0.92 | 0.74 | +20% |

[a] Control tube: addition of salt solution only.

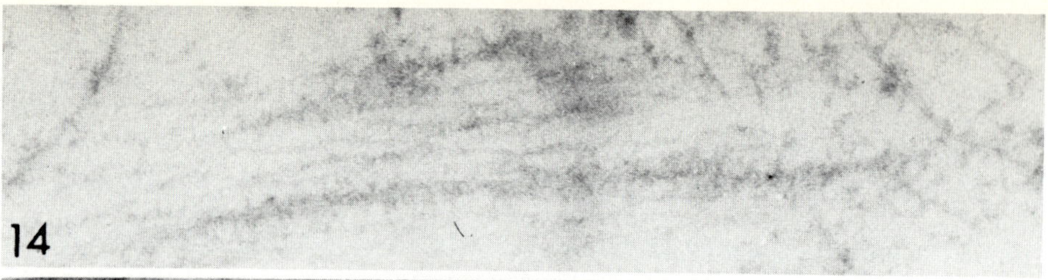


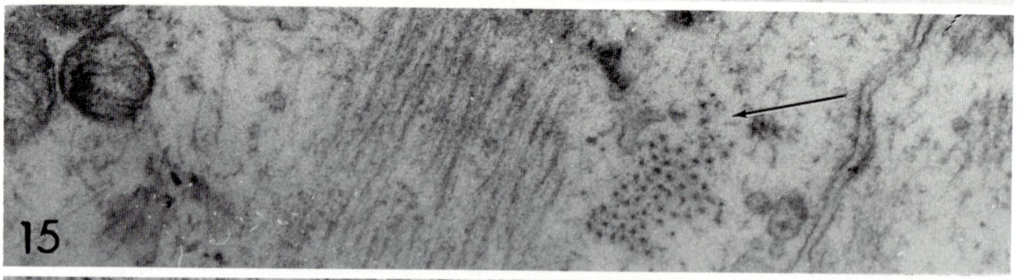


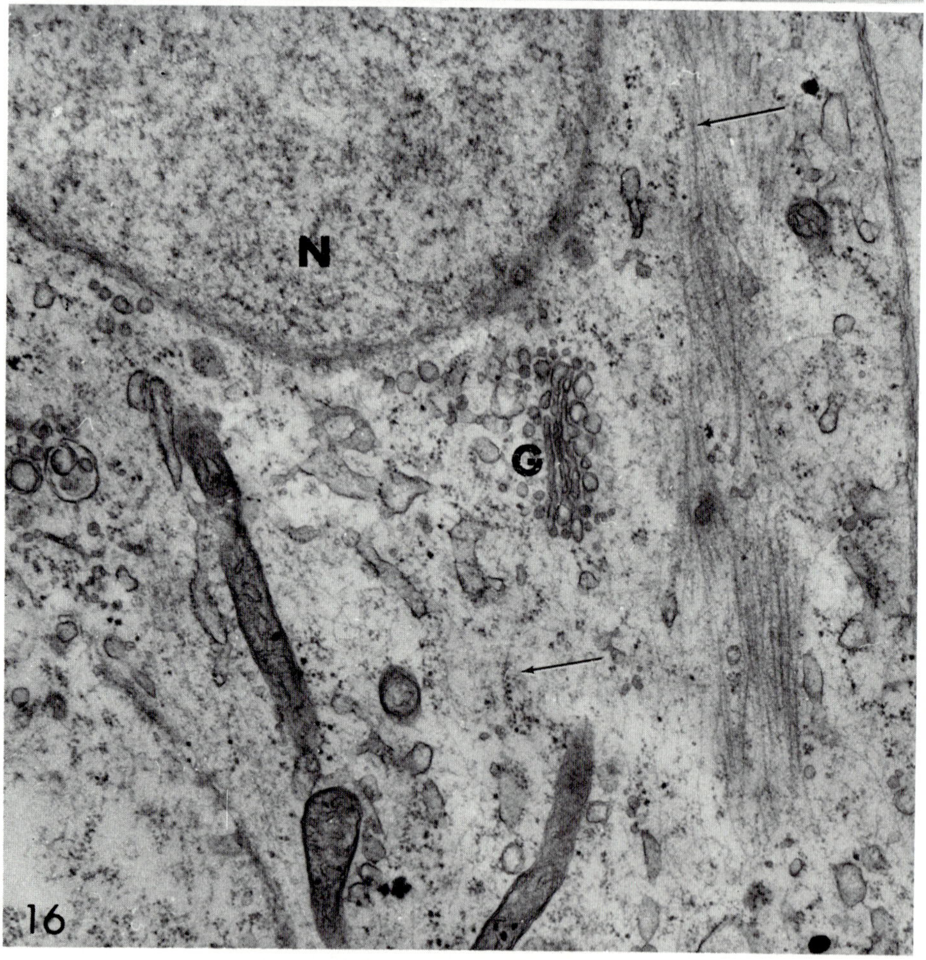


FIG. 14. Electron micrograph of a routinely embedded and sectioned myoblast. Only one thick myofilament is illustrated and it is surrounded (paralleled) by many thin ones. Phosphotungstic acid stain; × 144,000.

FIG. 15. Electron micrograph of an early myoblast showing a cross-section through early formed thick and thin filaments (*arrow*). There are approximately 40 thick filaments present and each is surrounded by a hexagonal array of thin filaments. This pattern is typical of mature functional muscle. Phosphotungstic acid stain; × 50,000.

FIG. 16. Electron micrograph of a myoblast from a stage 20, 3½-day chick embryo. Note the linear aggregation of thick and thin filaments to form a nonstriated myofibril. Polyribosomes (*arrows*) as well as elongate mitochondria are seen in the cytoplasm. Nucleus (*N*), Golgi (*G*). Many thin myofilaments are dispersed throughout the sarcoplasm but are not apparent because of the lead tartrate stain. × 20,750. (Reproduced from Allen and Pepe[2] by permission of The Wistar Press.)

studies of *in situ* myogenic material.[2, 13, 47, 54, 57] Considering that thin filaments interact with the cross bridges of thick filaments,[33] a hexagonal array of thin filaments surrounding the early thick ones would be inevitable. This hexagonal pattern, typical to adult muscle, was seen during very early stages of muscle differentiation,[2, 13] even when only a few thick filaments were present (Fig. 15). It was suggested that myofilament properties leading to aggregation may also be responsible for spatial arrangements.[13] As development proceeds, more thick and thin filaments come together in this hexagonal pattern and then, by a yet unknown mechanism, the myofilament packets become linearly arranged (Fig. 16). This linear arrangement forms a nonstriated myofibril. Membranous tubules can be detected at regular 1.5-$\mu$ intervals (A band size) along the nonstriated myofibril.[2] An invagination of the sarcolemma to form transverse tubules of the T system was recently demonstrated in fish extraocular muscle.[19] The tubules were shown to be in direct continuity both between individual myofibrils and with the sarcolemma.[1, 19] Since early myofibrils are often found near the sarcolemma,[1, 13, 32, 39, 57] this membranous continuity, or some modification thereof,[24] may be instrumental in developing myofibril allignment.[11] These tubules are illustrated in Figure 17. The period in which a myofibril is nonstriated during its assembly is very brief.[2, 10, 29] Typical Z lines then appear on or about the periodic tubules.[1, 2] Figure 18 illustrates a newly formed sarcomere (two Z lines) just prior to obtaining striations. Only after these Z lines are present can the normal banding pattern be detected in early formed myofibrils. Considering the widely accepted sliding filament model proposed for striated muscle contraction,[33] only after Z lines are present to afford attachment to the thin filaments[40] can a typical I band be formed. The thin filaments would slide toward the Z lines at either side of the A band, and an I band and H zone would be passively formed. Myofilament assembly of this type suggests the formation of cross-striations (banding patterns) just prior to the myofibril's ability to contract. This sequence is in agreement with fluorescence microscopy findings in both embryonic skeletal muscle[30] and in embryonic cardiac muscle.[31] In addition, results from fluorescence microscopy have indicated that new sarcomeres appear to be added meristematically at the ends of forming myofibrils, and that these new sections do not contain cross-striations.[29] Lengths of aligned thick and thin myofilament aggregates (A bands) interspersed periodically with tubules would give the early nonstriated appearance described in both fluorescent[29] and electron microscopy.[2]

## SUMMARY

Ultrastructural investigations have shown thin (actin) myofilaments to appear in developing chick somitic muscle 12 to 14 hours prior to thick (myosin) myofilaments. Simultaneously with the first appearance of myosin filaments, long polyribosomes, containing 60 to 65 helically arranged ribosomes, could be detected. They were morphologically correlated with myosin synthesis.

Immunochemical investigations combined with electron microscopic methods demonstrated the long polyribosomes to contain intact antigenic nascent myosin. The polyribosomal-myosin complex could be precipitated and isolated

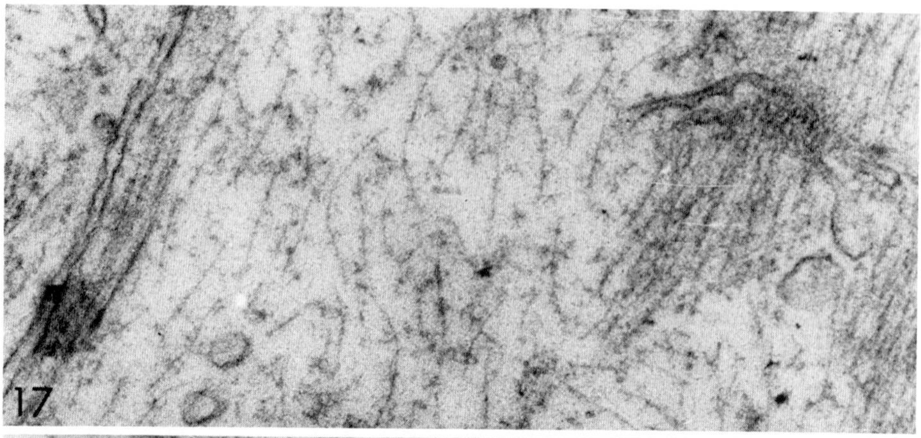


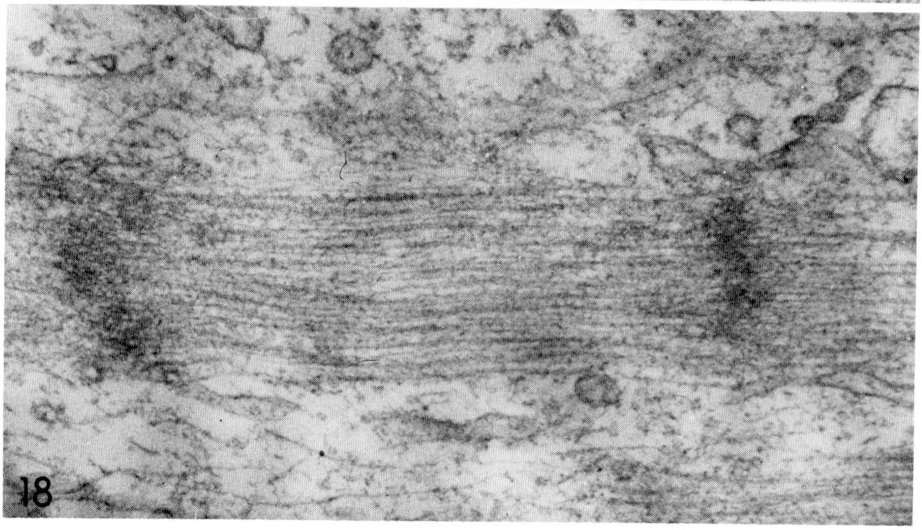


FIG. 17. Electron micrograph of a somatic myoblast from a stage 24 4-day chick embryo. Membranous tubules are seen surrounding the thick and thin myofilament aggregates at 1.5-$\mu$ intervals. Thin myofilaments are again randomly dispersed throughout the cytoplasm. These tubules appear to be instrumental in myofibril alignment. Phosphotungstic acid stain; × 50,000.

FIG. 18. Electron micrograph of a single sarcomere. Darkly staining material (Z lines) has appeared over or associated with the periodic tubules. This stage is found just prior to the visualization of normal banding patterns within the sarcomere. Phosphotungstic acid stain; × 69,000.

with antimyosin. Each of the three components was then separated and characterized.

Similarities between embryonic and mature myosin as well as embryonic and adult striated muscle were discussed. A possible method for the organization of newly synthesized myofilaments into functional striated myofibrils was presented.

## ACKNOWLEDGMENTS

Dr. Frank Pepe contributed the antimyosin and antitropomyosin used in these studies. Valuable assistance was given in the laboratory by Mr. Christopher Terrence and in the preparation of photomicrographs by Mr. Charles Abbott. Perseverance was provided through Christianity.

REFERENCES

1. Allen, E. R.: Unpublished observations, 1969.
2. Allen, E. R., and Pepe, F. A.: Ultrastructure of developing muscle cells in the chick embryo. *Am. J. Anat. 116:* 115–147, 1965.
3. Allen, E. R., and Terrence, C. F.: Immunochemical and ultrastructural studies of myosin synthesis. *Proc. Natl. Acad. Sci. U.S.A. 60:* 1209–1215, 1968.
4. Allen, D. W., and Zamecnik, P. C.: The effect of puromycin on rabbit reticulocyte ribosomes. *Biochim. Biophys. Acta 555:* 865–874, 1962.
5. Bergman, R. A. Observations on the morphogenesis of rat skeletal muscle. *Bull. John Hopkins Hosp. 110:* 187–201, 1962.
6. Cedergren, B., and Harary, I.: *In vitro* studies on single beating rat heart cells. *J. Ultrastruct. Res. 11:* 428–442, 1964.
7. Coons, A. H., Creech, H. J., and Jones, R. N.: Immunological properties of an antibody containing a fluorescent group. *Proc. Soc. Exp. Biol. Med. 47:* 200–202, 1941.
8. Draper, M. H., and Hodge, A. J.: Studies on muscle with the electron microscope. *Aust. J. Exp. Biol. Med. Sci. 27:* 465–484, 1949.
9. Duesberg, J.: Les chondriosomes des cellules embryonaires du poulet et leur role dans la geneses des myofibrelles avec guelgues observations sur le development des musculaires shises. *Arch. Zellforsch. 4:* 602–671, 1910.
10. Engle, W. K., and Horvath, B.: Myofibril formation in cultured skeletal muscle cells studied with antimyosin fluorescent antibody. *J. Exp. Zool. 144:* 209–223, 1960.
11. Ezerman, E. B., and Ishikawa, H.: Differentiation of the sarcoplasmic reticulum and T system in developing chick skeletal muscle *in vitro*. *J. Cell Biol. 35:* 405–420, 1967.
12. Finck, H., Holtzer, H., and Marshall, J. M., Jr.: An immunochemical study of the distribution of myosin in glycerol extracted muscle. *J. Biophys. Biochem. Cytol.* (*Suppl. 175*) *2:* 175–178, 1956.
13. Fischman, D. A.: An electron microscopic study of myofibril formation in embryonic chick skeletal muscle. *J. Cell Biol. 32:* 557–575, 1967.
14. Fiske, C. H., and SubbaRow, Y.: The colorimetric determination of phosphorus. *J. Biol. Chem. 66:* 375–400, 1925.
15. Fox, A. C., and Klein, M. D.: An immunochemical comparison of myosin from canine heart and skeletal muscle. *Biochim. Biophys. Acta 127:* 232–235, 1966.
16. Godchoux, W. S., Adamson, P., and Hebert, E.: Effects of cycloheximide on polyribosome function in reticulocytes. *J. Mol. Biol. 27:* 57–72, 1967.
17. Godlewski, E.: Die entwicklung des skelet-und herzmuskelgewebes der saugethiero. *Arch. Mikr. Anat. 60:* 111–156, 1902.
18. Godman, G. C.: Cell transformation and differentiation in regenerating striated muscle. In *Frontiers of Cytology*, ed. by Palay, S. L., pp. 381–416. New Haven, Yale University Press, 1958.
19. Goldstein, M. A.: A morphological and cytochemical study of sarcoplasmic reticulum and T system of fish extraocular muscle. *Z. Zellforsch. 102:* 31–39, 1969.
20. Hall, C. C., Jakus, A., and Schmitt, F. O.: An investigation of cross striations and myosin filaments in muscle. *Biol. Bull. 90:* 32–50, 1946.
21. Hamburger, V., and Hamilton, H.: A series of normal stages in the development of the chick embryo. *J. Morphol. 58:* 49–92, 1951.
22. Hanson, J., and Lowy, J.: The structure of F-actin and of actin filaments isolated from muscle. *J. Mol. Biol. 6:* 46–60, 1963.
23. Hay, E. D.: The fine structure of differentiating muscle in the salamander tail. *Z. Zellforsch. 59:* 6–34, 1963.
24. Heuson-Steinnon, J. A.: Morphogenese de la cellule musculaire striée étudiée au microscope electronique. *J. Microsc. 6:* 639–656, 1967.
25. Heywood, S. M., and Rich, A.: *In vitro* synthesis of native myosin actin and tropomyosin from embryonic chick polysomes. *Proc. Natl. Acad. Sci. U. S. A. 59:* 590–597, 1968.

26. Heywood, S. M., Dowben, R. M., and Rich, A.: The identification of polyribosomes synthesizing myosin. *Proc. Natl. Acad. Sci. U. S. A. 57:* 1002–1009, 1967.
27. Hibbs, R. C.: Electron microscopy of developing cardiac muscle in chick embryos. *Am. J. Anat. 99:* 17–52, 1956.
28. Hiramoto, R., Cairns, J., and Pressman, D.: Immunohistochemical studies on the antigenic relationship of some muscle tumors in normal skeletal muscle cells. *J. Natl. Cancer Inst. 27:* 937–947, 1961.
29. Holtzer, H.: Aspects of chondrogenesis and myogenesis. In *Synthesis of Molecular and Cellular Structure*, 19th Symposium of The Society of Development and Growth, ed. by Rudnick, D., pp. 35–87. New York, Ronald Press Co., 1961.
30. Holtzer, H., and Abbott, J.: Contraction of glycerinated embryonic myoblasts. *Anat. Rec. 131:* 417–428, 1958.
31. Holtzer, H., Abbott, J., and Cavanaugh, M. W.: Some properties of embryonic cardiac myoblasts. *Exp. Cell Res. 16:* 595–601, 1959.
32. Holtzer, H., Marshall, J. M., Jr., and Finck, H.: An analysis of myogenesis by use of fluorescent antimyosin. *J. Biophys. Biochem. Cytol. 3:* 705–724, 1957.
33. Huxley, H.: The double array of filaments in cross striated muscle. *J. Biophys. Biochem. Cytol. 3:* 631–648, 1957.
34. Huxley, H.: The contractile structure of cardiac and skeletal muscle. *Circulation 24:* 328–335, 1961.
35. Huxley, H.: Electron microscopic studies on the structure of natural and synthetic protein filaments from striated muscle. *J. Mol. Biol. 7:* 281–308, 1963.
36. Johnson, W., Jurand, J., and Hiramoto, R.: Immunological studies of tumors containing myosin. *Am. J. Pathol. 47:* 1139–1156, 1965.
37. Kesztyus, L., Nikodemusz, S., and Szilagyi, T.: Antigenic activity of myosin and actin. *Nature 163:* 136, 1949.
38. Kibo, Y., and Rich, A.: Induced enzymes formed on bacterial polyribosomes. *Proc. Natl. Acad. Sci. U. S. A. 51:* 111–118, 1964.
39. Klatzo, B., Horvath, B., and Emmart, E. W.: Demonstration of myosin in human striated muscle by fluorescent antibody. *Proc. Soc. Exp. Biol. Med. 97:* 135–140, 1958.
40. Knappeis, G. G., and Carlson, F.: The ultrastructure of the Z disc in skeletal muscle. *J. Cell Biol. 13:* 323–335, 1962.
41. Levi, G., and Chevremont, M.: Transformations structurales de elements des muscle sguelettique pendant leur croissance *in vitro. Arch. Biol. 52:* 523–551, 1941.
42. Lowey, S., Slayter, H. S., Weeds, A. G., and Baker, H.: Substructure of the myosin molecule I. Subfragments of myosin by enzymatic degradation. *J. Mol. Biol. 42:* 1–29, 1969.
43. Malkin, L. I., and Rich, A.: Partial resistance to nascent polypeptide chains to proteolytic digestion due to ribosomal shielding. *J. Mol. Biol. 26:* 329–346, 1967.
44. Manner, G., Kretsinger, R. H., Bernard, S., and Rich, A.: Polyribosomal synthesis of collagen. *Biochim. Biophys. Acta 134:* 411–429, 1967.
45. Mihalyi, E., and Szent-Gyorgyi, A. G.: Trypsin digestion of muscle proteins. *J. Biol. Chem. 201:* 189–196, 1953.
46. Nathans, D.: Puromycin inhibition of protein synthesis. Incorporation of puromycin into peptide chains. *Proc. Natl. Acad. Sci. U. S. A. 51:* 585–591, 1964.
47. Obinato, T., Yamamoto, M., and Maruyama, K.: The identification of randomly formed thin filaments in differentiating muscle cells of the chick embryo. *Dev. Biol. 14:* 192–213, 1966.
48. Obinato, T., Yamamoto, M., and Maruyama, K.: Morphological and biochemical studies on myofibrillar formation in developing chick embryos. *Sci. Pap. Coll. Gen. Educ. Univ. Tokyo 17:* 95–120, 1967.
49. Ogawa, Y.: Synthesis of skeletal muscle protein in early embryo and regenerating tissue of chick and triturus. *Exp. Cell Res. 26:* 269–274, 1962.
50. Pepe, F. A.: Some aspects of the structural organization of the myofibril as revealed by antibody-staining methods. *J. Cell Biol. 28:* 505–525, 1966.
51. Pepe, F. A.: Analysis of antibody staining patterns obtained with striated myofibrils in fluorescence microscopy and electron microscopy. In *Frontiers of Cytology*, ed. by Bourne, G. H., and Danielli, J. F., vol. 24, pp. 193–231. New York and London, Academic Press, 1968.

52. Pepe, F. A., and Finck, H.: The use of specific antibody in electron microscopy. *J. Biophys. Biochem. Cytol. 11:* 521–531, 1961.

53. Pepe, F. A., and Huxley, E. E.: Antibody staining of separated thin and thick filaments of striated muscle. In *Biochemistry of Muscle Contraction*, ed. by Gergeley, J., pp. 320–344. Boston, Little, Brown & Co., 1964.

54. Przybylski, R. J., and Blumberg, J. M.: Ultrastructural aspects of myogenesis in the chick. *Lab. Invest. 15:* 836–863, 1966.

55. Rich, A.: Polyribosomes. *Sci. Am. 209:* 44–53, 1963.

56. Samuels, A.: The immuno-enzymology of muscle proteins I. General features of myosin and 5′-adenylic acid deaminase. *Arch. Biochem. Biophys. 92:* 497–506, 1961.

57. Shafiq, S. A.: Electron microscopic studies on the indirect flight muscles of *Drosophila melanogaster*. II. Differentiation of myofibrils. *J. Cell Biol. 17:* 363–373, 1963.

58. Shy, G. M., Engel, W. K., Somers, J. E., and Wanko, T.: Nemaline myopathy: A new congenital myopathy. *Brain 86:* 793–810, 1963.

59. Slayter, H. S., and Lowey, S.: Substructure of the myosin molecule as visualized by electron microscopy. *Proc. Natl. Acad. Sci. U. S. A. 58:* 1611–1618, 1967.

60. Stanness, C. P.: The effect of cycloheximide and polyribosomes from hamster cells. *Biochem. Biophys. Res. Commun. 24:* 758–764, 1966.

61. Toselli, P. A., and Pepe, F. A.: The fine structure of the ventral intersegmental abdominal muscles of the insect *Rhodnius prolixus* during the molting cycle. *J. Cell Biol. 37:* 462–481, 1968.

62. Van Breeman, V. L.: Myofibril development observed with the electron microscope. *Anat. Rec. 113:* 179–195, 1952.

63. Waddington, C. H., and Perry, M. M.: Helical arrangements of ribosomes in differentiating muscle cells. *Exp. Cell Res. 30:* 599–600, 1963.

64. Warner, J. B., Rich, A., and Hall, C. E.: Electron microscopic studies of ribosomal clusters synthesizing hemoglobin. *Science 138:* 1399–1403, 1962.

65. Weed, I. G.: Cytological studies of developing muscle with special reference on myofibrils, mitochondria, Golgi material, and nuclei. *Z. Zellforsch. 25:* 516–590, 1936.

Chapter 5

# The Ultrastructure of Human Skeletal Muscle: Variations from Archetypal Morphology*

D. A. FISCHMAN, H. Y. MELTZER, AND R. W. POPPEI

---

Over the past 15 to 20 years, the application of coordinated electron microscopic, biochemical, and histochemical techniques for the study of skeletal muscle has resulted in a considerable advance in our understanding of the physiology of this tissue. In parallel with these investigations of normal muscle, and perhaps stimulated by them, there has been an extraordinary renewal of interest in the diseases of striated muscle, which has resulted in the description of a number of previously unrecognized myopathic disorders. Although there have been numerous studies of the ultrastructure of vertebrate muscle and an excellent recent review has been prepared by Pellegrino and Franzini-Armstrong,[57] our understanding of the normal deviations from prototypic morphology is quite incomplete. This is an especially acute problem for human biopsy material where tissue sampling, fixation artifact, and possible effects of medical intervention can substantially influence the interpretation of an electron microscopic study.

In preparation for this presentation and as part of our study of the structure of human muscle in patients with pathologic elevations in serum creatine phosphokinase activity,[46, 47] we have reviewed 59 publications† in the past 8 years concerning the ultrastructure of human muscle.[2, 4, 6, 9, 10, 13-18, 19b, 20, 22, 24, 25, 27, 29-31a, 32, 33, 36, 37, 40, 41, 44, 48, 49, 51, 53, 54a, 54b, 60, 63, 68, 69, 71, 73-79, 81, 83, 85-89, 91-95a, 98, 101] In this survey, we have recorded the total number of subjects biopsied and subdivided the cases into normal and abnormal categories as defined by the respective investigators. Although some papers may have been omitted, we believe this list is fairly representative of the existing literature and includes most of the important investigations concerned with normal human

---

* This research was supported in part by Grant 231-12-RD from the State of Illinois; United States Public Health Service Grants 16,127 and 18,396; and a grant from the Scottish Rite Foundation (to H.Y.M.); H.Y.M. is the recipient of a Research Scientist Development Award (KO MH 47,808) from the United States National Public Health Service; Science Foundation Grant GH-7591, Chicago and Illinois Heart Association Grant N 69-32; and United States Public Health Service Grant 1-RO1 HL-13505-02 (to D.A.F.).

† After completion of this manuscript, an additional electron microscopic study of muscle from 11 normal young adults has appeared.[68a]

muscle. From these 59 publications a total of 412 reported cases have been recorded, only 43 of which were considered normal by the authors. The largest series of normal biopsies which were examined by electron microscopy has been reported by Van Breeman[98] (10 biopsies), Shafiq, Gorycki, Goldstone, and Milhorat[81] (8 biopsies), Prineas and Ng[63] (5 biopsies), and Papadimitriou and Kakulas[54b] (5 biopsies). These four studies above contain data on more than half of all the normal subjects we have encountered in this literature search. Most of the other reports include one or two normal subjects or have omitted controls completely. In few of the studies is there an adequate description of the age or sex of the normal subjects and frequently no statement about the specific muscles biopsied is included in the reports. It is likely that a number of laboratories have accumulated normal material which has not been published, and this is somewhat unfortunate.

Our relative unease about the normal variations in structure of human muscle, and in particular, the variability in myofibrillar architecture, has prompted us to initiate an electron microscopic biopsy study of normal adults. Some of the results of this study are presented below.

## SUBJECTS

A total of 30 subjects have been biopsied to date. There were 15 men and 15 women ranging in age from 17 to 46, with a mean age of 25. None of these people exhibited any clinical signs of muscle disease, none had a personal or family history of any such disorder, and none had elevations in serum enzyme levels, such as creatine phosphokinase activity, which are typically observed in myopathic disorders.[48] Of these subjects, 20 were nonpsychotic psychiatric patients and 10 were normal volunteers. The psychiatric patients were part of a comparison group for our study of muscle structure in acute psychotic patients exhibiting elevations in serum creatine phosphokinase activity. All biopsies were performed after a 16-hour fast.

## METHODS

Skeletal muscle from either the vastus lateralis or belly of the gastrocnemius was obtained through an open biopsy performed under local xylocaine anesthesia. Since our experience has been more extensive with the thigh musculature, the results presented below will be confined to that muscle. A piece of muscle approximately 2 by 2 by 12 mm, oriented in parallel with the long axis of the myofibers, was clamped *in situ*,[52, 62] excised, and fixed while under extension in the muscle clamp. Fixation was performed by immersion for 2 hours in iced, 0.1 M phosphate-buffered (pH 7.4) 2.5% glutaraldehyde. Well fixed areas were dissected from the clamp and postfixed for 1 hour in cold, 0.1 M collidine-buffered (pH 7.4) 1% $OsO_4$. Following dehydration in ethanol, the tissue was embedded in Araldite. At least 10 blocks from each biopsy were sectioned longitudinally at 0.5- to 1-$\mu$ thickness, stained with alkaline toluidine blue, and a minimum of 100 to 150 different myofibers from each patient examined thoroughly by phase-contrast microscopy. Selected areas were subsequently thin sectioned, stained with uranyl acetate and lead citrate, and examined in an AEI EM 6B electron microscope.

## RESULTS

Before proceeding to a description of our results, a discussion of quality of fixation and possible artifacts of technique is in order. It has been our experience that care in the operating room during the surgical removal of the biopsy is amply rewarded in later study of the tissue. We fix the tissue in the operating room, immediately examine the biopsy under a dissecting microscope, and, if the muscle is contracted, damaged, or too covered with blood, a second, or if necessary, a third biopsy is performed. Inadequately stretched or poorly fixed muscle is extraordinarily difficult to interpret by either light or electron microscopy. We have utilized three criteria for judging the quality of fixation: (1) crispness and clarity of the sarcomere pattern in the light and electron microscope; (2) preservation of the sarcotubular system, especially the longitudinal terminal cisternae which are disrupted in poorly fixed areas; (3) absence of swollen or fragmented mitochondria. With the techniques now available, there is no reason to accept human material of inferior quality to that which is routinely obtained in animal studies.

## GENERAL MORPHOLOGY OF HUMAN MYOFIBER

The general ultrastructural features of the human striated muscle cell are familiar and need not be extensively reviewed here. This subject has been adequately covered in prior publications.[57, 59, 91] The basic structural characteristics of the human myofiber conforms to the general pattern common to all mammalian skeletal muscle; this is demonstrated in Figure 1. Such electron microscopic fields present no difficulty in interpretation. The elongated nucleus occupies a subsarcolemmal position, is frequently irregular in outline, and exhibits a prominent band of perinuclear chromatin. Accumulations of mitochondria are generally observed at the poles of the myonuclei, although the numbers of such mitochondria vary depending upon the fiber type.

The myofibrils exhibit the characteristic sarcomere pattern of vertebrate striated muscle; Z, A, I, and M bands are all clearly evident in Figure 1. Although Z bands are slightly wider in type I than type II fibers, the remaining characteristics of the myofibril, including the M band, are not clearly different in the different fiber types of mammalian muscle.[23, 81, 84] Pairs of mitochondria are generally found between the myofibrils on either side of the Z band in close association with one or two lipid droplets. The numbers of lipid droplets and mitochondria at this site vary widely. In type I fibers there are large mitochondria and numerous lipid droplets, whereas in type II fibers the lipid droplets are relatively rare and the mitochondria are small. A continuum in the amounts of mitochondria and lipid is frequently observed in which classification of fiber type is difficult, if not impossible, by electron microscopy alone. Glycogen granules (Fig. 2) are most frequently encountered in the sarcoplasm bordering the I band in close association with the terminal cisternae of the sarcoplasmic reticulum. Since the glycogen frequently obscures the details of the triad region, we have found it useful to routinely *en bloc* stain half of our material in aqueous 0.5% uranyl acetate (pH 4.5) before dehydration.[19a] It has been demonstrated that in addition to greatly accentuating membrane,

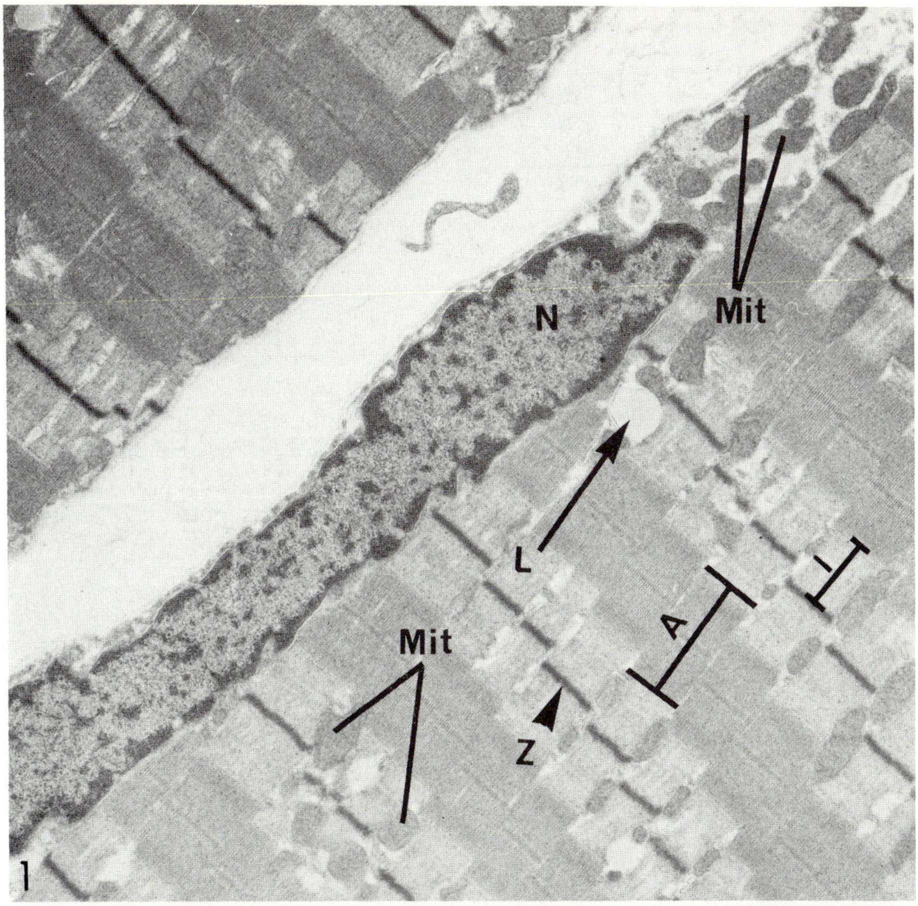


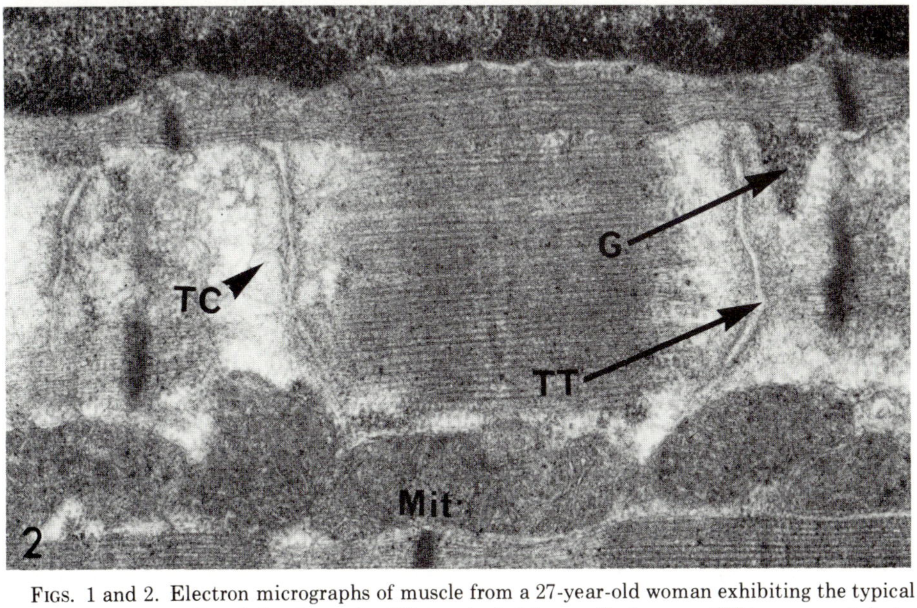


FIGS. 1 and 2. Electron micrographs of muscle from a 27-year-old woman exhibiting the typical structure of mammalian skeletal muscle. *TC*, terminal cisterna; *G*, glycogen; *TT*, transverse tubule; *Mit*, mitochondrion; *L*, lipid droplet; *N*, myonucleus; *Z*, Z band; *A*, A band; *I*, I band. Uranyl acetate and lead citrate stain; Fig. 1, × 10,000; Fig. 2, × 30,000.

ribosomal and myofibrillar contrast, this staining procedure causes the extraction of a particulate glycogen.[100] The details of the sarcotubular system are considerably clarified by such treatment (Figs. 3 and 4). The characteristic 400-Å periodicity of the I band, believed to result from the location of tropomyosin and troponin along the thin filaments, can be seen in Figure 4. The location of ribosomes in the interfibrillar space is readily apparent after glycogen extraction (Fig. 5). Some of the ribosomes definitely appear membrane bound in the regions of the triad. It has not been established whether such rough-surfaced membranes are continuous with the terminal or longitudinal cisternae of the sarcoplasmic reticulum or whether they comprise a separate membranous system. The disposition and amounts of ribosomal material in mature human muscle requires additional study. We have been impressed by the variability in ribosomal content between different myofibers and it would be of interest to relate such differences to the protein synthetic and catabolic conditions of such cells. This is a totally unexplored area of human or, for that matter, mammalian muscle research. In addition to this interfibrillar location, most of the remaining ribosomes of the myofiber are located in the perinuclear sarcoplasm, at the ends of each cell near the myotendon junction and in the subsynaptic cytoplasm of the myoneural junction. The structure of the mammalian myotendon junction has been well reviewed by Ishikawa,[31b] but the formation of myofibrils and the function of acetylcholinesterase at this site[42] remain poorly understood. Since Dr. Fardeau (see Chapter 17) has reviewed the structure of the myoneural junction in his contribution to this program, this subject will not be further discussed here.

## NORMAL VARIATIONS IN MYOFIBER STRUCTURE

### Z Band spreading

The interpretation of typical myofibrillar structure is not difficult. Problems arise when we consider variations from the usual pattern. One of the most frequently observed alterations of the myofibril, in almost every atrophic condition affecting skeletal muscle, is a spreading or smearing of Z bands. This subject has been reviewed by Engel,[14] but the interpretation of such Z band change remains controversial.[16] Shafiq, Gorycki, Goldstone, and Milhorat[81] have commented upon the fact that a focal loss of sarcomere pattern with Z band spreading can be observed in normal human muscle. We agree with this conclusion but would suggest that the degree or amount of Z band spreading in normal muscle is limited. Variations from this limit may be of use in detecting

---

FIG. 3. Micrograph of muscle from a 19-year-old man. The biopsy was stained *en bloc* with aqueous 0.5% uranyl acetate after $OsO_4$ fixation but prior to dehydration. This staining procedure extracts particulate glycogen, often permitting better visualization of the sarcoplasmic reticulum. Note the absence of ribosomes in this field, and compare with Figures 5, 6, 11, and 12. *TT*, transverse tubules; *SR*, sarcoplasmic reticulum. × 20,000.

FIG. 4. Micrograph of muscle from a 26-year-old woman illustrating the details of the triad and close association of mitochondria with the triadic elements. The M band and 400-Å periodicity of the I band are evident. *TT*, transverse tubule; *TC*, terminal cisterna; *M*, M band; *I*, I band. × 27,000.

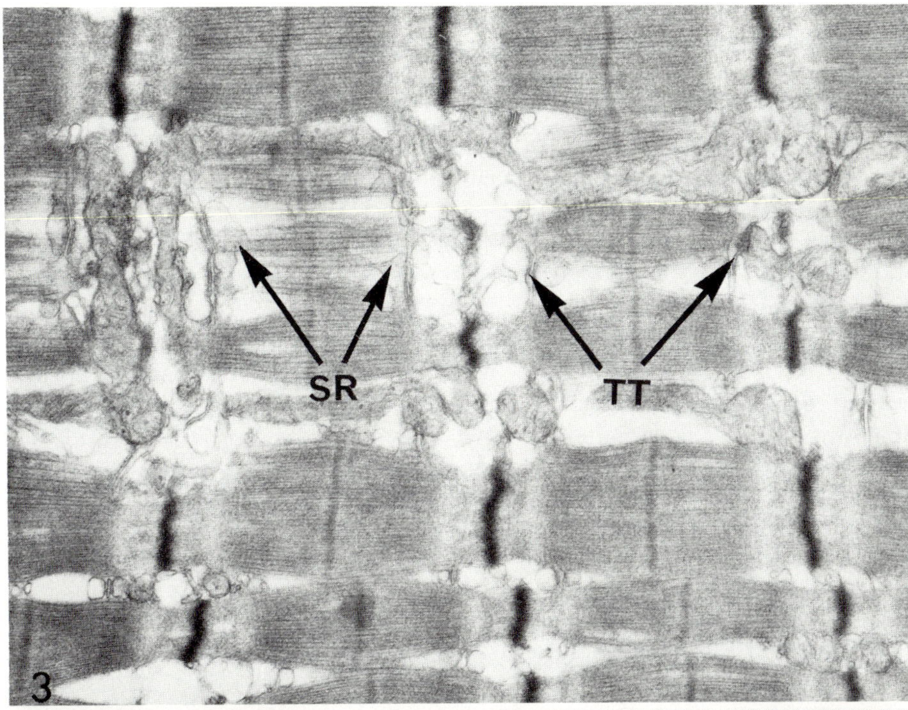


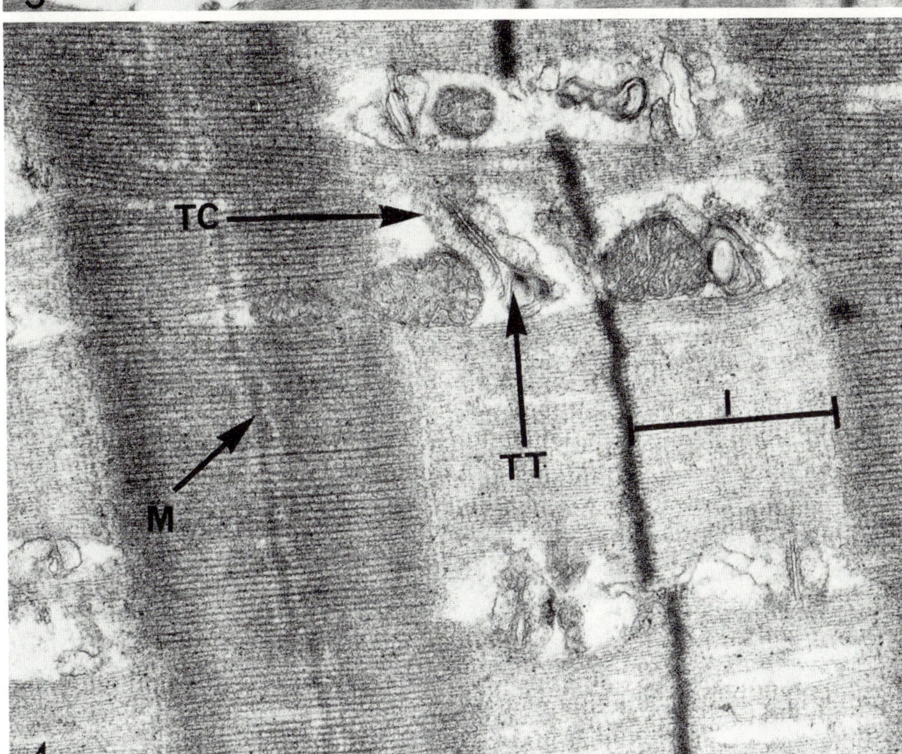


FIGS. 3 and 4

subtle myopathic alterations. In Figures 5 and 6, the spreading of Z bands in normal human muscle is demonstrated. The alterations are identical in morphology to what Engel[14] has described as early lesions of the myofibril in muscle disease. In agreement with his findings we have also noted a decrease in mitochondrial number in such locations of the cell.

If 0.5- to 1-$\mu$ thick sections of Araldite-embedded muscle are stained with toluidine blue and examined by phase contrast microscopy, these focal Z band irregularities can easily be visualized (Figs. 7 and 8). The *arrows* in these micrographs indicate darkly staining bodies which, in the electron microscope, have the appearance of widened Z bands as in Figures 5 and 6. In our experience, all normal muscle, if carefully searched, will exhibit sarcomere irregularities (Figs. 5 to 8). However, these changes are focal in nature, rarely extending over three to four consecutive sarcomeres or involving three to four adjacent myofibrils. Sometimes a fiber is detected in which relatively extensive myofibrillar disarray is evident (Figs. 9 and 10). These extensive alterations in myofibrillar structure are generally not found in greater than 1% of the muscle cells. An electron micrograph of the same biopsy as in Figure 10 is shown in Figure 11. Myofilaments are not arranged in a sarcomere pattern, the sarcotubular system is atypical, and mitochondria appear reduced in these regions of the myofiber. Such cytoplasmic regions bear resemblence to embryonic muscle[90] and sites of regeneration.[61] The interpretation of such micrographs must await further study. These could be regions of myofibrillar degeneration, regeneration, or normal turnover. As Pearson[55] has demonstrated, normal muscle exhibits focal regions of basophilia suggestive of regeneration.

A number of authors have stressed the relative nonspecificity of Z band spreading.[16, 70, 80] There seems no need to belabor the point that myofibrillar alteration can be manifested in only a limited number of ways. Apparently, the Z band, rather than being an inert, rigid, disclike structure, is probably the most labile structure of the myofibril. As such, it may be the most sensitive indicator of myofibrillar alteration, whether this be from a degenerative or regenerative situation. Our preliminary conclusion from quantifying the departure of the Z band from prototypic morphology in 30 control subjects is that, if Z band spreading covers more than four continuous sarcomeres in four adjacent fibrils and in greater than 2% of the myofibers, it tentatively should be considered beyond the limits of normal. Until more is understood about the structure,[21, 38, 66] composition,[96] and turnover of the Z band the interpretation of Z band spreading must remain a puzzling phenomenon.

### Lysosomal content of normal muscle

It has been demonstrated that the specific activities of several acid hydrolases increase in skeletal muscle undergoing atrophy secondary to denervation,[26, 58] vitamin E deficiency,[102] and hereditary dystrophy.[97] Pollack and Bird[58] have recently shown that membrane-enclosed acid hydrolases are found in normal muscle and the number of such proteolytic vesicles increases following denervation. Because of the possibility that such acid hydrolases are not myofiber in origin but are derived from connective tissue or blood cellular

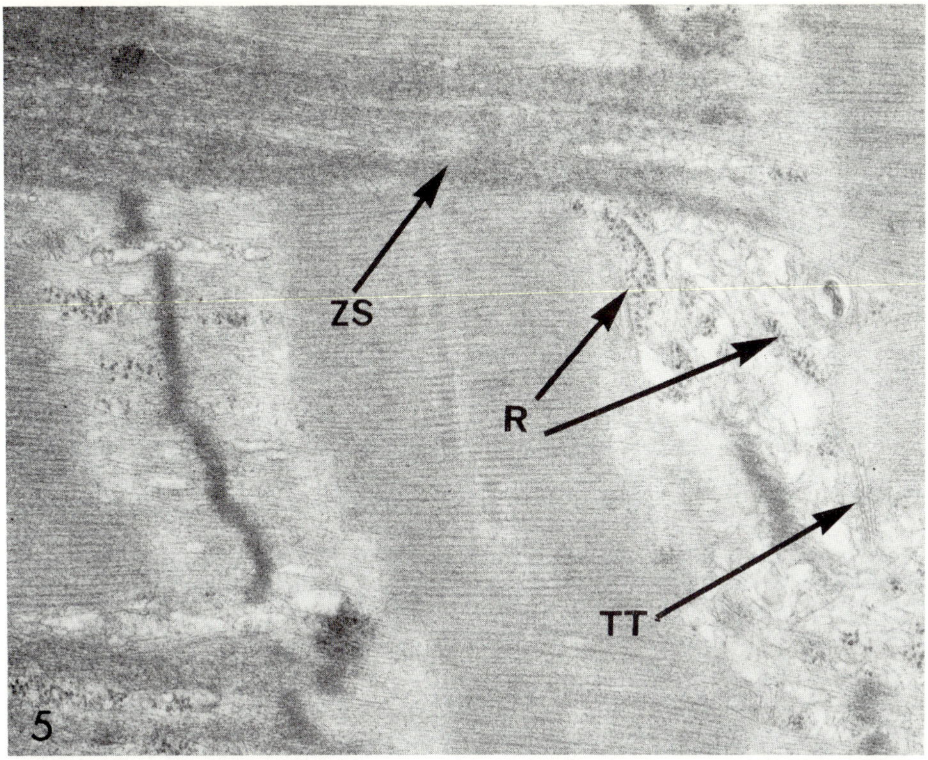


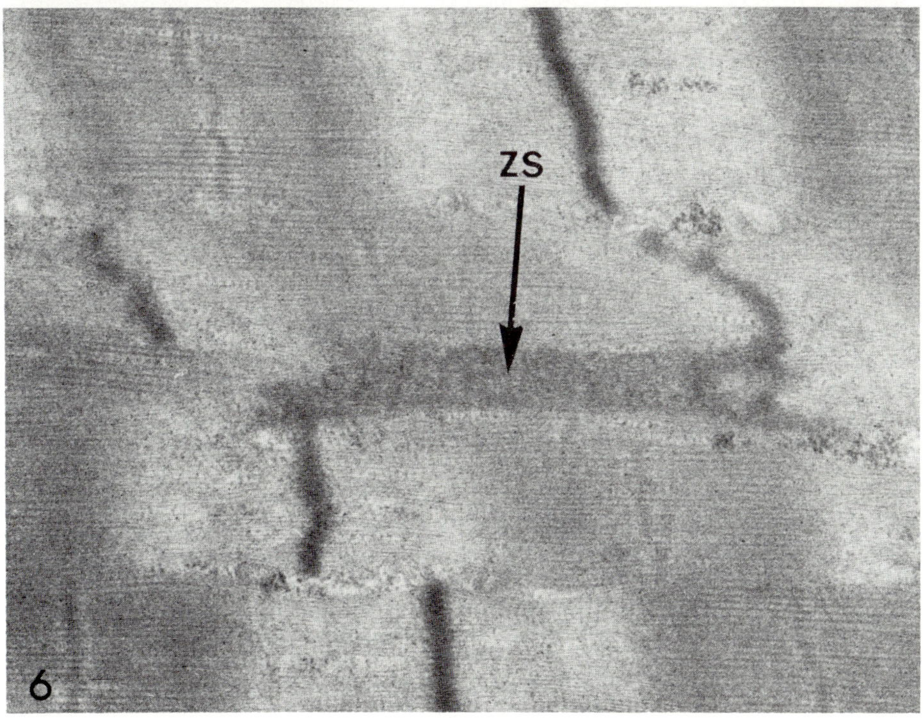


FIG. 5. Muscle from a 15-year-old girl which has been stained *en bloc* with aqueous 0.5% uranyl acetate. Ribosomes (*R*) attached to membranes are evident in the sarcoplasm surrounding the I band. Z band spreading (*ZS*) is evident at the *upper part* of the figure and, less clearly, in the *lower left corner*. *TT*, transverse tubule. × 26,000.

FIG. 6. An area of focal Z band spreading (*ZS*) in a 24-year-old woman. This region would present present the light micrographic appearance of Figures 7 and 8. × 28,000.

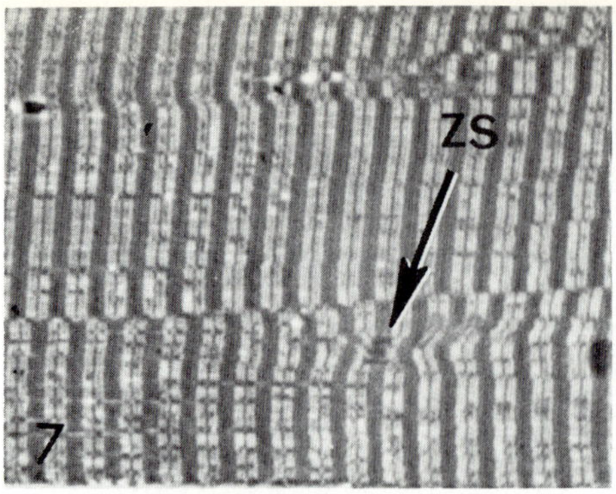


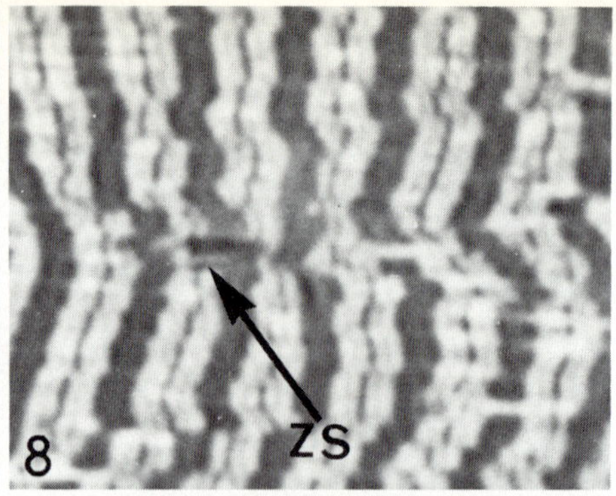


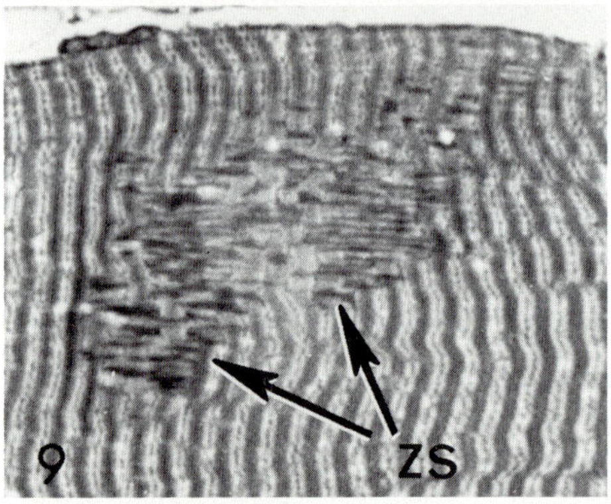


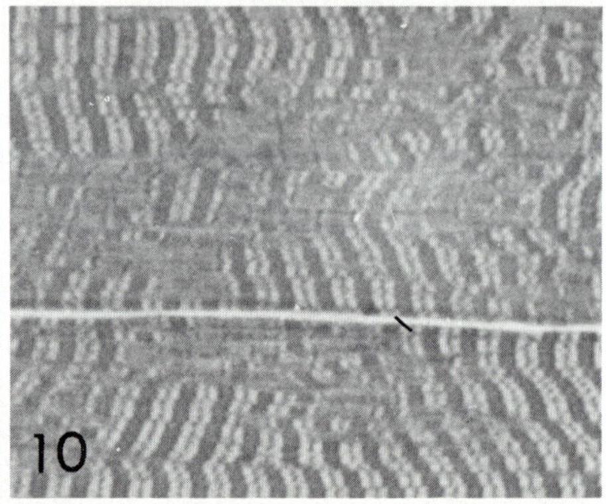


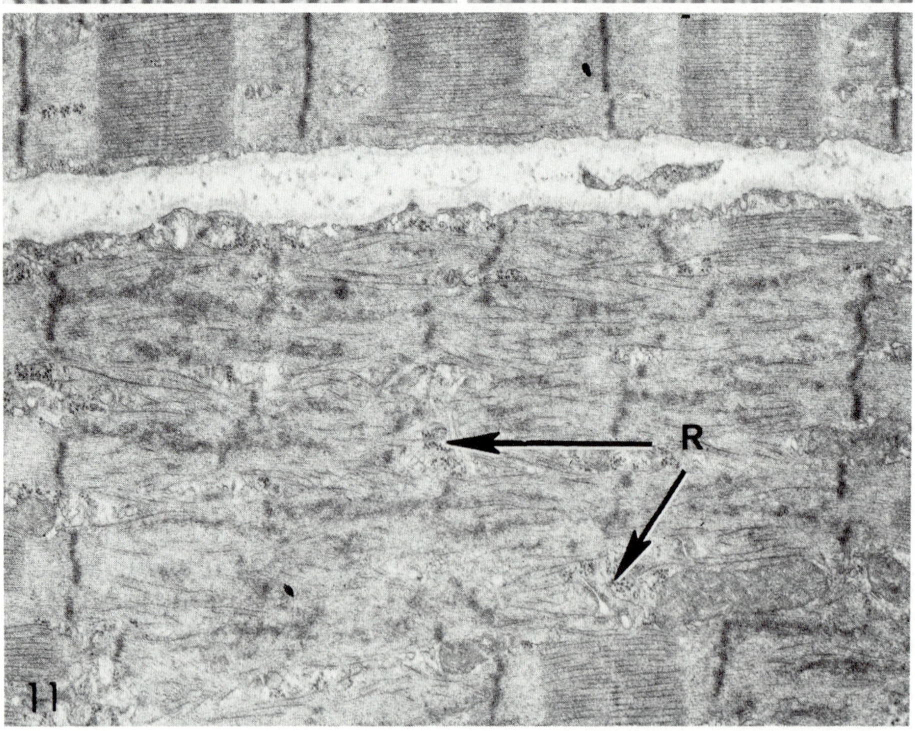


FIG. 7. Light micrograph of muscle from a 15-year-old girl. The *arrow* points to a focus of increased density which in the electron microscope is an area of Z band spreading (*ZS*). × 1,000.

FIG. 8. Z band spreading (*ZS*) in the muscle of a normal 24-year-old woman. × 2500.

FIG. 9. An area of extensive myofibrillar disruption and Z band spreading (*ZS*) in a 27-year-old woman. Such disruption of the usual sarcomere pattern has not been observed in greater than 1% of the muscle cells. ×1000.

FIGS. 10 and 11. Light and electron micrographs of muscle from a 17-year-old male exhibiting extensive fibrillar disarray with relatively little Z band spreading (compare with Fig. 9). Although the light micrograph (Fig. 10) might be interpreted as focally overcontracted sarcomeres, this can be ruled out by electron microscopy (Fig. 11). Fixation artifact is considered unlikely because of the well preserved transverse tubules and sarcoplasmic reticulum. *R*, ribosomes. × 13,000.

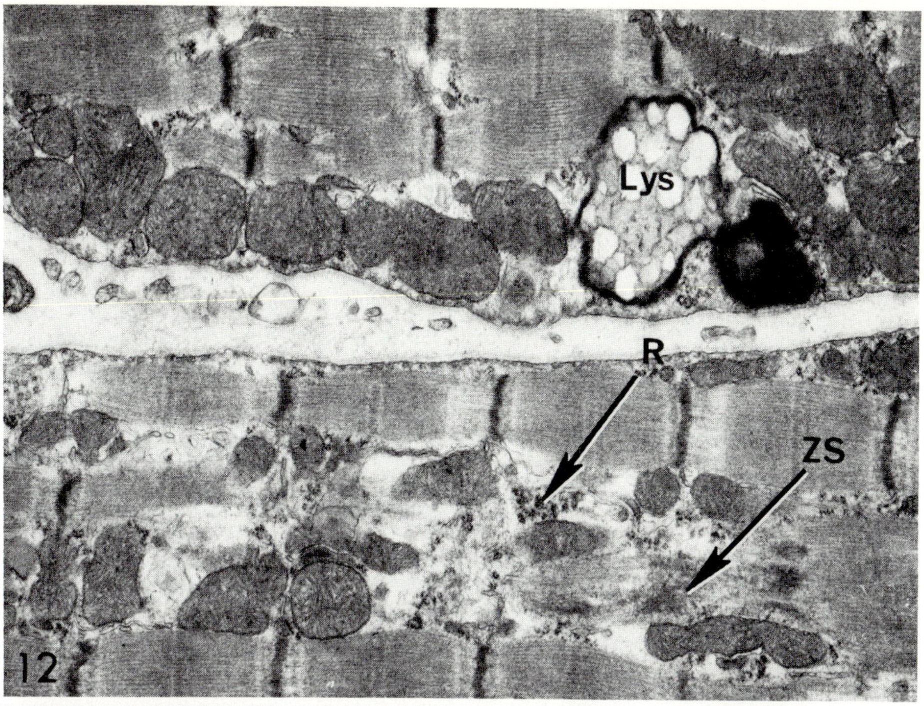


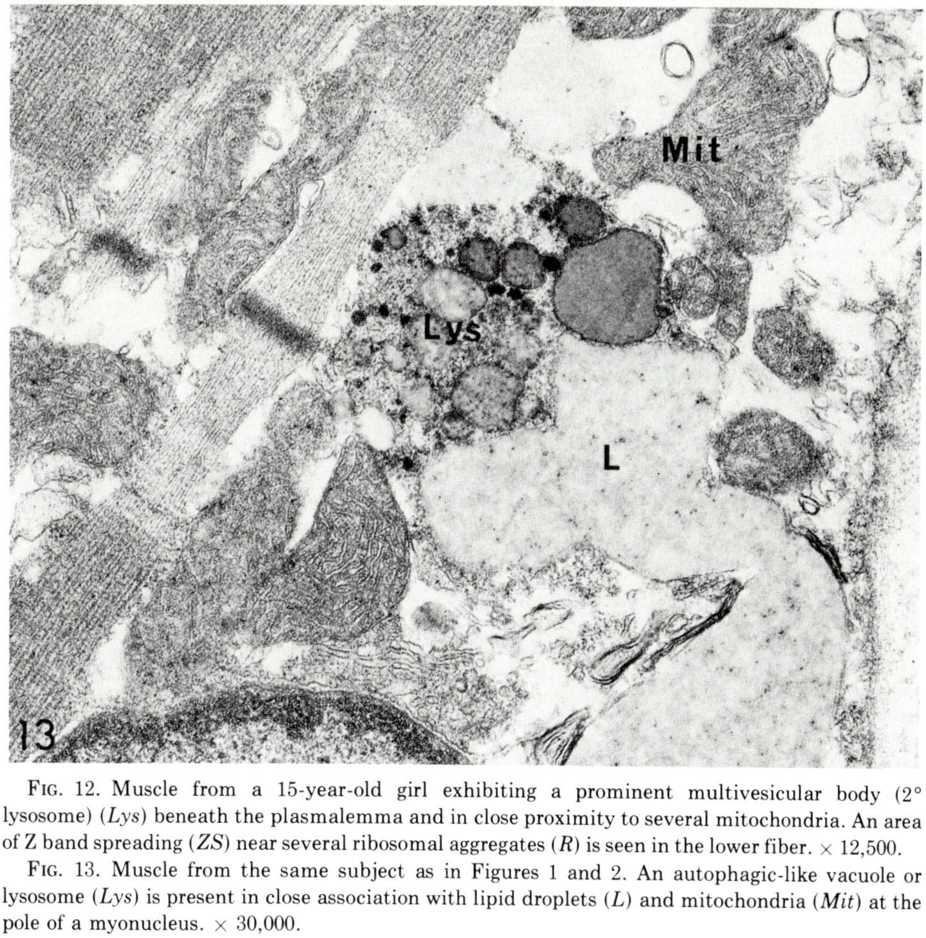


FIG. 12. Muscle from a 15-year-old girl exhibiting a prominent multivesicular body (2° lysosome) (*Lys*) beneath the plasmalemma and in close proximity to several mitochondria. An area of Z band spreading (*ZS*) near several ribosomal aggregates (*R*) is seen in the lower fiber. × 12,500.

FIG. 13. Muscle from the same subject as in Figures 1 and 2. An autophagic-like vacuole or lysosome (*Lys*) is present in close association with lipid droplets (*L*) and mitochondria (*Mit*) at the pole of a myonucleus. × 30,000.

elements, it is necessary to demonstrate, electron microscopically, that lysosomes are indeed located in muscle cells. Autophagic-like vacuoles have been described in the perinuclear sarcoplasm of normal muscle[28] and have been frequently observed in atrophic[56, 57] and myopathic conditions.[48] In our experience, although lysosomes are not plentiful in normal myofibers, structures which have all of the characteristics of an autophagic vacuole can be seen in almost every adult muscle cell, if carefully searched. As seen in Figures 12 and 13, pleomorphic membranous bodies are usually found in the subsarcolemmal cytoplasm and around the poles of the myonuclei. Generally, these autophagic vacuoles are closely associated with mitochondria (Fig. 12) and large lipid droplets (Fig. 13). Our impression is that they are more frequent in type I fibers.

The possible role of such vesicles in normal muscle metabolism is unclear. Although the turnover of myofibrillar proteins has been appreciated for some time[11] the proteolytic basis for this catoblism remains obscure. Recently, a soluble proteolytic system has been described in normal and atrophying muscle by Kohn[39] who has suggested a possible involvement in myofibrillar turnover. Clearly, the relationship of soluble proteases and membrane-enclosed proteases to myofibrillar protein catabolism requires further study.

### Satellite cell in normal muscle

Since the description of satellite cells by Mauro[45] a decade ago as mononucleated cells lying beneath the basement membrane of skeletal muscle, there has been continuing interest in the origin and role of such cells in the mature animal. Satellite cells have been described in association with both intrafusal[34, 35] and extrafusal myofibers.[45] They have been found in normal embryonic,[32] postnatal,[32, 50] and mature muscle of many vertebrate species including the frog,[45] alligator,[67] chicken,[64] mouse,[45] rat,[82] bat,[7] and man.[83] In human muscle, satellite cells have been observed in both normal and myopathic conditions.[83] The structure of this cell type has been amply reviewed by Mauro,[45] Muir, Kanji and Allbrook,[51] Ishikawa,[32] Reznik,[72] and Church.[7] Although the satellite cell is typically described as a spindle-shaped cell with an elongated nucleus surrounded by scanty cytoplasm, the ultrastructural characteristics of the nucleus and cytoplasm differ in many of these publications. For instance, in the micrographs of Ishikawa[32] and of Church[7] on embryonic satellite cells, the ample cytoplasm contains large numbers of free ribosomes, a moderately well developed granular endoplasmic reticulum and Golgi zone, and the nuclear chromatin is not very condensed. Similar appearing satellite cells are described in regenerating adult muscle by Allbrook[3] and Reznik.[72] Some authors have suggested that such satellite cells are identical in morphology to myoblasts and might be a later stage of such a cell.[4, 7, 8, 50] In contrast to this appearance, the satellite cell in adult, nonregenerating muscle usually contains much less cytoplasm with few mitochondria, few free ribosomes, and a poorly developed endoplasmic reticulum and Golgi zone. Nuclei have a more condensed chromatin than neighboring myonuclei and can frequently be distinguished from the muscle cell nuclei on this basis in histologic sections (see below). Thin 50- to 70-Å cytofilaments are frequently observed in satellite cells within embryonic

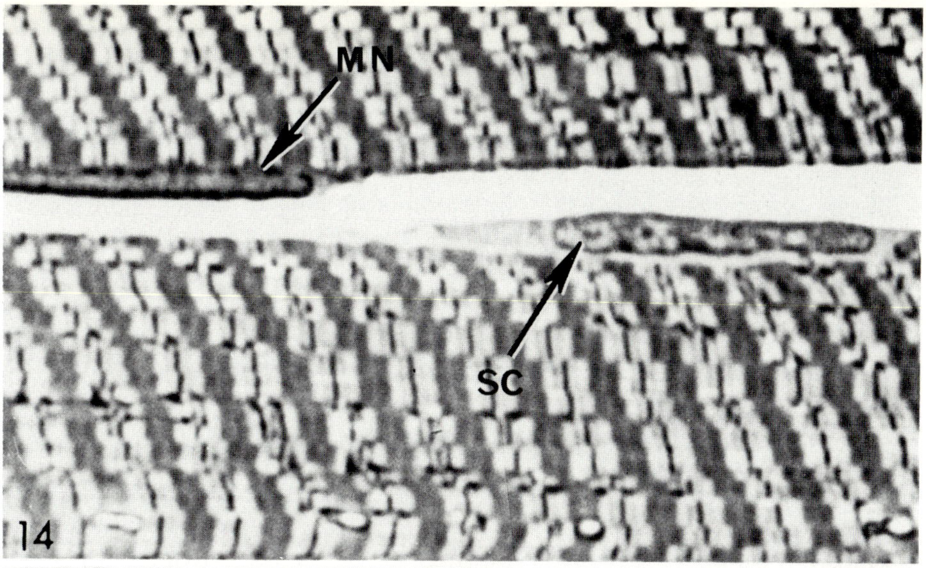


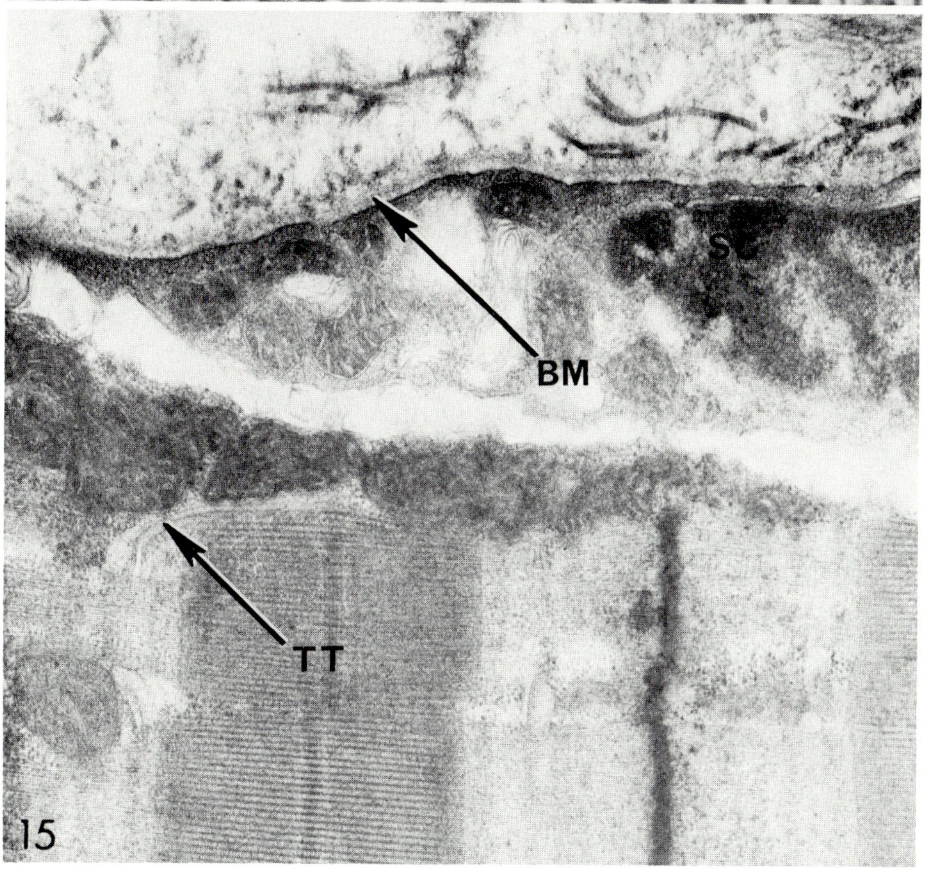


FIG. 14. A light micrograph of a toluidine blue stained 1-$\mu$ thick section examined under phase contrast microscopy. Muscle is from the same subject as Figure 13. Two elongated nuclei are visible in association with adjacent myofibers. The nucleus on the right is in a satellite cell; this has been verified by electron microscopy. Satellite cell nuclei can be recognized routinely in the light microscope by the pale staining gap between the nucleus and underlying myofibrils. *SC*, satellite cell; *MN*, myonucleus. × 2,500.

FIG. 15. Electron micrograph of a cell similar to that presented in Figure 14. Muscle is from a 24-year-old man. The satellite cell (*SC*) is seen beneath the basement membrane (*BM*) of the myofiber. *TT*, transverse tubule. × 26,000.

and regenerating muscle[83] but are not common in the satellite cell of the normal adult tissue.

The number of satellite cells in human muscle has not been systematically examined. Ishikawa[32] has stated that satellite cells are frequently observed in embryonic but rarely in adult human muscle. Based on his studies with rabbits and mice, Reznik[72] has stated that satellite cells are exceedingly rare in adult, uninjured muscle. In their study of 3-day-old mice, Muir, Kanji, and Allbrook[51] have estimated that satellite cells comprised 20% of the nuclei associated with the myofibers. Based on the figures of Church[7] one can calculate that satellite cells make up 12% of the nuclei on or in the myofibers of the web muscle in the 19-week-old bat embryo.

Based on the observations of Venable[99] that satellite cells can be visualized in 1-$\mu$ thick plastic-embedded sections, we have attempted a quantitation of satellite cell numbers in 10 normal adults (ages 19 to 23, five men and five women). Using phase contrast optics, two classes of nuclei can be identified along the myofibers (Fig. 14). One type of nucleus is directly applied to the myofibrils, while the other type of nucleus is separated from the myofibrils by a pale gap and appears to be within a short stretch of cytoplasm devoid of myofibrils. By electron microscopy we have verified that the latter group of nuclei are within satellite cells. In addition to this gap between myofibrils and the statellite cell nuclei, these latter nuclei generally stain more densely with toluidine blue and appear more heterochromatic than the muscle cell nuclei. Electron micrographs of three typical satellite cells in normal human muscle are shown in Figures 15 to 17. The intervening gap between satellite and muscle cells can vary markedly, even under identical conditions of tissue fixation.

In Figure 17 a satellite cell containing three autophagic-like vacuoles is presented. Such a cell would be difficult to differentiate on the basis of electron microscopy from a macrophage.

Using the light microscope we have estimated that satellite cells comprise 5 ± 0.5% of nuclei in or on the muscle fibers of the vastus lateralus in young adults. This figure is based on counts of 200 nuclei from each of 10 subjects (total of 2,000 nuclei examined). Studies are in progress to establish similar data for older age groups.

There is an extensive literature indicating that the number of myofibers in a muscle does not increase after the postnatal period and that subsequent muscle growth occurs by myofiber hypertrophy rather than proliferation.[1, 65] However, there now is good evidence that the number of nuclei in an adult myofiber is greater than that in the postnatal muscle cell.[12, 43] Since myonuclei do not divide or incorporate [$^3$H]thymidine[95b] there must be a recruitment of muscle nuclei from a source outside the muscle cells. Recently, Moss and Leblond[50] have shown one probable source to be satellite cells. Presumably, the satellite cell population represents, at least in part, a remnant of the myoblast population of the embryo. It has been demonstrated by a number of workers (see review by Betz, Firket, and Reznik[5]) that the number of myoblasts or satellite cells increases markedly within 48 to 72 hours after muscle injury. Reznik[72] has concluded that most of these cells arise by a pinching off of myonuclei with some

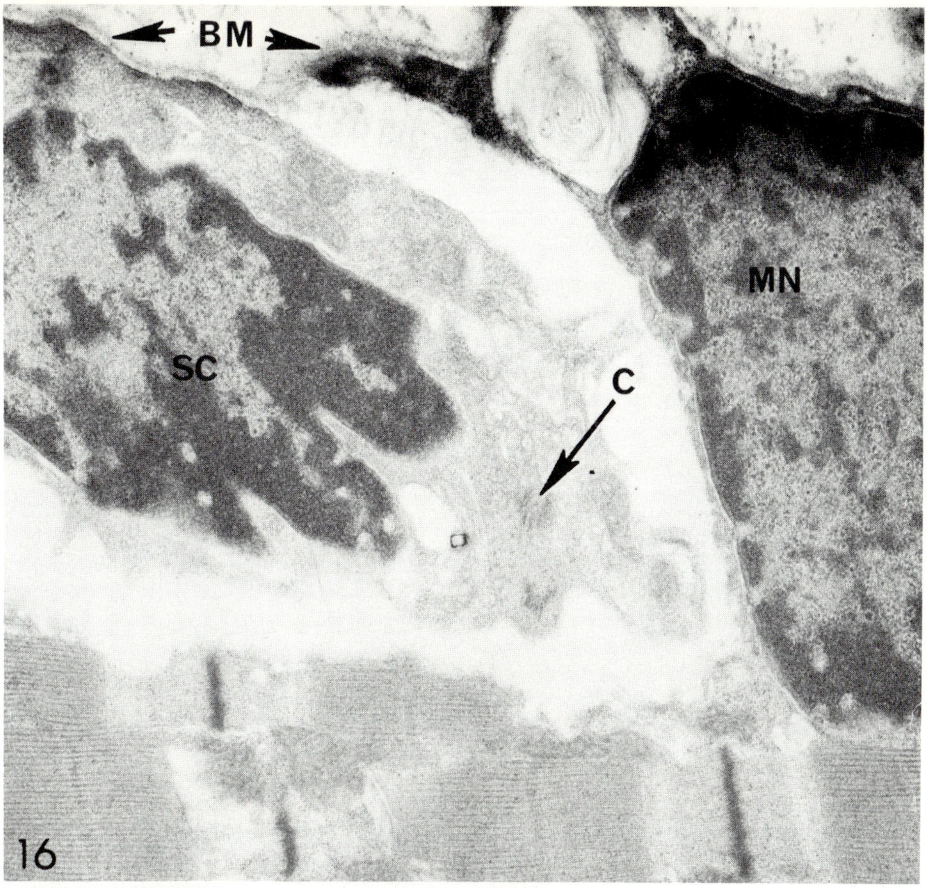


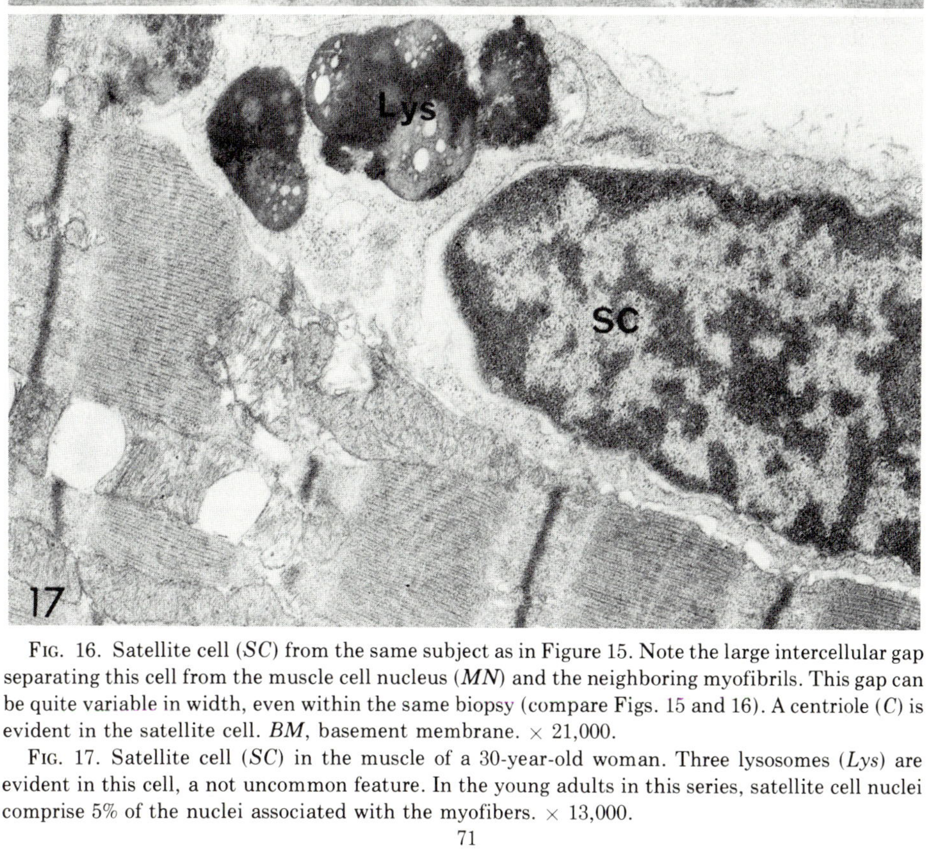


FIG. 16. Satellite cell (*SC*) from the same subject as in Figure 15. Note the large intercellular gap separating this cell from the muscle cell nucleus (*MN*) and the neighboring myofibrils. This gap can be quite variable in width, even within the same biopsy (compare Figs. 15 and 16). A centriole (*C*) is evident in the satellite cell. *BM*, basement membrane. × 21,000.

FIG. 17. Satellite cell (*SC*) in the muscle of a 30-year-old woman. Three lysosomes (*Lys*) are evident in this cell, a not uncommon feature. In the young adults in this series, satellite cell nuclei comprise 5% of the nuclei associated with the myofibers. × 13,000.

surrounding sarcoplasm. Other workers suggest that these myoblasts arise from the proliferation of preexisting satellite cells.[4, 8] Until quantitative measurements of satellite cell numbers and mitotic indices are correlated with the increase of myoblasts after injury, this question is likely to remain unanswered.

## ACKNOWLEDGMENTS

The authors acknowledge the assistance of Marshall Matz, M.D., and Miss Sarah Burnett in obtaining the biopsy specimens; the technical assistance of Mrs. George Burr and Miss Rita Yambot was of great help.

### REFERENCES

1. Adams, R. D., Denny-Brown, D., and Pearson, C. M.: *Diseases of Muscle, A Study in Pathology*, 2nd ed. New York, Hoeber, 1962.
2. Aleu, F. P., and Afif, A. K.: Ultrastructure of muscle in myotonic dystrophy. Preliminary observations. *Am. J. Pathol. 45:* 221–231, 1964.
3. Allbrook, D.: An electron microscopic study of regenerating skeletal muscle. *J. Anat. 96:* 137–152, 1962.
4. Allbrook, D., Baker, W. C., and Kilkcalydy-Willis, W. H.: Muscle regeneration in experimental animals and in man. *J. Bone Joint Surg. 48B:* 153–169, 1966.
5. Betz, E. H., Firket, H., and Reznik, M.: Some aspects of muscle regeneration. *Int. Rev. Cytol. 19:* 203–227, 1966.
6. Boehme, D., Themann, H., and Gold, J.: Structural and ultrastructural changes in striated human muscle caused by chronic ischemia. *Am. J. Pathol. 49:* 569–591, 1966.
7. Church, J. C. T.: Satellite cells and myogenesis; a study in the fruitbat web. *J. Anat. 105:* 419–438, 1969.
8. Church, J. C. T., Noronha, R. F. X., and Allbrook, D.: Satellite cells and skeletal muscle regeneration. *Br. J. Surg. 53:* 638–642, 1966.
9. Conen, P. E., Murphy, E. G., and Donohue, W. L.: Light and electron microscopic studies of myogranules in a child with hypotonia and muscle weakness. *Can. Med. Assoc. J. 89:* 983–986, 1963.
10. Cornog, J. L., Jr., and Gonatas, N. K.: Ultrastructure of rhabdomyoma. *J. Ultrastruct. Res. 20:* 433–450, 1967.
11. Dreyfus, J. C., Kruh, J., and Shapira, G.: Metabolism of myosin and life time of myofibrils. *Biochem. J. 75:* 574–478, 1960.
12. Enesco, M., and Puddy, D.: Increase in the number of nuclei and weight in skeletal muscle of rats of various ages. *Am. J. Anat. 114:* 235–253, 1964.
13. Engel, A. G.: Recent studies on neuromuscular disease. Late-onset rod myopathy (A new syndrome?). Light and electron microscopic observation in two cases. *Mayo Clin. Proc. 41:* 713–741, 1966.
14. Engel, A. G.: Pathological alterations of the Z disk. In *Exploratory Concepts in Muscular Dystrophy*, ed. by Milhorat, A. T., pp. 398–411. New York, Excerpta Medica Foundation, 1967.
15. Engel, A. G., and Gomez, M. R.: Nemaline (Z disc) myopathy. Observations on the origin, structure, and solubility properties of the nemaline structures. *J. Neuropathol. Exp. Neurol. 26:* 601–619, 1967.
16. Engel, W. K.: A critique of congenital myopathies and other disorders. In *Exploratory Concepts in Muscular Dystrophy and Related Disorders*, ed. by Milhorat, A., pp. 27–40. New York, Excerpta Medica Foundation, 1967.
17. Engel, W. K., Wanko, T., and Fenichel, G. M.: Nemaline myopathy. *Arch. Neurol. 11:* 22–39, 1964.
18. Engel, W. K., Foster, J. B., Hughes, B. P., Huxley, H. E., and Mahler, R.: *Brain 84:* 167–185, 1961.

19a. Farquhar, M. G., and Palade, G. E.: Cell junctions in amphibian skin. *J. Cell Biol. 26:* 263–291, 1965.
19b. Fidzianska, A., and Biczyskowa, W.: Electron microscopy in biopsy of the skeletal muscle. *Acta Med. Pol. 8:* 393–398, 1967.
20. Fisher, E. R., Cohn, R. E., and Danowski, T. S.: Ultrastructural observations of skeletal muscle in myopathy and neuropathy with special reference to muscular dystrophy. *Lab. Invest. 15:* 778–793, 1966.
21. Franzini-Armstrong, C., and Porter, K. R.: The Z disc of skeletal muscle fibrils. *Z. Zellforsch. 61:* 661–672, 1963.
22. Fulthorpe, J. J., Gardner-Medwin, D., Hudgson, P., and Walton, J. M.: Nemaline myopathy. *Neurology (Minneap.) 19:* 735–748, 1969.
23. Gauthier, G. F.: On the relationship of ultrastructural and cytochemical features of color in mammalian skeletal muscle. *Z. Zellforsch. 95:* 462–482, 1969.
24. Gonatas, N. K.: The fine structure of the rod-like bodies in nemaline myopathy and their relation to the Z-discs. *J. Neuropathol. Exp. Neurol. 25:* 409–421, 1966.
25. Gonatas, N. K., Perez, M. C., Shy, G. M., and Evangelista, I.: Central "core" disease of skeletal muscle. Ultrastructural and cytochemical observations in two cases. *Am. J. Pathol. 47:* 503–524, 1965.
26. Hajak, I., Gutmann, E., and Syrovy, I.: Proteolytic activity in denervated and reinnervated muscle. *Physiol. Bohemoslov. 13:* 32–38, 1963.
27. Heffernan, L. P., Rewcastle, N. B., and Humphrey, J. G.: The spectrum of rod myopathies. *Arch. Neurol. 18:* 529–542, 1968.
28. Holt, S. J.: Discussion. In *Ciba Foundation Symposium on Lysosomes*, ed. by de Reuck, A. V. S., and Cameron, M. P., p. 112. Boston, Little, Brown & Co., 1963.
29. Hopkins, I. J., Lindsey, J. R., and Ford, F. R.: Nemaline myopathy; a long-term clinicopathologic study of affected mother and daughter. *Brain 89:* 299–310, 1966.
30. Howes, E. L., Jr., Price, H. M., Pearson, C. M., and Blumberg, J. M.: Hypokalemic periodic paralysis-electromicroscopic changes in the sarcoplasm. *Neurology (Minneap.) 16:* 242–256, 1966.
31a. Hughes, J. T., and Brownell, B.: Ultrastructure of muscle in Werdning-Hoffmann disease. *J. Neurol. Sci. 8:* 363–379, 1969.
31b. Ishikawa, H.: The fine structure of myo-tendon junction is some mammalian skeletal muscles. *Arch. Histol. Jap. 25:* 275–296, 1965.
32. Ishikawa, H.: Electron microscopic observations of satellite cells with special reference to the development of mammalian skeletal muscles. *Z. Anat. Entwicklungsgesch. 125:* 43–63, 1966.
33. Johnson, A. G.: Alterations of the Z-lines and I-band myofilaments in human skeletal muscle. *Acta Neuropathol. 12:* 218–226, 1969.
34. Karlson, U., Anderson-Cedergren, E., and Ottoson, D.: Cellular organization of the frog muscle spindle as revealed by serial sections for electron microscopy. *J. Ultrastruct. Res. 14:* 1–35, 1966.
35. Katz, B.: The termination of the afferent nerve fibre in the muscle spindle of the frog. *Phil. Trans. R. Soc. 243:* 221–240, 1961.
36. Klinkerfuss, G. H.: An electron microscopic study of myotonic dystrophy. *Arch. Neurol. 16:* 181–193, 1967.
37. Klinkerfuss, G., Bleisch, V., Dioso, M. M., and Perkoff, G. T.: A spectrum of myopathy associated with alcoholism. II. Light and electron microscopic observation. *Ann. Intern. Med. 67:* 493–510, 1967.
38. Knappeis, G. G., and Carlsen, F.: The ultrastructure of the Z disc in skeletal muscle. *J. Cell Biol. 13:* 323–335, 1962.
39. Kohn, R. R.: A proteolytic system involving myofibrils and a soluble factor from normal and atrophying muscle. *Lab. Invest. 20:* 202–206, 1969.
40. Laguens, R.: Satellite cells of skeletal muscle in progressive muscular dystrophy. *Virchows Arch. (Pathol. Anat.) 336:* 564–569, 1963.
41. Lindsey, J. R., Hopkins, I. J., and Clark, D. B.: Pathology of nemaline myopathy. *Johns Hopkins Hosp. Bull. 119:* 378–406, 1966.

42. Lubinska, L.: Influence of denervation on acetylcholinesterase in developing fast and slow muscles of the rat. In *Exploratory Concepts in Muscular Dystrophy and Related Disorders*, ed. by Milhorat, A., pp. 168–174. New York, Excerpta Medica Foundation, 1967.
43. Mac Connachie, H. F., Enesco, M., and Leblond, C. P.: The mode of increase in the number of skeletal muscle nuclei in the postnatal rat. *Am. J. Anat. 114:* 245–251, 1964.
44. MacDonald, R. D., and Engel, A. G.: The cytoplasmic body; another structural anomaly of the Z disc. *Acta Neuropathol. (Berl.) 14:* 99–107, 1969.
45. Mauro, A.: Satellite cell of skeletal muscle fibers. *J. Biophys. Biochem. Cytol. 9:* 493, 1969.
46. Meltzer, H.: Creative kinase and aldolase in serum; abnormality common to acute psychoses. *Science 159:* 1368–1370, 1968.
47. Meltzer, H.: Muscle enzyme release in the acute psychoses. *Arch. Gen. Psychiatry 21:* 102–112, 1969.
48. Milhorat, A. T., Shafig, S. A., and Goldstone, L.: Changes in muscle structure in dystrophic patients, carriers and normal siblings seen by electron microscopy; correlation with levels of serum creatine phosphokinase (CPK). *Ann. N. Y. Acad. Sci. 138:* 246–292, 1966.
49. Mintz, G., González-Angulo, A., Fraga, A., and Zavala, B. J.: Ultrastructure of muscle in polymyositis. *Am. J. Med. 44:* 216–224, 1968.
50. Moss, F. P., and Leblond, C. P.: Nature of dividing nuclei in skeletal muscle of growing rats. *J. Cell Biol. 44:* 459–462, 1970.
51. Muir, A. R., Kanji, A. H. M., and Allbrook, D.: The structure of the satellite cells in skeletal muscle. *J. Anat. 99:* 435–444, 1965.
52. Namba, T., Shapiro, M. S., and Grob, D.: Laboratory suggestion; a simple clamp for skeletal muscle biopsy. *Am. J. Clin. Pathol. 50:* 250, 1968.
53. Neustein, H. B.: Fine structure of skeletal muscle in type III glycogenosis. *Arch. Pathol. 88:* 130–136, 1969.
54a. Nienhuis, A. W., Coleman, R. F., Brown, W. J., Munsat, T. L., and Pearson, C. M.: Nemaline myopathy; a histopathologic and histochemical study. *Am. J. Clin. Pathol. 48:* 1–13, 1967.
54b. Papadimitriou, J. M., and Kakulas, B. A.: Ultrastructure of skeletal muscle in muscular dystrophy, the carrier state and other human myopathies. *Proc. Aust. Assoc. Neurol. 5:* 87–94, 1968.
55. Pearson, C. M.: Incidence and type of pathologic alterations observed in muscle in a routine autopsy survey. *Neurology 9:* 757–766, 1959.
56. Pellegrino, C., and Franzini-Armstrong, C.: An electron microscope study of denervation atrophy in red and white skeletal muscle fibers. *J. Cell Biol. 17:* 327–349, 1963.
57. Pellegrino, C., and Franzinin-Armstrong, C.: Recent contributions of electron microscopy to the study of normal and pathological muscle. *Int. Rev. Exp. Pathol. 7:* 139–226, 1969.
58. Pollack, M. S., and Bird, J. W. C.: Distribution and particle properties of acid hydrolases in denervated muscle. *Am. J. Physiol. 215:* 717–722, 1968.
59. Price, H. M.: The skeletal muscle fiber in the light of electron microscope studies. A review. *Am. J. Med. 35:* 589–604, 1963.
60. Price, H. M., Gordon, G. B., Pierson, C. M., Myusat, T. G., and Blumberg, J. M.: New evidence for excessive accumulation of Z-band material in nemaline myopathy. *Proc. Natl. Acad. Sci. U. S. A. 54:* 1398–1406, 1965.
61. Price, H. M., Howes, E. L., and Blumberg, J. M.: Ultrastructural alterations in skeletal muscle fibers injured by cold. II. Cells of the sarcolemmal tube. Observations on discontinuous regeneration and myofibril formation. *Lab. Invest. 13:* 1279–1302, 1966.
62. Price, H. M., Howes, E. L., Jr., and Sheldon, D. B.: An improved biopsy technique for light and electron microscopic studies of human skeletal muscle. *Lab. Invest. 14:* 194–199, 1965.
63. Prineas, J., and Ng, R. C. Y.: Ultrastructural features of intracellular lipid in normal human muscle. *Neurology 17:* 1092–1098, 1967.
64. Przybylski, R. J., and Blumberg, J. M.: Ultrastructural aspects of myogenesis in the chick. *Lab. Invest. 15:* 836–864, 1966.
65. Rasch, P. J.: The problem of muscle hypertrophy. A review. *J. Am. Osteopath. Assoc. 54:* 525–528, 1955.

66. Reedy, M. K.: Discussion. *Proc. Roy. Soc. Lond. (Biol.) 160:* 458, 1964.
67. Reger, J. F.: The fine structure of iridial constrictor pupillae muscle of alligator mississippiensis. *Anat. Rec. 155:* 197–215, 1966.
68. Reger, J. F., and Craig, A. S.: Studies on the fine structure of muscle fibers and associated satellite cells in hypertrophic human deltoid muscle. *Anat. Rec. 162:* 483–500, 1968.
68a. Reske-Nielsen, E., and Harmsen, A.: Electron microscopic study of muscle biopsies from healthy young people. *Acta Pathol. Microbiol. Scand. (A) 80:* 449–467, 1972.
69. Resnick, J. S., and Engel, W. K.: Target fibers; structural and cytochemical characteristics and their relationship to neurogenic muscle disease and fiber types. In *Exploratory Concepts in Muscular Dystrophy*, ed. by Milhorat, A. T., pp. 255–265. New York, Excerpta Medica Foundation, 1967.
70. Resnick, J. S., Engel, W. K., and Nelson, P. G.: Changes in the Z disc of skeletal muscle induced by tenotomy. *Neurology (Minneap.) 18:* 737–740, 1968.
71. Rewcastle, N. B., and Humphrey, J. G.: Vacuolar myopathy. *Arch. Neurol. 12:* 570–582, 1965.
72. Reznik, M.: Origin of myoblasts during skeletal muscle regeneration. Electron microscopic observations. *Lab. Invest. 20:* 353–363, 1969.
73. Rose, A. L., Walton, J. N., and Pearce, G. W.: Polymyositis in an ultramicroscopic study of muscle biopsy material. *J. Neurol. Sci. 5:* 457–472, 1967.
74. Santa, T.: Fine structure of the human skeletal muscle in myopathy. *Arch. Neurol. 20:* 479–489, 1969.
75. Schmalbruch, H.: Crystalloids in human skeletal muscle fibers. *Naturwissenschaften 54:* 519, 1967.
76. Schmalbruch, H.: An atypical nuclear form in human skeletal muscle. *Virchows Arch. (Pathol. Anat.) 345:* 200–204, 1968.
77. Schotland, D. L.: An electron microscopic study of target fibers, target-like fibers, and related abnormalities in human muscle. *J. Neuropathol. Exp. Neurol. 28:* 214–228, 1969.
78. Schutta, H. S., Kelly, A. M., and Zacks, S. I.: Necrosis and regeneration of muscle in paroxysmal idiopathic myoglobinuria: electron microscopic observations. *Brain 92:* 191–202, 1969.
79. Shafiq, S. A., Dubowitz, V., Peterson, H. de C., and Milhorat, A. T.: Nemaline myopathy; report of a fatal case, with histochemical and electron microscopic studies. *Brain 90:* 817–828, 1967.
80. Shafiq, S. A., Gorycki, M. A., Asiedu, S. A., and Milhorat, A. T.: Tenotomy; effect on the fine structure of the soleus of the rat. *Arch. Neurol. 20:* 625–633, 1969.
81. Shafiq, S. A., Gorycki, M., Goldstone, L., and Milhorat, A. T.: Fine structure of fiber types in normal human muscle. *Anat. Rec. 156:* 283–301, 1966.
82. Shafiq, S. A., Gorycki, M. A., and Mauro, A.: Mitosis during postnatal growth in skeletal and cardiac muscle of the rat. *J. Anat. 103:* 135–141, 1968.
83. Shafiq, S. A., Gorycki, M. A., and Milhorat, A. T.: An electron microscopic study of regeneration and satellite cells in human muscle. *Neurology (Minneap.) 17:* 567–575, 1967.
84. Shafiq, S. A., Gorycki, M. A., and Milhorat, A. T.: An electron microscope study of fibre types in normal and dystrophic muscles of the mouse. *J. Anat. 104:* 281–293, 1969.
85. Shafiq, S. A., and Milhorat, A. T.: Some comments on the specificity of alterations in the fine structure in muscle disease. In *Exploratory Concepts in Muscular Dystrophy and Related Disorders*, ed. by Milhorat, A., pp. 268–273. New York, Excerpta Medica Foundation, 1967.
86. Shafiq, S. A., Milhorat, A. T., and Gorycki, M. A.: An electron microscope study of muscle degeneration and vascular changes in polymyositis. *J. Pathol. Bacteriol. 94:* 139–147, 1967.
87. Shafiq, S. A., Milhorat, A. T., and Gorycki, M. A.: Crystals in muscle fibers in patients with diabetic amyotrophy and neuropathy. *Neurology (Minneap.) 18:* 785–790, 1968.
88. Shafiq, S. A., Milhorat, A. T., and Gorycki, M. A.: Fine structure of human muscle in neurogenic atrophy. *Neurology (Minneap.) 17:* 934–948, 1967.
89. Shafiq, S. A., Milhorat, A. T., and Gorycki, M. A.: Giant mitochondria in human muscle with inclusions. *Arch. Neurol. 17:* 666–671, 1967.
90. Shimada, Y., Fischman, D. A., and Moscona, A. A.: The fine structure of embryonic chick skeletal muscle cells differentiated *in vitro*. *J. Cell Biol. 35:* 445–453, 1967.

91. Shy, G. M.: Advances in studies of human myopathies; chemical and morphological abnormalities in muscle disease. *Ann. N. Y. Acad. Sci. 138:* 232–245, 1966–67.

92. Shy, G. M., Engel, W. K., Somers, J. E., and Wanko, T.: Nemaline myopathy; a new congenital myopathy. *Brain 86:* 793–809, 1963.

93. Slotwiner, P., Song, S. K., and Anderson, P. J.: Skeletal muscle changes in leprosy; their relationship to changes in other neurogenic diseases affecting muscle. *J. Pathol. 97:* 211–218, 1969.

94. Slotwiner, P., Song, S. K., and Maker, H. S.: Myopathy resembling McArdle's syndrome. *Arch. Neurol. 20:* 586–598, 1969.

95a. Spiro, A. J., and Kennedy, C.: Hereditary occurrence of nemaline myopathy. *Arch. Neurol. 13:* 155–159, 1965.

95b. Stockdale, F., and Holtzer, H.: DNA synthesis and myogenesis. *Exp. Cell Res. 24:* 508, 1961.

96. Stromer, M. H., Hartshorne, D. J., Mueller, H., and Rice, R. V.: The effect of various protein fractions on Z and M line reconstitution. *J. Cell Biol. 40:* 167–178, 1969.

97. Tappel, A. L., Zalkin, H., Caldwell, A., Desai, I. D., and Shibko, S.: Increased lysosomal enzymes in genetic muscular dystrophy. *Arch. Biochem. Biophys. 96:* 340–346, 1962.

98. Van Breemen, V. L.: Ultrastructure of human muscle. I. Observations in normal striated muscle fibers. *Am. J. Pathol. 37:* 215–230, 1960.

99. Venable, J.: Morphology of cells of normal, testosterone-deprived, and testosterone-stimulated levator ani muscles. *Am. J. Anat. 119:* 271–302, 1966.

100. Vye, M. V., and Fischman, D. A.: The morphologic alteration of particulate glycogen by *en bloc* staining, with uranyl acetate. *J. Ultrastruct. Res. 33:* 278–291, 1970.

101. Zacks, S. I., Lipshutz, H., Markind, S., and Elliott, F. A.: Hyaline degeneration in severely atrophic muscle. Histochemical and electron microscopic observations. *Brain 91:* 463–472, 1968.

102. Zalkin, H., Tappel, A. L., Caldwell, A., Shibko, S., Desai, I. D., and Holiday, T. A.: Increased lysosomal enzymes in muscular dystrophy of vitamin E-deficient rabbits. *J. Biol. Chem. 237:* 2678–2682, 1962.

Chapter 6

# Histochemical Fiber Types in Clinical Disorders of Muscle*

PASQUALE A. CANCILLA AND M. ANTHONY VERITY

A vast literature has accumulated on the subject of enzyme histochemistry of muscle in the 10 years that these techniques have been applied to studies of muscle disorders. For example, histochemical fiber types have been defined,[6, 8] new disease entities have been described,[10, 11, 19, 20] and the role of the upper and lower motor neurons in maintaining the histochemical integrity of muscle fibers has been elucidated.[9] The subject has recently been comprehensively reviewed by Engel in an article in which he proposed a new approach to human neuromuscular disease based on the susceptibility of muscle fiber types,[9] and by Peter in this Symposium (see Chapter 1).

We have been using enzyme histochemistry to study biopsy tissue from selected patients with neuromuscular disorders. In this report, we will present our experiences with these techniques and indicate their value to others engaged in the morphologic and diagnostic aspects of muscle disease.

## MATERIALS AND METHODS

Muscle biopsies from diverse sites were examined from approximately 150 adults and children with muscle disease. In many patients, two different muscles were examined. The techniques that are used to obtain, preserve, and examine muscle have been described.[5] The stains used in this study and the appearance of the two basic types of muscle fibers are listed in Table 1. These two histochemical fiber types are regularly defined in stained sections of rapidly frozen tissue.[6, 8] Histochemical type I fibers are synonymous with red or slow fibers and are rich in oxidative enzymes. Histochemical type II fibers are synonymous with white or fast fibers and are rich in myophosphorylase and myosin ATPase. The myosin ATPase reaction is the least susceptible to change and it is the one recommended stain for identification of the two histochemical fiber types.[8, 9]

* This investigation was supported in part by United States Public Health Service Grants GM-15759 and RRWO238 and a grant from the Muscular Dystrophy Associations of America.

TABLE 1. HISTOCHEMICAL STAINS AND MUSCLE FIBER REACTIONS

| | Type I Fiber | Type II Fiber |
|---|---|---|
| DPNH diaphorase | Dark | Light |
| Succinic dehydrogenase | Dark | Light |
| Myophosphorylase | Light | Dark |
| Myosin ATPase | Light | Dark |

## RESULTS AND COMMENTS

The normal pattern of the two basic histochemical fiber types is shown in Figure 1. The two fiber types are present in about equal numbers, and they are distributed in a "checker-board" pattern (Figs. 1 and 2). This pattern is evident at birth, but in children the fibers are small and they gradually increase in size to reach adult proportions sometime after puberty.[4] In adult males, type II fibers are slightly larger than type I; in adult females, the opposite is true.[2] In addition to the stains for identification of fiber types, we use the Gomori method for demonstration of acid phosphatase.[13] Normal muscle has only a few lysosomes and is, therefore, low in stainable acid phosphatase. In contrast, inflammatory cells, particularly macrophages, and sometimes diseased muscle fibers, are rich in this enzyme and are easily identified by their prominent stain reaction for this enzyme (Fig. 3).

We have been able to recognize six abnormal histochemical patterns in the sections from our patients. These patterns are summarized in Table 2 and each will be discussed separately.

*Specific enzyme deficiency.* The best example of a disorder with a specific enzyme deficiency is McArdle's disease (myophosphorylase deficiency).[16, 18] In this disorder, insidious weakness begins in adult life. Microscopic sections of muscle reveal a vacuolar myopathy with increased glycogen that may suggest the diagnosis. However, the histochemical stains are definitive. All skeletal muscle fibers are grossly deficient in myophosphorylase and are unstained with this reaction, although myophosphorylase may be demonstrable in the media of adjacent blood vessels (Fig. 4). The case that is illustrated here was diagnosed with the histochemical reactions. Before these studies, the patient was thought to have had polymyositis. Other specific enzyme deficiencies could be detected by histochemical methods, but most are differentiated by more sensitive biochemical analyses.

*Fiber type grouping.* Fiber type grouping designates the aggregation of fibers of a single histochemical type in a muscle biopsy.[9] A group is generally quite large, specific, and well demarcated from adjacent fibers (Fig. 5). Neighboring fibers may display a normal histochemical pattern or may consist of a group of fibers of the opposite histochemical type. This pattern of fiber type grouping is specific for denervated muscle that has become reinnervated.[9] Recognition of this pattern in histochemical studies has allowed us to make a definite diagnosis in some cases which in routine studies is interpreted as normal or as nonspecific atrophy.

The normal histochemical pattern of striated muscle is determined by the

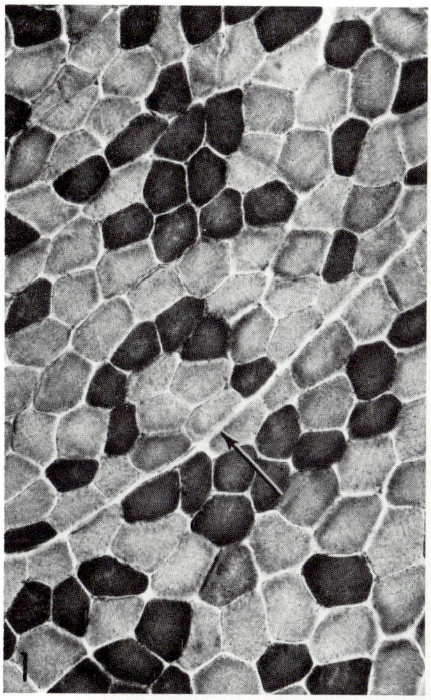

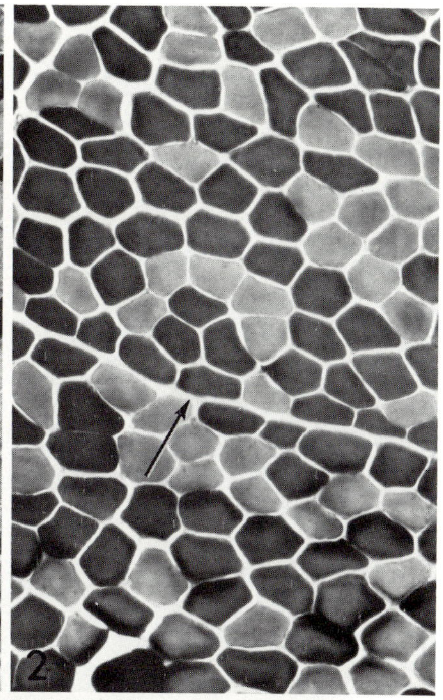

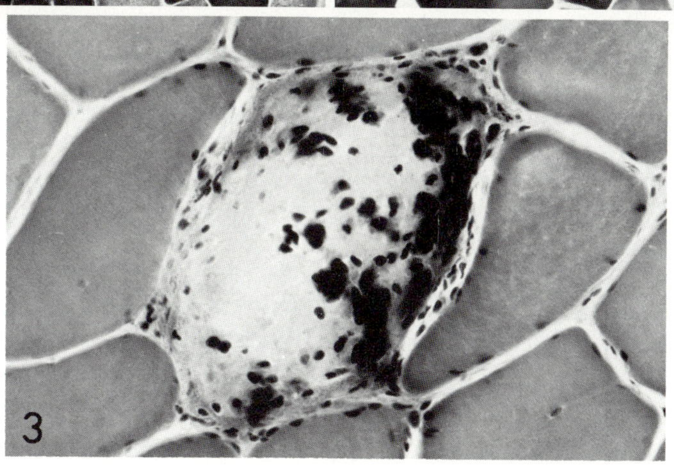

FIG. 1. Two histochemical fiber types are evident in this normal muscle stained for DPNH diaphorase. Type I fibers stain dark and type II fibers stain light with this technique. The *arrow* indicates a group of fibers to compare with an adjacent section in Figure 2. × 85.

FIG. 2. Two histochemical fiber types are evident in this normal muscle stained for myosin ATPase. Type I fibers stain light and type II fibers stain dark with this technique. The *arrow* indicates the same group of fibers as that illustrated in Figure 1. × 85.

FIG. 3. The muscle fiber in the center is necrotic and contains many intensely staining cells. Acid phosphatase; × 210.

type of innervation of the muscle. Experimental studies have shown that there are two types of nerve that innervate muscle.[12, 14] Type I fibers are determined by one type of nerve; type II fibers are determined by the other. If disease affects

TABLE 2. ABNORMAL HISTOCHEMICAL PATTERNS IN CLINICAL DISORDERS OF MUSCLE

| Designation | Fiber Type Involved | Examples |
|---|---|---|
| 1. Specific enzyme deficiency | Type I and II | McArdle's disease (myophosphorylase deficiency) |
| 2. Fiber type grouping | Type I and II | Denervation—reinnervation |
| 3. Generalized alteration | Type I and II | Dermatomyositis, muscular dystrophy, neurogenic atrophy |
| 4. Type I specific alteration | Type I | Myotonic dystrophy, nemaline myopathy, target fibers |
| 5. Type II specific alteration | Type II | Myasthenia gravis, disuse and cachectic atrophy, suprasegmental denervation |
| 6. Fiber predominance | Either Type I or Type II | Nemaline myopathy, denervation—reinnervation |

a lower motor neuron or its axon, the denervated muscle may become reinnervated by collateral sprouting of adjacent axons. The histochemical type of these muscle fibers will be determined by the innervating neuron. This leads to the formation of circumscribed groups of identically innervated fibers with production of the pattern of type grouping in properly stained sections.[15]

*Generalized alteration.* This pattern is probably the most frequently encountered and, from a diagnostic point of view, the least distinctive pattern (Fig. 6). Included in this group are cases of dermatomyositis, muscular dystrophy, and neurogenic atrophy. In all of these disease states the affected fibers are in different stages of involvement by the disease process and they vary in size.

In dermatomyositis there is a tendency for the fibers at the periphery of the fascicle to be smaller than the central fibers, but this change is apparent in sections stained with hematoxylin and eosin and enzyme histochemistry is not required to detect it[1] (Fig. 7). The acid phosphatase reaction has, by staining inflammatory cells, been invaluable in determining the extent of inflammation in a muscle fascicle in cases of dermatomyositis biopsied early in the course of the disease.

*Type I specific alteration.* There is a selective involvement of the histochemical type I fiber in the conditions included in this group. The alteration may be limited to atrophy of the type I fibers. In adults, this change occurs almost exclusively in myotonic dystrophy and is a distinctive feature of the condition[3] (Fig. 8).

Nemaline myopathy is another condition in which specific abnormalities of type I fibers are present.[9, 17] This is a nonprogressive disease in which distinctive rod-shaped structures accumulate in the region of the Z band.[19] The rods are demonstrated with the trichrome or phosphotungstic acid hematoxylin stain and are almost entirely limited to type I fibers. In some cases, type I fibers may be atrophic or predominant in the sections[9, 17] (Fig. 9).

Target fibers are seen in denervating conditions of muscle and are almost exclusively found as an abnormality of type I fibers.[7] As seen with histochemical stains, an affected fiber has an unstained central zone, increased reaction product in a middle zone, and a peripheral zone of normal stain intensity (Fig. 10). Target fibers have been found in a significant proportion of cases with denervation and are considered to be an almost specific abnormality.

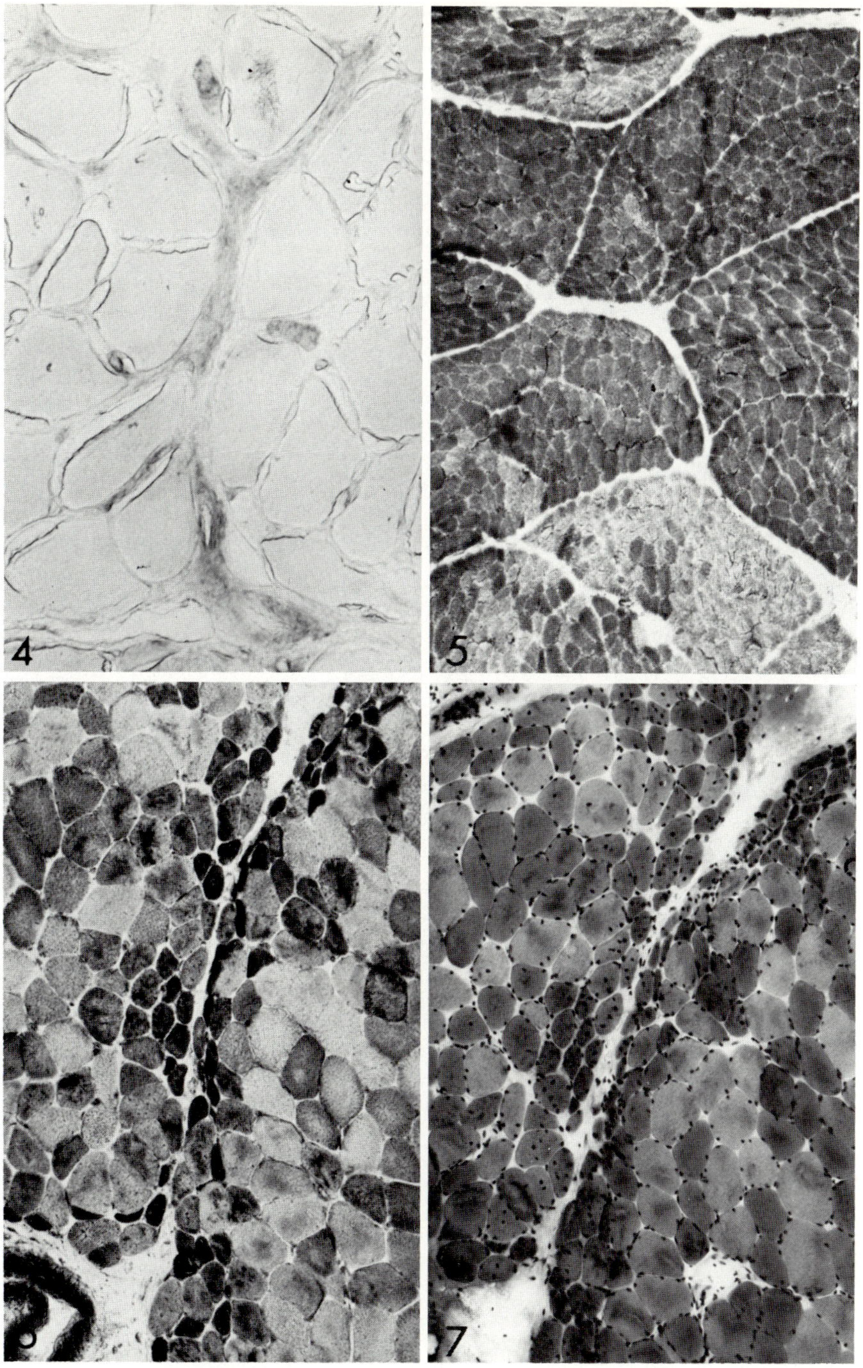

FIG. 4. Myophosphorylase deficiency. The muscle fibers are unstained while the media of the blood vessel in the center is stained. Myophosphorylase; × 150.

FIG. 5. Fiber type grouping. The "checkerboard" pattern of two histochemical fiber types has been replaced by groups of fibers with similar staining reactions. DPNH diaphorase; × 85.

FIG. 6. Dermatomyositis. There is considerable variation in the muscle fibers in these fascicles. Both histochemical fiber types are involved. Compare with Figure 7. DPNH diaphorase; × 85.

FIG. 7. Dermatomyositis. Marked variation in fiber size is evident. The staining qualities of the muscle fibers are such that two types of fibers comparable to the two histochemical fiber types are evident. This section is adjacent to the one illustrated in Figure 6. Frozen section. Hematoxylin and eosin stain; × 85.

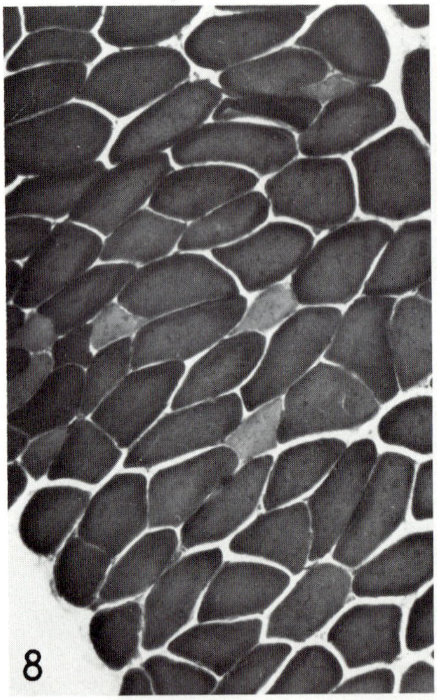

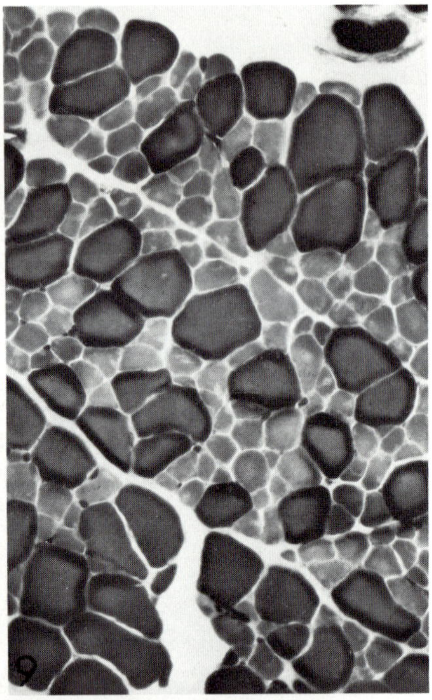

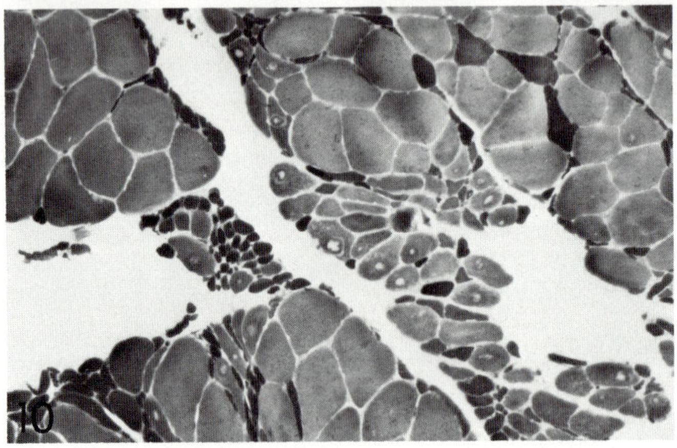

FIG. 8. Myotonic dystrophy. Atrophy of the light staining histochemical type I fibers. Myosin ATPase; × 85.

FIG. 9. Nemaline myopathy. Atrophy of the light staining histochemical type I fibers. Myosin ATPase; × 85.

FIG. 10. Target fibers in denervated muscle. Many of the light staining histochemical type I fibers in the center have alternating light and dark areas of reaction product. Myosin ATPase; × 85.

Type I fibers may be primarily involved in several rare diseases of muscle. These diseases are discussed in a recent review on the susceptibility of muscle fiber types.[9]

*Type II specific alteration.* There is a heterogeneous group of abnormal conditions which is included under this designation. These conditions have as their conspicuous and common feature involvement of type II fibers that usually appears as atrophy of that fiber type (Fig. 11). In the adult, the finding of type II atrophy lacks the diagnostic importance of type I fiber atrophy since it may be found in such diverse situations as myasthenia gravis,[3] disuse atrophy, cachexia, and suprasegmental denervation.[9] The reason for the diverse susceptibility of this fiber type is unknown, but a reasonable suggestion is that the muscle may atrophy because of a loss of trophic influence from the lower motor neuron.[9] Whatever the cause, type II fiber atrophy lacks diagnostic specificity.

*Fiber predominance.* In this situation, the histochemical studies are characterized by the identical staining reaction of nearly all muscle fibers rather than the usual pattern with two histochemical fiber types. This uniformity of muscle fiber staining may be found in a muscle biopsy with histologic evidence of denervation atrophy in which case it is most likely due to reinnervation (Fig. 12). In other instances, such as nemaline myopathy or a myopathy with lysis of fibrils in type I fibers, the reason for the predominance of one fiber type is unknown.

## DISCUSSION AND CONCLUSIONS

This presentation represents a distillation of our experience using enzyme histochemistry in a diagnostic service. It must be understood that we do not examine every muscle biopsy with these techniques. In many instances, the

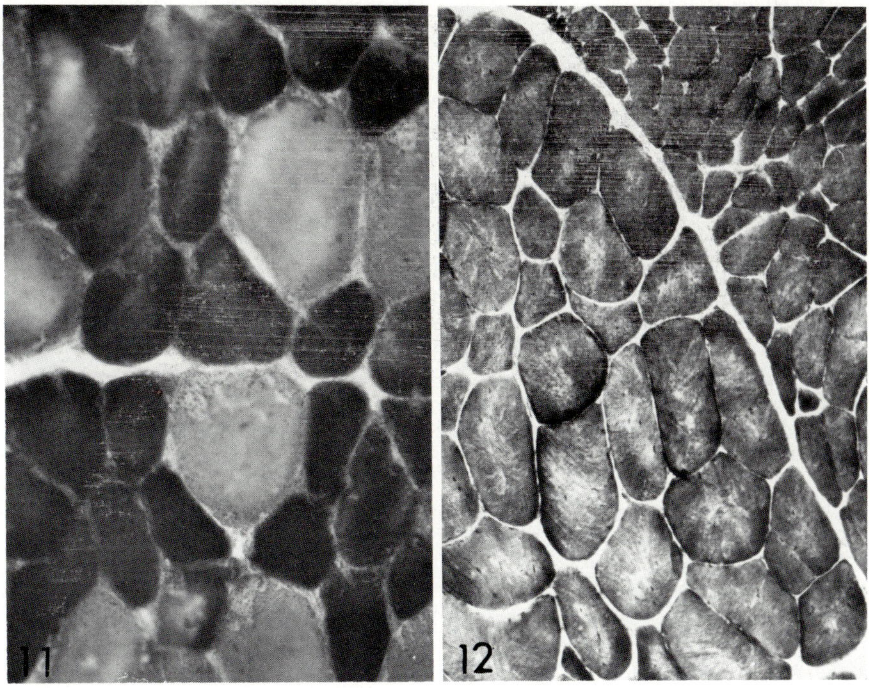


FIG. 11. Myasthenia gravis. Atrophy of the dark staining histochemical type II fibers. Myosin ATPase; × 210.

FIG. 12. Neurogenic atrophy with reinnervation. One histochemical fiber type predominates. DPNH diaphorase; × 85.

diagnosis is readily apparent clinically and the biopsy is done to substantiate the diagnosis. The usual methods of examination of muscle are all that are required in these cases.

For practical purposes, we can add to this group the cases of dermatomyositis, muscular dystrophy, and neurogenic atrophy with distinctive histology. From a diagnostic point of view, there is no need to proceed with histochemical analysis on a routine basis in these cases since the histochemical abnormalities are nonspecific and the diagnosis is readily apparent.

There remain, then, small but important groups of cases in which diagnostic information may be derived from enzyme histochemistry. Cases with the histochemical pattern of type grouping have been the most distinctive. This pattern of fiber type grouping is due to a process of denervation and reinnervation.[15] We have found this abnormality in some biopsies which have lacked other features of neurogenic disease; in these cases, its recognition has led to the proper classification of the disease process.

The definition of abnormalities restricted to a single fiber type has been of considerable diagnostic value. Cases with nemaline myopathy,[9, 17, 19] target fibers,[7] or other characteristic abnormalities need no further elaboration. Myotonic dystrophy has histologic features such as chains of sarcolemmal nuclei in the center of the fiber that may be found in other conditions, but the diagnosis is virtually assured when the change is accompanied by type I fiber atrophy.[3] Unfortunately, type II fiber atrophy lacks this specificity.[9]

Myophosphorylase deficiency (McArdle's disease) is another instance in which the use of enzyme histochemistry has led to a definite diagnosis. This disease can be detected clinically by the ischemic forearm test with lactic acid determinations; in these instances, the muscle biopsy and the histochemistry are done to corroborate the diagnosis.

Enzyme histochemistry achieves its most important application as a research technique. In carefully selected cases, enzyme histochemistry has augmented the information obtained from routine stains and has proved to be a useful method for the diagnosis of muscle disorders.

## ACKNOWLEDGMENTS

The authors would like to thank Margaret Hall, Laurel Reed, Joyce Brooks, Stephen Frommes, and Ethel Doucette for very valuable assistance.

### REFERENCES

1. Banker, B. Q.: Dermatomyositis of childhood. *Trans. Am. Neurol. Assoc. 87:* 11–13, 1962.
2. Brooke, M. H., and Engel, W. K.: The histographic analysis of human muscle biopsies with regard to fiber types, 1. Adult male and female. *Neurology (Minneap.) 19:* 221–233, 1969.
3. Brooke, M. H., and Engel, W. K.: The histographic analysis of human muscle biopsies with regard to fiber types. 3. Myotonias, myasthenia gravis and hypokalemic periodic paralysis. *Neurology (Minneap.) 19:* 469–477, 1969.
4. Brooke, M. H., and Engel, W. K.: The histographic analysis of human muscle biopsies with regard to fiber types. 4. Children's biopsies. *Neurology (Minneap.) 19:* 591–605, 1969.
5. Cancilla, P. A.: Techniques of muscle biopsy and staining methods with particular emphasis on commonly encountered artifacts. In *The Striated Muscle*, ed. by Pearson, C. M., and Mostofi, F. K. Baltimore, Williams and Wilkins, 1973.

6. Dubowitz, V., and Pearse, A. G. E.: A comparative histochemical study of oxidative enzyme and phosphorylase activity in skeletal muscle. *Histochemie 2:* 105–117, 1960.
7. Engel, W. K.: Muscle target fibers; a newly recognized sign of denervation. *Nature 191:* 389–390, 1961.
8. Engel, W. K.: The essentiality of histo- and cytochemical studies of skeletal muscle in the investigation of neuromuscular disease. *Neurology (Minneap.) 12:* 778–784, 1962.
9. Engel, W. K.: Selective and nonselective susceptibility of muscle fiber types. A new approach to human neuromuscular diseases. *Arch. Neurol. 22:* 97–117, 1970.
10. Engel, W. K., Foster, J. B., Hughes, B. P., Huxley, H. E., and Mahler, R.: Central core disease; an investigation of a rare muscle cell abnormality. *Brain 84:* 167–185, 1961.
11. Engel, W. K., Gold, G. N., and Karpati, G.: Type I fiber hypotrophy and central nuclei. A rare congenital muscle abnormality with possible experimental model. *Arch. Neurol. 18:* 435–444, 1968.
12. Guth, L.: "Trophic" Influences of Nerve on Muscle. *Physiol. Rev. 48:* 645–687, 1968.
13. Holt, S. J.: Factors governing the validity of staining methods for enzymes and their bearing upon the Gomori acid phosphatase technique. *Exp. Cell Res. (Suppl.) 7:* 1–27, 1968.
14. Karpati, G., and Engel, W. K.: Neuronal trophic function. A new aspect demonstrated histochemically in developing soleus muscle. *Arch. Neurol. 17:* 542–545, 1967.
15. Karpati, G., and Engel, W. K.: "Type grouping" in skeletal muscles after experimental reinnervation. *Neurology (Minneap.) 18:* 447–455, 1968.
16. McArdle, B.: Myopathy due to a defect in muscle glycogen breakdown. *Clin. Sci. 10:* 13–33, 1951.
17. Nienhuis, A. W., Coleman, R. F., Brown, W. J., Munsat, T. L., and Pearson, C. M.: Nemaline myopathy. A histopathologic and histochemical study. *Am. J. Clin. Pathol. 48:* 1–13, 1967.
18. Pearson, C. M., Rimer, D. G., and Mommaerts, W. F. H. M.: A metabolic myopathy due to absence of muscle phosphorylase. *Am. J. Med. 30:* 502–517, 1961.
19. Shy, G. M., Engel, W. K., Somers, J. E., and Wanko, T.: Nemaline myopathy; a new congenital myopathy. *Brain 86:* 793–810, 1963.
20. Spiro, A. J., Shy, G. M., and Gonatas, N. K.: Myotubular myopathy; persistence of fetal muscle in an adolescent boy. *Arch. Neurol. 14:* 1–14, 1966.

Chapter 7

# The Pathologic Interpretation of Muscle Histochemistry*

MICHAEL H. BROOKE

---

Until recently, the analysis of muscle biopsies has hinged upon whether the muscle showed the changes of denervation or of myopathy. The obligations of the pathologist were largely fulfilled if he could place the biopsy firmly in one of the two groups. With the advent of histochemistry, this classification has been extended and we are now able to recognize the changes seen in many different diseases. The purpose of this paper is to review some of these changes. It is not meant to be an exhaustive review of all muscle pathology. It is intended merely to highlight some of the diseases which we have encountered in a large and unselected series of biopsies.

The techniques which we use in the analysis of muscle biopsies have been described before.[18] Most of the information which we use in the interpretation of biopsies is derived from the following reactions: (1) modified trichrome stain[19]; (2) Verhoeff-Van Giesons stain[24]; (3) diphosphopyridine-reduced (DPNH) dehydrogenase[3]; (4) routine calcium method for adenosine triphosphatase (ATPase)[28]; (5) modified ATPase following preincubation at pH 4.3 and 4.6[9]; (6) amylophosphorylase[31]; (7) periodic acid-Schiff stain (PAS).[24]

The paper will be divided into three parts. Firstly, we would like to give a brief description of normal muscle; secondly, a description of the various abnormalities which we see in muscle disease will be given; thirdly, we will provide a description of changes seen in individual diseases. Many of these changes have been described previously, and excellent reviews of muscle pathology are available.[1, 12, 16, 22, 29, 32]

## NORMAL HUMAN MUSCLE

When normal human muscle is sectioned transversely, it can be seen to consist of individual fibers which are slightly polygonal in appearance, which are closely apposed to one another, and which are grouped into fascicles (Fig. 1). A fascicle is separated from its neighbor by a band of fibrous tissue of varying thickness. Using histochemical techniques various types of muscle fiber may be

* This work was supported by Grant 5 R01 NS 08435-02 NEUB from the National Institute of Neurological Diseases and Stroke.

differentiated.[13, 15] The exact number of fiber types present in muscle is somewhat problematical and the system of nomenclature of these types is, as yet, unresolved. We share the view that the ATPase reaction gives a clear differentiation which lends itself to the characterization of fibers more easily than do many of the other histochemical reactions. We do not necessarily wish to impose our own idiosyncrasies in regard to fiber typing upon others, but we would like to make one observation. In order for a system of fiber typing to have any value in regard to muscle pathology, it must be of practical use in the interpretation of biopsies. We have found the characterization of fiber types with the ATPase reaction to fulfill these criteria. The differentiation into type I muscle fibers, which are light, with the routine histochemical reaction for ATPase and type II muscle fibers, which are dark, has formed the basis for previous analyses of the changes of fiber size.[5-8] In order to evaluate fiber sizes, histograms are constructed from photographs of the ATPase reaction in each muscle biopsy. The lesser fiber diameter is measured and this is defined as the maximum distance which can be measured across the lesser aspect of the muscle fiber (Fig. 2). This measurement is used to avoid any artifact introduced by an oblique section of the muscle causing an artificially large measurement (Fig. 3). Each fiber type is measured independently and a histogram is constructed from measurements of 200 fibers in each biopsy.

In normal muscle, the values of fiber sizes depend to some extent upon whether the patient is male or female. In the male, type II fibers are larger than

---

FIG. 1. Normal human muscle. Muscle fibers of polygonal shape can be seen grouped into fascicles separated by fibrous tissue. Verhoeff-Van Gieson's stain; × 225.

FIG. 2. A diagram illustrating the measurement taken across the lesser fiber diameter, represented by the *dashed line*.

FIG. 3. In this diagram the lesser fiber diameter, represented by the *dashed line*, is shown to be the measurement least affected by obliquity of the plane of section or by kinking of the muscle fiber.

FIG. 4. Normal adult male muscle. In this muscle the type II fibers (dark) are slightly larger than the type I fibers (light) (ATPase (pH 9.4); × 125).

FIG. 5. Normal adult female muscle. The type II fibers are slightly smaller than the type I fibers (ATPase (pH 9.4); × 125).

FIGS. 6 to 10. Histochemical reactions of serial sections from normal human muscle. In all of these photographs, the fiber labeled *I* is a type I fiber. The fibers labeled *A* and *B* are type IIA and type IIB fibers, respectively. × 438.

FIG. 6. ATPase (pH 9.4). The type I fibers are light and the type II fibers, both A and B, are dark with this reaction.

FIG. 7. ATPase (pH 4.6 for 5 minutes preincubation). In this reaction, the type I fiber and the type IIB fibers are dark and the type IIA fiber is light.

FIG. 8. ATPase (pH 4.3 for 5 minutes preincubation). In this reaction, the type I fiber remains dark but both type IIA and B fibers are now light.

FIG. 9. DPNH dehydrogenase reaction. The type I fiber stains darkly with this reaction. The type IIB fiber is light, and the type IIA fiber is somewhat intermediate in staining characteristics.

FIGS. 10 and 11. Serial sections from a deltoid muscle biopsy. A type IIC fiber is labeled *C*. (× 125).

FIG. 10. ATPase (pH 9.4). The normal staining characteristics of type I and type II fibers can be seen.

FIG. 11. ATPase (pH 4.3 for 5 minutes preincubation). This photograph should be compared with Figure 8 as well as Figure 10. There are type II fibers which do retain some stain at pH 4.3. These are the type IIC fibers.

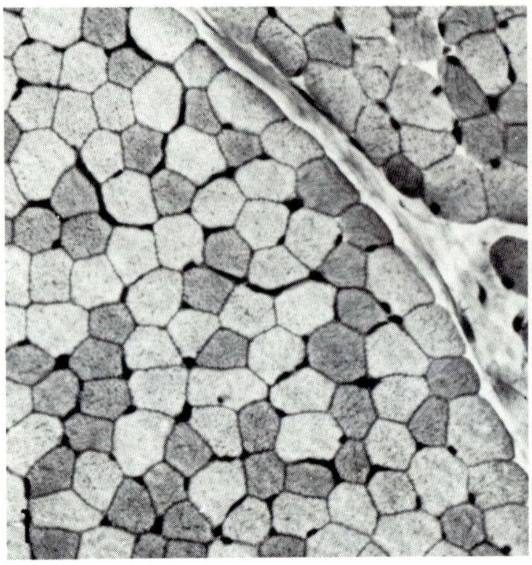

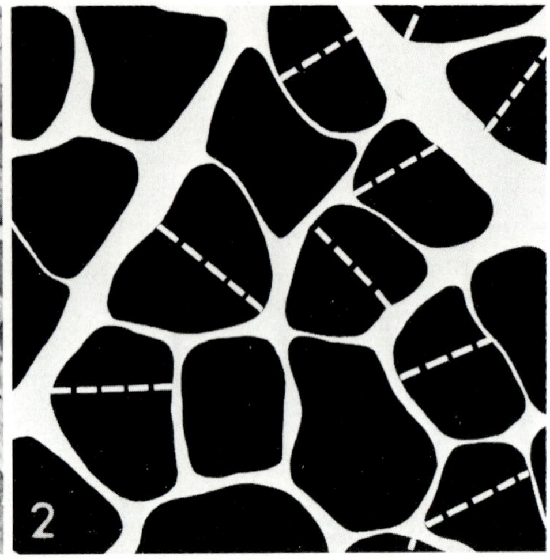

Effect of Obliquity and Kinking on Transverse Section

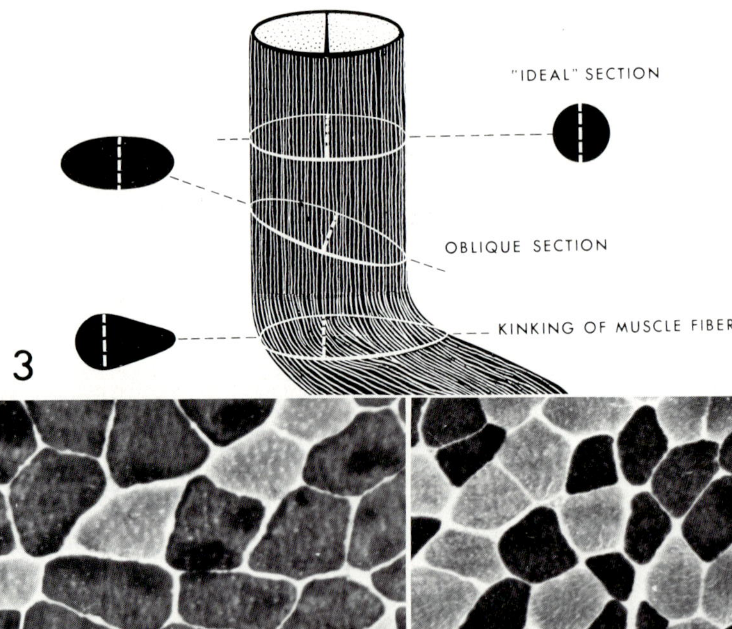


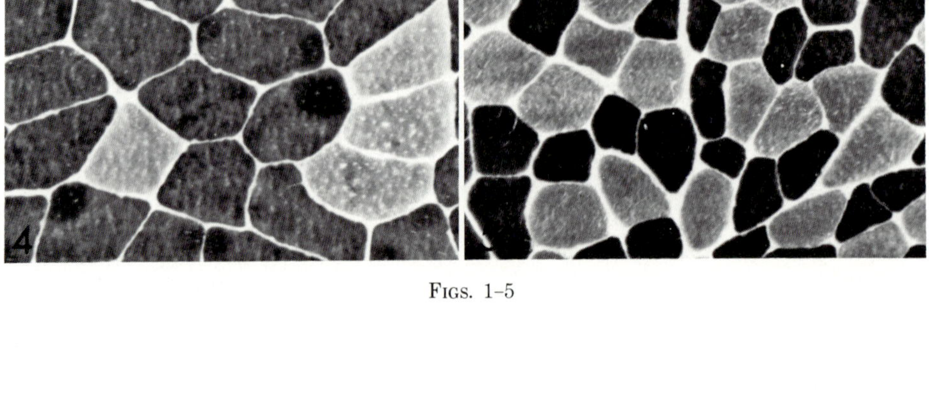

FIGS. 1–5

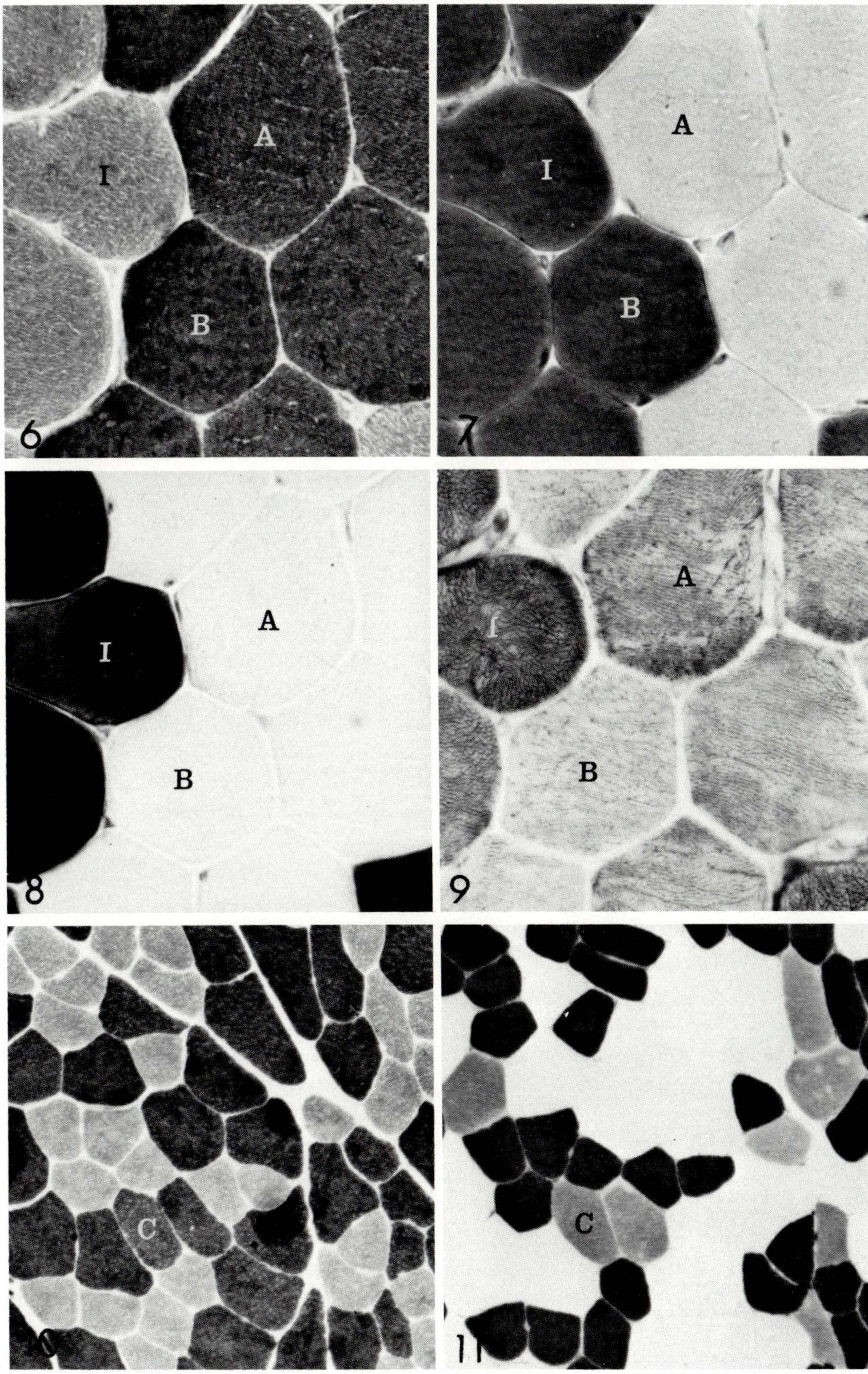


FIGS. 6–11

type I fibers, whereas in the female, the reverse is true (Figs. 4 and 5). The type I fibers in both sexes are the same size. Approximately two-thirds of the fibers in normal muscle are type II and one-third is type I. These different fiber types are distributed in a random mosaic pattern. Although the ATPase reaction shows clear differentiation into two fiber types, other histochemical reactions have suggested that three or more fiber types are present in normal muscle.[26] The existence of three fiber types is also indicated by electron microscopic studies.[27] Recently it has become apparent that the differentiation of fiber types with the histochemical reaction for ATPase is dependent upon the pH sensitivity of the enzyme system and that clear differentiation into three major and one minor fiber type may be obtained with the ATPase reaction.[10] Summarizing our results, we may say that the muscle fibers that we categorize as type I possess a myosin ATPase which resists, 5 minutes, preincubation at pH between 3.9 and 10.4. The development of the reaction is slightly impaired as one approaches the alkaline pH (*i.e.*, pH value greater than 9.0). What has appeared previously as a uniform group of type II fibers with the routine reaction at pH 9.4 may be differentiated into two further groups, type IIA and type IIB. The characteristic of type IIA fibers is that they possess a myosin ATPase which is resistant to preincubation between pH 4.9 and 10.8. These fibers are almost totally inhibited at pH values below 4.5. Type IIB fibers possess an ATPase which is resistant to preincubation from pH 4.5 to 10.8. The type IIB fiber is almost completely inhibited at pH 4.3. There is an additional fiber type which we have termed the type IIC fiber. These are not very numerous in most biopsies of the normal biceps and vastus, although we are not sure whether they may be seen in other human muscles. The mean diameter and variation of size of this fiber type is always identical to the IIA fiber. We have not, therefore, analyzed it separately. In some normal deltoid biopsies this fiber has been prominent. The characteristics of this type IIC fiber are that, although the stain in the fiber is inhibited at the same pH as the type IIA fibers, it is not inhibited to the same degree and the fiber can still be seen even below pH 4.3 as a relatively pale staining fiber. It should be stressed that these are not the same as the type IIB fibers. In our analysis we have used four different ATPase incubating conditions. Tissue sections have been preincubated at pH 10.4, 9.4, 4.6, and 4.3 before incubating them in the standard reaction for ATPase. When photographs of serial sections are taken, the various fiber types can be differentiated clearly and measured. The various fiber types are indicated in Figures 6 through 11. With the extension of the evaluation of fiber types to include the three main categories, type I, type IIA, and type IIB, it was found that each of these fiber types was present in approximately equal proportions in the biceps and vastus muscles. The type IIA fibers were always larger than the type IIB fibers. The values of mean diameters and fiber ratios are given in Table 1. It seems that there are slightly more type I fibers in the gastrocnemius and deltoid muscles than in the biceps and vastus muscles. We do not, however, have valid statistical results on the gastrocnemius and deltoid. Furthermore, in children the results are somewhat different. The sex difference in muscle fiber size does not seem to be present before the age of 12 and often does not appear until after

TABLE 1. AVERAGE DIAMETER AND PERCENTAGE OF FIBERS IN NORMAL HUMAN BIOPSIES[a]

| | Muscle Fiber Type | | | |
|---|---|---|---|---|
| | I | II | IIA | IIB |
| Male | 63 | 69 | 71 | 66 |
| Female | 57 | 51 | 57 | 47 |
| Percentage of fibers in total | 38% | 62% | 29% | 33% |

[a] The figures for the type I and type II fibers are calculated from biopsies in a previous study[5] together with 12 additional normal biopsies. The values for IIA and IIB fibers are calculated from six normal males and six normal females.

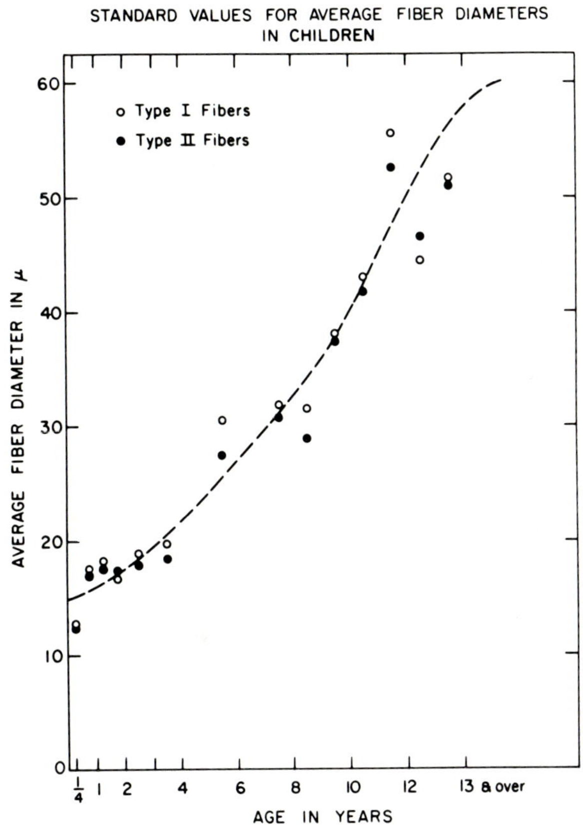


FIG. 12. Mean diameters of type I and type II fibers in normal children.

the age of 14. In most normal children under the age of 14, the type I and type II fibers are of roughly equal size. Their mean diameter is shown in an accompanying graph (Fig. 12).

In normal muscle the nuclei are situated close to the sarcolemma at the periphery of the fiber. Occasionally nuclei may be seen which are located within the fiber when seen on transverse section. In normal human muscle not more than 3% of the fibers display these internal nuclei[22] and furthermore, these fibers tend to be rather large.

The cytochemical architecture of normal fibers is relatively simple to

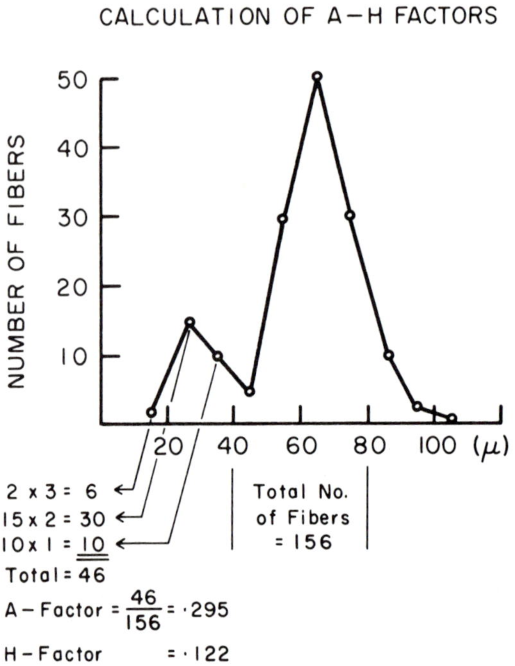


FIG. 13. The method of calculating atrophy and hypertrophy factors. The resulting number is multiplied by 1,000 and expressed as a whole number.

describe. The myofibrils which are seen on the histochemical reaction for myosin ATPase are separated from each other by aqueous sarcoplasm, mitochondria, the transverse tubular system, and the sarcoplasmic reticulum. All of these elements comprise a network lying between the myofibrils. This intermyofibrillar network is well demonstrated with the oxidative enzyme reaction such as DPNH dehydrogenase, with the phosphorylase reaction, and by the PAS stain. It is seen in these stains as a regular and fairly uniform net throughout the muscle fiber (Fig. 14). It should be noted that in very young children this network is coarsened and rather irregular (Fig. 15). This is termed the infantile network pattern. Additionally the oxidative enzyme reactions such as the DPNH dehydrogenase may be used to differentiate normal human fiber types. Type I fibers are dark with this reaction, type IIA fibers demonstrate an intermediate stain, and type IIB fibers are the lightest staining fibers (see Fig. 9). In addition to the muscle fibers, blood vessels, nerves, and muscle spindles may be seen on a muscle biopsy (Fig. 16). Since the main part of this paper will

FIG. 14. Normal human muscle. This illustrates the normal regular staining of the intermyofibrillar network pattern. DPNH dehydrogenase; × 885.

FIG. 15. The slightly more granular pattern of the infantile intermyofibrillar network. This biopsy was from a 2-year-old child. DPNH dehydrogenase; × 555.

FIG. 16. A bundle of nerve fibers can be seen running toward a muscle spindle which is on the right-hand side. Verhoeff-Van Gieson's stain; × 125.

FIGS. 17 and 18. These serial sections of a biopsy from a patient with limb girdle dystrophy illustrate fiber splitting. The *arrows* indicate the development of fibrous septa in one photograph that are not seen in the same muscle fiber on the other photograph. ATPase (pH 9.4); × 125.

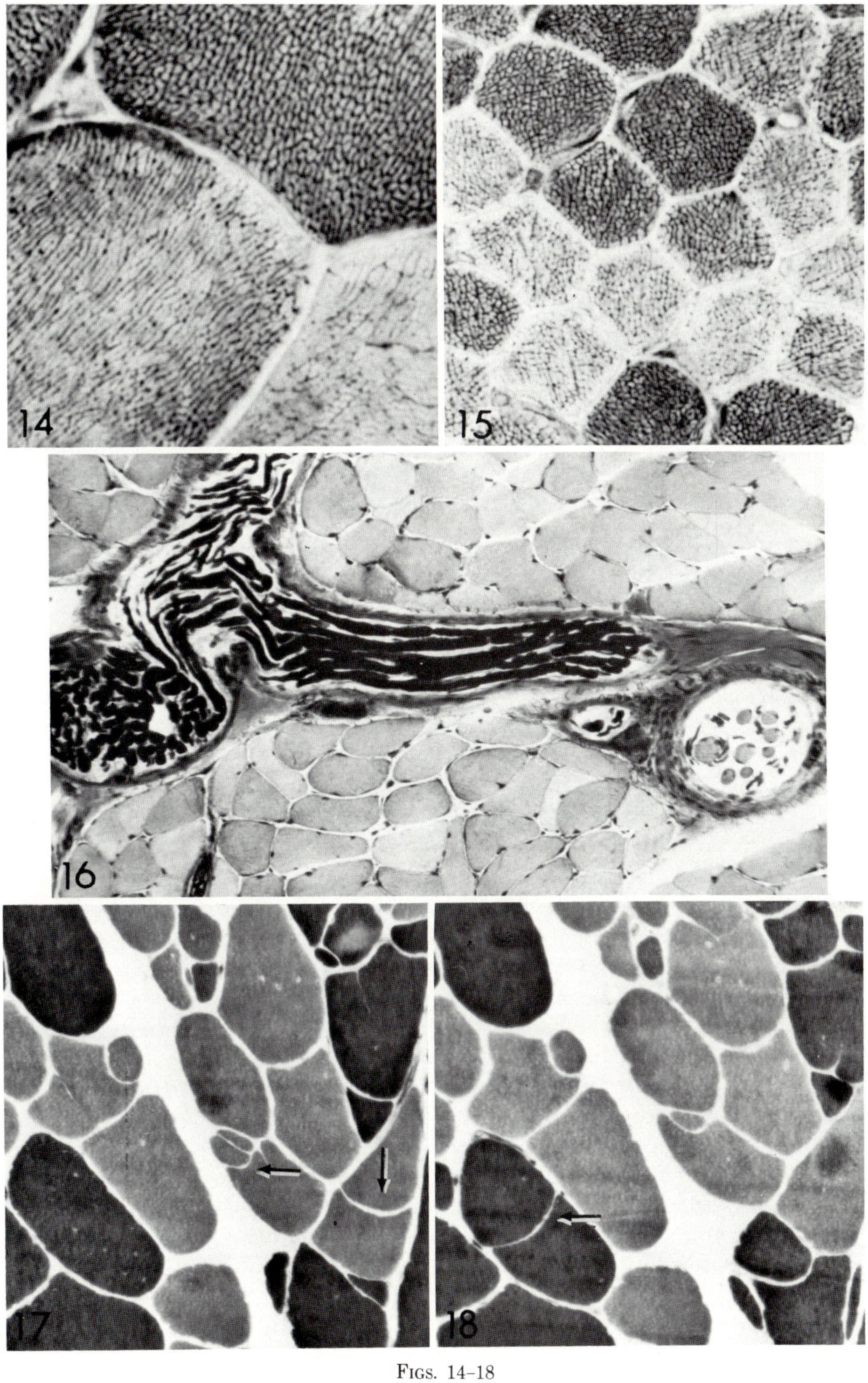


FIGS. 14–18

TABLE 2. MAXIMUM VALUES FOR NORMAL ATROPHY AND HYPERTROPHY FACTORS

| | Biceps | | | | Vastus | | | | Gastrocnemius | | | |
|---|---|---|---|---|---|---|---|---|---|---|---|---|
| | Type I | | Type II | | Type I | | Type II | | Type I | | Type II | |
| | Atrophy | Hypertrophy | Atrophy | Hypertrophy | Atrophy | Hypertrophy | Atrophy | Hypertrophy | Atrophy | Hypertrophy | Atrophy | Hypertrophy |
| Male ....... | 150 | 300 | 150 | 500 | 150 | 150 | 150 | 400 | 150 | 500 | 150 | 500 |
| Female ..... | 100 | 200 | 150 | 150 | 100 | 400 | 200 | 150 | 100 | 200 | 150 | 150 |

not be concerned with these last structures, we shall not describe them further here.

Finally, it should be remembered that in normal human muscle evidence of degeneration, regeneration, and cellular reactions is not prominent. However, mild trauma to a muscle whether from percussion or the introduction of an electromyogram needle or a hypodermic needle may result in findings of inflammation and degeneration.[17] Great care should, therefore, be taken to obtain a biopsy from a muscle which has not been subjected to such trauma.

## ILLUSTRATION OF PATHOLOGIC CHANGES

In order to describe the abnormalities seen in various muscle diseases, a vocabulary of pathologic change is needed. These are the building blocks of muscle pathology, and, although the changes themselves are similar, they are present in various combinations in individual muscle diseases. In describing these changes, they will be divided into various sections in the same way in which muscle biopsies are described when trying to evaluate the pathology. These sections are: (1) changes in fiber size, (2) changes in nuclei, (3) degeneration and regeneration, (4) cellular reactions, (5) changes in the cytochemical architecture, and (6) abnormalities of distribution.

### CHANGES IN FIBER SIZE

Changes in the size of fibers may be quite difficult to evaluate unless actual measurements are taken. The problem of comparing such measurements with the measurements of normal muscle led to the development of "atrophy-hypertrophy" factors. This system is simply a method of quantitating the reduction or increase in the size of fibers within a biopsy. Such quantitation has to take into account that the histograms derived from many abnormal biopsies do not lend themselves to the calculation of standard mean and deviation because of the skew or complex histogram which is obtained. In the male, normal fibers are considered to be those with lesser diameters between 40 and 80 $\mu$. To calculate atrophy factors, the number of fibers in the histogram with average diameters between 30 and 40 $\mu$ is counted. The number of fibers measuring between 20 $\mu$ and 30 $\mu$ is counted and doubled. The number of fibers measuring from 10 to 20 $\mu$ was counted and tripled, whereas the number measuring from 0 to 10 $\mu$ was multiplied by four. All of these resulting numbers are added. In order to express this value as a proportionate value for the total biopsy, this sum is divided by the total number of fibers in the biopsy. The resulting factor is multiplied by 1000 and expressed as a whole number. This number is referred to as the atrophy factor (Fig. 13). Hypertrophy factors may be calculated in a similar fashion. In the female the factors are calculated in the same way. The "normal" fibers in the female are assumed to occur with diameters between 30 and 70 $\mu$. Fibers outside of these limits are used in calculating atrophy-hypertrophy factors in the female. The normal maximum values for atrophy and hypertrophy factors for type I and type II fibers are shown in Table 2. If we are trying to determine whether there is an undue degree of atrophy of the type I fibers in a biopsy, we calculate the atrophy factor for type I fibers to ascertain whether this is above the normal threshold. Unless it is, we do not consider the type I fibers to be atrophied. By using this method, we may show nonspecific fiber atrophy in which all three fiber types are small; on the other hand, we may see examples of type I fiber atrophy, type IIA fiber atrophy, or type IIB fiber atrophy. We may also see selective hypertrophy of the various fiber types. The significance of some of the various forms of selective and nonselective atrophy will be discussed later. Changes in the size of fibers may take place in several ways. There may be atrophy, which results in a reduction in the size of the entire fiber. If all of the fibers in the biopsy are involved, this will result in a histogram which is shifted toward the left. However, in such conditions as the denervating illnesses, not all of the fibers are involved and the atrophic fibers constitute a separate population. Thus, the histogram reflects a population of relatively normal or even hypertrophied fibers together with small atrophic ones and presents a twin-peaked appearance. Small fibers may also be produced by the splitting of larger fibers (Figs. 17 and 18). This is particularly noticable in the limb girdle dystrophies, although it is not limited to this group. Very often this presents a histogram in which the average size of the fibers is normal but the fibers show a wide variability in size from very small to very large fibers.

---

FIG. 19. A biopsy from a patient with limb girdle dystrophy to show internal nuclei. Hematoxylin and eosin stain; × 125.

FIG. 20. Chronic neuropathy. A longitudinal section of muscle to show the clumps of pyknotic nuclei (formol-fixed, paraffin-embedded). Hematoxylin-Van Gieson's stain; × 300.

FIG. 21. Pale staining vesicular nuclei in a patient with Duchenne dystrophy (formol-fixed, paraffin-embedded). Hematoxylin and eosin stain; × 300.

FIG. 22. Tigroid nuclei from a patient with chronic familial peripheral neuropathy. These nuclei are marked by a heavily stippled appearance to the chromatin (formol-fixed, paraffin-embedded). Hematoxylin and eosin stain; × 620.

FIG. 23. A pale staining liquefied fiber from a patient with limb girdle dystrophy. Trichrome stain; × 125.

FIG. 24. A fiber undergoing phagocytosis in a patient with a chronic peripheral neuropathy. Hematoxylin and eosin stain; × 310.

FIG. 25. Perivascular reaction. The inflammatory response followed the course of the blood vessel which can be seen in cross-section at the right. This was from a patient with polymyositis. Trichrome stain; × 125.

FIG. 26. A lymphorrhage seen in a patient with polymyositis. Trichrome stain; × 225.

FIG. 27. Fibrosis. An increased amount of fibrous tissue can be seen separating the individual fibers in this patient with Duchenne dystrophy. Hematoxylin and eosin stain; × 310.

FIG. 28. Target fibers. A central unreactive zone surrounded by an intermediate zone of increased stain can be seen. ATPase (pH 4.65 for 5 minutes preincubation); × 225.

FIG. 29. Target fibers. DPNH dehydrogenase; × 310.

FIG. 30. Moth-eaten and whorled fibers in a patient with limb girdle dystrophy. The pattern of the intermyofibrillar network is distorted. DPNH dehydrogenase; × 310.

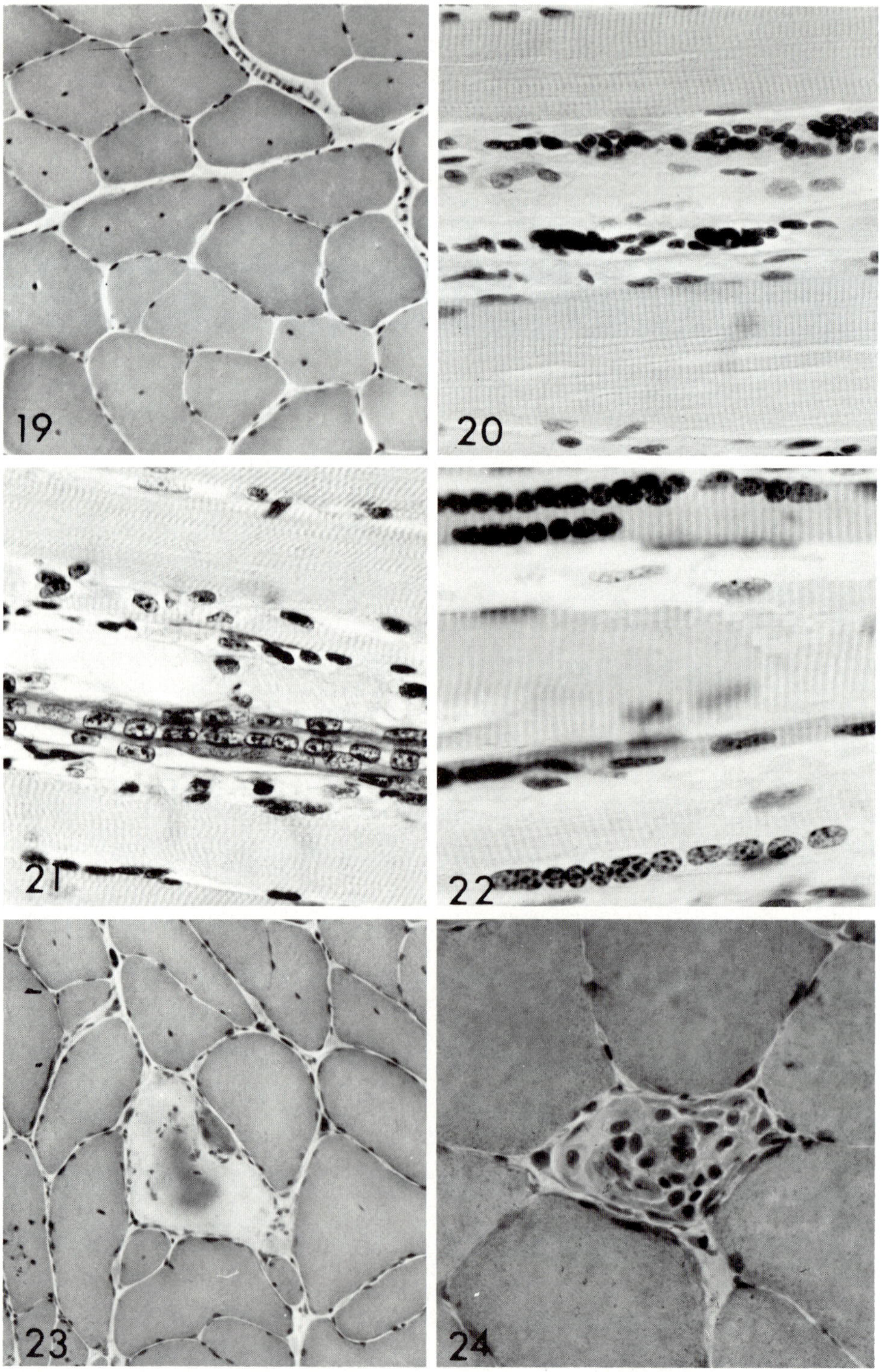


FIGS. 19–24

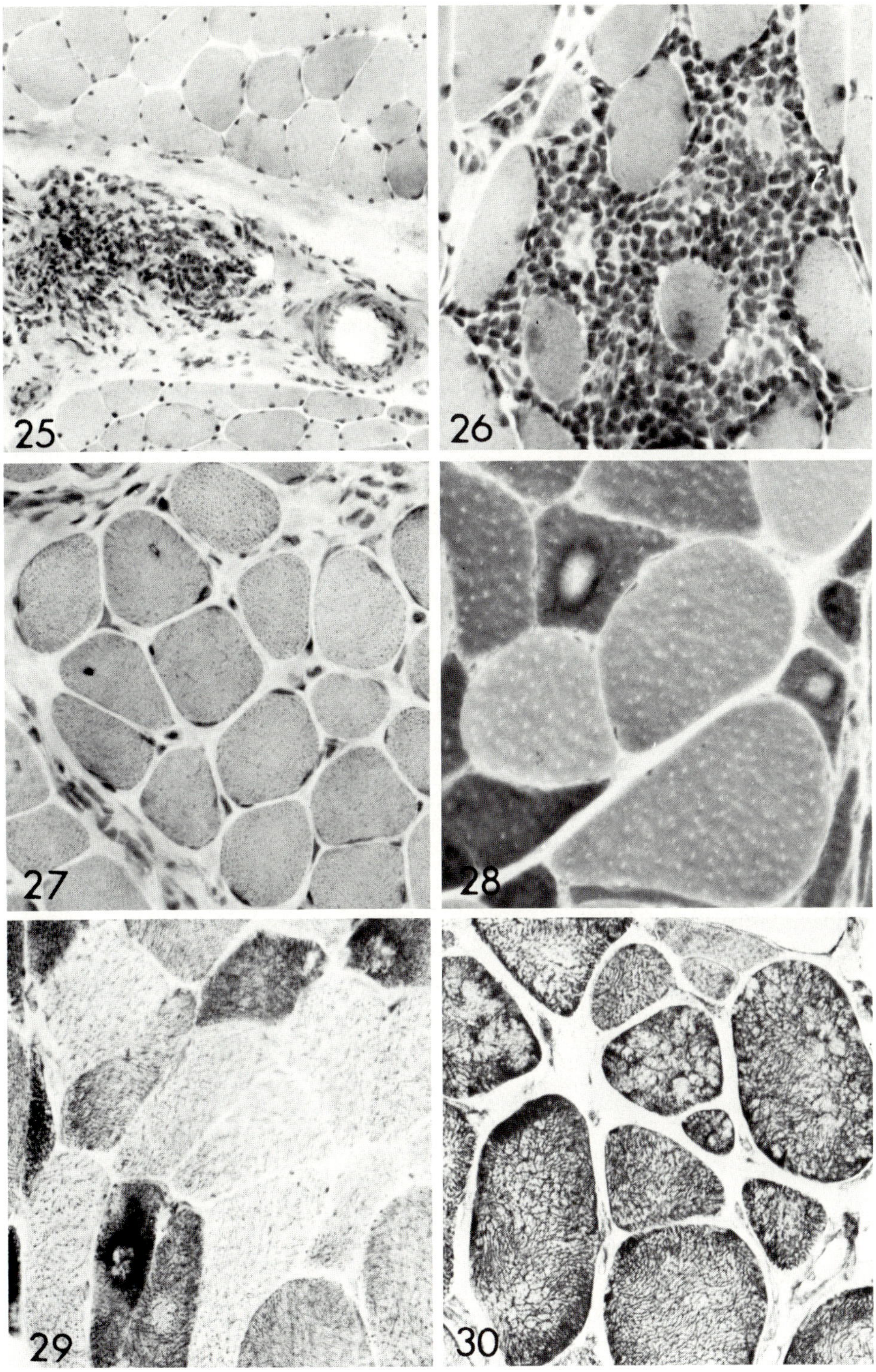


FIGS. 25–30

NUCLEAR CHANGES

In many conditions, the sarcolemmal nuclei migrate from their position at the periphery of the fiber and become quite numerous within the fiber itself. These are then referred to as internal nuclei and should only be looked for on transverse sections (Fig. 19). On a longitudinal section it is difficult to tell whether the nucleus is internal or otherwise since one may well be looking through myofibrils to see the nucleus at the periphery of the muscle fiber. With the progressive atrophy and disappearance of muscle fibers, very often the nuclei are left in clumps. Such pyknotic nuclear clumps are illustrated in Figure 20. The normal appearance of individual nuclei may be distorted in two major ways. The vesicular nucleus is characterized by a pale staining, ballooned appearance with a very prominent nucleolus (Fig. 21). When present in deeply basophilic fibers, these have been thought to indicate regeneration. Nuclei may also develop a heavy stippling of chromatin material within the nucleus and develop a "tigroid" appearance (Fig. 22). In general, vesicular nuclei are more prominent in myopathic conditions, whereas tigroid nuclei tend to be seen more frequently in neurogenic conditions.

DEGENERATION AND REGENERATION

There has been considerable discussion as to which changes represent regeneration and which degeneration. We do not wish to contribute to this controversy in the present paper but would like to describe certain changes under this heading. Some fibers may present a pale liquefied appearance with the routine stains (Fig. 23). Many of these fibers become filled with phagocytes (Fig. 24). Other fibers show marked basophilia especially with the hematoxylin and eosin stain (Fig. 21). A slightly different change noticed on the trichrome stain is the appearance of a granular red material within the fiber.

CELLULAR REACTIONS

In this context we consider those cellular reactions which are not confined to a single muscle fiber. Areas of phagocytosis may be quite extensive within a fascicle. Frequently, an inflammatory response is seen in association with blood vessels. This is the so-called perivascular reaction (Fig. 25). Another type of cellular reaction is a lymphorrhage (Fig. 26). This is a tight collection of small round cells which exists within relatively normal muscle. Another form of cellular reaction is fibrosis (Fig. 27). We only consider fibrosis to be abnormal when there is an increased amount of fibrous tissue between individual fibers in the fascicle, not when there is a large amount of fibrous tissue in the septa between the fascicles.

CHANGES IN ARCHITECTURE OF INDIVIDUAL FIBERS

In many cases it is necessary to have adequate histochemical preparations to detect changes within the fibers themselves. One of the earliest of these changes to be recognized was the target fiber.[14] This fiber, which is best seen with the oxidative enzyme stains, is a three-zoned fiber. The central zone is relatively unstained, whereas the intermediate zone stains rather heavily (Figs. 28 and

29). The outer zone may be normal. The significance of these target fibers will be discussed later. Frequently the regular ordered network lying between the myofibrils stains irregularly. The fibers then assume a moth-eaten appearance, frequently with a swirling of the intermyofibrillar network that looks like the eddying current of a stream. This is termed the "moth-eaten-whorled" fiber (Fig. 30). Tubular aggregates are areas of muscle fiber in which the normal architecture is replaced by a dense collection of abnormal tubules.[16] On histochemical stains these are well demonstrated with the oxidative enzyme DPNH dehydrogenase (Figs. 31 to 33). They then appear as intensely blue staining segments of the muscle fiber. With the ATPase reaction, these areas are not demonstrated since myofibrils are not present. Tubular aggregates are limited to IIB fibers in our experience. The presence within the muscle fiber of cytoplasmic bodies, which are frequently basophilic, is of unknown significance. These cytoplasmic bodies have been seen in many diseases although only rarely are they very numerous (Figs. 34 and 35). Another alteration in the normal architecture occurs when the normally oriented muscle fiber may be surrounded by a ring of myofibrils oriented at right angles to it (Fig. 36). This striated annulet or ring fiber is often associated with sarcoplasmic masses. Other abnormalities are seen in very rare conditions and would include the central cores (Fig. 37) as originally described by Shy and Magee[30] and the rods which are seen in several different conditions but are very numerous in nemaline myopathy[30] (Fig. 38). Mitochondrial abnormalities are only delineated in electron microscopic studies, but their presence may be suspected when granular red fibers are noted on the trichrome stain (Figs. 39 and 43).

### Abnormalities of distribution

We would now like to indicate those changes which can be found only by surveying large segments of the biopsy. Perhaps the best known is the grouped

---

Figs. 31 and 32. Serial sections of muscle from a patient with uncomplicated rheumatoid arthritis showing tubular aggregates. × 175.

Fig. 31. ATPase (pH 9.4).

Fig. 32. ATPase (pH 4.6 for 5 minutes preincubation).

Fig. 33. Tubular aggregates in a patient with Parkinson's disease. DPNH dehydrogenase; × 225.

Fig. 34. A large cytoplasmic body is seen within the muscle fiber. Trichrome stain; × 500.

Fig. 35. In this patient with acute dermatomyositis cytoplasmic bodies were seen in most of the type II muscle fibers as indicated by the *arrows*. Trichrome stain; × 555.

Fig. 36. A ring fiber in a patient with limb girdle dystrophy. Periodic acid-Schiff stain; × 555.

Fig. 37. Central core disease. The presence of the central cores of abnormally reacting muscle tissue can be seen in most of the muscle fibers. DPNH dehydrogenase; × 225.

Fig. 38. Dermatomyositis. At the tip of this muscle fiber are seen numerous rod bodies. Trichrome stain; × 500.

Fig. 39. Ragged red fibers seen in a patient with Kearn's and Sayre's ophthalmophlegia syndrome. The electron microscopic picture can be seen in Figure 43. Trichrome stain; × 125.

Fig. 40. Large group atrophy. To the left are fascicles consisting of very tiny fibers, whereas to the right the fibers are of more normal size. Trichrome stain; × 125.

Fig. 41. Amyotrophic lateral sclerosis. Small groups of atrophic fibers are scattered within more normal sized fibers. Hematoxylin and eosin stain; × 125.

Fig. 42. Perifascicular atrophy. The fibers around the edge of the fascicle are relatively atrophic compared to those in the center from a patient with polymyositis. ATPase (pH 9.4); × 50.

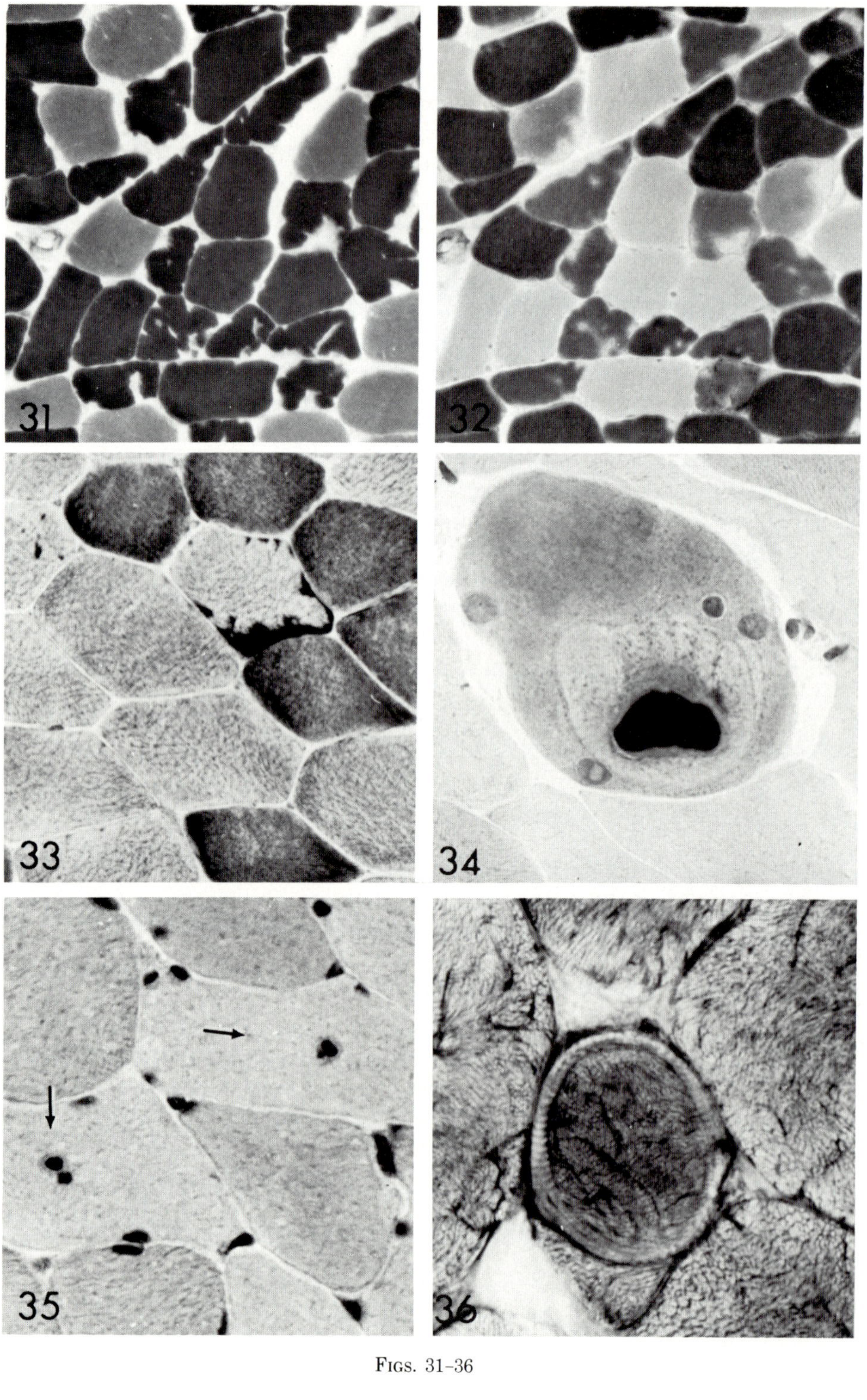


Figs. 31–36

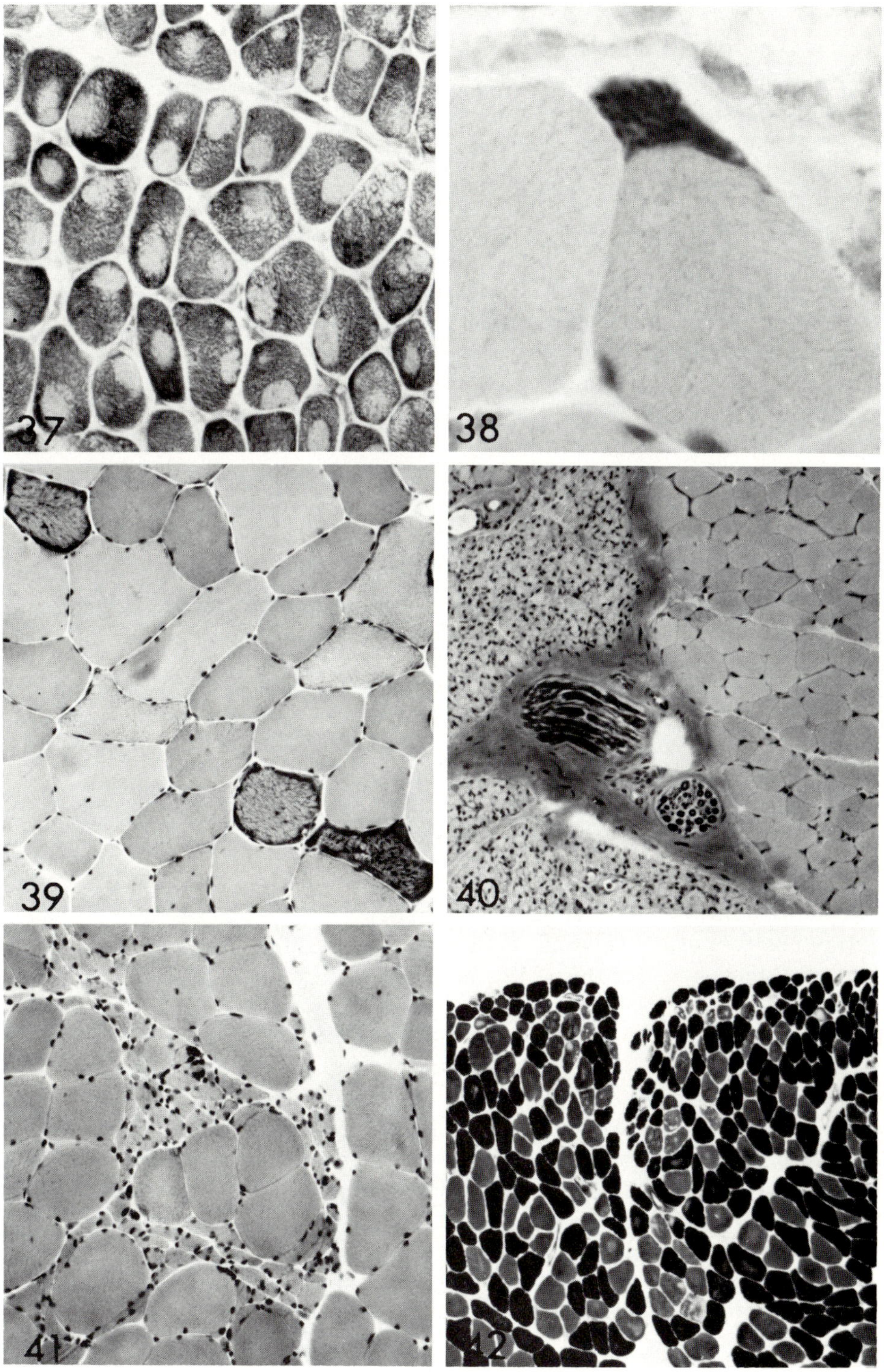


FIGS. 37–42

atrophy which we see in denervation. Large groups of atrophic fibers may be seen next to equally large groups of normal sized fibers. This has been termed variously geographic atrophy or large group atrophy (Fig. 40). We prefer the latter term to contrast it with the small group atrophy in which a few atrophic fibers are grouped together (Fig. 41). Another form of localized atrophy is the perifascicular atrophy seen in dermatomyositis or polymyositis.[2] In this change, the center of the fascicle is relatively normal but around the edge is a large area of atrophic fibers (Fig. 42). Histochemical abnormalities of distribution include type grouping and fiber type predominance. In type grouping the normal mosaic appearance of the fiber types is destroyed. Groups of one fiber type are seen next to groups of another fiber type (Fig. 44). The change described as fiber type predominance (Fig. 45) may be a reflection of fiber type grouping in some cases. It would be possible theoretically to take a biopsy from the middle of a very large type group. If this were the case, all of the fibers seen in the biopsy would be of one fiber type. However, certain diseases are so frequently associated with a predominance of one fiber type that it seems unlikely that type grouping is always the explanation of fiber type predominance. The opposite change from fiber type predominance is fiber type deficiency. In our evaluation of three fiber types, for example, we have noted that the type IIB fiber is sometimes absent, although type I and type IIA fibers are present in equal proportions. In order to evaluate these changes statistically, we say that fiber type predominance exists when more than 55% of the fibers in a given histogram are of one type (*i.e.*, I, IIA, or IIB). When type II fibers are considered without the subdivision into IIA and B, more than 80% of the fibers must be type II before predominance is said to occur. A fiber type deficiency exists when less than 10% of the fibers are of one type.

## CHANGES SEEN IN DISEASE

### "Disuse" and other causes of type II fiber atrophy

Reduction in size of type II fibers, and in particular of type IIB fibers, is a not uncommon finding. It has been noted that in normal female muscle the type II fibers are smaller than in the normal male.[5] It has also been speculated that this difference is due to the fact that male muscle is associated with more powerful voluntary activity. The larger type II fibers might be a primary phenomenon indicating an innate difference between male and female muscle or might be secondary to greater physical activity on the part of the male and might represent work hypertrophy. This issue is not settled, but it is of interest that mothers of children with diseases of sufficient severity that the child is unable to walk without assistance demonstrate larger type II fibers than are seen in biopsies of other women. It is not proven that mothers of severely handicapped children are performing more physical activity than other women but it seems a likely explanation in view of the fact that their children will need constant physical assistance, for example, lifting them in and out of wheelchairs. This would suggest that work hypertrophy results in type II fiber enlargement. At the other end of the scale, we have noticed type II fiber atrophy in many conditions in which the voluntary activity of the patient is limited. Thus, type II atrophy is noted in illnesses as diverse as pyramidal tract disease and prolonged bedrest. Periodic paralysis and myasthenia gravis are both diseases causing weakness in a patient and both are associated with type II fiber atrophy. As will be shown later, even mental retardation in a young child can be a cause of type II fiber atrophy, presumably because the intellectual handicap hampers motor development. In polymyalgia rheumatica, a desease in which the muscles are stiff and frequently movement is hampered because of muscle pain, type II atrophy is also seen. Perhaps, then, we should regard type II fiber atrophy as being due to "disuse" in its most general form. This disuse should be thought of as being produced by causes both physiologic and pathologic.

DISEASES OF LOWER MOTOR NEURON

The typical finding seen in denervation of muscle is an atrophy of the fibers which are deprived of their nerve supply. In most human diseases, the loss of innervation is a random process with the successive involvement of more and more muscle fibers. Thus, in early denervation one sees only scattered atrophic fibers which frequently assume a compressed angular configuration in the interstices between normal sized fibers. Frequently such fibers are very dark with the oxidative enzymes and are totally lacking in phosphorylase activity. Histographic analysis of the muscle biopsy shows, as might be expected, that the atrophy involves all three fiber types, and consequently the atrophy factors of all fiber types will be higher than the normal threshold value. As the severity of the denervation increases, the small angular fibers are found in increasingly large groups as will be discussed below. Two kinds of nuclear changes are seen in denervation. Quite early in denervation the nuclei frequently become tigroid when seen in paraffin-embedded fixed sections. In severe atrophy, pyknotic nuclear clumps are scattered throughout the biopsy, the remnants of groups of atrophic fibers. The changes of degeneration and regeneration are lacking in most neurogenic disease. Occasional liquefying fibers and basophilic fibers may be seen but are not a prominent part of the picture except in the chronic familial peripheral neuropathies.[4] Changes in the architecture of individual fibers are not common, but target fibers may be seen in diseases of the lower motor neuron. These fibers seem to be most numerous in our experience in patients recovering from denervation. Thus, target fibers may be numerous during the recovery phase of the peripheral neuropathy in acute porphyria or in Guillain-Barré's disease and are less numerous in the patients with amyotrophic lateral

---

FIG. 43. Electron microscopic picture of the biopsy shown in Figure 39. A large collection of abnormal mitochondria can be seen in the subsarcolemmal region.

FIG. 44. Type grouping. In this patient with a chronic familial peripheral neuropathy, clumps of type II fibers are seen next to groups of type I fibers. ATPase (pH 9.4); × 50.

FIG. 45. Type II fiber predominance. In this biopsy, from a child with mental retardation and pyramidal tract signs, type II fibers constituted more than 80% of the fibers in the biopsy. ATPase; × 50.

FIG. 46. Amyotrophic lateral sclerosis. In addition to the atrophy of both fiber types, type II fiber hypertrophy is present. ATPase (pH 9.4); × 125.

FIG. 47. Peripheral neuropathy. Comparing this figure with Figure 46, it can be seen that the type II fibers are not as large. ATPase (pH 9.4); × 125.

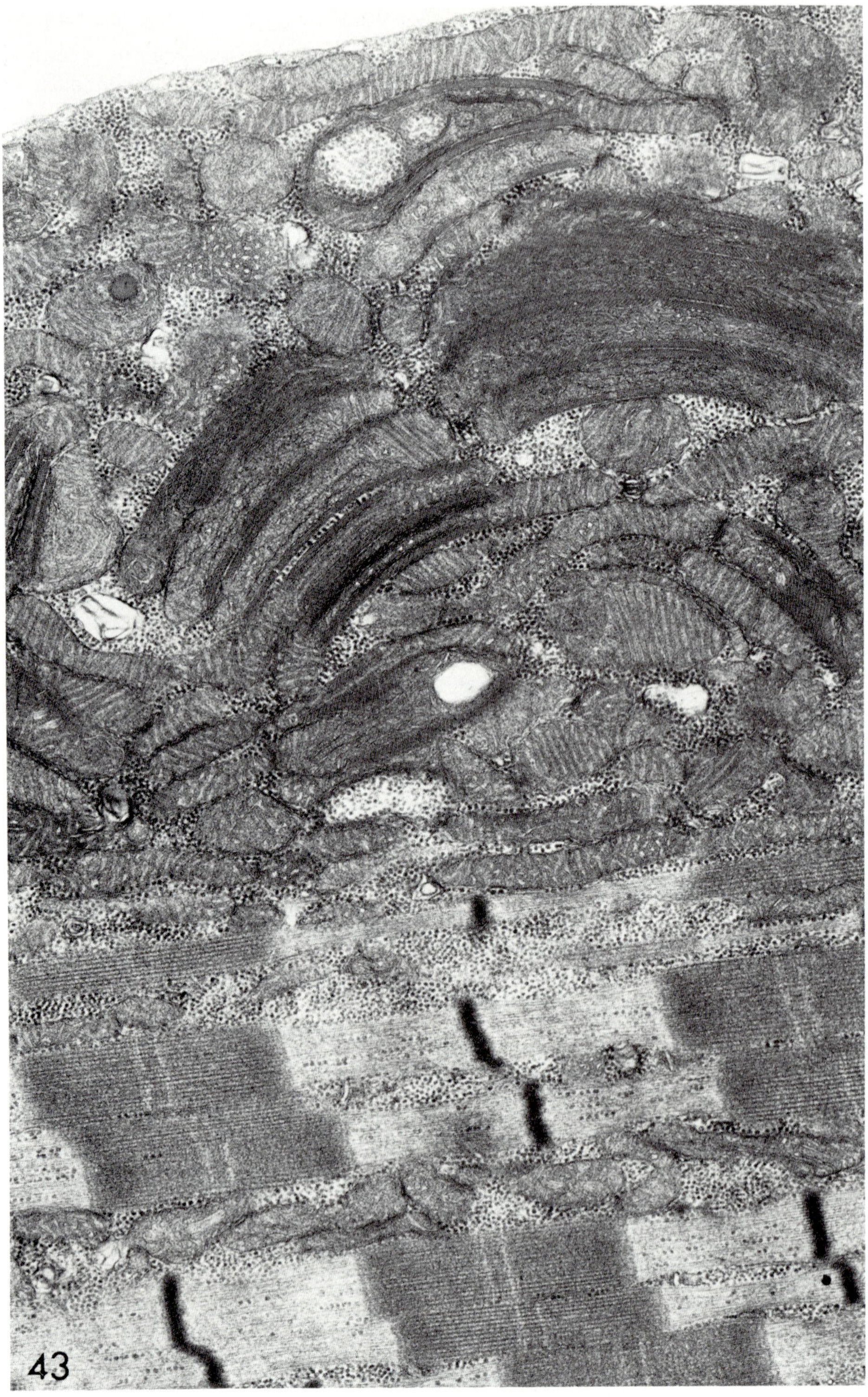


Fig. 43

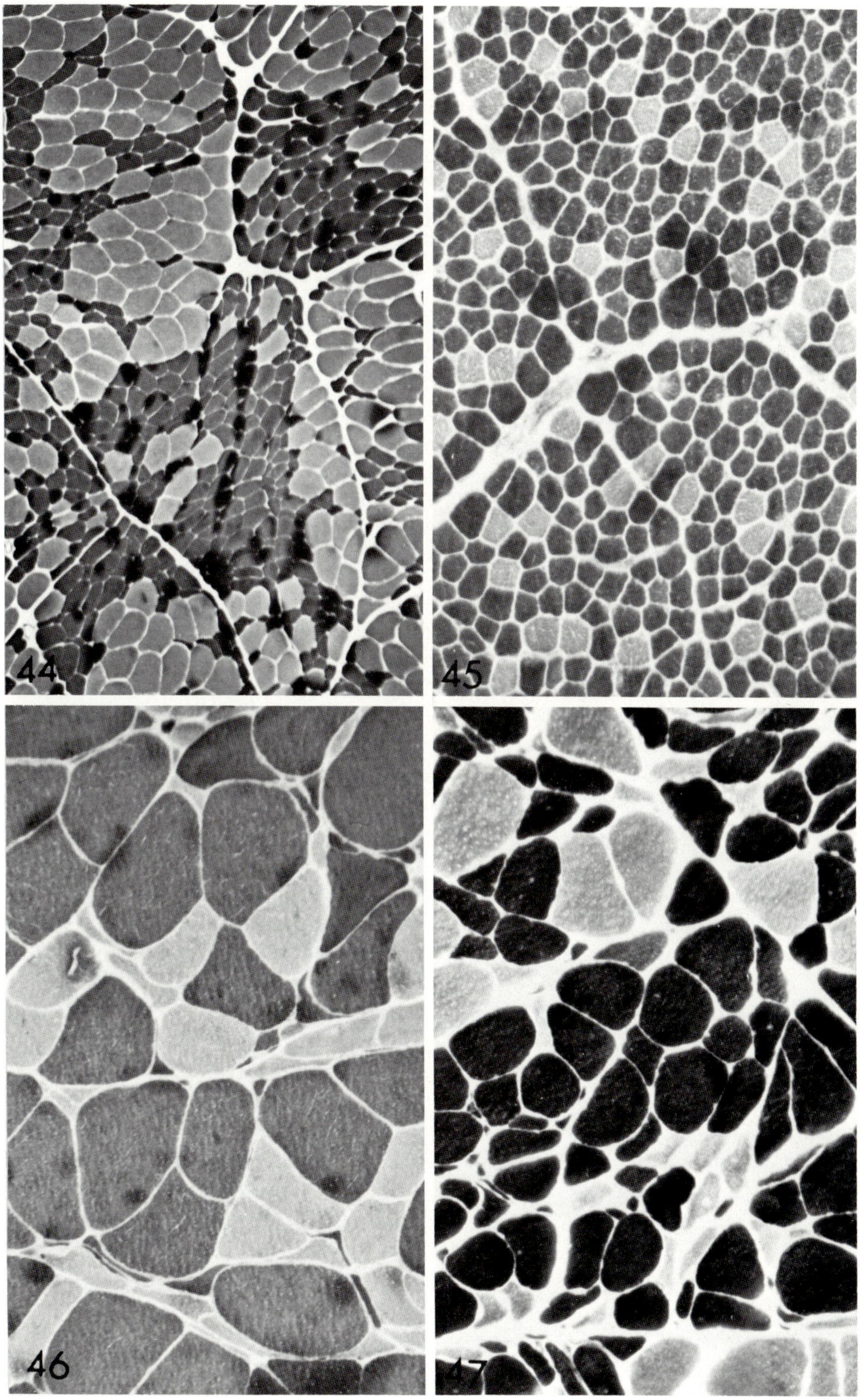


Figs. 44–47

sclerosis and chronic familial peripheral neuropathies. Others have suggested that the target fiber is a sign of reinnervation.[12] The abnormalities of distribution seen in patients with denervation are fairly straightforward. The atrophic fibers may appear in small groups and in large groups. The more chronic the neuropathy, the more likely are large groups of fibers to be seen. Another abnormality of distribution is the appearance of type grouping. Again, the more striking examples of type grouping are likely to be seen in chronic neurogenic conditions. Experimental evidence suggests that type grouping, although always associated with denervating disease, is indicative of reinnervation.[11, 23]

Once the biopsy has been established as belonging to the group associated with denervation, some effort should be made to place the biopsy in one of three further categories. The first category is that of amyotrophic lateral sclerosis. In this illness, particularly when only the lower motor neurons are involved, in addition to the changes described above, hypertrophy of the type II fibers is seen frequently (Fig. 46). This hypertrophy is evaluated by calculation of the hypertrophy factors for type II fibers. Type II fiber predominance is also seen more frequently in amyotrophic lateral sclerosis than in other illnesses. In chronic familial peripheral neuropathies, on the other hand, the type I fiber is frequently hypertrophied rather than the type II fiber (Fig. 44). Additionally, in the chronic familial peripheral neuropathies, changes other than those of simple denervation are seen. Fiber splitting may occur, cellular reactions are frequently present, basophilia and phagocytosis can be seen, and internal nuclei are usually plentiful. If the biopsy does not possess the characteristics of these two groups, we generally consider it to represent simple denervation (Fig. 47). The change of simple denervation is seen with many of the peripheral neuropathies such as in diabetes or in toxic neuropathies.

### Genetically determined dystrophies

In this group we would like to deal with four main categories. These are the pseudohypertrophic dystrophy of Duchenne, limb girdle dystrophy, facio-scapulo-humeral dystrophy, and myotonic dystrophy. Although these illnesses often commence during childhood, they will be seen during adult life and the biopsies can frequently be recognized. In all of the dystrophies, the pathologic emphasis shifts to involvement of almost all of the fibers in the biopsy. This is in contradistinction to the neurogenic diseases in which involvement of the various fibers is scattered. Changes in fiber size reflect this. There is a great variation in the size of fibers. This results in a spreading of the distribution curve of the histogram. The mean diameter of the muscle fibers is frequently normal, but the standard deviation of fiber size is greatly increased. All fiber types are involved in this abnormality, and frequently the changes in fiber size are due to the existence of very large fibers which then split into many smaller fibers. Such fiber splitting is much more common in the dystrophies than in any of the other diseases. The nuclear changes in the dystrophies are of two kinds. Internal nuclei are prominent, and on fixed paraffin-embedded sections the nuclei are frequently vesicular nuclei. The constant degeneration and attempted regeneration of muscle fibers is associated with basophilia. Necrotic fibers and fibers undergoing phagocytosis are present. Fibrous tissue is laid down within the fascicles, and it is in the dystrophies that fibrosis becomes most prominent. Changes in the architecture of individual fibers are also marked in the dystrophies as opposed to the neurogenic diseases. Ring fibers are often seen in the chronic myopathies, and, although this fiber has been described as pathognomonic of myotonic dystrophy, it is our experience that it is more common in the chronic limb gridle dystrophies. Probably the presence of many ring fibers is suggestive of any very long standing myopathy. It should be noted, though, that there may be numerous ring fibers in uncomplicated rheumatoid arthritis. The intermyofibrillar network is often distorted with the formation of motheaten and whorled fibers. Abnormalities of distribution are present in the generic dystrophies. Type IIB fiber deficiency is seen in a number of the biopsies. A consequence of the great numbers of split fibers may be an appearance that simulates type grouping. Thus, many fibers of the same type are seen clustered throughout the biopsy. These, however, have all been produced by splitting of one very large fiber and should not be considered to represent type grouping. We are very hesitant to make the diagnosis of type grouping in the presence of fiber splitting. We would now like to discuss the differentiating points between the various disease entities.

*Duchenne dystrophy*. In this illness, there is one abnormality which is not noted in the others. Very frequently with the routine ATPase reaction it is quite difficult to differentiate type I and type II fibers. This occurred in 11 of 12 biopsies in the present series. By poor differentiation we mean that at pH 9.4 the differentiation between types I and II fibers is not distinct (Fig. 48). If the sections are preincubated at a slightly higher pH, between 10 and 10.5, the differentiation between type I and type II fibers returns (Fig. 49). It is as if the

---

FIG. 48. Duchenne dystrophy. The distinction into types I and II fibers is unclear. ATPase (pH 9.4); × 125.

FIG. 49. A serial section to the photograph shown in Figure 48. The reaction for ATPase was carried out after preincubation at pH 10.5. There is now clear differentiation into type I (light) and type II (dark) fibers. × 125.

FIG. 50. Duchenne dystrophy. A biopsy from a 5-year-old boy who had two maternal uncles with the benign variety of Duchenne dystrophy. ATPase (pH 9.4); × 125.

FIG. 51. Duchenne dystrophy. A group of basophilic fibers. Hematoxylin and eosin stain; × 555.

FIG. 52. Duchenne dystrophy. The pattern of the intermyofibrillar network is relatively undisturbed compared to the changes seen in other myopathies. DPNH dehydrogenase; × 225.

FIG. 53. Facio-scapulo-humeral dystrophy. This photograph shows the marked inflammatory response frequently seen in this disease. Hematoxylin and eosin stain; × 125.

FIG. 54. Facio-scapulo-humeral dystrophy. Demonstrating the good differentiation of fiber types in the face of severe disease. ATPase (pH 9.4); × 125.

FIG. 55. Myotonic dystrophy. In this, an early case, the small atrophic fibers are seen to contain internal nuclei. Hematoxylin and eosin stain; × 225.

FIG. 56. Myotonic dystrophy. The same patient as is seen in Figure 55. Type I fiber atrophy together with type II fiber hypertrophy is pronounced. ATPase (pH 9.4); × 125.

FIG. 57. An advanced case of myotonic dystrophy. Many internal nuclei are seen in all fibers. Trichrome stain; × 125.

FIG. 58. Advanced myotonic dystrophy. This is the same patient as is seen in Figure 57. ATPase (pH 9.4); × 125.

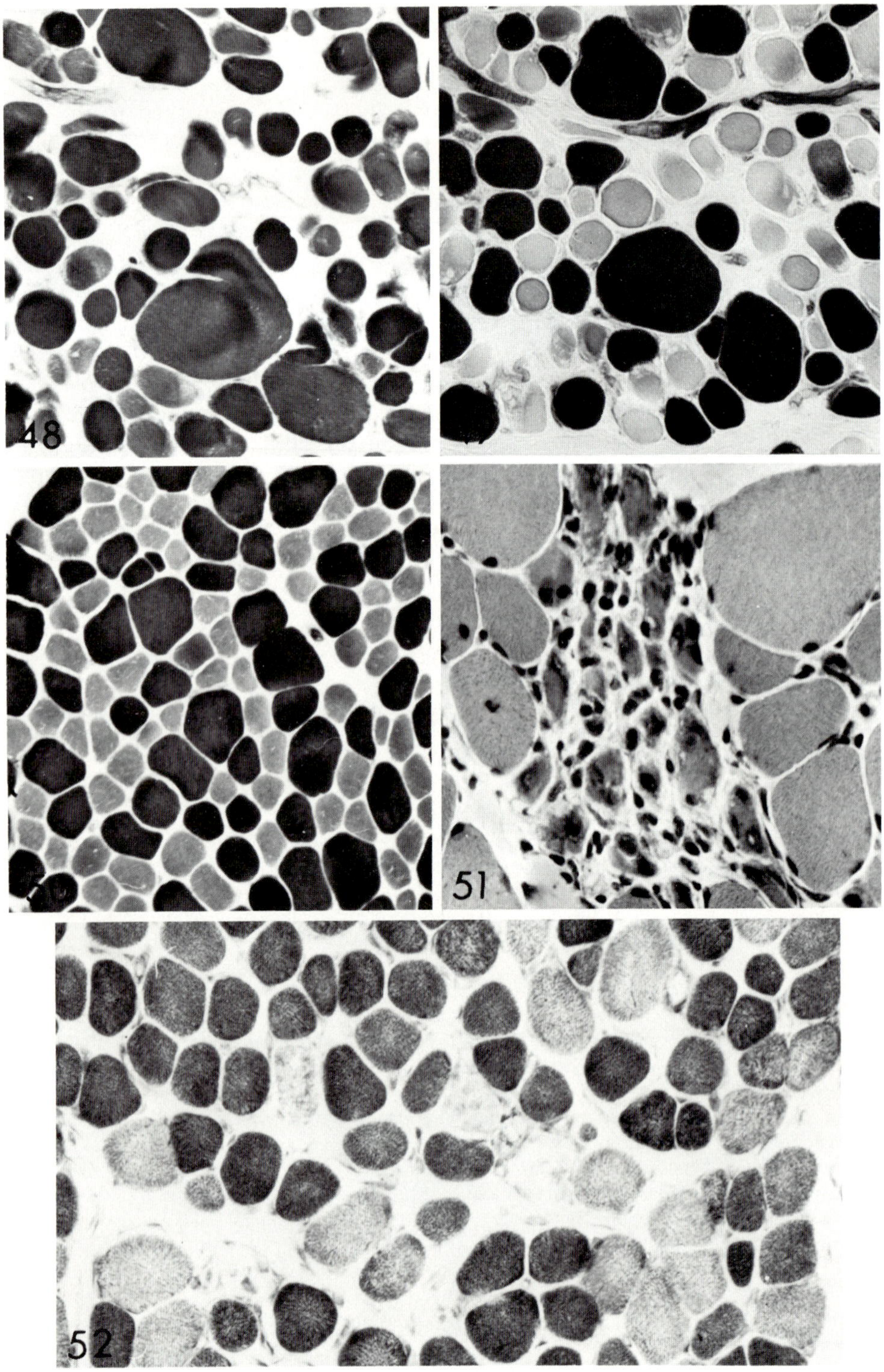


FIGS. 48–52

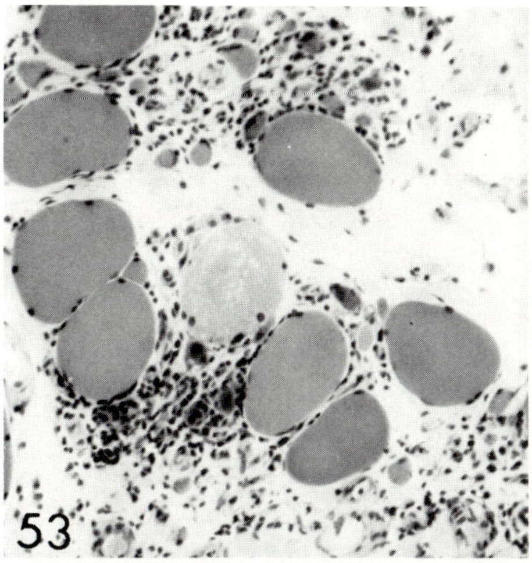
53

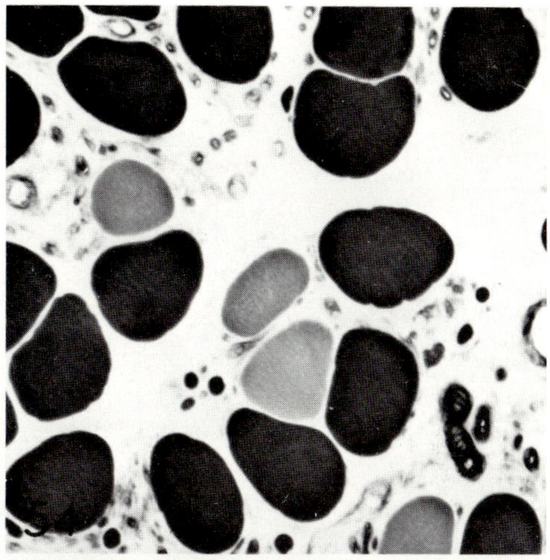

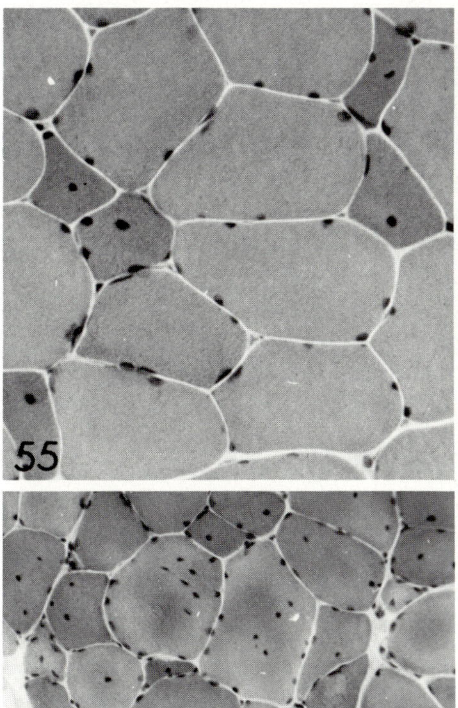
55

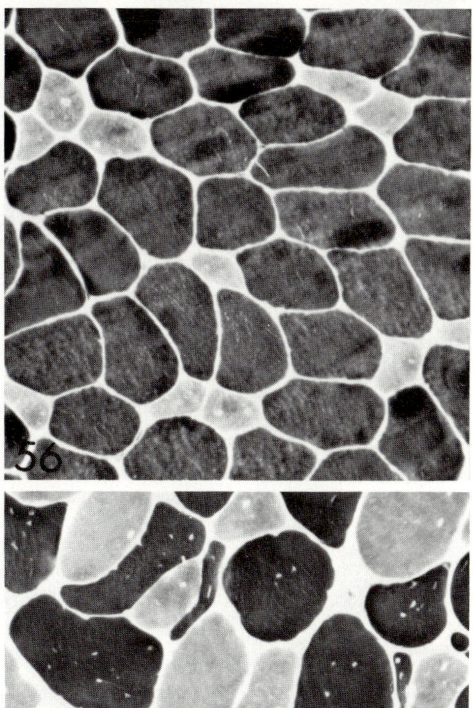
56

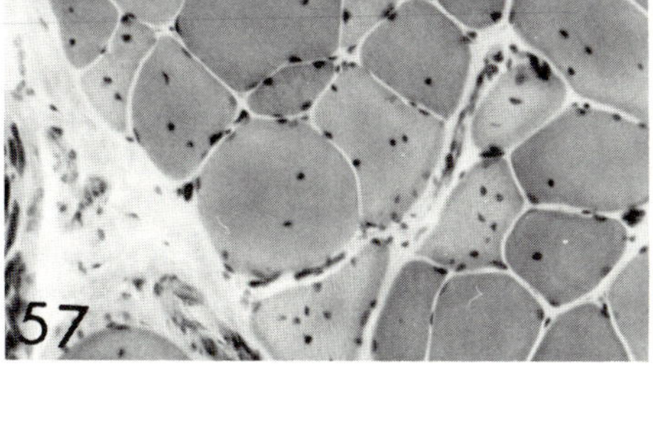
57

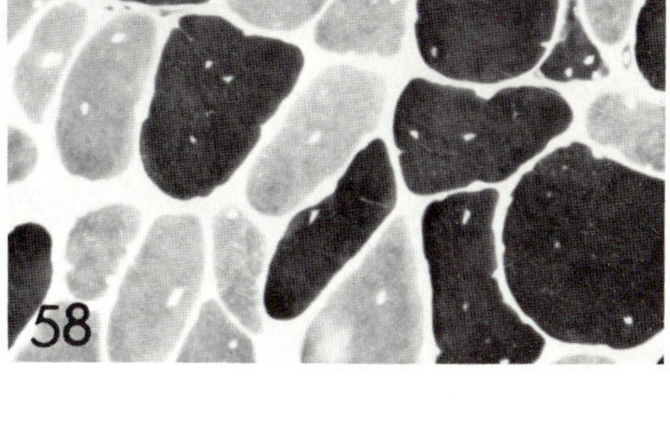
58

pH sensitivity of type I fibers in Duchenne dystrophy has in some way been altered. When the preincubation conditions are altered in order to demonstrate type I and type II fibers, it can be seen that, in all of the biopsies, there is a predominance of type I fibers, *i.e.*, more than 55% of the fibers in the biopsy consist of type I. In addition, in many biopsies there is a relative scarcity of type IIB fibers. Thus, in the biopsies in which type IIB fibers were looked for, there was 10% or less of the total number of fibers in 80% of the biopsies, even though approximately 30% of the fibers in these biopsies were IIA. An exception to the foregoing makes an interesting example. In a biopsy from a 5-year-old boy, the differentiation at pH 9.4 was quite clear (Fig. 50). Not only that but also there were a good number of type IIB fibers in spite of the type I fiber predominance. This 5-year-old boy had two maternal uncles with the benign variety of Duchenne dystrophy, and both were surviving at the ages of 27 and 32, respectively. The rest of the biopsy on this child showed the usual changes of respectively. The rest of the biopsy on this child showed the usual changes of Duchenne dystrophy with basophilia, phagocytosis, and a moderate number of internal nuclei. Basophilic fibers are a very prominent part of the Duchenne biopsy. Frequently, small groups of basophilic fibers are seen[18] (Fig. 51). Other differential points in the patient with Duchenne dystrophy are largely negative ones. The architectural changes are usually rather mild in comparison to the other dystrophies and it is rare to see a great number of moth-eaten fibers (Fig. 52). There may be some granular changes of the fibers but little else. Fiber splitting is not usually very evident in the patient with Duchenne dystrophy, and internal nuclei may only be moderately increased. Fibrosis on the other hand, is usually marked, especially in the severe disease. This gives a rather characteristic picture of many round muscle fibers embedded in a sea of fibrous tissue.

*Limb girdle dystrophy.* This entity is usually marked by a great degree of fiber splitting, perhaps more than is seen in any of the other diseases. Fiber differentiation is usually good, with type I and type II fibers being clearly present. There may, however, be type I fiber predominance in the biopsy. In the majority of our patients type IIB fiber deficiency is noted. Architectural changes are usually prominent in limb girdle dystrophy, and only a rare biopsy is seen without the moth-eaten whorled fibers which have been previously described.

*Facio-scapulo-humeral dystrophy.* Some of the patients with facio-scapulo-humeral dystrophy differ from those with other dystrophies in the degree of cellular response that is present.[4] This can, on occasions, be so marked that the diagnosis of an inflammatory myopathy is often considered (Fig. 53). However, in facio-scapulo-humeral dystrophy hypertrophic fibers are always seen whereas in polymyositis they are very rare. The hypertrophied fibers seen in facio-scapulo-humeral dystrophy are mostly type II (Fig. 54). Furthermore, in facio-scapulo-humeral dystrophy, the differentiation of fiber types with the routine ATPase is good, which may help to differentiate if from a Duchenne dystrophy. In mildly affected muscles of patients with facio-scapulo-humeral dystrophy, very often small angulated fibers are noted.[4]

*Myotonic dystrophy.* In the early stages, this disease is very easy to differentiate from the other dystrophies. In biopsies from the biceps muscle and most biopsies from the vastus muscle, type I fiber atrophy together with some type II fiber hypertrophy is pronounced (Figs. 55 and 56). As the disease progresses, this change becomes masked by the appearance of more general signs of myopathy (Figs. 57 and 58), although it can still be found by calculating atrophy-hypertrophy factors. Internal nuclei are profuse in myotonic dystrophy (Fig. 57) and occur to a degree seen in few other diseases, with the possible exception of chronic familial peripheral neuropathy and other rare diseases such as centronuclear myopathy.[25]

MYOPATHIES ASSOCIATED WITH COLLAGEN VASCULAR DISEASE AND RHEUMATOID DISEASES

Muscle may be involved in the various collagen vascular diseases. Perhaps the best known changes are seen in polymyositis and in dermatomyositis. In severe cases, all of the general findings of a myopathy may be present. There are, additional aspects which help differentiate this group of diseases from the dystrophies. There may be perivascular inflammatory responses (Fig. 25). Frequently there is type II fiber atrophy (Fig. 59) and this can often be seen in the histogram superimposed upon the myopathic changes of increased variability in the size of fibers. Lastly, perifascicular atrophy is often seen in this group of diseases[2] (Fig. 60). Fibers which have undergone a curious vacuolar change, as illustrated, are frequently seen in biopsies from patients with dermatomyositis or polymyositis (Fig. 61). Those biopsies from muscles which are only mildly involved usually demonstrate type II fiber atrophy, moth-eaten fibers, and less often perivascular inflammatory reactions. In addition to type II fiber atrophy, most patients also have a lesser degree of type I fiber atrophy. The type II atrophy involves both types IIA and B fibers. To the clinician, diseases such as lupus erythematosus, periarteritis nodosa, polymyositis, and dermatomyositis appear as distinct entities. In terms of muscle pathology, however, it is often difficult to distinguish differences in the pathology of the muscle. The preceding description applies to all of these diseases. The pathology in this group of patients merges with that of patients with rheumatoid diseases. Patients with polymyalgia rheumatica complain of severe incapacitating stiffness and pain in the muscles. The only abnormal laboratory finding is an elevated sedimentation rate. The response to steroids is dramatic. In our experience, the biopsy from patients with polymyalgia rheumatica represents a combination of type IIB fiber atrophy with moth-eaten whorled fibers (Figs. 62 to 64). A control group of patients, who were also complaining of aches, cramps, and pains but in whom there was no elevation of the sedimentation rate and no response to steroids, did not show these changes. The response of the muscle to steroid therapy is shown in Figures 65 and 66.

A quite different pathologic picture may be seen in patients with rheumatoid arthritis uncomplicated by any apparent muscle disease. In such patients, common finding is type I fiber atrophy together with some type IIA fiber atrophy (Fig. 67). Frequently ring fibers are seen (Fig. 68). The biopsy looks so abnormal

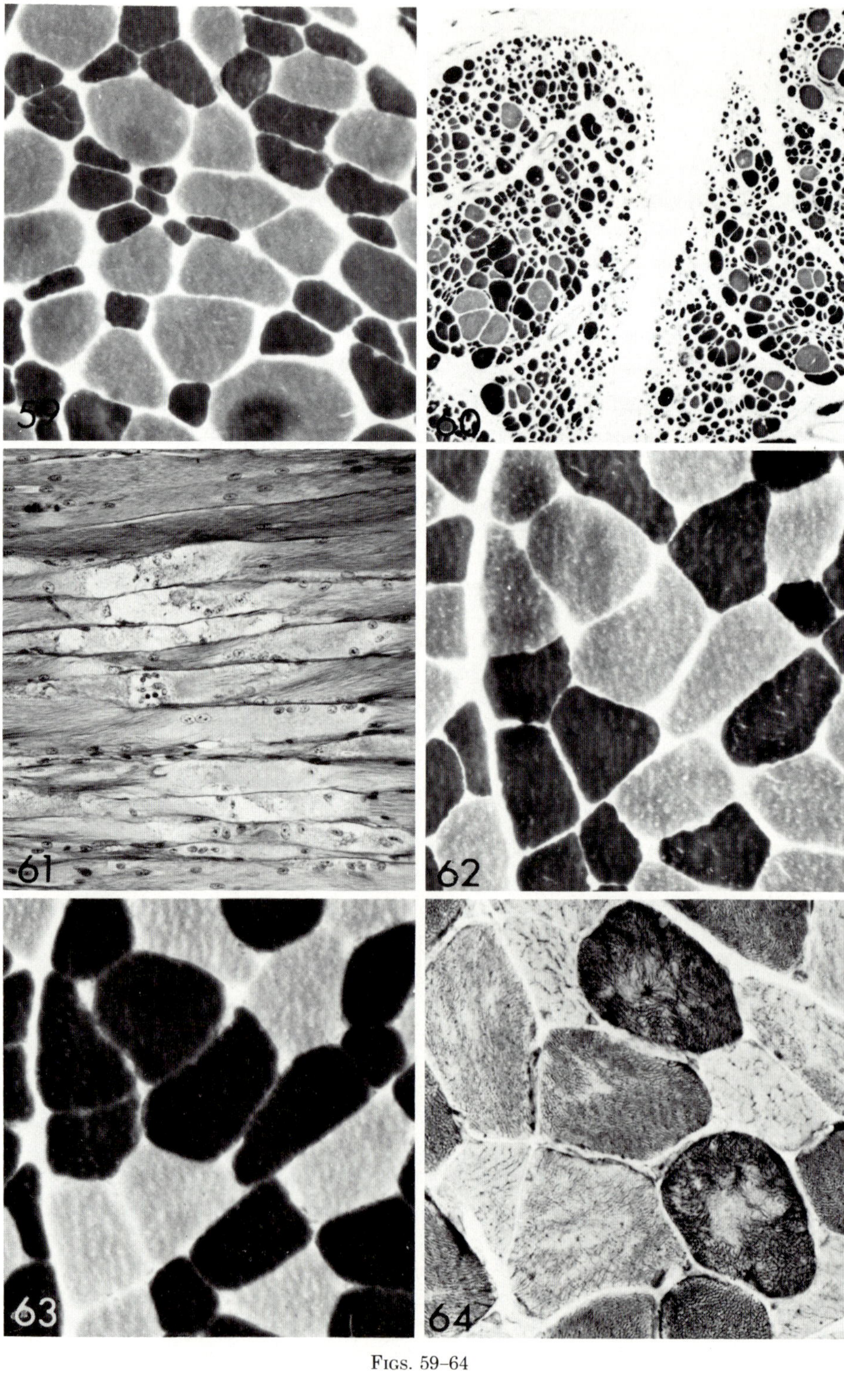


Figs. 59–64

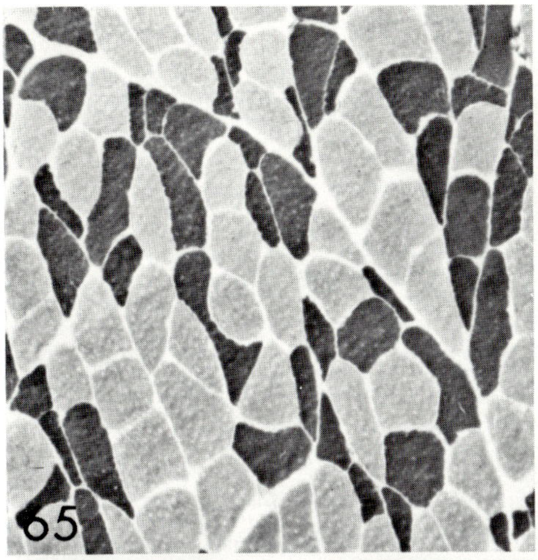
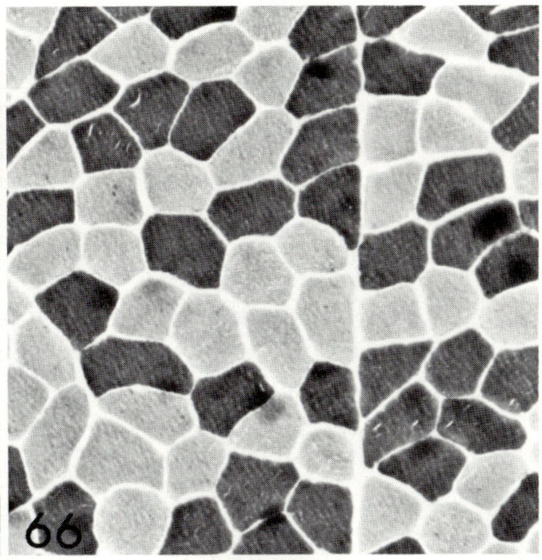
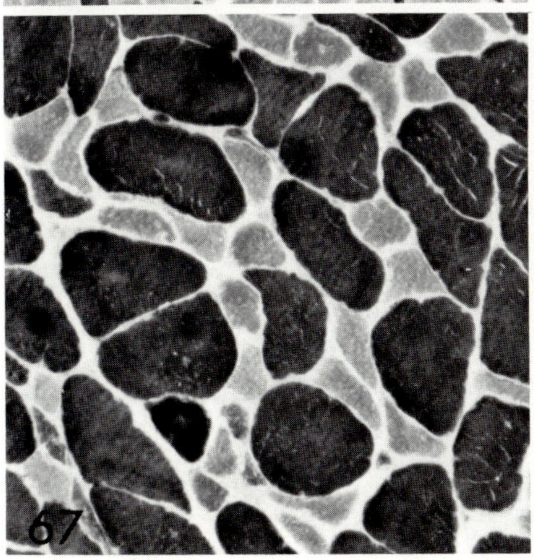
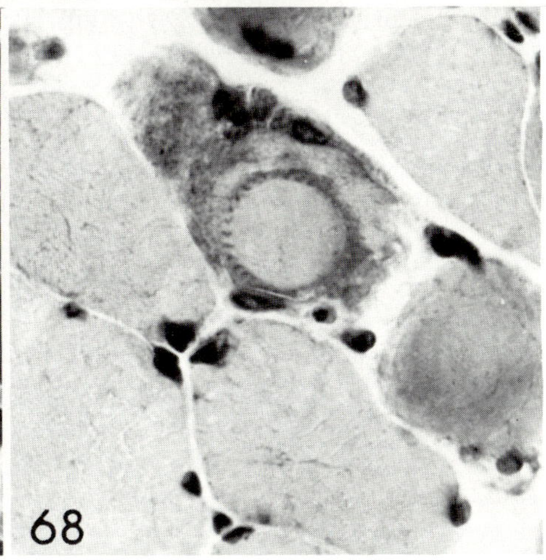

FIG. 59. Collagen vascular disease showing the type II fiber atrophy in a patient with periarteritis nodosa. ATPase (pH 9.4); × 225.

FIG. 60. Perifascicular atrophy. This patient had severe polymyositis. ATPase (pH 9.4); × 50.

FIG. 61. Dermatomyositis. A vacuolar change is seen in the muscle fibers (paraffin-embedded, Bouins-fixed). Trichrome stain; × 200.

FIGS. 62 and 63. These photographs are from serial sections of a patient with polymyalgia rheumatica to show the type IIB fiber atrophy in this condition. The type IIB fibers are identified because they are dark with both of these stains. × 225.

FIG. 62. ATPase (pH 9.4).

FIG. 63. ATPase (pH 4.6 for 5 minutes preincubation).

FIG. 64. Polymyalgia rheumatica showing the moth-eaten whorled fibers which occur in this condition. DPNH dehydrogenase; × 310.

FIG. 65. Polymyalgia rheumatica showing the type II fiber atrophy in a patient before treatment. ATPase (pH 9.4); × 125.

FIG. 66. The same patient as shown in Figure 65 following therapy with steroid for 1 year. Notice the increase in the size of type II fibers. ATPase (pH 9.4); × 125.

FIG. 67. Rheumatoid arthritis. In this patient with uncomplicated rheumatoid arthritis, type I fiber atrophy is present. ATPase (pH 9.4); × 125.

FIG. 68. Rheumatoid arthritis. Ring fibers and sarcoplasmic masses are frequently seen. Trichrome stain; × 555.

that it is difficult to believe that this does not represent primary pathology of the muscle. However, we have no other evidence of primary muscle involvement and we must, therefore, assume that such changes are secondary to involvement of the joint. In addition, in one patient with rheumatoid arthritis many fibers with tubular aggregates were seen.

## CHILDREN'S BIOPSIES

In general, the same principles of interpretation apply to children's biopsies, however, there is one important difference. We have found it easier to segregate the children's biopsies into five groups depending upon the fiber sizes rather than classifying the biopsies by disease entities.[8]

All of the children's biopsies fell into one of five groups. Group 1 consists of biopsies with type I and type II fibers having average diameters which were roughly equal; the difference between the diameters was less than 12% of the mean diameter of the larger group of fibers. Additionally, these biopsies had fibers with a normal distribution curve and a variability coefficient of less than 250. Group 2 biopsies consist of those in which the type II fibers are smaller than type I fibers by more than 12% of the diameter of the larger fibers and in which variability coefficient of the larger fiber type is normal. Group 3 biopsies are judged by the same criteria as group 2 biopsies except that in this case the type I fibers are smaller. Group 4 biopsies consist of those histograms which are "twin-peaked" in either fiber type. This indicates a population of abnormally small fibers and another of abnormally large fibers. Group 5 biopsies are those biopsies in which there is an abnormal variability in the size of the fibers indicated by a variability coefficient of greater than 250 for either fiber type and which do not fall into any of the other groups. Because this concept of pathologic classification is a unfamiliar one, we shall describe in detail the patients seen in these groups over the past year in our muscle clinic.

### Group 1 biopsies

There were eight biopsies in the present series which fell into group 1. Five of these were from patients with no evidence of neuromuscular disease (Fig. 69). The other three patients had retardation of the motor milestones; that is, they did not sit, stand, or crawl at the appropriate times. One of the patients had some degree of mental retardation. No specific diagnosis other than benign congenital hypotonia was entertained in these three patients. Our experience in the present series thus parallels our previous results and indicates that most normal children will have biopsies belonging to group 1 and that any biopsy which falls into the group 1 category is probably normal. In the six patients in whom analysis of types IIA and B was carried out, the types IIA were slightly larger than the types IIB in four of the patients. Both of the patients in which the types IIB were slightly larger than the types IIA were young (age 12 months and 17 months) and both had delayed motor milestones. There were no abnormalities in the biopsies belonging to group 1. In the younger patients, the intermyofibrillar network has an infantile pattern.

GROUP 2 BIOPSIES

In these patients, the type II fibers were smaller than type I (Figs. 70 and 71). There were seven patients in all, three of whom had both mental and motor retardation. Two patients had recurrent infections, one due to chronic pyelonephritis and the other secondary to hypogammaglobulinemia. One patient had been bedfast for 5 weeks with the diagnosis of a Guillain-Barré syndrome and one had an undiagnosed arthritic condition. This supported our previous impression on a larger number of patients that the group 2 biopsies were predominantly seen in children in whom the muscle problem was one of disuse. This disuse could be on the basis of mental retardation, *i.e.*, that the patient did not know how to use the muscles, or due to organic disease severe enough to cause the patient to spend much time in bed. Associated changes in these biopsies other than the variation of fiber size were not prominent. The ratios of the various fiber types were within normal limits in both group 1 and group 2.

GROUP 3 BIOPSIES

Group 3 consists of biopsies in which the type I fibers are smaller than the type II fibers (Figs 72 to 78). Very often predominance of type I fibers occurs. These are among the most interesting of the children's biopsies. It has been previously pointed out that the biopsies occurring in group 3 are from patients who tend to be mentally quite bright, in whom benign congenital hypotonia is the most frequent diagnosis.[8] Other patients with spinocerebellar degeneration of the predominantly ataxic variety were classified in this group as were patients with myotonic dystrophy. Many of the biopsies in the group present an appearance very similar to that illustrated. In our present series, there were 12 children with biopsies which were histographically group 3. One adult is also included because her clinical history was the same. The diagnosis entertained

---

FIG. 69. A group 1 biopsy in a child. ATPase (pH 9.4); × 125.

FIG. 70. A group 2 biopsy in a child showing the small type II fibers. ATPase (pH 9.4); × 310.

FIG. 71. This is a serial section to the photograph shown in Figure 70 showing that the small fibers are type IIB. ATPase (pH 4.6 for 5 minutes preincubation); × 310.

FIG. 72. A group 3 biopsy in a child showing the slightly smaller type I fibers with type I fiber predominance. ATPase (pH 9.4); × 310.

FIGS. 73 and 74. Serial sections from the biopsy of a floppy infant with no well established diagnosis. This shows, as in most of our group 3 biopsies, that the large fibers are type IIB. × 225.

FIG. 73. ATPase (pH 9.4).

FIG. 74. ATPase (pH 4.6 for 5 minutes preincubation).

FIGS. 75 and 76. Serial sections of a biopsy from a patient with an undiagnosed congenital nonprogressive myopathy. In addition to the group 3 pattern of the routine ATPase, it can be seen that there are no IIB fibers and that the large fibers are type IIA. × 175.

FIG. 75. ATPase (pH 9.4).

FIG. 76. ATPase (pH 4.6 for 5 minutes preincubation).

FIG. 77. This is from the same patient as Figures 75 and 76, and indicates the presence of moth-eaten whorled fibers and some ring fibers seen in the lower left hand corner. DPNH dehydrogenase; × 310.

FIG. 78. Type I fiber hypotrophy with internal nuclei. Hematoxylin and eosin stain; × 225.

FIG. 79. Werdnig-Hoffmann's disease. ATPase (pH 9.4); × 125.

FIG. 80. Werdnig-Hoffmann's disease. ATPase (pH 9.4); × 125.

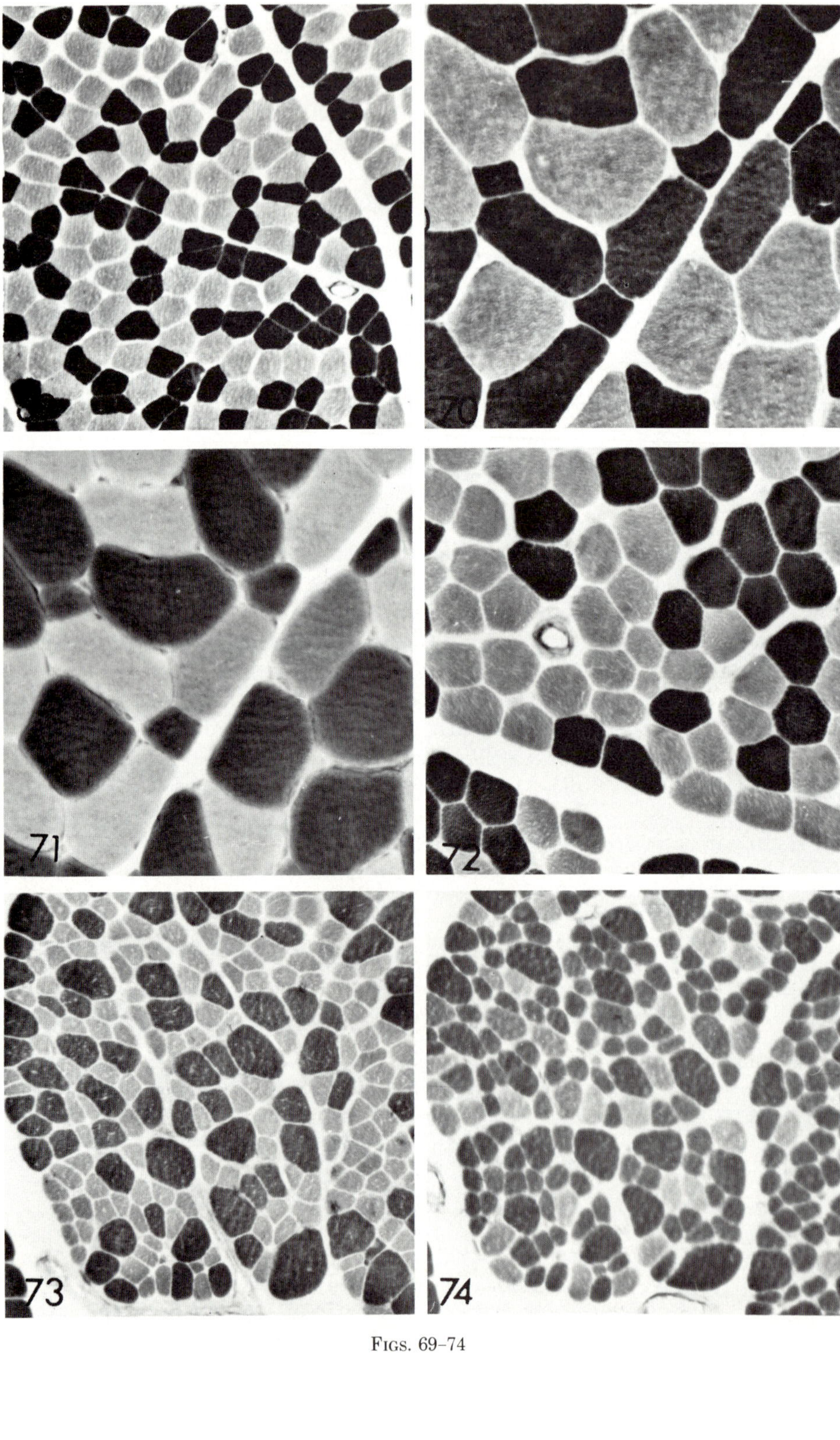


Figs. 69–74

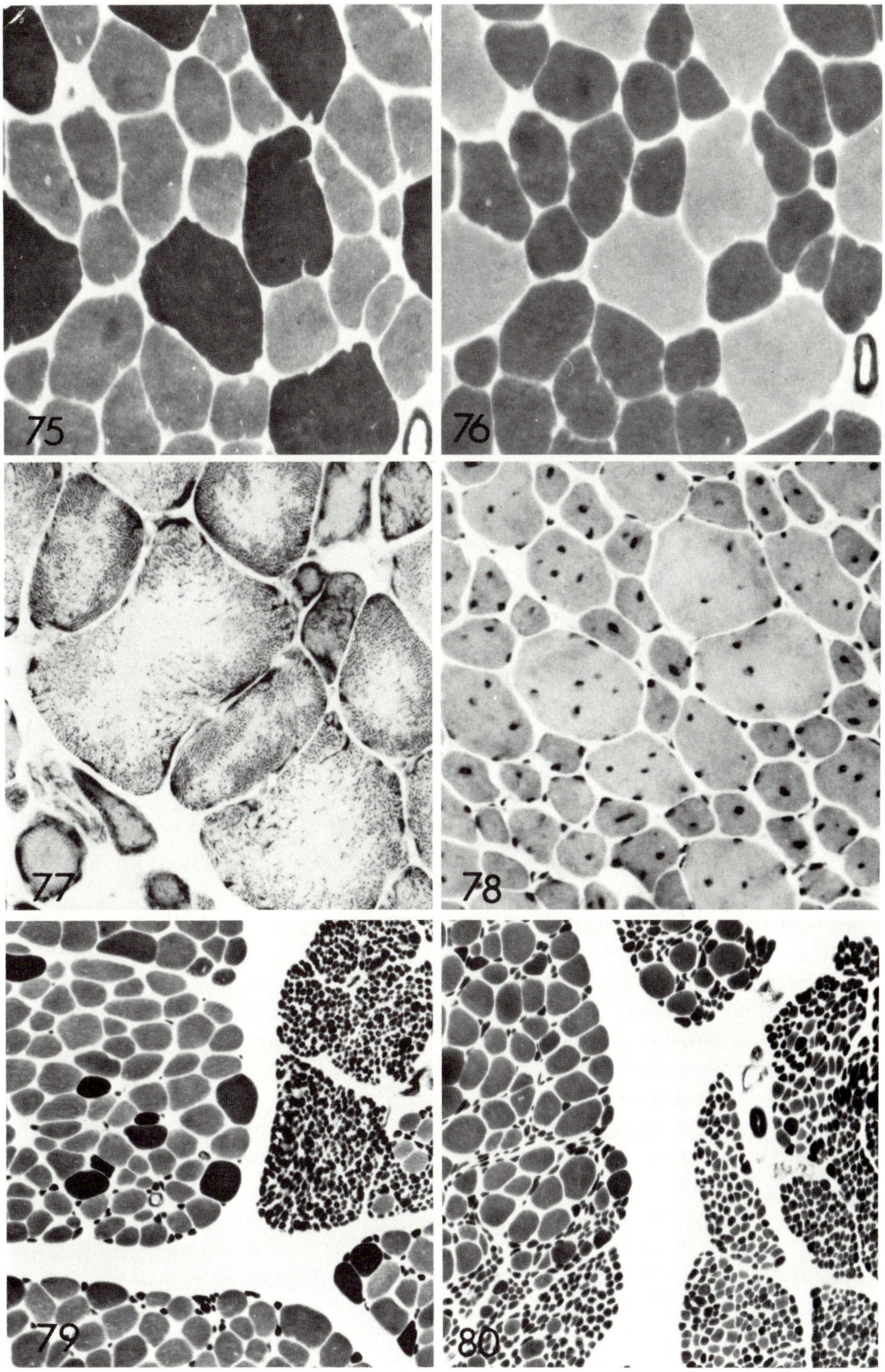


FIGS. 75–80

in these patients was not clear-cut. All of the children were hypotonic and had been so since the first year of life; none had any evidence of mental retardation. All of the patients were improving or had been stable for some time. One patient, who was biopsied at the age of 1, continued to lose motor function and when seen at the age of 2 was unable to sit. Clinically, she resembled a Werdnig-Hoffmann's disease except that there were no fasciculations seen. There was no evidence of denervation on electromyogram and the bulbar muscles were not involved. She also began to improve after 2 years of age. In some patients, there was a family history of benign congenital hypotonia. In all but one of the biopsies the largest fibers were type IIB. This is an unusual finding in normal biopsies of both adults and children in which the type IIB fiber is usually smaller than the type IIA fiber. Although we do not know what the normal fiber sizes are in the neonatal period, it is of interest that the two biopsies in the standard group (group 1) in which the type IIB fibers were slightly larger than were type IIA fibers were both from children who had delayed motor milestones and who were the youngest patients in the group. In addition to the changes in fiber size, type I predominance was seen in most of the biopsies. There did not seem to be any other abnormalities in the biopsies in the preceding patients.

One 13-year-old patient in the group deserves separate mention since, in addition to the findings of type I predominance and the changes in fiber size associated with group 3 biopsies, she also had many ring fibers, moth-eaten whorled fibers, and a number of internal nuclei (Fig. 78). During the mother's pregnancy she had a 45-pound weight gain and polyhydramnios. The baby was said to have clenched hands at birth and had early difficulty with feeding, being troubled with regurgitation of food. This passed off and the hands were also loose and unclenched by the age of 1 year. She sat at 8 months of age and walked without support at 22 months. However, she fell frequently and her arms were too weak to catch herself. She was a very asthenic child with a high arched palate and kyphosis. No fasciculations were seen; there was proximal weakness and decreased deep tendon reflexes. In addition to the weakness of the shoulders and the hips, the anterior tibials were also severely involved. There was neither electrical nor clinical evidence of myotonia nor of denervation and there was no evidence of progression of the disease within the last 10 years. Although differing from the previous biopsies in the presence of ring fibers and internal nuclei, the absence of type IIB fibers, and the hypertrophy of type IIA fibers, she nevertheless shared the good prognosis and relatively benign course of the previous patients.

Type I fiber hypotrophy with internal nuclei as described in Engel, Gold, and Karpati[20] is also included in the category of group 3 biopsies. Although the first patient died at the age of 18 months, we have seen a 22-year-old girl with this condition. Superimposed upon the basic change of fiber size and type I predominance as seen in group 3 biopsies was the presence of many internal nuclei in all fibers (Fig. 78) and a moth-eaten appearance of the smaller fibers. A sister of the patient had died in infancy from what was probably the same condition. The patient herself had experienced repeated bouts of pneumonia during her early childhood, but from the age of approximately 5 onward the disease had remained stationary and is at present improving slightly. Another nonprogressive myopathy with a relatively benign course is nemaline myopathy. The pathologic picture of the biopsy from some of these patients would fit the group 3 biopsy picture with superimposed changes of rods within the muscle fiber.[16] A third nonprogressive myopathy described originally by Shy and Magee[30] and others is central core disease. Three of our adult patients with central core disease demonstrate almost total predominance of type I fibers with only one or two type II fibers in the entire biopsy (Fig. 37). In these patients, the mean diameter of the type I fibers is below what would be expected for normal. They, therefore, demonstrate at least two of the properties of group 3 biopsies, namely, small type I fibers and type I predominance. Superimposed on this is the presence of central cores. We would like to suggest, therefore, that a biopsy which has the characteristics of group 3 biopsies may well be associated with a relatively benign course.

### Group 4 biopsies

These biopsies are characterized by the occurrence of twin peaks in the histogram, that is, there is a population of fibers larger than normal and smaller than normal. There were seven patients within this group. Five of these have Werdnig-Hoffmann's disease and two have chronic familial peripheral neuropathy. The biopsy of the patient with Werdnig-Hoffmann's disease is very characteristic.[21] There are many small round atrophic fibers in which occur islands of very much larger fibers (Figs. 79 to 82). These larger fibers are almost all type I, although occasional large type II fibers may be seen. There has been some discussion in the literature as to whether the large fibers seen in Werdnig-Hoffmann's disease are type I or type II. Some of the confusion has originated from the exact histochemical stain being used. We have always employed the ATPase stain for typing fibers, and with this stain the large fibers are mostly type I. When oxidative enzyme stains are used to type fibers, the large fibers may be of both types (Fig. 81). This is not necessarily true of the other neurogenic conditions. In Kugelberg-Welander's disease or juvenile spinal muscular atrophy, groups of large type II fibers as well as large type I fibers are seen. In the chronic familial peripheral neuropathy group in adults, characteristically there is hypertrophy of the type I fibers and this is seen in both of the patients in the present series.

### Group 5 biopsies

The group 5 biopsies consisted of those in which there was undue variability in the size of fibers. There were 16 biopsies in this category. All of the patients had some form of myopathy. This is comparable to the previous series in which 46 of 54 such biopsies were from patients with myopathy.[8] Twelve of the patients had Duchenne muscular dystrophy by clinical and laboratory criteria. One was a patient with limb girdle dystrophy, one with facio-scapulo-humeral dystrophy, one with congenital myopathy, and one with an undiagnosed myopathy manifested by myoglobinuria. The patients with Duchenne dystrophy presented a rather characteristic appearance, which has been described above.

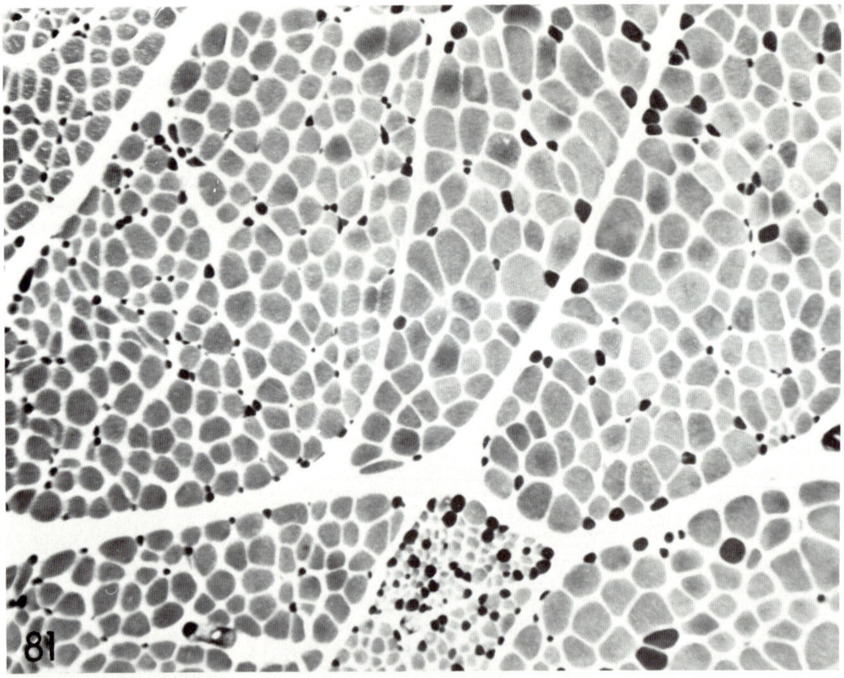

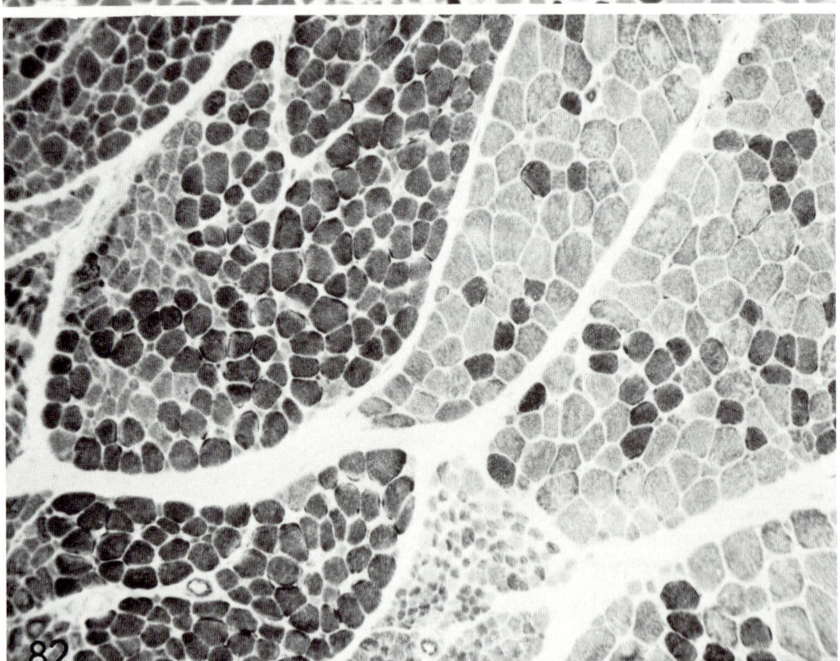

FIGS. 81 and 82. Serial sections from a patient with Werdnig-Hoffmann's disease to show that the large fibers are type I when characterized by the ATPase reaction but are both light and dark when characterized by the DPNH dehydrogenase reaction. × 125.

FIG. 81. ATPase (pH 9.4).

FIG. 82. DPNH dehydrogenase.

## SUMMARY

The conventional interpretation of muscle biopsy material has, in the past, depended upon findings seen in paraffin-embedded, formalin-fixed material. The changes seen in such material have been well recognized and have been useful in interpreting a given biopsy as either myopathic or neurogenic; they have occasionally enabled a more specific disease diagnosis to be made. With the advent of histochemistry, it has become apparent that many additional variables may now be recognized in muscle tissue. A logical approach to the interpretation of biopsy material is to describe the changes seen in a particular disease entity and to expand the original concept of neurogenic and myopathic diseases. This paper is concerned with the interpretation of both adult biopsies and children's biopsies. In the adult material, the changes seen in various groups of diseases have been described. In children's biopsies, it is pointed out that these may be divided into five groups depending upon the fiber size. Most of the normal biopsies occur in group 1. Group 2 biopsies occur in children whose motor difficulty stems from diseases which are frequently remote from the neuromuscular apparatus. The biopsies categorized as group 3 are frequently from children with various muscle diseases which usually are relatively benign. Group 4 biopsies are characteristic of the denervating diseases such as Werdnig-Hoffmann's and Kugelberg-Welander's. Group 5 biopsies are usually associated with the primary myopathies.

REFERENCES

1. Adams, R. D., Denny-Brown, D., and Pearson, C. M.: *Diseases of Muscle. A Study in Pathology*, 2nd ed. New York, Hoeber, 1962.
2. Banker, B. Q.: Dermatomyositis of childhood. *Trans. Am. Neurol. Assoc. 87:* 11–15, 1962.
3. Barka, T., and Anderson, P. J.: *Histochemistry*, p. 313, New York, Hoeber, 1963.
4. Brooke, M. H., and Engel, W. K.: The histologic diagnosis of neuromuscular diseases. A review of 79 biopsies. *Arch. Phys. Med. Rehabil. 47:* 99–121, 1966.
5. Brooke, M. H., and Engel, W. K.: The histographic analysis of human muscle biopsies with regard to fiber types. 1. Adult male and female. *Neurology (Minneap.) 19:* 221–233, 1969.
6. Brooke, M. H., and Engel, W. K.: The histographic analysis of human muscle biopsies with regard to fiber types. 2. Diseases of the upper and lower motor neurons. *Neurology (Minneap.) 19:* 378–393, 1969.
7. Brooke, M. H., and Engel, W. K.: The histographic analysis of human muscle biopsies with regard to fiber types. 3. Myotonias, myasthenia gravis, and hypokalemic periodic paralysis. *Neurology (Minneap.) 19:* 469–477, 1969.
8. Brooke, M. H., and Engel, W. K.: The histographic analysis of human muscle biopsies with regard to fiber types. 4. Children's biopsies. *Neurology (Minneap.) 19:* 591–605, 1969.
9. Brooke, M. H., and Kaiser, K. K.: Three myosin ATPase systems. *J. Histochem. Cytochem. 18:* 670–672, 1970.
10. Brooke, M. H., and Kaiser, K. K.: Muscle fiber types. How many and what kind. *Arch. Neurol. 23:* 369–379, 1970.
11. Dubowitz, V.: Pathology of experimentally reinnervated skeletal muscle. *J. Neurol. Neurosurg. Psychiatry 30:* 99–110, 1967.
12. Dubowitz, V.: *Developing and Diseased Muscle; Histochemical Study*, Symposium Research Monograph No. 2. London, Spastics International Medical Publications in association with William Heinman, 1968.
13. Dubowitz, V., and Pearse, A. G.: Reciprocal relationship of phosphorylase and oxidative enzymes in skeletal muscle. *Nature (Lond.) 85:* 701–702, 1960.

14. Engel, W. K.: Muscle target fibers, a newly recognized sign of denervation. *Nature (Lond.) 191:* 389–390, 1961.
15. Engel, W. K.: The essentiality of histo- and cytochemical studies of skeletal muscle in the investigation of neuromuscular disease. *Neurology (Minneap.) 12:* 778–794, 1962.
16. Engel, W. K.: Selective and nonselective susceptibility of muscle fiber types. A new approach to human neuromuscular diseases. *Arch. Neurol. 22:* 97–117, 1970.
17. Engel, W. K.: Focal myopathic changes produced by electromyographic and hypodermic needles. "Needle myopathy." *Arch. Neurol. 16:* 509–511, 1967.
18. Engel, W. K., and Brooke, M. H.: Muscle biopsy as a clinical diagnostic aid. In *Neurological Diagnostic Techniques*, ed. by Fields, W. S., pp. 90–146. Springfield, Illinois, Charles C Thomas Publishers, 1966.
19. Engel, W. K., and Cunningham, G. G.: Rapid examination of muscle tissue. An improved trichrome method for fresh-frozen biopsy sections. *Neurology (Minneap.) 13:* 919–923, 1963.
20. Engel, W. K., Gold, G. N., and Karpati, G.: Type I fiber hypotrophy and central nuclei. A rare congenital muscle abnormality with a possible experimental model. *Arch. Neurol. 18:* 435–444, 1968.
21. Fenichel, G. M., and Engel, W. K.: Histochemistry of muscle in infantile spinal muscular atrophy. *Neurology (Minneap.) 13:* 1059–1066, 1963.
22. Greenfield, J. G., Shy. G. M., Aword, E. C., and Berg, L.: *An Atlas of Neuromuscular Diseases*. London, E. S. Livingstone, 1957.
23. Karpati, G., and Engel, W. K.: Transformation of the histochemical profile of skeletal muscle by "foreign" innervation. *Nature (Lond.) 215:* 1509–1510, 1967.
24. Lillie, R. D.: *Histopathologic Technic and Practical Histochemistry*, 3rd ed. New York, McGraw-Hill, 1954.
25. Munsat, T. L., Thompson, L. R., and Coleman, R. F.: Centronuclear ("myotubular") myopathy. *Arch. Neurol. 20:* 120–131, 1969.
26. Ogata, T.: A histochemical study of the red and white muscle fibers. 1. Activity of the succinoxidase in muscle fibers. *Acta. Med. Okayama 12:* 228–231, 1958.
27. Padykula, H. A., and Gauthier, G. F.: Morphological and cytochemical characteristics of fiber types in normal mammalian skeletal muscle. In *Exploratory Concepts in Muscular Dystrophy and Related Disorders*, ed. by Milhorat, A. T., pp. 117–128. Amsterdam, Excerpta Medica Foundation, 1967.
28. Padykula, H. A., and Herman, E.: The specificity of the histochemical method for adenosine triphosphatase. *J. Histochem. Cytochem. 3:* 170–195, 1955.
29. Pearson, C. M., Coleman, R. F., Fowler, W. M., Jr., Mommaerts, W. F. H. M., Munsat, T. L., and Peter, J. B.: Skeletal muscle. Basic and clinical aspects and illustrative new diseases. *Ann. Intern. Med. 67:* 614–650, 1967.
30. Shy, G. M. and Magee, K. R.: A new congenital nonprogressive myopathy. *Brain 79:* 610–621, 1956.
31. Takeuchi, T.: Histochemical demonstration of branching enzyme (amylo-1,4 → 1,6-transglucosidase, in animal tissues. *J. Histochem. Cytochem. 6:* 208–216, 1958.
32. Walton, J. N.: *Disorders of Voluntary Muscle*. London, J. & A. Churchill, 1964.

## Chapter 8

# Cytochemical Properties of the External Lamina of Myofibers and Neuromuscular Junctions*

SUMMNER I. ZACKS, ATSUSHI SAITO, AND M. F. SHEFF

Although there has been extensive investigation of the terminal axon and the subneural apparatus of the neuromuscular junction by various methods, the intervening layer of amorphous material constituting the external lamina has been neglected. This material, which fills the synaptic gap between the axolemma of the terminal axon and the complex infolded structure of the subneural apparatus in vertebrate neuromuscular junctions is situated so that some of the chemical reactions of excitation may occur in this layer. Previous studies have revealed selective binding of metallic cations $Pb^{++}$, $Mn^{++}$, $Co^{++}$[27, 32] in unfixed muscle and periodic acid-Schiff (PAS) and ruthenium red staining of neuromuscular junction in fixed muscle. With the exception that organized collagen filaments are absent within the subneural apparatus, there are no fine structure differences between external membrane from innervated and noninnervated areas. The present study utilizes several cytochemical methods to investigate the properties of external lamina in innervated and noninnervated areas of mouse and rat striated muscle.

### MATERIALS AND METHODS

White Swiss mice weighing 20 to 25 g were used in the histochemical portion of the study, and white rats (200 to 300 g) were used to obtain muscle for sarcolemmal tube isolation.

#### Optical microscopy

*Lead staining.* Intact mouse thoraces with diaphragms were stained by the method of Sávay and Csillik[32] modified by use of buffers to obtain pH 4.5 to 5.6. After thorough washing, the intercostal muscles and diaphragms were soaked in 2% sodium sulfide solution, washed, dehydrated, cleared, and prepared as whole mounts.

*Histochemical methods for polysaccharides: Optical microscopy.* Mouse leg muscles were fixed for 24 hours in 10% formalin containing 2% calcium acetate, embedded in paraffin, sectioned at 5 $\mu$, and stained by a variety of histochemical methods.

Sections were stained with 0.02% azure A in acetate buffer at pH 1.5, 2.0, and 3.0 to demonstrate metachromasia of acid mucins,[34] by the periodate dimethylphenylenediamine method,[36] and by the combined aldehyde-fuchsin-alcian blue method of Spicer and Meyer.[35] Other sections were sulfated or methylated prior to staining with the aldehyde-fuchsin-alcian blue sequence.[35]

* This work was supported by Grant GB7367 from the National Science Foundation.

ELECTRON MICROSCOPY

*Lead staining*. Blocks of diaphragm and intercostal muscle after lead staining by the modified Sávay-Csillik method were treated with ammonium sulfide solution, washed, and postfixed in 6% glutaraldehyde or Dalton's chrome-osmium solution. The blocks were washed, dehydrated, embedded in Epon 812, and thin sectioned for electron microscopy.

*Ruthenium red*. Mice were anesthetized with intravenous Nembutal via the tail vein and perfused with 3% glutaraldehyde in cacodylate buffer (pH 7.2) or sacrificed by decapitation. Small blocks of intercostal and leg muscle were stained with ruthenium red according to the method of Luft.[22] Unfixed muscle was immersed in a solution of 2% glutaraldehyde containing 800 ppm of ruthenium red in 0.07 M cacodylate or 0.1 M phosphate buffers (pH 7.2) for 30 to 180 minutes. The blocks were rinsed several times with buffer and refixed with a solution containing 0.7% osmium tetroxide and 800 ppm of ruthenium red for 2 hours at 20°C. The blocks were washed, dehydrated, and embedded in Epon 812. Thin sections were cut with diamond knives and were examined unstained or stained with uranyl acetate and/or lead citrate in RCA EMU 3F or EMU 4 electron microscopes.

*Colloidal iron*. The colloidal iron method[26] was used for ultrastructural localization of acid mucopolysaccharides (AMPS). Leg or intercostal muscles were fixed for 12 hours in 10% formalin containing 2% calcium acetate or for 60 minutes in 6.25% buffered (pH 7.2) glutaraldehyde. Blocks of muscle were rinsed several times with cold buffer (pH 7.2) containing sucrose, and frozen sections were cut at 10 $\mu$. The sections were washed in distilled water and transferred to 12% acetic acid solutions for 5 minutes, after which they were stained with colloidal iron at pH 1.2. Some glutaraldehyde-fixed sections were washed and postfixed in osmium tetroxide. The blocks were washed, dehydrated, and embedded in Epon 812. Thin sections were prepared and examined with or without prior staining with lead citrate and/or uranyl acetate.

*Periodic acid-silver methenamine*. Small blocks of leg muscles fixed in 2.5% glutaraldehyde were washed several times with multiple changes of buffer at pH 7.4 for 16 to 24 hours, dehydrated with graded ethanol, and embedded in Epon 812-Araldite mixtures containing 12.5 ml of Epon 812, 27.5 ml of dodecenyl succinic anhydride, 7.5 ml of Araldite 502, 1.5 ml of dibutylphthalate, and 1 ml of tri-(dimethylaminomethyl)phenol. Thin sections were cut, floated on distilled water, and treated with 1% periodic acid and stained with silver methenamine according to the method of Rambourg and Leblond.[29] Some sections were also stained by the chromic-phosphotungstic acid method.[30] Control sections, not treated with periodic acid before methenamine silver staining, were also prepared.

*Chromic acid-phosphotungstic acid*. Glutaraldehyde-fixed muscle blocks were thoroughly rinsed with buffer and embedded in glycol methacrylate according to the method of Leduc and Bernhard.[18] Thin sections were stained with freshly prepared 1% phosphotungstic acid and 10% chromic acid for 2 to 5 minutes according to the methods of Rambourg, Hernandez, and Leblond.[30] Following a brief rinse in distilled water, the sections were collected on Formvar-coated grids and examined in the electron microscope. Thin sections on Formvar-coated copper or stainless steel grids were also stained with an aqueous solution of 1% phosphotungstic acid for 1 to 3 hours.

DENERVATION EXPERIMENTS

A group of mice were anesthetized with ether, and the right sciatic nerve was cut and ligated at mid-thigh level. Three weeks following operation, the mice were sacrificed and leg muscles were embedded and stained by the aldehyde fuchsin-alcian blue method[35] or by ruthenium red.[22] The former were examined in the optical microscope and the latter were prepared for examination in both optical and the electron microscopes.

ENZYME CYTOCHEMISTRY

The effect of three enzymes, neuraminidase (*Clostridium perfringens*, type VI, Sigma), hyaluronidase (bovine testes, type VI, Sigma), and collagenase (*Clostridium histolyticum*, type III, fraction A, Sigma), on the muscle basement lamina was tested under a variety of conditions. Hyaluronidase was applied in concentrations of 0.1 to 1 mg per ml at pH 6.0 in 0.1 M phosphate buffer with or without added 3% sucrose for 90 minutes to 24 hours at 37°C. Neuraminidase, 0.1 to 0.5 mg per ml, was used for 90 to 180 minutes. Collagenase was used in concentrations of 0.5 to 1 mg per ml in 0.1 M phosphate buffer (pH 7.0) for incubation periods of 20 minutes to 24 hours at 37°C.

Muscle was prepared in various ways before incubation in the enzyme solutions. Fresh, formalin-fixed, or glutaraldehyde-fixed whole muscle, 50- to 150-$\mu$ sections prepared with the Smith-Farquhar chopper or 20- to 50-$\mu$ frozen sections were stained with aldehyde-fuchsin-alcian blue or ruthenium red after incubation in the enzyme solutions. Tissue prepared in this way was examined in the optical or electron microscope when possible. Thin sections of muscle embedded in glycol-methacrylate were incubated on grids with the enzyme solutions and examined unstained or stained with ruthenium red, periodic acid-methenamine silver, or phosphotungstic acid-chromic acid methods in the electron microscope.

### Isolated sarcolemmal membranes

Rat sarcolemmal tubes were prepared by the method of McCollester.[24] Some pellets were resuspended in 4 to 6.0% buffered glutaraldehyde, dehydrated, and embedded in Epon 812 to follow the progress of myofiber emptying. Ohter pellets were resuspended with or without prior glutaraldehyde fixation in solutions of hyaluronidase or collagenase as used in the previous experiments, refixed in glutaraldehyde, or osmium 1% tetroxide, and embedded in Epon 812 for examination in the electron microscope. Still other pellets were incubated with either hyaluronidase or collagenase, stained with ruthenium red (800 ppm), postfixed in 1.7% osmium tetroxide, and then embedded in Epon 812 for electron microscopy.

In some experiments, sarcolemmal tube preparations were made from strips of muscle cut from the area of the "innervation band"[7] in either rat intercostal or diaphragm muscle. This yielded an "end-plate-rich" fraction; the remaining muscle yielded an "end-plate-poor" fraction.

Freshly prepared, unfixed sarcolemmal tubes were stained for acetylcholinesterase activity by naphthyl acetate[31] or thiolacetic acid[3] and examined in the optical microscope. Sarcolemmal tubes were also washed and resuspended in 4% glutaraldehyde after staining with thiolacetic acid, resedimented, and embedded in Epon 812 for electron microscopy. Control preparations were stained in the presence of $10^{-5}$ M eserine.

Fig. 1. Photomicrograph of mouse intercostal muscle illustrating the appearance of the centrally located "innervation band." The neuromuscular junctions are stained by the modified Sávay-Csillik method. The scale indicates 100 $\mu$.

Fig. 2. Photomicrograph of mouse intercostal muscle stained by the aldehyde fuchsin-alcian blue method after prior sulfation. The sarcolemma is deeply and selectively stained by aldehyde fuchsin. Note the clefts of the subneural apparatus delineated by aldehyde fuchsin staining (*arrow*). The scale indicates 10 $\mu$.

Fig. 3. Photomicrograph of mouse intercostal muscle stained by the aldehyde fuchsin-alcian blue method after prior sulfation. Aldehyde fuchsin staining material in the musculotendinous junction contrasts sharply with the lightly stained myofibers (*arrow*). The scale indicates 10 $\mu$.

Fig. 4. Photomicrograph (phase contrast) illustrating the deposition of lead in the subneural apparatus of the neuromuscular junction after staining with the Sávay-Csillik method (*arrow*). The scale indicates 30 $\mu$.

Fig. 5. Electron micrograph of mouse intercostal muscle neuromuscular junction after staining with the Sávay-Csillik lead method. Note the deposition of lead particles in the external lamina and in close proximity to the plasma membrane within the secondary synaptic clefts (*arrow*). Less intense lead staining is present in the external lamina of the muscle cell membrane at the periphery of the innervated area. The scale indicates 1 $\mu$.

Fig. 6. Electron micrograph of mouse neuromuscular junction stained with ruthenium red. The external lamina on the surface and within the primary and secondary synaptic clefts are filled with ruthenium red particles. The scale indicates 0.5 $\mu$.

Fig. 7. Electron micrograph of a fibroblast and a group of collagen filaments after staining with ruthenium red. The collagen fibrils appear as unstained white dots surrounded by electron-dense ruthenium red precipitate (*arrow*). The scale indicates 0.5 $\mu$.

Fig. 8. Electron micrograph illustrating myofiber external membrane stained with ruthenium red. Collagen filaments, unstained in the illustration (*arrow*), are surrounded by ruthenium red granules. The orifices of T-tubules near the external lamina also contain ruthenium red. The scale indicates 0.3 $\mu$.

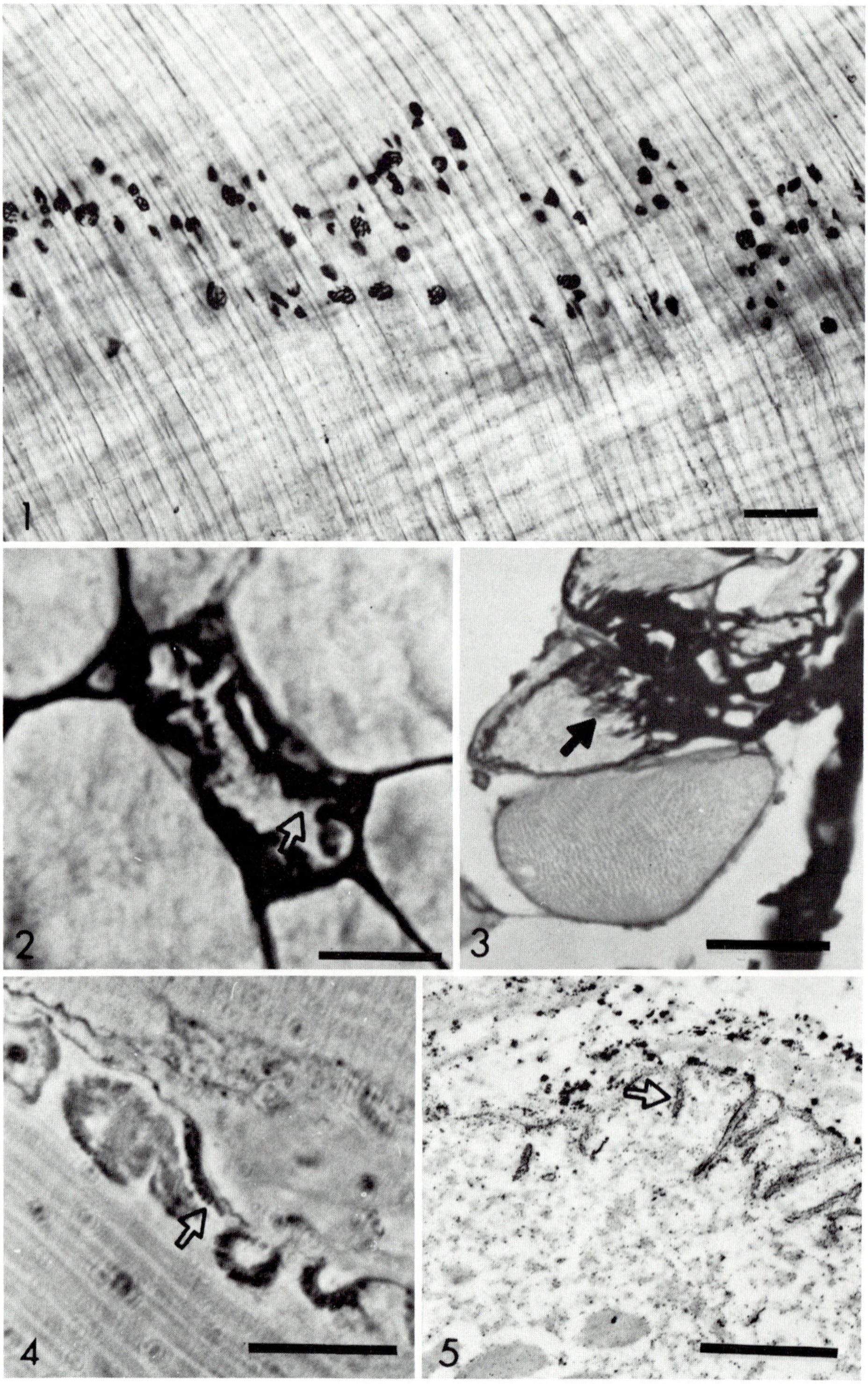

FIGS. 1–5

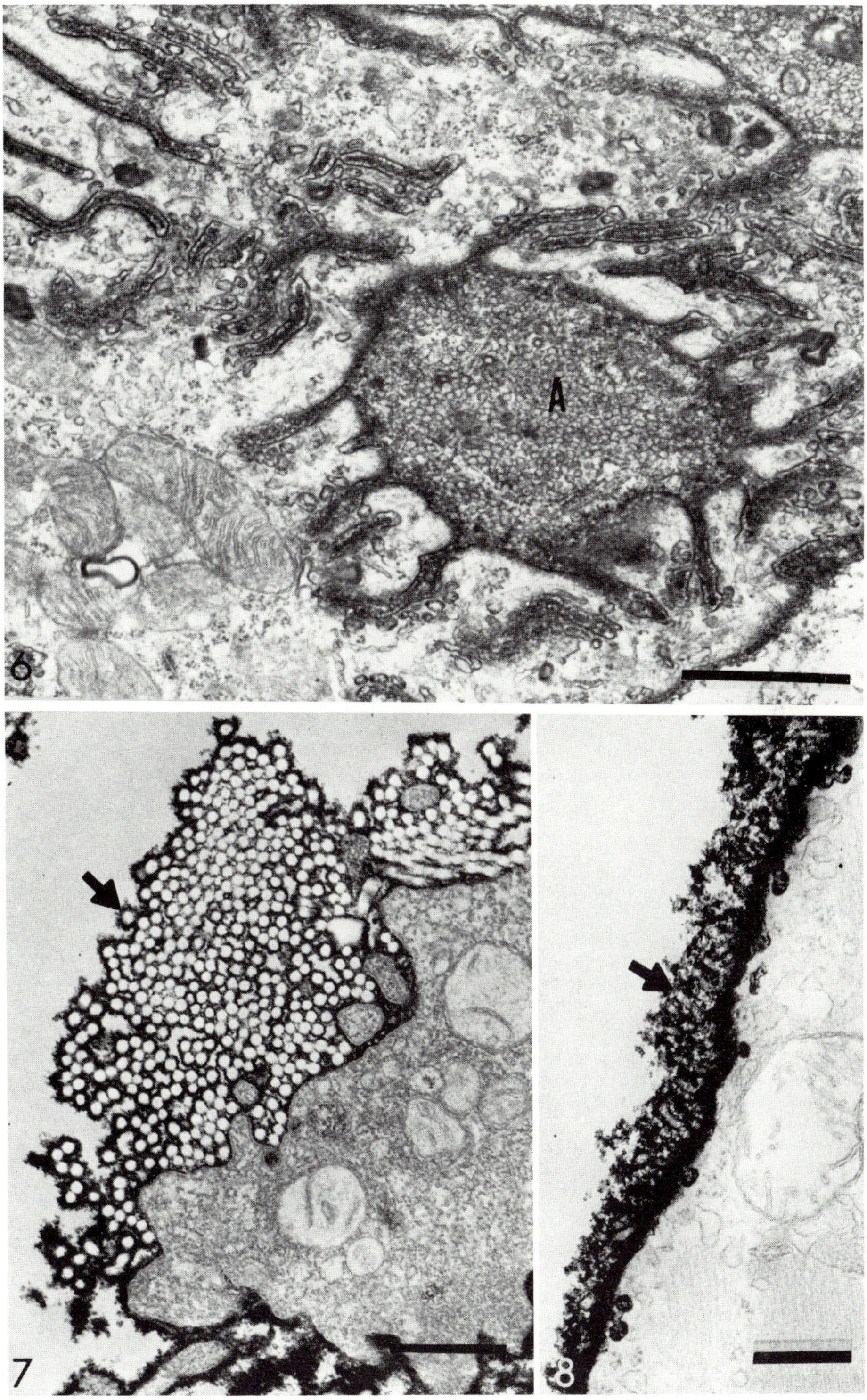


FIGS. 6–8

NITROGEN CONTENT OF ISOLATED SARCOLEMMAL TUBULES

Determinations of the nitrogen content of isolated sarcolemmal tubules were performed in an automatic Coleman Nitrometer model 29 and the percentage of nitrogen per mg dry weight was calculated.

## OBSERVATIONS

*Lead staining.* The cleared whole mounts revealed pale tan skeletal muscle with sharply delineated, black-stained motor end-plates lying within "innervation bands" (Fig. 1.) They closely resembled end-plates stained by histochemical methods designed to demonstrate acetylcholinesterase activity. Concentration of the end-plates in innervation bands permits the preparation of "end-plate-rich" and "end-plate-poor" fractions of sarcolemmal tubes.

*Azure A, Periodic acid phenylenediamine (PAPD), and alcian blue-aldehyde fuchsin.* The sarcolemma was colored pale blue when stained with azure A at pH 1.5 to 3.0, and pale grey after 24 hours staining with PAPD at pH 6.0.

The entire sarcolemma stained light blue when fixed muscle sections were stained by the alcian blue-aldehyde fuchsin (AB-AF) sequence. There was no staining of neuromuscular or musculotendinous junctions. When staining with aldehyde fuchsin preceded staining with alcian blue, the sarcolemma was uniformly stained red-violet. Again there was no staining of neuromuscular or musculotendinous junctions. Methylation prevented alcian blue staining of the sarcolemma, whereas after sulfation the sarcolemma was uniformly stained red-violet and, for the first time, the subneural apparatuses of neuromuscular junctions (Fig. 2) and musculotendinous junctions were stained red-violet (Fig. 3.).

Table 1 summarizes the staining characteristics of myofiber external lamina and subneural apparatuses after application of several histochemical staining procedures.

TABLE 1. CYTOCHEMICAL PROPERTIES OF EXTERNAL LAMINA OF MOUSE SKELETAL MUSCLE AND NEUROMUSCULAR JUNCTIONS

| Structure | Staining Reaction or Test | | | | | | | | | |
|---|---|---|---|---|---|---|---|---|---|---|
| | AF[a] | AB | Sulfation followed by AF | RR | Colloidal iron | Periodic acid-methenamine silver | Control methenamine silver | PTA-chromic acid | Collagenase incubation followed by RR | Hyaluronidase incubation followed by RR |
| Muscle myofiber external lamina | − | + | ++ | + | − | ++ | ± | + | + | + |
| Muscle cell membrane | | | | − | ++ | − | | | | |
| Subneural apparatus | − | − | ++ | + | − | ++ | ± | ++ | + | + |
| Collagen fibrils | | | | − | | ++ | + | − | | |
| Collagen matrix | | | | ++ | | ++ | − | | | |
| Axolemma | | | | − | ++ | − | | − | | |

[a] The abbreviations used are: AF, aldehyde fuchsin; AB, alcian blue; RR, ruthenium red; PTA, phosphotungstic acid.

ELECTRON MICROSCOPY

*Lead staining.* In the phase contrast and electron microscopes, electron-dense lead particles were deposited in the external lamina of the myofibers and in greater quantities within the subneural apparatus (Figs. 4 and 5). Selectivity of binding within the subneural apparatus was maximal in the pH range 4.0 to 5.6.

*Ruthenium red.* After staining with ruthenium red (RR), opaque deposits of the stain were observed in the outer cell layers of the muscle blocks. The neuromuscular junctions were easily located and resembled end-plates stained by techniques used to demonstrate acetylcholinesterase activity. In the electron microscope, sections not further stained with lead citrate or uranyl acetate demonstrated well defined membrane structures, and dense deposits of RR were present in the muscle external lamina. External lamina within the subneural apparatus of neuromuscular junctions was also uniformly and intensely stained with RR (Fig. 6). Ruthenium red staining of external lamina covering noninnervated areas of myofibers was identical to the staining of innervated areas except for the presence of collagen fibers surrounded by a coating of RR staining material in the noninnervated areas (Figs. 7 and 8).

There also was increased electron density in the muscle surface membranes of the synaptic and coated vesicles. Intense RR staining was present in the external lamina of interstitial capillaries and Schwann and perineural epithelial cells. The intracellular space between the Schwann cell processes and the terminal axons was occasionally filled with RR, and some smooth coated vesicles in the sole plate sarcolemma contained RR.

In some experiments, we added 3% dimethylsulfoxide to the enzyme solutions in an attempt to increase penetration. Subsequent attempts to stain external lamina with ruthenium red were unsuccessful, possibly due to methylation of carboxyl groups.

*Colloidal iron.* After staining with colloidal iron, there was a thin coating of electron-dense iron particles covering the outer surface of the myofiber external lamina and in some areas, iron particles coated the external leaflet of the myofiber cell membrane. Few particles were present within the myofiber external lamina. In innervated areas, a thin layer of iron particles covered the axolemma and the myofiber cell membrane within the primary and secondary synaptic clefts of the subneural apparatus (Fig. 9). Few iron particles were present in the external lamina within the subneural apparatus. A narrow layer of iron particles also coated outer mitochondrial membranes, endothelial cells, and Schwann cell processes.

*Periodic acid-silver methenamine.* Control sections from glutaraldehyde-fixed skeletal muscle without prior phosphotungstic acid (PTA) oxidation stained with silver methenamine demonstrated intense silver staining of nuclei and collagen filaments. Moderate numbers of coarse silver granules were also present within the external lamina coating myofibers and filling the subneural apparatus. After PTA, there was intense staining of nuclei and collagen as before, and, in addition, there was greatly increased deposition of fine silver granules within the external lamina covering myofibers and within the subneural apparatuses (Fig. 10).

*Phosphotungstic acid-chromic acid.* When thin sections of glycol methacrylate-embedded muscles were stained with a mixture of phosphotungstic and chromic acids at low pH, uniform staining of the external lamina of collagen fibers, myofibers, endothelial cells, Schwann and perineural epithelial cells occurred. There was also diffuse staining of the external lamina within the synaptic clefts (Fig. 11). This method did not stain glycogen but ribosomes, filaments, and tubular structures within the terminal axons were weakly stained. The membranes of synaptic vesicles in terminal axons were also stained by phosphotungstic acid and resembled the synaptic vesicles stained by this technique in the rat cerebral cortex.[29]

*Phosphotungstic acid.* Glutaraldehyde-fixed muscle embedded in glycol methacrylate and stained with phosphotungstic acid demonstrated intense staining of both the external lamina within the subneural apparatus and covering adjacent noninnervated muscle. The appearance was similar to PTA-chromic acid-stained preparations.

### Denervated neuromuscular junctions

Denervated neuromuscular junctions examined 3 to 8 weeks after nerve section and stained by the AB-AF method or RR showed no alterations in the staining of external lamina of innervated or noninnervated myofibers (Fig. 12).

### Enzyme cytochemistry

*Optical microscopy.* Staining with AB-AF sequences or RR was unaffected by incubation of fresh or fixed muscle blocks or chopper sections in solution of neuraminidase or collagenase for up to 24 hours at 37°C. Incubation for 3 hours with solutions of hyaluronidase (0.5 mg per ml) slightly reduced AB staining of the sarcolemma, particularly in fresh or frozen sections. No alterations in staining occurred after incubation with neuraminidase or collagenase.

*Electron microscopy.* Fresh blocks of muscle, incubated with hyaluronidase (1 mg per ml) for 24 hours at 37°C in balanced salt solutions containing tetracycline (1 mg per ml) showed slight reduction of external lamina, and fixed

---

FIG. 9. Electron micrograph of mouse intercostal neuromuscular junction stained by colloidal iron. Note the iron particles attached to the external surface of the myofiber external membrane (*open arrow*) and similar particles in close relationship to the myofiber plasma membrane within the primary and secondary synaptic clefts of the subneural apparatus (*closed arrow*). External membranes of mitochondria (*double arrow*) are also coated with iron particles. The scale indicates 0.5 $\mu$.

FIG. 10. Electron micrograph of mouse intercostal muscle neuromuscular junction stained by the methenamine silver method. Note the dense aggregates of silver covering collagen filaments (*open arrow*) and the diffuse deposition of silver granules in the primary and secondary synaptic clefts of the subneural apparatus (*closed arrows*). The scale indicates 1.0 $\mu$.

FIG. 11. Electron micrograph of mouse intercostal neuromuscular junction embedded in glycol methacrylate and stained with the phosphotungstic acid-chromic acid method. Note the granular precipitate in the muscle surface external lamina and within the subneural apparatus. The scale indicates 0.5 $\mu$.

FIG. 12. Electron micrograph of mouse leg muscle neuromuscular junction 3 weeks following denervation. Staining with ruthenium red shows persistence of external lamina covering the myofibers and within the subneural apparatus. The scale indicates 1 $\mu$.

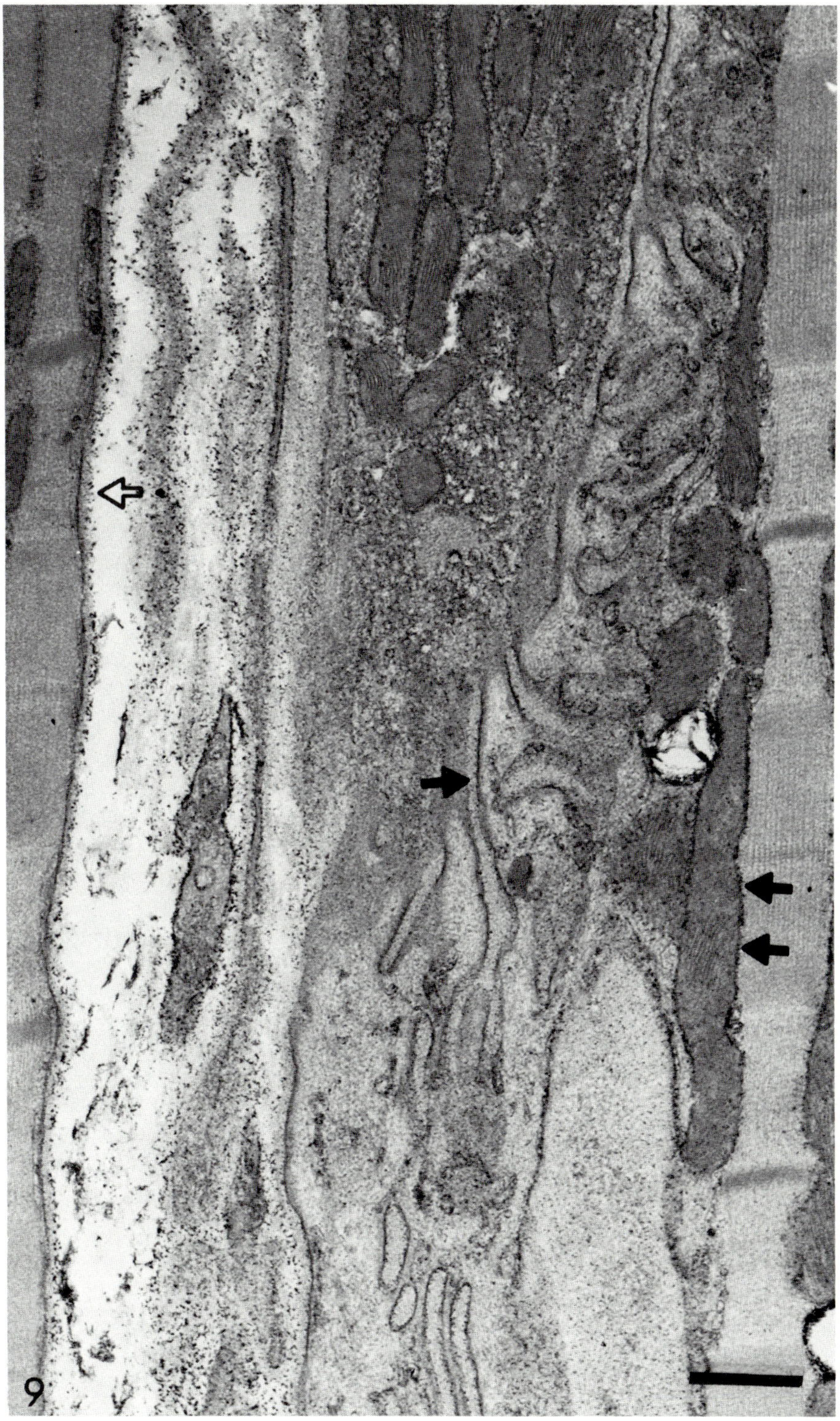


FIG. 9

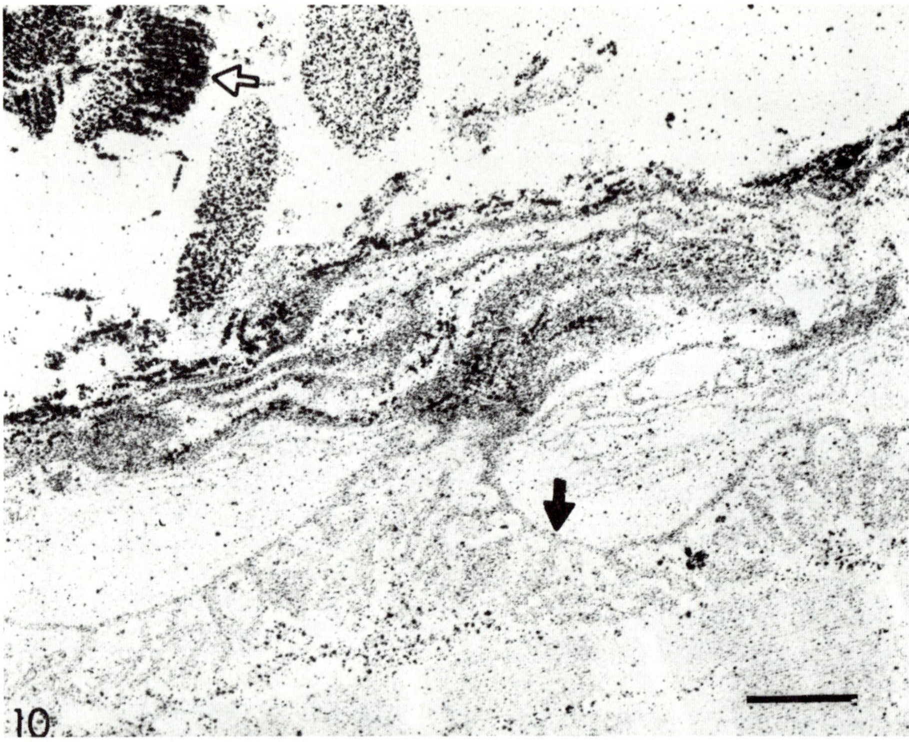


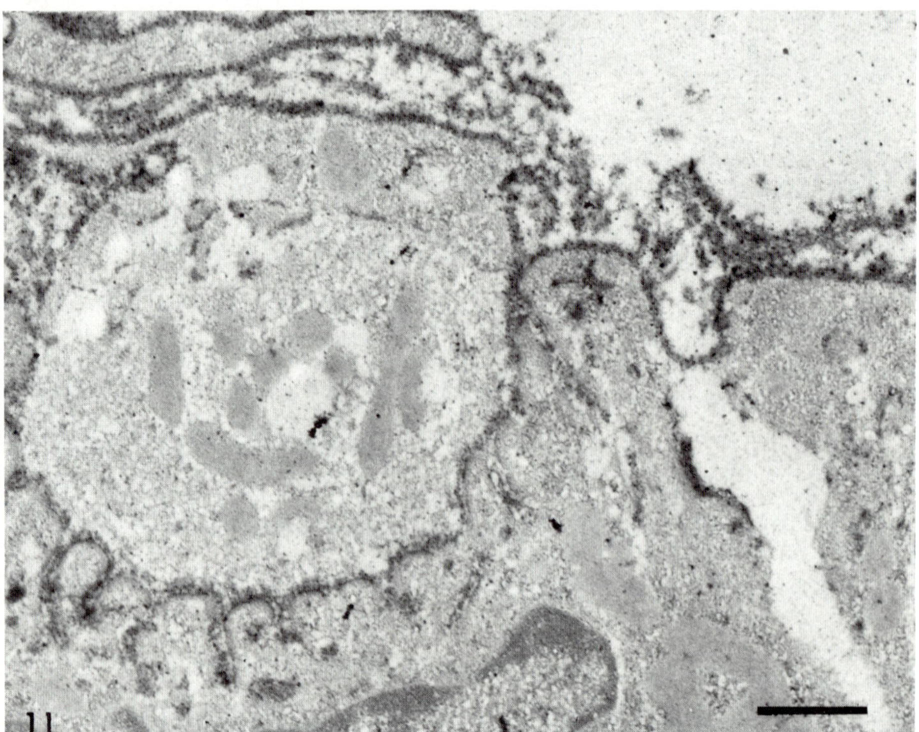


Figs. 10 and 11

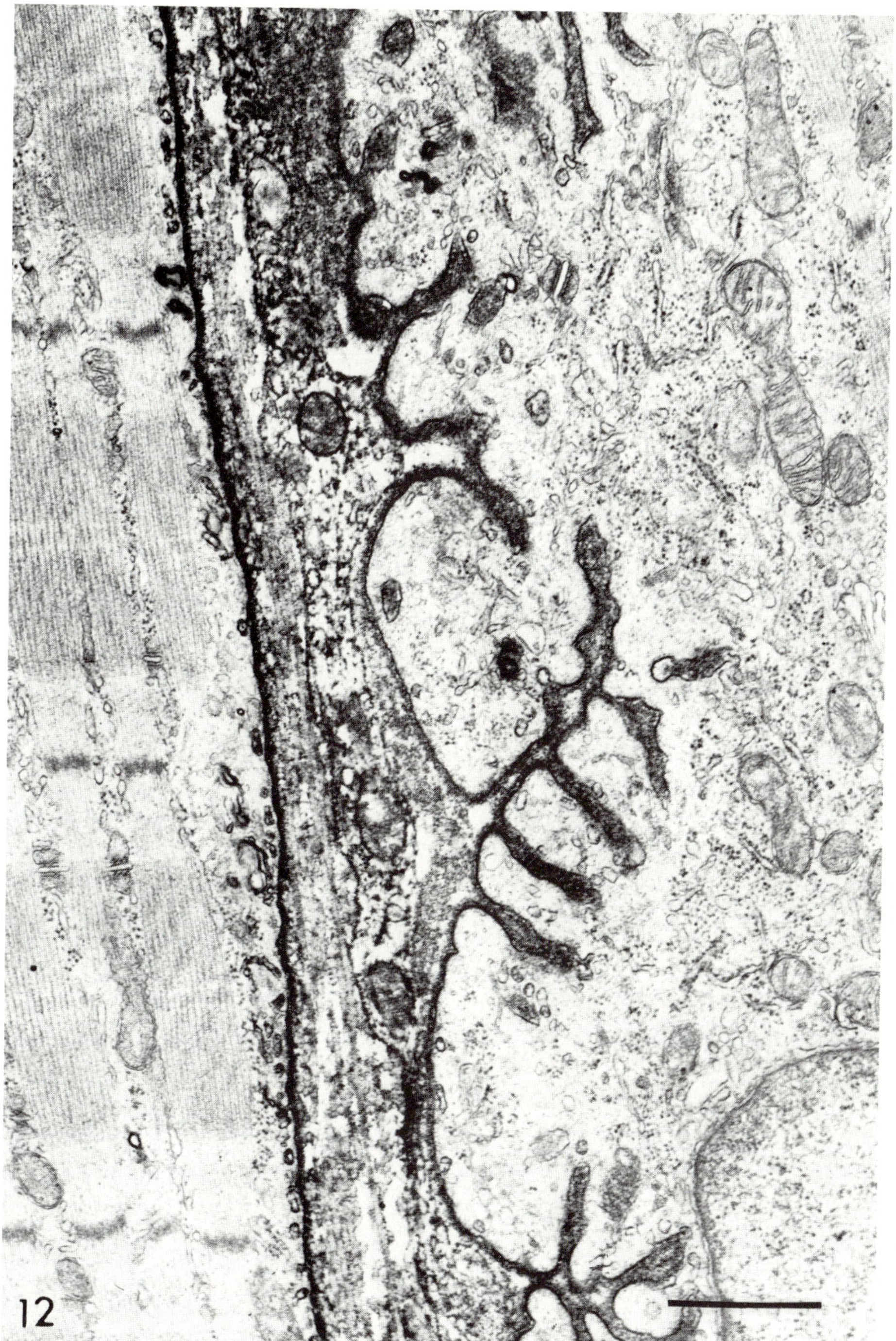


FIG. 12

sections demonstrated no significant reduction in the width or density of this layer (Fig. 13). Similarly, thin sections embedded in glycol methacrylate, incubated in hyaluronidase and subsequently incubated with PTA-chromic acid or PTA-methenamine silver, failed to show significant extraction of the myofiber basement lamina.

Fresh muscle blocks, frozen or fixed sections, incubated with large quantities (1 mg per ml) of collagenase known to hydrolyze gelatin all demonstrated persistence of myofiber external lamina (Fig. 14) and organized collagen filaments (Fig. 15). There was also no alteration in subsequent RR staining after incubation with hyaluronidase (Fig. 16) or collagenase (Fig. 17).

### Experiments with sarcolemmal tubes

Successful sarcolemmal tube preparations contained approximately 90% empty tubes (Fig. 18).

In the electron microscope, isolated sarcolemmal membranes consisted of fragmented or partially coiled muscle plasma membranes covered by an amorphous mass of external lamina that appeared less dense than in fixed sections (Fig. 19). The ends of occasional filaments, recognized as collagen by their periodic banding, were embedded in the external lamina. RR staining of sarcolemmal tubes demonstrated electron-dense granules of the dye in the myofiber external lamina (Fig. 20). When pellets of isolated sarcolemmal tubes were incubated with hyaluronidase or collagenase solutions as in experiments with tissue sections and stained with RR, there was no reduction of RR staining when the sarcolemmal tubes were examined in either the optical or electron microscopes.

Acetylcholinesterase activity was demonstrated in neuromuscular junctions on isolated sarcolemmal tubes from "end-plate-rich" sarcolemmal tube preparations (Fig. 21). Thin sections from pellets of tubes stained by the thiolacetic acid method revealed occasional discrete membrane patches covered with the histochemical reaction product. Some structures closely resembled subneural apparatuses (Fig. 22). These consisted of paired membranes separated by gaps of approximately 500 Å. The membranes were covered with lead granules resembling subneural apparatuses in sections stained by this method.

### Composition of Isolated Sarcolemmal Tubes

Isolated empty rat sarcolemmal tubes contain 78.4% protein, 14.2% lipid, and 1.8% carbohydrate.

## DISCUSSION

The cytochemical data obtained in the present study suggest that the external lamina covering skeletal myofibers is a moderately acidic glycoprotein that is either mixed with or contains as an integral part components with free carboxyl groups. This material covers the myofibers in innervated as well as noninnervated areas. There is also a layer of probable acid mucopolysaccharide attached to the external leaflet of the muscle cell membrane, a "cell coat,"[29] that stains differently than the external lamina. Absence of metachromasia after azure A staining at pH 2.5 and failure to stain with the dimethylphenylenediamine method after periodate oxidation, colloidal iron, or aldehyde fuchsin rule out the presence of significant quantities of acid mucopolysaccharide within the external lamina.[34, 36] Although the myofiber external lamina stains intensely with ruthenium red, this reaction does not appear to be due to AMPS as suggested by Luft.[21-23] However, the presence of acid groups within the myofiber external lamina is suggested by the binding of metallic cations by unfixed external lamina,[27] staining with alcian blue,[38] and blockade of alcian blue staining by methylation, a reaction which blocks carboxyl groups in sialic, uronic acids in weakly acidic AMPS and mucoproteins. Furthermore, staining with methenamine silver without prior oxidation with periodate suggests residual glutaraldehyde in the sections despite washing.[30] Yamada[37] has pointed out that alcain blue stains hydroxyl as well as carboxyl groups.

Other cytochemical evidence suggesting the myofiber external lamina is composed of glycoprotein include the positive PAS reaction and similar procedures such as periodate-methenamine silver and periodate-chromic acid methods. Induction of metachromasia and aldehyde fuchsin staining in the external lamina by sulfation,[35] also suggests that a neutral mucopolysaccharide or a glycoprotein is present. Thus, the intense ruthenium red staining of the external lamina is probably not due to AMPS. Ruthenium red has been demonstrated to stain the narrow AMPS coating that covers the surface of the external leaflet of the myofiber cell membrane, [14, 21] but in the present study this staining was probably obscured by the large amount of RR deposited in the external lamina of both innervated and noninnervated areas of myofibers.

Data obtained by incubating fresh or fixed muscles in solutions containing neuraminidase, hyaluronidase or collagenase under a wide variety of conditions, with or without subsequent staining, revealed slight reduction of alcian blue staining only after incubation of fresh muscle with hyaluronidase or collagenase. Electron microscopic studies revealed no significant change in the myofiber external lamina following incubation of thin sections on grids with these enzymes with or without subsequent staining to demonstrate the basement lamina. When glycol methacrylate embedding was used to facilitate penetration of the enzymes into thin sections[18] there also was no extraction of the myofiber external lamina. Failure to remove myofiber external lamina with a preparation of collagenase demonstrated to lyse gelatin was unexpected since Kono, Ka-

FIG. 13. Electron micrograph of mouse intercostal neuromuscular junction extracted with hyaluronidase illustrating persistence of external lamina. The scale indicates 0.5 $\mu$.

FIG. 14. Electron micrograph of mouse intercostal neuromuscular junction following incubation with collagenase illustrating persistence of external lamina. The scale indicates 0.5 $\mu$.

FIG. 15. Electron micrograph of banded collagen filaments from mouse intercostal muscle incubated with collagenase. There is no evidence of digestion of the fibrils. The scale indicates 0.5 $\mu$.

FIG. 16. Electron micrograph of mouse intercostal neuromuscular junction following incubation with hyaluronidase and subsequent staining with ruthenium red. Note the persistence of the external lamina covering the myofibers and similar material within the subneural apparatus. The scale indicates 0.5 $\mu$.

FIG. 17. Electron micrograph of mouse intercostal neuromuscular junction following incubation with collagenase and subsequent staining with ruthenium red. Note the persistence of the myofiber external membrane and the similar material within the subneural apparatus. White dots indicating collagen filaments are present in the external lamina. The scale indicates 0.5 $\mu$.

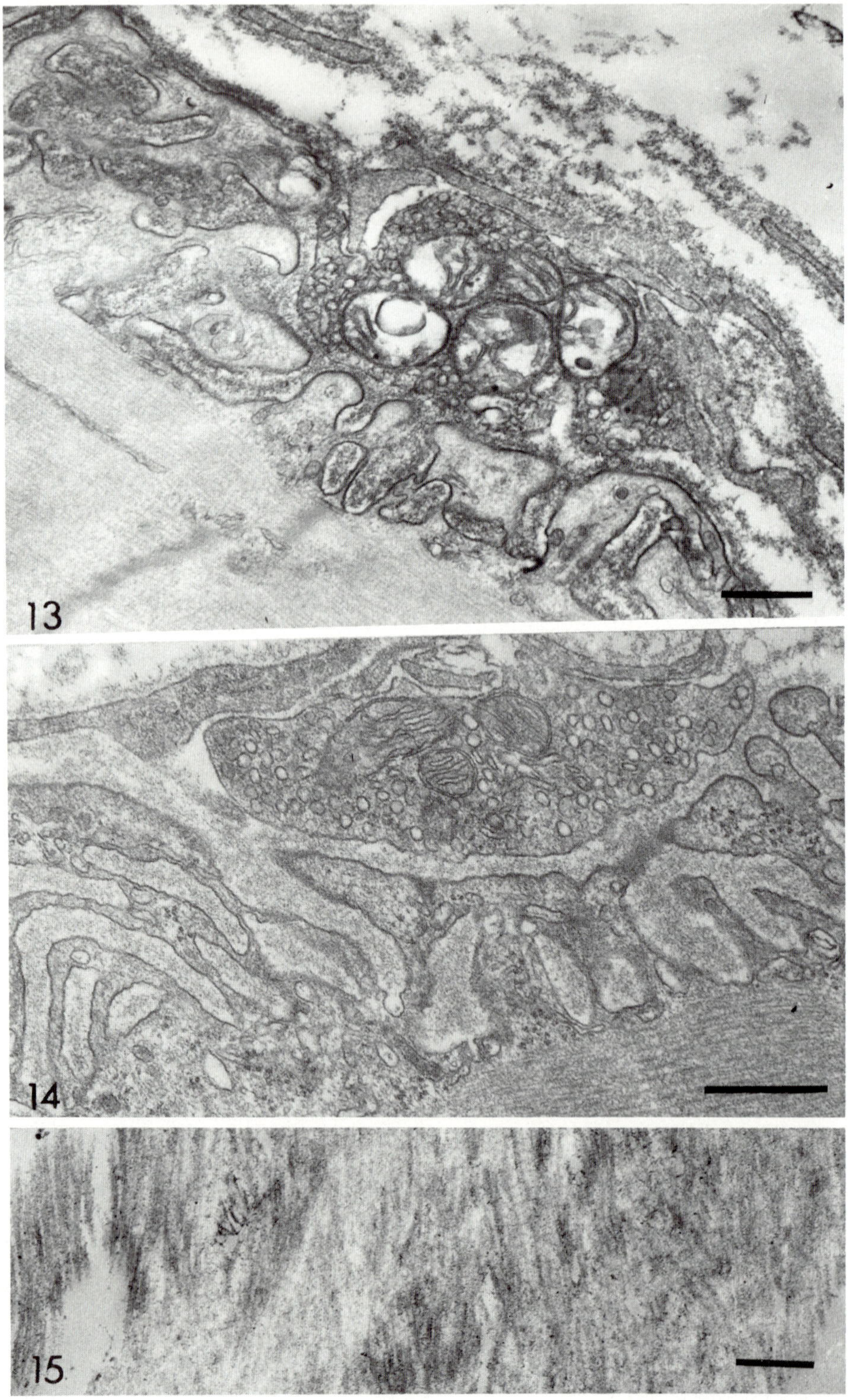


FIGS. 13–15

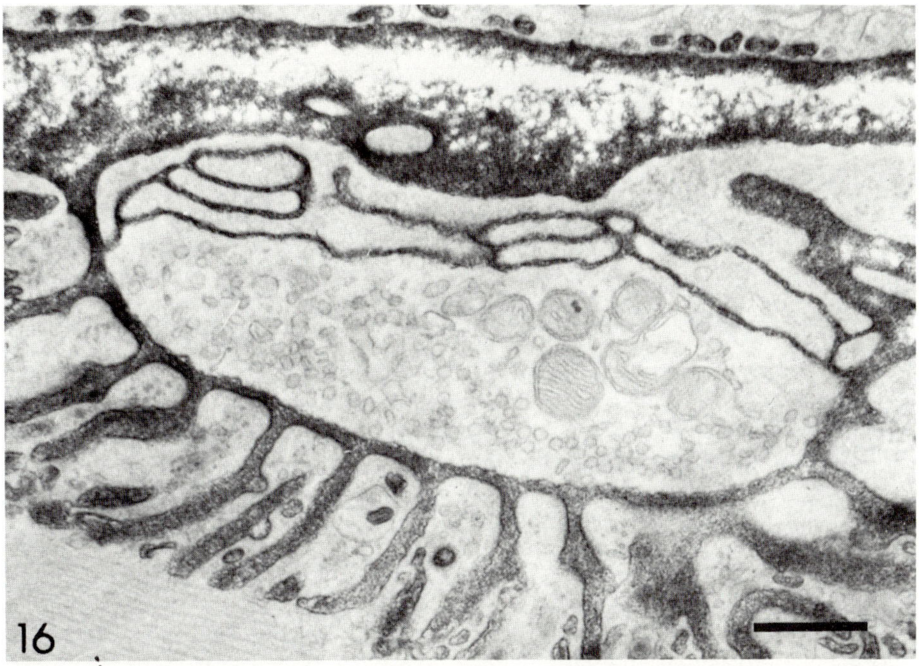


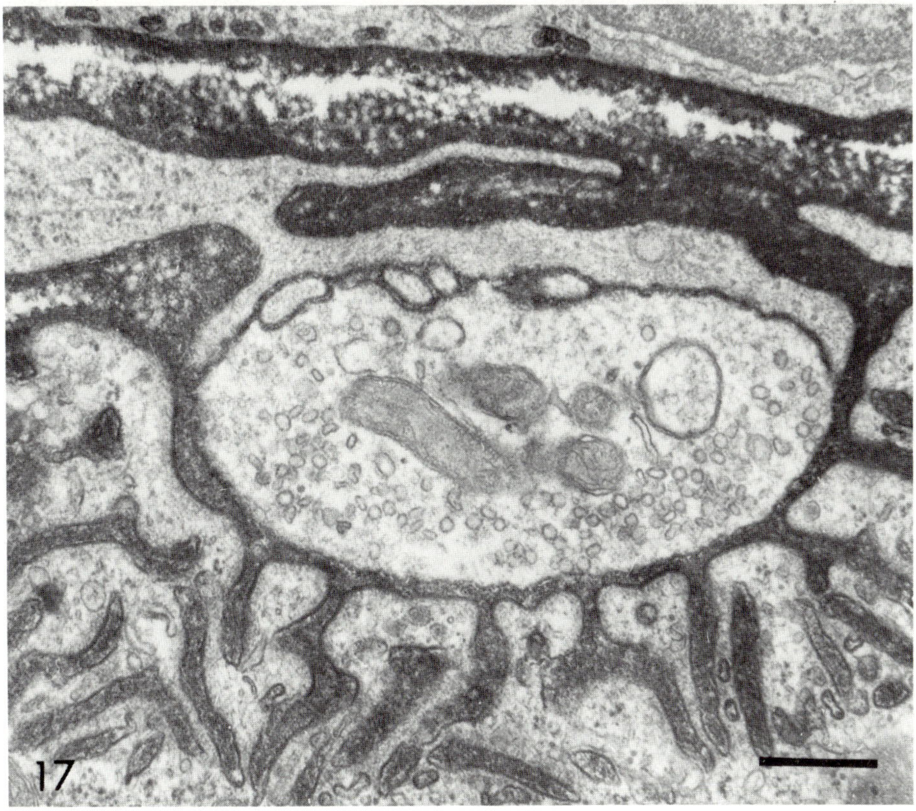


FIGS. 16 AND 17

kuma, Homa, and Fukuda[16] reported that an unspecified collagenase preparation did remove rat myofiber external lamina. In our preparations, organized collagen fibrils, recognized by their banding, resisted incubation with high concentrations of collagenase for up to 24 hours at 37°C. Furthermore, collagen-like "reticulin," that resembles collagen but lacks its typical organization, resists tryptic but not collagenase digestion.[2]

Rat sarcolemma contains acetylneuraminic acid, glucose, glucosamine, glucuronic acid, and xylose according to Kono and Collowick.[15] Abood, Kurahasi, Brunngraber, and Koketsu[1] have reported that myofiber external lamina contains 0.9 ± 0.2% polysaccharide and 60 ± 3% protein calculated on a dry weight basis. Abood *et al.*[1] concluded that an amino acid analysis of the sarcolemma excluded the presence of large amounts of collagen. In the present experiments, the failure of collagenase to remove the myofiber external lamina may be due to the absence of the Gly-Pro-X-Gly-Pro-X sequence consisting of 3 to 6 residues as occurs in gelatin that is required for collagenase proteolysis.[2] On the other hand, failure of neuraminidase to prevent alcian blue staining of the myofiber external lamina suggests that the free carboxyl groups in sialic acids were not responsible for alcian blue staining or that these groups were not accessible to the enzyme.

The acidic glycoprotein covering the myofibers appears to differ from similar appearing material in the glomerular external lamina. Lazarow and Speidel[17] found that collagenase solubilized 75% of isolated glomerular external lamina, and amino acid "fingerprinting" revealed the presence of a protein resembling collagen containing a high concentration of proline and hydroxyproline. This material contained 3.4% carbohydrate and was also partially solubilized by sialidase. Other investigators working with glomerular external membrane were Bruchhausen and Merker,[5] who found 28% collagen and 59.6% noncollagen protein but no mucopolysaccharides containing uronic acids. Misra and Berman[25] concluded that glomerular external lamina is a sialic acid-glycoprotein complex with collagen. Preliminary nitrogen determinations of approximately 14% from isolated rat sarcolemmal tubes in present study suggests that myofiber external lamina contains a considerable percentage of protein and could not be a mucopolysaccharide. However, its content of hexosamine nitrogen is unknown.

Our present hypothesis is that myofiber external lamina is a loosely structured acidic glycoprotein or carbohydrate-protein complex that is permeable to both horseradish peroxidase (40 Å) and ferritin (100 Å). It appears that the myofiber external lamina is negatively charged under the conditions of fixation and staining used.

Another difference between myofiber and glomerular external lamina is the restricted permeability of glomerular external lamina to ferritin.[9] Capillary external lamina, however, is permeable to ferritin.[6]

In addition to the external lamina, the present cytochemical data indicate the presence of a "cell coat" that was revealed by colloidal iron staining and suggested by lead staining of fresh muscle. The "cell coat" was described by Fawcett[10] as a layer of amorphous material containing fine filaments that lay on the surface of myofibers, Schwann cells, and fibroblasts. Bennett[4] coined the term "glycocalyx" to describe the PAS staining material coating many kinds of cells. Although in the earlier work the "cell coat" was not clearly differentiated from the external lamina, more recent studies by Rambourg and Leblond[29] and Kelly[14] have emphasized differences between the "cell coat" and the overlying external lamina. The "cell coat" demonstrated by colloidal iron staining is a thin layer attached to the external leaflet of the myofiber cell membrane. It was present in both innervated and noninnervated areas of myofibers. This material, thought to be AMPS, was not specifically studied in the present investigation. The specificity of colloidal iron staining has been doubted, but at pH 2 much of the artifactual deposition is minimized.[8] Lead deposition also occurs in the "cell coat" area when unfixed muscle is soaked in high concentrations of lead nitrate (1 mg per ml). If it was stained in ruthenium red and periodate methenamine silver preparations, the "cell coat" was probably obscured by the large amount of histochemical reaction product within the overlying external lamina.

The only cytochemical differences between innervated and noninnervated areas of the myofiber are the selective binding of metallic cations,[27] absence of alcian blue staining, and the presence of acetylcholinesterase activity in the subneural apparatus. Metallic cation binding and alcian blue staining are primarily properties of the external lamina, whereas acetylcholinesterase activity appears to be situated in the "cell coat" or within the cell membrane. The localization of acetylcholinesterase activity is particularly well shown in isolated membranes from "end-plate-rich" sarcolemmal tube preparations which tend to confirm previous histochemical observations.[19]

The failure of alcian blue to stain the subneural apparatus of neuromuscular junctions and the selective binding of divalent cations indicates that there may be differences in the ionization or accessibility of binding sites in external lamina within the subneural apparatus as compared to noninnervated areas. The failure of the external lamina within the subneural apparatus to stain with alcian blue may result from inaccessibility of dissociating groups or possibly interference from other proteins.[28, 33]

A possible explanation for differences in staining of external lamina in innervated and noninnervated areas is that, within the subneural apparatus, the external lamina is composed of the conjoined contributions of the overlying Schwann cell in conjunction with the myofiber external lamina. Possibly

FIG. 18. Photomicrograph of emptied sarcolemmal tube contrasted with a portion of a still unemptied tube (*arrow*). The scale indicates 100 $\mu$.

FIG. 19. Electron micrograph of sarcolemmal tubes illustrating the low density external lamina (*open arrow*) and the fragmented, denser muscle surface membranes (*closed arrow*). The scale indicates 0.5 $\mu$.

FIG. 20. Electron micrograph of isolated sarcolemmal tube stained with ruthenium red. The scale indicates 0.5 $\mu$.

FIG. 21. Photomicrograph of sarcolemmal tube preparations stained by the azo method for acetylcholinesterase activity. The darkly stained objects are neuromuscular junctions. The scale indicates 100 $\mu$.

FIG. 22. Electron micrograph of membranes isolated from end-plate-rich sarcolemmal tube preparations after staining with the thiolacetic method for acetylcholinesterase. Note the deposition of lead particles on the surface of the membranes. The scale indicates 1 $\mu$.

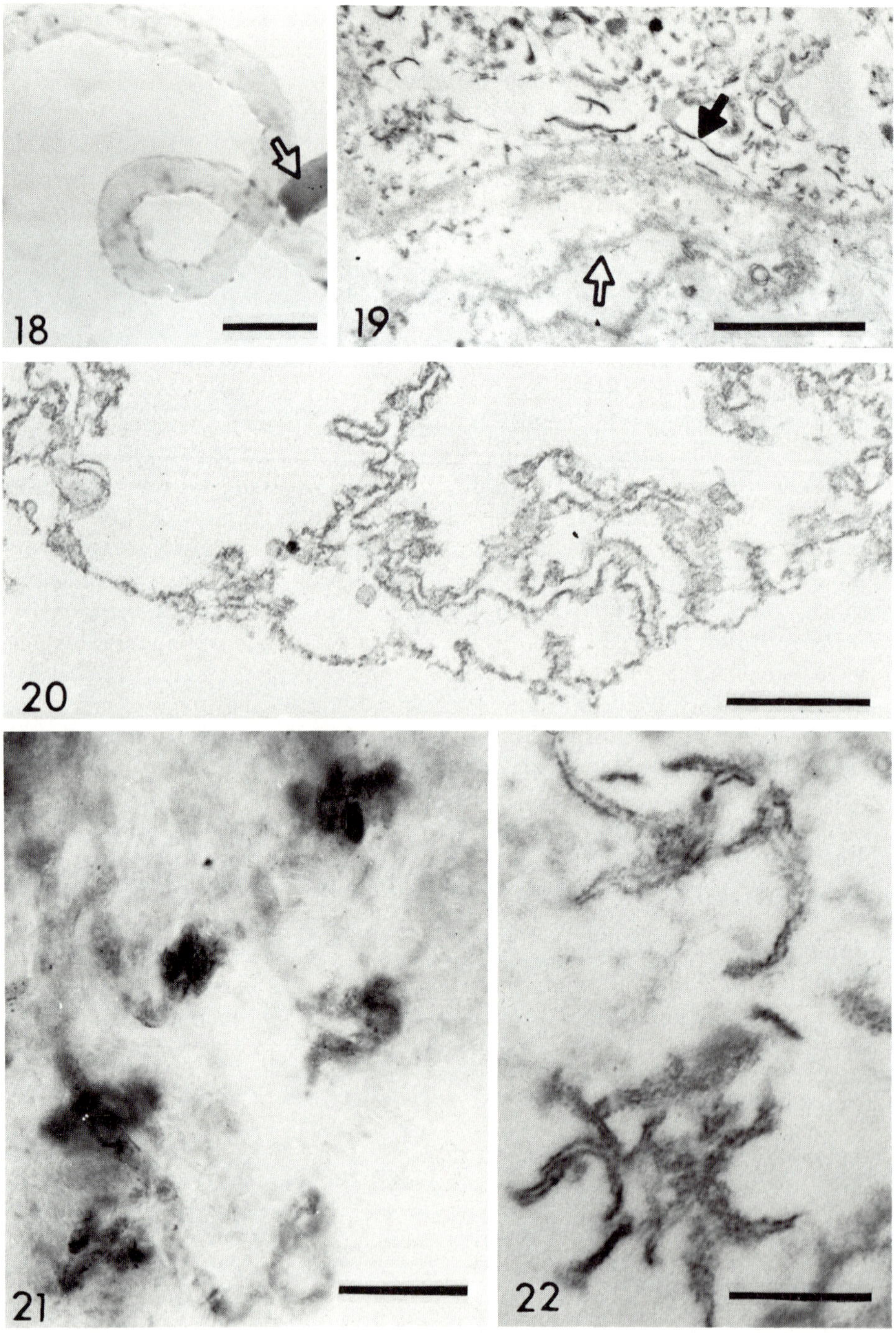


Figs. 18–22

related to this phenomenon is the absence of organized collagen filaments within the subneural apparatus. An analogous situation occurs in the musculotendinous junction where myofiber external lamina joins the periosteum. This area has acetylcholinesterase activity, binds lead, stains with PAS, and has a separate peak of acetylcholine sensitivity.[13]

Aldehyde fuchsin staining after prior sulfation is a useful method for the study of neuromuscular junction in formalin-fixed, paraffin-embedded tissue. In thin sections, individual clefts of the subneural apparatus can be resolved in the optical microscope.

Another glycoprotein present in close proximity to the muscle external lamina is the argentophilic, PAS-positive, ruthenium red staining reticulun[11, 12, 20] in which mature organized collagen fibers are embedded. In the electron micrographs prepared from muscles stained with methenamine silver, these materials stained more intensely than the external lamina, probably due to the presence of free aldehyde groups. Similarly, after ruthenium red staining, collagen filaments were surrounded by layers of amorphous, deeply staining material. The present cytochemical data are insufficient to distinguish reticulin associated with collagen fibers from the external lamina material. Only silver staining appears more intense in the reticulin.

## SUMMARY

A variety of cytochemical methods was employed to investigate the properties of the external lamina covering innervated and noninnervated areas of myofibers. The myofiber is also covered by a thin layer of material external to the myofiber plasma membrane that stains with colloidal iron and appears to be identical to the "cell coat" described by other investigators. External to this, the external lamina had the staining properties of an acidic glycoprotein containing acid groups that were not sufficiently dissociated to produce metachromasia or aldehyde fuchsin staining. Following sulfation, both subneural apparatus and musculotendinous junctions were sharply delineated by aldehyde fuchsin-stained basement lamina. Major differences between the subneural apparatus of neuromuscular junction and noninnervated areas on the myofiber surface were the absence of alcian blue staining in the subneural apparatus, selective binding of metallic cations by unfixed neuromuscular junctions, and the presence of acetylcholineaterase activity in the vicinity of the external leaflet of the muscle plasma membrane. Experiments with fixed or unfixed sections and isolated sarcolemmal tube preparations have shown that neither hyaluronidase, neuraminidase, or collagenase removes appreciable amounts of the external lamina from either innervated or noninnervated areas. The nature of the substances responsible for the staining reactions is discussed.

## ACKNOWLEDGMENTS

The technical assistance of Miss Marilyn Salscheider, Mrs. Eileen Kershaw, and Mr. Marion Rhoades is acknowledged.

REFERENCES


1. Abood, L. G., Kurahasi, K., Brunngraber, E., and Koketsu, K.: Biochemical analysis of isolated bullfrog sarcolemma. *Biochim. Biophys. Acta 112:* 330–339, 1966.
2. Bailey, A. J.: The nature of collagen. Extracellular and supporting structures. In *Comparative Biology*, ed. by Florkin, M., and Stotz, E. H. vol. 26 B, chap. 5. Amsterdam, Elsevier, 1968.
3. Barrnett, R. J., and Palade, G. E.: Enzymatic activity in the M-bands. *J. Biophys. Biochem. Cytol. 6:* 163–169, 1959.
4. Bennett, H. S.: Morphological aspects of extracellular polysaccharides. *J. Histochem. Cytochem. 11:* 15–23, 1963.
5. Bruchhausen, F. v. u., and Merker, H. J. Morphologischer und Chemischer aufbau Isolierter. Basalmembranen aus der Nierenrinde der Ratte. *Histochemie 8:* 90–108, 1967.
6. Bruns, R. R., and Palade, G. E.: Studies on blood capillaries. II. Transport of ferritin molecules across the wall of muscle capillaries. *J. Cell Biol. 37:* 277–299, 1968.
7. Cöers, C., and Woolf, A. L.: *The Innervation of Muscle: A Biopsy Study*. Oxford, Blackwell Scientific Publications, 1959.
8. Curran, R. C.: The histochemistry of mucopolysaccharides. *Int. Rev. Cytol. 17:* 149–212, 1964.
9. Farquhar, M. G.: Glomerular permeability investigated by electron microscopy. In *Small Blood Vessel Involvement in Diabetes Mellitus*, ed. by Siperstein, M. D., Colwell, A. R., Sr., and Meyer, K., pp. 31–38. Washington, D.C., American Institute of Biological Sciences, 1964.
10. Fawcett, D. W.: *Modern Development in Electron Microscopy*, ed. by Seigel, B. M., p. 257. New York, Academic Press, 1964.
11. Glegg, R. E., Eidenger, D., and Leblond, C. P.: Some carbohydrate components of reticular fibers. *Science 118:* 614–616, 1953.
12. Irving, E. A., and Tomlin, S. G.: Collagen, reticulum and their argyrophilic properties. *Proc. Roy. Soc. Lond. (Biol.) 142:* 113–135, 1954.
13. Katz, B., and Miledi, R.: Further observations on the distribution of acetylcholine-reactive sites in skeletal muscle. *J. Physiol. 170:* 379–388, 1964.
14. Kelly, D. E.: Fine structure of desmosomes, hemidesmosomes, and an adepidermal globular layer in developing newt epidermis. *J. Cell Biol. 28:* 51–72, 1966.
15. Kono, T., and Collowick, S. P.: Isolation of skeletal muscles cell membrane and some of its properties. *Arch. Biochem. Biophys. 93:* 520–533, 1961.
16 Kono, T., Kakuma, F., Homa, M., and Fukuda, S.: The electron microscopic structure and chemical composition of the isolated sarcolemma of the rat skeletal muscle cell. *Biochim. Biophys. Acta 88:* 155–176, 1964.
17. Lazarow, A., and Speidel, E.: The chemical composition of the glomerular basement membrane and its relationship to the production of diabetic complications. In *Small Blood Vessels Involvement in Diabetes Mellitus*, ed. by Siperstein, M. D., Colwell, A. R., Sr., and Meyer, K., pp. 127–150. Washington, D. C., American Institute of Biological Sciences, 1964.
18. Leduc, E. H., and Bernhard, W.: Recent modification of the glycol methacrylate embedding procedure. *J. Ultrastruct. Res. 19:* 196–199, 1967.
19. Lehrer, G. M., and Ornstein, L.: A diazo coupling method for the electron microscopic localization of cholinesterase. *J. Biophys. Biochem. Cytol. 6:* 399–406, 1959.
20. Little, K., and Kramer, H.: Nature of reticulin. *Nature* (Lond.) *170:* 499–500, 1952.
21. Luft, J. H.: Ruthenium red staining of the striated muscle cell membrane and the myotendinal junction. In *Proceedings of the 6th International Congress for Electron Microscopy*, ed. by Uyeda, R. vol. 2, pp. 65–66. Tokyo, Maruzen Co., 1966.
22. Luft, J. H.: Ruthenium red and violet. I. Chemistry, purification, methods of use for electron microscopy and mechanism of action. *Anat. Rec. 171:* 347–368, 1966.
23. Luft, J. H.: Fine structure of nerve and muscle cell permeability to ruthenium red. *Anat. Rec. 154:* 379, 1966.
24. McCollester, D. L.: A method for isolating skeletal-muscle cell-membrane components. *Biochim. Biophys. Acta 57:* 427–437, 1962.
25. Misra, R. P., and Berman, L. B.: Studies on glomerular basement membrane. II. Isolation and chemical analysis of diseased glomerular basement membrane. *Lab. Invest. 18:* 131–138, 1968.

26. Mowry, R. W.: Improved procedure for the staining of acidic polysaccharide by Müller's colloidal (hydrous) ferric oxide and its combinations with the Feulgen and the periodic acid-Schiff reactions. *Lab. Invest. 7:* 567–576, 1958.
27. Nakamura, T., Mamba, T., and Grob, D.: Motor endplate reactivity to divalent metal ions. Histochemical studies. *J. Histochem. Cytochem 15:* 276–284, 1967.
28. Palladini, G., and Lauro, G.: Observations sur la significativité de la coloration aux Alcians pour les mucopolysaccharides. *Histochemie 16:* 15–22, 1968.
29. Rambourg, A., and Leblond, E. P.: Electron microscope observations on the carbohydrate-rich cell coat present at the surface of cells in the rat. *J. Cell Biol. 32:* 27–54, 1967.
30. Rambourg, A., Hernandez, W., and Leblond, C. P.: Detection of complex carbohydrates in the Golgi apparatus of rat cells. *J. Cell Biol. 40:* 395–414, 1969.
31. Ravin, H. A., Zacks, S. I., and Seligman, A. M.: The histochemical localization of acetylcholinesterase in nervous tissue. *J. Pharmacol. Exp. Ther. 107:* 37–53, 1953.
32. Sávay, G., and Csillik, B.: Lead reactive substances in myoneural synapses. *Nature* (*Lond.*) *181:* 1137–1138, 1958.
33. Scott, J. E., Quintarelli, G., and Delovo, M. C.: The chemical and histochemical properties of alcian blue. I. The mechanism of alcian blue staining. *Histochemie 4:* 73–85, 1964.
34. Spicer, S. S.: Histochemical differentiation of sulfated rodent mucins. In *Proceedings of the First International Congress of Histochemistry and Cytochemistry*, p. 92. New York, Pergamon Press, 1960.
35. Spicer, S. S., and Meyer, D. B.: Histochemical differentiation of acid mucopolysaccharide by means of combined aldehyde fuchsin-alcian blue staining. *Am. J. Clin. Pathol. 33:* 453–460, 1960.
36. Spicer, S. S., and Jarrels, M. H.: Histochemical reaction of an aromatic diamine with acid groups and periodate engenered aldehydes in mucopolysaccharides. *J. Histochem. Cytochem. 9:* 368–379, 1961.
37. Yamada, K.: Effects of oxidation upon alcian blue staining of acid mucopolysaccharides. *Nagoya J. Med. Sci. 26:* 217–220, 1964.
38. Zacks, S. I., and Blumberg, J. M.: Observations on the fine structure of mouse and human muscular junctions. *J. Biophys. Biochem. Cytol. 10:* 517–528, 1961.

Chapter 9

# Ultrastructural Pathologic Characteristics of the Skeletal Muscle Fiber: An Introductory Survey*

HAROLD M. PRICE

## INTRODUCTION

### Background and Objectives

Almost as soon as electron microscopy began to influence our knowledge of normal striated muscle, investigations were initiated to gain further insight into the pathology of human muscle disorders. As a consequence, the literature on the ultrastructure of diseased muscle cells has mushroomed at almost logarithmic proportions. Although significant gains have resulted from these studies, some rash unsubstantiated communications have also evolved. At first, it seemed as if each new subcellular irregularity was a significant clue to the etiology of a particular muscle disorder. In some cases undue emphasis was placed on ultrastructural alterations within the confines of a few injured fibers while the over-all histopathologic picture was ignored. As subsequent studies of various myopathies showed that many of the so-called unique ultrastructural patterns were relatively nonspecific changes, disparaging comments were heard about the usefulness of electron microscopy. These extreme positions evidently are the irrational growing pains of the rapid application of an exciting new technology to the field of cellular pathology. It should be recognized that electron microscopy has an important role to play in histopathologic studies and that we have barely begun to utilize its potential in defining the macromolecular components of diseased muscle fibers.

This paper is a brief survey of the main ultrastructural reactions of injured skeletal muscle fibers. We hope that it will stimulate some thought on future work as well as serve as a reasonable account of what has passed. Emphasis is given to observations of human tissues and selected citings from other mammalian studies that appear to augment our understanding of the changes encountered in man. The descriptions are based primarily on tissues which have been fixed with osmium tetroxide or fixed with glutaraldehyde (or paraformal-

* This work was mainly supported by a grant from the Muscular Dystrophy Associations of America.

dehyde) and then post-osmicated, embedded in an epoxy resin, and stained with lead, phosphotungstic acid, and/or uranyl solutions.

## Some Advantages and Disadvantages of Electron Microscopy

Before enumerating features of the organelles in diseased skeletal muscle fibers, some comments are offered on the benefits and limitations of electron microscopy as well as the results that may be realistically anticipated from the use of this methodology to aid in the understanding of muscle disorders. More detailed discussions on many of the concepts expressed in the next few paragraphs may be found in a number of excellent references.[22, 41, 65, 86, 134, 136, 142]

Although the electron microscope reveals structural details and differences not visible by light microscopy, it is important to remember that the increased magnification and resolution gained with electron optics is associated with a marked restriction in the amount of tissue that can be effectively examined. This can create sampling and interpretation problems, which cannot be too strongly overemphasized, when dealing with the morphologic variables of pathologic material or the artifacts that may be caused by tissue processing. One must maintain a proper perspective of the role of the diseased cells as structural-functional units within a tissue and, in turn, within the organism as a whole. A good understanding of basic pathologic reactions at the light microscopic level and of the clinical manifestations of the various muscle disorders is necessary in order to evaluate effectively the significance of ultrastructural changes.

On the other hand, the new spectrum of structural features seen at the subcellular level may also improve the diagnostic accuracy of the histopathologist. Anatomic pathology is a discipline based on the premise that biologic behavior can usually be reliably predicted and classified from structural changes. However, the experienced pathologist is well aware that it is unusual to be able to match a specific structural alteration with a specific disorder and that the morphologic modulations of tissues are limited, being largely governed by the structural, functional, and developmental characteristics of a specific tissue as well as the nature of the injurious agent. Identical morphologic alterations frequently accompany different disorders and alone are nonspecific. A histodiagnosis results from being able to recognize various morphologic changes and then fitting them into a meaningful scheme based on an understanding of the dynamics of normal and abnormal biologic processes. Electron microscopy provides the means to obtain knowledge of structural reaction patterns at a more finite dimensional level of organization. Thus, when integrated with observations of gross and light microscopic proportions, the additional subcellular information enlarges our perspective of structural changes and should improve our ability to differentiate and classify disease entities.

Because electron microscopy exceeds the resolving capability of light microscopy and makes it possible to characterize cellular structures in proportions more compatible with macromolecular biochemical concepts, investigators have hoped that it would aid in localizing where specific physical-chemical determinations could be most effectively utilized to uncover the etiology of various muscular disturbances. For example, it would be of interest to know whether in a specific muscular disorder, the primary or earliest pathologic changes involve the sarcoplasmic reticulum, the myofilaments, or the mitochondria, thereby providing a clue as to which cell fraction may best profit from biochemical investigation. Although at first glance this approach may seem reasonable, it may also be misleading in that the initial functional insult may have widespread repercussions. The subcellular structure first observed to show a change may in fact be far removed from the basic defect. Therefore, before we are able to pinpoint a specific etiologic factor, we will probably need a broader understanding of the basic molecular control and reactive mechanisms that work within injured muscle fibers.

Many of the anticipated results from this technology are dependent on newer methods of tissue preparation as well as a better understanding of current ones. The effects of fixation and embedding methods are not well understood and are difficult to check adequately by independent methods. In the injured cell, this may be particularly critical, since the deleterious factor responsible for altering a specific cell structure may also make the organelle more susceptible to the stresses of fixation and embedding. It is important to employ various fixatives and embedding procedures as well as to correlate findings independently when possible with other physical methods of investigation, such as x-ray defraction, electron probe, scanning electron microscopy, and polarizing, interference, and phase microscopy. The wider application of histochemical, autoradiographic, and immunochemical methods at the fine structural level also gives indications of eventually improving our insight into abnormal cell mechanisms. In addition, detailed characterizations of the isolated filaments, membranes, and nuclear components of the normal and diseased muscle fibers are yet to be fully expounded. More information is needed on the normal distinguishing structural features of the different fiber types in man. Some of this knowledge of the fiber types has been acquired from work in lower animals, but preliminary information on higher mammals, especially primates, indicates the presence of perhaps more subtle differences.

It should also be noted that electron microscopists' preoccupation with fixation tends to make them even more discerning in the evaluation of conventional light microscopic preparations. There seems to be an improvement in the ability to distinguish between "real" "artifactual" changes. With this experience, it behooves us to wonder why some people still advocate the use of fixatives like Bouin's solution when there is good evidence from electron microscopic studies that buffered glutaraldehyde or formaldehyde solutions are better fixatives for muscle tissue.

## INJURED MUSCLE FIBER

### Surface of Muscle Fiber

#### Normal aspects

In thin sections viewed by the electron microscope, the muscle cell is seen to be bounded by a plasma membrane (*sarcolemma*) which at very high magnifications displays the triple-layered unit membrane structure characteristic of most, if not of all, cell membranes.[88, 129] Other membranes form various intracellular compartments and organelles. Many of these membranous cellular constituents are discussed below under separate topic headings. The outer surface of the sarcolemma is covered by an extraneous coat (*external lamina* or *basement membrane*) of material that contains a substantial amount of polysaccharides (Fig. 1).[88, 129] One function of the external lamina is thought to be as a microskeleton, helping to maintain the shape and stability of cells. Thus, this external coating generally does not appear to encompass those cells in which mobility is an important functional trait, such as myoblasts, macrophages, and fibroblasts. The fibroblast-like cells, however, which become an intricate part of the rigid framework of a muscle spindle capsule, generally have an elaborate external lamina.[106]

### Interruptions in continuity of sarcolemma

Large breaks in the continuity of the sarcolemma are seen (Figs. 2 and 3) following a variety of necrotizing injuries such as rhabdomyolysis,[23, 114] acute polymyositis,[101, 118] zones of acute degeneration in early Duchenne dystrophy,[59, 74, 117] animal dystrophies,[11, 15] ischemia,[128] and a variety of experimental injuries.[6, 19, 49, 91, 132] Unlike other cells,[135] this change in the multinucleate muscle fiber is not necessarily lethal. Frequently, at least some regenerative activity ensues. In other disturbances, such as a form of myopathy associated with alcoholism,[67] no detectable structural change in the plasma membrane has yet been noted, but a defect allowing increased permeability is postulated because of evidence of prominent increased intracellular edema. In most chronic atrophic disorders, as emphasized by several groups of workers,[77, 101, 102, 139] the sarcolemma appears to be relatively unchanged when examined with current electron microscopic techniques, except for some degree of uniform thickening. The amount of thickening may vary with the method of fixation and section thickness.

### Branched fibers and incomplete lateral fusion

Split or branched fibers have been noted in various myopathies but are particularly frequent in muscular dystrophy.[11] Incomplete fusion of the lateral margins of developing muscle fibers has been postulated as a cause of fiber branching in muscular dystrophy.[11] In early Duchenne dystrophy incomplete lateral fusion may also be seen without overt branching (Fig. 4). Fiber branching may also be the result of exteriorization of a part of the large intracellular vacuoles that persist in the chronic myopathic phase of hypokalemic periodic paralysis (Fig. 10).[53]

### Irregularities of surface contour

In chronic myopathic conditions where there is an apparent reduction of fiber volume, irregularities in the surface of the muscle fiber may take several forms.[14, 30, 53, 56, 57, 143] Small regular surface undulations may be seen (Fig. 44), or irregular broad or narrow based surface projections (Figs. 7 and 12) can occur which may or may not be associated with deep infoldings (Fig. 12) of the fiber

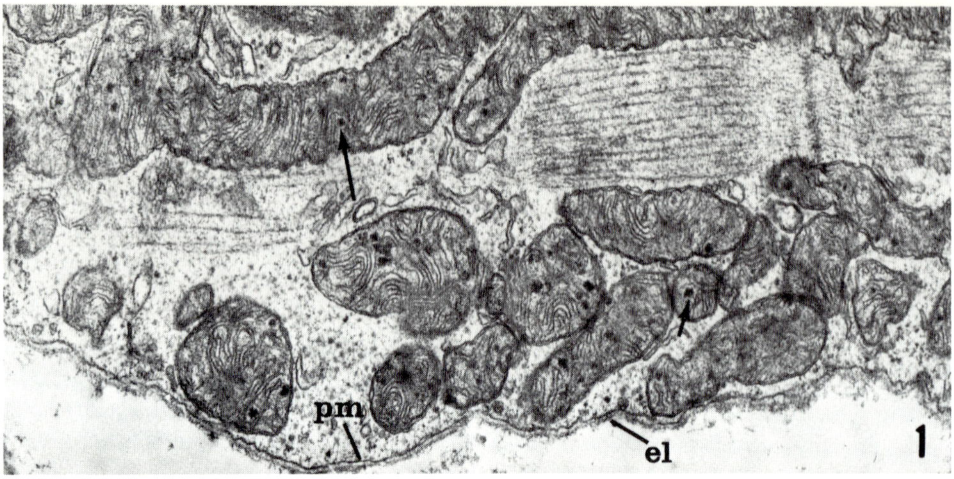


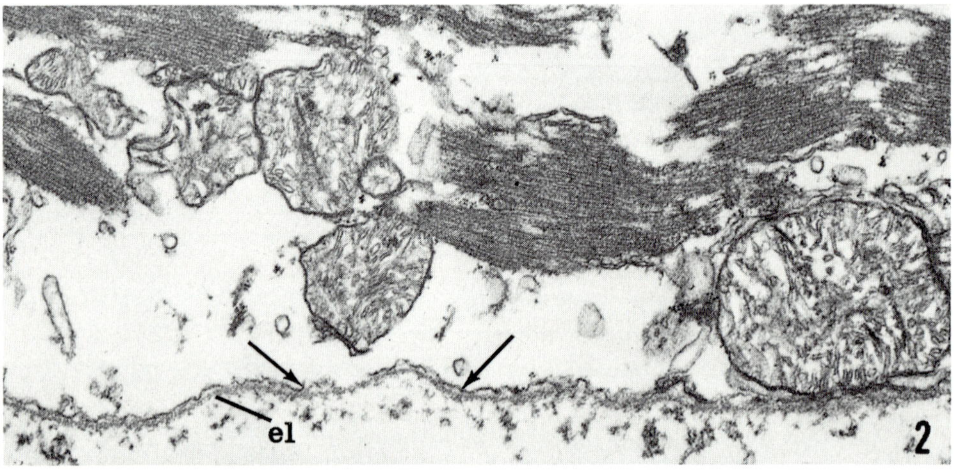


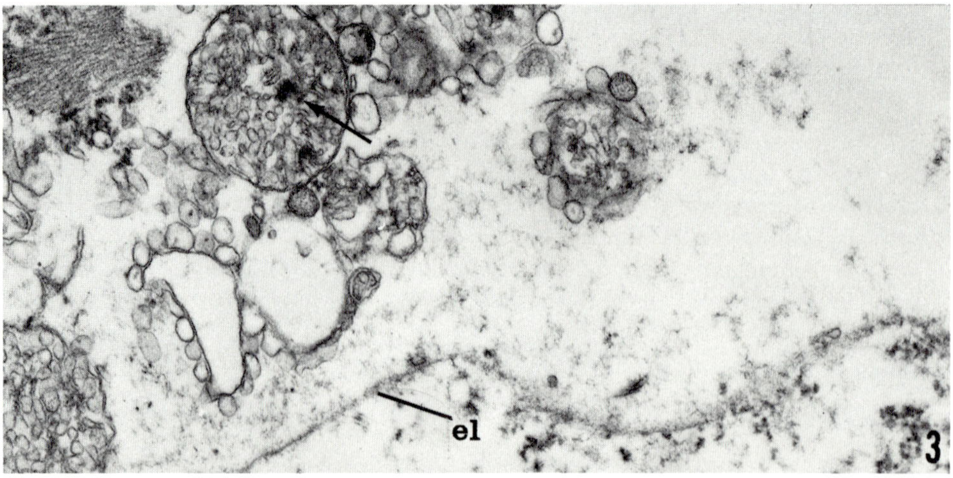


FIG. 1. Longitudinal section showing small area at margin of a normal skeletal muscle fiber. The sarcolemma (*pm*) is separated by a thin translucent region from the external lamina (*el* or basement membrane. A typical marginal distribution of mitochondria is evident. Within the mitochondria are small dense matrix granules (*arrows*). The A and I bands and the Z line of two sarcomeres in a marginal myofibril can be seen. In sectioning, the plane of section has partially grazed the surface of

surface. The subsarcolemmal regions of these surface projections may contain clusters of mitochondria and glycogen or other particulate sarcoplasmic components. These changes should not be confused with the junctional folds of the motor end-plate, or with the narrow surface projections and deep infoldings at the ends of normal fibers which terminate either within a muscle (Fig. 5) or at the myotendon junction.

#### External laminal (or sarcolemmal) tube

In contrast to the sarcolemma, the external lamina seems relatively resistant to acute injurious agents, and, with adjacent endomysial collagen fibrils, it forms the so-called "sarcolemmal tube" (Figs. 3 and 8).[6, 92] This tube forms a framework within which regeneration of the injured segment of a muscle fiber can take place.

#### Dissociation of external lamina

Along the margins of some injured fibers, strands of the external lamina can be seen which are completely dissociated from a plasma membrane surface. In certain instances, they may represent ghostlike remnants of collapsed necrotic fibers,[92] but in other cases they appear to be true reduplicated strands of this external coating material (Fig. 11). Another type of dissociation may be in the form of small repetitious pleats[67, 113] (Fig. 6).

### Myofibrils

#### Normal aspects[37, 61, 88]

Myofibrils occupy most of the interior of the muscle fiber and are the basic contractile units of this cell. They are best conceived as closely packed, nonbranching,* elongated, cylindrical structures approximately 1 $\mu$ in diameter† that extend the length of the muscle fiber and exhibit a repetitious pattern of light and dark bands. The band pattern is a consequence of the regular overlapping arrangement of numerous *thick* and *thin filaments* making up the myofibril. The extent to which these myofilaments overlap depends on the state of muscle contraction and results in changes in the length of the I band

* Branched myofibrils are normally seen in mammalian extraoccular muscles and intrafusal fibers, as well as in the limb muscles of lower animal forms.

† No finite limits of normality have been established for the diameter of human muscle myofibrils. This is an approximate determination based on personal observations and suggestions from other publications.

---

the myofibril resulting in a picture that could be misinterpreted as a break in the continuity of the myofibril (compare with Fig. 46). × 31,600.

Fig. 2. Margin of damaged fiber 15 minutes after cold injury. The plasma membrane shows evidence of degeneration and fragmentation (*arrows*). The external lamina (*el*) appears to be intact. The mitochondria are swollen. Intracellular edema probably accounts for the clear spaces between degenerating myofibrillar elements. × 22,100.

Fig. 3. Twenty-four hours after cold injury. Only the external lamina (*el*) defines the margin of the damaged fiber; *the plasma membrane is absent.* Large dense bodies are seen within mitochondria (*arrow*). Vesiculation of the cristae is evident. × 22,200.

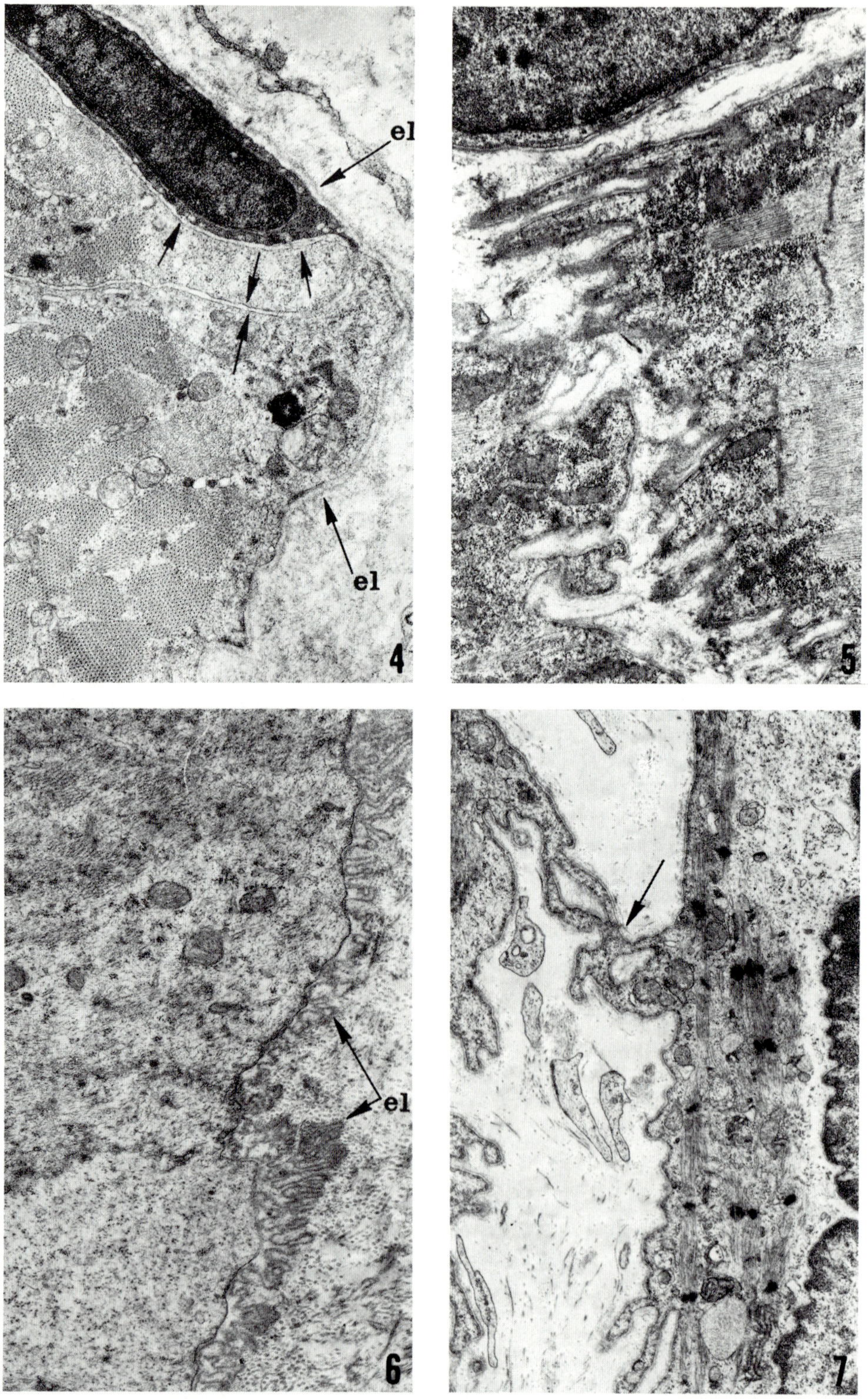


FIG. 4. Cross-section through a fiber from a case of early Duchenne dystrophy. Within a single external laminal (*el* or basement membrane) lining, there are portions of separate plasma membrane-lined cells. At the top is a satellite cell with its nucleus and dense cytoplasm separated

and H zone while the A band seems to retain essentially the same length. With these structural features in mind, it is obvious that the appearance of myofibrils in a given electron micrograph is dependent on (1) the plane of section through the muscle fiber—longitudinal, oblique, or cross-section—and (2) the state of muscle activity at the time of fixation—contracted, relaxed, or stretched. Over stretching results in extensive lengthening of the I band and sometimes small myofilament separation gaps in the I band near the A and I band junction. Marked contraction causes obliteration of the I band and double overlapping of the thin filaments in the highly shortened sarcomeres (Fig. 14). Oblique sections distort sarcomere length and the features of the H zone and M line. Because of the cylindrical shape of the myofibrils, care should be taken in assessing changes in the diameter of these contractile units in a random longitudinal section; cross-sections or serial longitudinal sections are preferable. Another normal variable in longitudinal sections that sometimes leads to misinterpretations is the elongated sinuous configurations that myofibrils may assume in fixed preparations. A given thin section passing through the base or apex of the curves of a wavy myofibril may result in a micrograph depicting an apparently fragmented myofibril (Figs. 1 and 46). The solutions used for fixing the tissues also influence the features of the myofibrils.[61] As corroborated by x-ray defraction, glutaraldehyde or formaldehyde seem to be more effective fixatives for the myofilamentous proteins than osmiun tetroxide.

I BAND AND Z LINE

The I band region is frequently the initial myofibrillar site to exhibit morphologic variation. A spectrum ranging from total disintegration of I band and Z line filaments to an organized proliferation of Z line substance in the form of rods has been seen in a variety of muscle disturbances (Figs. 17–24). Table 1 summarizes the pathologic changes associated with these myofibrillar components.

*Rod formation.* Table 2 lists the main morphologic features of rods. Although there is general agreement that rods originate from the Z line, it is difficult to explain the discrepancy in the spacings of the normal Z line and the rods, and also the periodicity reported by different investigators as recorded in Table II. As recently discussed,[27, 28, 44] several variables probably contribute to these differences. Among these are disparities in planes of sectioning relative to the long and short axes of the rods and also the presence of a complex repeating unit, which may be further complicated by an irregular growing pattern

---

by its own plasma membrane (*arrows*) from the adjacent muscle fiber. There is internal branching or incomplete lateral fusion of the muscle fiber as indicated by the separate plasma membranes (*arrows*) which separate two myofibril-bearing portions of sarcoplasm. × 19,400.

FIG. 5. Longitudinal section showing narrow surface projections and deep infoldings at the ends of two opposing muscle fibers. × 10,100.

FIG. 6. Longitudinal section in a zone of regeneration. The external lamina (*el*) is dissociated from underlying plasma membrane and forms a number of small repetitious pleats. × 17,000.

FIG. 7. Longitudinal section showing margins of two atrophic muscle fiber elements with irregular surface projections. There appears to be fusion (*arrow*) between several of the opposing surface projections. × 10,500.

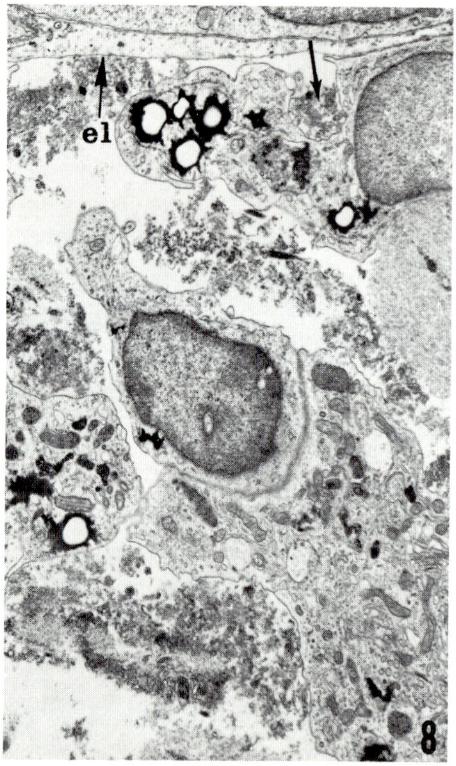


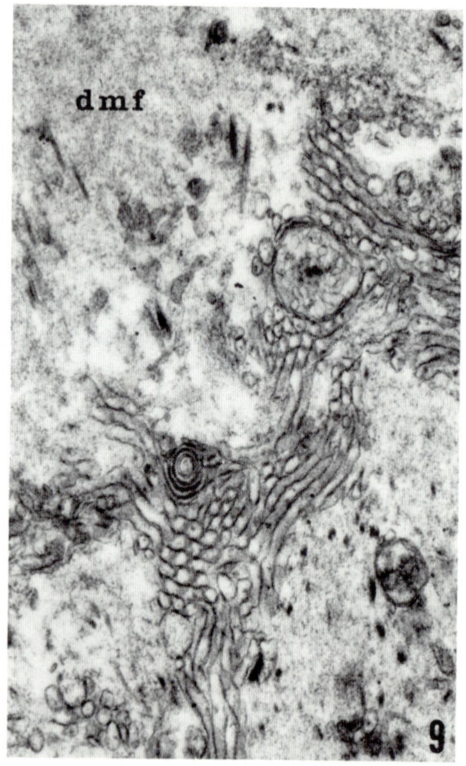


FIG. 8. Margin of a sarcolemmal tube, formed by the external lamina (*el*), in an injured fiber. The plasma membrane in this portion of the injured fiber has completely degenerated. Within the sarcolemmal tube are several macrophages and degenerating sarcoplasmic debris. *Arrow* indicates site of possible phagocytosis along margin of a macrophage. × 6100.

FIG. 9. Degenerating mitochondria and a form of tubular aggregates in a muscle fiber acutely injured by cold. Granular degenerating myofibrillar elements (*dmf*) are also evident. × 21,900.

whereby several adjacent rods fuse at slightly different degrees of orientation. In addition, the interpretation of closely related linear structures is influenced by the fact that the depth of field in the electron microscope exceeds section thickness. Thus, an entire section is visible in one plane, resulting in superimposition of images. Other factors may be compression during tissue sectioning and varying shrinkage effects of tissue processing. Another interesting structural feature of the rods that is not well understood is the role of the thin I band filaments. Except for the apparent structural continuity of the thin filaments and the rods, no further details about this relationship can be clearly

---

FIG. 10. Longitudinal section showing muscle fiber splitting in a long standing case of hypokalemic periodic paralysis. The presence of an intracellular vacuole (*V*) suggests the possibility that in this disease enlargement and coalescence of these internal vacuoles may lead to a process of exteriorization resulting in fiber branching. *el*, external lamina; *pn*, sarcolemma. × 9000.

FIG. 11. Longitudinal section showing marked reduplication of the external lamina around dense extracellular particles (electron micrograph made from blocks loaned by Dr. William Morningstar, Veterans Administration Hospital, Cleveland, Ohio). × 38,300.

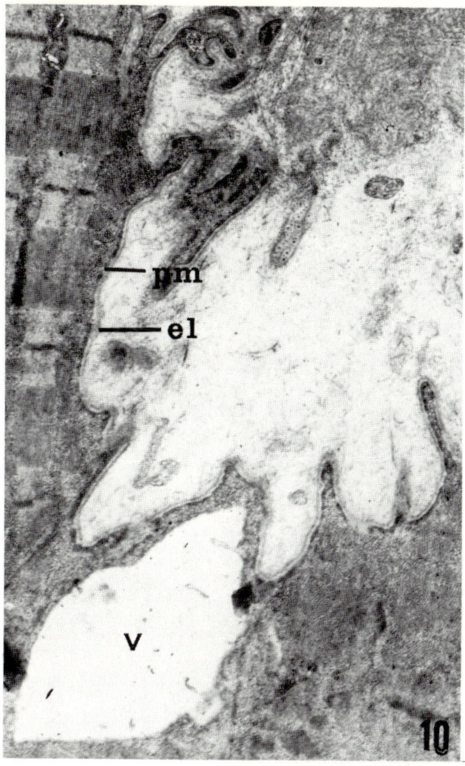


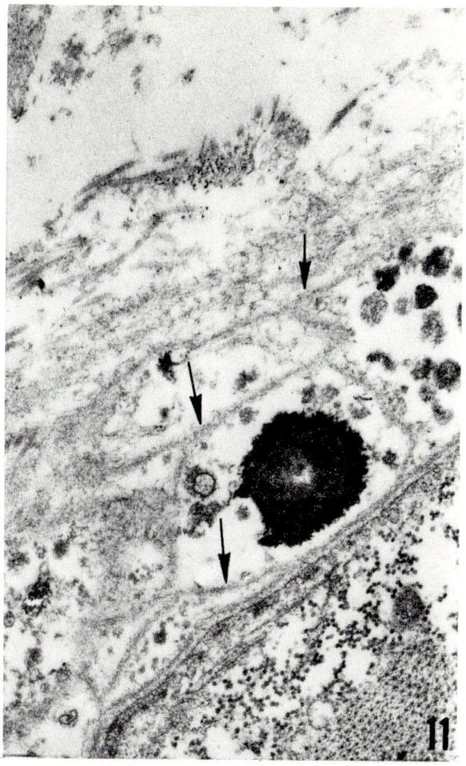


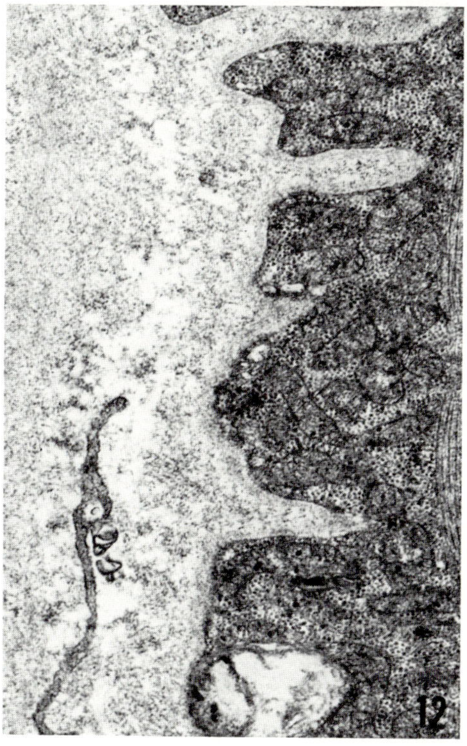


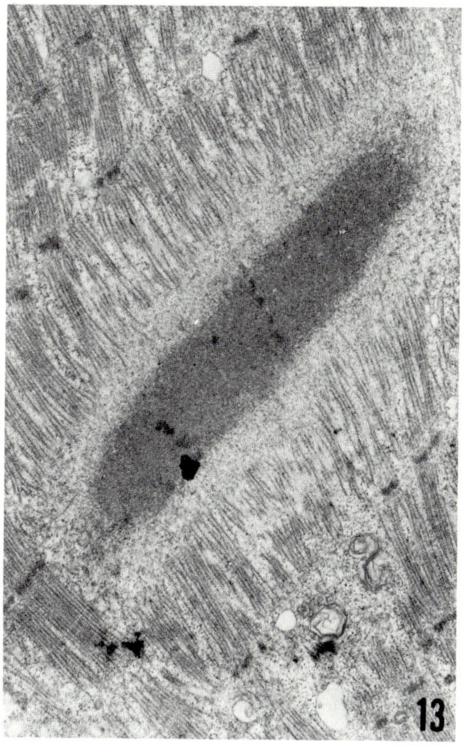


FIG. 12. Irregular surface infoldings along margin of moderately atrophic fiber. The subsarcolemmal pads contain mitochondria and numerous glycogen granules. × 19,600.

FIG. 13. "Cytoplasmic body" in degenerating muscle fiber. × 13,000.

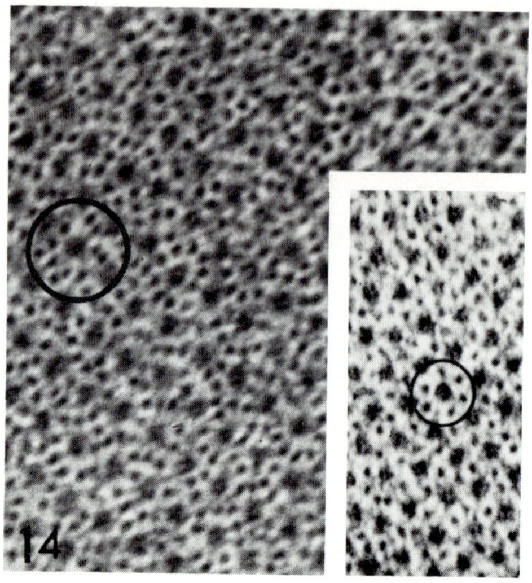


FIG. 14. Cross-section through A band region of markedly contracted muscle fiber. The overlapping of thin filaments results in apparent increase in the number of thin filaments surrounding each thick filament. Compare to cross-section of relaxed fiber (*inset*). × 170,000. *Inset*, × 130,000.

discerned. At first becuase of their structural configuration, the rods were thought to consist of tropomyosin B[90]; however, subsequent evidence from chemical extraction and immunochemical studies indicates that this is unlikely.[28]

*Pathologic significance of rods.* Rods are the principal structural finding in nemaline myopathy; however, as evident in Table 1, they are *not* pathognomonic of this disorder. Instead, they are best conceived as part of a clinical-pathologic syndrome. In nemaline myopathy the rods are the *dominant* tissue change, and, in contrast to Z line streaming, there is usually no decrease in mitochondria. This appears to be an accumulative or storage process rather than a degenerative change. Other alterations such as fragmentation of adjacent myofibrils and displacement of organelles seem to be secondary to the excessive amounts of Z line substance. Rods have also been found within intrafusal fibers[121] and within nuclei.[63]

*Z line streaming.* Streaming and disintegration of the Z line may be confused with rod formation. This change is more irregular than rods and lacks the transverse and longitudinal periodicity of rods. It is frequently associated with an apparent decrease or absence of adjacent mitochondria and has been observed in a variety of disorders (see Table 1).

### Central myofibrillar degeneration

Selective degeneration of the myofibrils in the center of muscle fibers is manifested in light microscopic preparations by either a three-zone (target) or two-zone (central core) pattern. Although the light microscopic findings suggest two types of pathologic alterations, by electron microscopy the central myofibrillar changes appear to be more of a difference in the extent of a single type of degenerative process rather than two distinct entities (Figs. 16, 18, and 19).

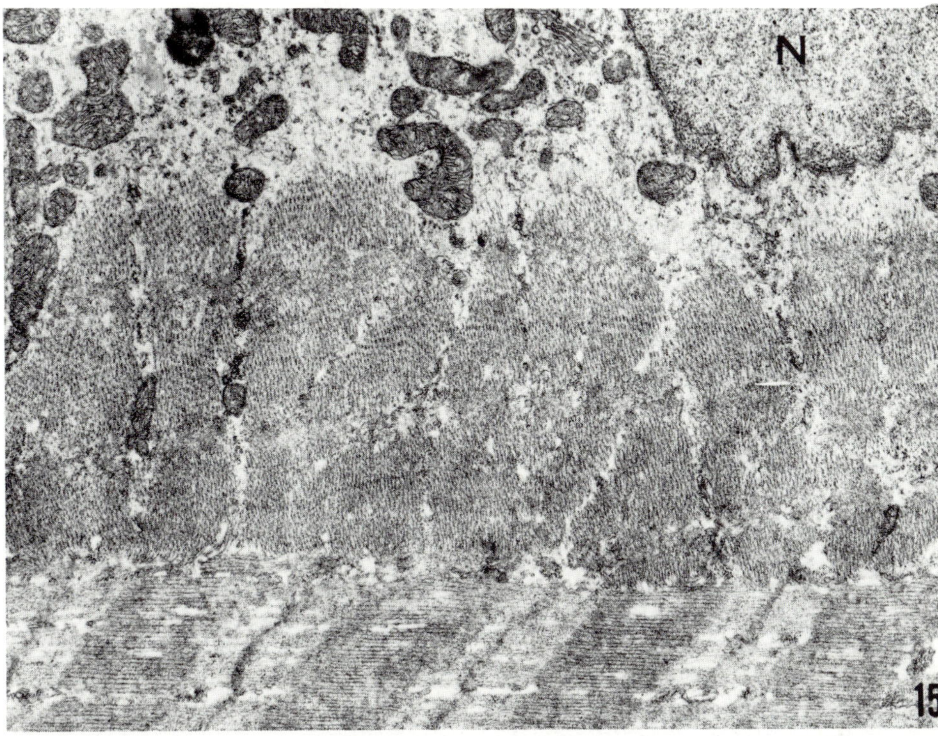


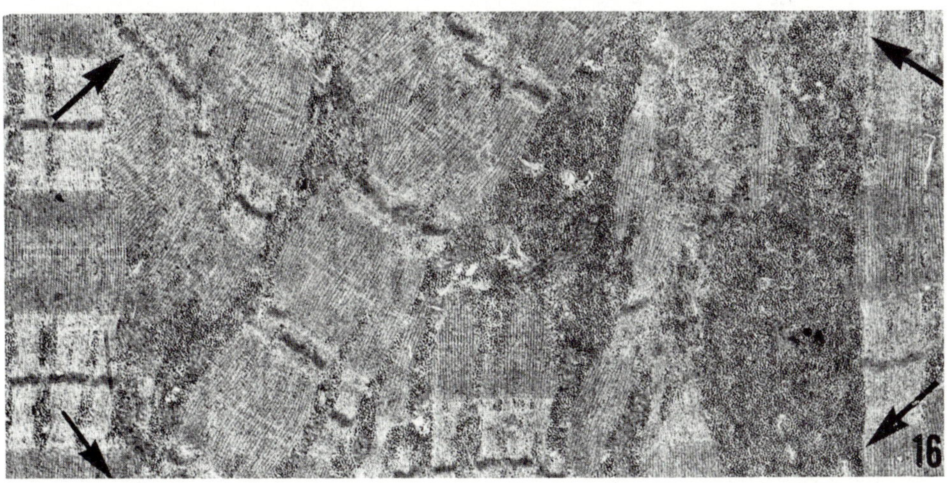


FIG. 15. Longitudinal section through a portion of a muscle fiber showing "ringbinden effect." Lower one-fourth of picture shows normally oriented myofibrils. The myofibrils at periphery of fiber, in the middle half of the picture, are oriented in a position almost at a right angle to the direction of the normal myofibrils. × 13,000.

FIG. 16. Central disorganization of myofibrils with increased glycogen deposition from a case of long-standing hypokalemic paralysis. *Arrows* mark margins of normally oriented myofibrils bordering central zone of degeneration. × 13,500.

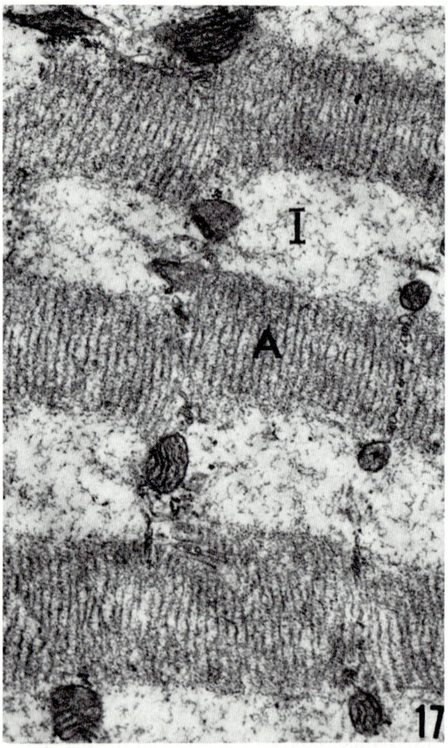


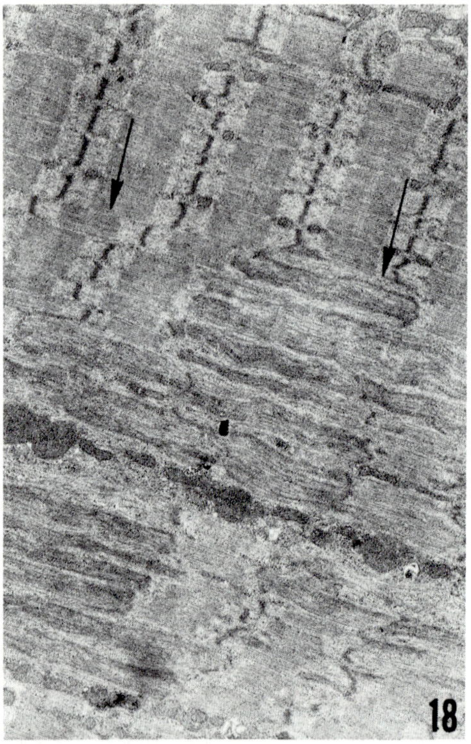


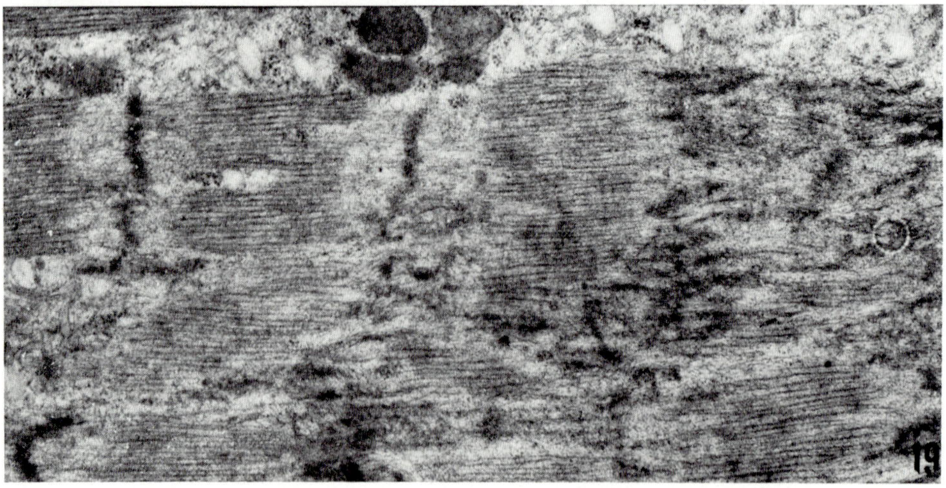


FIG. 17. Dissolution of thin filaments which makeup I band (*I*) and extend into A band (*A*) area, and also disintegration of Z line substance, in diaphragm muscle of rat after plasmocid injection. × 22,000.

FIG. 18. Early central myofibrillar degeneration with Z line streaming. As degeneration progresses mitochondrial elements will probably drop out entirely. *Arrows* indicate normal peripheral myofibrillar elements. × 8600.

FIG. 19. Central zone of degeneration and Z line streaming. Within zones of degeneration there is a lack of mitochondria and sarcoplasmic reticulum. The few mitochondrial profiles in upper portion of the picture are adjacent to zone of normal myofibrils. × 22,000.

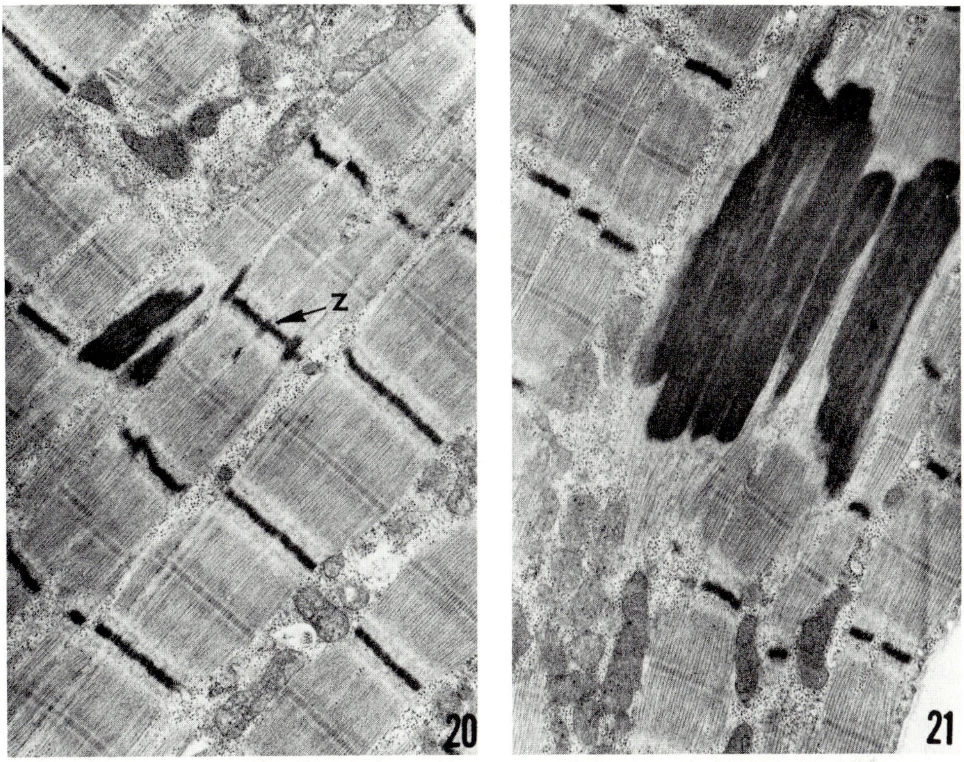


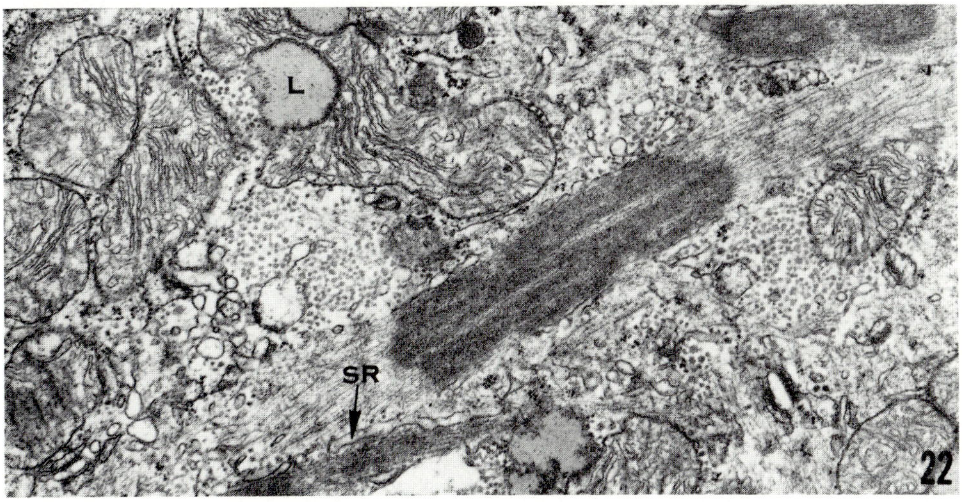


FIG. 20. Longitudinal section from case of rod myopathy showing focal expansion of Z band (*Z*) material and a well developed rod. × 13,900.

FIG. 21. Abnormal accumulation of rod (or Z line) material from case of rod myopathy with displacement of normal myofibrillar and organelle components. × 14,600.

FIG. 22. Rod structures in a rhabdomyoma. Note intimate association of sarcoplasmic reticulum (*SR*) components with developing rods. Lipid droplets (*L*) normally are frequently associated with mitochondria. (Electron micrographs made from biopsy material obtained from Dr. J. Smith, University of Vermont.) × 26,000.

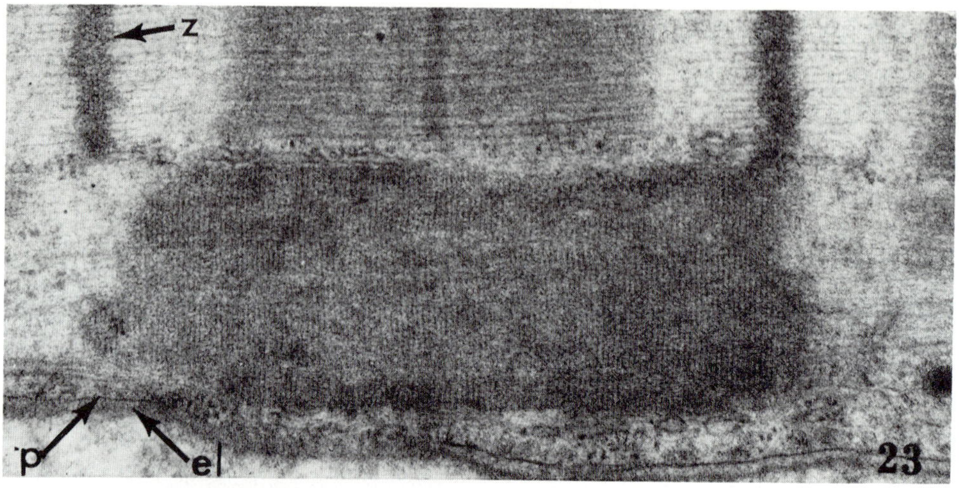


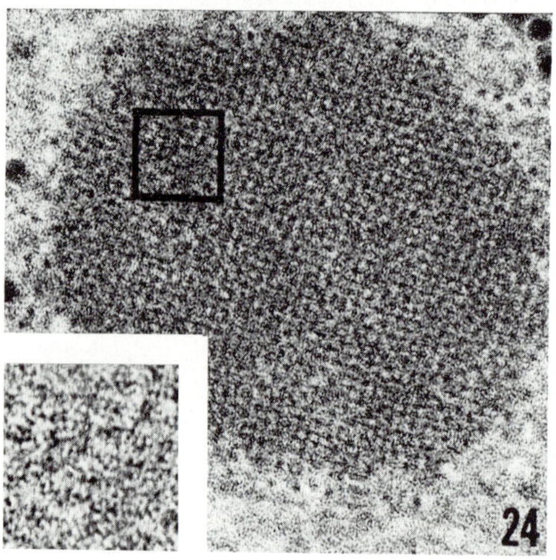


FIG. 23. Subsarcolemmal rod showing periodicity of approximately 180 Åoriented perpendicular to the direction of myofibril. Note periodicity in adjacent Z bands (*Z*). *P*, plasma membrane; *el*, external lamina. × 46,000.

FIG. 24. Cross-section through a rod. The rod lattice consists of compressed squares with approximately 90-Å sides (measured from center to center of dense 34-A filaments forming squares). Dots of 60 A appear to occupy the corners of most of the squares. *Inset* is higher power micrograph of a region of this rod showing that the filaments forming the rod lattice may in turn be composed of even smaller twisted filamentous components. × 140,000. *Inset*, × 325,000.

*Target fibers.*[32, 95, 109] On cross-section, these fibers have three concentric zones which are easily detected with phosphotungstic acid-hematoxylin or trichrome stains but are barely discernible with hematoxylin and eosin. If the intermediate zone is incompletely developed, the fibers are referred to as *targetoid*. Cytochemical methods reveal no reaction for mitochondrial enzymes, phosphorylase, or myofibrillar ATPase in the central zone. These reactions may show increased activity in the intermediate zone and are often normal in the outer zone. Electron microscopy of the central zone reveals severe disorganization of the myofibrils, almost no membranous elements (sarcoplasmic reticu-

TABLE 1. PATHOLOGIC REACTIONS OF I BAND AND Z LINE[a]

| Reaction Pattern | Muscle Disorder | Example Reference |
|---|---|---|
| 1. Total disappearance of I band filaments with Z line | Plasmocid poisoning (rats) | Price *et al.*[93] |
| | Central core disease | Gonatas *et al.*[46] |
| | Myotonic dystrophy | Johnson[64] |
| | Denervation atrophy | Johnson[64] |
| 2. Fragmentation | Myotonic dystrophy | Aleu and Afifi[5] |
| | Ischemia | Stenger *et al.*[128] |
| | Cold injury | Price *et al.*[91] |
| | Myopathy associated with alcoholism | Klinkerfuss *et al.*[67] |
| 3. Z line streaming and disintegration | Central core disease | Engel *et al.*[34] |
| | Target fibers | Resnick and Engel[95] |
| | Vitamin E deficiency (rat) | Howes *et al.*[54] |
| | Acute denervation (rat) | Pellegrino and Franzini[87] |
| | Motor neuron disease | Roth *et al.*[102] |
| | Chronic human myopathies | Howes *et al.*[53]; Fisher, *et al.*[40] |
| | Steroid myopathy (rat) | Tice and Engel[135] |
| | Emetine treatment | Duane and Engel[24] |
| | Acute psychotic patients with elevated creatine phosphokinase | Fischman[38a] |
| 4. Cytoplasmic body (see Fig. 13) | | MacDonald and Engel[71] |
| 5. Rod formation | Congenital rod myopathy | See Table 2 |
| | Rod myopathy of later life | Engel, A. G.,[29] Engel, W. K., and Resnick,[33]; Hopkins, *et al.*[52] |
| | Tenotomy | Shafiq *et al.*[115] |
| | Heart muscle (cat) | Fawcett[36] |
| | Acute psychotic patients with elevated creatine phosphokinase | Fischman[38a] |

[a] Updated from table originally published by Engel.[28] The ultrastructural pathologic changes of the I band Z line regions are summarized along with the various disorders in which they have been seen. A reference example is listed with each; however, other publications have further documented most of these observations.

lum, T system, and mitochondria), markedly reduced glycogen particles, and an abundant amorphous electron-dense material that resembles disorganized Z line substance (Fig. 19). In the intermediate zone the myofibrils are often narrowed and there is some disorganization of the Z substance; the sarcotubular membranes are present, but may form small stacks of flattened vesicles; and the mitochondria may seem relatively numerous. The outer zone generally appears normal.

*Central core fibers.*[32, 95] On cross-section, these fibers have two distinct zones which have histochemical and ultrastructural characteristics like the central and outer zones of the target fibers. In one report,[46] central core changes were associated with almost complete absence of I band and Z line constituents (see Table 1). In hypokalemic periodic paralysis,[53] a central zone of myofibrillar

TABLE 2. MORPHOLOGIC CHARACTERISTICS OF RODS[a]

| Feature | Dimension | Reference |
|---|---|---|
| Origin in the Z disk | | Price *et al.*[90]<br>Gonatas *et al.*[44, 45]<br>Engel and Gomez[27]<br>Engel[29]<br>Afifi *et al.*[4] |
| Square filamentous array in transverse section | Squares with 60- to 70-Å sides<br>Squares with 95- to 110-Å sides<br><br>Squares with approximately 90-Å sides | Gonatas[45]<br>Engel and Gomez[27]<br>Engel[29]<br>Price (Chapter 9) |
| Structural continuity with thin filaments | | Price *et al.*[90]<br>Gonatas *et al.*[44, 45]<br>Engel and Gomez[27]<br>Engel[29] |
| Periodic lines perpendicular to long axis | 125 Å<br>145 Å<br><br>180 to 200 Å<br>175 Å<br>140 to 170 Å<br>140 to 200 Å | Conen *et al.*[18]<br>Shy *et al.*[123]<br>Engel[28]<br>Price *et al.*[90]<br>Gonatas *et al.*[44]<br>Gonatas[45]<br>Engel and Gomez[27]<br>Engel[29] |
| Periodic lines parallel to long axis | 80 to 100 Å<br>120 to 170 ÅK<br>80 to 120 Å | Price *et al.*[90]<br>Gonatas[45]<br>Engel and Gomez[27]<br>Engel[29] |

[a] Updated from table originally published by Engel.[28] Subsequent publications on rod structures have generally confirmed the above observations; some of these are listed in the bibliography.

disorganization and increased glycogen stores has also been seen (Fig. 16).

Both target and central core changes occur almost exclusively in *type I* fibers.[32, 95] Target changes usually occur along the entire longitudinal extent of a fiber, whereas central core changes may be patchy. In the latter, the changes may also be seen to extend to the surface of the fiber. Target and targetoid fibers are thought to be associated primarily with denervating disorders.[32] Central core changes are found as the primary pathologic change in a congenital myopathy called central core disease; because of the similarity of the lesion to target changes, a neurogenic factor in the pathogenesis of this myopathy has been proposed.[32] Several recent papers, however, seriously question the validity of the relationship of these central degenerative changes and denervation.[1, 75]

PERIPHERAL MYOFIBRILLAR DEGENERATION[68, 112]

Peripheral (subsarcolemmal) myofibrillar alterations have been found to take several forms. They have been most frequently reported in myotonic dystrophy, but they have also been seen in other human myopathic disorders and experimental animal material.

In one form, they may consist of peripheral circumferential ring myofibrils oriented in a plane at right angles to the muscle fiber and to the long axis of adjacent nuclei. This change was recognized long ago in early light microscopic studies of muscle and has been termed "ringbinden," "striated annulets," and "anneux des fibrilles" (Fig. 15). Some workers consider it to be an artifact related to tissue handling, whereas other accept it as a significant pathologic change.

Another peripheral change consists of subsarcolemmal sarcoplasmic masses (or "subsarcolemmal pads") which may contain only a few disarrayed degenerating myofibrillar components, some scattered sarcotubular components, varying numbers of mitochondria, glycogen, and sarcolemmal nuclei.

#### Transverse amorphous degeneration

Degeneration may be so complete that there is hardly a semblance of a myofibrillar pattern or cross-striations (Fig. 2). Only filamentous fragments and/or irregular accumulations of dense Z line-like substance remain which may phase into amorphous finely granular deposits (Figs. 8 and 9). These are ultrastructural manifestations of the granular degeneration or coagulative necrosis seen by light microscopy. Usually these necrotic changes traverse the width of the fiber and, depending on the extent of the injurious insult, may involve the total length of the fiber. Among the diseases and experimental conditions in which this extensive architectural destruction has been encountered are Duchenne dystrophy,[58, 74, 76, 104, 133, 140] polymyositis,[101, 118] idiopathic rhabdomyolysis,[23, 114] cold[91] or crush[6] injuries, ischemia,[10, 128] vitamin E deficiency,[54] animal dystrophies,[7, 11, 15] steroid myopathy[1, 2, 19, 100] and with *Clostridium perfringens* $\alpha$-toxin.[132]

#### Regenerative changes[38, 92, 117]

In isolated electron micrographs, regenerative and degenerative alterations may at times be difficult to differentiate. In many cases both processes may be going on simultaneously. As a rule, groups of myofilaments associated with polyribosomes are probably zones of regeneration (Fig. 42). See Chapter 10 for further presentation of the morphologic features associated with regeneration.

### T System and Sarcoplasmic Reticulum

#### Normal aspects[81, 88]

In addition to the myofibrils, the T system and sarcoplasmic reticulum (SR) are two other organelles unique to striated muscle. They are membrane-limited tubulous structures that form a complex network throughout the muscle fiber and play an important role in activation and deactivation of muscle contraction. Adjacent to each myofibril, they form a repetitive tubular pattern called a *triad* which consists of a pair of dilated *terminal cisternae* of the SR separated by a slender *transverse tubule of the T system*. In human and other mammalian skeletal muscle, there are two triads per sarcomere, one near each A-I band junction. This contrasts to mammalian cardiac muscle with only one triad per sarcomere, at the Z line. The SR is probably homologous to smooth surface endoplasmic reticulum of other cells and seems to be a separate intracellular fluid compartment with no connections to the sarcoplasmic matrix, the extracellular space, or the T system. Frequently electron-dense amorphous material may be seen in the SR, especially the lateral cisternae. Slight dilatation of the lateral cisternae can be seen in otherwise normal muscle and may be a normal physiologic variation or a result of tissue processing. The membranes forming the T system appear to be confluent with the plasma membrane, and there seems to be a functional continuity between the space of the T system and the extracellular space.

VACUOLES[12, 31, 53, 56, 70, 78, 82, 95, 96, 124]

Relatively large vacuoles have been seen in almost all myopathies including familial periodic paralysis (Figs. 10, 38 and 39), lupus erythematosus, polymyositis, chloroquine myopathy, and myotonic dystrophy. Most studies have found evidence indicating that these vacuoles are dilated portions of the SR; however, the T system may also be involved in some cases.[32] These should be distinguished from autophagic vacuoles which often contain recognizable constituents (Fig. 41). Vacuolar myopathies are discussed in more detail in Chapter 16.

AGGREGATES OF LINEAR TUBULES

In several cases of periodic paralysis[8, 32] (Figs. 33 and 34), neurogenic atrophy,[113] and trichinosis[99] abnormal aggregates of long straight single membrane-limited tubules have been seen. A tubular aggregate of questionable similarity has also been observed in degenerating muscle fibers (Fig. 9). They usually are found to have either a solid core of dense amorphous material (Figs. 33 and 34) or an inner hollow sleeve. Occasionally, the tubules appear to form triadlike structures and have been compared to components of the SR. In

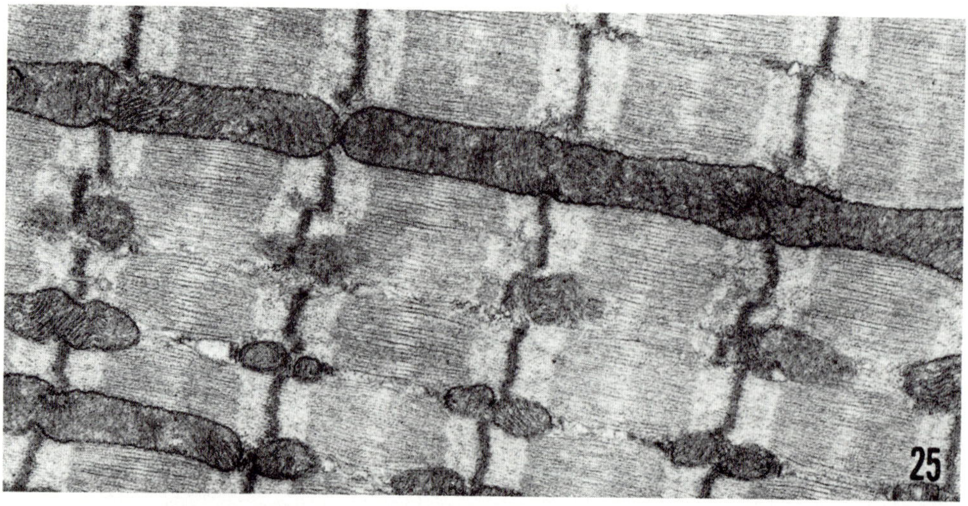


FIG. 25. A longitudinal section; mitochondria between the myofibrils may normally extend several sarcomeres in length. × 17,200.

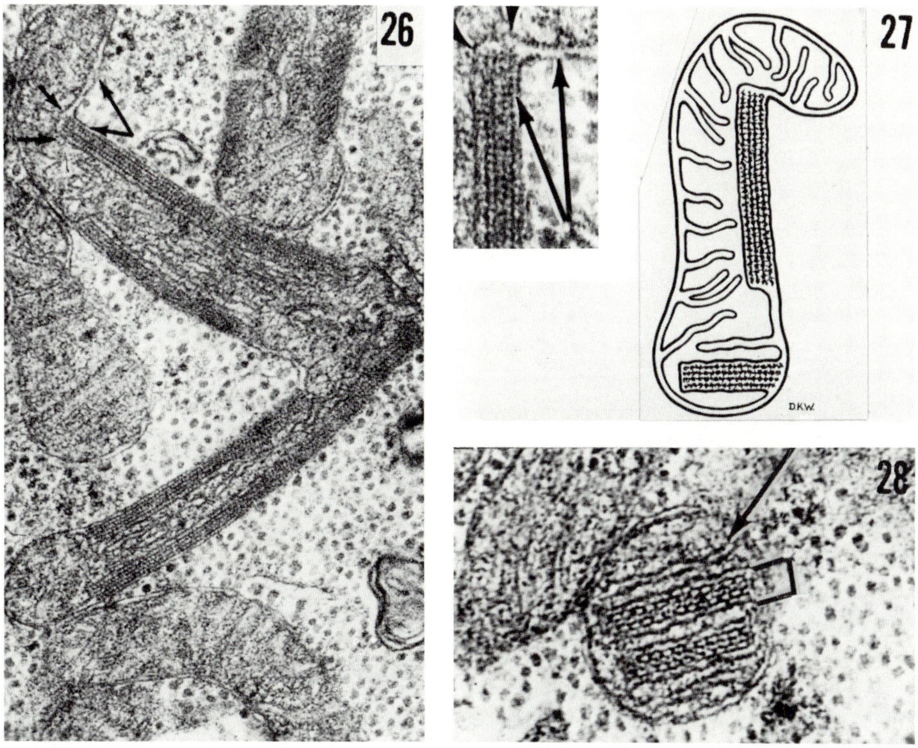

FIG. 26. Abnormal mitochondrion with paracrystalline inclusion in space between inner and outer mitochondrial membranes. The *arrows* show point of focal widening of this space to accommodate the abnormal inclusion. × 50,400.

FIG. 27. Diagram summarizing typical locations in outer mitochondrial chamber where abnormal paracrystalline structures have been found. Compare with Figs. 26 and 28. *Inset* is enlargement of area indicated by *arrows* in Fig. 26. × 100,000.

FIG. 28. Abnormal mitochondrion having paracrystalline inclusion within widened intracristae space. The bracket indicates margins of widened intracristae space. An *arrow* points to a normal crista. × 72,500.

experimental neurogenic atrophy, the intertubular matrix material contained glycogen particles or had a spiral (or herringbone) pattern.[113] Histochemically, the regions containing these tubular aggregates in periodic paralysis have been found to show a high activity of DPNH-tetrazolium reductase and are thought to be primarily confined to type II fibers.[32] This tubular change has been compared to the drug-induced proliferation of the endoplasmic reticulum which has been seen in the liver;[32] however, such a drug-dependent relationship has not yet been shown in muscle. It has also been pointed out that similar tubular structures have been demonstrated in virus-infected cells.[8]

AGGREGATES OF SHORT BUDDING TUBULES

Not to be confused with the aggregates of linear tubules are groups of narrower short budding tubules (or "honeycombs") which have been seen in hyperkalemic periodic paralysis,[72] neurogenic atrophy,[87] trichinosis,[99] myotonic

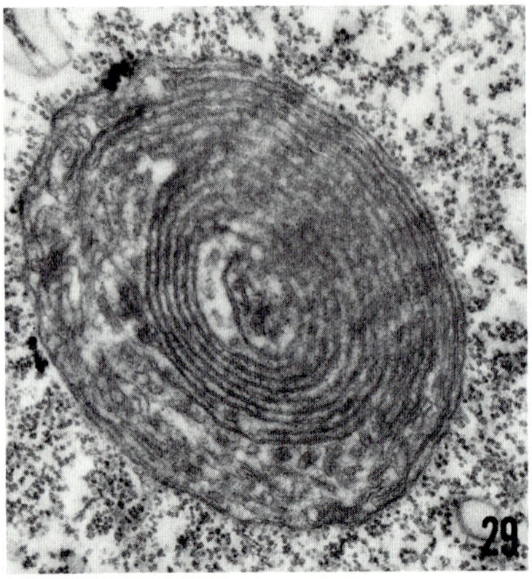


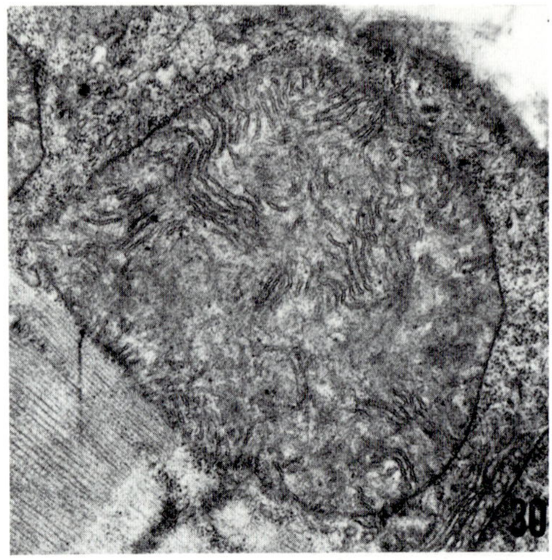


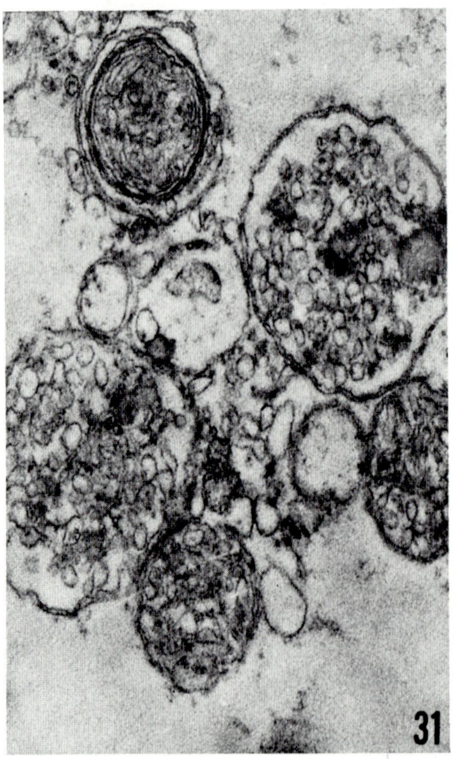


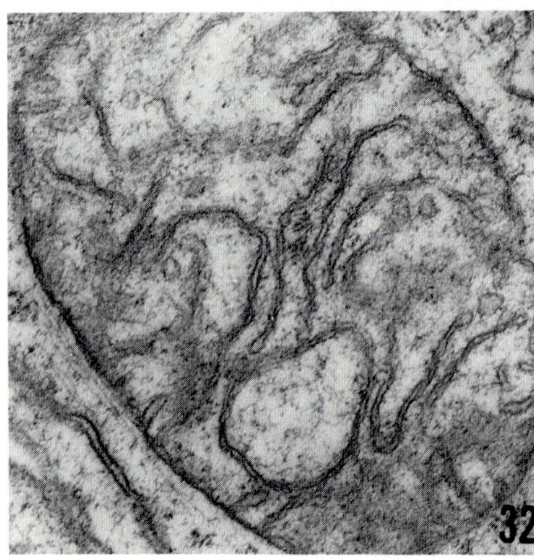


FIG. 29. Enlarged mitochondrion with closely packed concentrically arranged cristae. × 29,300.

FIG. 30. Enlarged mitochondrion with short, blunted cristae and increased dense matrix material. × 28,000.

FIG. 31. Mitochondria in acutely injured cell with vesiculation of cristae, large dense deposits, and focal fragmentation of the outer mitochondrial membranes. × 27,600.

FIG. 32. Enlarged mitochondrion in regenerating muscle fiber. Note marked variation in pattern of cristae. × 27,000.

dystrophy[111] (Fig. 35), a rhabdomyoma (Fig. 36), a late onset glycogenosis,[26] and polymyositis.[82] They resemble budding T system elements seen in chick skeletal muscle tissue culture preparations.[35] In all of these conditions one factor appears to be constant—namely, a lack of normal neural influence because of a direct injury to the nerve, focal damage of the muscle fiber which may prevent continuous normal neural influence across the fiber surface, the uncontrolled growth of muscle in the form of a neoplasm, or growth in an artificial environment devoid of neural elements.

## Golgi System

In normal fibers the Golgi complex is an inconspicuous member of the sarcoplasm in the region of the nucleus. It is well developed in injured cells, particularly in zones of regeneration. In acid maltase deficiency the Golgi system was thought to contribute to the membranes limiting autophagic vacuoles.[25, 26]

## Mitochondria

### Normal aspects[88]

In skeletal muscle fibers mitochondria tend to be distributed in the interfibrillar sarcoplasm as isolated profiles symmetrically positioned near the Z line or aligned end-to-end in a single column running parallel to adjacent myofibrils (Fig. 25). Other clusters of mitochondria can be seen adjacent to the plasma membrane frequently associated with the nuclei of a muscle fiber or in the vicinity of a motor end-plate.

The number of mitochondria varies in muscle fibers, being more abundant in *type I* fibers than in *type II* fibers. The shape, size, and number of mitochondria as well as of the cristae may be a reflection of normal variations in metabolic activity. The cristae are often perpendicular to the long axis of the mitochondria, but they may run parallel and sometimes even form a localized concentric pattern.

### Abnormal mitochondria

A number of abnormal mitochondrial patterns has been reported in injured skeletal muscle fibers. This discussion is limited to those most commonly seen in human disorders (see Table 3). None of these mitochondrial alterations seem to be distinctive for a specific disease; however, in some disorders they are the dominant pathologic finding, being present in excessive numbers (pleoconial) and/or increased in size (megaconial). In other disturbances similar abnormal mitochondria are an incidental finding and not the primary pathologic change. Detailed combined biochemical and structural studies of cell fractions containing these mitochondria may shed more light on their role in the diseased cell. These deviations from normal mitochondrial architecture may be a compensatory change to some non-mitochondrial-induced abnormality in the metabolic milieu; or on the other hand they could be a primary inherent reproducible defect in mitochondrial structure which has hampered normal muscle function. It

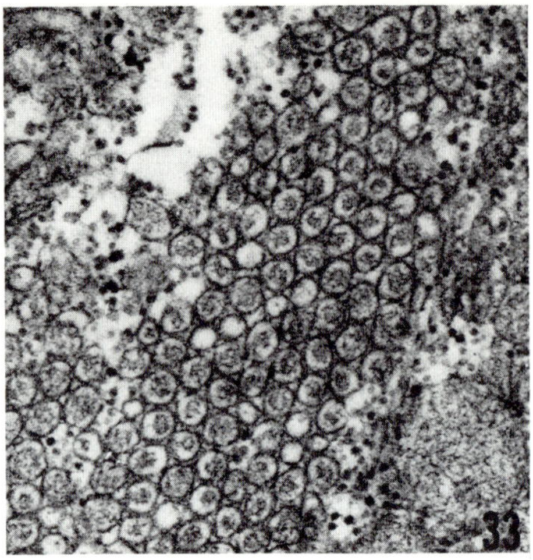


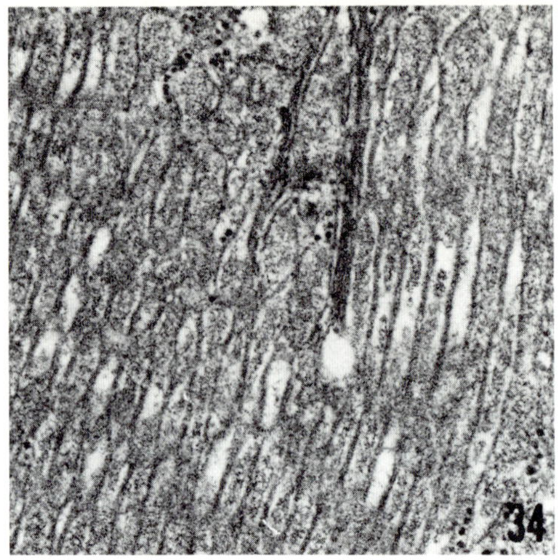


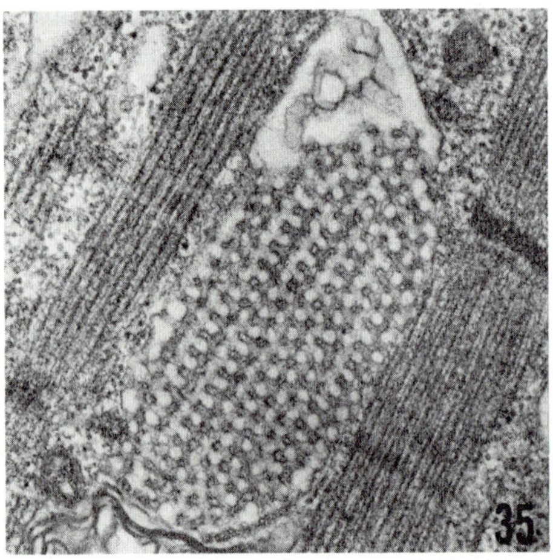


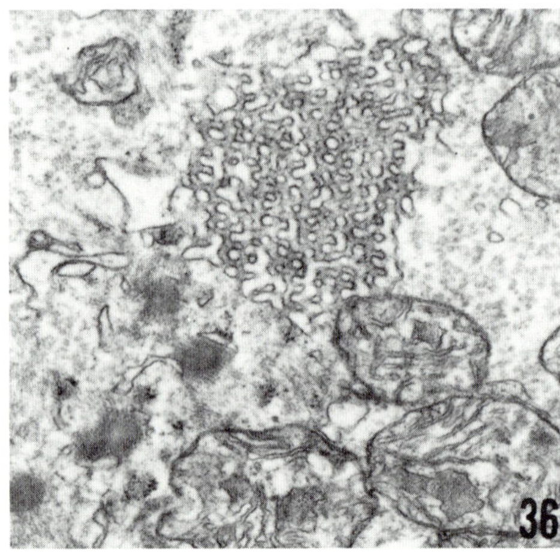


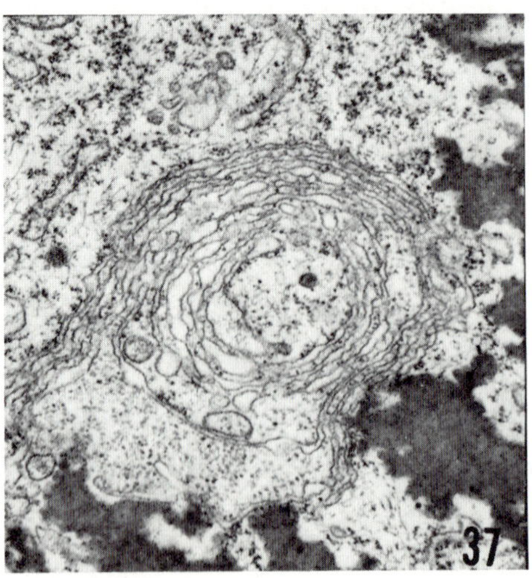


Fig. 33. Cross-section through tightly packed tubular aggregates. The cores of the tubules contain finely granular election dense material. (Electron micrograph made from blocks loaned by Dr. K. Meyer, Armed Forces Institute of Pathology, Washington, D. C.) × 50,300.

Fig. 34. Slightly oblique section through tightly packed tubular aggregates. The cores appear to contain electron-dense finely granular material. × 39,200.

is also important to keep in mind when analyzing correlative structural and functional studies that the metabolic functions of mitochondria seem to be dependent not only on the presence of specific enzymes but also on the proper sequential arrangement of these enzymes within mitochondrial membranes. Thus, it is theoretically possible to have a dysfunctional mitochondrial condition that may be related to variations in the membrane-linked enzymatic assembly lines of the mitochondria rather than an actual deficiency of a specific enzyme.

*Paracrystalline inclusions.* These (Figs. 26 to 28) have been observed in the outer mitochondrial chamber (which includes the space between the outer and inner mitochondrial membranes and the intracristae space) in a number of clinically different muscle disorders (Table 3). The affected mitochondria are usually enlarged. In several cases the change seems to be confined to the mitochondria of *type I* fibers. These inclusions appear to have a regular periodic pattern. One report[119] has likened these inclusions to those seen in the mitochondria associated with yolk protein formation in amphibian oocytes.[139] Similar mitochondria have also been reported recently in liver cells of women under increased steroid effect such as contraceptive therapy, pregnancy, and trophoblastic growths.[47] In this latter group, the affected mitochondria could represent a compensatory structural modulation to an increased demand on liver metabolism in the breakdown of steroids.

*Hypercristae.* Hypercristae, or increased numbers of mitochondrial cristae, have also been seen in several different clinical muscular disturbances (Table 3). In some cases, the increased numbers of closely packed cristae assume a linear pattern and in others a concentric arrangement (Fig. 29). These mitochondria are usually enlarged and seem to be present in increased numbers. In one patient, a definite hypermetabolic state similar to hyperthyroidism was demonstrated.[69] In several cases, the inner mitochondrial chamber contained dense inclusions and also glycogen-like particles.[20]

Other changes such as a reduction in the number of mitochondria or in the number of cristae, variations in density of mitochondrial matrix material (Fig. 30), fusion of cristae, and mitochondrial swelling have been found to varying degrees in muscle cell disorders.

In the early phases of acute injuries large dense matrix bodies are often seen, and they are soon followed by vesiculation of the cristae (Figs. 3 and 31).

In regions of regeneration, it is not unusual to see large mitochondria with irregular cristae (Fig. 32), increased numbers of relatively small mitochondrial profiles, or other mitochondrial changes.[97]

## Lysosomes and Autophagic Vacuoles

Cell-fractionation studies have demonstrated the presence of a subcellular fraction in normal skeletal muscle tissues having the biochemical and physical

---

Fig. 35. Short budding tubular structure from patient with myotonic dystrophy. × 20,000.

Fig. 36. Short budding tubular structure in rhabdomyoma. × 28,800.

Fig. 37. Circularly arranged tubular structure and dense lipid droplets in regenerating muscle fiber. × 24,000.

TABLE 3. ABNORMAL MITOCHONDRIA WITH HUMAN MYOPATHIES

| Clinical Diagnosis | Sex | Age | Family History | Abnormal Mitochondria | | | Remarks | Reference |
|---|---|---|---|---|---|---|---|---|
| | | | | Structural features[a] | Incidence | Fiber type | | |
| Hypermetabolism of non-thyroid origin | F | 35 | No | a, b, d, e | Frequent | ? | Respiratory studies of isolated mitochondria revealed loosely coupled state of oxidative phosphorylation | Luft *et al.*[69] |
| Megaconial myopathy | F | 8 | Yes | a, d, e | Frequent | I? | Increased lipid deposits in fibers | Shy *et al.*[123] |
| Pleoconial myopathy | M | 8 | Yes | b, d, e | Frequent | ? | | Shy *et al.*[123] |
| Mitochondrial myopathy | M | 20 | No | a, d, e | Frequent | I | Increased lipid deposits in fibers | Price *et al.*[89] |
| Progressive ocular muscle dystrophy | M<br>F<br>M | 20<br>51<br>38 | No<br>No<br>No | a, b, d, e | Frequent | ? | | Zintz and Villiger[144] |
| Diffuse muscular weakness and growth retardation | F<br>F | 15<br>8 | Sisters | a, b, d, e | Frequent | ? | Intramitochondrial glycogen deposits | D'Agostino *et al.*[20] |
| Autophagic glycogenosis of late onset | M | 46 | Yes | a, d | Frequent | I | Great variety of other ultrastructural alterations | Engel and Dale[26] |
| Myopathy with abnormal mitochondria | M | 17 | No | b, d, e | Frequent | I | Biochemical studies indicated loosely coupled state of oxidative phosphorylation | van Wijngaarden *et al.*[137] |
| Proximal myopathy with abnormal mitochondria | F | 27 mos. | No | b, d, e, f | Frequent | I? | Biochemical studies indicated loosely coupled state of oxidative phosphorylation; intramitochondrial glycogen | Hulsmann *et al.*[60] |
| Failure to thrive and persistent lactic acidemia | M | 4 | ? | a, d, e, f | Frequent | I | Thought to be defect in pyruvate metabolism; intramitochondrial glycogen | Pardo *et al.*[83] |
| Moderate lack of muscular endurance and abnormal muscle mitochondria | ? | 34 | ? | b, d, e | Frequent | I | | Engel[32] |

TABLE 3. (*continued*)

| | | | | | | | | |
|---|---|---|---|---|---|---|---|---|
| Mitochondrial myopathy | M | 39 | No | a, d, e | Frequent | I | Clinical evidence of cardiac involvement | Fisher and Danowski[39] |
| Dermatomyositis | M | 48 | No | a, d | ? | ? | | Shafiq *et al.*[119] |
| Polymyositis | F | 42 | No | a, d | ? | ? | | Shafiq *et al.*[119] |
| Spinal muscular atrophy | F | 53 | Yes | a, d | ? | ? | | Shafiq *et al.*[119] |
| Normal (?) muscle in nephrotic syndrome | M | 25 | No | a, d | ? | ? | | Shafiq *et al.*[119] |
| Progressive distal myopathy | ? | Adult | (?) | a | Infrequent | ? | | Gruner[50] |
| Hypothyroid myopathy | M | 61 | No | a | Infrequent | ? | | Norris and Panner[79] |
| Refsum's disease with Hurler's syndrome | F | 21 | No | a, c, d | Infrequent | ? | Abnormal mitochondria also within liver | Gonatas *et al.*[43] |
| Amyotrophic lateral sclerosis | F | 45 | ? | a | Infrequent | ? | | DeRecondo *et al.*[21] |
| Chronic polymyositis with Raynaud's syndrome | M | 66 | No | a, b, d | Infrequent | ? | Myxovirus-like structures in muscle | Chou[16] |
| Progressive spinal muscular atrophy | M | 60 | No | a, d | ? | ? | | Sato and Tsubaki[105] |
| Type V glycogenosis | F | 8 | No | a, d | ? | ? | | Sluga *et al.*[125] |
| | F | 25 | No | | | | | |
| Myopathy with intramitochondrial inclusions | M | 37 | No | a, d | ? | ? | | Vye[138] |

[a] Key to structural features:

a, paracrystalline inclusions within space between inner and outer mitochondrial membranes and/or the intracristae space.

b, hyperplasia of cristae (either with linear or concentric pattern).

c, fusion of cristae.

d, enlarged mitochondria.

e, apparent increase in number of mitochondria.

f, other change.

properties of lysosomes, but the intracellular origin of this lysosomal fraction has not been definitely established.[13] Morphologic characterization of the lysosome at the light and electron microscopic levels has been attempted by histochemistry using Gomori's lead phosphate technique for acid phosphatase. In normal skeletal muscle, this enzyme has been found to be localized mainly to the walls of the T tubules, within the immediate adjacent terminal cisternae of the sarcoplasmic reticulum, and in round bodies which frequently have a paranuclear location and structural features resembling lipofuscin.[48, 56, 84] These histochemical observations and recent biochemical evidence have led several workers to suggest that the lysosomal hydrolases in normal muscle are contained primarily within the sarcotubular system.[13, 84]

It has been postulated that during focal degradation of cell constituents, either in response to pathologic stress or normal physiologic need, the lysosomal enzymes may in some way be transferred to the forming autophagic vacuoles[56] (Fig. 41). These single membrane-bound autophagic vacuoles ("cytosomes" or "cytosegresomes") resembling lysosomes of other cells have been seen in various myopathic disorders such as acute polymyositis,[101, 115] preclinical muscular dystrophy,[56, 58] and vitamin E deficiency.[54] Pompe's disease may be a lysosomal disorder since it is a deficiency of the lysosomal enzyme acid maltase.[14, 126] Vacuoles containing glycogen, and large amounts of free sarcoplasmic glycogen, are frequently seen in the skeletal muscle of patients with this disease; but similar vacuoles have also been seen in other disorders (Figs. 38 and 40). Much more work is required on the role of lysosomes in muscle.

### Satellite Cells (A Modified Form of Cytoplasmic Segregation?)

These are mononucleated myoblast-like cells which lie separately beneath the external lamina adjacent to an adult muscle fiber. Several hypotheses have been proposed as to their function and origin, chiefly that they are myoblasts with a potential to contribute to the postnatal growth and/or repair of the main multinucleated skeletal muscle cell.[73] These concepts are discussed in more detail in Chapter 10. It has been noted that relatively increased numbers of these mononuclear cells may be seen at the site of injury in mature muscle fibers, whereas the chances of finding satellite cells in uninjured mature muscle are markedly reduced.[98] In a recent ultrastructural study[98] of regenerating muscle it has been suggested that myoblasts are probably pinched off from the injured muscle sarcoplasm as mononucleated cells which pass through a satellite stage. Plasma membranes appeared to form by a process of progressive coalescence of vesicles and clefts and eventually isolated a normal appearing muscle nucleus and its surrounding sarcoplasm from the degenerative portion of the cell.[98] This seems to this reviewer to be an extention of the concept of cytoplasmic segregation which is a characteristic response of cells to focal injury. Normal cells appear to have the capability to form membrane-limited compartments around damaged portions of cytoplasm, thus protecting the viable part of the cell from the unfavorable influence of the degenerative zone. Within these membrane-limited areas cytoplasmic degradation may take place through the effects of hydrolytic (or lysosomal) enzymes.[55, 136] In case of muscle we have

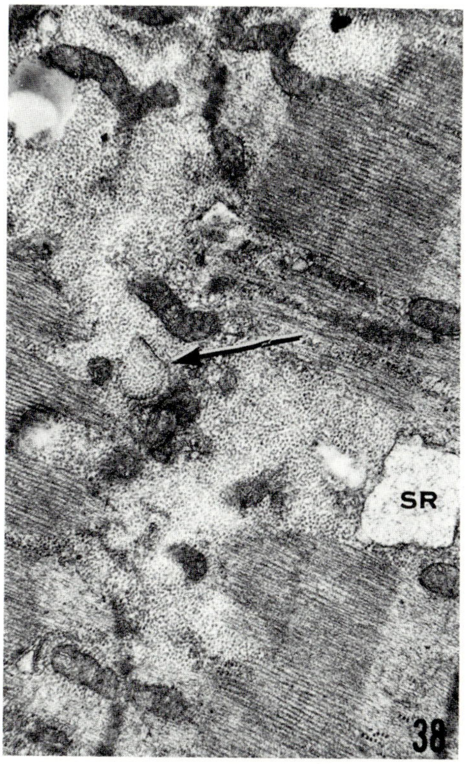


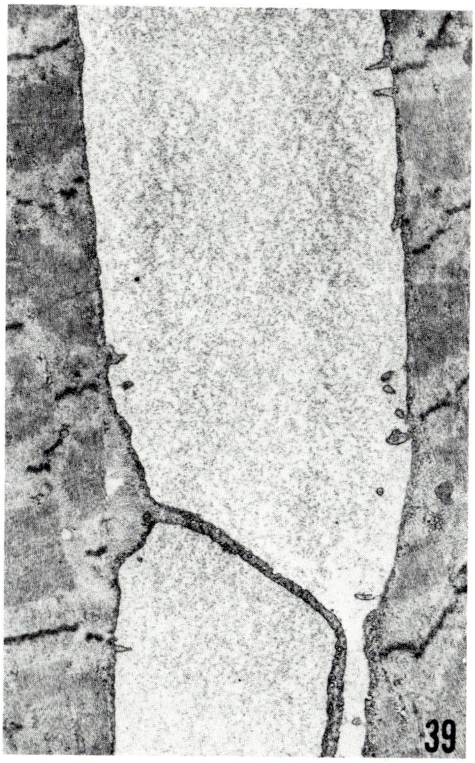


FIG. 38. Early dilatation of sarcoplasmic reticulum (*SR*) and abundant glycogen accumulation in patient with hypokalemic periodic paralysis. *Arrow* indicates partial membrane formation around focal accumulation of glycogen particles. × 20,300.

FIG. 39. Large vacuoles containing finely granular material within center of fiber from case of hypokalemic periodic paralysis. × 10,000.

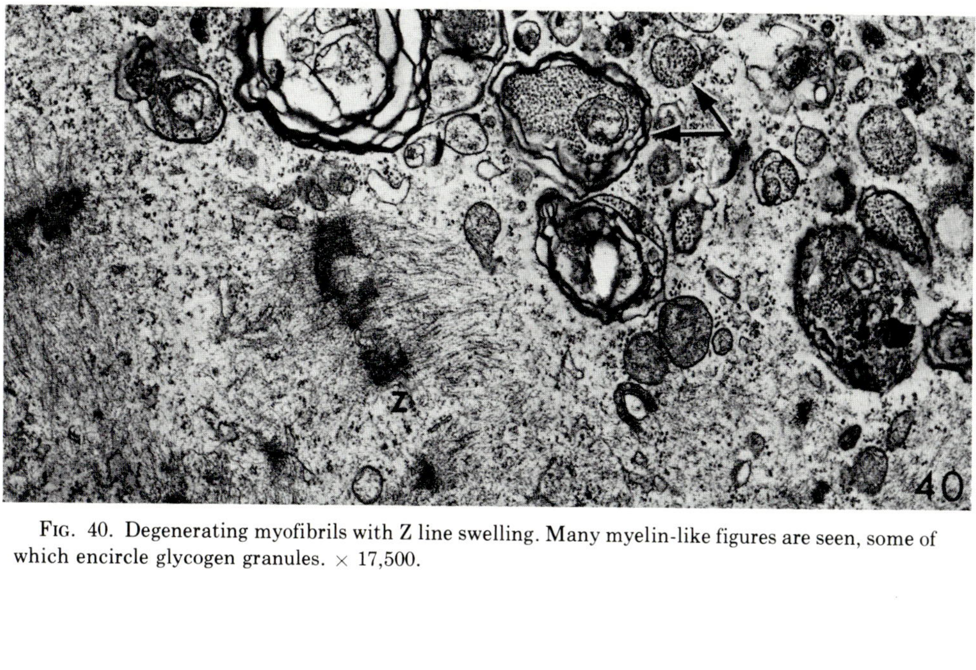


FIG. 40. Degenerating myofibrils with Z line swelling. Many myelin-like figures are seen, some of which encircle glycogen granules. × 17,500.

the unusual situation of a cell which is normally multinucleated and in which the principle of cytoplasmic segregation seems to be maintained, but it is modified so that a viable nucleus and sarcoplasm are surrounded by a membrane to separate them from the nonviable injured elements. The degenerative debris is then removed by wandering macrophages; and the viable newly formed "satellite cell," which has a nucleus, may begin to initiate protein production and repair.

## Sarcoplasmic Particulate Components—Normal and Abnormal

### Glycogen

In routine electron microscopic preparations glycogen particles generally have a slight irregular outline and a diameter ranging from 150 to 350 Å. In skeletal muscle, they occur singly ($\beta$ particles) without forming rosette forms ($\alpha$ particles) as seen in the liver.[9, 37] The number of glycogen particles in routine electron micrographs of normal muscle varies to some extent, depending on the metabolic demands of the cell. Increased numbers of these particles have been particularly noticeable in the periodic paralyses[53, 56] and the glycogen storage diseases.[14, 26, 42, 56, 110] As noted above, in at least some of these conditions, membrane-bound pockets of glycogen particles may be seen.

### Lipid inclusions[32, 37, 88]

These are generally droplets of triglycerides of fatty acids and they are found normally in skeletal muscle fibers. These droplets are usually intimately associated with mitochondria (Fig. 22). They are relatively more abundant in the *type I* fibers. The fixation of lipid particles is usually difficult. When fixed with osmium tetroxide, these triglyceride droplets should be blackened by the reduction of osmium and rendered insoluble to the alcohol and other reagents used for dehydration and embedding of the tissues; however, this is not always the case, particularly in tissue fixed with glutaraldehyde. There appears to be a fair amount of variation in the degree of fixation of lipids, which is difficult to control. Often, failure of the fixative to penetrate adequately results in extraction of lipid from the center of the droplets during tissue processing, which causes either a clear center surrounded by a dense band of blackened lipid or complete extraction of the lipid leaving only a thin continuous rim of osmophilic material at the interface between the droplet and the sarcoplasm. As indicated earlier, in several disorders associated with abnormal mitochondria a substantial increase in the numbers of lipid droplets appears to be present in the *type I* fibers.

### Lipopigments

Cytoplasmic lipid pigment deposits have been described under a variety of titles such as "lipofuscin," "lipochrome pigment," "aging pigment," "ceroid," and "wear and tear pigment." Some of these terms probably overlap in meaning, but all of these lipopigments most likely represent different stages in the oxidation of lipid material.[85] Chemical analyses indicate that these pigments consist of autooxidized lipids in association with a protein moiety.[131]

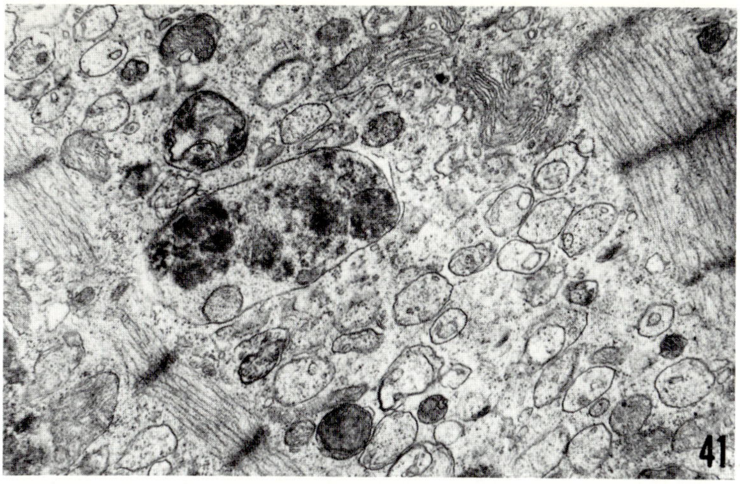


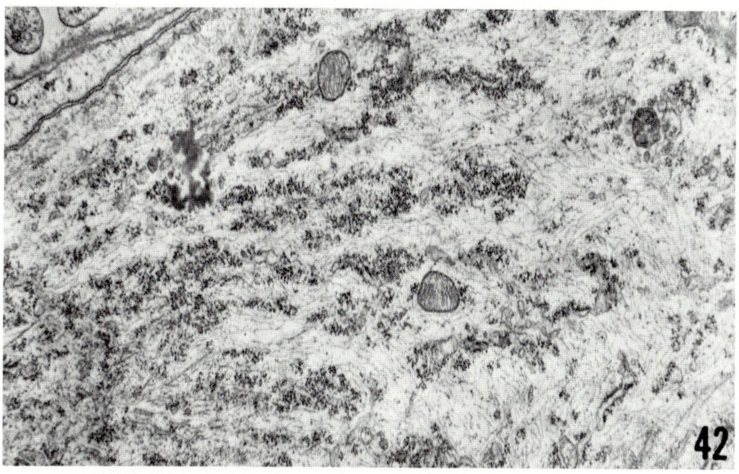


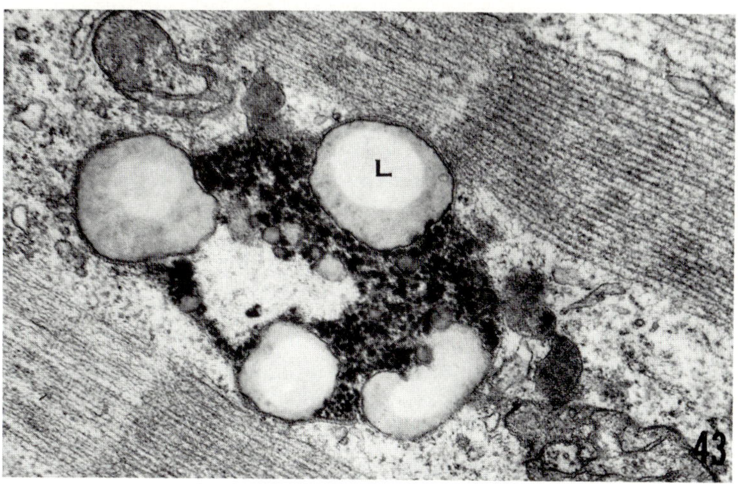


FIG. 41. Autophagic vacuoles. × 17,200.

FIG. 42. Aggregates of ribosomes (polyribosomes) and fine filaments in sarcoplasm from a zone of regeneration. × 17,200.

FIG. 43. Electron microscopic features of lipofuscin body. Note lipid inclusions (*L*) within this membrane-bound structure. × 40,600.

By electron microscopy these pigment bodies are bound by a single membrane and have a variable density pattern consisting of large round pale bodies (resembling lipid droplets) intermixed with dense granular material (Fig. 43). Increased numbers of these lipopigment deposits are seen in various chronic myopathic disturbances. In vitamin E-deficient animals they also contain membrane fragments or myelin-like figures.[54, 103] It is possible that they are terminal forms of autophagic vacuoles as they may contain acid phosphatase[48, 80] and other acid hydrolase activities.[13, 80, 130]

#### Ribosomes

These 150- to 200-Å particles, composed of protein and RNA, are relatively sparse in normal mature skeletal muscle fibers. However, ribosomes are abundant in developing and regenerating muscle fibers (Fig. 42). Due to the RNA the net charge of the ribosome is strongly negative.[22] For this reason they bind cations and basic dyes, thereby imparting a basophilic hue to the sarcoplasm of regenerating muscle in hematoxylin and eosin preparation. It is generally accepted that ribosomes directly participate in protein production.[22, 136] In regenerating muscle (or myoblasts), where protein production is intended for intracellular structures, the ribosomes are distributed either in clusters (polyribosomes) or as isolated particles. This contrasts with the membrane-bound ribosomes seen within cells such as fibroblasts which make protein for extracellular use.

#### Subsarcolemmal crystals[116]

Recently unique subsarcolemmal crystals were found in the muscle fibers from two patients with diabetic amyotrophy and neuropathy. The crystals did not appear to be related to any sarcoplasmic structures and lacked peripheral membranes. They had a periodic pattern of dense parallel bands approximately 120 Å wide separated by spaces 150 Å wide. The chemical properties of these crystals are not known.

#### Viruslike structures[17]

Myxovirus-like structures were demonstrated in both the sarcoplasm and the muscle fiber nuclei in a case of chronic polymyositis. Abnormal mitochondria were also present (see Table 3).

### Nuclei

#### Normal aspects

The oval nuclei lie adjacent to the plasma membrane with their long axes parallel to the longitudinal orientation of the myofibrils. The appearance of the chromatin in the interphase nucleus varies with fixation and embedding techniques.[86, 136] After osmium tetroxide fixation, the chromatin is diffusely distributed throughout the nucleoplasm, whereas, following aldehyde fixation and osmium tetroxide postfixation, the chromatin tends to be aggregated along the nuclear membrane and around the nucleus. By serial sectioning, nucleolar structures can be found in all sarcolemmal nuclei, even in mature fibers whose nuclei often appear to have a dense homogenous chromatin pattern by light microscopy. A nuclear envelope separates the cytoplasm from the nucleoplasm and is composed of two membranes which display an occasional nuclear pore. These pores seem to be more frequent in nuclei of regenerating muscle.[88]

INJURED CELLS

In acutely injured cells the chromatin material becomes markedly clumped along the nuclear membrane and around the nucleolus even in osmium tetroxide-fixed tissue.[91, 136] Many of these nuclei are associated with zones of regeneration. The nuclei may be increased in number and may be found within the center of the fiber, sometimes aggregated in clumps. On the other hand, in myotonic dystrophy[141] and centronuclear (or myotubular) myopathy[66, 77, 122, 127] most of the nuclei appear to be in a central location and are associated with the sarcoplasmic features of regeneration.

With nuclear activation the nuclear surface is often thrown into folds of varying degrees of complexity (Figs. 44 to 47). *Pseudoinclusions* result from deep invaginations of the nuclear surface by which portions of cytoplasm, due to the plane of section, appear to be intranuclear in location (Fig. 47). *True inclusions* such as myxovirus-like and rodlike structures have been discussed above. Glycogen particles have also been seen within sarcolemmal nuclei.[108]

## CONCLUDING COMMENTS

Critical consideration has been given to the ultrastructural aspects of myopathology. Electron microscopy has increased the precision by which we can view the pathologic reactions within injured muscle fibers. The vague myofibrillar, sarcolemmal, and sarcoplasmic alterations seen with light microscopy can now be translated into variations of particular organelles. Myofibrillar changes are described in terms of the myofilamentous components that are normally responsible for the myofibrillar banded pattern. Variations in the surface of the fiber can be specifically related to the plasma membrane and the external lamina. Satellite cells can be identified, and myoblasts are more easily differentiated from macrophages and fibroblasts. The role of the SR and T system in vacuole formation is perceptible. In addition, structural modulations of mitochondria, and of the SR and T system, such as aggregates of linear tubules and short budding tubules, have been recognized.

As with conventional microscopy, muscle at the ultrastructural level manifests a limited number of responses to injury. Only a combination of individual morphologic features and their apparent quantitative relationship can be used to formulate diagnostic patterns. To date, the primary influence of ultrastructural studies has been on the nosology of congenital and metabolic muscle disturbances such as nemaline (or rod) myopathy and the group of disorders characterized primarily by the presence of abnormal mitochondria. Also, a single report, which needs confirmation, has suggested that one form of polymyositis may be related to and recognized by the presence of myxovirus-like structures.

For the pathologist using electron microscopy, the sampling and interpretive problems inherent with this technique must be reemphasized. Because the specimen examined is relatively small and many observations are made on a few cells, misleading conclusions may result unless large numbers of blocks are examined and correlated with light microscopic inspection. It is sometimes difficult to decide what is a normal physiologic reaction and what is pathologic or even artifactual. We do not know what the tolerable levels of abnormality are

FIG. 44. Centrally positioned nuclei in fibers of patient with familial centronuclear myopathy. Note fine undulations of fiber surface. × 4700.

FIG. 45. Same case as Fig. 44. Symmetrical deep infoldings of nucleus (*arrows*) may suggest a phase of amitotic division. Dense deposits represent tightly packed aggregates of glycogen particles. × 11,400.

FIG. 46. Irregular undulations of surface of muscle nucleus. The significance of this change is not understood, but it is frequently seen in otherwise normal fibers. Apparent fragmentation of myofibrils in this slightly oblique section is *not* an abnormal finding (see Fig. 1 and text). × 5700.

FIG. 47. Pseudo-intranuclear inclusion (*pni*) due to deep invagination of nuclear surface with apparent engulfment of adjacent sarcoplasm. × 6700.

FIG. 48. Some widening of basement lamina around capillary in skeletal muscle biopsy from patient with polymyositis and no clinical evidence of diabetes mellitus. This finding supports observations in muscle biopsies from four patients with polymyositis by Gonzalez-Angulo, A., Fraga, A., and Mintz, G. (*Amer. J. Med. 45:* 873–879, 1968). In addition, the author has observed capillary basement membrane thickening in muscle biopsies from patients with muscular dystrophy. × approximately 11,000.

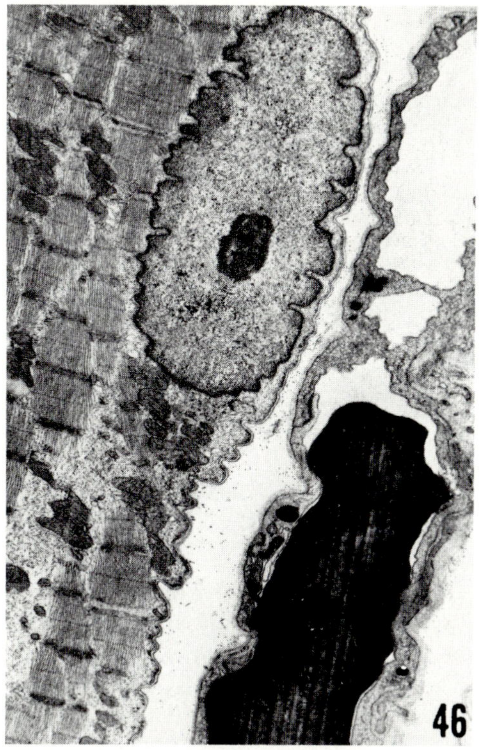

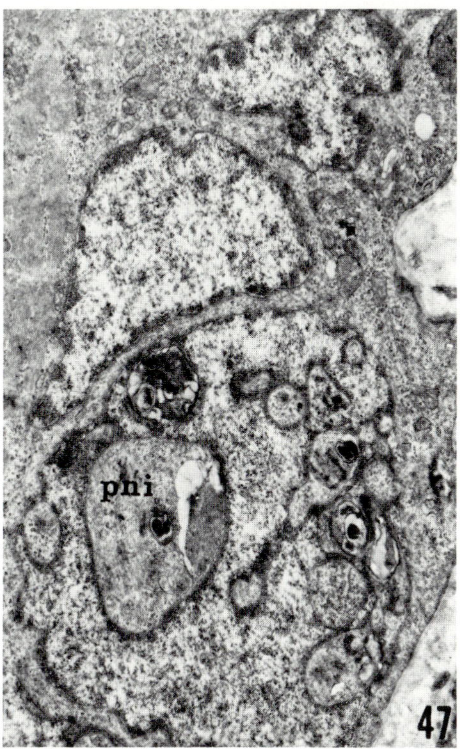

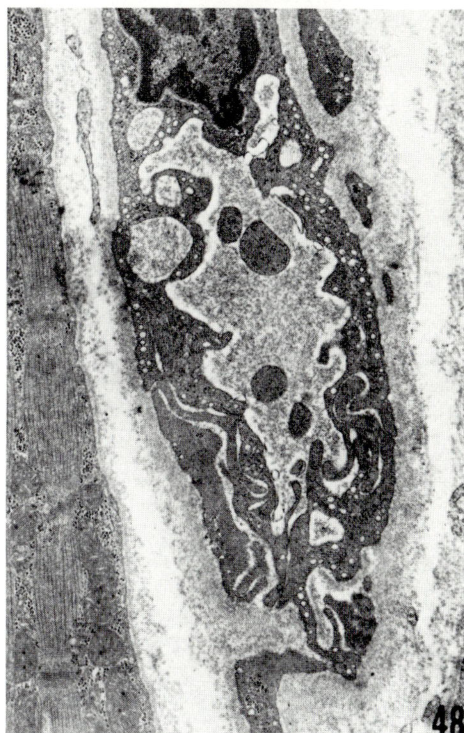


FIGS. 45–48

or what changes represent subclinical disease. What is the significance of an occasional rod or abnormal mitochondria? We need more information on the variations of normal as well as the abnormal.

In conclusion, a meaningful number of electron microscopic observations of injured muscle fibers has been accumulated; however, it is constantly necessary to critically evaluate data keeping in mind the technique limitations and the total pathobiologic process being studied. It is important to correlate ultrastructural findings with other physical and chemical methods of investigation. The wider application of cytochemical, immunochemical, and autoradiographic procedures at the fine structural level should also improve our insight into abnormal cells.

## ACKNOWLEDGMENTS

I am indebted to Drs. Gerald Gordon and Edward L. Howes, Jr., for several electron micrographs from past collaborative efforts. I also appreciate the cooperation of Drs. William Morningstar, James Smith, and Karl Meyers for blocks or tissues from which some electron micrographs were made. Finally, I also wish to express my gratitude to Mr. Robert Florida and Mrs. Carol Raber for their generous technical assistance.

### REFERENCES

1. Adams, R. D.: The giant muscle fiber. Its place in myopathy. In *Modern Neurology*, ed. by Locke, S. Boston, Little, Brown & Co., 1969.
2. Afifi, A. K., Bergman, R. A., and Harvey, J. C.: Steroid myopathy. Clinical, histologic and cytologic observations. *Johns Hopkins Med. J. 123:* 158–173, 1968.
3. Afifi, A. K., Aleu, F. P., Goodgold, J., and Mackay, B.: Ultrastructure of atropic muscle in amyotrophic lateral sclerosis. *Neurology (Minneap.) 16:* 475–481, 1966.
4. Afifi, A. K., Smith, J. W., and Zellweger, H.: Congenital nonprogressive myopathy. Central core disease and nemaline myopathy in one family. *Neurology (Minneap.) 15:* 371–381, 1965.
5. Aleu, F. P., and Afifi, A. K.: Ultrastructure of muscle in myotonic dystrophy. *Am. J. Pathol. 45:* 221–232, 1964.
6. Allbrooks, D.: An electron microscopic study of regenerating skeletal muscle. *J. Anat. 96:* 137, 1962.
7. Banker, B.Q.: A phase and electron microscopic study of dystrophic muscle. I. The pathological changes in the two week old Bar Harbor 129 dystrophic mouse. *J. Neuropathol. Exp. Neurol. 26:* 259–275, 1967.
8. Bergman, R. A., Afifi, A. K., Dunkle, L. M., and Johns, R. J.: Muscle pathology in hypokalemic periodic paralysis with hyperthyroidism. I. High resolution light microscopic study of a case. *Johns Hopkins Med. J. 126:* 88–99, 1970.
9. Biava, C.: Identification and structural forms of human particulate glycogen. *Lab. Invest. 12:* 1179–1197, 1963.
10. Boehme, D., Themann, H., and Gold, J.: Structural and ultrastructural changes in striated human muscle caused by chronic ischemia. *Am. J. Pathol. 49:* 569–591, 1966.
11. Bray, G. M., and Banker, B. Q.: An ultrastructural study of degeneration and necrosis of muscle in the dystrophic mouse. *Acta Neuropathol. (Berl.) 15:* 34–44, 1970.
12. Brody, I. A., and Dudley, A. W.: Thyrotoxic hypokalemic periodic paralysis. *Arch. Neurol. 21:* 1, 1969.
13. Canonico, P. G., and Bird, J. W. C.: Lysosomes in skeletal muscle tissue. Zonal centrifugation evidence for multiple cellular sources. *J. Cell Biol. 45:* 321, 1970.
14. Cardiff, R. D.: A histochemical and electron microscopic study of skeletal muscle in a case of Pompe's disease (glycogenosis II). *Pediatrics 37:* 249–259, 1966.

15. Caulfield, J. B.: Electron microscopic observations on the dystrophic hamster muscle. *Ann. N. Y. Acad. Sci. 138:* 151–159, 1966.
16. Chou, S. M.: "Megaconial" mitochondria observed in a case of chronic polymyositis. *Acta Neuropathol. (Berl.) 12:* 68, 1969.
17. Chou, S. M.: Myxovirus-like structures in a case of human chronic polymyositis. *Science 158:* 1453, 1967.
18. Conen, P. E., Murphy, E. G., and Donohue, W. L.: Light and electron microscopic studies of "myogranules" in a child with hypotonia and muscle weakness. *Can. Med. Assoc. J. 89:* 983–986, 1963.
19. D'Agostino, A. N., and Chiga, M.: Cortisone myopathy in rabbits. A light and electron microscopic study. *Neurology (Minneap.) 16:* 257–263, 1966.
20. D'Agostino, A. N., Ziter, F. A., Rallison, M. L., and Bray, P. F.: Familial myopathy with abnormal muscle mitochondria. *Arch. Neurol. 18:* 388–401, 1968.
21. DeRecondo, J., Fardeau, M., and Lapresle, J.: Etude au microscope electronique des leisons musculaires d'atrophie neurogene par atteinte de la corne anterieure. *Rev. Neurol. (Paris) 114:* 169, 1966.
22. DeRobertis, E. D. P., Nowinski, W. W., and Saez, F. A.: *Cell Biology*. Philadelphia, W. B. Saunders Co., 1965.
23. Douglas, R. M., Fewings, J. D., Casley-Smith, J. R., and West, R. F.: Recurrent rhabdomyolysis precipitated by alcohol. A case report with physiological and electron microscopic studies of skeletal muscle. *Aust. Ann. Med. 15:* 251–261, 1966.
24. Duane, D. D., and Engel, A. G.: Emetine myopathy. *Neurology (Minneap.) 18:* 274, 1968.
25. Engle, A. G.: Acid maltose deficiency of adult life. *Trans. Am. Neurol. Assoc. 94:* 250, 1969.
26. Engle, A. G., and Dale, A. J.: Autophagic glycogenosis of late onset with mitochondrial abnormalities. Light and electron microscopic observations. *Mayo Clin. Proc. 43:* 233, 1968.
27. Engel, A. G., and Gomez, M. R.: Nemaline (Z disk) myopathy. Observations on the origin, structure, and solubility properties of the nemaline structures. *J. Neuropathol. Exp. Neurol. 26:* 601–619, 1967.
28. Engel, A. G.: Pathologic reactions of the Z-disk. In *Exploratory Concepts in Muscular Dystrophy and Related Disorders*, ed. by Milhorat, A. T., International Congress Series No. 147. Amsterdam, Excerpta Medica Foundation, 1966.
29. Engel, A. G.: Late-onset rod myopathy (a new syndrome?). Light and electron microscopic observations in two cases. *Mayo Clin. Proc. 41:* 713–741, 1966.
30. Engel, A. G.: Electron microscopic observations in thyrotoxic and corticosteroid-induced myopathies. *Mayo Clin. Proc. 41:* 785–796, 1966.
31. Engel, A. G.: Electron microscopic observations in metabolic myopathies (three types of hypokalemic periodic paralysis, thyrotoxic myopathy, and steroid-induced myopathy). *J. Neuropathol. Exp. Neurol. 24:* 141, 1965.
32. Engel, W. K.: Selective and nonselective susceptibility of muscle fiber types. *Arch. Neurol. 22:* 97, 1970.
33. Engel, W. K., and Resnick, J. S.: Late onset rod myopathy. A newly recognized, acquired, and progressive disease. *Neurology (Minneap.) 16:* 308, 1966.
34. Engel, W. K., Foster, J. B., Hughes, B. P., Huxley, H. E., and Mahler, R.: Central core disease. An investigation of a rare muscle cell abnormality. *Brain 84:* 167–185, 1961.
35. Ezerman, E. B., and Iskikawa, H.: Differentiation of the sarcoplasmic reticulum and T-system in developing chick skeletal muscle in vitro. *J. Cell Biol. 35:* 405–420, 1967.
36. Fawcett, D. W.: The sporadic occurrence in cardiac muscle of anomalous Z-bands exhibiting a periodic structure suggestive of tropomyosin. *J. Cell Biol. 36:* 266–270, 1968.
37. Fawcett, D. W.: *An Atlas of Fine Structure*. Philadelphia, W. B. Saunders Co., 1966.
38. Firket, H.: Ultrastructural aspects of myofibril formation in cultured skeletal muscle. *Z. Zellforsch. Mikrosk. Anat. 78:* 313, 1967.

38a. Fischman, D. A.: Personal communication.

39. Fisher, E. R., and Danowski, T. S.: Mitochondrial myopathy. *Am. J. Clin. Pathol. 51:* 619, 1969.
40. Fisher, E. R., Cohn, R. E., and Danowski, T. S.: Ultrastructural observations of skeletal muscle in myopathy and neuropathy with special reference to muscular dystrophy. *Lab. Invest. 15:* 778–793, 1966.

41. Frost, J. K.: *The Cell in Health and Disease*. Baltimore, Williams and Wilkins Co., 1969.
42. Garancis, J. D.: Type II glycogenosis. A biochemical and electron microscopic study. *Am. J. Med. 44:* 289, 1968.
43. Gonatas, N. K., Evangelista, J., and Martin, J.: A generalized disorder of nervous system, skeletal muscle, and heart resembling Refsum's disease and Hurler's syndrome. II. Ultrastructure. *Am. J. Med. 42:* 169–178, 1967.
44. Gonatas, N. K., Shy, G. M., and Godfrey, E. H.: Namaline myopathy. The origin of nemaline structures. *New Engl. J. Med. 274:* 535–539, 1966.
45. Gonatas, N. K.: The fine structure of the rod-like bodies in nemaline myopathy and their relation to the Z-discs. *J. Neuropathol. Exp. Neurol. 25:* 409–421, 1966.
46. Gonatas, N. K., Perez, M. C., Shy, G. M., and Evangelista, I.: Central "core" disease of skeletal muscle. Ultrastructural and cytochemical observations in two cases. *Am. J. Pathol. 47:* 503–524, 1965.
47. Gonzalez-Angulo, A., Aznar-Ramos, R., Bautista O'Fariel, J., and Martinez-Manautou, J.: The fine structure of liver cells in women under steroid effect. Contraceptive therapy, normal pregnancy and trophoblastic growths. In *The Muscle*, ed. by Pearson, C. M. Baltimore, Williams and Wilkins Co., 1971.
48. Gordon, G. B., Price, H. M., and Blumberg, J. M.: Electron microscopic localization of phosphatase activities within striated muscle fibers. *Lab. Invest. 16:* 422–435, 1967.
49. Grossman, I. W., Heitkamp, D. H., and Sacktor, B.: Morphologic and biochemical effects of *Clostridium perfringens* alpha toxin on intact and isolated skeletal muscle mitochondria. *Am. J. Pathol. 50:* 77–88, 1967.
50. Gruner, J. E.: Sur quelques anomalies mitochondriales observes au cours d'affections musculaires variees. *C. R. Soc. Biol. (Paris) 157:* 181–182, 1963.
51. Heffernan, L. P., Rewcastle, N. B., and Humphrey, J. G.: The spectrum of rod myopathies. *Arch. Neurol. 18:* 529, 1968.
52. Hopkins, I. J., Lindsey, J. R., and Ford, F. R.: Nemaline myopathy. A long-term clinicopathologic study of affected mother and daughter. *Brain 89:* 299–310, 1966.
53. Howes, E. L., Jr., Price, H. M., Blumberg, J. M., and Pearson, C. M.: Hypokalemic periodic paralysis. Electromicroscopic changes in the sarcoplasm. *Neurology (Minneap.) 16:* 242–256, 1966.
54. Howes, E. L., Jr., Price, H. M., and Blumberg, J. M.: The effects of a diet producing lipochrome pigment (ceroid) on the ultrastructure of skeletal muscle in the rat. *Am. J. Pathol. 45:* 599–631, 1964.
55. Hruban, A., Spargo, B., Swift, H., Wissler, R. W., and Kleinfeld, R. G.: Focal cytoplasmic degradation. *Am. J. Pathol. 42:* 657–683, 1963.
56. Hudgson, P., and Pearce, G. W.: Ultramicroscopic studies of diseased muscle. In *Disorders of Voluntary Muscle*, ed. by Walton, J. N., 2nd ed. London, J. & A. Churchill Ltd., 1969.
57. Hudgson, P., Gardner-Medwin, D., Worsfold, M., Pennington, R. J. T., and Walton, J. N.: Adult myopathy from glycogen storage disease due to acid maltase deficiency. *Brain 91:* 435–462, 1968.
58. Hudgson, P., Pearce, G. W., and Walton, J. N.: Pre-clinical muscular dystrophy. Histopathological changes observed on muscle biopsy. *Brain 90:* 565, 1967.
59. Hudgson, P., Gardner-Medwin, D., Fulthoyce, J. J., and Walton, J. N.: Nemaline myopathy. *Neurology (Minneap.) 17:* 1125, 1967.
60. Hulsmann, W. C., Bethlem, J., Meijer, A. E. F. H., Fleury, P., and Schellers, J. P. M.: Myopathy with abnormal structure and function of muscle mitochondria. *J. Neurol. Neurosurg. Psychiatry 30:* 519, 1967.
61. Huxley, H. E.: Electron microscope studies of natural and synthetic protein filaments from striated muscle. *J. Mol. Biol. 7:* 281, 1963.
62. Isch, F., Juif, J. G., Sacrez, R., and Thiebaut, F.: Glycogenose musculaire, a forme myopathique par deficit en maltase acide. *Pediatrie 21:* 71–86, 1966.
63. Jenis, E. H., Lindquist, R., and Lister, R. C.: New congenital myopathy with crystalline intranuclear inclusions. *Arch. Neurol. 20:* 281, 1969.
64. Johnson, A. G.: Alterations of the Z-lines and I-band myofilaments in human skeletal muscle. *Acta Neuropathol. (Berl.) 12:* 218, 1969.

65. King, D. W.: *Ultrastructural Aspects of Disease*. New York, Hoeber Medical Division, Harper & Row, 1966.
66. Kinoshita, M., and Cadman, T. E.: Myotubular myopathy. *Arch. Neurol. 18:* 265, 1968.
67. Klinkerfuss, G., Bleisch, V., Marcone, M. D., and Perkoff, G. T.: A spectrum of myopathy associated with alcoholism. II. Light and electron microscopic observations. *Ann. Intern. Med. 67:* 493–510, 1967.
68. Lapresle, J., and Fardeau, M.: Les desorganisations spatiales des myofibrilles, des sarcomeres et des myofilaments dars les zones peripheriques de fibres musculaires pathologiques etudiees en microscopie electronique. *Acta Neuropathol. (Berl.) 10:* 105, 1968.
69. Luft, R., Ikkos, D., Palmieri, G., Frnster, C., and Afzelius, B.: A case of severe hypermetabolism of nonthyroid origin with a defect in the maintenance of mitochondria respiratory control. A correlated clinical, biochemical, and morphological study. *J. Clin. Invest. 41:* 1776, 1962.
70. MacDonald, R. D., Rewcastle, N. B., and Humphrey, J. G.: Myopathy of hypokalemic periodic paralysis. An electron microscopic study. *Arch. Neurol. 20:* 565, 1969.
71. MacDonald, R. D., and Engel, A. G.: The cytoplasmic body. Another structural anomaly of the Z-disk. *Acta Neuropathol. (Berl.) 14:* 99, 1969.
72. MacDonald, R. D., Rewcastle, N. B., and Humphrey, J. G.: The myopathy of hyperkalemic periodic paralysis. *Arch. Neurol. 19:* 274, 1968.
73. Mauro, A.: Satellite cell of skeletal muscle fibers. *J. Biophys. Biochem. Cytol. 9:* 493, 1961.
74. Milhorat, A. T., Shafiq, S. A., and Goldstone, L.: Changes in muscle structure in dystrophic patients, carriers and normal siblings seen by electron microscopy; correlation with levels of serum creatinine phosphokinase (CPK). *Ann. N. Y. Acad. Sci. 138:* 246–292, 1966.
75. Mittelbach, F.: Die Begleitmyopathie Bei Neurogenen Atrophien. *Monogr. Neurol. Psychiat. (Ber.) 113:* 1–72, 1966.
76. Molbert, E.: Das electronenmikroskopische Bild des Skelettmuskels bei dystrophia musculorum progressive Erb. *Naturwissenschaften 41:* 186, 1960.
77. Munsat, T. L., Thompson, L. R., and Coleman, R. S.: Centronuclear ("myotubular") myopathy. *Arch. Neurol. 20:* 120, 1969.
78. Norris, F. H., Panner, B. J., and Stormont, J. M.: Thyrotoxic periodic paralysis. *Arch. Neurol. 19:* 88, 1968.
79. Norris, F. H., Jr., and Panner, B. J.: Hypothyroid myopathy. Clinical, electromyographical, and ultrastructural observations. *Arch. Neurol. 14:* 574–589, 1966.
80. Novikoff, A. B.: Lysosomes and related particles. In *The Cell*, ed. by Bracket, J., and Mirsky, A. E., Vol. 2. New York, Academic Press, 1962.
81. Page, S.: Structure of the sarcoplasmic reticulum in vertebrate muscle. *Br. Med. Bull. 24:* 170–173, 1968.
82. Palmeiro, J., Behrend, R. C., and Wechsler, W.: Electronenmikroskopische Befunde an der Skeletmuskulatur bei Polymyositis. *Acta Neuropathol. (Berl.) 7:* 26–43, 1966.
83. Pardo, V., Bejar, R. I., Cleveland, W. W., Nyhan, W. L., and Toshyuki, A.: Multiple mitochondrial abnormalities associated with an error in lactate and pyruvate metabolism. *Lab. Invest. 20:* 599, 1969.
84. Pearce, G. W.: The sarcolemma and sarcotubular systems in normal and dystrophic muscle. *Res. Muscular Dystrophy Proc. Symp. 3:* 146, 1965.
85. Pearse, A. G. E.: Pigments. In *Histochemistry, Theoretical and Applied*, 2nd ed., pp. 632–680. Boston, Little, Brown & Co., 1960.
86. Pease, D. C.: *Histological Techniques for Electron Microscopy*, 2nd ed. New York, Academic Press, 1964.
87. Pellegrino, C., and Franzini, C.: An electron microscope study of denervation atrophy in red and white skeletal muscle fibers. *J. Cell Biol. 17:* 327–349, 1963.
88. Price, H. M.: Ultrastructure of the skeletal muscle fibre. In *Disorders of Voluntary Muscle*, ed. by Walton, J. N., Chapter 3. London, J. & A. Churchill, 1969.
89. Price, H. M., Gordon, G. R., Munsat, T. L., and Pearson, C. M.: Myopathy with atypical mitochondria in type I skeletal muscle fibers. A histochemical and ultrastructural study. *J. Neuropathol. Exp. Neurol. 26:* 475–497, 1967.
90. Price, H. M., Gordon, G. B., Pearson, C. M., Munsat, T. C., and Blumberg, J. M.: New

evidence for excessive accumulation of Z-band material in nemaline myopathy. *Proc. Natl. Acad. Sci. U.S.A. 54:* 1398, 1965.

91. Price, H. M., Howes, E. L., Jr., and Blumberg, J. M.: Ultrastructural alterations in skeletal muscle fibers injured by cold. I. Acute degenerative changes. *Lab. Invest. 13:* 1264–1278, 1964.
92. Price, H. M., Howes, E. L., Jr., and Blumberg, J. M.: Ultrastructural alterations in skeletal muscle fibers injured by cold. II. Cells of the sarcolemmal tube. Observations on "discontinuous" regeneration and myofibril formation. *Lab. Invest. 13:* 1279–1302, 1964.
93. Price, H. M., Pease, D. C., and Pearson, C. M.: Selective actin filament and Z-band degeneration induced by plasmocid. *Lab. Invest. 11:* 549, 1962.
94. Prineas, J. W., Hall, R., Barwick, D. D., and Watson, A. J.: Myopathy associated with pigmentation following adrenalectomy for Cushing's syndrome. *Q. J. Med. 37:* 63, 1968.
95. Resnick, J. S., and Engel, W. K.: Target fibers. Structural and cytochemical characteristics and their relationships to neurogenic muscle disease and fiber types. In *Exploratory Concepts in Muscular Dystrophy and Related Disorders*, p. 255. Amsterdam, Excerpta Medica Foundation, 1967.
96. Rewcástle, N. B., and Humphrey, J. G.: Vacuolar myopathy. Clinical, histochemical, and microscopic study. *Arch. Neurol. 12:* 570–582, 1965.
97. Reznick, M., and Hansen, J. L.: Ultrastructure of mitochondria in degenerating and regenerating skeletal muscle. Annual Meeting of the International Academy of Pathology, San Francisco, March, 1969.
98. Reznick, M.: Origin of myoblasts during skeletal muscle regeneration. Electron microscopic observations. *Lab. Invest. 20:* 353, 1969.
99. Ribas-Mujal, D., and Rivera-Pomar, J. M.: Biological significance of the early structural alterations in skeletal muscle fibers infected by *Trichenella spiralis*. *Virchows Arch.* (*Pathol. Anat.*) *345:* 154, 1968.
100. Ritter, R. A.: Effect of cortisone on the structure and strength of skeletal muscle. *Arch. Neurol. 17:* 403, 1967.
101. Rose, A. L., Walton, J. N., and Pease, G. W.: Polymyositis. An ultramicroscopic study of muscle biopsy material. *J. Neurol. Sci. 5:* 457, 1967.
102. Roth, R. G., Graziani, L. J., Terry, R. D., and Scheinberg, L. C.: Muscle fine structure in the Kugelberg-Welander syndrome (chronic spinal muscular atrophy). *J. Neuropathol. Exp. Neurol. 24:* 444–454, 1965.
103. Rumery, R. E., and Hampton, J. C.: Microscopic and submicroscopic observations in skeletal muscle from vitamin E-deficient rats. *Anat. Rec. 133:* 1, 1959.
104. Santa, T.: Fine structure of the human skeletal muscle in myopathy. *Arch. Neurol. 20:* 479, 1969.
105. Sato, T., and Tsubaki, T.: Electron microscopic studies on the muscles in neuromuscular diseases. *Clin. Neurol. 8:* 3, 1968.
106. Scalzi, H., and Price, H.: The arrangement and sensory innervation of the intrafusal fibers in the feline muscle spindle. *J. Ultrastruct. Res. 36:* 375, 1971.
107. Schmalbruch, H.: Lyse und Regeneration von Fibrillen in der normaler menschlichen Skeletmuskulatur. *Virchows Arch.* (*Pathol. Anat.*) *344:* 159, 1968.
108. Schmalbruch, H. Eine atypische Kernform in der menschlichen Skeletmuskulatur. *Virchows Arch.* (*Pathol. Anat.*) *345:* 200, 1968.
109. Schotland, D. L.: An electron microscopic study of target fibers, target-like fibers, and related abnormalities in human muscle. *J. Neuropathol. Exp. Neurol. 28:* 214, 1969.
110. Schotland, D. L., Spiro, D., Rowland, L. P., and Carmel, P.: Ultrastructural studies of muscle in McArdle's disease. *J. Neuropathol. Exp. Neurol. 24:* 629–644, 1965.
111. Schotland, D. L.: Ultrastructural abnormalities in myotonic dystrophy including an unusual T-system alteration. *J. Neuropathol. Exp. Neurol. 27:* 109, 1968.
112. Schotland, D. L., Spiro, D., and Carmel, P.: Ultrastructural studies of ring fibers in human muscle disease. *J. Neuropathol. Exp. Neurol. 25:* 431–442, 1966.
113. Schrodt, G. R., and Walker, S. M.: Ultrastructure of membranes in denervation atrophy. *Am. J. Pathol. 49:* 33–51, 1966.

114. Schutta, H. S., Kelley, A. M., and Zacks, S. I.: Neurosis and regeneration of muscle in paroxysmal idiopathic myoglobinuria. Electron microscopic observations. *Brain 92:* 191, 1969.
115. Shafiq, S. A., Gorycki, M. A., Asiedu, S. A., and Milhorat, A. T.: Tenotomy effect on the fine structure of the soleus of the rat. *Arch. Neurol. 20:* 625, 1969.
116. Shafiq, S. A., Milhorat, A. T., and Gorycki, M. A.: Crystals in muscle fibers in patients with diabetic amyotrophy and neuropathy. *Neurology (Minneap.) 18:* 785, 1968.
117. Shafiq, S. A., Milhorat, A. T., and Gorycki, M. A.: An electron microscopic study of regeneration and satellite cells in human muscle. *Neurology (Minneap.) 17:* 567–574, 1967.
118. Shafiq, S. A., Milhorat, A. T., and Gorycki, M. A.: An electron microscopic study of muscle degeneration and vascular changes in polymyositis. *J. Pathol. Bacterol. 94:* 139, 1967.
119. Shafiq, S. A., Milhorat, A. T., and Gorycki, M. A.: Giant mitochondria in human muscles with inclusions. *Arch. Neurol. 17:* 666, 1967.
120. Shafiq, S. A., Milhorat, A. T., and Gorycki, M. A.: Fine structure of human muscle in neurogenic atrophy. *Neurology (Minneap.) 17:* 934, 1967.
121. Shafiq, S. A., Dubowitz, V., Peterson, H. de C., and Milhorat, A. T.: Nemaline myopathy. Report of a fatal case, with histochemical and electron microscopic studies. *Brain 90:* 817, 1967.
122. Sher, J. H., Rimalovski, A. B., Athanassiades, T. J., and Aranson, A. M.: Familial centronuclear myopathy; a clinical and pathological study. *Neurology (Minneap.) 17:* 727–742, 1967.
123. Shy, G. M., Gonatas, N. K., and Perez, M.: Two childhood myopathies with abnormal mitochondria. I. Megaconial myopathy. II. Pleconial myopathy. *Brain 89:* 133–158, 1966.
124. Shy, G. M., Wanko, T., Rowley, P. T., and Engel, A. G.: Studies in familial periodic paralysis. *Exp. Neurol. 3:* 53, 1961.
125. Sluga, E., Aertelberger, F., and Mosen, K.: Über eine progressive Myopathie mit Muskelphosphorylosemangel und Riesenmitochondrien. *Wien. Klin. Wochenschr. 79:* 917, 1967.
126. Smith, J., Zellweger, H., and Afifi, A. K.: Muscular form of glycogenosis, type II (Pompe). *Neurology (Minneap.) 17:* 537–549, 1967.
127. Spiro, A. J., Shy, G. M., and Gonatas, N. K.: Myotubular myopathy. Persistence of fetal muscle in an adolescent boy. *Arch. Neurol. 14:* 1–14, 1966.
128. Stenger, R. J., Spiro, D., Scully, R. E., and Shannon, J. M.: Ultrastructural and physiologic alterations in ischemic skeletal muscle. *Am. J. Pathol. 40:* 1, 1962.
129. Stoeckenius, W., and Engelman, D. M.: Current models for the structure of biological membranes. *J. Cell Biol. 42:* 613, 1969.
130. Straus, W.: Lysosomes, phagosomes, and related particles. In *Enzyme Cytology*, ed. by Roodyn, D. E., p. 239. New York, Academic Press, 1967.
131. Strehler, B. L., and Mildvan, A. S.: Studies on the chemical properties of lipofuscin age pigment. In *Biological Aspects of Aging*, ed. by Shock, N. W., pp. 174–181. New York, Columbia University Press, 1962.
132. Strunk, A. W., Smith, C. W., and Blumberg, J. M.: Ultrastructural studies on the lesion produced in skeletal muscle fibers by crude type A *Clostridium perfringens* toxin and its purified alpha fraction. *Am. J. Pathol. 50:* 89, 1967.
133. Tetsuji, S.: Fine structure of the human skeletal muscle in myopathy. *Arch. Neurol. 20:* 479, 1969.
134. Threadgold, L. T.: *The Ultrastructure of the Animal Cell.* Oxford, Pergamon Press, 1967.
135. Tice, L. W., and Engel, A. G.: The effects of glucocorticoids on red and white muscles in the rat. *Am. J. Pathol. 50:* 311–333, 1967.
136. Trump, B. F., and Ericsson, J. L. E.: Some ultrastructural and biochemical consequences of cell injury. In *The Inflammatory Process*, ed. by Zweifach, B. W., Grant, L., and McCluskey, R. T., Chapter 2, New York, Academic Press, 1965.
137. van Wijngaarden, G. K., Bethlem, J., Meijer, T. H., Hiilsmann, W., and Feltkamp, C. A.: Skeletal muscle disease with abnormal mitochondria. *Brain 90:* 577, 1967.
138. Vye, M. W.: A myopathy with intramitochondrial inclusions. *Amer. J. Clin. Pathol 55:* 130–131, 1971.

139. Ward, R. T.: The origin of protein and fatty yolk in *Rana pipiens*. II. Electron microscopical and cytochemical observations of young and mature oocytes. *J. Cell Biol. 14:* 309, 1962.

140. Wechsler, W.: Comparative electron microscopic studies on various forms of muscle atrophies and dystrophies in animals and man. *Ann. N. Y. Acad. Sci. 138:* 113–137, 1966.

141. Wechsler, W., and Hager, H.: Elektronenmikroskopische Untersuchungen bei myotonischen Muskeldystrophie. *Arch. Psychiatr. Nervenkr. 201:* 668, 1961.

142. Weiss, P.: From cell to molecule. In *The Molecular Control of Cellular Activity*, ed. by Allen, J. M. New York, McGraw-Hill Book Co., 1962.

143. Zellweger, H., Brown, B. I., McCormick, W. F., and Tu, J. B.: A mild form of muscular glycogenosis in two brothers with alpha-1,4-glucosidase deficiency. *Ann. Pediatr. (Paris) 205:* 413–437, 1965.

144. Zintz, V. R., and Villiger, W.: Elektronenmikroskopische Befunde bei 3 Fällen von chronisch progressive okulärer Muskeldystrophie. *Ophthalmologica 153:* 439, 1967.

Chapter 10

# Current Concepts of Skeletal Muscle Regeneration

M. REZNIK

For the pathologist examining muscle tissue from spontaneous and chronic neuromuscular disorders, the diagnosis of the disease depends on complex histologic, histochemical, or ultrastructural alterations that can be variously interpreted. In most of these cases, however, the basic morphologic abnormality of the muscle tissue is constituted from a picture of coexisting degenerating and regenerating myofibers.[1, 7, 32, 33, 45, 56] The process of muscle regeneration, or its failure, has become a more important problem since its role increasingly appears as an intriguing factor in the evolution of many skeletal muscle disorders.

Theoretically, the future of any altered striated muscle is likely to be related to: (1) the origin and extent of tissue necrosis or myofiber atrophy, (2) the nature and intensity of the inflammatory reaction, (3) the rapidity and effectiveness of myofiber regeneration, and (4) the innervation of the regenerated myofibers and the nature of the physiologic activities in which they become involved. Despite intensive efforts to study pathologic reactions of human skeletal muscle,[4, 32, 33, 62, 84, 96] this material does not appear suitable to elucidate the repair and regeneration of damaged skeletal muscle fibers because the muscle fragment is often obtained by a single biopsy in the course of a chronic disease. Therefore, it seems appropriate to refer to experimental research that provides possible evaluation of degeneration and regeneration of skeletal muscle fibers. It is hoped that a more complete knowledge of induced pathologic conditions in animals will eventually provide a basis for clarifying some aspects of human neuromuscular disorders.

In this work, most of the findings have been obtained from cold-injured gastrocnemius muscle of normal adult mice, acute ischemic tibialis anterior muscle of normal adult rabbits, or partially cross-sectioned muscles of the two species. After partial cross-section, the area of muscle necrosis is limited to a small margin around the wound; there is, however, a gap in the bulk of the muscle created by the retraction of the interrupted myofibers. After local cold injury or acute ischemia, the bulk of the damaged muscle seems intact, but three areas are visible inside the muscle (Fig. 1): (1) a necrotic zone, (2) an undamaged portion, and (3) an intermediate segment which includes the margin of both the distal part of the undamaged myofibers and the proximal zone of the degenerating fibers. This intermediate region is the most interesting because it is in this field that the process of regeneration is the most prominent.

Whatever the method of damaging the skeletal muscle (physical or chemical injury, ischemia, infection), regenerative phenomena begin while degenerative changes are still in progress and before phagocytosis of the destroyed parts is

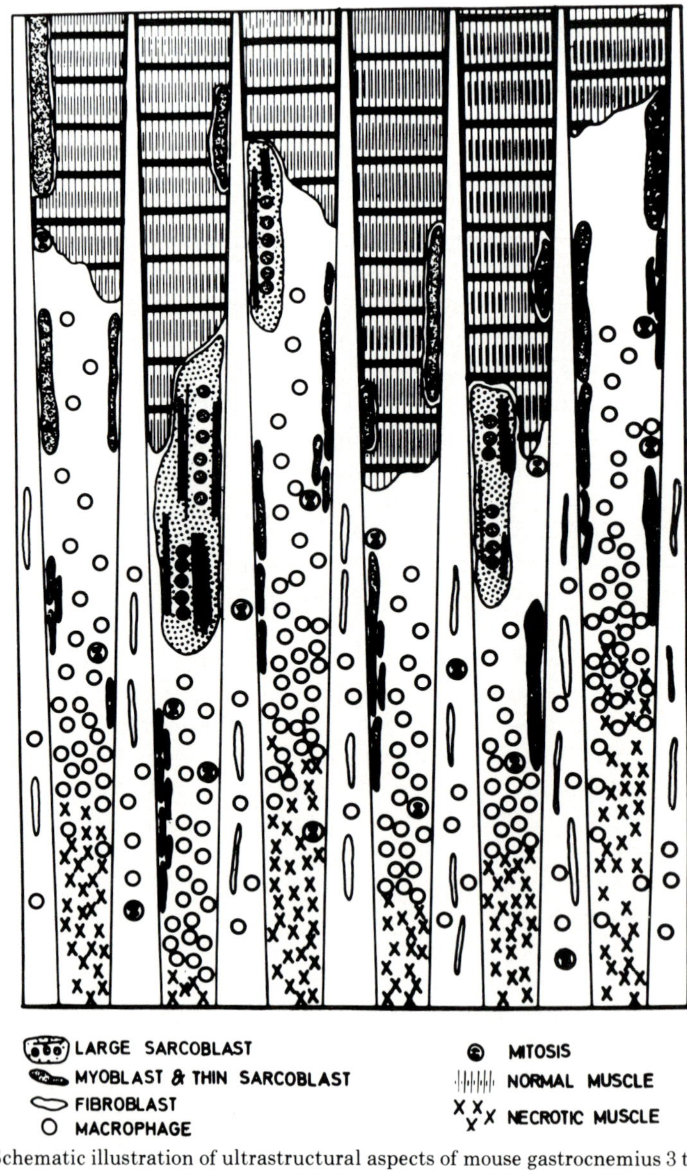


FIG. 1. Schematic illustration of ultrastructural aspects of mouse gastrocnemius 3 to 5 days after cold injury. Three areas are visible (from *top* to *bottom*): (1) an undamaged portion, (2) an intermediate segment where muscle regeneration is prominent; (3) a necrotic region (see text for details).

completed. Cellular heterogeneity is an obvious histologic observation in injured muscle, as well as in most of the spontaneous muscular disorders.

With the intent of providing a distinct picture of the different pathologic modifications that occur between the normal stage and the stage of the regenerated skeletal muscle, the four major points previously mentioned (necrosis, inflammation, regeneration, and innervation) will be discussed separately.

## NECROSIS AND DEGENERATIVE CHANGES

Necrosis of muscle fibers has been induced by a large variety of physical agents, drugs, experimental infections, or diet disturbances (see References 1 and 63 for review). Depending upon which noxious factor is used, one or several muscles can be damaged. In any muscle, lesions can be scattered among different myofibers, or localized on a group of adjacent myofibers.

*Histologic alterations of muscle fibers* which result from exposure to any of the damaging physical agents, chemical products, infections, or immunologic reaction vary from minimal swelling of the sarcoplasm to total necrosis of the muscle fiber. Myonuclei rapidly become pyknotic and loose their staining quality. The necrotic sarcoplasm commonly appears faintly eosinophilic and floccular.

Most authors believe that a series of pathologic changes can be observed in muscle undergoing necrosis regardless of the cause.[1, 94] It is believed that the natural tension within the muscle causes part of the fragmentation of the necrotic myofiber. The fiber usually ruptures at the transition between necrotic and unaffected portions of the muscle, with retraction of the content of the sound part of the fiber away from the necrotic portion. This reaction is responsible for the so-called "discoid" degeneration, which results in empty stretches of sarcolemmal tubes. It has been shown that some motor end-plates are included in the acute degenerative changes.[11]

Many muscle fibers immediately bordering the necrotic mass rapidly lose their bulk. In some of these fibers, the surviving striated substances stain poorly. Myonuclei are often intact, even enlarged, and may migrate into the center of the fiber.

*Electron microscopic observations* of skeletal muscle altered in different experimental conditions[63, 66, 70] provide new insight into the structural changes in damaged myofibers.

1. In the intermediate region of the muscle that suffers mild injury, different ultrastructural alterations can be grouped into three categories: (1) formation of abnormal membranous structures, (2) modification of mitochondria, and (3) change of the Z band.

By comparison with normal myofibers (Fig. 2), abnormal membranous structures consist of irregular loops and coils which are found in some myofibers. The intensification of this process leads to the formation of whorls of densely packed membranes; they are present both among the myofibrils and under the plasmalemma, often near a myonucleus (Fig. 3). The origin of these bodies from the sarcotubular system and from mitochondria is possible. Some of these

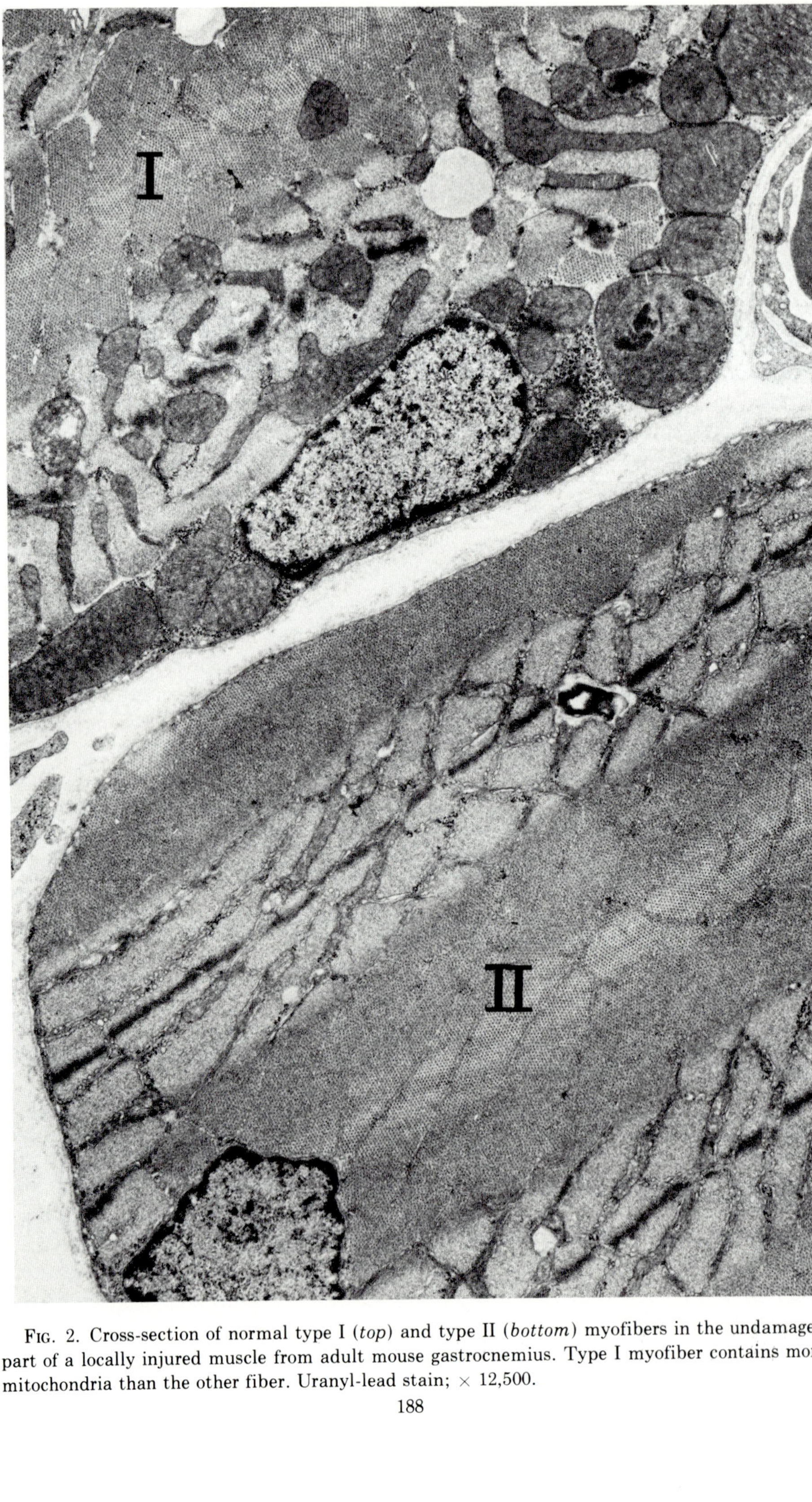


FIG. 2. Cross-section of normal type I (*top*) and type II (*bottom*) myofibers in the undamaged part of a locally injured muscle from adult mouse gastrocnemius. Type I myofiber contains more mitochondria than the other fiber. Uranyl-lead stain; × 12,500.

elements can be an expression of autophagy and development of lysosomes as seen in other tissues.[20] Similar spheromembranous bodies are found in skeletal muscle of animals suffering vitamin E deficiency[41] or those treated with vincristine.[6]

Less frequently, disruption and rearrangement of sarcoplasmic tubules are visible; these modifications result in the development of honeycomb-like structures (Fig. 4) that can occupy a large portion of the intermyofibrillar space.

Mitochondria are often enlarged and more densely packed in myofibers that are in the vicinity of tissue necrosis.[78] Mitochondria in these muscle fibers can be increased three to five times in size and are found between myofibrils or beneath the plasmalemma, where they are grouped in large clusters that can dislocate the myofibrillar networks. The mitochondria have many densely packed cristae displaying various patterns (Figs. 5 and 6). By analogy with changes in other regenerating tissues observed by electron microscopy,[55] it is suggested that the increase in size and number of mitochondria in skeletal muscles of rabbits or mice is correlated with a process of cellular repair after injury. Metabolic disturbances are also a possible factor since such changes in mitochondria are observed in steroid-induced myopathy[19] and after thyroidectomy or treatment with thyroxine.[36]

Some Z lines appear irregularly enlarged and thickened in partially damaged myofibers after cold injury (Fig. 6). Such a modification of the Z lines is also induced after tenotomy,[85] although it can be observed in clinically healthy animals.[61]

2. In myofibers suffering acute degenerative alteration, the first morphologic modifications, at about 6 hours after injury, are mitochondrial swelling, dilatation of the sarcotubular system, and disappearance of cytoplasmic granules.[70] In the early stages of degeneration, the mitochondria are slightly enlarged and they possess many cristae. These mitochondria contain different types of inclusions, such as dense osmophilic bodies, whorled myelin-like membranes, and rodlike structures (Fig. 7). In more advanced stages of degeneration, mitochondria are rounded and fragmented, and they tend to disappear from the necrotic myofibers (Fig. 8).

The nuclei become pyknotic (Fig. 8) with marked clumping of chromatin at the periphery and disruption of their outer membrane. Simultaneously, most of the sarcoplasmic tubules are fragmented. Plasma membranes also disappear, but the basement membrane often persists to constitute the wall of the so-called sarcolemmal tube (Fig. 9). It should be remembered that the term "sarcolemma" was originally defined as an elastic sheath of the muscle fiber that could be demonstrated in teased preparations examined with the light microscope.[18] Observed with the electron microscope, this sheath is normally composed of: (1) the plasma membrane of the muscle cell, or plasmalemma, not visibly different from the limiting membrane of other living cells; (2) an extracellular dense membrane, or basement membrane, of approximately 500-Å thickness, and separated from the plasmalemma by a continuous gap of 500 to 1,500 Å. In pathologic myofibers, when the muscle cell and its plasma membrane are destroyed, the sarcolemmal sheath appears to be comprised of the persistent basement membrane and an outer associated mesh of fine collagen fibrils.

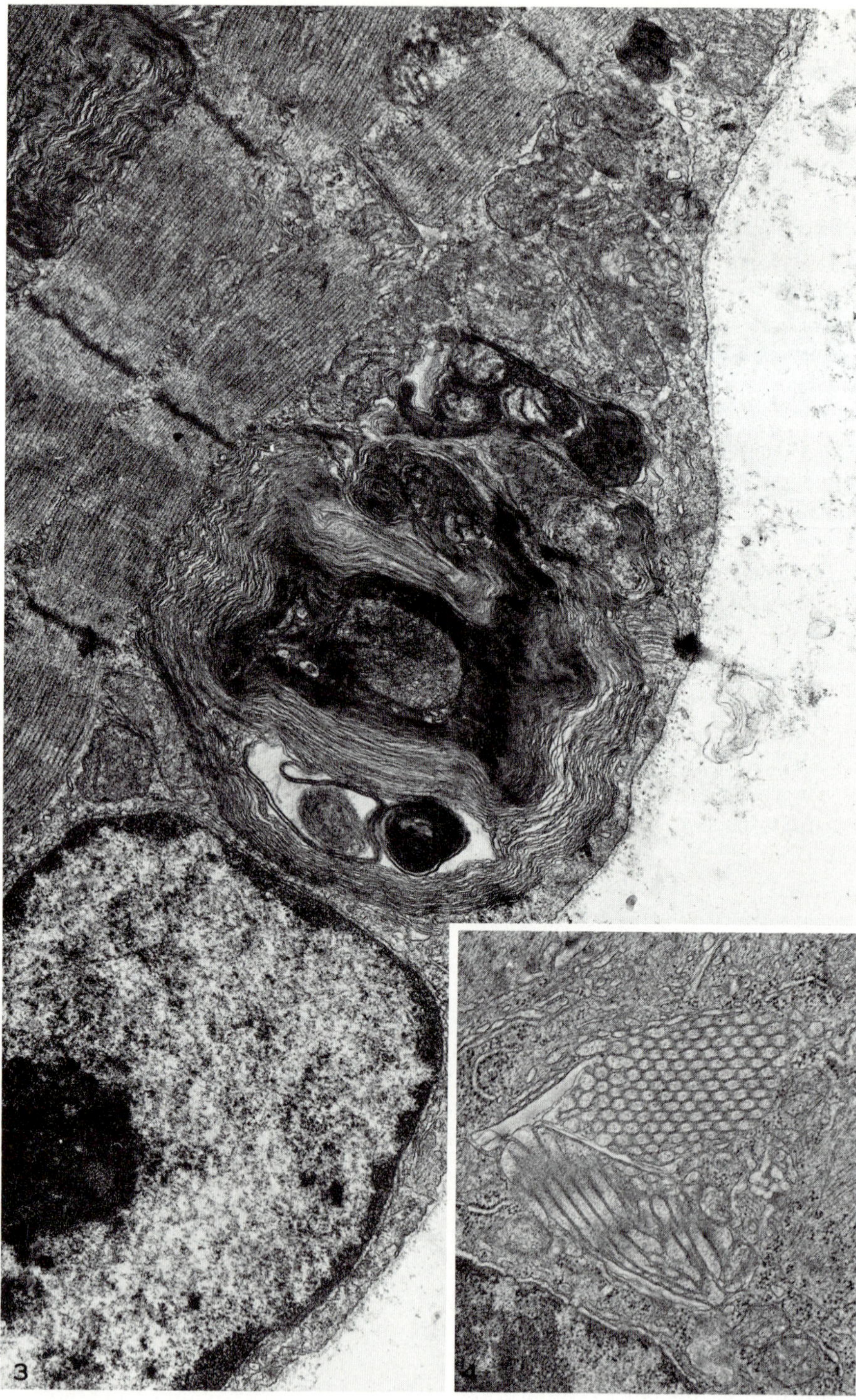

FIG. 3. Myofiber near a site of cold injury (mouse gastrocnemius). Heterogeneous bodies composed of whirled membranes, dense material and granular structures near a normal myonucleus and also between myofibrils. Uranyl-lead stain; × 11,800.

FIG. 4. Honeycomb-like structure in a partially altered myofiber near a site of cold injury (mouse gastrocnemius). Uranyl-lead stain; × 18,000.

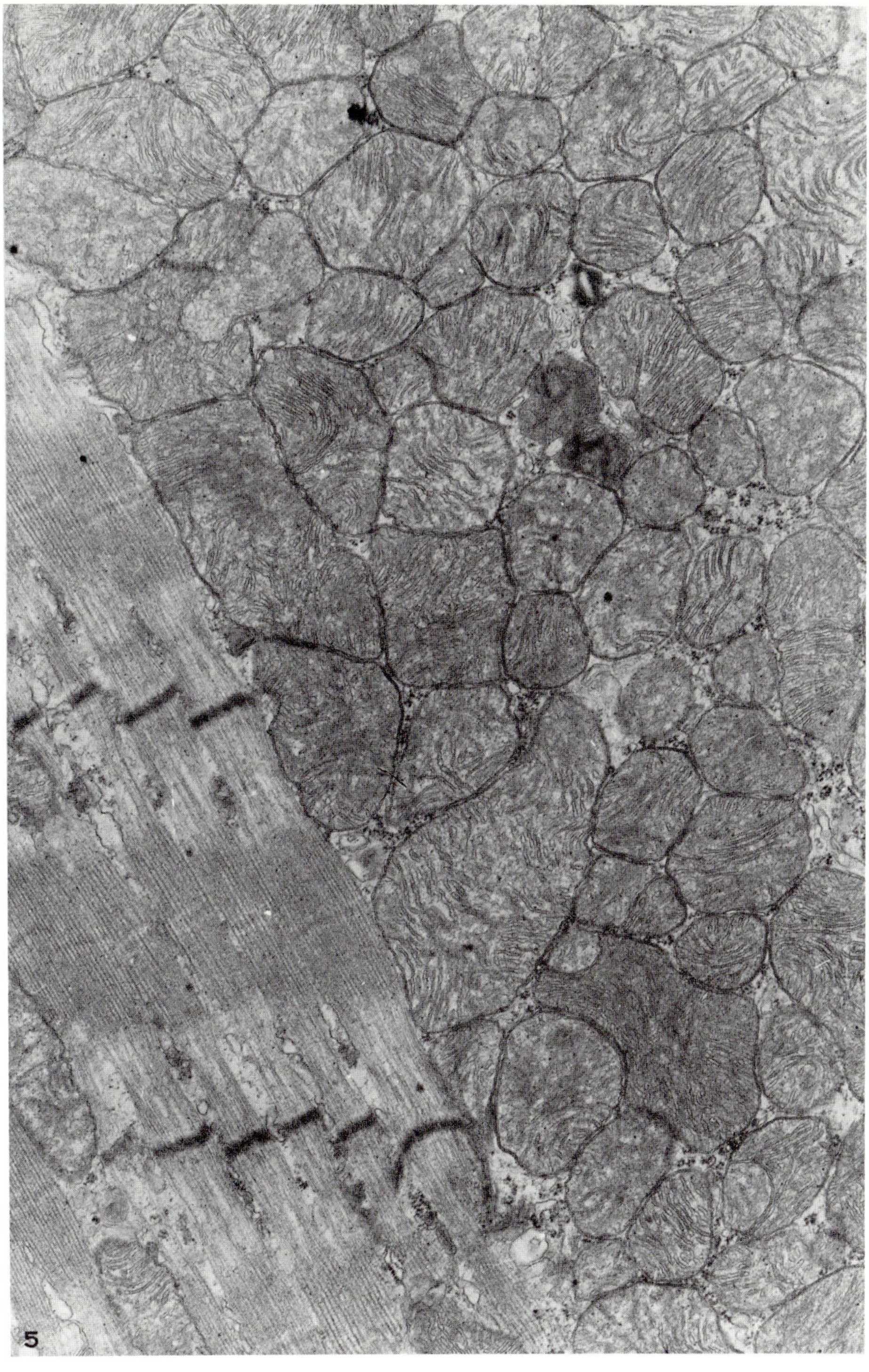


FIG. 5. Subsarcolemmal accumulation of enlarged mitochondria in a myofiber near an area of necrosis 2 days after cold injury. Compare with normal myofibers in Figure 2 (mouse gastrocnemius). Uranyl-lead stain; × 16,000.

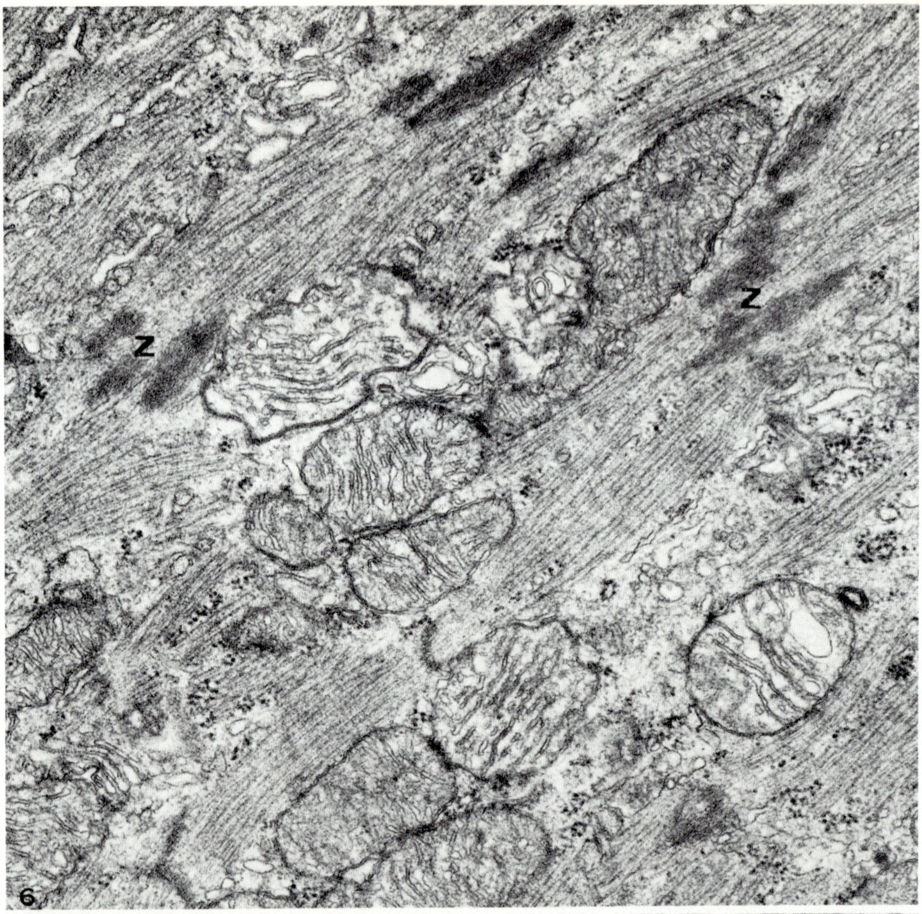


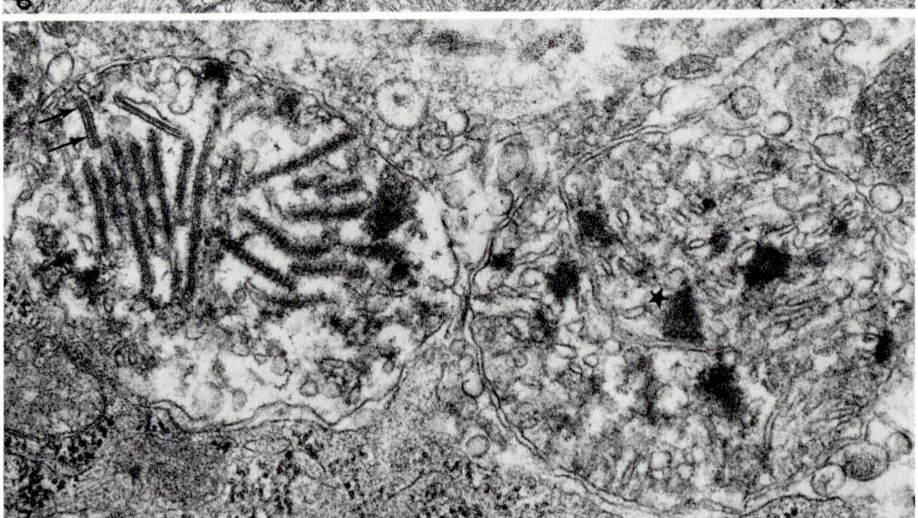


FIG. 6. Partially altered myofiber 4 days after cold injury (mouse gastrocnemius). Mitochondria are swollen and irregular in shape. Myofibrils are disarrayed and Z lines (*Z*) appear abnormal. The sarcotubular system seems irregular. Uranyl-lead stain; × 26,000.

FIG. 7. Degenerating myofiber in cold-injured mouse gastrocnemius. Example of abnormal mitochondria with irregular dense inclusions (★) or rodlike elements (*arrows*). Uranyl-lead stain; × 78,000.

Fig. 8. Necrotic myofiber in rabbit tibialis anterior muscle 2 days after acute ischemia. Pyknotic nucleus with clumped chromatin. Disappearance of mitochondria and sarcotubular structures. Alterations of Z bands but partial preservation of the myofibrillar network. Uranyl-lead stain; × 21,000.

The main and earliest feature of myofibrillar alteration is loss of the Z lines (Fig. 8), while the whole A band is unaltered.[70] Later, there is a fragmentation of the I bands, often in their middle portion, and a subsequent disruption of the myofibrillar network (Fig. 10).

*Histochemical reactions*, which are usually used for fiber typing and are considered as stains of some sarcoplasmic organelles,[22, 23, 80] provide a good correlation with most of the ultrastructural alterations described above.[76]

After the myofibrillar ATPase reaction, many necrotic myofibers are still uniformly stained at nearly the same intensity as the undamaged or partially damaged myofibers. The persistence of the latter reaction is very likely linked with the relative preservation of A bands in the damaged fibers.[76] Necrotic myofibers are not stained after the phosphorylase reaction or after histochemical reaction for four tested oxidative enzymes (lactate, succinate, malate, and reduced disphosphopyridine nucleotide dehydrogenase) which are likely linked with mitochondria. In a few necrotic myofibers, however, it is possible to find occasional small dots that are scattered in the remnant sarcoplasm.[89]

On the other hand, many preexisting myofibers immediately bordering the area of tissue necrosis present a strong intensification of their staining properties after reactions for the tested oxidative enzymes (lactate, succinate, malate, and DPNH). Such a modification of the latter histochemical reactions is likely

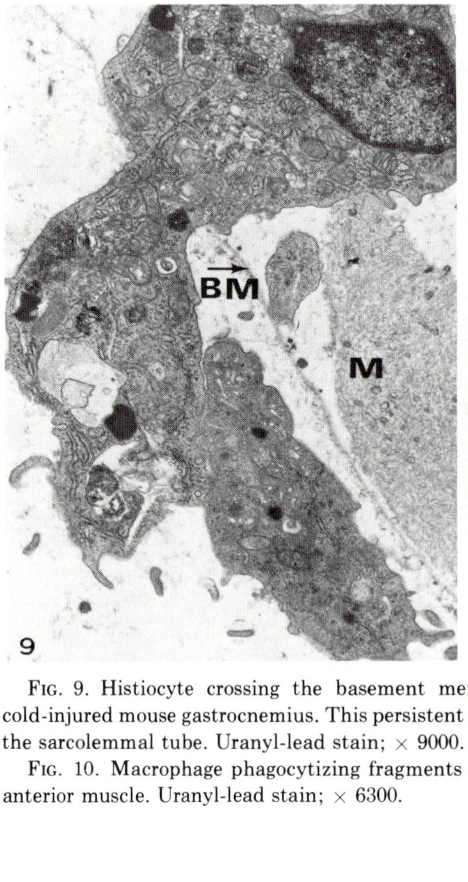


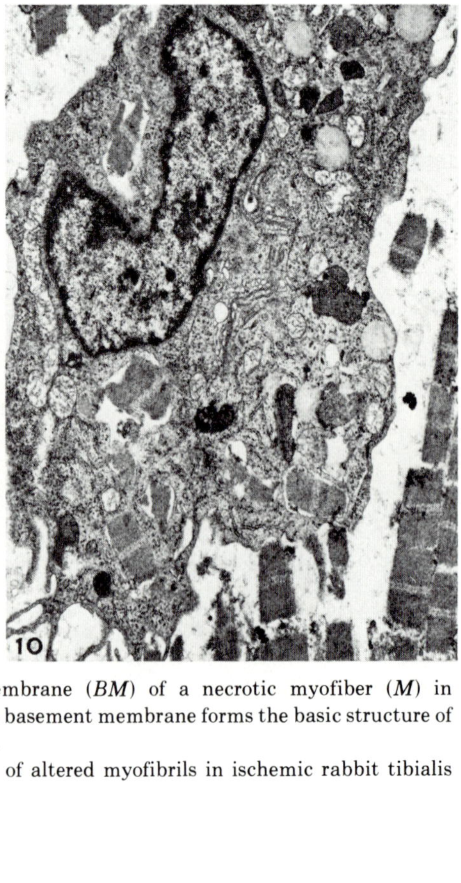


FIG. 9. Histiocyte crossing the basement membrane (*BM*) of a necrotic myofiber (*M*) in cold-injured mouse gastrocnemius. This persistent basement membrane forms the basic structure of the sarcolemmal tube. Uranyl-lead stain; × 9000.

FIG. 10. Macrophage phagocytizing fragments of altered myofibrils in ischemic rabbit tibialis anterior muscle. Uranyl-lead stain; × 6300.

related to the presence of collections of enlarged mitochondria in these myofibers.

## INFLAMMATORY REACTION

Whatever the noxious factor, experimental muscle necrosis is invariably followed by an inflammatory reaction. In the acute stage, the inflammatory reaction starts a few hours after the injury, and 12 to 24 hours later it is marked by the presence of a considerable number of various mononucleated cells. It is chiefly in the intermediate region between the center of the necrotic tissue and the area of partially damaged myofibers that the proliferation of mononucleated cells is the most prominent. These cells increase in number rapidly during the following 3 to 4 days, after which they are progressively less numerous. Simultaneously, the inflammatory cells move from the periphery toward the interior of the necrotic tissue.

*Light microscopic observations*, with the usual histologic techniques, do not allow differentiation of all types of mononucleated cells which are present in the injured muscle. However, three major categories can be recognized.[1, 10, 25, 34] (1) Leukocytes appear early around the vessels and in the interstitial spaces. In aseptic surgical procedures, polymorphonuclear leukocytes are not numerous and disappear very soon from the damaged muscle. Lymphocytes are usually limited to the perivascular spaces. (2) Rounded cells, with an eosinophilic cytoplasm containing various cellular debris, are recognized as macrophages. Some of these cells can appear a few hours after the injury, but they are more abundant 2 to 4 days later. Although these cells can be found anywhere in the injured muscle, a great many macrophages are always packed in the sarcolemmal sheath of necrotic myofibers, mainly at the periphery of the damaged tissue. They phagocytize the necrotic muscle material, degenerated blood cells, and myelin fragments, all of which are later removed from the area. During such a process, the edge of the necrotic area is removed first and the center later. In large infarcts, as after acute ischemia in rabbits, necrotic myofibers which are not phagocytized after 2 to 3 weeks can be encapsulated by fibrous tissue and persist in the muscle. (3) Fusiform basophilic cells appear 1 or 2 days after injury. At the beginning, most of these cells are observed in necrotic myofibers, lying beneath the surface of the sarcolemmal sheath. Many of these cells seem to be surviving fragments of the degenerated muscle fibers. They exhibit mitotic division and become rapidly more abundant until the appearance of regenerating myofibers (see below). At that time, 4 to 10 days after injury in the material of this study, these mononucleated cells become less numerous and disappear from the sarcolemmal tube.

When the necrotic process is limited mainly to the sarcoplasmic structures, such as after plasmocid intoxication,[1] Coxsackie virus infection,[35] or nutritional deficiency,[45] mononucleated cells disappear while regeneration of myofibers is in progress. If the injury disrupts the endomysial structure of the muscle, such as after cross-section, crushing, or a graft of minced muscle fragments,[16] the proliferation of fusiform mononucleated cells is more prominent. In the latter condition, formation of new myofibers is scanty and irregular; on the contrary, fibrous tissue becomes more abundant and forms irregular large scars.

Contrary to observations in lower vertebrates in which an almost normal limb can regenerate from the blastema (see Hay[38] and Lentz[51] for review), it seems that in higher vertebrates, the speed of regenerative process of muscle competes with the rate of connective scar development.[16, 62] Unfortunately, in adult mammals, it seems that the proliferation of connective tissue often prevails over the development of other tissues.

*Electron microscopic observations* confirm most of the previous description but help to differentiate the nature of the cells which appear in the damaged muscle. This aspect will be discussed in the following chapter.

## MUSCLE REGENERATION

As previously mentioned, the process of skeletal muscle regeneration in normal adult animals occurs: (1) after a stage during which mononucleated cells are particularly abundant, (2) before phagocytosis of the necrotic material is completed, and (3) simultaneously with the development of connective tissue.

Cellular heterogeneity resulting from these simultaneous modifications remains one of the major difficulties for direct recognition of the origin of the new myofibers. Various experimental studies, however, have produced specific data upon which new concepts of skeletal muscle regeneration can be proposed.

*Histologic aspects* of skeletal muscle regeneration have been described for many decades and extensively reviewed more recently.[1, 4, 10, 16, 25, 34, 46, 53, 91]

Light microscopic observations with conventional techniques support the concepts that the process of muscle repair can occur in two different ways (Fig. 11): (1) "continuous" and (2) "discontinuous" regeneration.

In the so-called continuous form, regeneration is supposed to take place by budding from the extremities of preexisting myofibers, at the site of their separation from the necrotic sarcoplasma. Three to 4 days after cross-section or crushing, for example, the distal extremities of the interrupted myofibers present various cytologic changes. Surviving myonuclei become enlarged and vesicular, with one or more prominent nucleoli. They migrate to the center of the fibers. The striation of the sarcoplasma becomes less apparent, or absent, and the cytoplasm surrounding the nuclei is intensively basophilic. By the 5th to 10th days, ribbons of nonstriated sarcoplasm lie between the necrotic segments. In the larger ribbons, when basophilia is reduced, some striated myofibrils are observed.

If the distance between the two sides of the wound is short, sprouts from each border become interdigitated and apparently restore continuity between the ends of interrupted muscle fibers. If an obstruction is met, such as a wall of fibrous tissue, the sprouts seem to be deflected or stopped by it.

In the so-called discontinuous form, regeneration is supposed to proceed from mononucleated cells (myoblasts). The source of myoblasts during regeneration of adult skeletal muscle is a controversial subject. Theoretically, they could originate in various ways (Fig. 11), which are not necessarily exclusive. Whatever their mode of formation, myoblasts are probably a fraction of the fusiform basophilic cells found along the inner surface of the basement

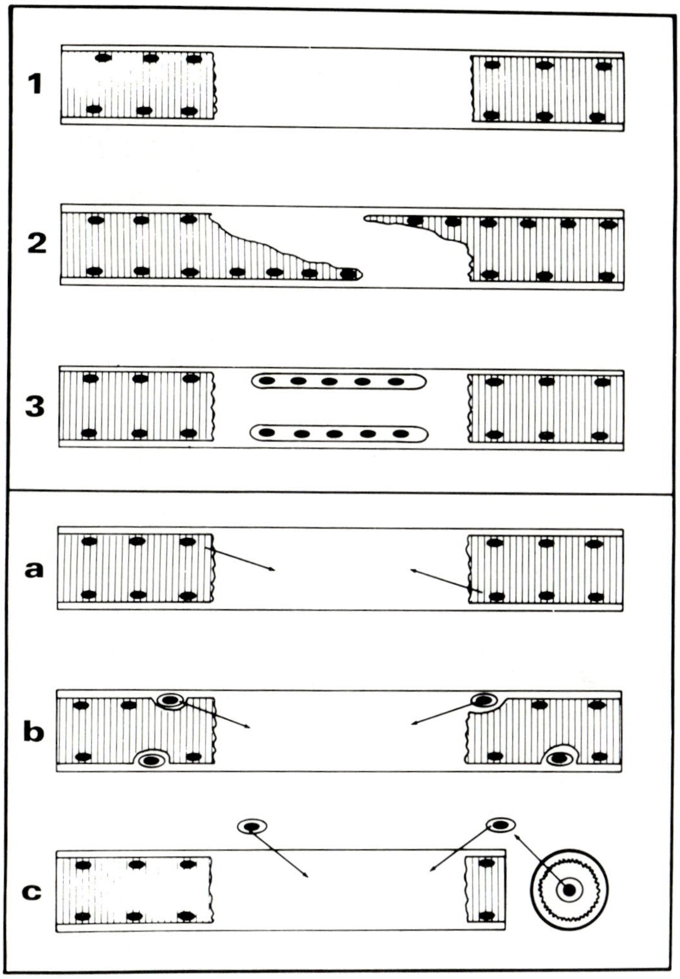


FIG. 11. Theoretical modifications in locally injured skeletal myofiber (from *top* to *bottom*). *1*, no regeneration; *2*, continuous regeneration by budding from the extremities of the preexisting myofiber; *3*, discontinuous regeneration by proliferation of isolated cells. These cells can originate in various ways: *a*, by segregation from the damaged myofiber; *b*, by activation of satellite cells preexisting in the sarcolemmal sheath; *c*, by metaplasia from mesenchymal cells or circulating cells.

membrane of degenerated muscle fibers (Figs. 12 and 13). Intense mitotic activity is present in these cells. Three to 4 days after cold injury or acute ischemia, isolated, thin, long basophilic ribbons with numerous large nuclei in single file become prominent (Fig. 14). This multinucleated basophilic cytoplasm, which eventually transforms into an adult striated muscle fiber, can take several forms during its development and has been variously named "myotube," "early muscle fiber," or "sarcoblast." To avoid confusion, it seems appropriate to reserve the name "myotube" for the early muscle fiber formed during embryonic myogenesis, and to use the term "sarcoblast" for the early stage of the regenerating myofiber in adult animals.

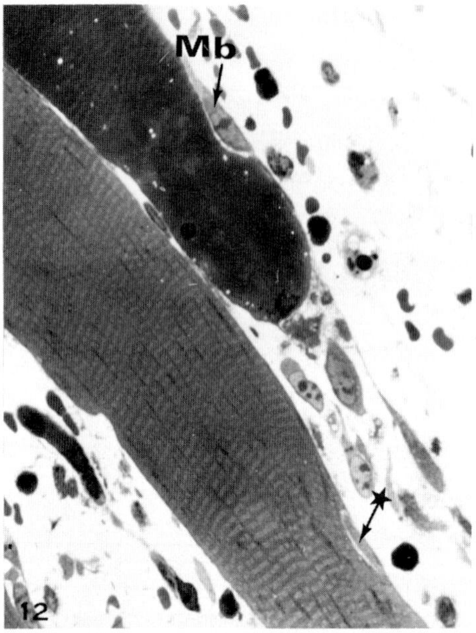


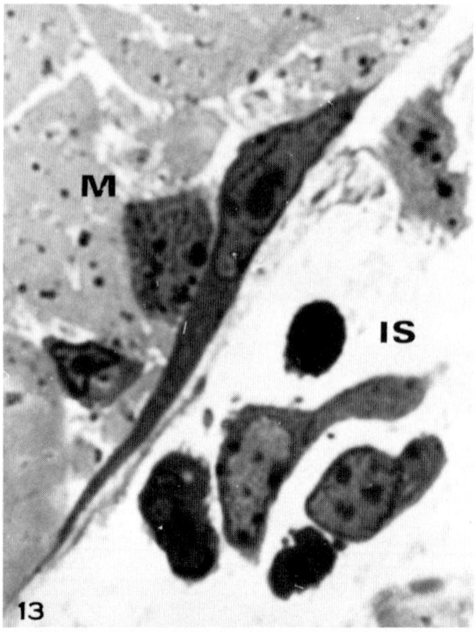


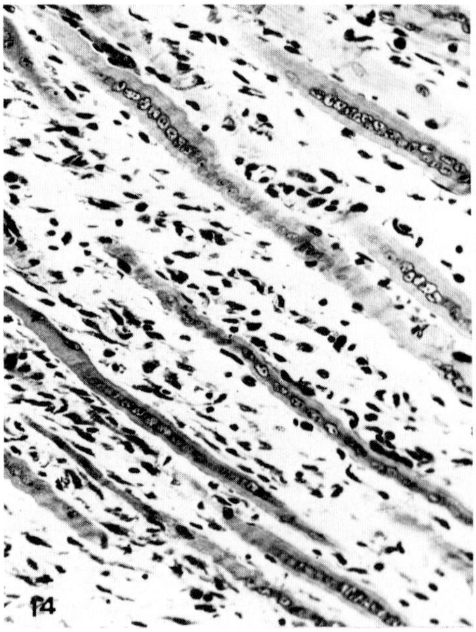


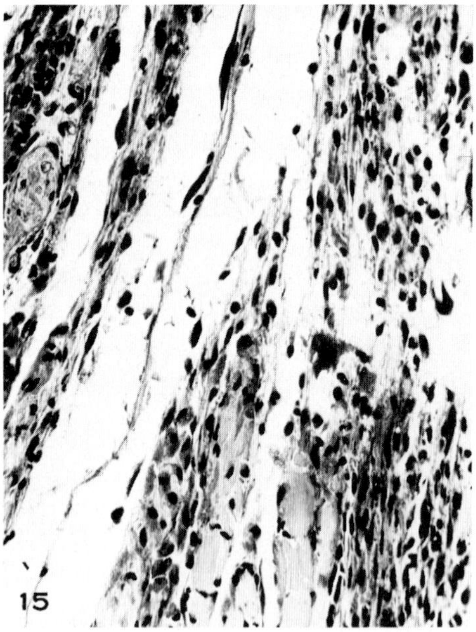


FIG. 12. Satellite cell or presumptive myoblast (*Mb*) beneath the basement membrane of a partially altered myofiber. A presumptive myoblast (★) is also visible in the normal fiber 68 hours after cold injury (mouse gastrocnemius). Toluidine blue stain; × 600.

FIG. 13. Same condition as in Figure 12. A fusiform satellite-like cell or presumptive myoblast is seen beneath the basement membrane of a necrotic myofiber (*M*). Histiocytes and fibroblasts are seen in the interstitial space (*IS*). Toluidine blue stain; × 1240.

FIG. 14. Longitudinal section of mouse gastrocnemius 5 days after cold injury. Numerous multinucleated sarcoblasts are growing in parallel rows. Hematoxylin and eosin stain; × 220.

FIG. 15. Contralateral gastrocnemius of the same animal as in Figure 14, injured simultaneously. This muscle was locally irradiated (1500 R treatment) 2 days after cold injury and removed at the same time as the nonirradiated muscle shown in Figure 14. No sarcoblasts are visible, although inflammatory cells are numerous. Hematoxylin and eosin stain; × 220.

MYOBLASTS

Myoblasts cannot be readily recognized in sections by usual histologic techniques, although they are more easily identified when they are dissociated and growing *in vitro*.[50]

Electron microscopic investigations do not provide unquestionable criteria for distinguishing early myoblasts from all other mesenchymal cells. However, ultrastructural characteristics of myoblasts emerge from study during embryonic myogenesis,[5, 21, 29, 39, 68] development of muscle *in vitro*,[24, 28, 44, 86] or skeletal muscle regeneration *in vivo*.[2, 17, 38, 51, 67, 73, 74, 83, 84]

Two to 3 days after injury, cells recognized as myoblasts are fusiform mononucleated cells often lying along the inner side of the basement membrane (Fig. 16). The single nucleus is large and in a central position; it presents dispersed chromatin, with one or two prominent nucleoli. The cytoplasm is moderately abundant, with very few or no pseudopodic extensions; it is packed with ribosomes in various arrangements, and contains a poorly developed endoplasmic reticulum with rough membranes. A small Golgi apparatus and centrosome are sometimes discovered near the nucleus. Mitochondria are often small and scanty. Few lipid droplets are seen at one pole of the cell. Wavy thin filaments are randomly distributed in the cytoplasm of most of the myoblasts.

These ultramicroscopic features differentiate myoblasts from other mononucleated cells found in injured skeletal muscle. Macrophages are found inside and outside the sarcolemmal sheath of necrotic myofibers (Figs. 9 and 10); they contain a great many mitochondria, various amounts of endoplasmic reticulum, well developed Golgi apparatus, and large quantities of vesicles with necrotic fragments. These cells are very irregular in shape and possess a great number of pseudopodic extensions, often encompassing cellular debris (Fig. 10). Fibroblasts are found only outside the basement membrane of myofibers; they contain few mitochondria and a large endoplasmic reticulum bordered by particles of ribosomes. They also exhibit many cytoplasmic extensions that can extend far from the main cellular body. These ultrastructural aspects are schematically illustrated in Figure 17.

It is clear that, during embryonic myogenesis, all myoblasts come from mononucleated cells derived from early mesenchyme. During regeneration of mature muscle, the origin of myoblasts is less evident. The major difficulty lies in the uncertainty of identifying myoblasts at an early stage of their existence and tracing their evolution in fixed material. However, there is now strong evidence that myoblasts arise locally in injured muscle fibers. This assertion is based on: (1) the effects of local x-ray irradiation,[69, 75] (2) autoradiographic study after injection of tritiated thymidine,[71, 72, 95] and (3) electron microscopic observations.[73, 74]

1. Local x-ray irradiation (1500 R treatment) given 1 or 2 days after injury completely inhibits the muscle regeneration that should have occurred in the following days. There is an inflammatory reaction however, but very few, if any, basophilic mononucleated cells are discovered in the damaged muscle. Sarcoblasts are not apparent 5 days (Fig. 15) or even 2 weeks after injury. Similarly, if the muscle is irradiated before injury, the regeneration is also inhibited. An

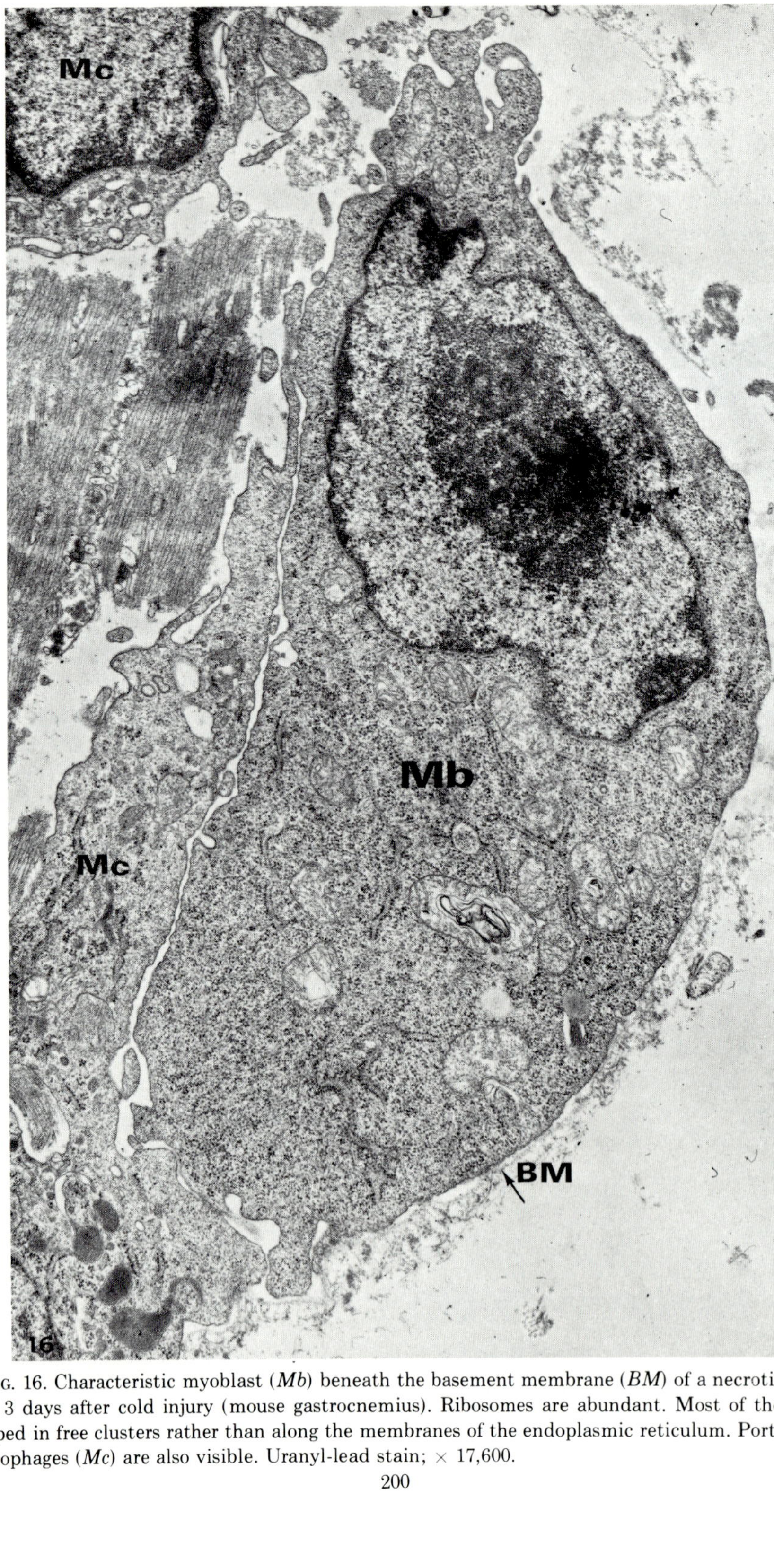


FIG. 16. Characteristic myoblast (*Mb*) beneath the basement membrane (*BM*) of a necrotic myofiber 3 days after cold injury (mouse gastrocnemius). Ribosomes are abundant. Most of them are grouped in free clusters rather than along the membranes of the endoplasmic reticulum. Portions of macrophages (*Mc*) are also visible. Uranyl-lead stain; × 17,600.

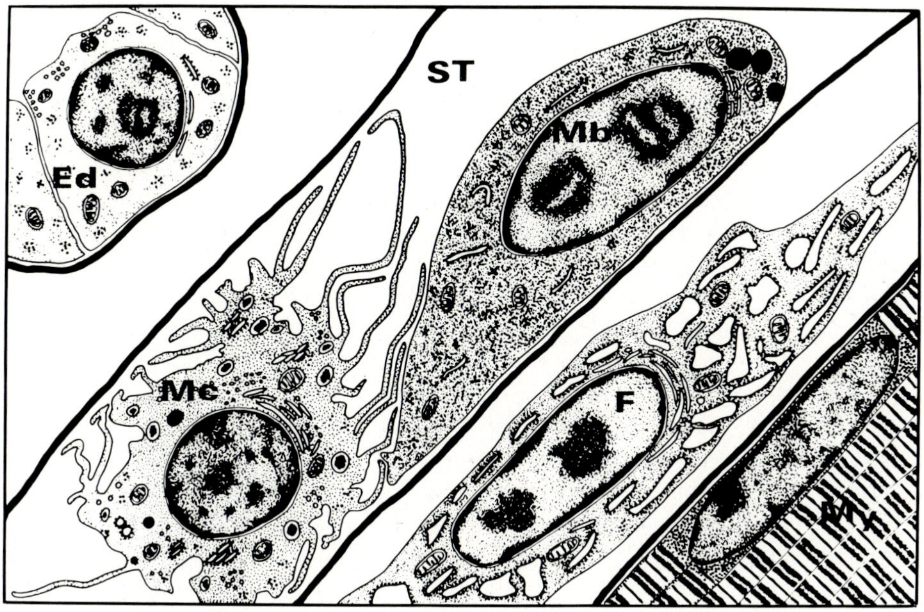


FIG. 17. Schematic ultrastructure of principal cells in mammal regenerating skeletal muscle. A fibroblast (*F*) is in the interstitial space, near a normal myofiber (*My*). A myoblast (*Mb*) and a macrophage (*Mc*) are seen in a sarcolemmal tube (*ST*). A blood vessel (*Ed*) is also illustrated (see text for explanation).

inflammatory reaction with leukocytes or macrophages is observed in the irradiated muscle after injury, but very few, if any, basophilic mononucleated cells are found in the sarcolemmal tubes.[75]

These results suggest that the myoblasts needed for muscle regeneration are formed locally in the damaged muscle and, further, that these cells originate from intramuscular structures that can be altered by previous irradiation. As opposed to inflammatory cells, myoblasts do not appear to be a part of the cell population brought to the muscle from the blood stream.

2. A sequential autoradiographic study of the incorporation of tritiated thymidine during skeletal muscle regeneration also supports the assumption that myoblasts are formed locally in the damaged muscle[71, 72] and are not derived from circulating cells. If tritiated thymidine is injected into the animal after the muscle is injured, a great number of mononucleated cells are labeled (Fig. 24), and labeled nuclei are found in sarcoblasts (see below). On the contrary, if tritiated thymidine is injected before muscle injury, labeled mononucleated cells and inflammatory cells are found in the damaged muscle, but sarcoblasts that form later do not contain labeled nuclei.[95] Such experimental results also would exclude the concept that myoblasts could be circulating cells formed outside the injured muscle.[9] Therefore, the origin of myoblasts must be discovered inside the muscle itself.

3. Electron microscopic research provides three main hypotheses for the possible origin of myoblasts during skeletal muscle regeneration.

The first hypothesis is based on studies of the regenerating salamander limb in which strips of nucleated sarcoplasm arising from the fragmentation of sectioned muscle fibers are considered capable of becoming myoblasts.[38, 51] Other authors[17, 84] support the second theory that the precursors of myoblasts are preexisting cells such as the muscle satellite cells described by Mauro in 1961.[54] A third possibility is supported by recent investigations during regeneration of cold-injured muscle.[73, 74] According to the latter theory, early myoblasts can originate in two ways. (1) In damaged myofibers they are single cells that probably survive necrosis of the sarcoplasm from which they are derived (Fig. 16). (2) In the partially damaged myofibers that are in the vicinity of tissue necrosis, it is possible to observe ultramicroscopic features that can be sequentially assembled so as to suggest that myoblasts are pinched off from the sarcoplasm. Membranes isolate a previously normal myonucleus and its surrounding envelope of cytoplasm. In a first step, vesicles and clefts are seen in the sarcoplasm surrounding myonuclei (Figs. 18 and 19). Very often there is an increase in the number of small dense particles in the sarcoplasm around these myonuclei. The clefts or planes of cleavage appear between the normal myofibrils and the sarcoplasm containing these particles which are ribosomes or ribosome-like elements (Figs. 18 and 19). It is suggested that these clefts and planes of cleavage are connected with preexisting sarcotubular structures of the normal myofiber (Fig. 19). At a later stage, the separation of the mononucleated cell is almost complete (Fig. 20), although cytoplasmic bridges are still visible between the developing cell and the sarcoplasm of the myofiber (Fig. 21). In the course of such a process, presumptive myoblasts pass through a satellite cell stage.[73, 74] These cells extend longitudinally under the basement membrane and sometimes overlap each other in a single cavity of a myofiber (Fig. 22).

After they are isolated as mononucleated cells which exhibit the ultrastructural aspect of myoblasts, these cells become more numerous by mitotic division. Thereafter, many myoblasts are seen at the edge of the partially injured myofibers (Fig. 23).

### Sarcoblasts

The formation and multinucleation of sarcoblasts, in which no sign of nuclear division is ever visible, do not remain the puzzling problems of previous investigators (see Betz *et al.*[10, 11] for review). Supported by elegant research *in vitro*,[15, 27, 57, 60, 98] the current concept is that multinucleation of sarcoblasts during muscle regeneration *in vivo* occurs by cytoplasmic fusion of myoblasts that have divided at least one before fusing.

1. Local x-ray irradiation, given before or immediately after injury (see above), completely inhibits muscle regeneration by interfering with the formation and multiplication of myoblasts. When a similar single local irradiation (1500 R treatment) is made 4 to 5 days after injury at a time when many sarcoblasts are already formed, muscle regeneration progresses as usual and appears identical in the irradiated muscle and in the contralateral nonirradiated muscle of the same animal.[69] If irradiation is applied 3 days after injury, at a time when myoblasts are still dividing and sarcoblasts are supposed to develop, muscle regeneration is completely inhibited in some animals or

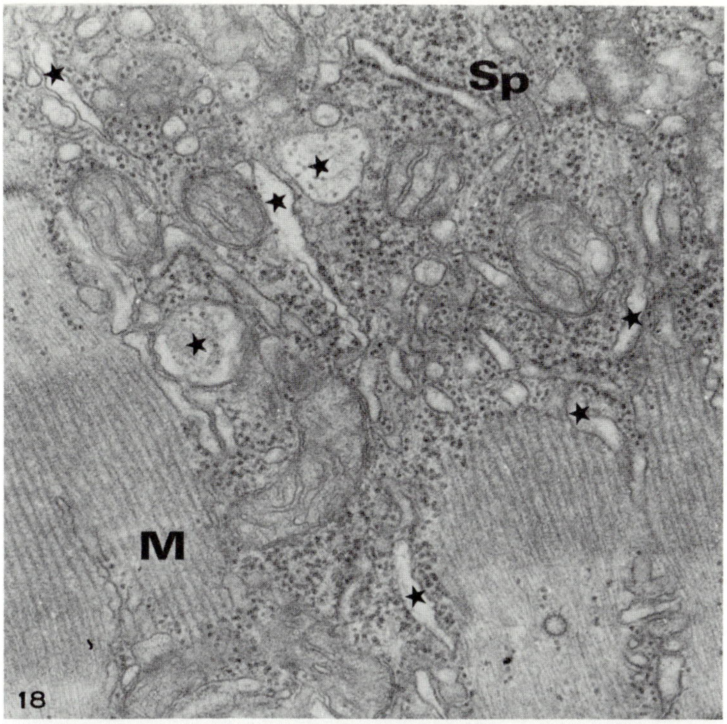


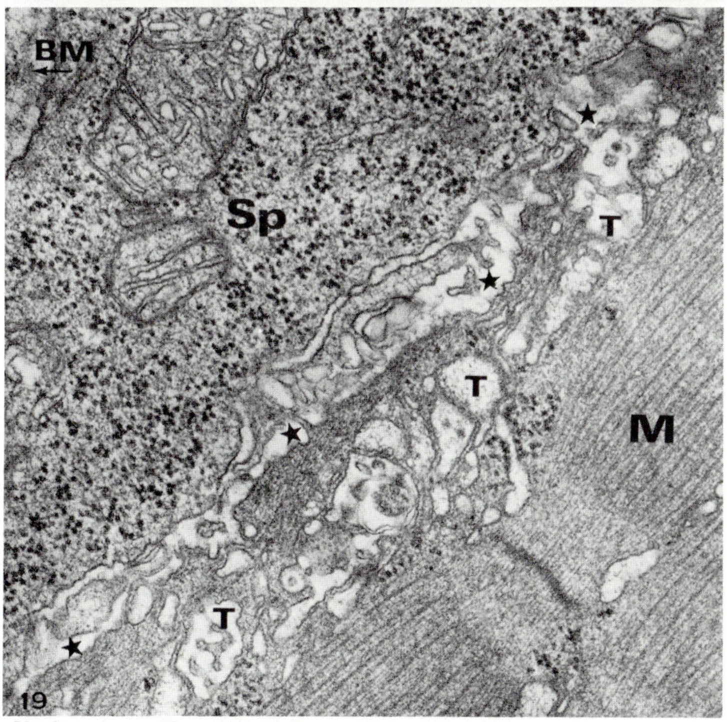


FIG. 18. Unaltered myofiber (*M*) 68 hours after cold injury (mouse gastrocnemius). Many vesicles and clefts (★) are visible between preexisting myofibrils and an area of perinuclear sarcoplasm (*Sp*) filled with ribosome-like particles. Uranyl-lead stain; × 38,000.

FIG. 19. Unaltered myofiber (*M*) in cold-injured mouse gastrocnemius. A portion of sarcoplasm (*Sp*) filled with ribosomes lies beneath the basement membrane (*BM*) and is separated from preexisting myofibrils by enfolded clefts and planes of cleavage (★) that seem to be connected with dilated portions of the sarcotubular system (*T*). Uranyl-lead stain; × 38,000.

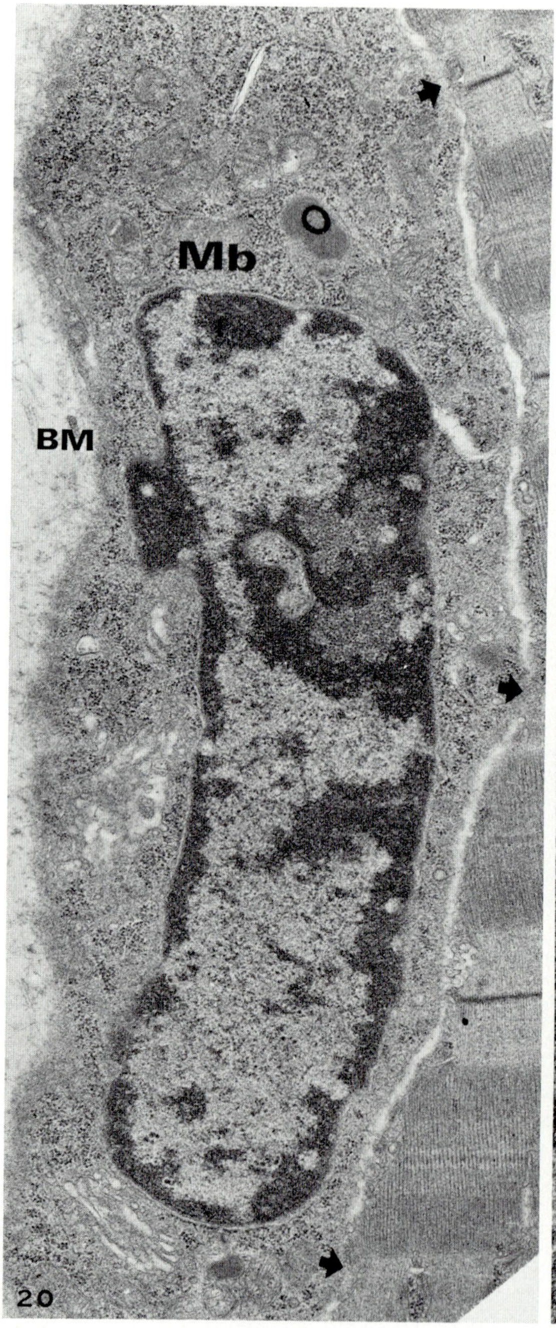


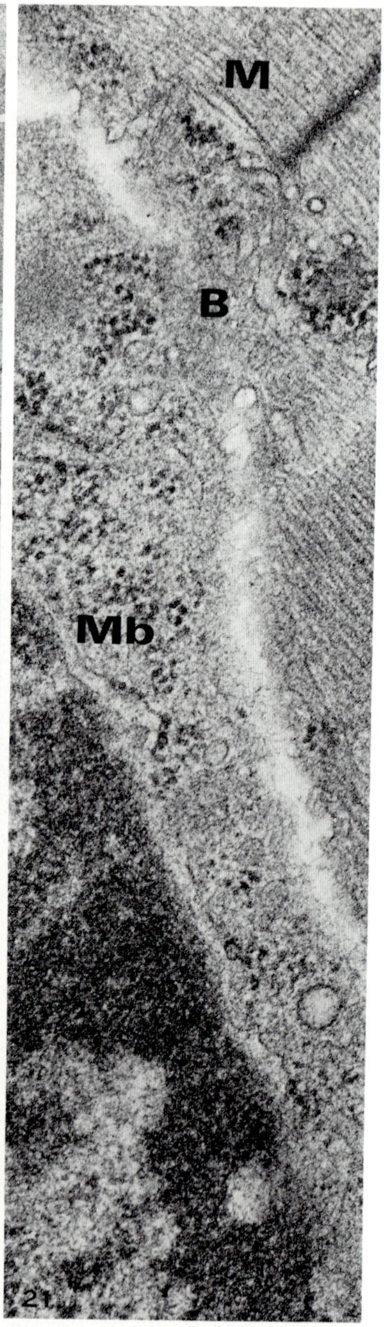


FIG. 20. Prominent satellite cell or presumptive myoblast (*Mb*) beneath the basement membrane (*BM*) of an unaltered myofiber 60 hours after cold injury (mouse gastrocnemius). This mononucleated cell (*Mb*) is still partially connected with the sarcoplasm of the myofiber (*arrows*). Uranyl-lead stain; × 24,000.

FIG. 21. Detail of Figure 20 (*central arrow*) demonstrating a cytoplasmic bridge (*B*) between the sarcoplasm (*M*) and the cytoplasm of the presumptive myoblast (*Mb*). Uranyl-lead stain; × 102,000.

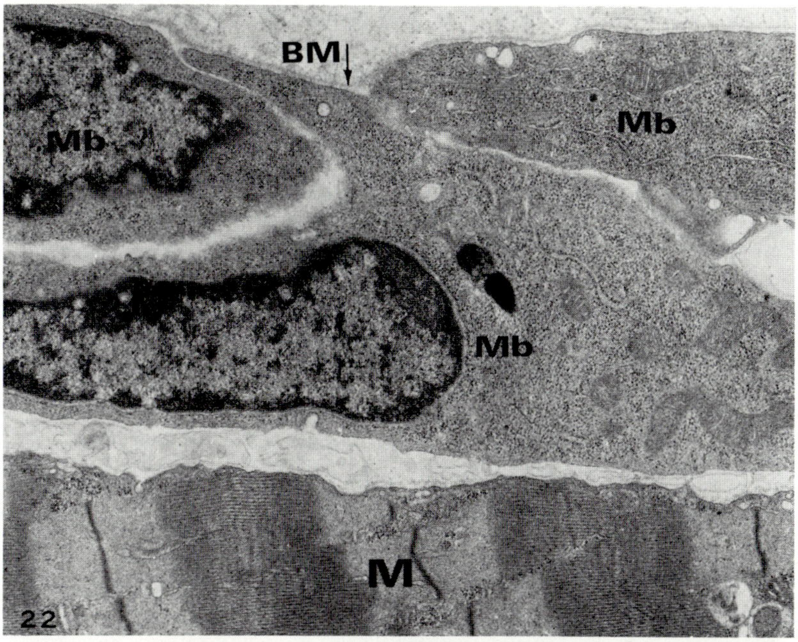


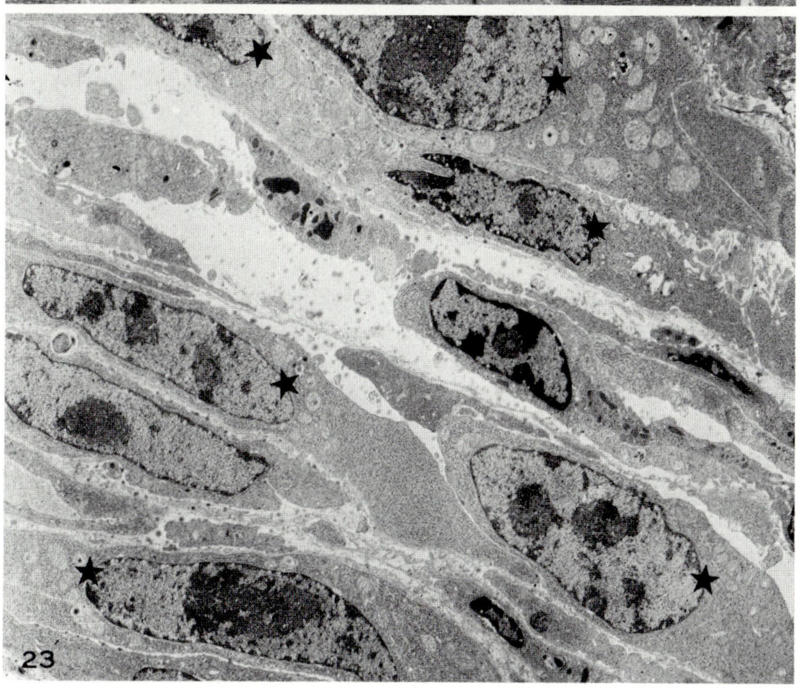


FIG. 22. Overlapping myoblasts (*Mb*) between the basement membrane (*BM*) and the sarcoplasm in an unaltered portion of a myofiber (*M*) 3 days after cold injury (mouse gastrocnemius). Uranyl-lead stain; × 12,000.

FIG. 23. Sarcolemmal tubes filled with overlapping myoblasts (★) 78 hours after cold injury (mouse gastrocnemius). Uranyl-lead stain; × 4000.

partially reduced in others. In the latter group, nuclei of varying size can be found in some of the developing sarcoblasts; similar alterations of size are described in nuclei derived from mitotic division occurring in the presence of mitotic inhibitors.[8]

These experiments demonstrate that mitotically dividing myoblasts are necessary for the multinucleation and early development of sarcoblasts. When sarcoblasts are already formed, 4 to 5 days after injury, it would seem that their further development is not impaired by the local irradiation and, therefore, no longer requires the presence of dividing cells.[69]

2. Light and electron microscopic autoradiography provide qualitative and quantitative data regarding the incorporation of myoblasts into sarcoblasts.[71, 72] Results depend on the interval of time between muscle injury and the moment when the labeled precursor of DNA is injected into the animal. From such sequential autoradiographic studies some important conclusions can be pointed out. It is only when the tritiated thymidine is injected *after* the injury that labeled nuclei are found in sarcoblasts (see above). In damaged muscles observed 1 to 3 hours after a single injection of the radioactive compound, only mononucleated cells are labeled (Figs. 24 and 26), including myoblasts and so-called satellite cells. It is not until 6 to 8 hours after the injection of tritiated thymidine that labeled nuclei are discovered in sarcoblasts (Figs. 25 and 27). These findings demonstrate that the labeled nuclei in the sarcoblasts are derived from the fusion of recent mitotically divided mononucleated cells. Moreover, if tritiated thymidine is given to the animal by several appropriate injections, developing sarcoblasts are found to contain more than 90% of labeled nuclei (Fig. 25). Such figures indicate that almost all nuclei inside the sarcoblasts result from postmitotic myoblasts. These investigations also underline the abrupt formation of multinucleated sarcoblasts 3 to 4 days after injury. Later, sarcoblasts continue to form, but they do so at a slower rate and incorporate fewer mononucleated cells.[71] Such results are in agreement with findings on myotube formation *in vitro*.[12]

At this point, it should be mentioned that when myoblasts are cultured in presence of 5-bromodeoxyuridine (BUdR, a thymidine analogue), they continue to divide but fail to form myotubes.[13, 90] This action of BUdR is reversible in that, after removal of the drug, myoblasts begin to form myotubes. Under these circumstances, more muscle fibers may be obtained than would normally be expected from the original population of myoblasts.[59] This apparent enhancement of muscle development *in vitro* is probably a result of an increase in the number of myoblasts which did not withdraw from the mitotic cycle and kept dividing before fusing into a multinucleated cell.

3. Ultrastructural investigation of regenerating skeletal muscle *in vivo* confirms that myoblasts have a tendency to become aggregated and intermingled (Figs. 23 and 28). It is also possible to observe pictures indicative of cellular fusion. In many areas, reciprocal segments of unit membranes of two adjacent myoblasts disappear, and cytoplasmic bridges are visible between the two cells (Figs. 28 and 29). Cytoplasmic fusion can exist simultaneously between one myoblast and a few other adjacent myoblasts. Thereafter, multinucleated

sarcoblasts are present in the sarcolemmal tubes (Fig. 32). More than one sarcoblast can be discovered in the same sarcolemmal tube, thus giving the false image of a longitudinal splitting of myofibers.

The morphologic appearance of sarcoblasts changes during various stages of maturation into well differentiated muscle fibers.

At the beginning of their existence, sarcoblasts possess large nuclei which occupy the central axis of the cytoplasm (Fig. 32). It is in the cytoplasm of these multinucleated sarcoblasts that myofilaments and myofibrils become visible.

*Myofibrillogenesis* has been widely debated for many decades, but it is now obvious that the optical microscope gives a view of only the latest stages of myofibril organization. Ultrastructural aspects of myofibrillogenesis have been extensively studied in various conditions of muscle development,[10, 28, 29, 40, 49, 97] but there are still many questions about the time, the site, and the mode of appearance and organization of the myofibrils.

In regenerating muscles examined in this work, the ultrastructural appearance of myofibrillogenesis exists only in sarcoblasts, but it is asynchronous from one sarcoblast to another and even in the same sarcoblast.

On morphologic grounds, myofibrillogenesis starts with the appearance of thin (50 to 60 Å) and thick (100 to 130 Å) myofilaments which are, respectively, supposed to be actin and myosin filaments.[42,43] These two types of myofilaments are detectable almost simultaneously, and they are soon associated into irregular bundles (Fig. 30). Thin and thick myofilaments alternate regularly in these bundles which are randomly scattered in the cytoplasm (Figs. 30 and 31), though they are more frequent at the periphery of the multinucleated cell. At first, these bundles of thin and thick filaments have a uniform aspect, without apparent banding. Later, they become thicker and elongated, usually surrounded by clusters of polysomes. They are oriented in the long axis of the cell. At the same time, A and I bands become visible, followed by Z line formation in the middle of the I bands (Fig. 32).[49] Myofibrils become progressively more numerous, and they occupy a major portion of the cytoplasm while nuclei remain in central position (Figs. 32 and 33). As differentiation progresses, the myofibrils in one sarcoblast tend to become aligned with each other, while being parallel to the main axis of the cell. Therefore, the characteristic cross-striated pattern of skeletal muscle is progressively established (Fig. 34).

During maturation of sarcoblasts and development of myofibrils, ribosomes and polysomes are progressively reduced in amount. The role of microtubules for the orientation and alignment of myofibrils may certainly be important but probably is not essential.[97] It is also apparent that the sarcotubular system is developed simultaneously with the formation of myofibrils and the maturation of new myofibers,[24,82] but detailed results are not yet available during muscle regeneration.

### Regenerated myofibers

Regenerated myofibers are the ultimate stage of sarcoblast maturation, but this final form still requires several progressive changes.

1. Morphologically, the increase in size of regenerating myofibers is linked

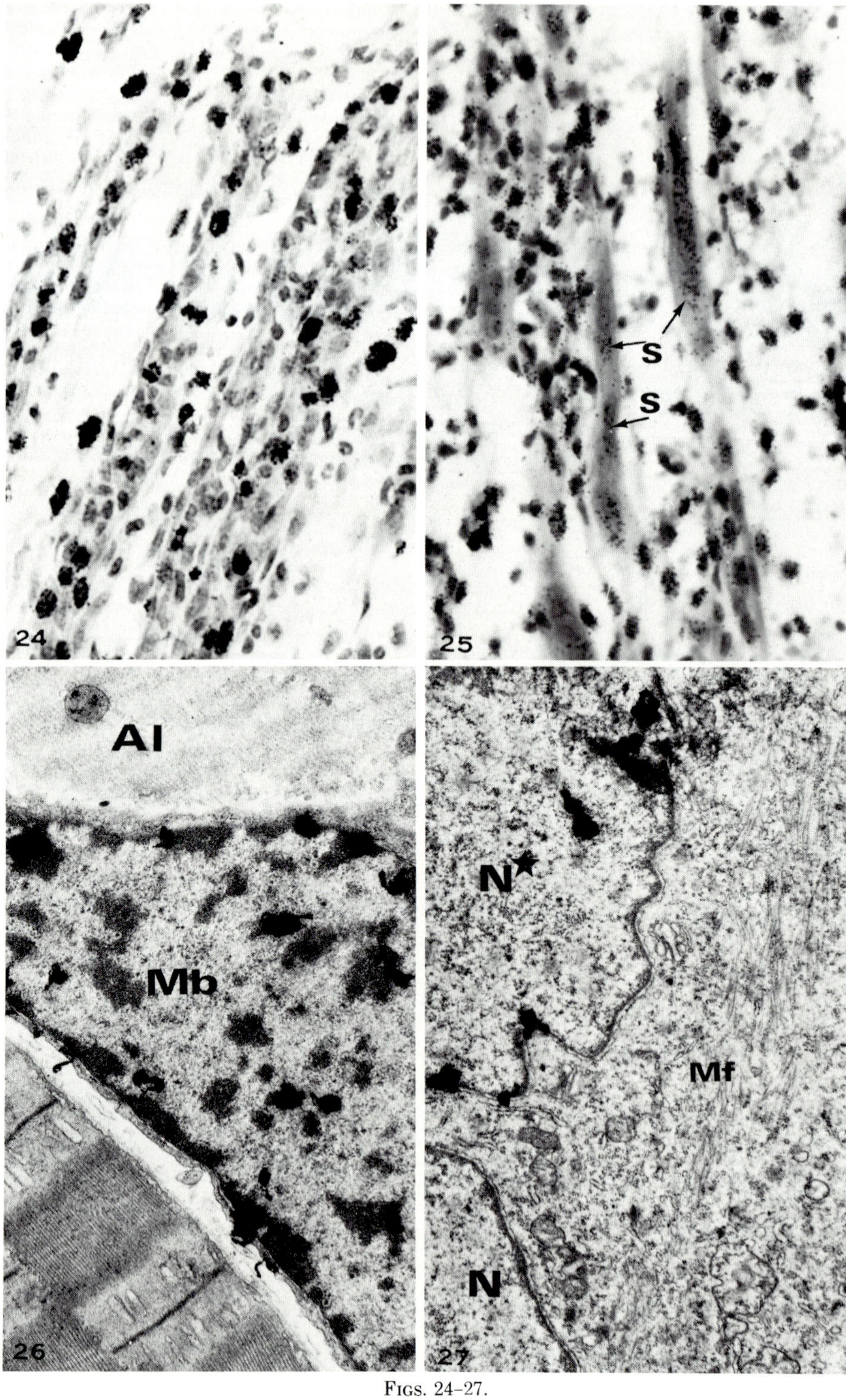


FIGS. 24–27.

with continuous formation of myofibrils which become more numerous (Figs. 33 and 34) but also more elongated. Among many other questions, the way in which myofibrils are limited in thickness but continue to grow at their free end is not yet known. The nuclei which were located in single file in the axis of the sarcoblast separate from each other and migrate slowly to the periphery of the sarcoplasm.

2. Histochemical investigations, as those described above, provide additional information on the process of skeletal muscle regeneration and maturation, as demonstrated in the study of injured muscle.[70, 81, 89, 92]

As already mentioned in this work, developing sarcoblasts exhibit a very strong basophilic cytoplasm which is related to their content of ribosomes. At an early stage of their development, sarcoblasts are well recognized with the myofibrillar ATPase reaction (Figs. 35 and 36); a few days after the injury, only foci of cytoplasm are stained, rendering a dark rim around the unstained center of the sarcoblasts. On serial cross-sections, it is possible to observe that the unstained portions of the sarcoblasts, after the myofibrillar ATPase reaction (Fig. 36), correspond to the nuclear region (Fig. 37), or that part of cytoplasm darkly stained with most of the tested oxidative enzyme reactions (lactate, succinate, malate, DPNH) (Fig. 38). At the same time, sarcoblasts have no phosphorylase activity.

One or 2 weeks after the injury, the regenerated myofibers with central nuclei (Fig. 39) are intensively stained with the tested oxidative enzyme reactions; the end product of these reactions produces a very irregular lattice of dots and cross bars through the sarcoplasm (Fig. 40). The myofibrillar ATPase reaction, however, gives a more diffuse and homogeneous staining of the new myofibers, while still leaving the nuclear region unstained (Fig. 41). The phosphorylase reaction is very irregular but, when present, is stronger in the larger regenerated myofibers (Fig. 42).

One month after the injury, two types of fiber population are more readily recognized among the regenerated myofibers, while most of them still contain central nuclei.

The activity of acetylcholinesterase follows an interesting pattern in sarcoblasts and regenerated myofibers.[10] A few days after injury (ischemia, section), a diffuse acetylcholinesterase activity is detected in many sarcoblasts (Fig. 43). The first sarcoblasts demonstrating such an enzymatic reaction seem to be in continuity with relatively intact muscle fibers. However, almost all of the sarcoblasts progressively exhibit the same diffuse reaction. The pattern of

FIG. 24. Histoautoradiography of mouse gastrocnemius 3 hours after a single injection of 4 to 5 $\mu$Ci per g of tritiated thymidine given 66 hours after cold injury. Many mononucleated cells are labeled. Hematoxylin stain; × 420.

FIG. 25. Histoautoradiography of mouse gastrocnemius 80 hours after cold injury and several injections of tritiated thymidine. The great majority of nuclei inside the sarcoblasts (*S*) are labeled. Hematoxylin stain; × 420.

FIG. 26. Typical early myoblast (*Mb*), in an altered myofiber (*Al*), labeled 3 hours after a single injection of 4 to 5 $\mu$Ci per g of tritiated thymidine given 66 hours after cold injury (mouse gastrocnemius). Uranyl-lead stain; × 12,600.

FIG. 27. Sarcoblast with labeled (*N*✱) and unlabeled (*N*) nuclei, 24 hours after a single injection of 4 to 5 $\mu$Ci per g of tritiated thymidine given 66 hours after cold injury. Myofilaments (*Mf*) are visible (mouse gastrocnemius). Uranyl-lead stain; × 11,000.

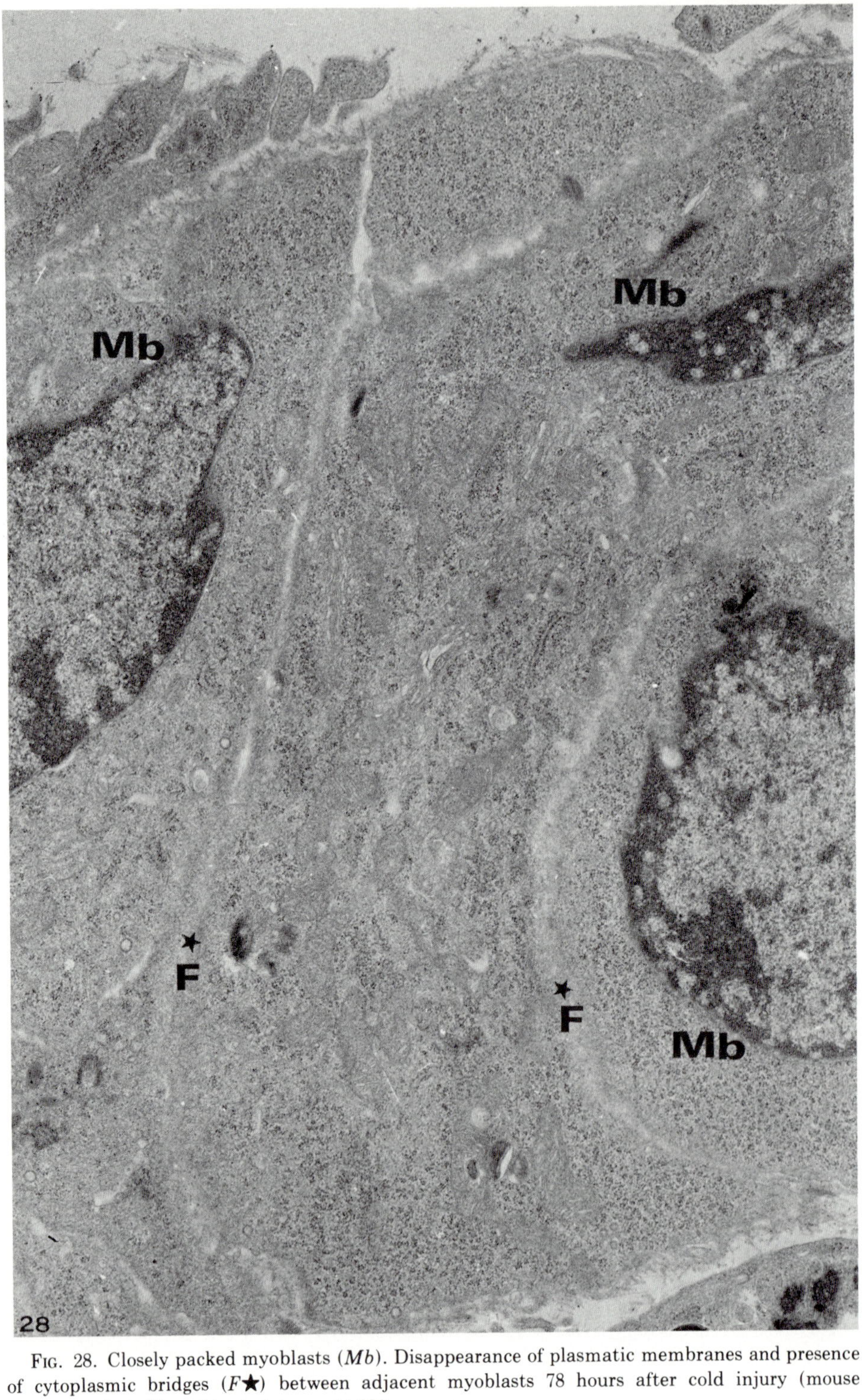


FIG. 28. Closely packed myoblasts (*Mb*). Disappearance of plasmatic membranes and presence of cytoplasmic bridges (*F*★) between adjacent myoblasts 78 hours after cold injury (mouse gastrocnemius). Uranyl-lead stain; × 14,000.

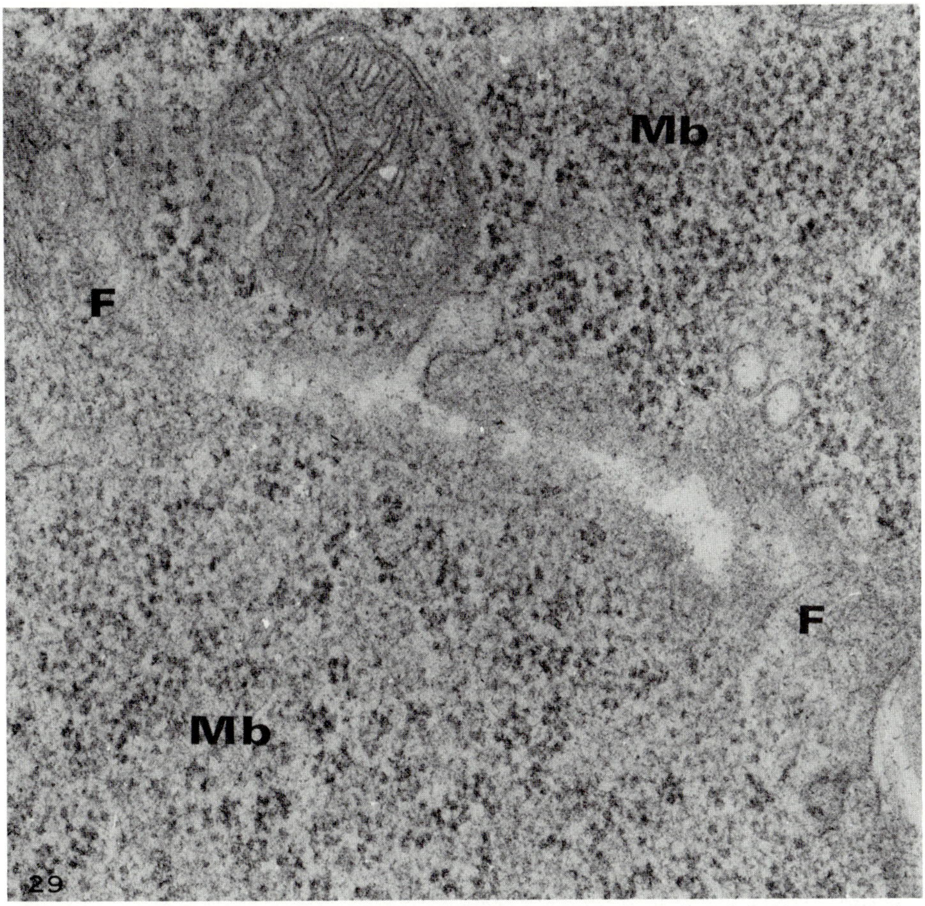


FIG. 29. Similar condition as in Figure 28. Details of an area of cytoplasmic bridges and fusion (*F*) between two adjacent myoblasts (*Mb*). Uranyl-lead stain; × 60,000.

histochemical reaction changes when sarcoblasts mature. After 2 to 3 weeks, the reaction is no longer diffuse but appears to be located in small irregular foci scattered along the new fibers (Fig. 44). The acetylcholinesterase reaction is also more intense where the extremity of the regenerated myofiber comes in contact with fibrous tissue.[10]

Without going into details, one should be reminded that similar histochemical reactions are observed during embryonic myogenesis in higher vertebrates before and after the formation of motor end-plates,[26,31] as well as when embryonic muscle develops *in vitro*.[77] These findings, however, are different from observations in lower vertebrates in which it appears that the nerve endings seem to have an inductive effect on the development of cholinesterase activity in regenerating muscle.[52]

The ultimate structure and histochemical type of these regenerated myofibers in adult animals will depend upon their future innervation. This remark introduces the complex problem of the nerve-muscle interrelationship during skeletal muscle regeneration.

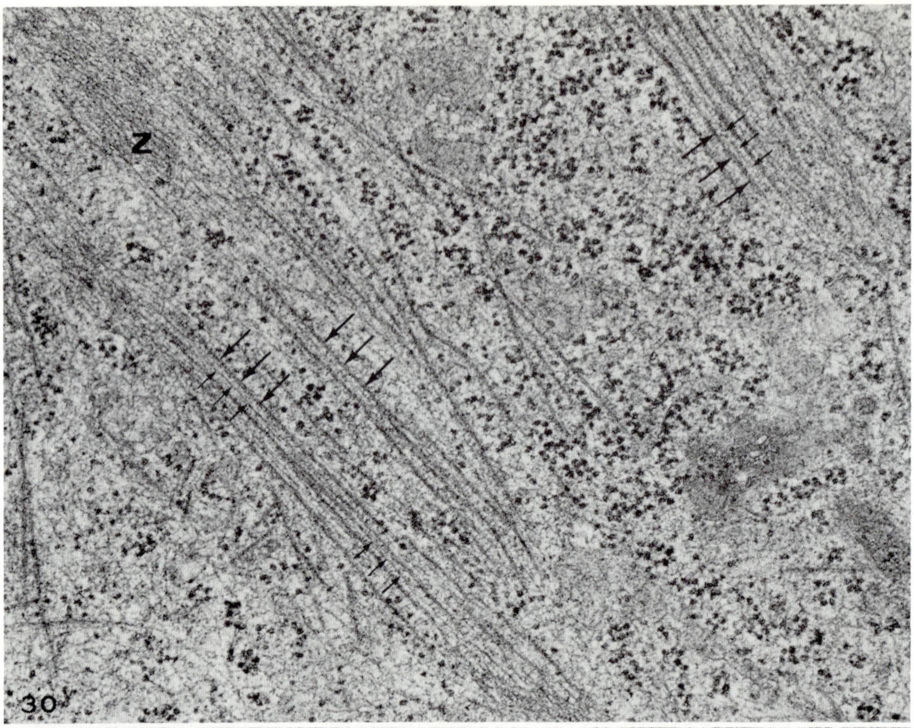


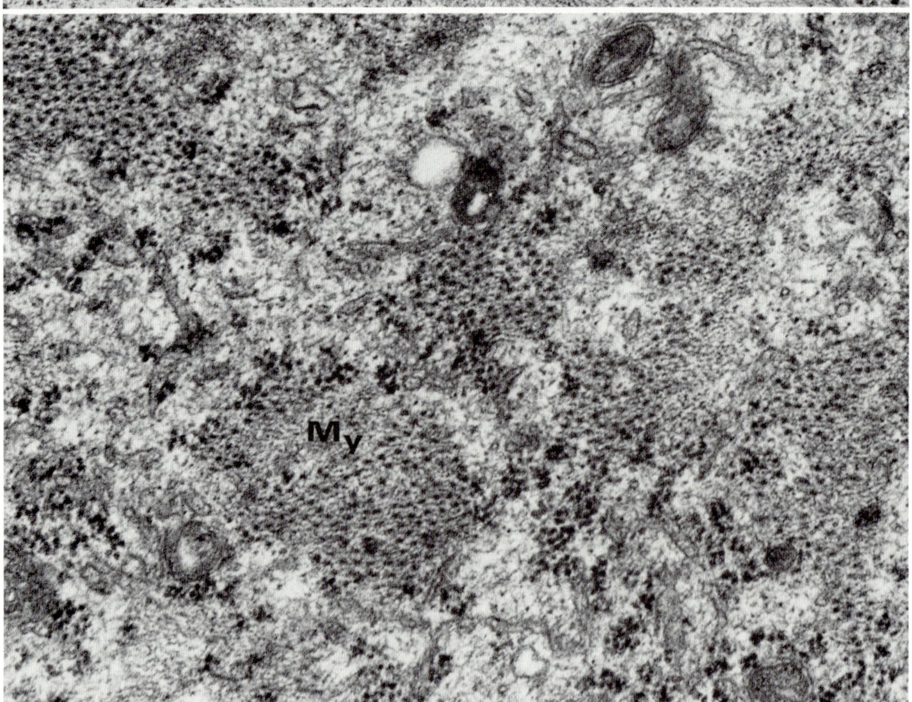


FIG. 30. Early appearance of myofibril organization in longitudinal section of sarcoblast 78 hours after cold injury (mouse gastrocnemius). Ribosomes are grouped in clusters. Thin (*thin arrows*) and thick (*thick arrows*) myofilaments are partially oriented in bundles. Z band material (*Z*) is suggested in some areas. Uranyl-lead stain; × 60,000.

FIG. 31. Cross-section of sarcoblast 4 days after cold injury (mouse gastrocnemius). Bundles of thin and thick myofilaments (*My*) in hexagonal array are distributed at random in the cytoplasm. Uranyl-lead stain; × 45,000.

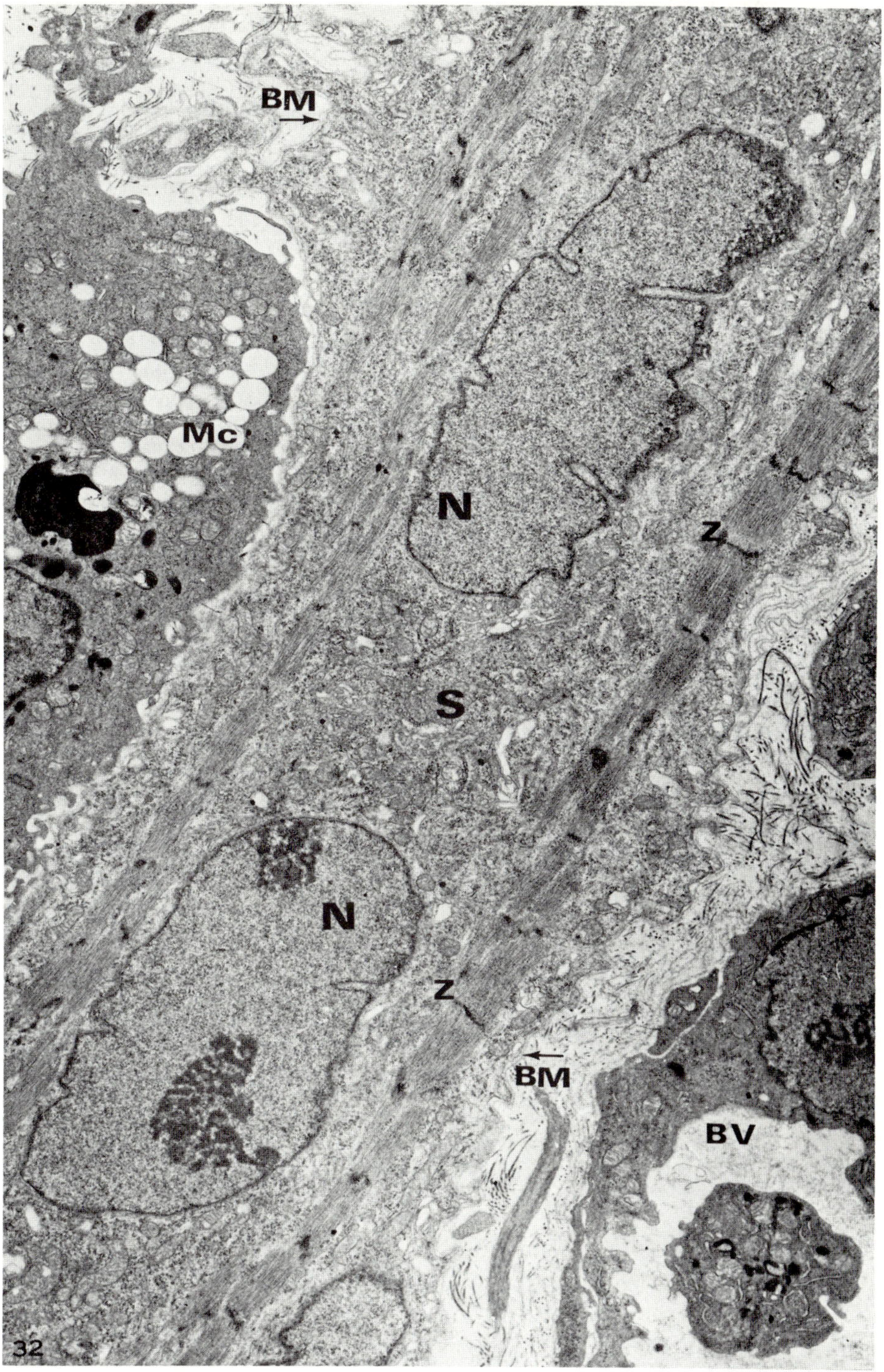


FIG. 32. Longitudinal section of a sarcoblast (*S*) with central nuclei (*N*). Myofibrils with Z lines (*Z*) are aligned at the periphery of the cell. Histiocytes and macrophages (*Mc*) are visible outside the sarcolemmal sheath (*BM*). A blood vessel (*Bv*) is present. Rabbit tibialis anterior muscle 6 days after acute ischemia. Uranyl-lead stain; × 8400.

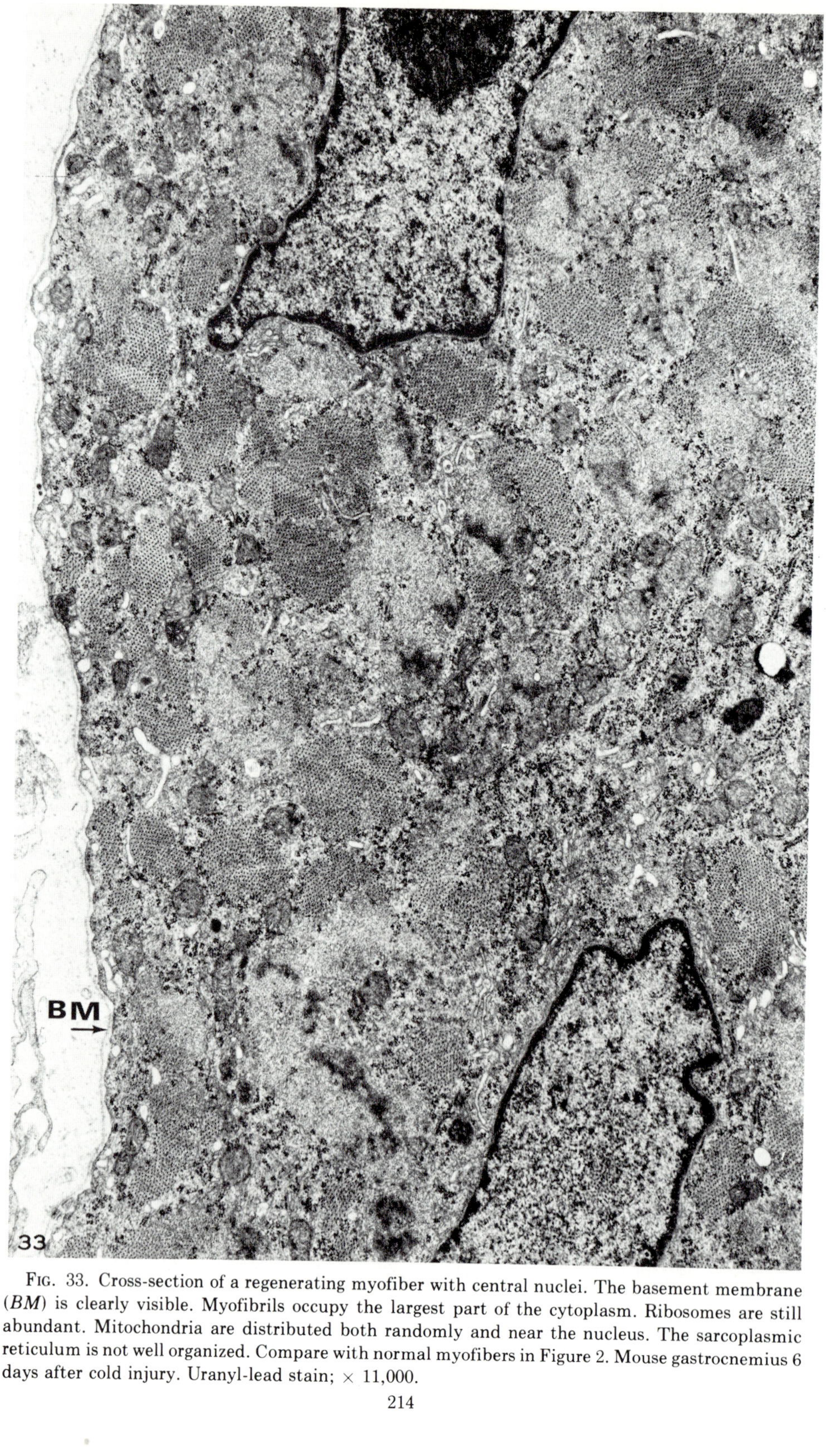


FIG. 33. Cross-section of a regenerating myofiber with central nuclei. The basement membrane (*BM*) is clearly visible. Myofibrils occupy the largest part of the cytoplasm. Ribosomes are still abundant. Mitochondria are distributed both randomly and near the nucleus. The sarcoplasmic reticulum is not well organized. Compare with normal myofibers in Figure 2. Mouse gastrocnemius 6 days after cold injury. Uranyl-lead stain; × 11,000.

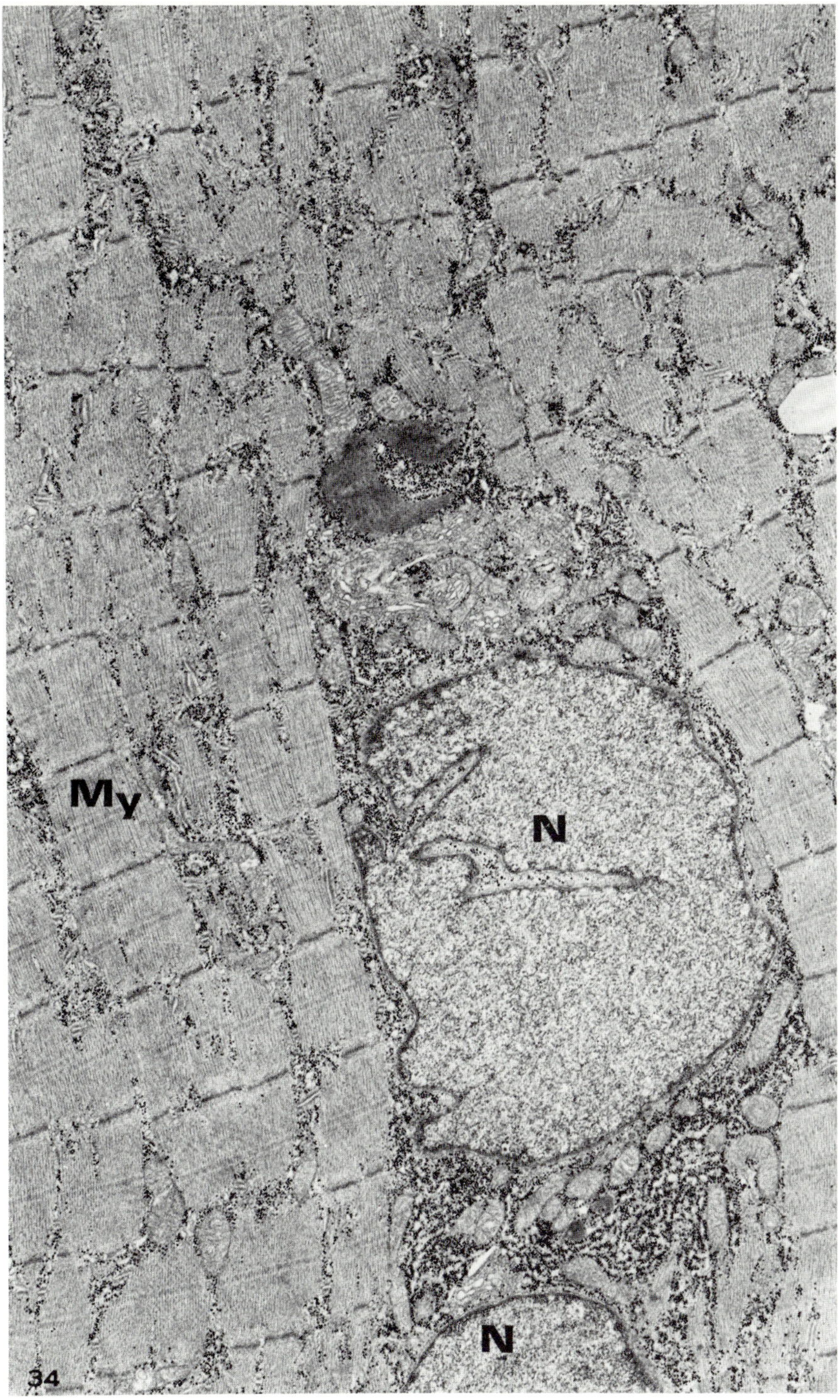


FIG. 34. Longitudinal section of a regenerated myofiber in rabbit tibialis anterior muscle 14 days after acute ischemia. Mitochondria and various vesicles are seen near central nuclei (*N*). The myofibrillar network (*My*) exhibits a more regular pattern than in previous stages. Uranyl-lead stain; × 10,500.

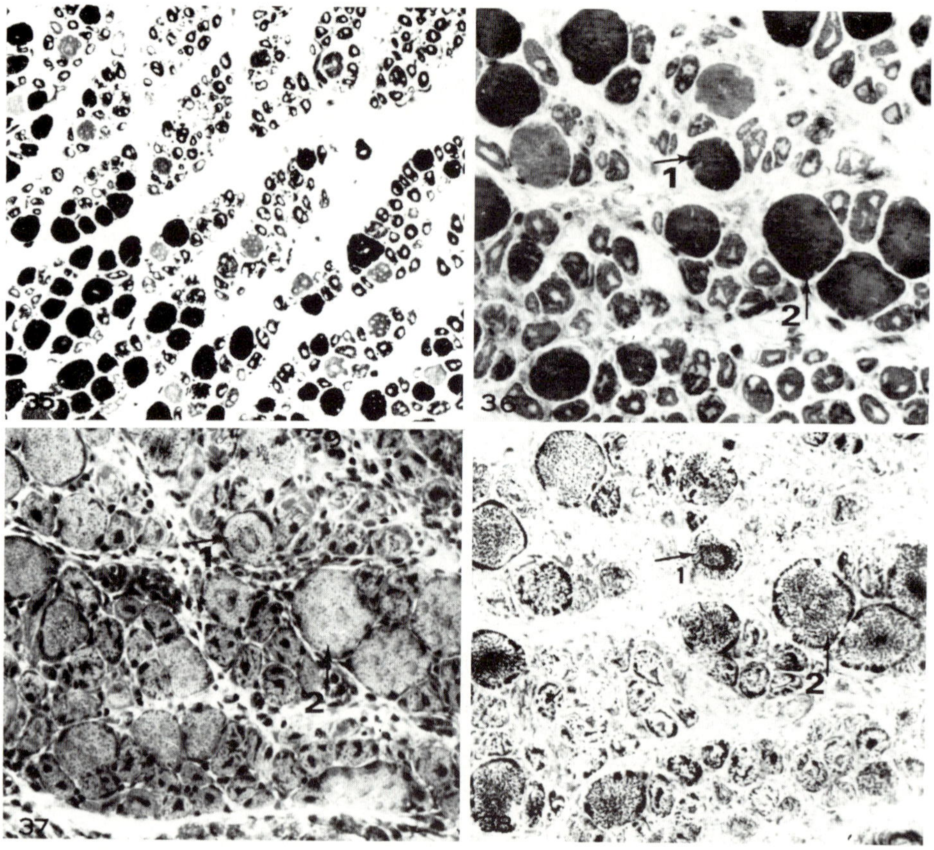


FIG. 35. ATPase reaction on a cross-section of mouse gastrocnemius 4 days after cold injury. Preexisting large myofibers exhibit a normal uniform aspect. Sarcoblasts have a smaller diameter and present a dark rim around a clear center. × 75.

FIG. 36. Details of the same muscle as in Figure 35. Dark and light large fibers are the preexisting myofibers. Sarcoblasts are small with the end product of the enzymatic reaction forming a dark rim around a clear center. Compare with Figures 37 and 38 (two fibers are labeled for reference). ATPase reaction; × 200.

FIG. 37. Serial cross-section of the same area as in Figure 36. The dark areas of the sarcoblasts in this figure are nuclei or clusters of mitochondria, and they correspond to the unstained regions with the ATPase reaction, as shown in Figure 36 (two fibers are labeled for reference). Modified Gomori trichrome stain; × 200.

FIG. 38. DPNH reaction on a serial cross-section of the same area as in Figures 36 and 37. Intense reaction is present in most of the large myofibers. Sarcoblasts exhibit large and irregular dark dots similar to those viewed in Figure 37 (two fibers are labeled for reference). × 200.

## EFFECT OF INNERVATION ON MUSCLE REGENERATION

The single fact that, *in vitro*, dissociated mononucleated myoblasts are capable of multiplying actively to form multinucleated myotubes and to mature into differentiated myofibers presenting spontaneous contraction proves that innervation is not necessary for the initial steps of muscle development. Similar conditions exist during the early stages of normal embryonic myogenesis, and partially in the course of skeletal muscle regeneration in adult animals.

Despite extensive research in this particular field (see References 1, 10, 16, and 99 for review), the influence of the motor innervation on the skeletal muscle is largely unexplained, and the effect of innervation on muscle regeneration *in vivo* remains unclear. The fundamental observations are that: (1) denervation induces muscle atrophy, and, when myofibers are atrophic, regenerative processes are considerably impaired; (2) when denervation is recent, muscle regeneration is almost normal, but without an efficient innervation, regenerated myofibers undergo degenerative changes. The first assumption is very well illustrated by a great number of experiences in different species (see References 1, 10, and 63 for review).

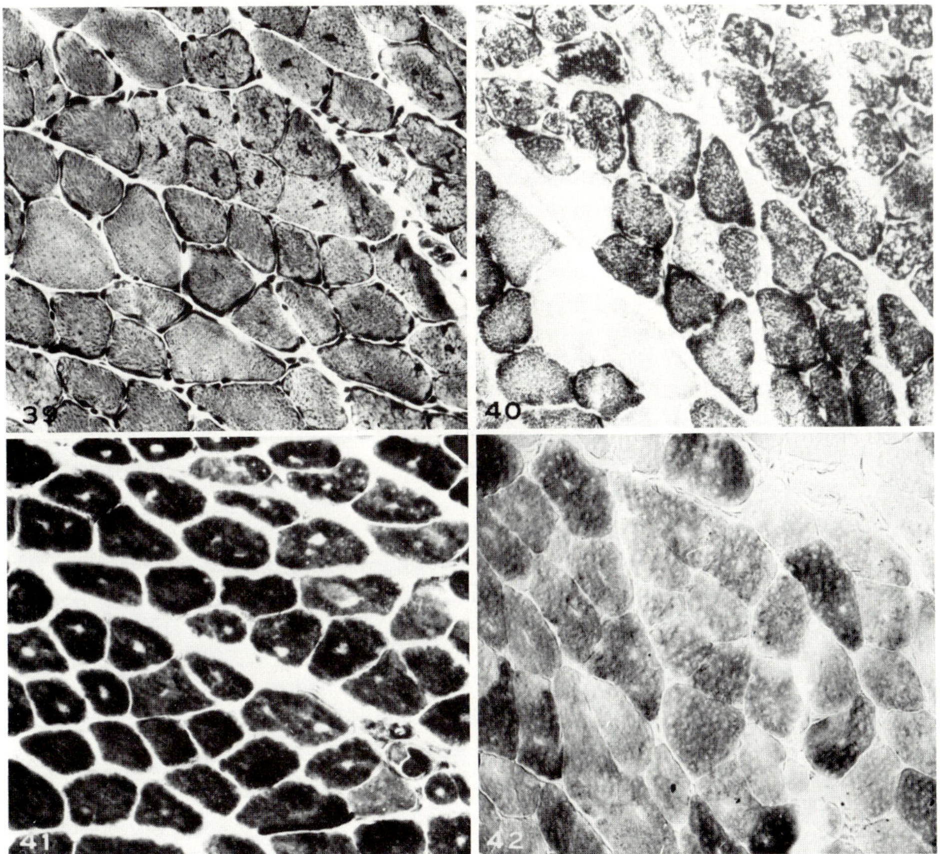


FIG. 39. Near preexisting large myofibers, regenerated myofibers are recognizable by their smaller diameters and central nuclei; 12 days after cold injury (mouse gastrocnemius). Modified Gomori trichrome stain; × 200.

FIG. 40. DPNH reaction on a similar area as in Figure 39. Normal type I (dark) and type II (light) myofibers are present (*bottom left*). An intense reaction is visible in almost all of the regenerated myofibers (*top right*). × 200.

FIG. 41. ATPase reaction on a similar area as in Figures 39 and 40. Diffuse reaction in normal and regenerated myofibers; the latter still present a clear center corresponding to the central nuclei. × 200.

FIG. 42. The phosphorylase reaction is very irregular in regenerated myofibers in a similar area as in Figures 39 and 40; 12 days after cold injury (mouse gastrocnemius). × 200.

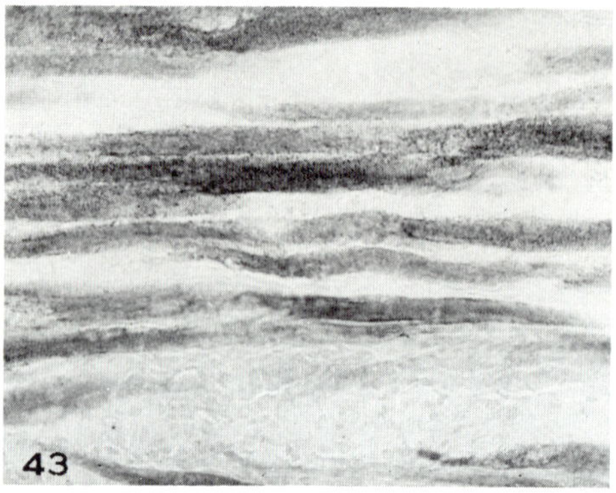

FIG. 43. Diffuse acetylcholinesterase reaction in sarcoblasts. Rabbit tibialis anterior muscle 7 days after acute ischemia. × 200.

FIG. 44. Two foci of intense enzymatic reaction in the same regenerated myofiber 25 days after acute ischemia (rabbit tibialis anterior muscle). Acetylcholinesterase reaction; × 520.

On the contrary, little is known about the reinnervation of regenerated muscle fibers.[3, 52, 91] At the present time, recent research on the embryologic development of motor end-plates[26, 47] or the formation of motor end-plates *in vitro*[14, 58, 64, 93] is of little use for the elucidation of the problems now under discussion.

The mechanisms and the effects of additional innervation or reinnervation of preexisting muscle fibers in adult animals are extensively investigated. Results from these studies are interesting and can illuminate some aspects of the innervation of regenerated myofibers.

For example, there is evidence that spontaneous reinnervation of preexisting neuromuscular junctions occurs 5 to 6 weeks after nerve section under optimal conditions.[81] The regenerating axons return to old subneural apparatuses, but apparently normal innervation of some motor end-plates and restoration of normal acetylcholinesterase activity do not occur before 15 to 30 weeks after the operation. In these types of experiments, there is little tendency to form new neuromuscular junctions, thus confirming the current concept that myofibers are resistant to multiple innervation.

When a motor nerve is cut and reimplanted, or when a foreign nerve is implanted into previously denervated muscle, a new innervation seems to develop near the point of entry of the transplanted nerve.[37,81] It is not until 6 to 8 weeks after nerve implantation that new motor end-plates develop, as shown by acetylcholinesterase activity, silver impregnation of nerves, and ultramicroscopic investigations. The presence of ectopic motor end-plates in normally end-plate-free portions of the muscle would indicate that these are, in fact, new endings.

The earliest ultramicroscopic evidence of motor end-plate formation during embryonic development,[48] neoinnervation of previously denervated muscle,[81] or regenerated myofibers in amphibian[52] occurs when focal areas of thickening are observed in muscle membranes opposite axon sprouts. These are then followed by primary and secondary cleft development.

Actually, however, it appears too early to clearly differentiate a newly formed motor end-plate from a preexisting neuromuscular junction. The problem of identification of new motor end-plates is magnified in injured muscle, in which myofibers at varying stages of development are present.

In addition to physiologic evidence, the demonstration of reinnervation of regenerated myofibers would require a number of morphologic criteria: (1) proof that it is a regenerated muscle fiber, (2) ultrastructural confirmation of a motor end-plate, (3) presence of nerve endings in the new motor end-plate, and (4) demonstration of acetylcholinesterase activity. The interpretation of histologic sections and electron micrographs is difficult because it is not always possible to ascertain whether or not the regenerating fiber is in continuity with the normal part of a preexisting muscle fiber and whether this fiber is innervated or not.

Despite the actual difficulties in analyzing the progressive state of innervation (or denervation) of individual regenerated myofibers *in vivo*, current concepts on the nerve-muscle relationship can be put together to consider potential neuromuscular alterations in damaged muscle (Fig. 45).

When injured myofibers undergo acute pathologic alteration, necrosis extends mainly to the aneural part of the sarcoplasm, but some motor end-plates are also involved in the degenerative changes.[10,11]

As previously indicated in this review, mononucleated myoblasts appear to be segregated from the syncytium of the partially damaged myofiber. Further evolution of the persistent part of the myofiber, however, is certainly related to its state of motor innervation. If the motor end-plate is destroyed by the necrotic process, whatever its origin, the remnant myofiber will behave like a denervated

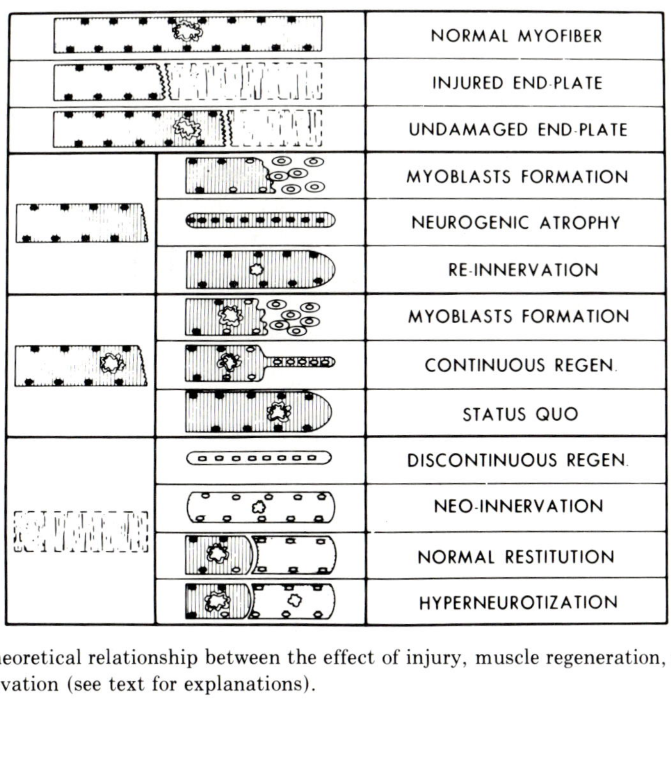


FIG. 45. Theoretical relationship between the effect of injury, muscle regeneration, and the role of motor innervation (see text for explanations).

muscle fiber. Unless it is rapidly reinnervated, this myofiber will become atrophic (neurogenic atrophy). On the contrary, if the partially damaged myofiber remains correctly innervated, it can either persist unchanged (*status quo*) or could, eventually, be the basic element of active repair such as a type of continuous regeneration.

In necrotic areas, myoblasts will proliferate and fuse to form multinucleated sarcoblasts (discontinuous regeneration) that can mature into new myofibers. These latter modifications can occur without necessity of immediate innervation, as observed during muscle development *in vitro*. However, the ultimate evolution and stability of the regenerated myofibers will depend upon their future innervation.

If regenerated myofibers are not effectively innervated, it is obvious that they will not be able to maintain their state of differentiation and will become atrophic like any other noninnervated muscle fibers *in vivo*. In theory, a regenerated myofiber can become innervated either by fusing with a preexisting innervated myofiber (*restitutio ad integrum*), or by receiving its own innervation (neoinnervation). Eventually, a state of hyperneurotization can be created if a preexisting innervated myofiber fuses with a regenerated myofiber which has been recently innervated. These various alternatives may have no significance after a single injury of a normal muscle. In that circumstance, the regenerative process can restore an almost normal muscle and the abnormality of some individual fibers can escape the scrutiny of any investigator. Nevertheless, whatever the apparent completeness of muscle regeneration, it is clear that repeated injuries, or episodes of degeneration, will reduce the population of preexisting normal muscle fibers and alter the ratio of motor innervation. In such skeletal muscles, there will be more persistent pathologic lesions, signs of neurogenic atrophy, and evidence of an exhaustion of the regenerative capacity. Secondary increase of connective tissue or fatty infiltration can become prominent.

It may be reasonable to postulate that the final morphologic aspect of skeletal muscles suffering repeated cycles of degenerative and regenerative changes will be similar to the heterogenous fiber pattern observed in most of the chronic progressive muscular dystrophies.

## EFFECT OF DRUGS ON MUSCLE REGENERATION

Medical enhancement of skeletal muscle regeneration is an obvious goal for research and has already attracted various investigators. Although many questions remain open, the effect of some medications on muscle regeneration *in vivo* have been suggested.

The administration of the lathyrogen aminoacetonitrile[30] or the local injection of heparin[79] may improve muscle regeneration by reducing the development of connective scar tissue. On the other hand, it is possible that a drug such as ethylenedinitramine stimulates muscle regeneration by increasing protein synthesis in developing myofibers.[65]

Steroids appear to produce more versatile effects on muscle regeneration.[88] Catabolic steroid reduces inflammatory reactions and phagocytosis, thus leading to the persistance of a large amount of necrotic material in crushed muscles. Thereafter, sarcoblasts are fewer than in control animals. On the contrary, anabolic steroids seem to accelerate both phagocytosis of dead muscle and the growth of new sarcoblasts which appear larger and more numerous than in control animals.

These contributions to the study of experimental muscle regeneration point the way to further investigations. No doubt, workers in this field will be able to gather more facts so that the process of muscle regeneration will be better understood and may be oriented to improve the course of some neuromuscular disorders.

## CONCLUSIONS AND SUMMARY

From this review of research using new techniques of investigation, it appears that: (1) the response of the muscle fiber to a variety of noxious agents is relatively uniform, and (2) some older misconceptions about skeletal muscle regeneration have been dispelled.

In the field of degenerative processes, two areas are considered. (1) Necrotic myofibers are characterized by rapid disappearance of Z lines, rupture of I bands with relative preservation of A bands, fragmentation of sarcotubular system, mitochondria which can exhibit various inclusions, pyknosis of myonuclei, and disruption of plasmalemma but usual persistence of basement membrane. These morphologic changes are correlated with varying histochemical patterns including persistence of myofibrillar ATPase reaction, loss of phosphorylase, and oxidative enzymes (lactate, succinate, malate, and DPNH) reactions. (2) at the periphery of the precedent area, morphologic changes are mainly enlargement of Z lines and presence of mitochondria which are increased in size and in number. Histochemical tests indicate significant increase of staining after oxidative enzyme reactions (lactate, succinate, malate, and DPNH).

In the field of regenerative processes, three successive yet coordinated stages are recognized: (1) the presence of a pool of proliferating mononucleated myoblasts which are segregated from preexisting myofibers. The stem cells either survive acute degenerative changes or are pinched off from the muscle sarcoplasm as mononucleated cells that then pass through a satellite cell state; (2) abrupt formation of multinucleated sarcoblasts by cytoplasmic fusion of myoblasts which have divided at least once and have left the mitotic cycle before fusing. It is in sarcoblasts that myofibrillogenesis becomes apparent; (3) regenerated myofibers result from maturation of sarcoblasts which became progressively thicker and contain larger amounts of sarcoplasm, myofibrils, mitochondria and an organized sarcotubular system. Cytoplasmic basophilia fades out simultaneously with maturation into new myofibers. Histochemical patterns of regenerated myofibers include dark irregular staining with the tested oxidative enzyme reactions, early appearance of myofibrillar ATPase reaction extending from the periphery to the center of developing myofibers, late appearance of diffuse phosphorylase reaction, and diffuse and then progressively localized acetylcholinesterase activity.

Innervation seems an important factor for maintaining the state of differentiation of normal and regenerated skeletal muscle fibers, but its action and morphologic changes remain difficult to analyze in sequential detail. Theoretical considerations suggest that exhaustion of regenerative capacity and failure of muscle restoration, with increase of connective tissue and signs of neurogenic atrophy, can be secondary to repeated acute degenerative alterations whatever their origin.

Medical enhancement of skeletal muscle regeneration remains a challenge for further research.

## REFERENCES


1. Adams, R. D., Denny-Brown, M. D., and Pearson, C. M.: *Diseases of Muscle. A Study in Pathology*, 2nd ed. New York, Hoeber Medical Division, Harper and Row, 1962.
2. Allbrook, D. B.: An electron microscopic study of regenerating skeletal muscle. *J. Anat. 96:* 137–152, 1962.
3. Allbrook, D. B., and Aitken, J. T.: Re-innervation of striated muscle after acute ischemia. *J. Anat. 85:* 376–390, 1951.
4. Allbrook, D. B., Baker, W. C., and Kirkcaldy-Willis, W. H.: Muscle regeneration in experimental animals and in man. *J. Bone Joint Surg. 48B:* 153–169, 1966.
5. Allen, E. R., and Pepe, F. A.: Ultrastructure of developing muscle cells in the chick embryo. *Am. J. Anat. 116:* 115–146, 1965.
6. Anderson, P. J., Song, S. K., and Slotwiner, P.: The fine structure of spheromembranous degeneration of skeletal muscle induced by vincristine. *J. Neuropathol. Exp. Neurol. 26:* 15–24, 1967.
7. Banker, B. Q.: A phase and electron microscopic study of dystrophic muscle. I. The pathological changes in the two week old Bar Harbor 129 dystrophic mouse. *J. Neuropathol. Exp. Neurol. 26:* 259–275, 1967.
8. Bassleer, R., Collignon, P., and Matagne-Dhoosche, F.: Effets cytologiques du myleran sur le muscle strié de l'embryon de poulet in ovo et de la queue de l'axolotl en régénération. *Arch. Biol. (Liege) 74:* 79–94, 1963.
9. Bateson, R. G., Woodrow, D. F., and Sloper, S. C.: Circulating cell as a source of myoblasts in regenerating injured mammalian skeletal muscle. *Nature (Lond.) 213:* 1035–1036, 1967.
10. Betz, E. H., Firket, H., and Reznik, M.: Some aspects of muscle regeneration. *Int. Rev. Cytol. 19:* 203–222, 1966.
11. Betz, E. H., and Reznik, M.: La régénération des fibres musculaires striées dans différentes conditions expérimentales. *Arch. Biol. (Liege) 75:* 567–599, 1964.
12. Bischoff, R., and Holtzer, H.: Mitosis and the processes of differentiation of myogenic cells in vitro. *J. Cell Biol. 41:* 188–200, 1969.
13. Bischoff, R., and Holtzer, H.: Inhibition of myoblast fusion after one round of DNA synthesis in 5-bromodeoxyuridine. *J. Cell Biol. 44:* 134–150, 1970.
14. Bornstein, M. B., Iwanami, H., Lehrer, G. M., and Breitbart, L.: Observations on the appearance of neuromuscular relationship in cultured mouse tissues. *Z. Zellforsch. Mikrosk. Anat. 92:* 197–206, 1968.
15. Capers, C. R.: Multinucleation of skeletal muscle *in vitro. J. Biophys. Biochem. Cytol. 7:* 559–566, 1960.
16. Carlson, B. M.: Regeneration of the completely excised gastrocnemius muscle in the frog and rat from minced muscle fragments. *J. Morphol. 125:* 447–471, 1968.
17. Church, J. C. T., Noronha, R. F. X., and Allbrook, D. B.: Satellite cells and skeletal muscle regeneration. *Br. J. Surg. 53:* 638–642, 1966.
18. Clara, M.: Was ist das Sarkolemm? *Anat. Anz. 110:* 41–51, 1961.
19. D'Agostino, A. N., and Chiga, M.: Morphologic changes in cardiac and skeletal muscle induced by corticosteroids. *Ann. N. Y. Acad. Sci. 138:* 73–81, 1966.
20. De Duve, C., and Wattiaux, R.: Functions of lysosomes. *Annu. Rev. Physiol. 28:* 435–492, 1966.
21. Dessouky, D. A., and Hibbs, R. G.: An electron microscope study of the development of the somatic muscle of the chick embryo. *Am. J. Anat. 116:* 523–566, 1965.

22. Engel, W. K.: The essentiality of histo- and cytochemical studies of skeletal muscle in the investigation of neuromuscular disease. *Neurology* (*Minneap.*) *12:* 778–794, 1962.
23. Engel, W. K., and Brooke, M. H.: Muscle biopsy as a clinical diagnosic aid. In *Neurological Diagnostic Techniques*, ed. by Fields, W. S., pp. 90–147. Springfield, Illinois, Charles C Thomas, 1965.
24. Ezerman, E. B., and Ishikawa, H.: Differentiation of the sarcoplasmic reticulum and T system in developing chick skeletal muscle in vitro. *J. Cell Biol. 35:* 405–420, 1967.
25. Field, E. D.: Muscle regeneration and repair. In *Structure and Function of Muscle*, ed. by Bourne, G. H., vol. 3, pp. 139–170. New York, Academic Press, 1960.
26. Filogamo, G., and Gabella, G. The development of neuromuscular correlations in vertebrates. *Arch Biol.* (*Liege*) *78:* 9–60, 1967.
27. Firket, H.: Recherches sur la synthèse des acides désoxyribonucléiques et la préparation de la mitose dans des cellules cultivées in vitro. *Arch. Biol.* (*Liege*) *69:* 1–166, 1958.
28. Firket, H.: Ultrastructural aspects of myofibril formation in cultured skeletal muscle. *Z. Zellforsch. Mikrosk. Anat. 78:* 313–327, 1967.
29. Fischman, D. A.: An electron microscope study of myofibril formation in embryonic chick skeletal muscle. *J. Cell Biol. 32:* 557–575, 1967.
30. Galluci, V., Novello, F., Margreth, A., and Aloisi, M.: Biochemical correlates of discontinuous muscle regeneration in the rat. *Brit. J. Exp. Pathol. 47:* 215–277, 1966.
31. Gerebtzoff, M.: *Cholinesterases*. Pergamon Press, New York, 1959.
32. Gilbert, R. K., and Hazard, J. B.: Regeneration in human skeletal muscle. *J. Pathol. Bacteriol. 89:* 503–512, 1965.
33. Gilbert, R. K., Chin, B., and Hawk, W. A.: The incidence of necrosis of muscle fibers in Duchenne type muscular dystrophy. *Am. J. Pathol. 43:* 107–122, 1963.
34. Godman, G. C.: Cell transformation and differentiation in regenerating striated muscle. In *Frontiers in Cytology*, ed. by Palay, S. L., pp. 381–416. New Haven, Yale University Press, 1958.
35. Godman, G., Bunting, H., and Melnick, S. L.: The histopathology of Coxsackie virus infection in mice. I. Morphologic observations with four different viral types. *Am. J. Pathol. 28:* 223–257, 1952.
36. Gustafsson, R., Tata, J. R., Lindberg, O., and Ernster, L.: The relationship between the structure and activity of rat skeletal muscle mitochondria after thyroidectomy and thyroid hormone treatment. *J. Cell Biol. 26:* 555–578, 1965.
37. Gwyn, D. G., and Aitken, J. T.: The formation of new motor end-plates in mammalian skeletal muscle. *J. Anat. 100:* 111–126, 1966.
38. Hay, E. D.: Electron microscopic observations of muscle dedifferentiation in regenerating Amblystoma limbs. *Dev. Biol. 1:* 555–585, 1959.
39. Hay, E. D.: The fine structure of differentiating muscle in the salamander tail. *Z. Zellforsch. Mikrosk. Anat. 59:* 6–34, 1963.
40. Heuson-Stiennon, J. A.: Morphogenèse de la cellule musculaire striée au microscope électronique. I. Formation des structures fibrillaires. *J. Microsc. 4:* 657–678, 1965.
41. Howes, E. L., Price, H. M., and Blumberg, J. M.: The effects of a diet producing lipochrome pigment (ceroid) on the ultrastructure of skeletal muscle in the rat. *Am. J. Pathol. 45:* 599–631, 1964.
42. Huxley, H. E.: The double array of filaments in cross striated muscle. *J. Biophys. Biochem. Cytol. 3:* 631–648, 1957.
43. Huxley, H. E.: Electron microscope studies on the structure of natural and synthetic protein filaments from striated muscle. *J. Mol. Biol. 7:* 281–308, 1963.
44. Ishikawa, H., Bischoff, R., and Holtzer, H.: Mitosis and intermediate-sized filaments in developing skeletal muscle. *J. Cell Biol. 38:* 538–555, 1968.
45. Kakulas, B. A., and Adams, R. D.: Principles of myopathology as illustrated in the nutritional myopathy of the Rottnest Quokka. *Ann. N. Y. Acad. Sci. 138:* 90–101, 1966.
46. Kaspar, U., Wiesmann, U., and Mumenthaler, M.: Necrosis and regeneration of the tibialis anterior muscle in rabbit. I. Histological changes following the ligation of the tibialis anterior artery. *Arch. Neurol. 21:* 363–372, 1969.

47. Kelly, A. M., and Zacks, S. I.: The histogenesis of rat intercostal muscle. *J. Cell Biol. 42:* 135–153, 1969.
48. Kelly, A. M., and Zacks, S. I.: The fine structure of motor end-plate morphogenesis. *J. Cell Biol. 42:* 154–169, 1969.
49. Kelly, D. E.: Myofibrillogenesis and Z-band differentiation. *Anat. Rec. 163:* 403–426, 1969.
50. Konigsberg, I. R.: Clonal analysis of myogenesis. *Science 140:* 1273–1284, 1963.
51. Lentz, T. L.: Cytological studies of muscle dedifferentiation and differentiation during limb regeneration of the Newt Triturus. *Am. J. Anat. 124:* 447–480, 1969.
52. Lentz, T. L.: Development of the neuromuscular junction. I. Cytological and cytochemical studies on the neuromuscular junction of differentiating muscle in the regenerating limb of the Newt Triturus. *J. Cell Biol. 42:* 431–443, 1969.
53. Levander, G.: The regeneration of striated muscle. In *Induction Phenomena in Tissue Regeneration*, Chapter 5, pp. 88–104. Baltimore, Williams and Wilkins Co, 1964.
54. Mauro, A.: Satellite cell of skeletal muscle fibers. *J. Biophys. Biochem. Cytol. 9:* 493–495, 1961.
55. McMinn, R. M. H.: The cellular morphology of tissue repair. *Int. Rev. Cytol. 22:* 63–145, 1967.
56. Milhorat, A. T., Shafiq, S. A., and Goldstone, L.: Changes in muscle structure in dystrophic patients, carriers and normal siblings seen by electron microscopy; correlation with level of serum creatine phosphokinase (CPK). *Ann. N. Y. Acad. Sci. 138:* 246–292, 1966.
57. Mintz, B., and Baker, W. W.: Normal mammalian muscle differentiation and gene control of isocitrate dehydrogenase synthase. *Proc. Nat. Acad. Sci. U.S.A. 58:* 592–598, 1967.
58. Nakai, J.: The development of neuromuscular junctions in cultures of chick embryo tissues. *J. Exp. Zool. 170:* 85–106, 1969.
59. Nameroff, M.: Personal communication, 1969.
60. Nameroff, M., and Holtzer, H.: Contact-mediated reversible suppression of myogenesis. *Dev. Biol. 19:* 380–396, 1969.
61. Orci, L., Forssman, W. G., Matter, A., Pictet, R., and Rouiller, C.: Phasenkontrastoptische und Ultrastrukturelle Untersuchungen über Degenerationsformen der Skeletmuskelfasern von Laboratoriumstieren und von Menschen. *Z. Zellforsch. Mikrosk. Anat. 84:* 24–43, 1968.
62. Pearson, C. M.: Pathology of human muscular dystrophy. In *Muscular Dystrophy in Man and Animal*, ed. by Bourne, G. H., and Golarz, M. N., pp. 2–45. Basel, S. Karger, 1963.
63. Pellegrino, C., and Franzini-Armstrong, C.: Recent contributions of electron microscopy to the study of normal and pathological muscle. *Int. Rev. Exp. Pathol. 7:* 139–226, 1969.
64. Peterson, E. R., and Crain, S. M.: Re-innervation of denervated mammalian skeletal muscle *in vitro. Anat. Rec. 160:* 408–409, 1968.
65. Pietsch, P., and Harrold, J.: Enhanced muscle regeneration and increased strength in myopathic animals treated with ethylenedinitramine. *Nature (Lond.) 213:* 601–603, 1967.
66. Price, H., Howes, E. L., and Blumberg, J. M.: Ultrastructural alterations in skeletal muscle fibers injured by cold. I. The acute degenerative changes. *Lab. Invest. 13:* 1264–1278, 1964.
67. Price, H. M., Howes, E. L., and Blumberg, J. M.: Ultrastructural alterations in skeletal muscle fibers injured by cold. II. Cells of sarcolemmal tube. Observations on "discontinuous" regeneration and myofibril formation. *Lab. Invest. 13:* 1279–1302, 1964.
68. Przybylski, R. J., and Blumberg, J. M.: Ultrastructural aspects of myogenesis in the chick. *Lab. Invest. 15:* 836–863, 1966.
69. Reznik, M.: Influence d'une irradiation locale unique sur le muscle strié en régénération. *Arch. Biol. (Liege) 78:* 91–106, 1967.
70. Reznik, M.: Aspect ultrastructural de la dégénérescence du muscle strié ischémié. *Ann. Anat. Pathol. (Paris) 12:* 209–222, 1967.
71. Reznik, M.: Etude autoradiographique du muscle strié en régénération in vivo. *J. Embryol. Exp. Morphol. 19:* 283–297, 1968.
72. Reznik, M.: Thymidine-3H uptake by satellite cells of regenerating skeletal muscle. *J. Cell Biol. 40:* 568–571, 1969.
73. Reznik, M.: Origin of myoblasts during skeletal muscle regeneration. Electron microscopic observations. *Lab. Invest. 20:* 353–363, 1969.
74. Reznik, M.: Satellite cells, myoblasts, and skeletal muscle regeneration. In *Myogenesis and*

*Skeletal Muscle Regeneration*, ed. by Mauro, A., Shafig, S. A., and Milhorat, A. T., pp. 133–156, Amsterdam, Excerpta Medica Foundation, 1970.

75. Reznik, M., and Betz, E. H.: Influence de l'irradiation locale préalable sur les capacités régénératrices du muscle strié squelettique. *Pathol. Eur. 2:* 69–80, 1967.
76. Reznik, M., and Engel, K. W.: Ultrastructural and histochemical correlations of experimental muscle regeneration. *J. Neurol. Sci. 11:* 167–185, 1970.
77. Reznik, M., and Firket, H.: Activité cholinesterasique précoce du muscle strié d'embryon de poulet en régénération "in vitro." *C. R. Soc. Biol. 158:* 1168–1170, 1965.
78. Reznik, M., and Hansen, J. L.: Mitochondria in degenerating and regenerating muscle skeletal muscle. *Arch. Pathol. 87:* 601–608, 1969.
79. Riva, D.: Experimental research into the prevention of hyperplasia of connective scar tissue in torn muscle by use of heparin. *J. Sports Med. Phys. Fitness 6:* 253–258, 1966.
80. Romanul, F. C. A.: Enzymes in muscles. *Arch. Neurol. 11:* 355–368, 1964.
81. Saito, A., and Zacks, S. I.: Fine structure of neuromuscular junctions after nerve section and implantation of nerve in denervated muscle. *Exp. Mol. Pathol. 10:* 256–273, 1969.
82. Schiaffino, S., and Margreth, A.: Coordinated development of the sarcoplasmic reticulum and T system during postnatal differentiation of rat skeletal muscle. *J. Cell Biol. 41:* 855–875, 1969.
83. Shafiq, S. A., and Gorycki, M. A.: Regeneration in skeletal muscle of mouse. Some electron microscope observations. *J. Pathol. Bacteriol 90:* 123–127, 1965.
84. Shafiq, S. A., Gorycki, M. A., and Milhorat, A. T.: An electron microscopic study of regeneration and satellite cells in human muscle. *Neurology (Minneap.) 17:* 567–575, 1967.
85. Shafiq, S. A., Gorycki, M. A., Asiedu, S. A., and Milhorat, A. T.: Tenotomy. Effect on the fine structure of the soleus of the rat. *Arch. Neurol. 20:* 625–633, 1969.
86. Shimada, Y., Fischman, D. A., and Moscona, A. A.: The fine structure of embryonic chick skeletal muscle cells differentiated *in vitro. J. Cell Biol. 3:* 445–453, 1967.
87. Shimada, Y., Fischman, D. A., and Moscona, A. A.: Formation of neuromuscular junctions in embryonic cell cultures. *Proc. Nat. Acad. U. S. A. Sci. 62:* 715–721, 1969.
88. Sloper, J. C., and Pegrum, G. D.: Regeneration of crushed mammalian skeletal muscle and effects of steroids. *J. Pathol. Bacteriol. 93:* 47–63, 1967.
89. Smith, B.: Histochemical changes in muscle necrosis and regeneration. *J. Pathol. Bacteriol. 89:* 139–143, 1965.
90. Stockdale, F. E., Okasaki, K., Nameroff, M., and Holtzer, H.: 5-Bromodeoxyuridine. Effect on myogenesis *in vitro. Science 146:* 533–535, 1964.
91. Studitsky, A. N.: Dynamics of the development of myogenic tissue under conditions of explanation and transplantation. In *Cinematography in Cell Biology*, ed. by Rose, G. G., pp. 171–200. New York, Academic Press, 1963.
92. Susheela, A. K., Hudgson, P., and Walton, J. N.: Histological and histochemical studies of experimentally induced degeneration and regeneration in normal and dystrophic mouse muscle. *J. Neurol. Sci. 9:* 423–442, 1969.
93. Veneroni, G., and Murray, M. R.: Formation *de novo* and development of neuromuscular junctions *in vitro. J. Embryol. Exp. Morphol. 21:* 369–382, 1969.
94. Veress, B., Kerenyi, T., Hüttner, I., and Jellinek, H.: The phases of muscle necrosis. *J. Pathol. Bacteriol. 92:* 511–517, 1966.
95. Walker, B. E.: The origin of myoblasts and the problem of dedifferentiation. *Exp. Cell Res. 30:* 81–92, 1963.
96. Walton, J. N., and Adams, R. D.: The response of the normal, the denervated and the dystrophic muscle cell to injury. *J. Pathol. Bacteriol. 72:* 273–298, 1956.
97. Warren, R. H.: The effect of colchicine on myogenesis *in vivo* in *Rana pipiens* and *Rhodnius prolixus* (Hemiptera). *J. Cell Biol. 39:* 544–555, 1968.
98. Yaffeé, D., and Feldman, M.: The formation of hybrid multinucleated muscle fibers from myoblasts of different genetic origin. *Dev. Biol. 11:* 300–317, 1965.
99. Zelena, J.: The effect of denervation on muscle development. In *The Denervated Muscle*, ed. by Gutman, E., pp. 103–126. Prague, Czechoslovak Academy of Sciences, 1962.

Chapter 11

# Ultrastructural Aspects of the Normal Mammalian Muscle Spindle*

HOWARD A. SCALZI† AND HAROLD M. PRICE

Muscle spindles are the most complex sensory receptors found within skeletal muscle. By light microscopy, the mammalian muscle spindle consists of a group of specialized (intrafusal) muscle fibers, sensory and motor nerves, and blood vessels enclosed within a fusiform capsule.[3-5, 11, 14] Recent electron microscopic studies have improved our knowledge of the structural relationships between the various intrafusal fibers and the sensory innervation of this receptor.[1, 2, 6-10, 11a, 11b, 12-14] This review discusses some of these current ultrastructural concepts of mammalian muscle spindle morphology (Figs. 1 to 11).

## CAPSULE

The fusiform capsule which separates the constituents of the muscle spindle from the other components of the muscle can be arbitrarily divided into an outer and inner capsule (Figs. 1 and 2). The outer capsule is composed of concentric sheets of flattened cells and collagen which are often continuous with the sheath of Henle of the nerves innervating the spindle (Fig. 2). Depending upon the species of animal, each layer of cells may be bounded by an external lamina (basement membrane). Junctional complexes, *i.e.*, desmosomes, zonula occludens, or zonula adherents, are present in areas where two adjacent sheets of cells are approximately 200 to 300 Å apart.[10] These complexes have not been found between the cells of a given sheet.[10] Small arterioles, sensory and motor nerves, and capillaries can penetrate the capsule (Figs. 2 and 3).

An inner capsule, composed of one to three layers of cells, surrounds the intrafusal fibers and is well developed in the mid or sensory region but is not well developed in the polar or motor regions of the receptor (Figs. 1 and 2). The inner capsule inserts into the innermost layer of the outer capsule (Fig. 2). The cells of the inner capsule are more ovoid and contain more cytoplasm than the flattened cells of the outer capsule (Figs. 2, 4, and 5). A periaxial space of variable magnitude, containing collagen and flocculent precipitate, is interposed be-

* This research was supported by funds from the Muscular Dystrophy Associations of America and from the Health Research and Service Foundation of Pittsburgh.

† United States Public Health Service Postdoctoral Fellow (12-F02-NB-2Y272-02).

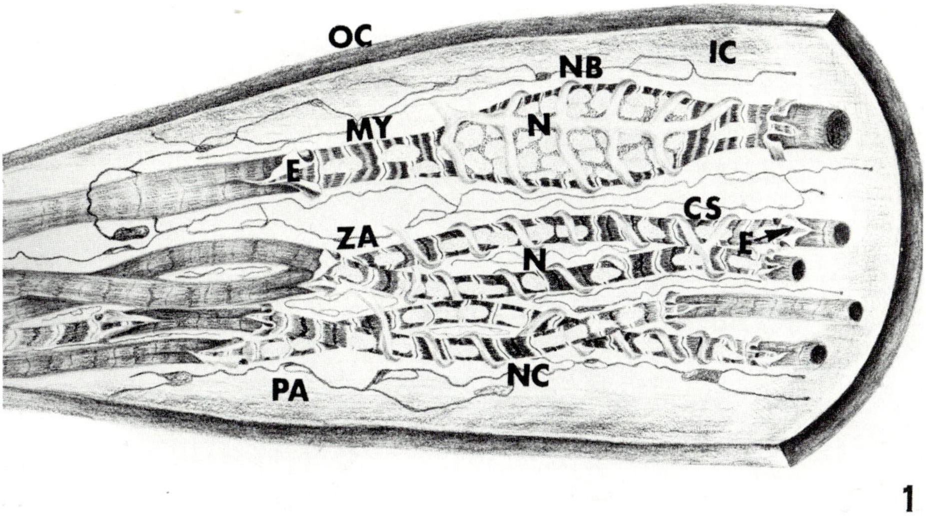


FIG. 1. The mammalian muscle spindle is composed of nuclear-bag (*NB*) and nuclear-chain fibers (*NC*) surrounded by a fusiform outer capsule (*OC*). This drawing depicts a portion of the OC region of the feline muscle spindle. The primary sensory innervation (*N*) of the nuclear-bag fibers may not always have an annulo-spiral configuration. Here the nerve (*N*) appears to form a "cage" around the fiber. The nuclear-chain fibers form zones of close apposition (*ZA*) and are surrounded by a continuous external laminal sac (*E*). Sensory nerves which innervate one fiber may send branches across (*CS*) to adjacent nuclear-chain fibers. Cross-sensory innervation (*CS*) of nuclear-chain fibers has also been observed in rat[8] and sheep.[13] *MY*, myotube region; *PA*, periaxial space.

tween the outer and inner capsules or outer capsule and intrafusal fibers depending upon the region of the receptor under examination (Figs. 1 to 4).

## INTRAFUSAL FIBERS

### NUCLEAR-BAG FIBERS

Two types of intrafusal fibers, large nuclear-bag fibers (22 $\mu$) and smaller nuclear-chain fibers (11 $\mu$), have been identified in most mammalian muscle spindles. The one or two nuclear-bag fibers extend the length of the receptor.[4] From its cytologic appearance, these fibers can be divided symmetrically at the nuclear-bag into a central zone filled with nuclei and into peripheral myotube and polar regions (Fig. 1). In the poles, at the ends of the nuclear-bag fibers, myofibrils are closely packed and the elongate nuclei found in this region have either a central or peripheral orientation. Scattered clumps of mitochondria, Golgi elements, granular endoplasmic reticulum, and centrioles are found near the nuclei. The transition of a fiber crowded with myofibrils in the polar regions to the less compact myotube region is gradual. There is a progressive increase in the number of central nuclei with the typical perinuclear cytoplasmic constituents. The junction of the myotube and equatorial nuclear-bag region is more abrupt. A very thin zone of peripheral myofibrils is present in the nuclear-bag region with its unique cluster of closely packed nuclei (Figs. 1, 3, 4, 5, and 6). The myofibrils of these fibers do not possess a distinct M band (Fig. 6).

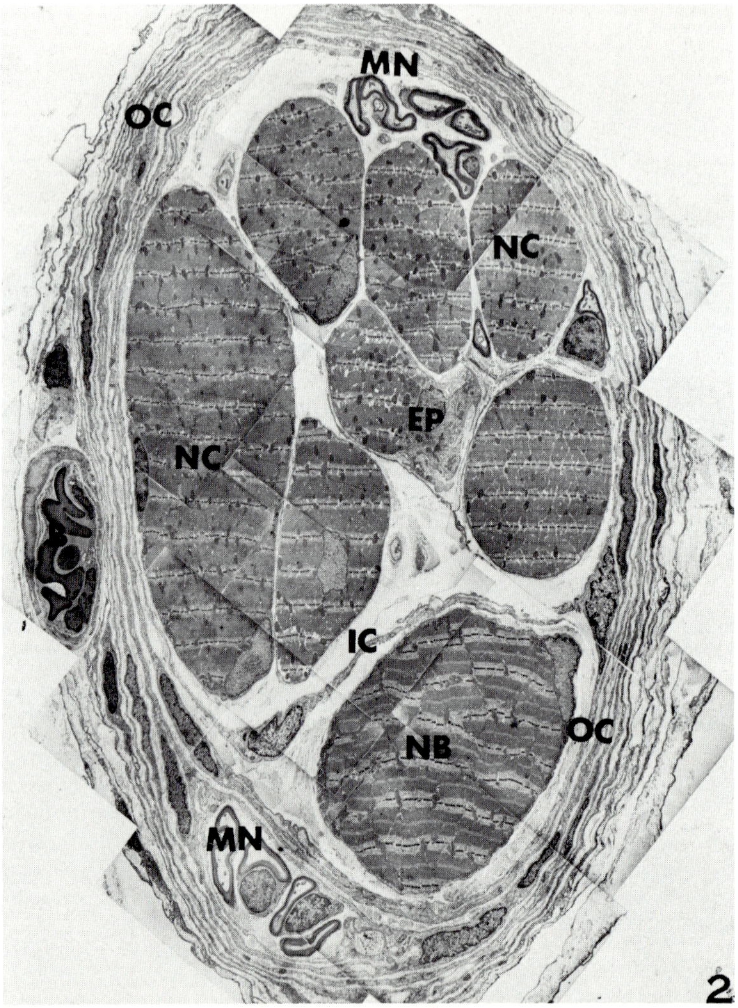


FIG. 2. A montage of electron micrographs reveals the arrangement of the intrafusal fibers in the motor region of the muscle spindle. The seven nuclear-chain fibers appear to be separated from the nuclear-bag fiber by a cellular extension of the inner layer of the outer capsule (*OC*). Blood vessels (*B*) and myelinated nerves (*MN*) can be found in the capsule. *EP*, motor end-plate on a nuclear-chain fiber; IC, inner capsule. × 2000.

Although classic light microscopic studies have indicated that the primary sensory innervation at the nuclear-bag region of the fibers, supplied by the group Ia sensory nerves, was as an "annulo-spiral," recent electron microscopic observations show that the primary innervation is more variable.[12, 14] Many times, it also has a "rib-cage-like" orientation (Figs. 1, 3, and 4). This conclusion was arrived at by the examination of numerous longitudinal sections through a nuclear-bag fiber (Figs. 3 and 4). A loss of symmetry around the bag region which would be consistent with a tightly wound coil was noted as were large Y-shaped nerves and interconnections between the "coils" (Figs. 3 and 4). Secondary sensory innervations, supplied by group II sensory nerves, may be

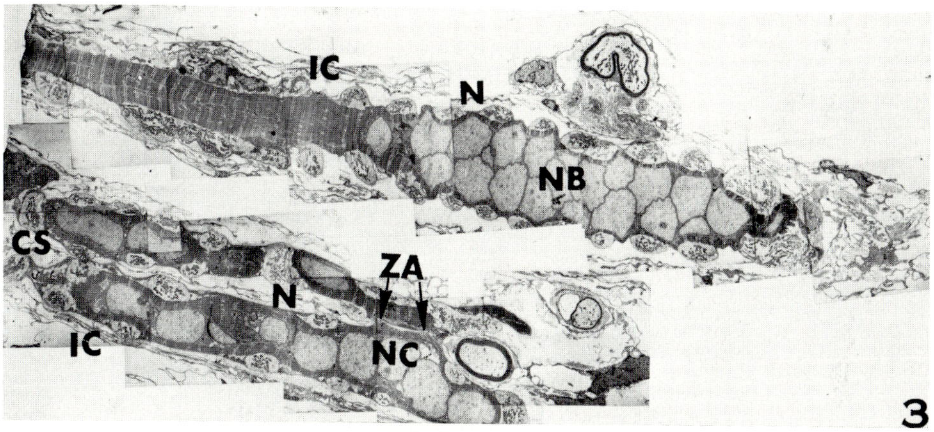


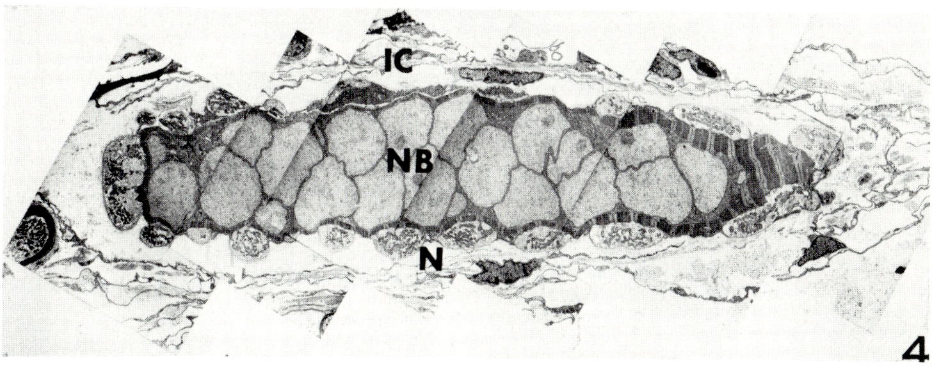


FIG. 3. The characteristic arrangement of the nuclei of the nuclear-bag (*NB*) and nuclear-chain fibers (*NC*) is depicted in this longitudinal section through the expanded central (sensory) region of a muscle spindle. The sensory innervation (*N*) of both types of intrafusal fibers can be compared. The sensory nerves encompassing the nuclear-bag fiber are suggestive of a spiral (coil) symmetry. *IC*, inner capsule; *ZA*, zones of apposition; *CS*, cross-sensory innervation. × 1000.

FIG. 4. The spiral symmetry of the sensory nerve (*N*) around the nuclear-bag fiber (*NB*) as suggested in Fig. 3 is not present at all depths of the same fiber. Other patterns of innervation are indicated. *IC*, inner capsule. × 1800.

localized along the myotube region of the fiber (Fig. 3). Both types of sensory nerves are found in shallow troughs with approximately 50% of the surface area of the nerve in contact with the muscle (Fig. 6). These nerves penetrate the external lamina (basement membrane) of the muscle fiber and lie immediately adjacent to the sarcolema. The motor innervation of the nuclear-bag fibers, on the other hand, is localized toward the myotube and polar regions of these fibers. Trail endings (terminaison en ligne) and plate endings (terminaison en plaque) have been identified (Figs. 9 and 10). The motor nerves, unlike the sensory nerves, are separated from the plasma membrane of the muscle by the external lamina (Fig. 10).

NUCLEAR-CHAIN FIBERS

Mammalian muscle spindles possess three to nine nuclear-chain fibers. These small diameter (11 $\mu$) muscle fibers may or may not, depending on the species,

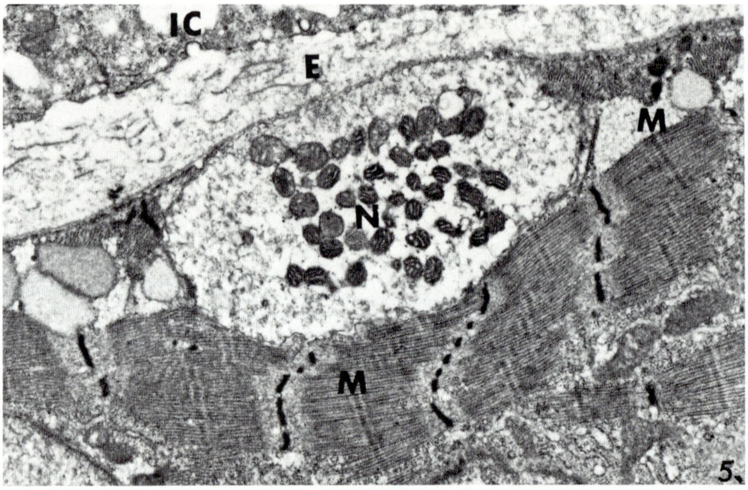


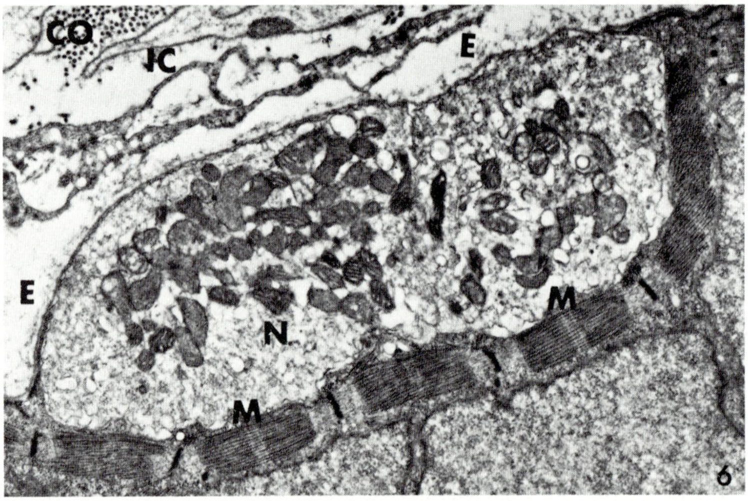


FIG. 5. The sensory nerves (*N*) innervating a nuclear-chain fiber lie in a deep trough with approximately 75% of the nerve in contact with the muscle. The external lamina (*E*) of the muscle covers the upper aspect of the nerve but none is found between nerve and muscle. Another distinguishing characteristic of the nuclear-chain fiber is the appearance of a distinct M line (*M*) in the myofibrils. *IC*, inner capsule. × 11,800.

FIG. 6. The sensory innervation (*N*) of the nuclear-bag fibers lies in shallow troughs. The M line (*M*) is less distinct. *CO*, collagen; *E*, external lamina; *IC*, inner capsule. × 15,700.

extend the length of the muscle spindles.[4] They can insert into the capsule. The nuclear-chain fibers are characterized by a chainlike alignment of nuclei within the mid-region of the fiber, a distinct M band in the myofibrils, and more interfibrillar sarcoplasm (Figs. 1, 3, and 4). These fibers form zones of close apposition where the external lamina which surrounds the adjacent fibers is continuous and the opposing plasma membranes are separated by only a 20- to 80-mμ space (Fig. 7). The zones of close apposition were initially described in cats by Adal[1] as the phenomenon of pairing. Subsequently, Scalzi and Price[13, 14] demonstrated that all of the nuclear-chain fibers in a given muscle spindle

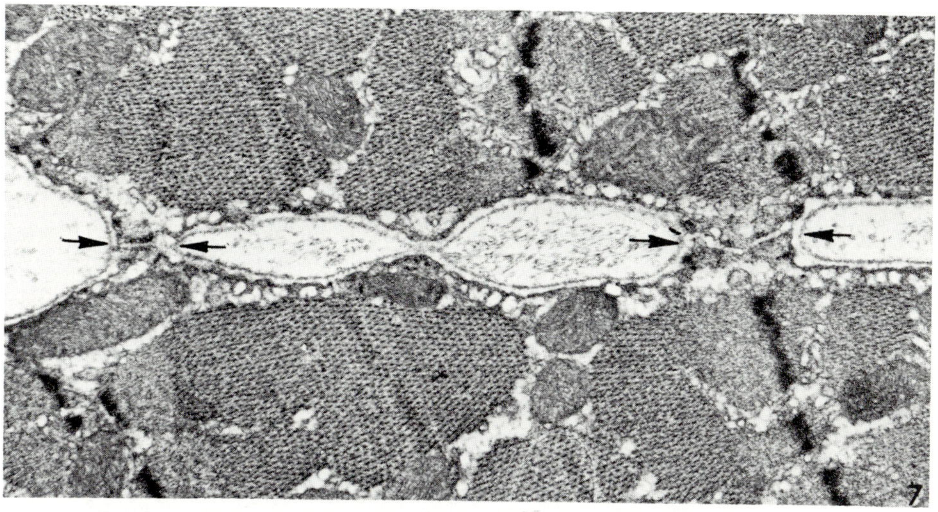


FIG. 7. At the zones of close apposition the external lamina which surrounds the nuclear-chain fiber is continuous (*arrows*), and the adjacent plasma membranes are approximately 20 m$\mu$ apart. × 22,500.

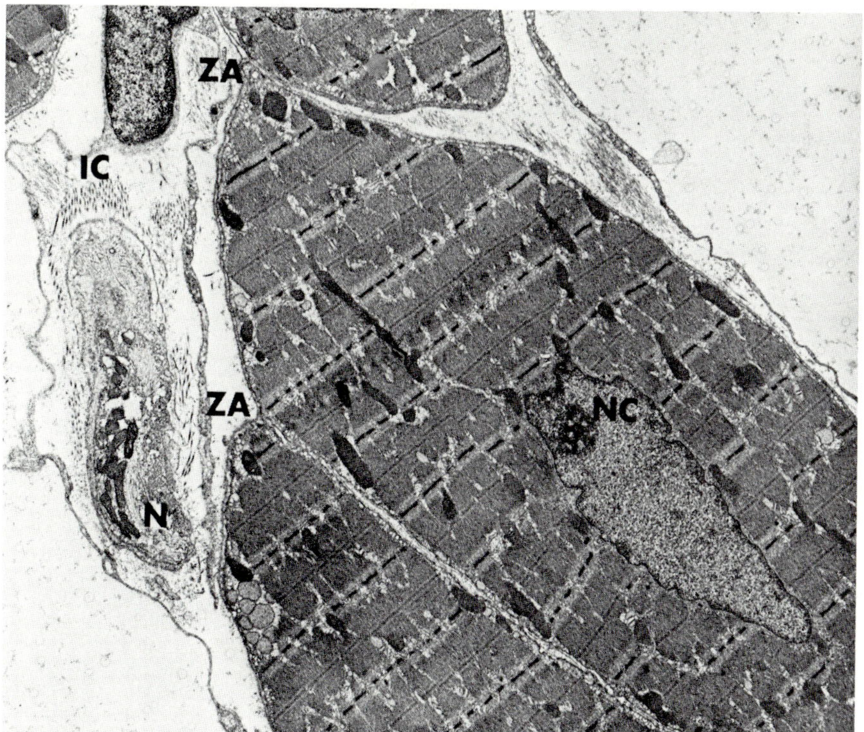


FIG. 8. In this micrograph three nuclear-chain fibers (*NC*) form zones of close apposition (*ZA*) with each other. All nuclear-chain fibers are thought to form zones of close apposition with each other and to be enclosed in a common external laminal sac. *N*, sensory nerve; *IC* inner capsule. × 8200.

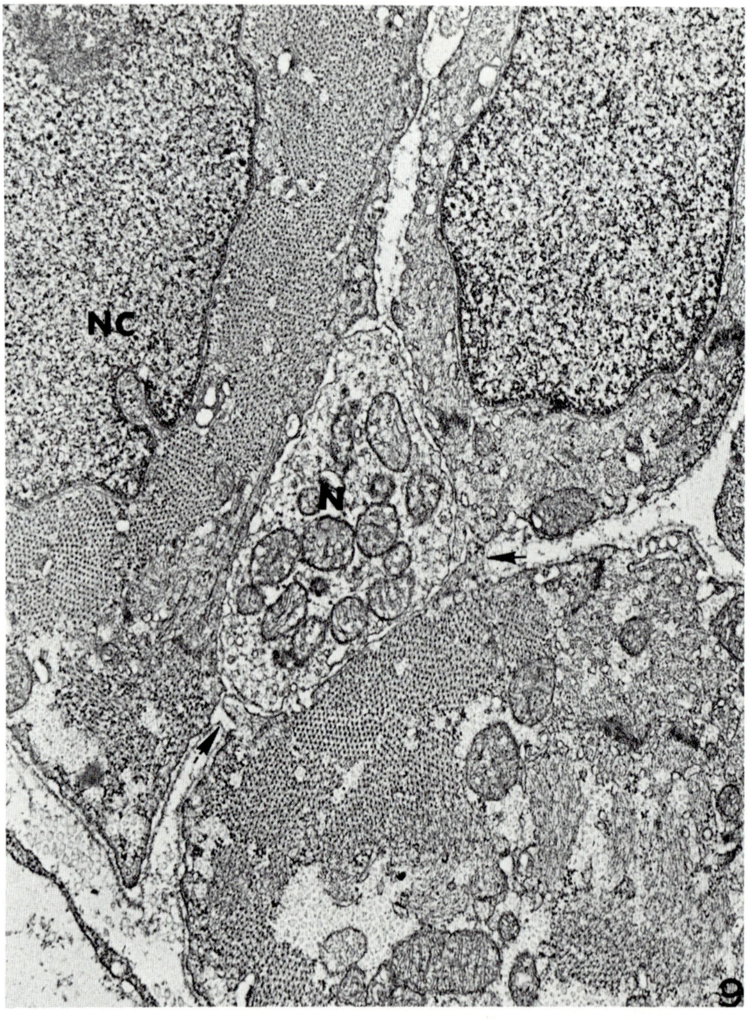


FIG. 9. In the newborn kitten one sensory nerve (*N*) can innervate three adjacent nuclear-chain fibers (*NC*). *Arrows* indicate site of continuity of the external lamina. × 19,400.

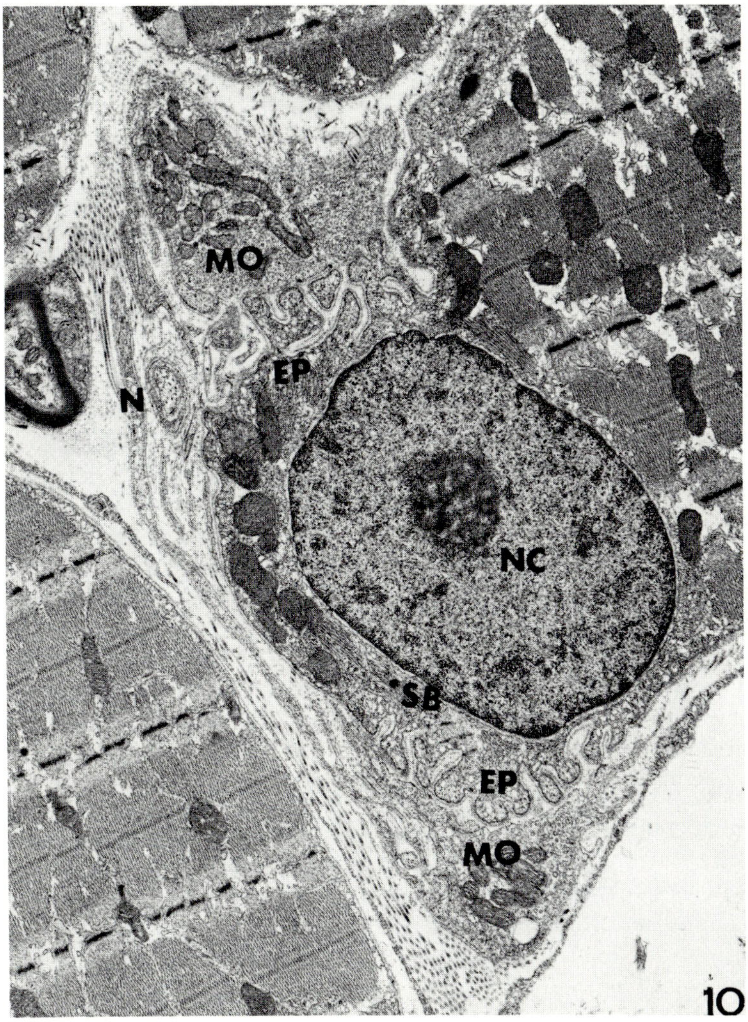


FIG. 10. A section through the motor region of this nuclear-chain fiber (*NC*) reveals that the motor end-plates (*EP*) are similar to those found in extrafusal fibers[2]; however, these motor end-plates do not markedly indent the surface of the fibers as do extrafusal end-plates. This type of ending has been called a $P_2$ plate. *MO*, motor nerve; *N*, small unmyelinated nerve; *SB*, striated body. × 9300.

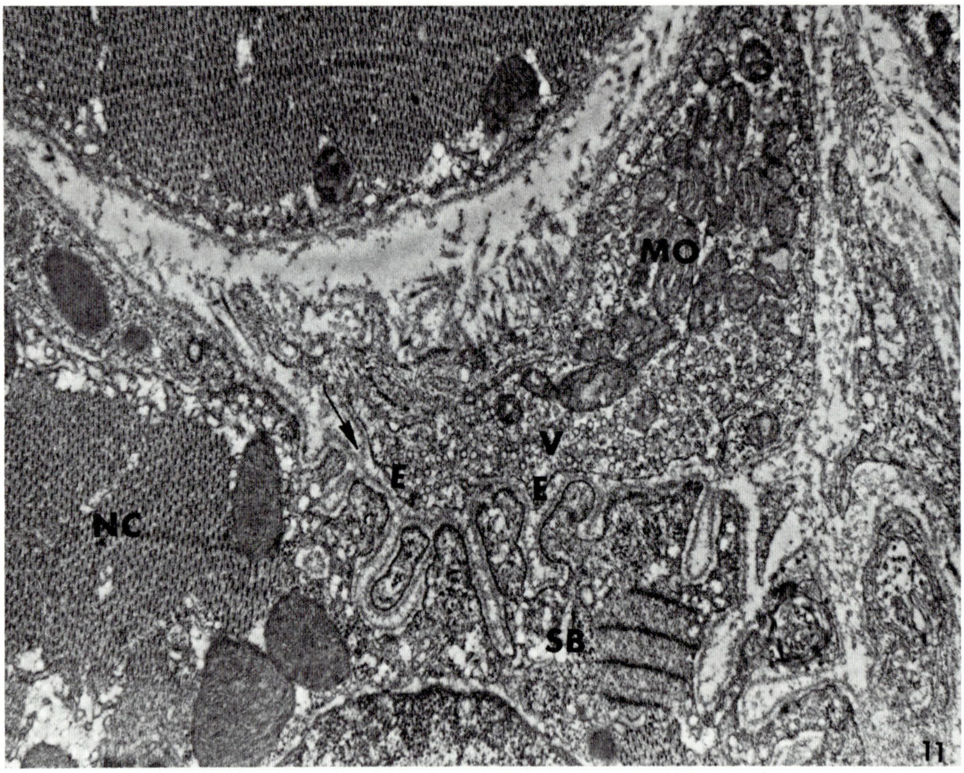


FIG. 11. The characteristics of a $P_2$ end-plate are depicted in this micrograph. A single external lamina (*E*) is present between the nerve (*MO*) and muscle. *Arrows* demonstrate areas of continuity of the external lamina of the muscle and nerve. *V*, synaptic vesicles; *NC*, nuclear-chain fibers; *SB*, striated bodies. × 21,600.

formed zones of close apposition with each other and that a single external lamina sac appears to enclose the group of nuclear-chain fibers (Figs. 1 and 8). Some of the zones of close apposition contain sensory nerve twigs which cross between adjacent fibers (Fig. 1 and 9). These sensory nerves innervating the nuclear-chain fibers lie in deeper troughs than those innervating nuclear-bag fibers, with approximately 75% of the surface of the nerve in contact with the muscle (Fig. 5). The motor endings of nuclear-chain fibers are similar to those of nuclear-bag fibers (Figs. 9 to 11).

## REFERENCES

1. Adal, M. N.: The fine structure of the sensory region of cat muscle spindle. *J. Ultrastruct. Res. 26:* 332–353, 1969.
2. Adal, M. N. and Barker, D. J. The fine structure of cat fusimotor endings. *Physiologist 192:* 50–52, 1967.
3. Boyd, I. A. The structure and innervation of the nuclear bag muscle fiber system and the nuclear chain muscle fiber system in mammalian muscle spindles. *Phil. Trans. Roy. Soc. Lond. B 245:* 81, 1962.
4. Bridgman, C. F., Shumpert, E., and Eldred, E.: Insertions of intrafusal fibers in muscle spindles of the cat and other mammals. *Anat. Rec. 164:* 391–402, 1969.

5. Cooper, S.: Muscle spindles and other muscle receptors. In *The Structure and Function of Muscle*, ed. by Bourne, G. H., vol. I. Chapter XI. New York, Academic Press, 1960.

5a. Corvaja, N., Marinozzi, U., and Pompeiano, O.: Muscle spindles in the lumbrical muscle of the adult cat. Electron microscopic observations and functional considerations. *Arch. Ital. Biol. 107*, 365–543, 1969.

5b. Corvaja, N., and Pompeiano, O.: The differentiation of two types of intrafusal fibers in rabbit muscle spindles. *Pflugers Arch. Ges. Physiol. 317*, 187–197, 1970.

6. During, V. M., and Andres, K. H.: On the fine structure of muscle spindle of mammals. *Anat. Anz. 124:* 566–573, 1969.

7. Gruner, J. E.: The fine structure of the human neuromuscular spindle. *Rev. Neurol. (Paris) 104:* 490–507, 1961.

8. Hennig, G.: Nerve endings of rat muscle spindle studied by electron and phase contrast microscopy. *Z. Zellforsch Mikrosk. Anat. 96:* 275–294, 1969.

9. Landon, D. N.: In *Control and Innervation of Skeletal Muscles*, ed. by Andrews, B. L., p. 96, E. & S. Livingston, London, 1966.

10. Merrillees, N. C. R.: The fine structure of muscle spindles in the lumbrical muscles of the cat. *J. Biophys. Biochem. Cytol., 7:* 725–742, 1960.

10a. Ovalle, W. K., Jr.: Fine structure of rat intrafusal muscle fibers. The Polar region. *J. Cell Biol.* 83–103, 1971.

10b. Ovalle, W. K., Jr.: Motor nerve terminals on rat intrafusal muscle fibers, a correlated light and electron microscopic study. *J. Anat. 111*, 239–252, 1972.

10c. Ovalle, W. K., Jr.: Fine structure of rat intrafusal muscle fibers. The equatorial region. *J. Cell Biol.*, 52:382–396, 1972.

10d. Ovalle, W. K., Jr. and Smith, R. S.: Histochemical identification of three types of intrafusal muscle fibers in the cat and monkey based on the myosin ATP-ase reaction. *Can. J. Physiol. Pharmacol. 50*, 195–202, 1972.

11. Ruffini, A.: On the minute anatomy of the neuromuscular spindles of the cat and on their physiological significance. *J. Physiol. 23:* 190–208, 1898.

11a. Scalzi, H. A.: Observation of the secondary sensory innervation of nuclear-chain fibers from the feline muscle spindle. *J. Cell Biol. 55*, 228a, 1972.

11b. Scalzi, H. A., and Price, H. M.: Electron-microscopic observations of the sensory region of the mammalian muscle spindle in research. In *Muscle Development and the Muscle Spindle*, ed. by Banker, B. Q., Przybylski, R. J., Van Der Meulen, J. P., and Victor. Excerpta Medica, International Congress Series. No. 240, 1972.

12. Scalzi, H. A., and Price, H. M.: Ultrastructure of the sensory region of the mammalian muscle spindle. *J. Cell Biol. 43:* 124a, 1969.

13. Scalzi, H. A., and Price, H. M.: The arrangement and sensory innervation of the intrafusal fibers in the feline muscle spindle. *J. Ultrastruct. Res. 36*, 375–390, 1971.

14. Scalzi, H. A., and Price, H. M.: Ultrastructure of muscle spindles from the extraocular muscles of sheep. *J. Cell Biol. 47*, 180a, 1970.

Chapter 12

# THE INFLUENCE OF THE NERVOUS SYSTEM ON MUSCLE STRUCTURE AND FUNCTION

VICTOR DUBOWITZ

Every once in a while someone does an experiment which is so simple in its design and so fundamental in its importance that it influences thinking for many years to follow. I think this is true of the fascinating cross-innervation experiment of Buller, Eccles, and Eccles.[4] They sectioned the nerves to the slow soleus and the fast flexor hallucis longus (FHL) muscles in the cat, crossed the nerves over, and then approximated the proximal stumps to the distal stumps of the opposite muscle. After reinnervation they found that stimulation of the original soleus nerve now innervating FHL produced a slow contraction in that muscle, while stimulation of the nerve of FHL now supplying soleus produced a fast contraction in that previously slow muscle (Fig. 1).

They postulated that this was due to a chemical influence from the nervous system, passing down the axon of the peripheral nerve and finding its way into the muscle fiber where it exerted its influence.

In a further but somewhat limited experiment, they showed that isolation of the segment of cord innervating the muscles concerned, by transection and deafferentation of the cord, inhibited the effects of cross-union. They suggested that the influence might in fact be supranuclear, thus coming from the central nervous system rather than being purely localized to the peripheral nerve.

Physiologists pursued this experimental model and similar studies on cross-union were subsequently reported in the rat,[5, 34] the guinea pig,[32] the chick,[25, 27] and the frog.[30]

## HISTOCHEMICAL STUDIES

Since the contractile properties of muscle must be closely correlated with the chemical character of the fibers involved, this experimental model was obviously an interesting one for the application of other techniques for the assessment of muscle function.

*Fiber types*. By enzyme histochemical methods one can distinguish different fiber types within skeletal muscle.[13, 14, 21, 33, 38] There is still controversy on the

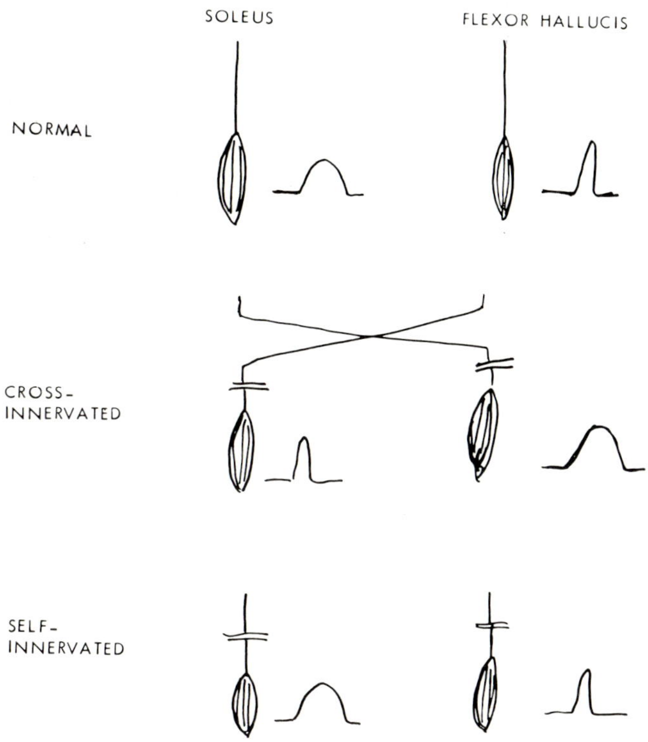


FIG. 1. Diagramatic representation of change in twitch contraction of the slow soleus muscle and fast flexor hallucis after cross-innervation.

nomenclature of fiber types and how many different types there are. I have used a simple subdivision into two types; type 1 fibers are rich in oxidative enzymes and poor in phosphorylase and myosin ATPase, and type 2 fibers are rich in phosphorylase and ATPase and poor in oxidative enzymes. I also recognized the existence of intermediate fibers.[14] Engel[21] has also adopted this two-fiber type classification, but he prefers the ATPase reaction as the main yardstick.

A two-fiber type classification is particularly useful in dealing with human muscle, both normal and diseased. In animals, many workers have preferred a three-fiber classification.[38]

The slow soleus muscle of animals such as the cat, rabbit, guinea pig, and rat consists mainly or entirely of type 1 fibers (Fig. 2), while fast muscles such as flexor hallucis and flexor digitorum comprise a mixture of fibers with a predominance of type 2 fibers (Fig. 3).

*Cross-innervation.* Following cross-innervation of the soleus and flexor digitorum muscles in the cat and rabbit, we demonstrated a conversion in fiber type.[8] The flexor digitorum acquired large plaques of type 1 fibers indistinguishable from normal soleus muscle (Fig. 4), while the soleus developed areas of type 2 fibers. The changeover in the flexor digitorum longus seemed to be more extensive than that in soleus. In a similar study in the rat and cat, Romanul and Van der Meulen[34] showed a transformation of fiber types of equal

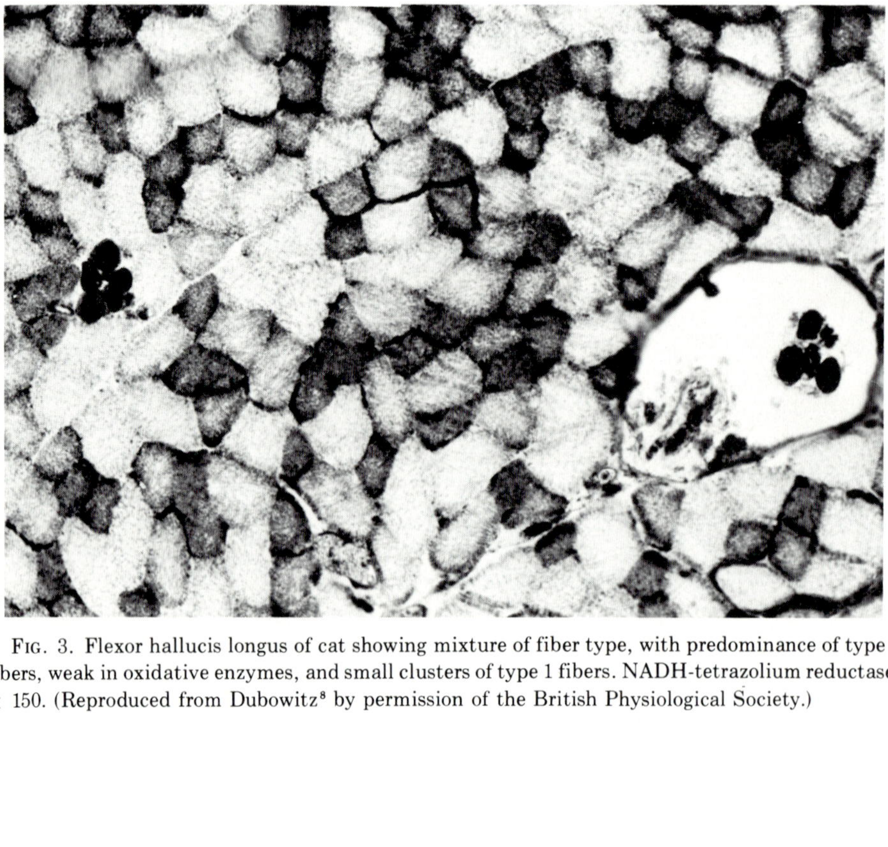

FIG. 3. Flexor hallucis longus of cat showing mixture of fiber type, with predominance of type 2 fibers, weak in oxidative enzymes, and small clusters of type 1 fibers. NADH-tetrazolium reductase; × 150. (Reproduced from Dubowitz[8] by permission of the British Physiological Society.)

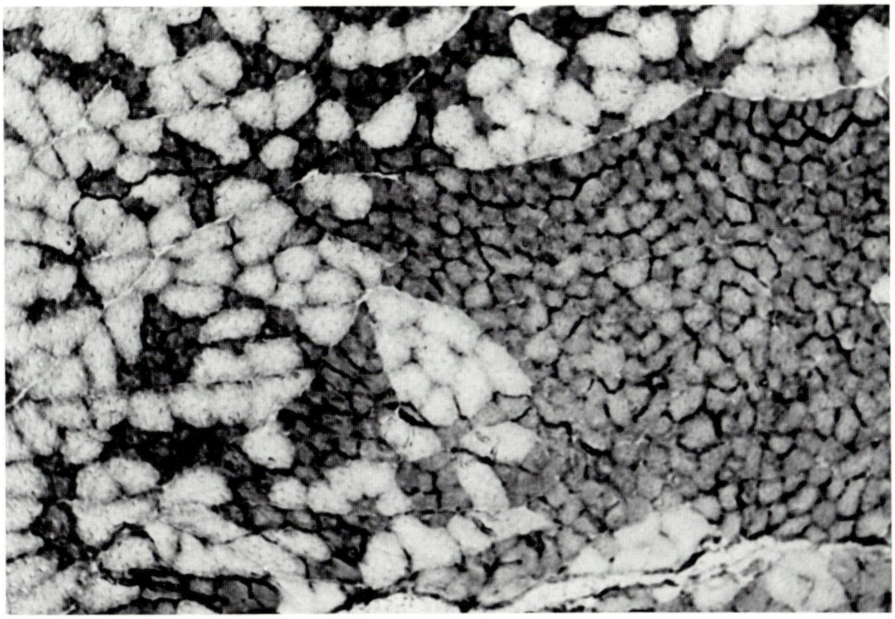

FIG. 4. Flexor hallucis longus of cat after cross-innervation by nerve from soleus. Shows large area of type 1 fibers resembling normal soleus. NADH-tetrazolium reductase; × 80. (Reproduced from Dubowitz[8] by permission of the British Physiological Society.)

extent in both the soleus and flexor digitorum muscles of these animals. Guth, Watson, and Brown[26] repeated the experiment on the soleus and gastrocnemius of the rat and obtained a more marked transformation in soleus than in gastrocnemius. In the guinea pig, Karpati and Engel[28] studied the effects on the soleus muscle of three different procedures: implantation of the peroneal nerve after denervation of soleus; sectioning and resuturing after rotation of the sciatic nerve; and crushing of the sciatic or post-tibial nerves. Each of these procedures resulted in the appearance of a variable proportion (5 to 40%) of type 2 fibers in the soleus, which is normally a purely type 1 fiber muscle in the guinea pig. Yellin[40] sectioned and resutured the sciatic nerve in the rat and, on assessment 6 months later, found islands of one fiber type in the soleus, the plantaris, and gastrocnemius, which were either alien to that muscle or present in larger clusters than normal.

These experiments have clearly shown that the fiber type in skeletal muscle can be changed under the influence of alien innervation.

*Developing muscle.* Buller, Eccles, and Eccles[3] observed that all muscles in the kitten were slow at birth and that fast muscles such as flexor hallucis changed their contractile properties in the course of the first few weeks after birth. Histochemical studies have shown a similar pattern.[6, 10] In many animals such as the rat and rabbit the muscles are initially of uniform fiber type and differentiate after the first week or two. Other animals such as the guinea pig may already be differentiated at birth. It is possible that a similar nervous influence is responsible for the maturation of normal muscle during develop-

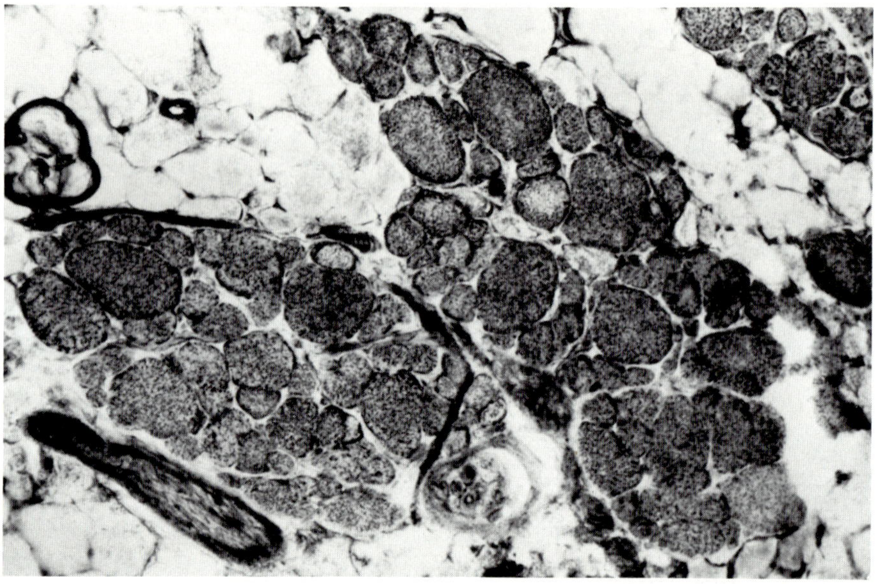

FIG. 5. Flexor digitorum longus muscle of rabbit 5 weeks after cross-innervation. Note variation in fiber size and excess of perimysial adipose tissue, giving appearance resembling a myopathy. NADH-tetrazolium reductase; × 150. (Reproduced from Dubowitz[9] by permission of the British Medical Association.)

ment as in the change in fiber type after cross-innervation. This is supported by the observation of Karpati and Engel[29] that section of the sciatic nerve in the newborn rat may delay the maturation into fiber types in the muscle.

*Structural changes*. In the cross-innervation experiment above, most animals were studied 6 months or more after operation, by which time the muscle fibers were histologically normal and the over-all architecture resembled normal muscle. In a series of rabbits, however, a number of animals were studied for a relatively short period after cross-innervation. These muscles showed some striking structural abnormalities[9] (Figs. 5 and 6). Many showed a marked variation in fiber size and shape and an excess of perimysial adipose tissue, a pattern strikingly reminiscent of muscular dystrophy. Many fibers showed aggregations of mitochondria or central zones devoid of mitochondria, as judged by mitochondrial enzyme stains. Morphologic abnormalities took various forms. The most common was "snake-coils" which showed some bizarre appearances. Other fibers seemed to be subdivided into separate components and a few resembled ringbinden. These abnormal fibers were often clustered together. It was suggested that these changes might be a feature of denervated muscle in the early phases of reinnervation. It is of interest that Karpati and Engel[29] were able to produce similar structural changes in the skeletal muscle of the rat after cordotomy, suggesting that the influence of the central nervous system on the lower motor neuron may be of importance in relation to cyto-architectural changes in muscles as well as the histochemical pattern. In these experiments it is difficult to assess the possible role of disuse or immobility on the structure of the muscle fiber.

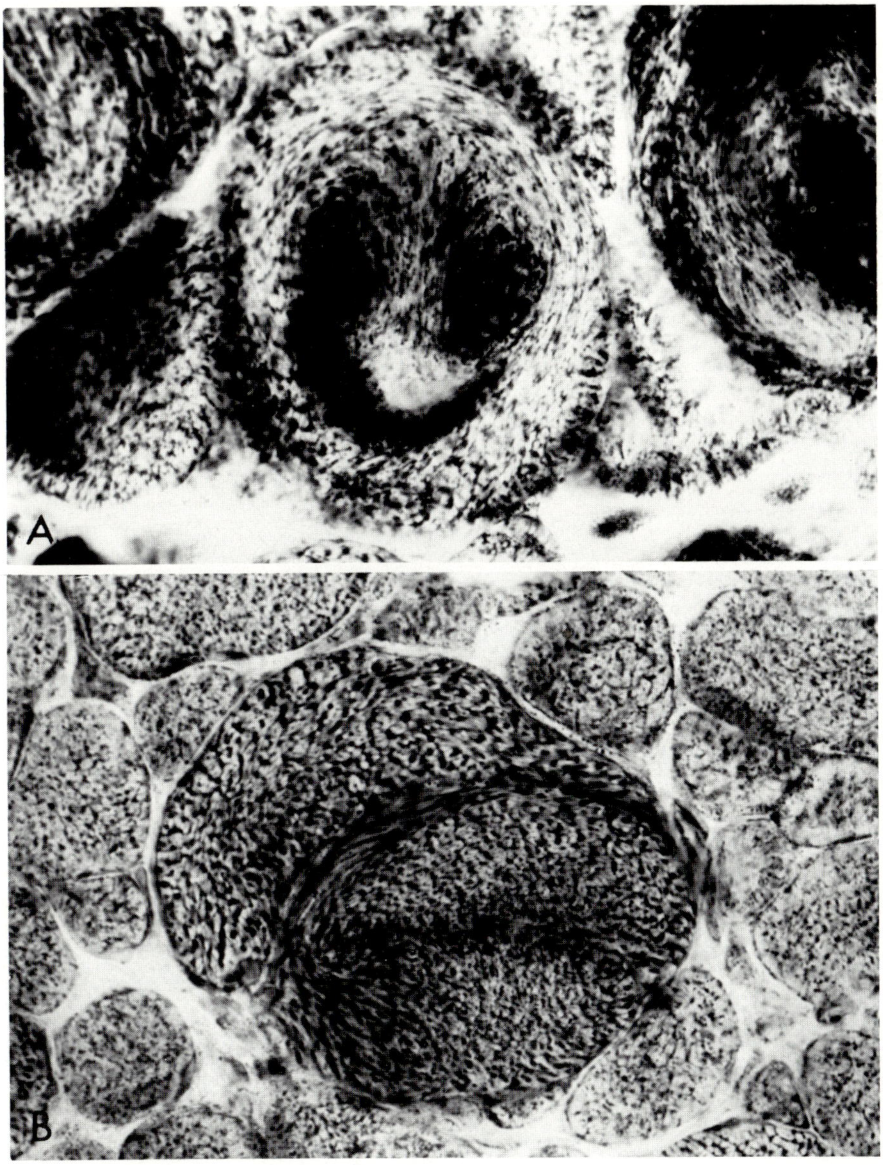


FIG. 6. Rabbit soleus (*a*) and flexor digitorum (*b*) 5 weeks after denervation, showing bizarre structural changes and distribution of enzyme activity. NADH-tetrazolium reductase; × 500. (Reproduced from Dubowitz[9] by permission of the British Medical Association.)

## OBSERVATIONS IN HUMAN DISEASE

These experimental data have helped to clarify a number of aspects of muscle pathology.

*Neurogenic atrophies.* One of the striking features in various syndromes of neurogenic atrophy, such as infantile spinal muscular atrophy, is that the clusters of normal-sized or enlarged fibers seen in histologic sections (Fig. 7) often show a uniformity of histochemical fiber type[7, 16] (Fig. 8). This would suggest that these apparently normal fibers are not simply unaffected fibers, since they would then be expected to have the checkerboard pattern of two fiber types of normal muscle. The uniform pattern in these fibers is best explained on

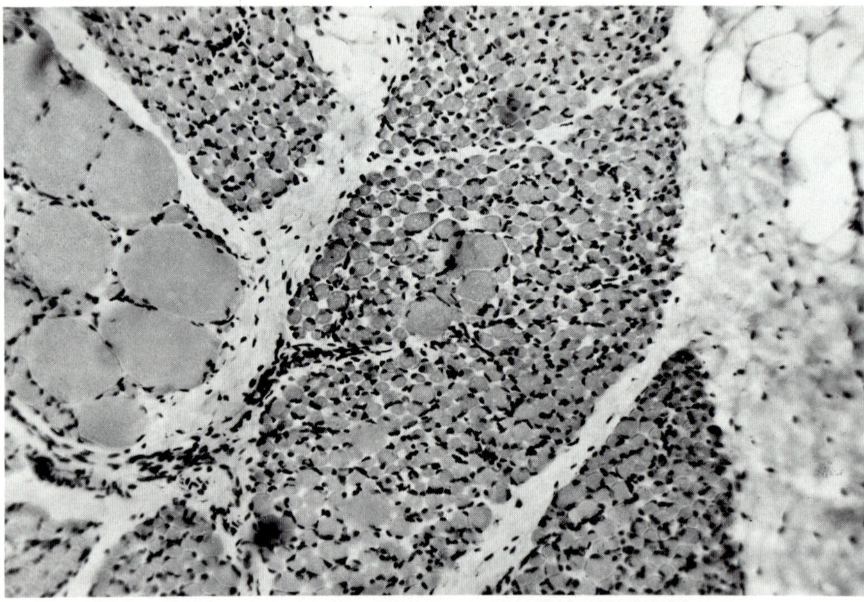

FIG. 7. Spinal muscular atrophy (severe form); aged 2 years; quadriceps. Note uniform atrophy of large groups of fibers and clusters of normal looking fibers. Hematoxylin and eosin stain; × 150.

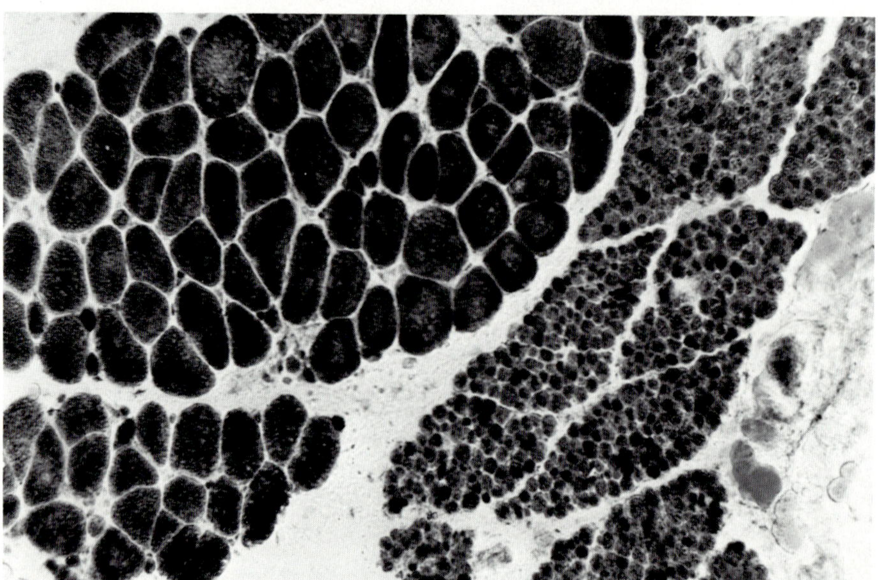

FIG. 8. Spinal muscular atrophy; same muscle as shown in Figure 7. Note uniformity of enzyme activity in large fibers, suggesting reinnervation by nerves of one type. Atrophic fibers show mixed fiber pattern. NADH-tetrazolium reductase; × 130.

the basis of reinnervation of denervated fibers, presumably from collateral sprouting of surviving nerves, which produce muscle fibers of one type. Moreover, Engel[22] has observed fiber type grouping in neurogenic atrophies, even in the absence of the usual histologic changes of denervation, and Morris[31] has recently shown a correlation between collateral sprouting of nerve terminals, as demonstrated by methylene blue staining, and grouping of fibers of one histochemical type.

Another interesting histochemical pattern was the selective atrophy of type 1 fibers only in the muscle of a child with metachromatic leucodystrophy, which was the clue to an associated peripheral nerve involvement.[12] Changes of this type can only be recognized by histochemical methods and are not apparent on routine histologic staining.

In an adult with benign proximal neurogenic atrophy (Kugelberg-Welander syndrome) I have recently observed on histochemical study structural changes comparable to those seen in the rabbit muscle during reinnervation. Routine histologic stains showed group atrophy of fibers interspersed between normal-sized or enlarged fibers which appeared histologically normal (Fig. 9). Histochemical enzyme stains, however, showed bizarre distribution of oxidative enzymes and associated "snake coils" and other changes (Fig. 10). I have not observed similar changes in cases of severe infantile muscular atrophy of the Werdnig-Hoffmann type, and they may thus be a feature only of the more benign and long standing cases where there is more active reinnervation.

*Myopathies*. Specific fiber type atrophy may also be a feature of some other neuromuscular disorders. Thus in dystrophia myotonica there is a tendency to

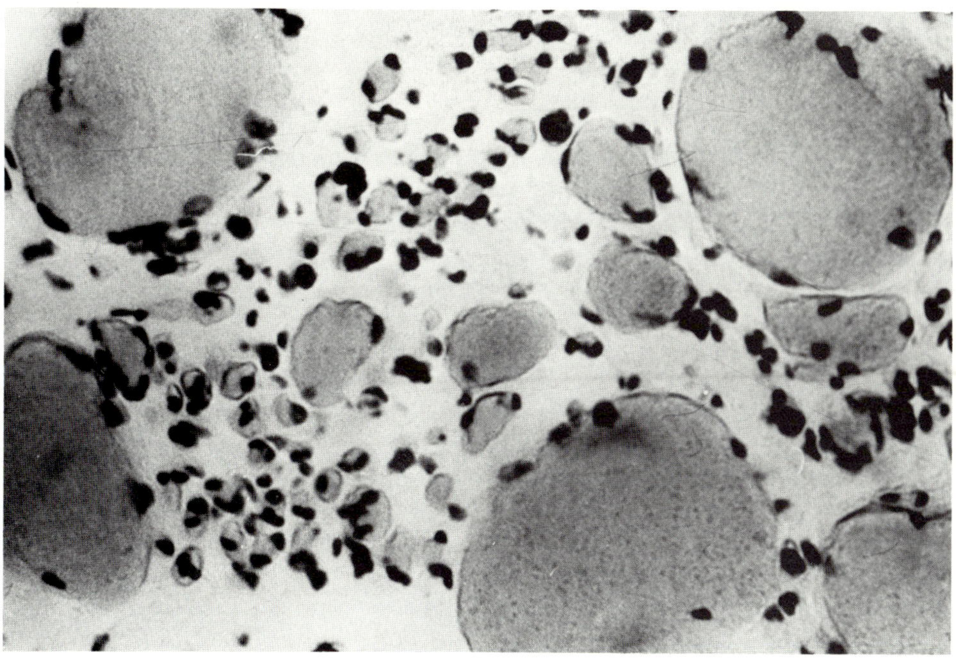

FIG. 9. Spinal muscular atrophy (mild form); aged 21 years. Shows clusters of atrophic fibers and isolated normal looking fibers. Hematoxylin and eosin stain; × 360.

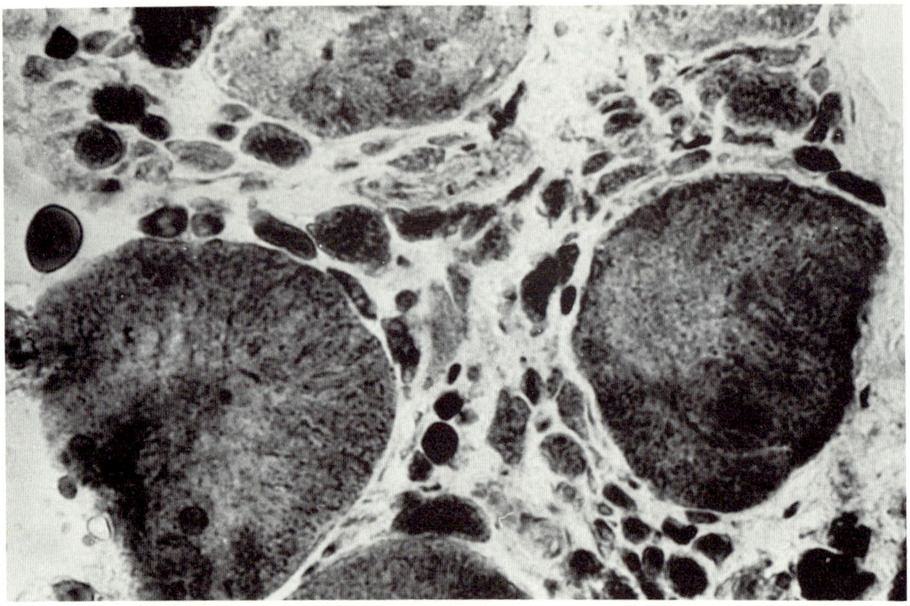

FIG. 10. Spinal muscular atrophy (mild form); same muscle as shown in Figure 9. Note bizarre appearance of the large fibers; compare with Fig. 6.

type 1 fiber atrophy. In addition, a number of specific disease entities have a selective involvement of one fiber type only. Central core disease of muscle, in which a "core" within the fiber is devoid of mitochondrial as well as sarcoplasmic enzymes, selectively affects type 1 fibers. In the first case studied histochemically the muscle appeared undifferentiated and composed of one fiber type only, approximating to type 1 fibers with a high oxidative enzyme content.[15] In a subsequent case I studied,[17] the muscle showed a normal subdivision into two fiber types, but only type 1 fibers were affected. I have recently had the opportunity of studying another typical example of central core disease affecting both mother and son.[19] The mother's muscle was practically undifferentiated (99% type 1 fibers) and every type 1 fiber had a single, centrally placed, central core (Fig. 11). The muscle of the 4-year-old boy, in contrast, showed a normal subdivision into fiber types (37% type 1, 63% type 2), and only 3% of his type 1 fibers had central cores, which were single but eccentrically placed (Fig. 12).

In the light of our observation on the presence of central cores in the reinnervated muscle fibers of rabbits following the cross-innervation experiment,[9] and the observation by Engel of "target fibers" resembling central cores in the muscles of patients with long standing neurogenic atrophies,[20] one is tempted to look upon central core disease as a disorder determined by a neural influence rather than being a primary myopathy. Recently Karpati and Engel[29] were able to demonstrate the presence of extensive core formation in the fibers of the guinea pig soleus muscle 2 weeks after cordotomy but not at a later stage. This once again suggests the possibility of a close correlation of central and peripheral neural factors in this disorder.

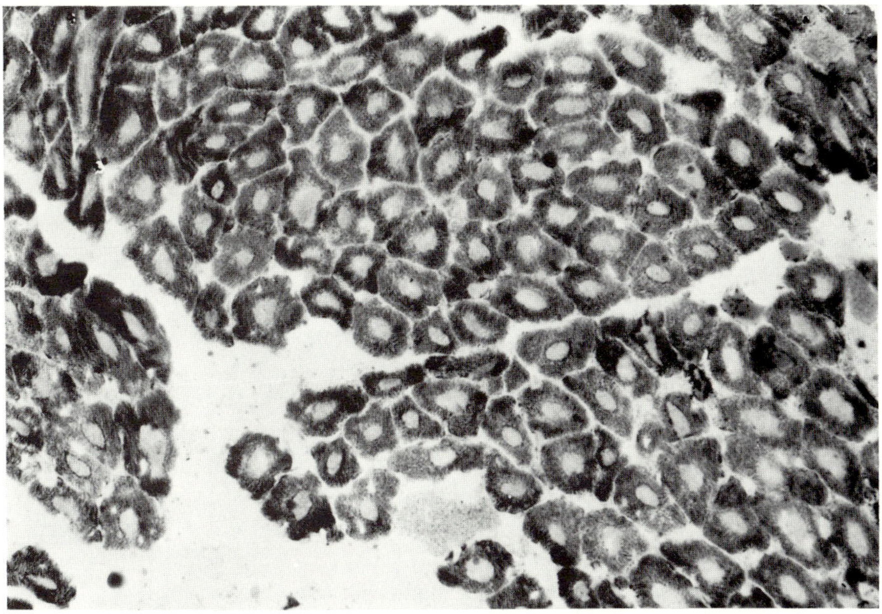

FIG. 11. Central core disease; aged 29 years; rectus femoris. Note uniformity of enzyme activity (all type 1 fibers) and centrally placed core in each fiber. NADH-tetrazolium reductase; × 180. (Reproduced from Dubowitz and Roy[19] by permission of The Macmillan Company.)

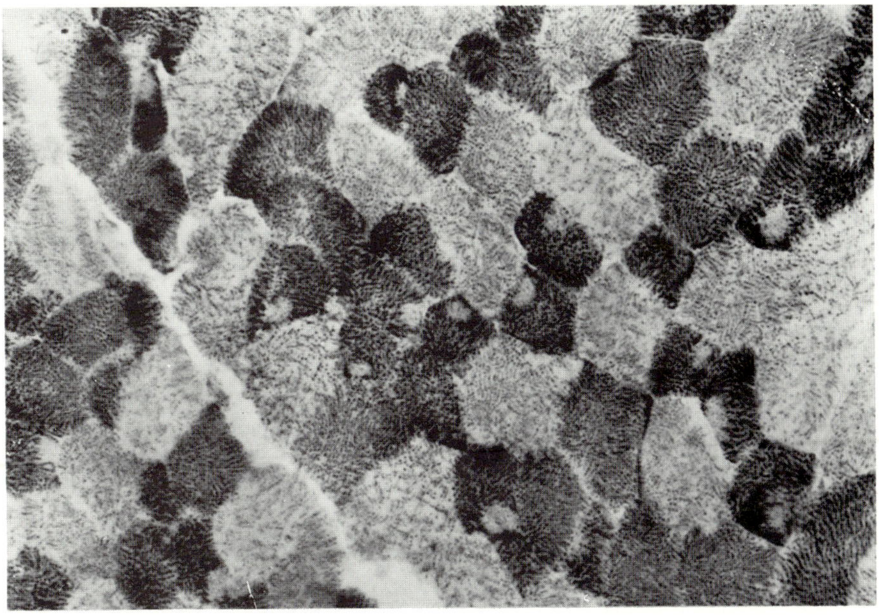

FIG. 12. Central core disease; aged 4 years, son of above case (Fig. 11); rectus femoris. Shows normal subdivision into type 1 (darker) and type 2 fibers. Eccentric central cores are present in some type 1 fibers. NADH-tetrazolium reductase; × 470. (Reproduced from Dubowitz and Roy[19] by permission of The Macmillan Company.)

In myotubular myopathy, another of these so-called congenital myopathies, there have recently been reports of specific fiber type involvement. The initial reports on myotubular myopathy contained no reference to localization to one or other fiber type,[1, 36, 37] but recently cases have been recorded with atrophy of type 1 fibers with associated central nuclei and myotube-like structure.[2, 23, 39] The pathologic picture is strikingly reminiscent of the picture observed by Engel and Karpati[24] in rats at 21 days after neonatal denervation, where one fiber type (in that case, however, type 2 fibers) was small, with central nuclei, and resembled myotubes, whereas the type 1 fibers were normal in size and appearance. Once again the type-specific involvement, and either atrophy or lack of maturation of the fibers, suggests a neural influence in myotubular myopathy.

In Duchenne muscular dystrophy a fair proportion (approximately 30 to 40%) of cases has associated intellectual impairment, suggesting that this may be a more general disease of the nervous system rather than only a "primary" myopathy.[11] Other forms of muscular dystrophy are not associated with intellectual impairment but dystrophia myotonica particularly when occurring in childhood, tends to have associated intellectual retardation. Rosman and Kakulas[35] reported structural changes in the nervous system in a small series of autopsy specimens from cases of muscular dystrophy and dystrophia myotonica. In contrast, in a recent study of 21 cases of classical Duchenne dystrophy, we were unable to find any significant histologic abnormality.[18]

It is still possible though that there may be a biochemical abnormality within the central nervous system which in some way alters the normal neural influence on skeletal muscle and leads to a "myopathic" degeneration.

## CONCLUSIONS

The nervous system, both central and peripheral, seems to exert an important influence on the structure and functional activity of skeletal muscle. It is probably in large part responsible for the normal process of maturation of the muscle and its differentiation into specific fiber types. Moreover, striking abnormalities in structure may also result from the denervation-reinnervation process in the experimental animal and seem to have direct counterparts in human disease. A number of disorders looked upon as myopathies, or "primary" diseases of muscle, preferentially affect one specific fiber type within the muscle. Because of the dependence of the fiber types on the innervation, it is possible that these disorders may well have a neural basis as well.

The pathogenesis of the muscular dystrophies still remains obscure and we are still groping for possible clues. It is possible that if we grope a little more centrally within the nervous system, we may not have to search so long for the answer to the so-called "primary" disorders of muscle.

### REFERENCES

1. Bethlem, J., Meijer, A. E. F. H., Schellens, J. P. M., and Vroom, J. J.: Centronuclear myopathy. *Eur. Neurol. 1:* 325–333, 1968.
2. Bethlem, J., van Wijngaarden, G. K., Meijer, A. E. F. H., and Hülsmann, W. C.: Neuromuscular disease with type 1 fiber atrophy, central nuclei, and myotube-like structures. *Neurology (Minneap.) 19:* 705–710, 1969.

3. Buller, A. J., Eccles, J. C., and Eccles, R. M.: Differentiation of fast and slow muscles in the cat hind limb. *J. Physiol. 150:* 399–416, 1960.
4. Buller, A. J., Eccles, J. C., and Eccles, R. M.: Interactions between motoneurones and muscles in respect of the characteristic speeds of their responses. *J. Physiol. 150:* 417–439, 1960.
5. Close, R.: Effects of cross-union of motor nerves to fast and slow skeletal muscles. *Nature (Lond.) 206:* 831–832, 1965.
6. Dubowitz, V.: Enzyme histochemistry of skeletal muscle. I. Developing animal muscle. II. Developing human muscle. *J. Neurol. Neurosurg. Psychiatry 28:* 516–524, 1965.
7. Dubowitz, V.: Enzyme histochemistry of skeletal muscle. III. Neurogenic muscular atrophies. *J. Neurol. Neurosurg. Psychiatry 29:* 23–28, 1966.
8. Dubowitz, V.: Cross-innervated mammalian skeletal muscle. Histochemical, physiological and biochemical observations. *J. Physiol. 193:* 481–496, 1967.
9. Dubowitz, V.: Pathology of experimentally re-innervated skeletal muscle. *J. Neurol. Neurosurg. Psychiatry 30:* 99–110, 1967.
10. Dubowitz, V.: *Developing and Diseased Muscle. A Histochemical Study*. London, Spastics International Medical Publications, 1968.
11. Dubowitz, V.: Chemical and structural changes in muscle. The importance of the nervous system. In *Some Inherited Disorders of Brain and Muscle*, ed. by Allan, J. D., and Raine, D. R., pp. 32–43. Edinburgh, E. & S. Livingstone Ltd., 1969.
12. Dubowitz, V.: The role of histochemistry in muscle disease. In *Modern Trends in Neurology*, ed. by Williams, D. 5th Ed. pp. 189–208. London, Butterworths, 1970.
13. Dubowitz, V., and Pearse, A. G. E.: Reciprocal relationship of phosphorylase and oxidative enzymes in skeletal muscle. *Nature (Lond.) 185:* 701–792, 1960.
14. Dubowitz, V., and Pearse, A. G. E.: A comparative histochemical study of oxidative enzyme and phosphorylase activity in skeletal muscle. *Histochemie 2:* 105–117, 1960.
15. Dubowitz, V., and Pearse, A. G. E.: Oxidative enzymes and phosphorylase in central core disease of muscle. *Lancet 2:* 23–24, 1960.
16. Dubowitz, V., and Pearse, A. G. E.: Enzymic activity of normal and dystrophic human muscle. A histochemical study. *J. Pathol. Bacteriol. 81:* 365–378, 1961.
17. Dubowitz, V., and Platts, M.: Central core disease of muscle with focal wasting. *J. Neurol. Neurosurg. Psychiatry 28:* 432–437, 1965.
18. Dubowitz, V., and Crome, L.: The central nervous system in Duchenne muscular dystrophy. *Brain 92:* 805–808, 1969.
19. Dubowitz, V., and Roy, S.: Central core disease of muscle. Clinical, histochemical and electron microscopic studies of an affected mother and child. *Brain 93:* 133–146, 1970.
20. Engel, W. K.: Muscle target fibers. A newly recognized sign of denervation. *Nature (Lond.) 191:* 389–390, 1961.
21. Engel, W. K.: The essentiality of histo- and cytochemical studies of skeletal muscle in the investigation of neuromuscular disease. *Neurology (Minneap.) 12:* 778–784, 1962.
22. Engel, W. K.: Histochemistry of neuromuscular disease. Significance of muscle fiber types, "neuromuscular diseases," In *Proceedings of the VIII International Congress of Neurology*, vol. 2, pp. 67–101. Amsterdam, Excerpta Medica Foundation, 1965.
23. Engel, W. K., Gold, G. N., and Karpati, G.: Type 1 fiber hypotrophy and central nuclei. A rare congenital muscle abnormality with possible experimental model. *Arch. Neurol. 18:* 435–444, 1968.
24. Engel, W. K., and Karpati, G.: Impaired skeletal muscle maturation following neonatal neurectomy. *Dev. Biol. 17:* 713–723, 1968.
25. Feng, T. P., Wu, W. Y., and Yang, F. Y.: Selective reinnervation of a "slow" or "fast" muscle by its original motor supply during regeneration of mixed nerve. *Sci. Sin. 14:* 1717–1720, 1965.
26. Guth, L., Watson, P. K., and Brown, W. C.: Effects of cross-reinnervation on some chemical properties of red and white muscles of rat and cat. *Exp. Neurol. 20:* 52–69, 1968.
27. Hnik, P., Jirmanova, I., Vyklicky, L., and Zelena, J.: Fast and slow muscles of the chick after nerve cross-union. *J. Physiol. 193:* 309–325, 1967.
28. Karpati, G., and Engel, W. K.: Neuronal trophic function. A new aspect demonstrated histochemically in developing soleus muscle. *Arch. Neurol. 17:* 542–545, 1967.
29. Karpati, G., and Engel, W. K.: Correlative histochemical study of skeletal muscle after

suprasegmental denervation, peripheral nerve section, and skeletal fixation. *Neurology (Minneap.) 18:* 681–692, 1968.

30. Miledi, R., and Orkand, P.: Effect of a "fast" nerve on "slow" muscle fibres in the frog. *Nature (Lond.) 209:* 717–718, 1966.
31. Morris, C. J.: Human skeletal muscle fibre type grouping and collateral re-innervation. *J. Neurol. Neurosurg. Psychiatry 32:* 440–444, 1969.
32. Robbins, N., Karpati, G., and Engel, W. K.: Histochemical and contractile properties in the cross-innervated guinea pig soleus muscle. *Arch. Neurol. 20:* 318–329, 1969.
33. Romanul, F. C. A.: Enzymes in muscle. I. Histochemical studies of enzymes in muscle fibres. *Arch. Neurol. 11:* 355–368, 1964.
34. Romanul, F. C. A., and Van der Meulen, J. P.: Slow and fast muscles after cross innervation. *Arch. Neurol. 17:* 387–402, 1967.
35. Rosman, N. P., and Kakulas, B. A.: Mental deficiency associated with muscular dystrophy. A neuropathological study. *Brain 89:* 769–787, 1966.
36. Sher, J. H., Rimalovski, A. B., Athanassiades, T. J., and Aronson, S. M.: Familial centronuclear myopathy. A clinical and pathological study. *Neurology (Minneap.) 17:* 727–742, 1967.
37. Spiro, A. J., Shy, G. M., and Gonatas, N. K.: Myotubular myopathy. *Arch. Neurol. 14:* 1–14, 1966.
38. Stein, J. M., and Padykula, H. A.: Histochemical classification of individual skeletal muscle fibers of the rat. *Am. J. Anat. 110:* 103–123, 1962.
39. van Wijngaarden, G. K., Fleury, P., Bethlem, J., and Meijer, A. E. F. H.: Familial "myotubular" myopathy. *Neurology (Minneap.) 19:* 901–908, 1969.
40. Yellin, H.: Neural regulation of enzymes in muscle fibers of red and white muscle. *Exp. Neurol. 19:* 92–103, 1967.

## Chapter 13

# The Histopathology of Neurogenic Muscular Atrophy

JON D. DORMAN

In the past 20 years significant advances have been made in both the theoretical and empirical understanding of neurogenic muscular atrophy through animal experimentation, more frequent performance of muscle biopsy, and particularly through application of the techniques of enzyme histochemistry to muscle. In this review the histopathologic features of neurogenic muscular atrophy are discussed in the context of current knowledge of the anatomy and physiology of the motor unit.

### THE NORMAL MOTOR UNIT

The motor unit is the final common pathway for all muscular activity and consist of (1) the motor neuron cell body, (2) the primary motor axon passing through the ventral root into peripheral nerve, (3) the secondary and terminal branchings of the axon, (4) the neuromuscular junctions, and (5) the muscle fibers. An action potential initiated in the motor neuron cell body passes down the parent axon into all of the secondary and terminal axonal fibers and produces a twitch response in all of the muscle fibers of that motor unit.

The number of muscle fibers in single human motor units varies significantly such that muscles capable of finer movements have fewer muscle fibers innervated by one axon. In normal human muscles the following ratios of muscle fibers to motor axons were determined: platysma-25, first lumbrical-107, and gastrocnemius-1934.[9] Such figures indicate that the primary motor axon undergoes considerable branching en route to the muscle fibers. Although occasional branching occurs within peripheral nerve trunks it is near and within muscle where the terminal arborization occurs into dozens or hundreds of terminal axons. All muscle fibers are innervated by a single terminal axon, but a proportion of terminal axons innervate more than one muscle fiber. Coërs and Woolf, using methylene blue stains of intramuscular nerves, calculated the average terminal innervation ratio in normal human limb muscles to be 1.09 muscle fibers to 1.0 terminal axon.[5]

Within a given mammalian muscle motor units vary in physiologic function. Individual cat and rat motor units studied by stimulation of single ventral root axons have been shown to vary significantly in the tension developed in twitch contraction, speed of contraction,[18] and fatiguability.[8] This variability in motor unit performance reflects differing characteristics of the motor unit components. Motor neuron cell bodies vary in size and excitability[16] and in enzyme activity determined histochemically.[3] The primary motor axons have variable diameters and their conduction velocities are directly proportional to the axon diameter. The end organs of the motor unit, the muscle fibers, are now also known to vary significantly in enzymatic and biochemical properties. The classification of muscle fiber types is discussed in detail elsewhere in this volume. In this review red muscle fibers high in oxidative enzyme content and low in myofibrillar ATPase and phosphorylase content will be referred to as *type I*, and white fibers with the opposite histochemical characteristics are referred to as *type II*.

One important question bearing on the histopathology of neurogenic muscular atrophy is whether muscle fibers of a single motor unit are biochemically and histochemically uniform, or whether each motor neuron innervates a mixed collection of types I and II muscle fibers. An elegant experiment by Edström and Kugelberg[8] provides a strong evidence for the histochemical uniformity of muscle fibers within a given motor unit. These authors isolated single ventral root fibers supplying rat tibialis anterior and were able to identify histochemically the muscle fibers innervated by that motor axon by depleting the fibers of glycogen with repetitive axon stimulation and mapping their territory with frozen sections of muscle stained for glycogen or phosphorylase. The muscle fiber type of stimulated fibers was then identified histochemically on a serial section, and it was found that motor units were 98% histochemically uniform. Another important observation in this study is that the distribution of muscle fibers for a given motor unit is widely scattered and interspersed with muscle fibers belonging to other motor units, with no more than a few muscle fibers of the same motor unit lying adjacent to one another.

The uniformity of muscle enzyme activity within fibers of a given motor unit implies that the motor neuron has a critical influence on muscle enzyme activity and physiologic properties. This inference is supported by a variety of cross-innervation experiments in which biochemical, histochemical, or physiologic properties of muscle have been altered by surgically changing the nerve supply. These studies, reviewed extensively by Guth[10] consistently show that muscle fibers tend to adopt properties inherent to a particular motor neuron. The trophic influence of nerve not only maintains muscular bulk and contractility but in some fashion "imprints" muscle fibers with specific characteristics.

## PATHOLOGIC REACTIONS OF MOTOR NEURON

The lower motor neuron in its long course from spinal cord through peripheral nerve to muscle is subject to pathologic alteration by an enormous variety of disorders. In the comprehensive Classification of Neuromuscular Disorders prepared through the World Federation of Neurology[4] several pages were needed merely to list the recognized disorders of anterior horn cell, motor nerve root, peripheral nerve, and neuromuscular junction. All of these entities may cause neurogenic muscular atrophy by interfering with the normal trophic effect of nerve upon muscle. When the anatomic continuity of the motor neuron has been disrupted by any cause, the fate of the muscle fibers is similar; either they undergo progressive atrophy and, eventually, degeneration, or the atrophic process is reversed at some point by reinnervation by the same or a neighboring axon.

There are two general mechanisms by which the function of motor axons may be impaired without loss of their anatomic continuity. In the presence of a peripheral nerve *conduction block*, as can be experimentally induced by continuous infusion of local anesthetic[11] or light compression,[6] the distal effects on muscle morphology or biochemistry are significantly less than effects seen after nerve section. The implication of these studies and others of a similar nature is that motor axons have a trophic effect by some mechanism not requiring the conduction of action potentials. In human nerve pathology a conduction block without disruption of the neuromuscular contact, as in a mild nerve entrapment syndrome, could similarly cause muted denervation (? disuse) effects. Conduction block could also occur when a neurogenic disorder causes *segmental demyelination* of nerve fibers. In segmental demyelination of peripheral nerve, Schwann cells degenerate focally leaving the axon anatomically intact but subject to a conduction block. Segmental demyelination has been observed in several human neuropathies.[1]

A second mechanism by which partial denervation effects could theoretically be induced is by impairment of *axoplasmic flow*. The normal bulk passage of axoplasm from the neuron cell body into axons is undoubtedly important to axonal maintainance and might also be carrying trophic substances for transynaptic delivery to muscle fibers. Slowing of axoplasmic flow was found in experimental acrylamide neuropathy,[15] and this mechanism may in the future prove to be significant in human neurogenic disorders.

In denervating disorders unaffected terminal axons may react, presumably to a chemical signal, by sprouting and reinnervating nearby denervated muscle fibers.[5] By this process of *collateral reinnervation* more muscle fibers are added to the reinnervating motor unit. By the neural "imprinting" of specific biochemical characteristics onto the reinnervated muscle fibers, the histochemical reactivity of all muscle fibers reinnervated by the same axon will become identical.[13] Enzyme histochemical stains of reinnervated muscle will then reveal *fiber-type grouping*, focal aggregations of muscle fibers which have been converted to the same fiber type by their common reinnervating terminal axon (see Figs. 1 and 2).

## HISTOPATHOLOGIC FEATURES OF DENERVATED MUSCLE

*Stages of denervation.* The specific histopathologic features seen in denervated muscle depend upon which point is sampled in the longitudinal history of that neurogenic atrophy. In the earliest stages fiber atrophy is the hallmark, in a pattern that depends on the severity and distribution of the neural involvement. Following nerve section in the opossum, there was a 70% reduction of the average fiber cross-sectional area over the first 60 days,[17] and in human nerve

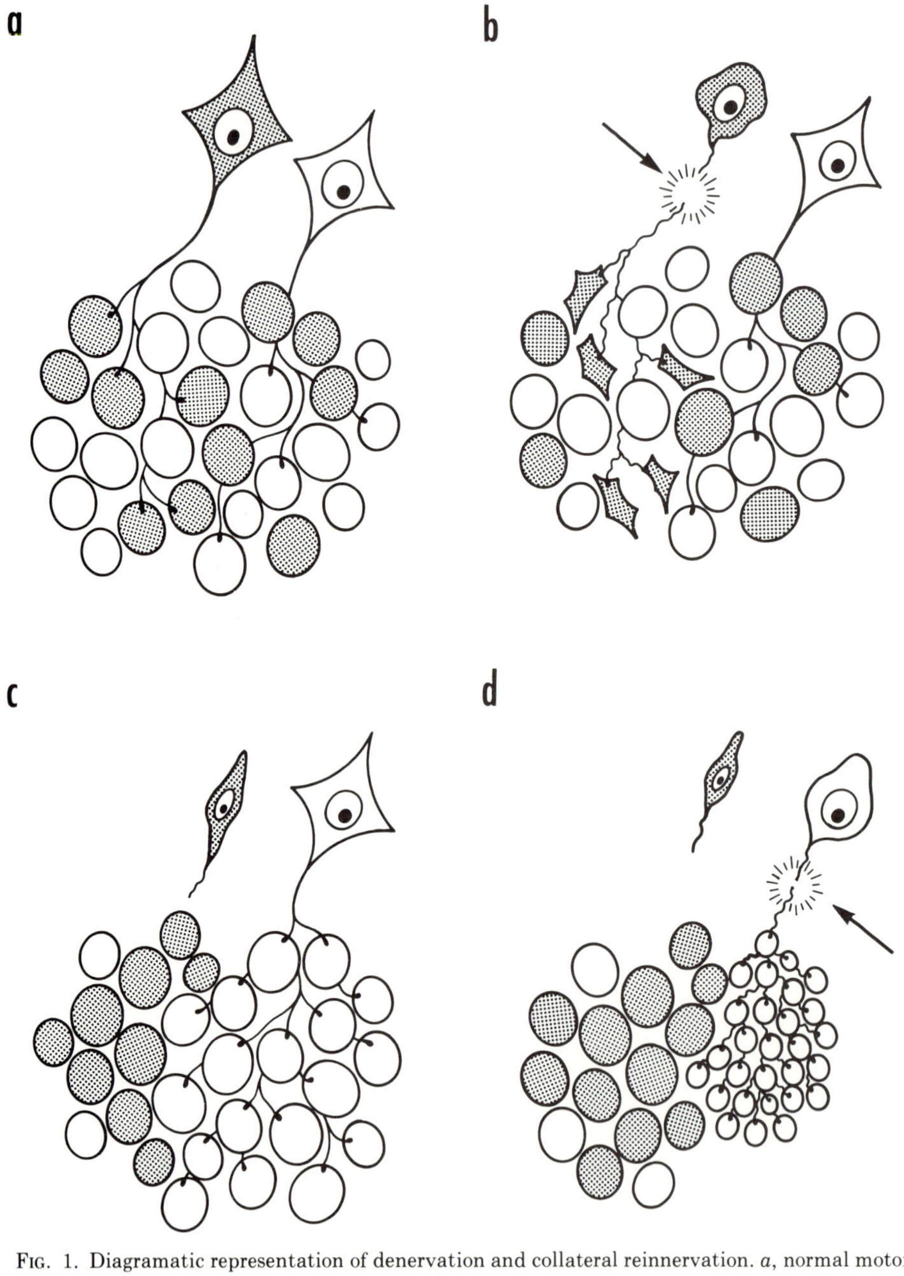


Fig. 1. Diagramatic representation of denervation and collateral reinnervation. *a*, normal motor units with neurons imprinting specific characteristics on their scattered muscle fibers. *b*, the initial effect of denervation is muscle fiber atrophy and angulation within that motor unit. *c*, fiber-type grouping. By collateral reinnervation from the "white" neuron denervated muscle fibers regain normal bulk and are imprinted with "white" characteristics. *d*, grouped atrophy occurs when the reinnervating neuron becomes affected by disease.

injuries progressive loss of muscle bulk is well known clinically. Following denervation the atrophic fibers continue to shrink and may persist for months or years with a cross-sectional diameter of less than 10 $\mu$. If reinnervation occurs, the denervated muscle fibers regain normal bulk and are imprinted with the histochemical pattern of the reinnervating motor neuron. When collateral reinnervation has occurred, histochemical stains will reveal fiber-type grouping. A further stage then occurs when the pathologic process extends to motor neurons which had sprouted terminally to provide collateral reinnervation. When these later axons become involved, grouped atrophy of muscle fibers occurs. When muscle has remained denervated for many months or years the final end-stage appearance consists of scattered nuclear debris, possibly in association with tiny strands of denervated sarcoplasm, and replacement by connective tissue and fat. In chronic neurogenic disorders such as peroneal muscular atrophy, the muscle may manifest all of these stages concurrently.

*Changes in fiber size and contour.* By light microscopy the most essential characteristic and often the only abnormality of the denervated fiber is atrophy. The diagnosis of fiber atrophy in a biopsy depends on a knowledge of the normal fiber size spectrum for skeletal muscle. In fresh frozen material stained for myofibrillar ATPase Brooke and Engel[2] found by histographic analysis that the normal male biceps brachii and vastus lateralis have a modal diameter of 60 $\mu$ with nearly all fibers falling in the range of 40 to 80 $\mu$. Females have an average fiber size diameter 10 $\mu$ smaller than males, and fixed, paraffin-embedded muscle will have smaller fibers than fresh frozen specimens. In judging whether small fibers are denervated the contour of the fiber may be of further help since denervated fibers tend to assume a sharply angulated contour in the early and middle phases of denervation (Fig. 2,*b* and *d*). Chronically denervated fibers, on the other hand, often become rounded (Fig. 5). One note of caution in the interpretation of fiber contour is that tissue handling and processing, both in frozen and fixed sections, may artifactually affect fiber contour toward either greater angulation or rounding. Fiber hypertrophy is also seen in denervated muscle to a variable extent. Presumably this mainly represents a compensatory hypertrophy of nondenervated fibers, but in experimental denervation, the denervated muscle fibers themselves undergo a transitory hypertrophic phase.[11]

*Distribution of atrophic fibers.* Following a complete disruption of innervation the early findings will be a generalized fiber atrophy similar to but more severe than that seen in simple disuse atrophy, often with greater atrophy of type II fibers (Fig. 3). With partial denervation the initial distribution of atrophic fibers follows a random and scattered pattern in accordance with the normal anatomy of the motor unit (Fig. 2*b*). Small groups of 2 to 10 atrophic and angulated fibers are observed when a few fibers from one motor unit lie together or when fibers from two or three adjacent motor units are affected simultaneously.

Atrophy of large groups or entire fascicles of fibers of mixed histochemical types may be observed when denervation is relatively severe, effectively eliminating the local availability of healthy terminal axons for collateral reinnervation. The perimysium acts as a barrier to prevent ingrowth of axon sprouts from adjacent fasicles.[13] Following collateral reinnervation and subse-

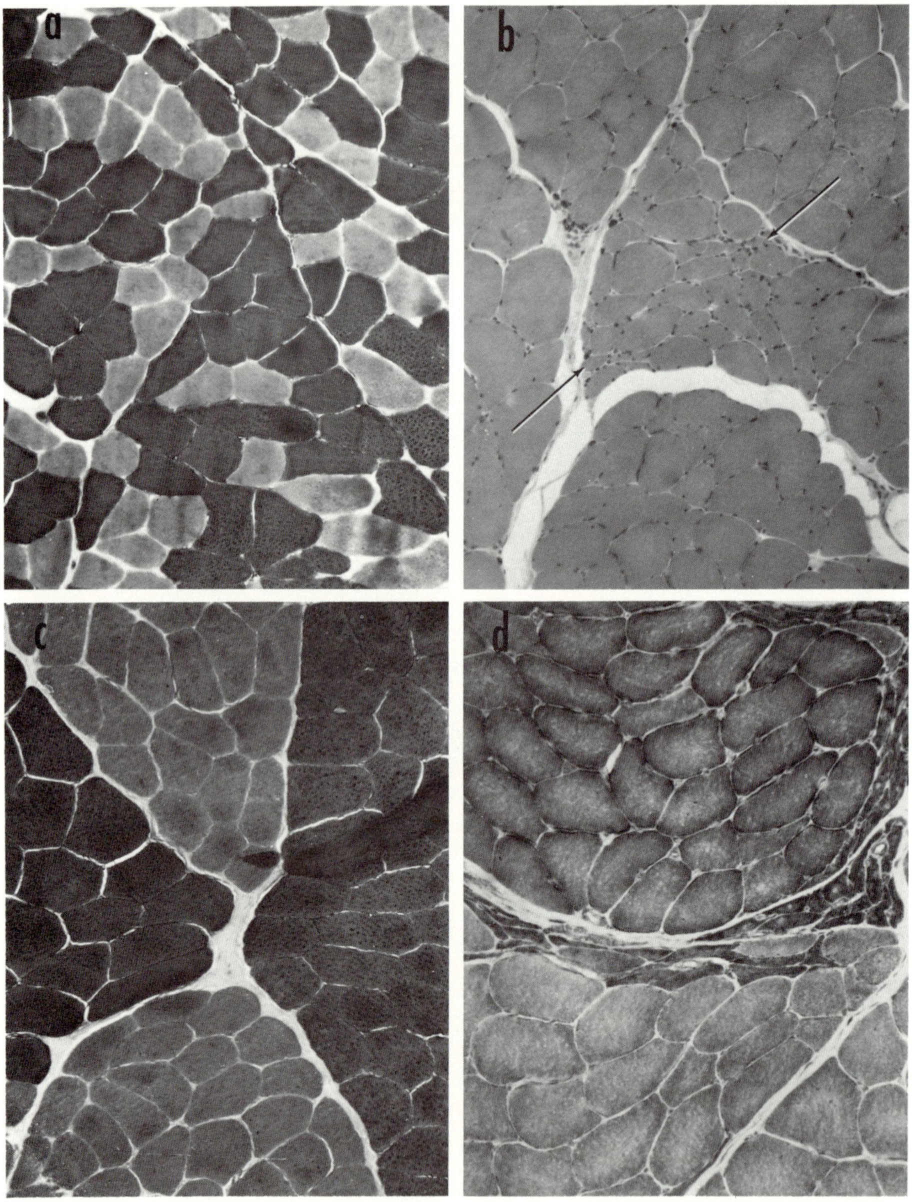


FIG. 2. Muscle sections illustrating stages of denervation and collateral reinnervation diagrammed in Fig. 1. *a*, normal mosaic pattern of types I and II fibers. Frozen; myofibrillar ATPase; × 68. *b*, early denervation showing scattered small angulated fibers. Nutritional neuropathy. Frozen; hematoxylin and eosin stain; × 68. *c*, fiber-type grouping and restoration of muscle bulk. Peroneal muscular atrophy. Frozen; myofibrillar ATPase; × 68. *d*, grouped atrophy of histochemically uniform fibers. Peroneal muscular atrophy. Frozen; NADH diaphorase; × 68.

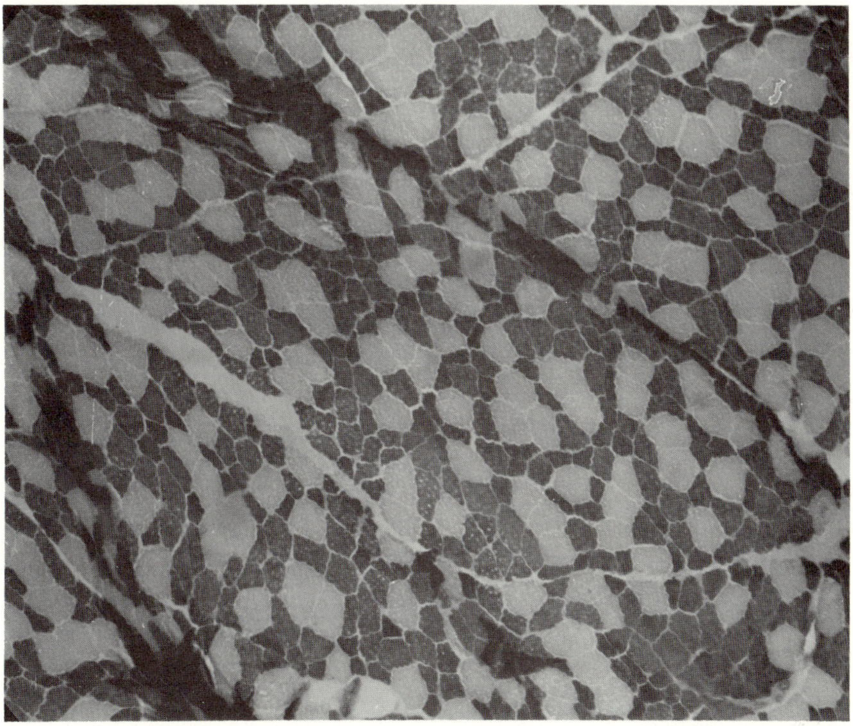

FIG. 3. Acute denervation in Guillain-Barré syndrome. Diffuse atrophy greater in type II (dark) fibers. Frozen; myofibrillar ATPase; × 100.

quent spread of disease to the reinnervating neuron, grouped atrophy of histochemically identical fibers may be observed in which 20 or more fibers up to an entire fascicle may become denervated simultaneously by loss of their common motor neuron (Fig. 2*d*). The pattern of grouped atrophy is commonly observed in chronic motor neuron disorders, such as Werdnig-Hoffmann disease (Fig. 4) or Kugelberg-Welander disease, and in chronic neuropathies.

*Fiber type grouping.* The loss of the normal, random, mosaic pattern of fiber types and tendency for muscle fibers of the same histochemical type to be clumped together provide strong evidence that the muscle has undergone previous denervation followed by reinnervation. The muscle fibers may be of normal size if the collateral reinnervation has remained intact (Fig. 2*c*). Coërs and Woolf[5] found evidence of collateral reinnervation commonly in motor neuron disease, poliomyelitis, and peroneal muscular atrophy but infrequently in neuropathies associated with diabetes, porphyria, and nutritional deficiency. Although no numerical standards are presently available to establish the normal range of fiber-type grouping in borderline cases, in many instances the fiber-type grouping is quite obvious without applying any numerical standards. On occasion an entire biopsy specimen may be composed of a single fiber type. In some patients fiber type preponderance with an even distribution of the minor fiber type may be observed. This finding, when prominent, may be

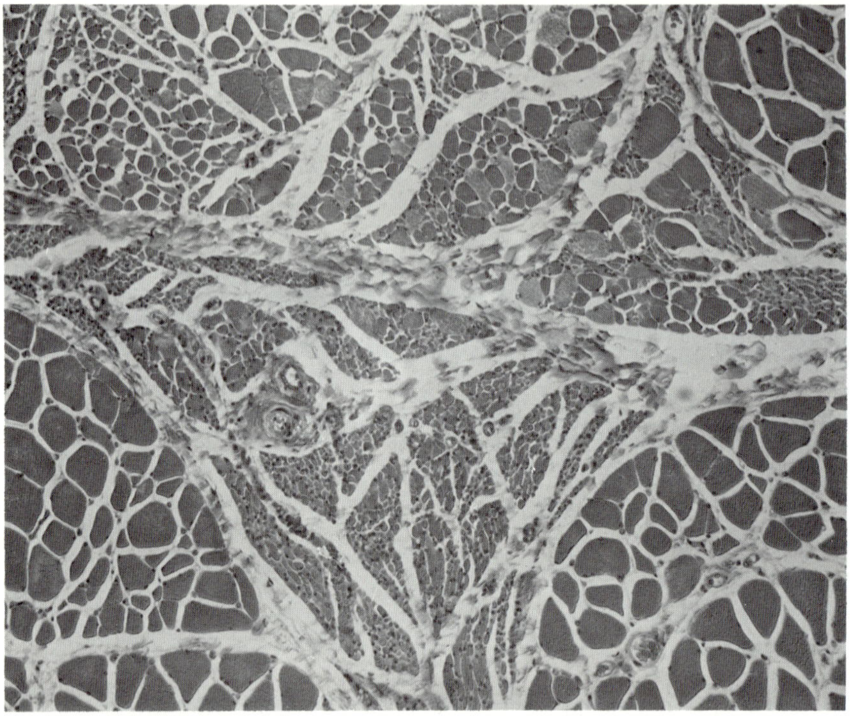

FIG. 4. Werdnig-Hoffmann disease. Large group and fascicular atrophy. Paraffin-embedded; hematoxylin and eosin stain; × 100.

compatible with denervation followed by reinnervation, but it is not diagnostic thereof and may be observed in nondenervating disorders or possibly normal individuals.

*Morphologic changes in denervated fibers.* By light microscopy, particularly if only hematoxylin and eosin stained material is examined, the morphologic changes within denervated fibers are relatively scattered and mild. *Basophilia* of atrophic fibers occurs in modest frequency (Fig. 5) and may be occasionally observed in normal sized fibers, particularly in chronic denervation. The occurrence of *target fibers* is diagnostically more helpful since this change seldom occurs in disorders without a neurogenic component. When fully developed targets consist of a central understained zone surrounded by a rim of hyperstaining, frequently only the more-or-less round area of central understaining is seen (Fig. 6). Target fibers are most often seen in type I fibers and are best observed in the NADH diaphorase stain of fresh frozen material (Fig. 6). However, they are seen in other histochemical stains and less frequently in paraffin-embedded material. The phosphotungstic acid-hematoxylin stain is the best preparation for demonstrating target fibers in fixed material. Another diagnostically helpful finding seen on the NADH diaphorase stain is *dark angulated fibers* (Fig. 7) with a coarser pattern of granules than normal. As with target fibers this change is not invariably present in denervated muscle. Denervated fibers also may have decreased phosphorylase activity histochemically.

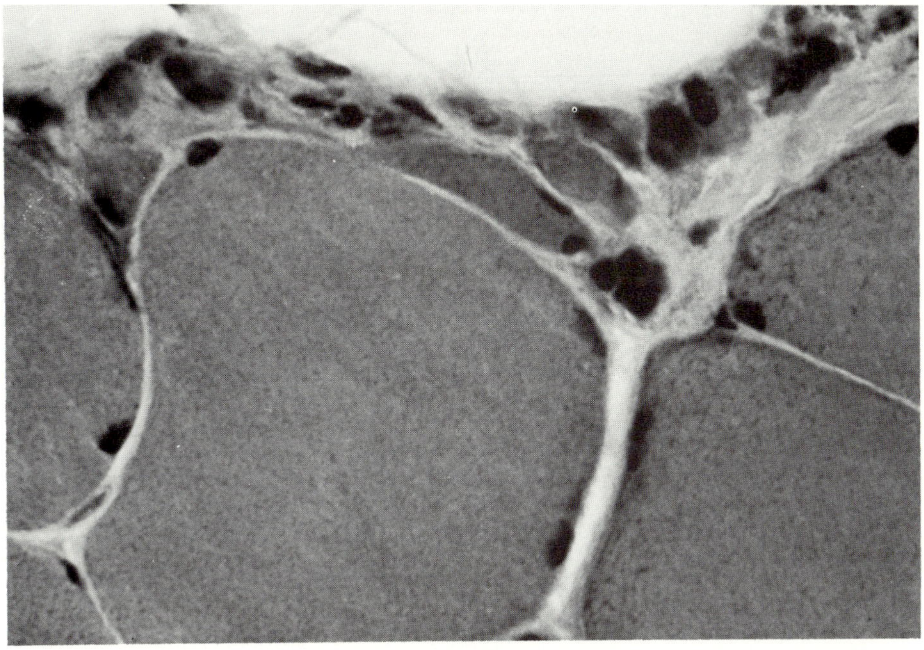

FIG. 5. Chronic denervation. Note rounding and basophilia of atrophic fibers and pyknotic nuclear clumping. Frozen; hematoxylin and eosin stain; × 304.

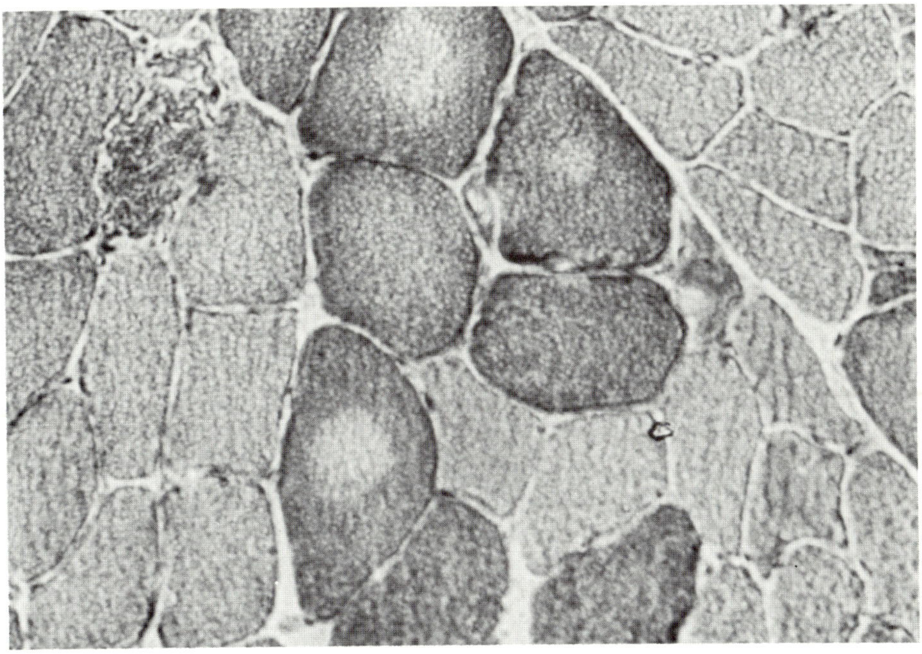

FIG. 6. Uremic neuropathy. Target changes in some type I fibers. Frozen; NADH diaphorase; × 76.

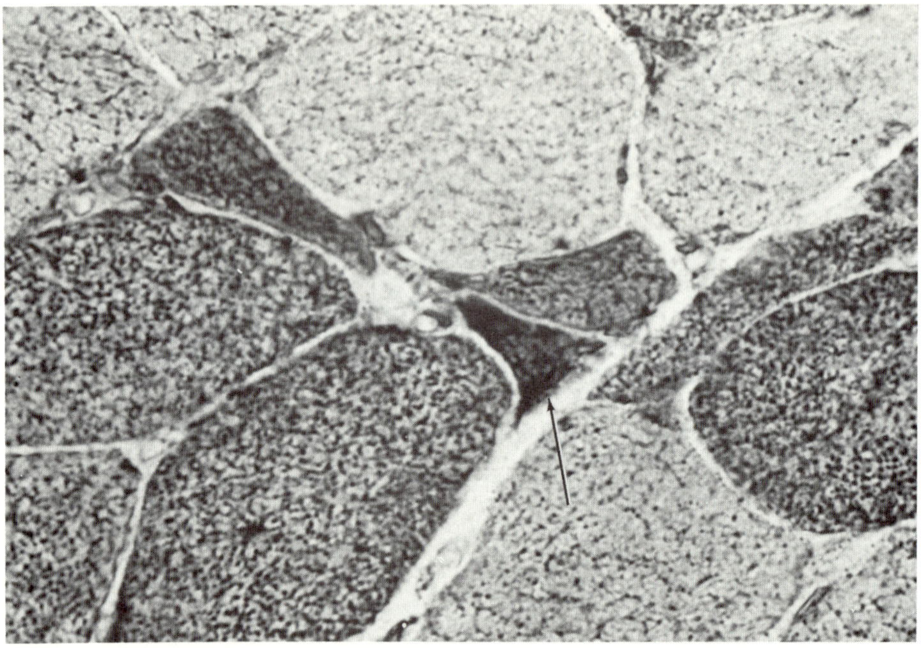

FIG. 7. Nutritional neuropathy. Atrophic angulated fiber darkly stained in NADH diaphorase reaction. Frozen; × 304.

In early denervation *nuclear* changes are minimal, but some fibers may show nuclear activation with enlargement and decreased chromatin staining. In chronic denervation sarcolemmal nuclei frequently condense to form pyknotic nuclear clumps lying within very atrophic fibers (Fig. 5). This nuclear alteration is very suggestive of a neurogenic disorder when it is prominent, but nuclear clumps may occur occasionally in myopathic disorders. Ultimately the sarcoplasm and sarcolemmal membrane may degenerate and be reabsorbed, leaving only pyknotic nuclear clumps lying in the connective tissue.

End-stage denervation, consisting of nuclear clumps, very atrophic degenerated fibers, and increased connective tissue and fat, may be indistinguishable from end-stage myopathic muscle. Such nondiagnostic findings should be avoided by selecting muscles for biopsy which are not completely or severely paralyzed.

In chronic neurogenic muscular atrophy "myopathic" changes in muscle fibers may be observed—including hyalin, granular or liquifactive necrosis of fibers with or without phagocytosis by mononuclear cells, rounding of the fiber contours, fiber splitting, internal migration of sarcolemmal nuclei, and increase in endomysial connective tissue (Fig. 8). These changes are commonly seen in chronic familial neuropathies such as peroneal muscular atrophy, chronic proximal spinal muscular atrophy (Kugelberg-Welander disease), or patients with a previous history of poliomyelitis. In carcinomatous neuromyopathy and collagen vascular diseases neurogenic muscular atrophy may co-exist with primary myopathic alterations.

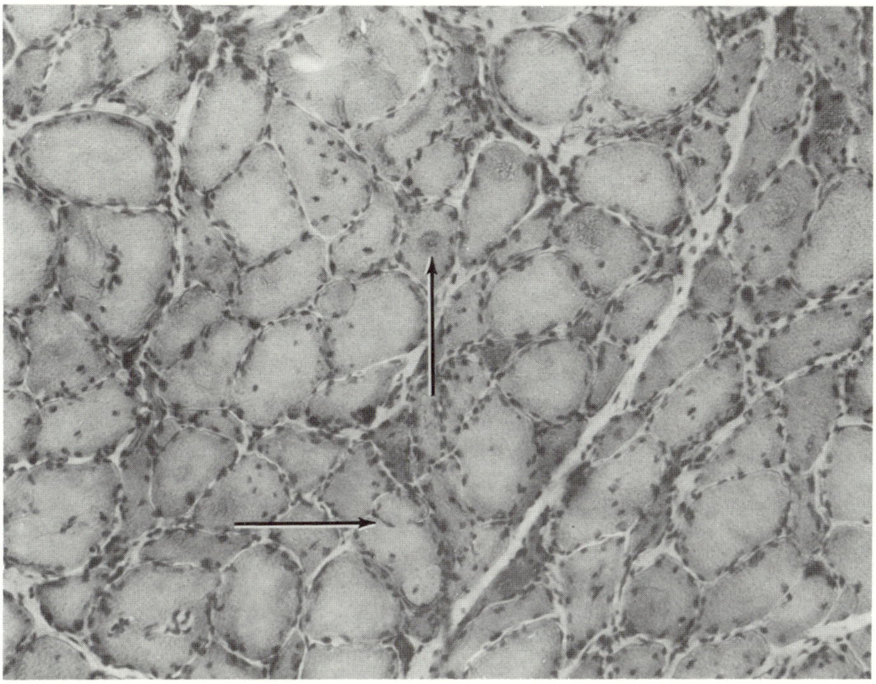

FIG. 8. Chronic denervation in a diabetic patient demonstrating "myopathic" changes of fiber splitting (*horizontal arrow*) and internal migration of sarcolemmal nuclei. Also note denervation changes of target fibers (*vertical arrow*), scattered atrophic and angulated fibers, and pyknotic nuclear clumps. Frozen; hematoxylin and eosin stain; × 100.

*Connective tissue, blood vessels, and nerves.* Intramuscular connective tissue and blood vessels are unchanged in the early and middle phases of a denervating disorder. In areas of significant atrophy however, condensation of endomysial connective tissue gives the appearance of fibrosis (Fig. 5). In an end-stage muscle a nonspecific thickening of blood vessels may be observed. In neurogenic disorders intramuscular nerve bundles may be observed to have increased collagen or axonal drop out.

*Denervation or disuse atrophy?* Disuse atrophy may histopathologically be associated with atrophic and angulated fibers, potentially causing difficulty in distinguishing disuse atrophy from early neurogenic muscular atrophy. Disuse atrophy regularly affects type II fibers to a more severe degree, and it may therefore mimic the pattern of atrophy seen in early, severe denervation which may also preferentially affect type II fibers (compare Figs. 3 and 9). The NADH diaphorase stain may help to differentiate the two conditions by revealing scattered very dark fibers and possibly target fibers in the denervated cases. Ultrastructurally the degenerative changes seen in disuse atrophy resemble those found in denervation,[14] suggesting that the type II atrophy of disuse may result from an altered neural influence.

Since there are differences in motor neurons as discussed above, it may be

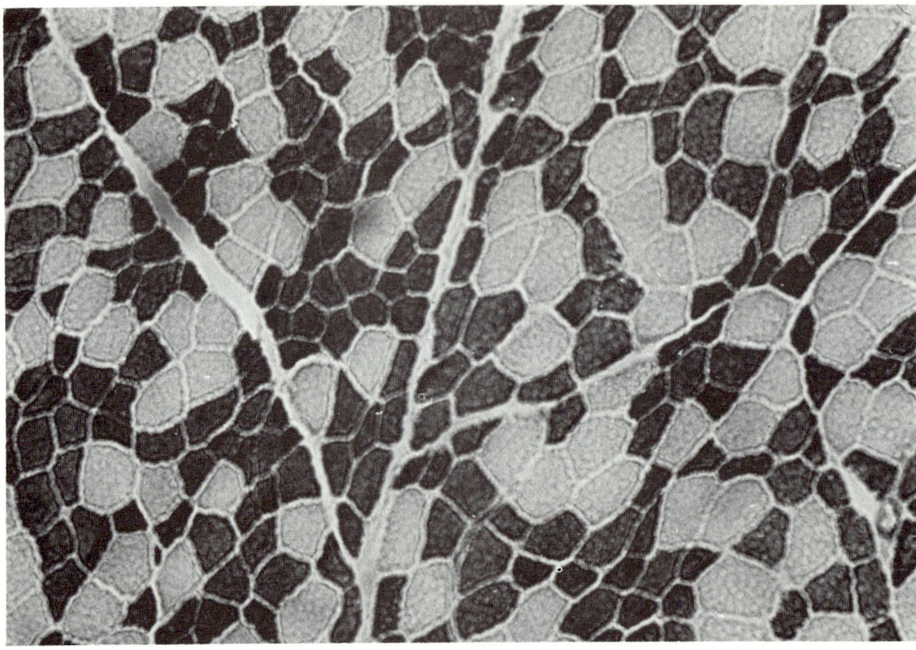

FIG. 9. Disuse atrophy, with greater effect on type II (dark) fibers. Upper motor neuron lesion. Frozen; ATPase; × 76.

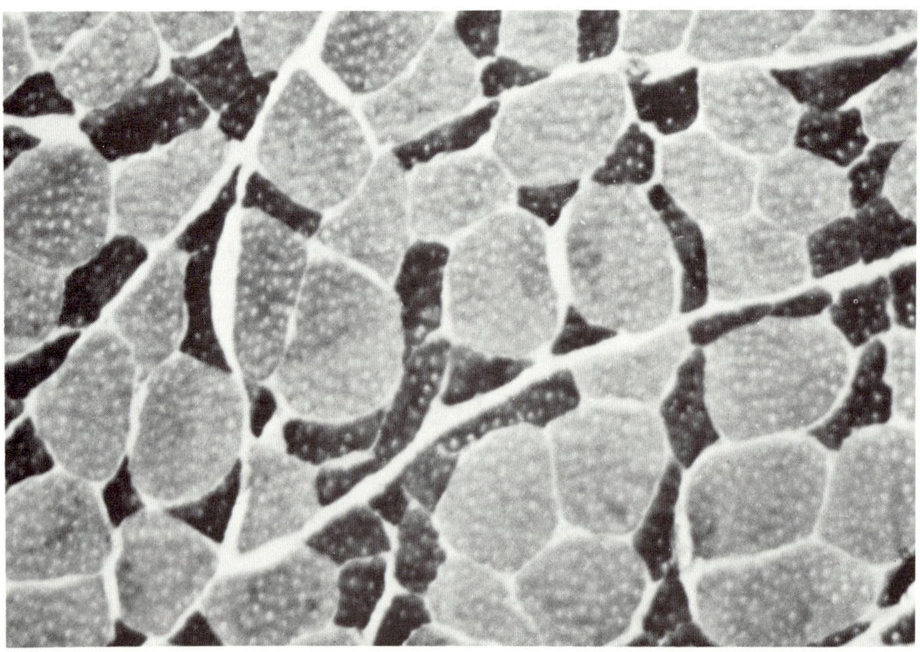

FIG. 10. Neuropathy of unknown etiology with preferential type II (dark) fiber involvement. Frozen; ATPase; × 76.

anticipated that some neurogenic disorders could preferentially affect one population of motor neurons, which would be reflected in the muscle by greater involvement of one fiber type. Figure 10 depicts a myofibrillar ATPase stain of a biceps brachii muscle from a young woman with a severe neuropathy of unknown etiology. This muscle had fibrillation potentials by electromyographic examination, implying denervation, and the biopsy revealed atrophic angulated fibers which were nearly exclusively type II. Did this patient have a type II neuropathy?

*Trophic influence of nerve on muscle.* It is still an unresolved issue whether the trophic effects that motor neurons have on muscle are mediated exclusively by patterns of acetylcholine stimulation at the neuromuscular junctions[7] or whether trophic substances are trans-synaptically supplied to muscle. The evidence for trophic substances is indirect and circumstantial for mammals, although it is more strongly supported in the snail.[12] If future research supports the existence of trophic compounds supplied by axoplasmic flow to muscle fibers, some histopathologic changes now regarded as "myopathic" may eventually fall within the spectrum of neurogenic muscular atrophy.

## REFERENCES


1. Bots, G.: Pathology of nerves. In *Handbook of Clinical Neurology: Vol. 7, Diseases of Nerves*, ed. by Vinken, P. J., and Bruyn, G. W. New York, American Elsevier Publishing Co., 1970.
2. Brooke, M. H., and Engel, W. K.: The histographic analysis of human muscle biopsies with regard to fiber types. 1. Adult male and female. *Neurology* (Minneap.) *19:* 221–233, 1969.
3. Campa, J. F., and Engel, W. K.: Histochemistry of motor neurons and interneurones in the cat lumbar spinal cord. *Neurology* (*Minneap.*) *20:* 559–568, 1970.
4. Classifications of the neuromuscular disorders. *J. Neurol. Sci. 6:* 165–177, 1968.
5. Coërs, C., and Woolf, A. *The Innervation of Muscle.* Oxford, Blackwell Scientific Publications, 1959.
6. Denny-Brown, D. E., and Brenner, C.: Paralysis of nerve induced by direct pressure and by tourniquet. *Arch. Neurol. Psychiat. 51:* 1–26, 1944.
7. Drachman, D.: Is acetylcholine the trophic neuromuscular transmitter? *Arch. Neurol. 17:* 206–218, 1967.
8. Edström, L., and Kugelberg, E.: Histochemical composition, distribution of fibers and fatiguability of single motor units. Anterior tibial muscle of the rat. *J. Neurol. Neurosurg. Psychiatry 31:* 424–433, 1968.
9. Feinstein, B., Lindegaard, B., Nyman, E., and Wohlfart, G.: Morphologic studies of motor units in normal human muscles. *Acta Anat.* (*Basel*) *23:* 127–142, 1955.
10. Guth, L.: "Trophic" influences of nerve on muscle. *Physiol. Rev. 48:* 645–687, 1968.
11. Gutmann, E.: Evidence for the trophic function of the nerve cell in neuromuscular relations. In *The Effect of Use and Disuse on Neuromuscular Functions*, ed. by Gutmann, E., and Hník, P., p. 29. New York, American Elsevier Publishing Co., 1963.
12. Kerkut, G.: Biochemical aspects of invertebrate nerve cells. In *Invertebrate Nervous Systems*, ed. by Wiersma, C., p. 26. Chicago, University of Chicago Press, 1967.
13. Kugelberg, E., Eström, L., and Abbruzzese, M.: Mapping of motor units in experimentally reinnervated rat muscle. Interpretation of histochemical and atrophic fiber patterns in neurogenic lesions. *J. Neurol. Neurosurg. Psychiatry 33:* 319–329, 1970.
14. Mendell, J., and Engel, W.: The fine structure of type II muscle fiber atrophy. *Neurology* (*Minneap.*) *21:* 358–365, 1971.

15. Pleasure, D. E., Mishler, K. C., and Engel, W. K.: Axonal transport of proteins in experimental neuropathies. *Science 166:* 524–525, 1969.
16. Somjen, G., Carpenter, D., and Henneman, E.: A precise correlation between the sizes of motor neurons and their reflex excitabilities. *Fed. Proc. 23:* 111, 1964.
17. Sunderland, S.: *Nerves and Nerve Injuries*, p. 233. Edinburgh & London, E. & S. Livingstone Ltd., 1968.
18. Wuerker, R. B., McPhedran, A. M., and Henneman, E.: Properties of motor units in a heterogenous pale muscle (m. gastrocnemius) of the cat. *J. Neurophysiol. 28:* 85–99, 1965.

Chapter 14

# Progressive Muscular Dystrophy: Structural Alterations in Various Stages and in Carriers of Duchenne Dystrophy*

JOHN N. WALTON

The investigation of neuromuscular disease with the aid of muscle biopsy is a technique which has been practiced for more than a century. Duchenne in 1868[13] was among the first to use classical anatomopathologic techniques, both macroscopic and microscopic, to study human muscular disease. He obtained muscle specimens for examination under the microscope with the aid of a transcutaneous trocar. Erb in 1883[19] described a number of important pathologic criteria upon which the diagnosis of muscular dystrophy could be based, criteria which in many respects are equally valid today. Despite the enormous technical progress of the last 15 years, it is unfortunate that certain pathologists have been either unable or unwilling to study the progressive changes which occur at different stages of various diseases of the neuromuscular apparatus. Furthermore, the study of muscle obtained at autopsy, although of considerable importance and interest in identifying the extent of the disease process, is of little or no value in identifying the successive changes within voluntary muscle which occur at different stages of a particular disease. Such autopsy observations, too, have the disadvantage that many of the abnormalities observed in the muscle specimens are often due to secondary factors related to immobilization of the patient. This is why it is particularly important to utilize the biopsy technique in order to study the changes occurring in voluntary muscle at the earliest possible stage of the illness. It must, of course, be admitted that multiple biopsy is only rarely needed or justifiable, although the technique of single biopsy has the important disadvantage that it can only demonstrate the abnormalities which are present in a small part of the muscle from which the biopsy has been taken at one instant in time. Successive biopsies taken at short intervals from the same area of an affected muscle have helped us to learn more of the response which the diseased muscle shows to injury, while successive biopsies obtained at longer intervals have proved to be of particular value in de-

* This work was aided by grants from the Muscular Dystrophy Associations of America, the Muscular Dystrophy Group of Great Britain, and the Medical Research Council.

picting the morphologic changes which accompany the evolution of the disease process.

Other authors contributing to this symposium have dealt in some detail with the question of biopsy technique. I would say first that in studying cases of progressive muscular dystrophy it is important to select a muscle which is moderately affected by the disease process at an early stage (*e.g.*, biceps brachii in the upper limb and quadriceps femoris in the lower) and not one such as deltoid or gastrocnemius which is comparatively spared. Gastrocnemius has the further disadvantage of being a muscle which is frequently exposed to minor trauma, and it is well recognized that even apparently trivial injury, such as a blow with a tendon hammer, may produce areas of focal necrosis and inflammatory cell reaction in an otherwise normal muscle. Similarly, biopsy of a muscle recently examined electromyographically must be avoided. It is our policy to carry out electromyographic examination in muscles on the right side of the body and to biopsy, as a routine, those on the left to avoid such error. It is also our general policy, particularly in children and in many adults, to carry out biopsy under general anesthesia as the procedure is more comfortable for the patient, the operation can be more leisurely and precise, the possibility of artifact resulting from injection of local anesthetic into the muscle is avoided, and in our hands complications of general anesthesia have been negligible. Clearly there are circumstances, particularly in patients whose respiratory reserve is seriously impaired, when local anesthesia is to be preferred. I do not propose to dwell upon the technique of biopsy and upon fixation and preparation of the muscle specimen in detail, as this has already been dealt with by other contributors. I will content myself with saying that in general the technique described by Price *et al.*[44] has proved to be admirable in practice. We now prefer to study both frozen sections prepared in a cryostat and sections obtained from formalin-fixed material embedded in paraffin, making sure that the blocks are carefully prepared so that well aligned transverse and longitudinal sections of muscle are available for examination. Fixation of the specimen in an acid fixative must not be carried out too quickly after its removal as this has a tendency to produce contraction artifact, and it is often wise to stretch the specimen slightly on a card or in a clamp and to allow it to remain exposed to the air for 10 minutes or so prior to being plunged into the fixative. Without question, care taken in the handling, fixation, and sectioning of specimens following careful operative removal is the factor of greatest importance in allowing one to obtain meaningful biopsy specimens.

Other contributors have been invited to consider the information which has accrued in recent years from the histochemical and ultrastructural studies of normal and diseased muscle, and I shall therefore confine myself to observations based upon frozen and paraffin-embedded material stained with conventional stains and examined under the light microscope. For such routine histopathologic studies my colleagues and I prefer to examine sections stained with hemalum and eosin, which best demonstrate changes in the staining characteristics of the sarcoplasm and in the nuclei, with Mallory's phosphotungstic acid hematoxylin, which gives admirable definition of the myofibrils and of their striations, and with a trichrome stain such as picro-Mallory, which also gives good myofibrillar definition but is of particular value in identifying infiltrating connective tissue. On occasions we use special stains such as azur B before and after predigestion with ribonuclease if we are particularly anxious to study the changes of muscle regeneration as identified by the presence of RNA in the sarcoplasm of regenerating muscle fibers.

## PROGRESSIVE MUSCULAR DYSTROPHY: HISTOPATHOLOGIC CHARACTERISTICS

Before describing and discussing the pathologic changes occurring in skeletal muscle in the various forms of muscular dystrophy at different stages of the illness, it is important first to consider briefly the classification of this group of disorders. I do not propose to consider the changes observed in cases of myotonic dystrophy (myotonia atrophica); I shall confine my attention to the pure muscular dystrophies, that group of genetically determined primary degenerative myopathies in which current investigative evidence suggests that the disease process is essentially one of the muscle cell itself. These disorders have been classified by Walton and Gardner-Medwin[53] as follows:

a. X-linked muscular dystrophy
    Severe (Duchenne type)
    Benign (Becker type)
b. Autosomal recessive muscular dystrophy
    Limb-girdle types
    Childhood muscular dystrophy (except Duchenne)
    Congenital muscular dystrophies
c. Facio-scapulo-humeral muscular dystrophy
d. Distal muscular dystrophy
e. Ocular muscular dystrophy
f. Oculopharyngeal muscular dystrophy

After dealing with some general considerations, I propose to describe the types of abnormality identified by muscle biopsy in muscle specimens obtained from patients suffering from each of these individual subvarieties of muscular dystrophy.

### Some general considerations

Since Erb in 1883[19] suggested that the principal pathologic criteria for the histologic diagnosis of muscular dystrophy consisted of (1) rounding of the outline of muscle fibers in transverse section with hypertrophy of some fibers and atrophy of others, (2) random variation in fiber size and shape, (3) the central migration of sarcolemmal nuclei in many fibers, (4) fiber splitting, and (5) extensive infiltration with fat and connective tissue, these have been the criteria upon which many pathologists have based a histologic diagnosis of muscular dystrophy (Fig. 1). It was generally accepted that inflammatory cell infiltration was absent in the affected muscle, and although a number of authors in the 1930's and 1940's referred to loss of striations and proliferation of sarcolemmal nuclei, the occurrence of actual necrosis of muscle fibers was not

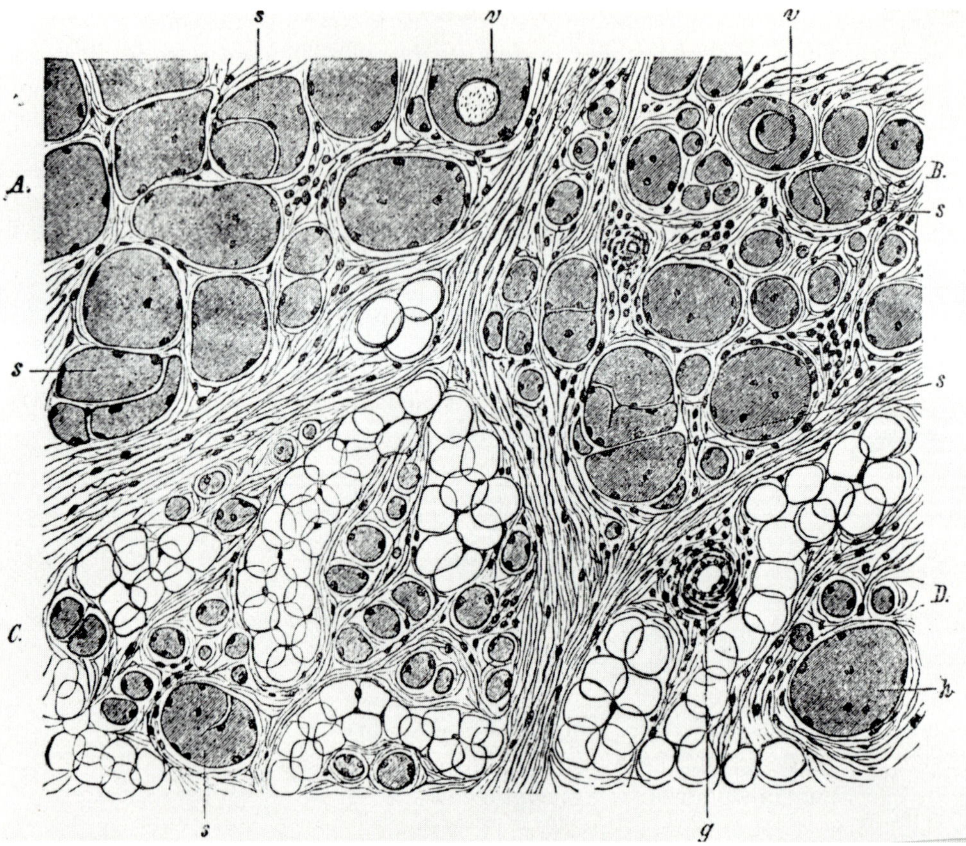


FIG. 1. A diagrammatic representation of the histopathologic changes occurring in voluntary muscle in cases of muscular dystrophy. (Reproduced from Erb[19] by permission of J. F. Bergman Publishers.)

commented upon specifically, and Adams, Denny-Brown and Pearson[3] in the first edition of their book published in 1953 went so far as to suggest that evidence of spontaneous regenerative activity was not seen in dystrophic muscle and that the presence of such changes strongly suggested an active necrotizing myopathy such as polymyositis. I must confess in the middle 1950's to having made more than once a diagnosis of polymyositis based upon the presence of profuse regenerative activity in a muscle biopsy section, in the face of abundant clinical evidence to indicate that the affected children from whom the biopsies were taken were suffering from typical Duchenne type muscular dystrophy.

The limited value of the macroscopic examination of dystrophic muscle must also be mentioned. Clinical observation has indicated clearly that muscular dystrophy, unlike many other myopathic afflictions, is in the early stages a curiously selective disorder in that certain muscles become weak at an early stage of the disease and others are comparatively spared. While the pattern varies in the different varieties, it can be said that in general terms in the Duchenne, limb-girdle and facio-scapulo-humeral cases the muscles most often affected in the upper limbs are the pectorals, serrati, biceps, and brachi-

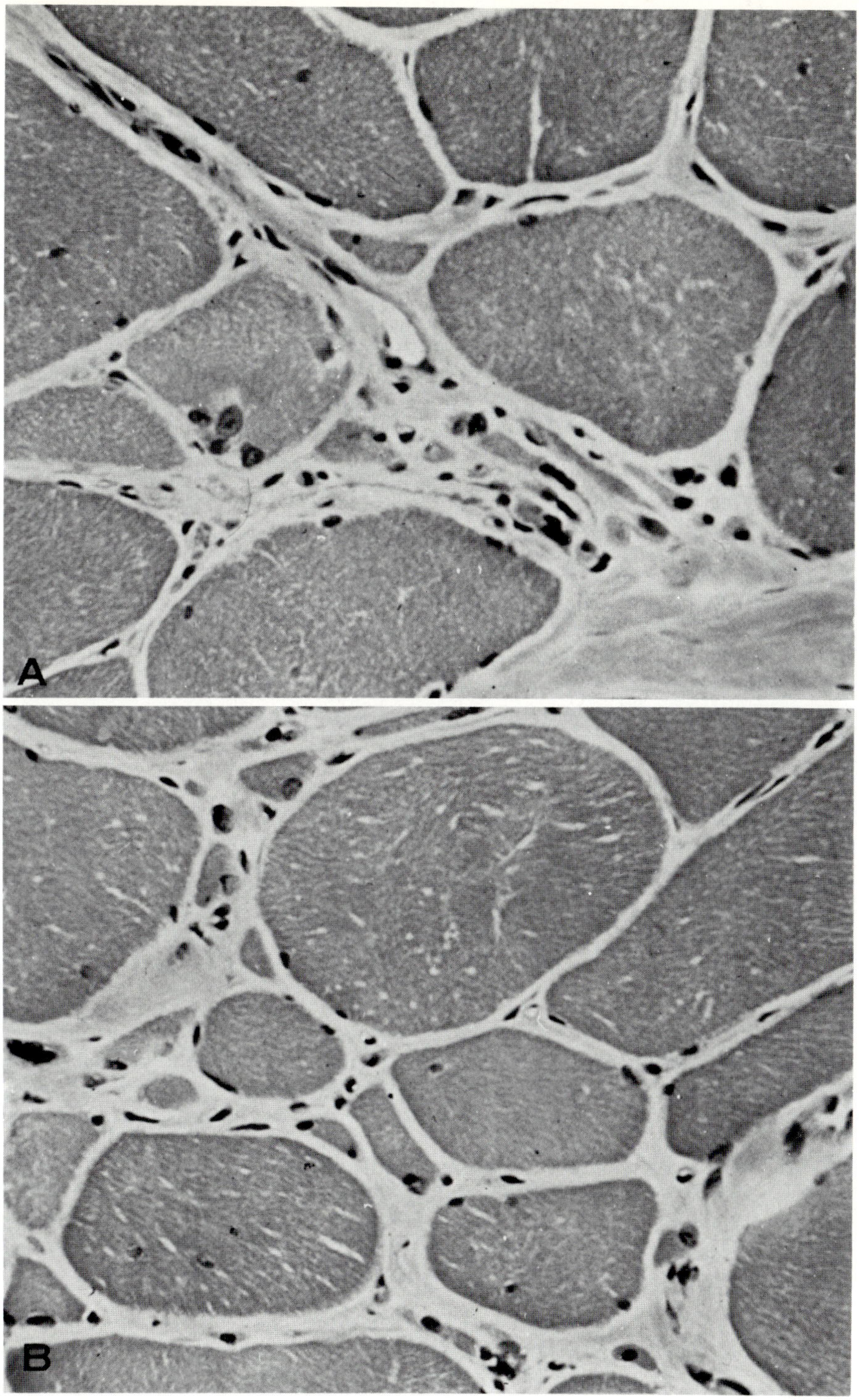


FIG. 2. Transverse sections of a quadriceps biopsy obtained from a youth of 20 years suffering from benign spinal muscular atrophy (the Kugelberg-Welander syndrome). Cryostat sections; hemalum and eosin stain; × 960. *A*, in the center of the section there is a small group of uniformly atrophic denervated fibers. The other muscle fibers show an abnormal degree of variation in size and shape, and one to the left of center shows a minute area of necrosis in which an active sarcolemmal nucleus is seen. *B*, these appearances are more suggestive of so-called disseminated neurogenic atrophy, but there is also evidence of secondary myopathic change with marked variation in fiber size and some central nuclei.

oradiales, while deltoid is almost invariably unaffected in a comparative sense. Similarly, in the lower limbs, the quadriceps and/or hamstrings and anterior tibial muscles tend to become weak comparatively early, but gastrocnemius remains strong until a comparatively late stage in the illness. Morphologic studies such as those carried out by Erb[19] and more recently by Bonsett[7] have confirmed that the gross macroscopic appearance of dystrophic muscles, which become excessively pale and slender and tend to float in water, unlike normal muscle, owing to the very large quantity of fat they contain, remains somewhat selective even at postmortem in that certain muscles are plainly more severely damaged than others, although in the terminal stages virtually every skeletal muscle in the body is abnormal. It is, however, important to make the point that the well recognized clinical phenomenon of enlargement of muscles, which is seen particularly well in cases of the Duchenne and Becker varieties of muscular dystrophy, particularly in the calves but sometimes in deltoids and the lateral vasti, has often been described as being due to the phenomenon of pseudohypertrophy and has been attributed to massive infiltration of the affected muscles by fat. While fatty infiltration has certainly been demonstrated, particularly at autopsy and less often by biopsy in such cases, it is important to recognize that the evidence obtained from biopsy material suggests that in these particular muscles there is often a true hypertrophy of the muscle fibers which accounts for the enlargement of the muscle, and infiltration with fat and connective tissue plays only a minor role. There is some evidence to indicate that this hypertrophy is an attempt at a compensatory process occurring as a response to weakness in other muscle groups, although it seems unlikely that this can be the entire explanation for a phenomenon which is at times so striking. I must, however, stress that the term "pseudohypertrophic" is now virtually outmoded.

To return to a consideration of the histopathologic changes observed in dystrophic muscle, more recent evidence, fully described and reviewed by Pearce and Walton,[40] Pearce,[38] Mumenthaler,[36] Adams,[1, 2] Engel,[18] Mastaglia and Kakulas,[32] and Bell and Conen,[5] has indicated that the primary pathologic process in all varieties of muscular dystrophy appears to be one of necrosis with active degeneration of the muscle cell to which there is a secondary regenerative response of varying extent and efficacy. Vacuolar change is sometimes seen in some necrotic fibers, but it is not a prominent histologic feature in any form of muscular dystrophy. In this respect the primary abnormality observed in dystrophic muscle does not differ significantly from that seen in many other forms of myopathy. It is the combination of necrosis with regeneration, hypertrophy, atrophy, fat, and connective tissue infiltration and the secondary effects upon the muscle of all of these phenomena, together with the tempo of the interplay of these processes, which define, and at the same time distinguish, the changes observed in the various clinical varieties of the disease.[51] One may conclude that, while the classical pathologic descriptions of dystrophic muscle as given by Erb[19] and others describe the changes which one may expect to find in affected voluntary muscle in cases of muscular dystrophy of all clinical types at a late stage of the affliction, recent work has plainly demonstrated that the patterns of pathologic change and its tempo vary greatly from one form of the disease to another, and some of the important variations will be discussed below. It should further be pointed out that much recent work has been done upon so-called secondary myopathic change in chronically denervated muscle and this has been commented upon by Dr. Dorman (see Chapter 13). Among others, Drachman, Murphy, Nigam, and Hills[10] have shown that in patients with chronic partial denervation of skeletal muscle, whether due to an acute illness such as poliomyelitis occurring some years before or to chronic spinal muscular atrophy, the continuing interplay of denervation and reinnervation processes, together with the additional trauma and strain imposed upon a weakened muscle which is attempting to produce physical effort beyond its capacity, may lead to the development of hypertrophic as well as atrophic fibers scattered randomly throughout a muscle section (Fig. 2*A*), separated by infiltrating fat and connective tissue, and accompanied by central nucleation, fiber splitting, vacuolation, necrosis, and phagocytosis. In other words, these so-called myopathic changes, which are commonly accompanied by an abnormal effluence of creatine kinase from damaged muscle into the serum, are indistinguishable in virtually every respect from those appearances observed in an advanced case of muscular dystrophy. It is only when one discovers in such a sample small groups of uniformly atrophic fibers, indicating, or at least suggesting, denervation (Fig. 2*B*), that one may obtain a clue as to the nature of

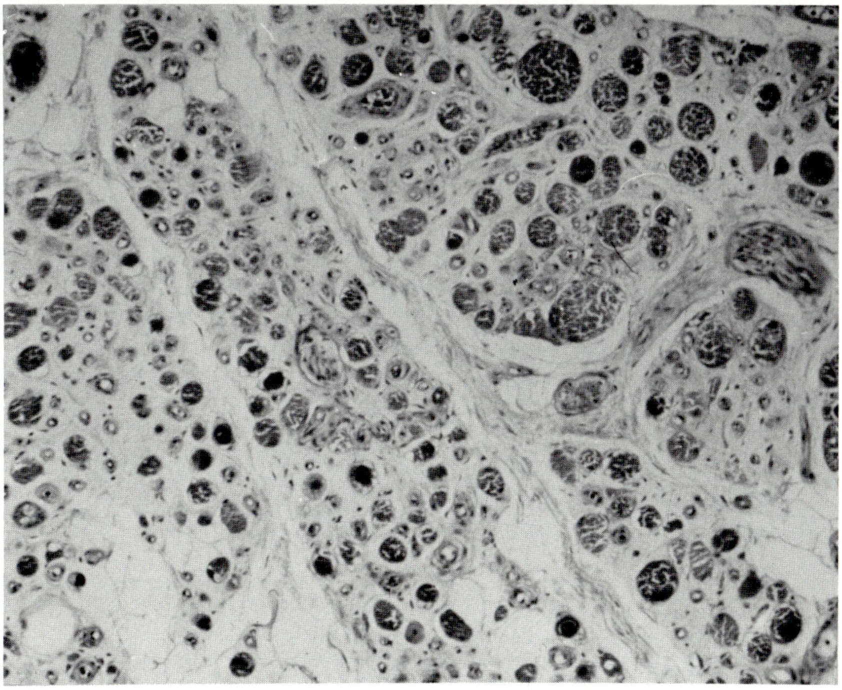

FIG. 3. Transverse section of a tibialis anterior biopsy obtained from a man of 40 years with long standing distal muscular dystrophy. Paraffin section; hemalum and eosin stain; × 120. The small, rounded muscle fibers show considerable variation in size and are embedded in a background of dense connective tissue with a small number of fat cells.

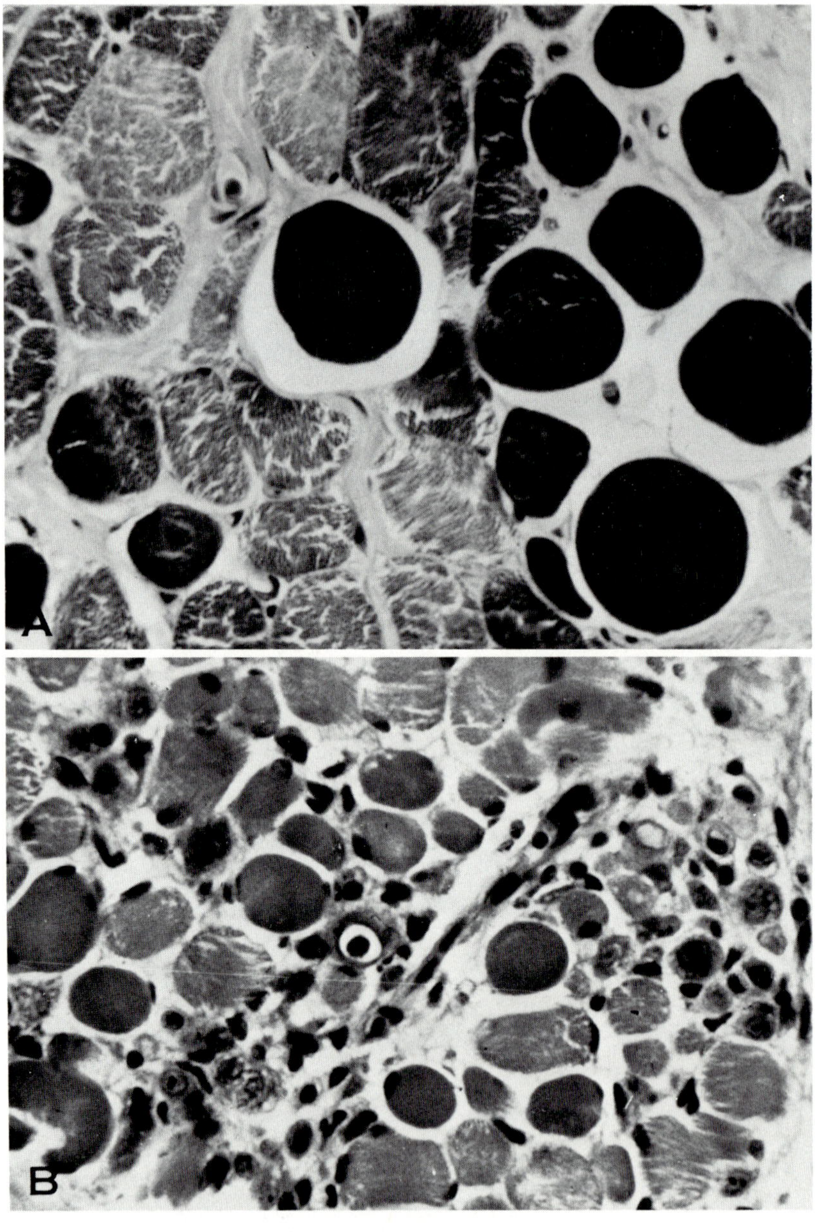


FIG. 4. *A*, transverse section of a quadriceps biopsy from a male child aged 4 years suffering from the Duchenne type muscular dystrophy. Paraffin section; hemalum and eosin stain; × 360. Many darkly staining necrotic and hyalinized muscle fibers are seen. *B*, transverse section of a quadriceps biopsy from an infant aged 3 months with preclinical muscular dystrophy of the Duchenne type. Paraffin section; hemalum and eosin stain; × 360. Many muscle fibers are hyalinized and necrotic but many others show basophilic stippled sarcoplasm and large vesicular nuclei with prominent nucleoli, indicating widespread regenerative activity. (Reproduced from Hudgson, Pearce, and Walton[23] by permission of the Macmillan Company.)

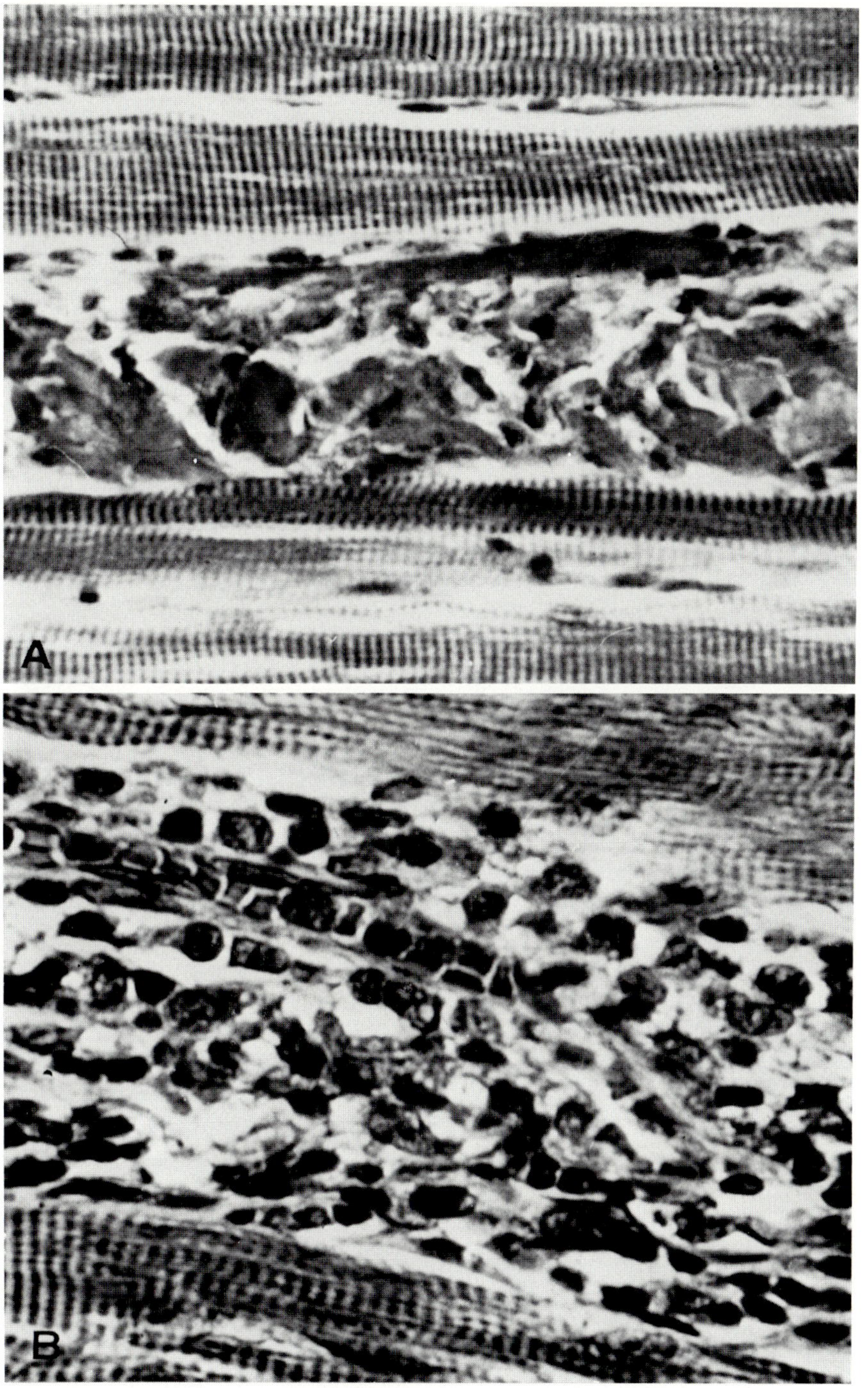


FIG. 5. *A*, longitudinal section of a quadriceps biopsy from a male child aged 5 years with muscular dystrophy of the Duchenne type. Paraffin section; picro-Mallory stain; × 540. The fiber in the center of the section is undergoing necrosis with early invasion by phagocytes. (Reproduced from Brain *et al.*[7a] by permission of Oxford University Press.) *B*, longitudinal section of a quadriceps biopsy from a male child aged 6 years with muscular dystrophy of the Duchenne type. Paraffin section; picro-Mallory stain; × 720. A necrotic segment of a muscle fiber has been completely invaded by phagocytes.

the primary disease process, a clue which is often given considerable support by the results of electromyographic studies. However, in the terminal stages it may be virtually impossible to distinguish between a long standing myopathic process on the one hand and long standing denervation atrophy on the other (Fig. 3), as it is now clear that the final end-stage of muscle degeneration observed in such chronic processes is virtually identical, and these terminal changes are accompanied by abnormalities in the serum enzymes and on electromyography which are equally nonspecific.

### X-LINKED MUSCULAR DYSTROPHIES

*Duchenne type muscular dystrophy*. It was Pearson[41] who first reported studies of muscle biopsies obtained from preclinical cases of muscular dystrophy of the Duchenne type. These were young male infants whose elder brothers were established cases of Duchenne dystrophy and in whom estimation of the serum enzymes had clearly indicated, despite the absence of any clinical abnormality, that they were suffering from this form of muscular dystrophy in a preclinical stage. Pearson[41] and subsequently Hudgson, Pearce, and Walton,[23] who also examined a series of cases of this type, showed that in such infants there were very large numbers of rounded, hyalinized, and necrotic muscle fibers scattered throughout the sections which they studied (Fig. 4), and many other necrotic fibers were observed to be undergoing phagocytosis (Fig. 5*A*). The hyalinized necrotic fibers were seen best in transverse sections of muscle; these changes have been regarded by many as being artifactual, but I am now satisfied that this appearance indicates genuine necrosis, and in fact a large number of such fibers scattered randomly throughout a transverse section of muscle when examined with the low power objective of a microscope is in my opinion an important hallmark of early muscular dystrophy of the Duchenne type. If one follows these hyalinized fibers, which are easy to identify in longitudinal section, either through serial transverse sections or longitudinally, one will invariably find within the affected fiber one or more areas of focal necrosis with infiltration by phagocytes (Fig. 5*B*). Sometimes as many as a third of all of the muscle fibers within such a section are seen to be undergoing necrosis, but yet at this early stage of the disease process there is little, if any, evidence of the random variation in fiber size, fiber splitting, fiber hypertrophy, and central nucleation (with small pyknotic nuclei) which is so common a feature of the dystrophic process at a late stage. Equally characteristic and striking in sections obtained from these patients is evidence of widespread muscle fiber regeneration (Fig. 6). Often large clumps of fibers are seen to have basophilic and coarsely stippled cytoplasm and to contain numerous large vesicular nuclei with prominent nucleoli. In longitudinal section these may be seen to be present in buds adjacent to areas of muscle fiber necrosis, or else a number of such nuclei within muscle spindle cells can be observed amongst the phagocytes which are still devouring necrotic sarcoplasm. Mastaglia and Kakulas[33] have made a particular study of the regenerative activity which occurs in the early stages of Duchenne muscular dystrophy and have found that this is chiefly embryonal in type with the formation of mononucleate myoblasts

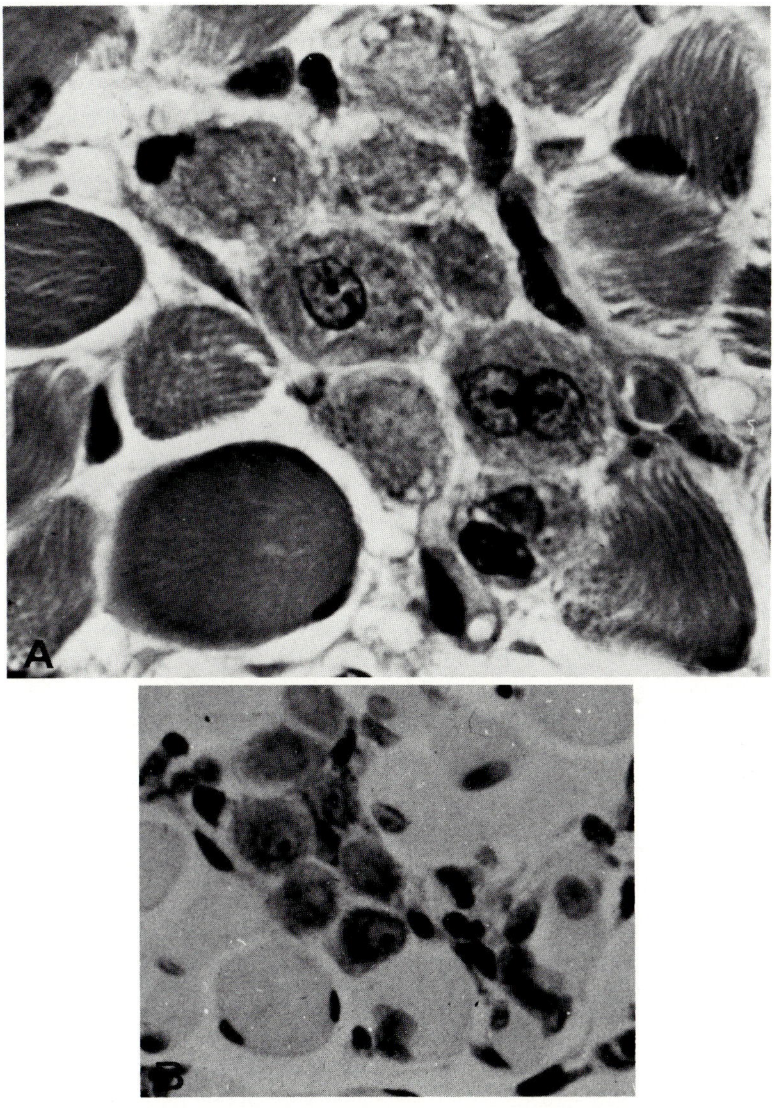

FIG. 6. *A*,transverse section of a quadriceps biopsy from an infant aged 3 months suffering from preclinical muscular dystrophy of the Duchenne type. Paraffin section; hemalum and eosin stain; × 960. The group of fibers seen in the center shows characteristic features of regeneration. (Reproduced from Brain *et al.*[7a] by permission of Oxford University Press.) *B*, a group of fibers similar to those seen in *A*, from the same biopsy, azur B stain; × 360. The cytoplasm of the regenerating fibers stains deeply for RNA. (Reproduced from Hudgson, Pearce, and Walton[23] by permission of the Macmillan Company.)

in necrotic segments of muscle fibers (Fig. 7). They have shown that the cells divide by mitosis and appear to fuse to form multinucleated syncytia, but they also demonstrated "continuous regeneration" taking the form of muscle bud formation at the surviving ends of fibers which had undergone segmental necrosis. In the very early stages of the disease they found that regeneration appeared to proceed to completion with the formation of fully mature muscle fibers. As the disease progressed, however, there appeared to be an arrest in the regenerative sequence so that many regenerating fibers showed degenerative features, a finding in common with that described by Hudgson, Pearce, and Walton.[23] It is thus apparent that in the earliest stages of muscular dystrophy of the Duchenne type the muscle is undergoing an active process of necrosis and that pari passu with this process there is an active attempt being made to repair the damage by regenerative activity which seems to become progressively less efficient and less widespread as the disease progresses. Similar observations have been reported by Walker and Drager[50] who studied 11 cases of muscular dystrophy, though they do not specify whether all of their patients were suffering from dystrophy of the Duchenne type. Using the technique of taking a second biopsy from the same area of muscle 4 to 7 days after the first, they found evidence of regeneration with frequent mitoses and the formation of myotubes in both 4- and 7-day repeat biopsies in 5 of the 11 patients they studied.

If the disease process is studied by muscle biopsy in children under the age of 5 years, the appearances are in general those described above. As the child becomes progressively older, however, necrosis of muscle fibers continues, but fewer and fewer fibers appear effectively to be replaced by regenerative activity. Later biopsies demonstrate less and less active necrosis, and regeneration is often absent or extremely difficult to find. It is in these later stages that secondary changes become more clearly apparent. Thus it seems that many surviving fibers undergo hypertrophy with rounding of their outline and an increase in their myofibrillar population, and in such fibers it is common to discover that there are many central nuclei. Whether these nuclei are present in the center of the fibers in order to produce the new myofibrils necessary for the fiber to undergo hypertrophy it is impossible to say. Concurrently, splitting of such fibers is commonly observed and there has been much speculation as to whether this process is due to direct invasion of the hypertrophied fiber by a connective tissue septum with its nuclei, or whether this appearance is due to the end-stage of a regenerative process in which two myotubes and newly formed fibers arise from a single muscle bud after a period of regeneration, so that a broad fiber, if followed in longitudinal section (Fig. 7), gives the appearance of branching. The significance of this change is therefore obscure, but an appearance suggesting partial or complete splitting of fibers is commonly seen in transverse sections (Fig. 8). At this stage, too, it is generally observed that between the random admixture of excessively large, normal-sized but rounded, and excessively atrophic fibers strands of connective tissue proliferate, so that virtually every muscle fiber has a sheath of connective tissue of variable thickness and many such fibers are separated by fat cells. The source of the infiltrating fat has not been determined, but there is no evidence that it arises as a result of metaplasia from damaged muscle. Finally, one reaches an end-stage where virtually the whole muscle is occupied by large amounts of fat and thick strands of connective tissue surrounding and separating remnants of functionally incompetent muscle fibers of varying shapes and sizes, as well as scattered nuclear debris in small and pyknotic clumps, representing the remnants of those

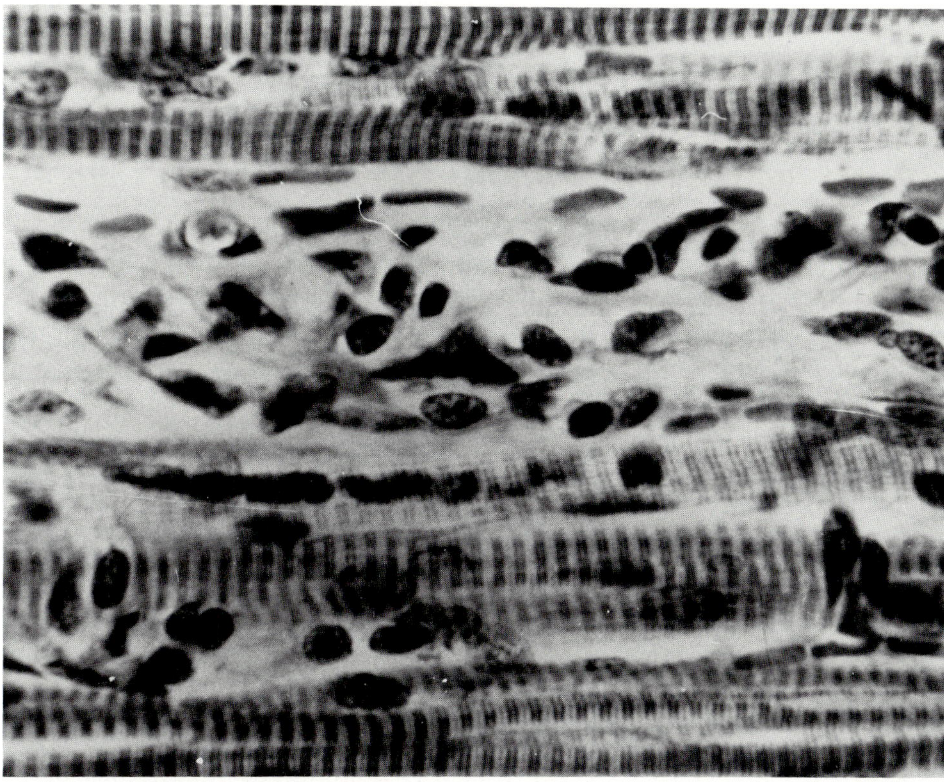

FIG. 7. Longitudinal section of a quadriceps biopsy from a male child aged 4 years with muscular dystrophy of the Duchenne type. Paraffin section; picro-Mallory stain; × 640. In the center there is a necrotic segment of muscle fiber containing numerous phagocytes but also several muscle spindle cells indicating "embryonal regeneration" (see text). The fibers above and below show appearances suggestive of longitudinal branching or splitting, with large and active muscle nuclei often lying alongside or between the branches.

fibers whose myofibrils and sarcoplasm have virtually disappeared completely. As described above, this end-stage is virtually indistinguishable from that of any other form of muscular dystrophy and even from some varieties of chronic denervation atrophy.

So far as the other structures seen within muscle biopsy sections are concerned, the blood vessels remain remarkably normal throughout and no specific abnormalities have been described. In the late stages they appear to increase somewhat in number but this is simply the result of concentration and of a reduction in the over-all size of the affected muscle. Similarly, one may have the impression that the muscle spindles are increased and that they are relatively spared by the disease process. This is true to some extent, as normal spindles may be observed even in severely dystrophic muscle. However, as pointed out by Lapresle and Milhaud,[28] Cazzato and Walton,[8] and Patel, Lalitha, and Dastur,[37] some pathologic reactions may be observed in certain spindles in severely dystrophic muscle. Cazzato and Walton[8] in particular found that in some cases of muscular dystrophy the capsule of the spindle was

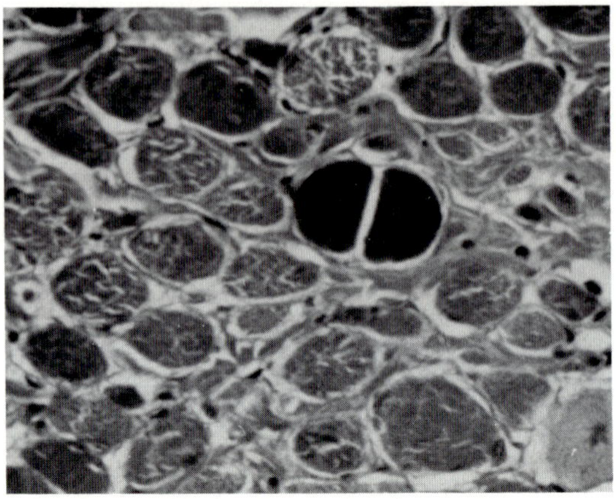

FIG. 8. Transverse section of a quadriceps biopsy from a male child aged 5 years with muscular dystrophy of the Duchenne type. Paraffin section; hemalum and eosin stain; × 240. The hyalinized fiber in the center shows an appearance suggestive of fiber splitting.

thickened, that there was marked edematous swelling and atrophy of both the nuclear-bag and nuclear chain fibers (Fig. 9). In general, however, these changes occurred relatively late in the disease process and were much more severe in cases of congenital muscular dystrophy (*vide infra*) (Fig. 10) than in those of the Duchenne type, although even in Duchenne type cases occasional spindles were observed which were very abnormal.

No consistent abnormalities in the intramuscular nerves have been reported in muscle biopsy material from cases of muscular dystrophy, though Coërs and Woolf[9] have pointed out that intravital staining of the muscle end-plates with methylene blue, followed by biopsy of the end-plate zone, may demonstrate changes in many of the end-plates examined in dystrophic muscle and abnormal degrees of proliferation of the terminal intramuscular nerves. It seems likely, however, that these changes are secondary to the abnormalities in the muscle fibers rather than the reverse.

*Becker type muscular dystrophy*. This X-linked variety of muscular dystrophy tends to be of later onset and runs a much more benign course than does the Duchenne type, which it otherwise resembles clinically. Walton and Gardner-Medwin[52] described the case of a young man of 19 years, two of whose maternal uncles suffered from this form of muscular dystrophy and in whose case virtually no clinical abnormality apart from diffuse muscle hypertrophy could be demonstrated. Examination of a quadriceps biopsy taken from this patient (Fig. 11) demonstrated a uniform hypertrophy of all of the muscle fibers but no other pathologic changes could be demonstrated. The muscle fibers in general measured in diameter and area almost twice the known upper limit for fibers in the rectus femoris in a young man of this age, and the authors concluded that there may be an initial phase of true muscular hypertrophy in some patients suffering from this variety of the disease.

Our own experience and that of others indicates that subsequently when muscular weakness becomes apparent the histologic changes differ only in

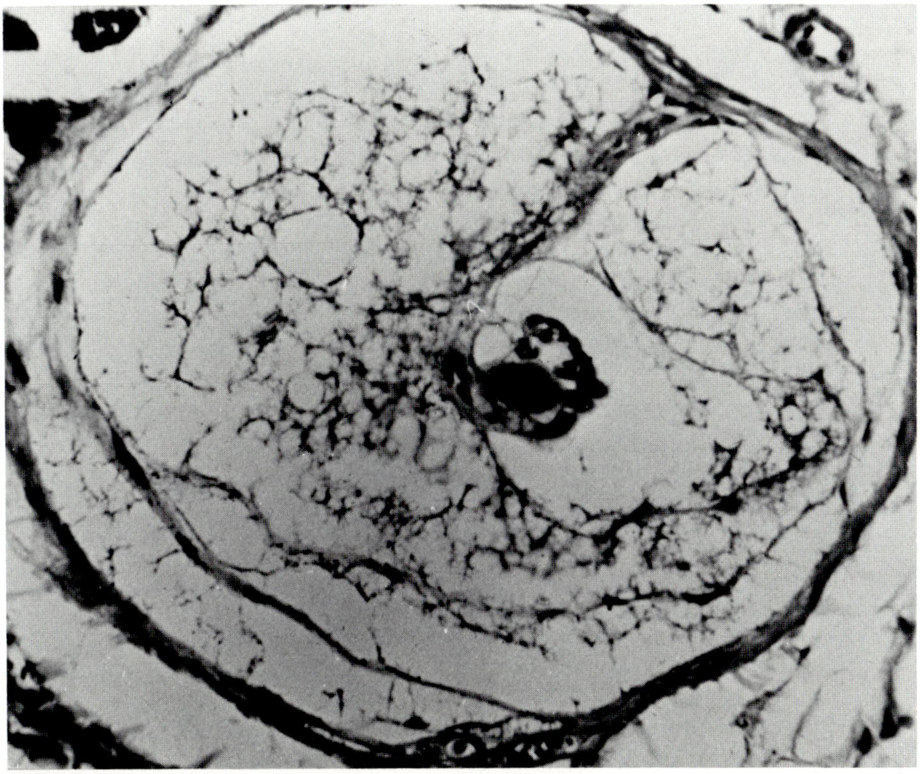

FIG. 9. A grossly swollen and edematous muscle spindle showing marked atrophy of the intrafusal fibers. Limb-girdle muscular dystrophy, quadriceps biopsy. Paraffin section; phosphotungstic acid-hematoxylin stain; × 338. (Reproduced from Cazzato and Walton[8] by permission of Elsevier Publishing Co.)

severity and ubiquity from those of the Duchenne type. Scattered areas of fiber necrosis and phagocytosis are seen, and so too is regeneration, but this is comparatively scanty. Even at a stage when such changes are apparent, one usually finds a considerable degree of variation in fiber size and shape with central nucleation, typical of so-called myopathic changes but with comparatively little infiltration by connective tissue and fat occurring until very much later (Fig. 12). In these cases, too, the ultimate appearances are identical with those observed in any form of dystrophy.

*Carrier detection.* Many attempts have been made within the last few years to identify those clinically unaffected female carriers of the X-linked genes responsible for the Duchenne and Becker types of muscular dystrophy who are likely to pass the disease on to their sons. At the present time estimation of the serum creatine kinase is successful in identifying 60 to 70% of carriers of the Duchenne type gene but is much less successful in identifying carriers of the Becker variety.[17] In an attempt to improve carrier detection, several authors have studied the histologic changes observed in muscle biopsy specimens obtained from carriers. Emery,[15] Mannucci, Ideo, Cao, and Macciotta,[3] Pearson and Kar,[43] and Johnston, Wilkinson, Withycombe, and Raymond[24]

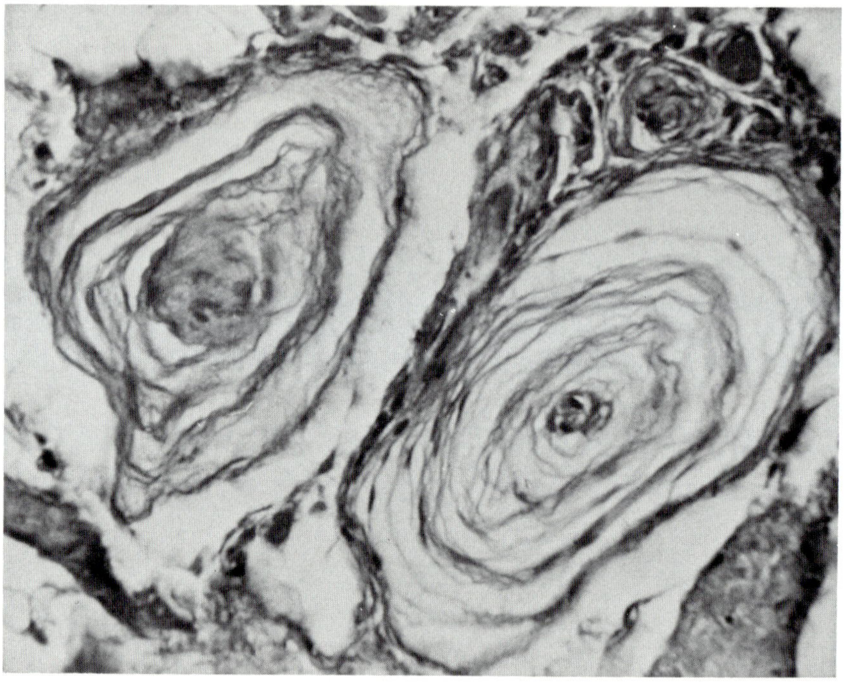

FIG. 10. Two grossly abnormal muscle spindles seen in a quadriceps biopsy from a child aged 3 years with congenital muscular dystrophy; there is marked proliferation of fibrous tissue within the capsules and the intrafusal muscle fibers are grossly atrophic. Paraffin section; picro-Mallory stain; × 615. (Reproduced from Cazzato and Walton[8] by permission of Elsevier Publishing Co.)

found that in such muscle biopsy specimens there was a reduction in the concentration of lactate dehydrogenase-5 isoenzyme, but this change appears to be relatively nonspecific as a similar abnormality has been demonstrated in denervated muscle. A number of authors have, however, demonstrated histologic abnormalities, as revealed by light microscopy, in muscle biopsy specimens obtained from carriers of the gene responsible for the severe Duchenne type,[12, 14, 16, 27, 30, 39, 42, 45, 47, 48] and changes have also been observed in specimens examined with the electron microscope.[25, 27, 34, 47] Changes are, however, very variable in different carriers. Sometimes clear-cut areas of focal necrosis and phagocytosis (Fig. 13) with some regenerative activity have been observed,[39] but in other cases the changes are minimal and nonspecific, consisting of no more than moderate variation in fiber size and shape with occasional central nuclei. These changes have been accounted for on the basis of the Lyon hypothesis[29]; it has been suggested that in a proportion of the nuclei incorporated into muscle cells in the female carrier, the paternal X chromosome may be inactivated with the result that that part of the cell whose metabolism is controlled by the nucleus in question becomes dystrophic. If the Lyon hypothesis were to apply fully in this situation, one would expect that approximately one-half of all the nuclei in the muscle cells of a female carrier would be dystrophic, and certainly abnormalities of this magnitude have not been revealed by histologic study. Thus one may conclude that if unequivocal histologic abnormalities of the type described, which cannot be accounted for by focal trauma to the muscle or by other extraneous factors, are found in a muscle biopsy obtained from a putative female carrier of this X-linked gene, these histologic changes give considerable support to the collateral evidence obtained from studies of the serum creatine kinase and from quantitative electromyography,[20] and it may be reasonably assumed that the female in question is a carrier. Negative findings in a muscle biopsy examination can, however, in no sense be taken to exclude the carrier state with any degree of certainty.

AUTOSOMAL RECESSIVE MUSCULAR DYSTROPHY

*Limb-girdle muscular dystrophy.* This is the least clearly defined of the commonly occurring varieties of muscular dystrophy, and our experience, in common with that of others, has recently indicated that many patients so diagnosed in the past are in fact suffering from relatively benign spinal muscular atrophy (the Kugelberg-Welander syndrome) in which, because of the relative chronicity of the disease process, secondary myopathic change may dominate the histologic findings. There can, however, be no doubt that cases of true limb-girdle dystrophy, in which the disease begins in the shoulder girdle

FIG. 11. Transverse section of a quadriceps biopsy from a very early case of the Becker type muscular dystrophy, patient aged 19 years, taken when there was clinical evidence of diffuse muscular hypertrophy but none of muscle weakness. Paraffin section; hemalum and eosin stain; × 360. The muscle fibers are morphologically normal but have an average transverse diameter of 130 $\mu$. (Reproduced from Walton and Gardner-Medwin[52] by permission of the Pitman Medical Publishing Company, Ltd.)

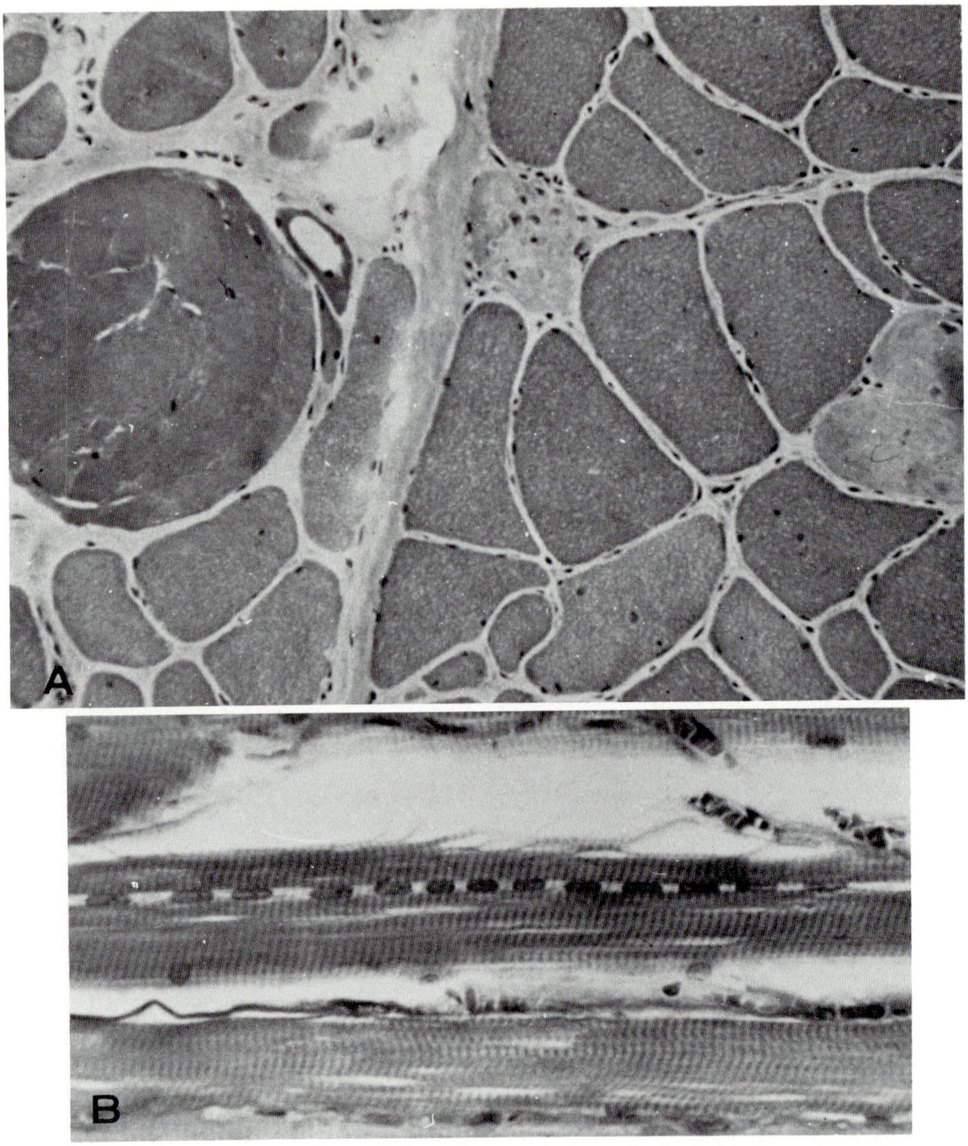


FIG. 12. *A*, transverse section of a quadriceps biopsy from a man of 28 years with the Becker type muscular dystrophy. Cryostat section; hemalum and eosin stain; × 540. Two fibers are seen to be necrotic and to be undergoing phagocytosis; the grossly enlarged fiber to the left is also partially necrotic; the other fibers show many central nuclei with random variation in size and shape. *B*, longitudinal section from the same biopsy as in *A*. Paraffin section; picro-Mallory stain; × 480. The fiber in the center contains a chain of enlarged nuclei; the appearances show some resemblance to those of a fetal myotube and probably indicate one phase of regeneration.

muscles or alternatively in the pelvic girdle muscles, do occur, that the condition is inherited by an autosomal recessive gene, and that it runs a clinical course which is much more benign than that of the severe Duchenne variety. Previous reports on the histopathologic changes observed in muscle biopsy

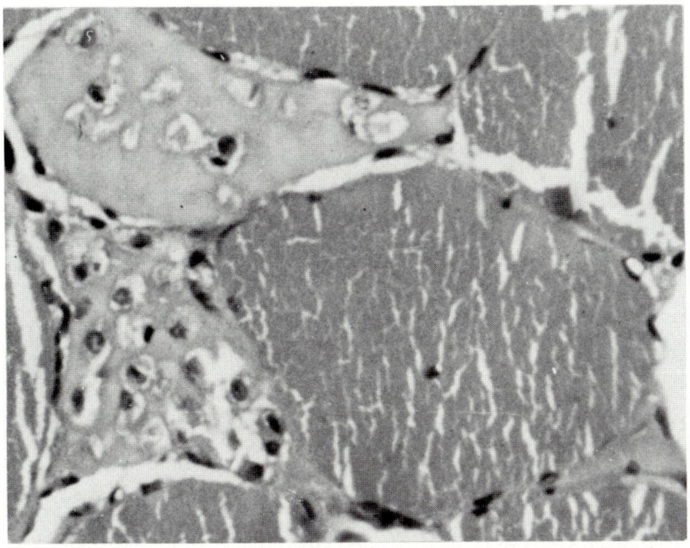

FIG. 13. Transverse section from a quadriceps biopsy from a woman aged 36 years who had two affected sons and a brother who had died of the Duchenne type muscular dystrophy; she was therefore classed as a definite carrier. Paraffin section; hemalum and eosin stain; × 640. Two fibers to the left are undergoing necrosis and phagocytosis, and the other fibre contains a central nucleus.

sections from such cases[1, 2, 4, 18, 21, 36, 40] and our own recent experience have indicated that the changes observed in these cases lack specificity. By the time that most such patients come to muscle biopsy the disease as a rule is moderately advanced and little information is available concerning the pathologic changes observed in the early stages. The characteristic appearances observed are those of variation in fiber size and shape, central nucleation, fiber splitting, and marked infiltration by fat and connective tissue, as described above (Fig. 14), although in the majority of cases some patchy areas of focal fiber necrosis and phagocytosis and occasional evidence of sparse regenerative activity are seen. Our own studies have failed to demonstrate any pattern of histologic change which is in any way specific for this clinical variety of the disease. We have, however, been impressed by the finding in some such cases of frequent ringbinden (Fig. 15). So-called ringbinden or striated annulets have been described as occurring particularly in combination with pale staining, homogeneous peripheral masses of sarcoplasm in muscle sections obtained from patients with myotonic dystrophy,[46] but in our experience these changes are relatively rare in muscle biopsies obtained from such cases, although we have observed this type of change frequently in muscle obtained at autopsy (Fig. 16). There has been considerable controversy in the past about the significance and etiology of ringbinden, although Morris[35] and Bethlem and Van Wijngaarden[6] and others have demonstrated that large numbers of such striated annulets seen in transverse section encircling the central myofibrils of a muscle fiber which are cut transversely may readily be produced experimentally by tenotomy and are commonly seen in normal muscles which have no tendinous insertion. In several cases of limb-girdle muscular dystrophy we have seen these structures,

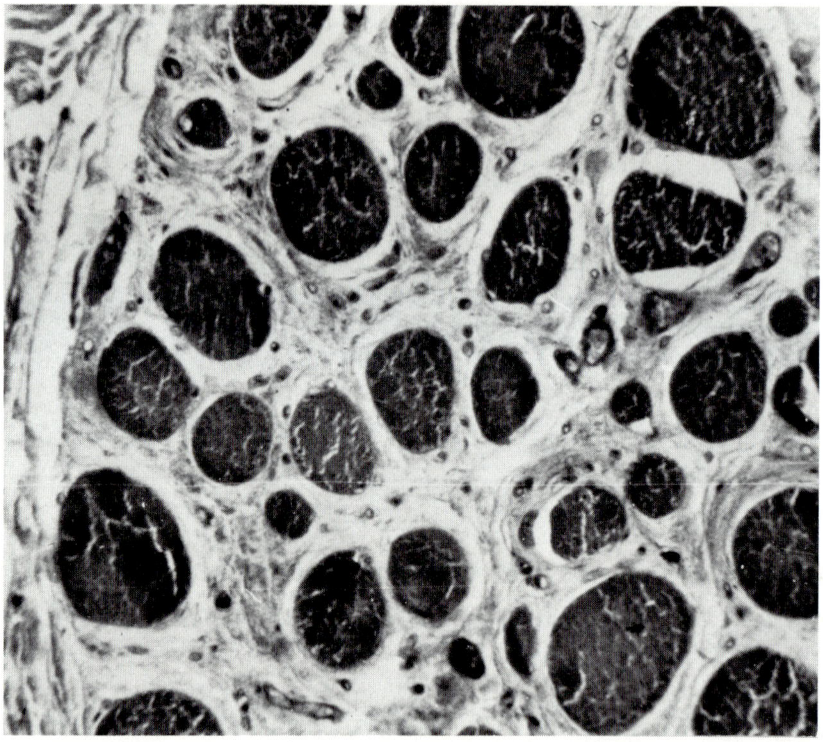

FIG. 14. Transverse section of a biopsy of biceps brachii from a moderately advanced case of limb-girdle muscular dystrophy. Paraffin section; hemalum and eosin stain; × 360. The fibers are abnormally rounded, show a marked variation in size and shape, and are embedded in a dense background of connective tissue. (Reproduced from Brain *et al.*[7a] by permission of Oxford University Press.)

sometimes with, but more often without, sarcoplasmic masses, and in two such cases we found very large numbers of ringbinden clearly apparent in both longitudinal and transverse sections in muscle biopsy material embedded in paraffin (Fig. 15), but these appearances were totally absent in another portion of the same biopsy which had been frozen immediately on removal and from which sections had been cut in a cryostat (Fig. 17). This and other observations lead to the irresistible conclusion that ringbinden are usually the result of fracture of peripherally situated myofibrils within certain muscle fibers. Because of the elasticity of the myofibrils which fracture, they then wind in a spiral around the remaining myofibrils of the fiber which remain in continuity. Between these ringbinden and the intact sarcolemma, pale staining, homogeneous masses of sarcoplasm (sarcoplasmic masses) may then be formed (Fig. 16). The reason why sarcoplasmic masses were not seen in the material from our two cases of limb-girdle muscular dystrophy referred to above was probably because the fracture of the peripheral myofibrils, giving rise to ringbinden, occurred after removal of the muscle biopsy. When such fracture occurs *in vivo* (as, for instance, in the irritable myotonic muscle of patients with myotonia dystrophica) this process is more likely to be followed by the formation of sarcoplasmic

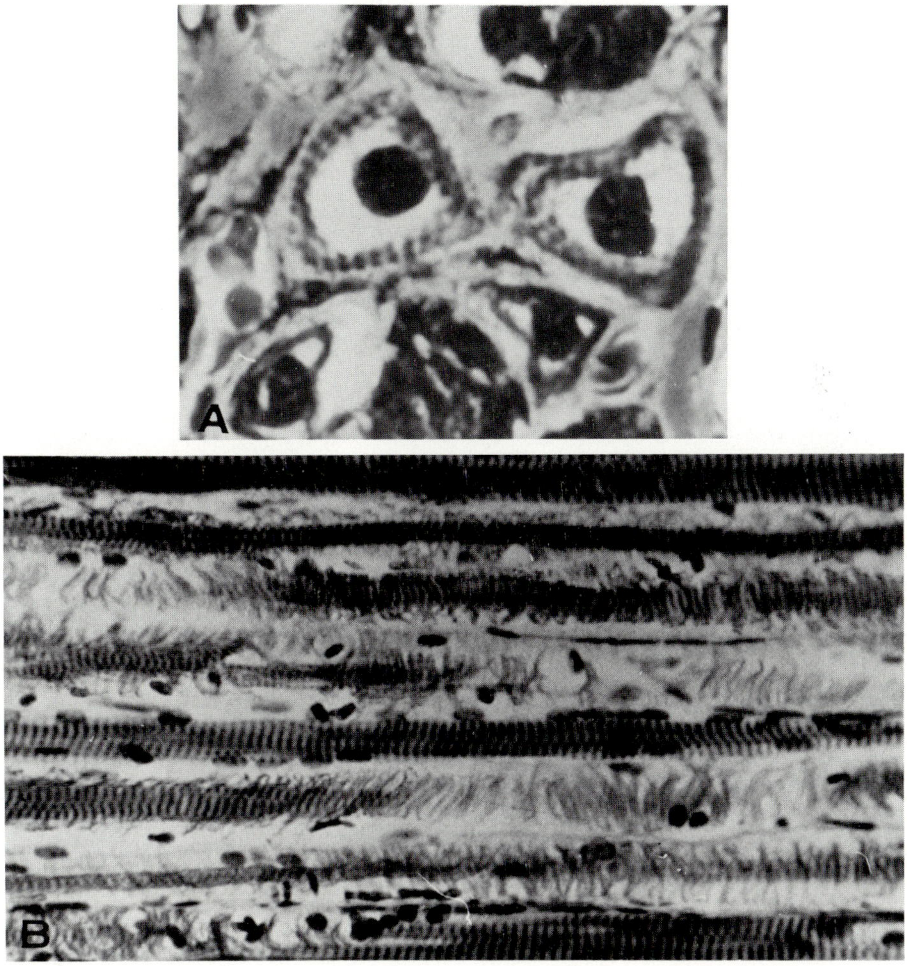


FIG. 15. *A*, transverse section of a quadriceps biopsy from a moderately advanced case of limb-girdle muscular dystrophy. Paraffin section; phosphotungstic acid-hematoxylin stain; × 640. The sparse central myofibrils, cut transversely, are surrounded by striated annulets (ringbinden). *B*, longitudinal section from the same biopsy as that illustrated in *A*. Paraffin section; phosphotungstic acid-hematoxylin stain; × 240. The fractured peripheral myofibrils are seen winding in dense spirals around those central myofibrils which remain in continuity.

masses which fill in the space remaining between the ringbinden and the intact sarcolemma. Thus it would be my personal view that ringbinden developing *in situ* in muscle during life depend either upon an unusual degree of irritability of the muscle fibers such as is present as a result of myotonia, or upon functional or anatomic separation of the muscle, in whole or in part, from its tendinous attachment. Our experience, however, indicates that large numbers of ringbinden may occur as an artifact developing either during fixation or embedding; I do not believe that the finding of this type of change in some cases of limb-girdle muscular dystrophy casts any light upon the pathogenesis of this particular variety of the disease.

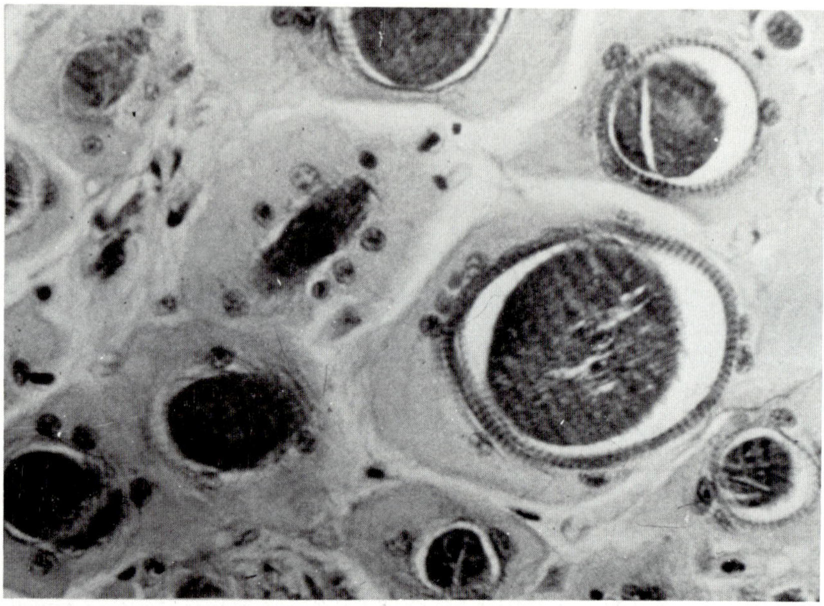

FIG. 16. Transverse section of iliopsoas obtained at autopsy from a case of dystrophia myotonica. Paraffin section; phosphotungstic acid-hematoxylin stain; × 480. There are ringbinden and large, amorphous, peripheral sarcoplasmic masses in practically every muscle fiber. (Reproduced from Brain *et al.*[7a] by permission of Oxford University Press. Figure lent by G. W. Pearce.)

*Childhood muscular dystrophy (except Duchenne).* A number of cases are on record in which a disorder resembling muscular dystrophy of the Duchenne type has affected male and female members of the same sibship and has therefore been presumed to be the result of autosomal recessive inheritance. These cases are rare and many patients so diagnosed have been found on further investigation to be suffering from chronic spinal muscular atrophy. In those cases in which all investigations indicate that the muscular weakness and wasting is the result of a dystrophic process, histologic changes are identical with those of moderately advanced Duchenne type dystrophy.

*Congenital muscular dystrophy.* This group of cases remains nosologically unsatisfactory. It has been recognized for many years that in some children who are born with widespread muscular weakness, hypotonia, and contractures of skeletal muscle, and who on clinical grounds are diagnosed, therefore, as suffering from arthrogryposis multiplex congenita, the pathologic changes in skeletal muscle may be strongly suggestive of a dystrophic process. In many others pathologic evidence indicates that the muscular weakness and the resultant contractures are due to a spinal muscular atrophy beginning in fetal life, whereas in some few cases this clinical picture may well result from a primary disorders of tendons and connective tissue. While doubt has been cast upon the existence of an entity which may be identified by the title of "congenital muscular dystrophy," there can be no doubt that cases occur which cannot be described in any other way.[22] In many such cases the muscular weakness, although severe and sufficient to prevent the child from ever standing or walking, is relatively nonprogressive; electromyography gives clear indication of a myopathic process but yet the serum enzymes may be normal. Histologic examination of muscle biopsy sections obtained from such cases usually demonstrates some variation in fiber size and shape with rounding of the surviving fibers and gross atrophy of others, together with occasional features of focal necrosis, little or no regeneration, and above all a massive infiltration with connective tissue and lesser amounts of fat (Fig. 18). While some may postulate that this clinical picture could result from a long arrested fetal spinal muscular atrophy, one can only conclude that in certain fatal cases the population of anterior horn cells has been found to be normal at autopsy and in the great majority of those so diagnosed no electrophysiologic or histopathologic evidence of denervation atrophy of muscle is recognized. Thus for the present no better term than "congenital muscular dystrophy," despite the confusion engendered by its use, is possible for the identification of these cases. Zellweger, Afifi, and McCormick[55] have suggested that many cases previously diagnosed as examples of benign congenital nonprogressive myopathy occurring in children who are late in standing and in walking, who demonstrate widespread muscular weakness and hypotonia from birth, and in whom the disease process appears to be relatively nonprogressive, are probably suffering from congenital muscular dystrophy in a mild form. It is true that cases of this nature are not uncommon and we ourselves have recognized more than a dozen such cases within the last 7

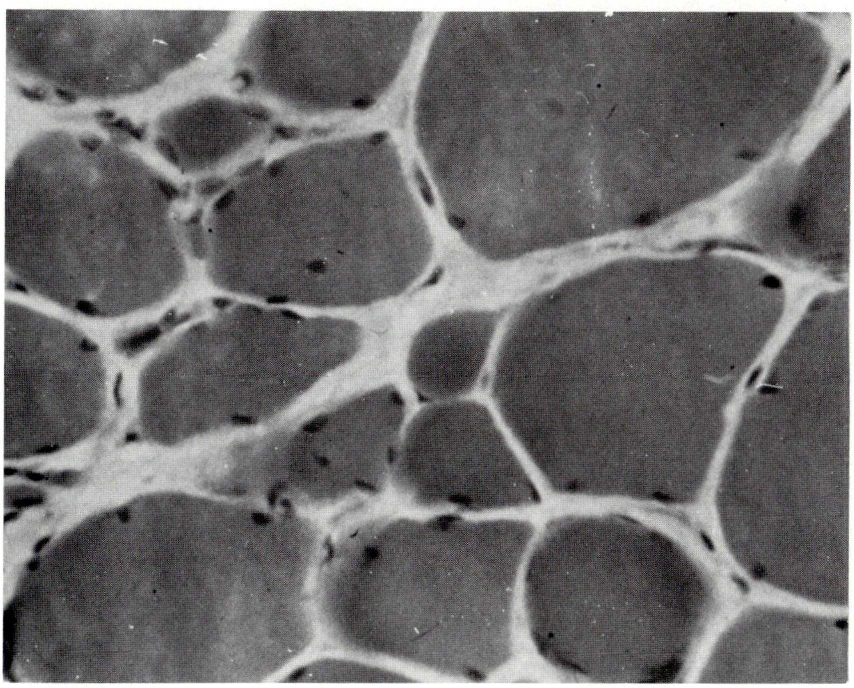

FIG. 17. Transverse section of a quadriceps biopsy from a moderately advanced case of limb-girdle muscular dystrophy. Cryostat section; hemalum and eosin stain; × 640. There is marked variation in size and shape of the muscle fibers but no ringbinden are seen. This section was cut from the same biopsy as that illustrated in Fig. 15.

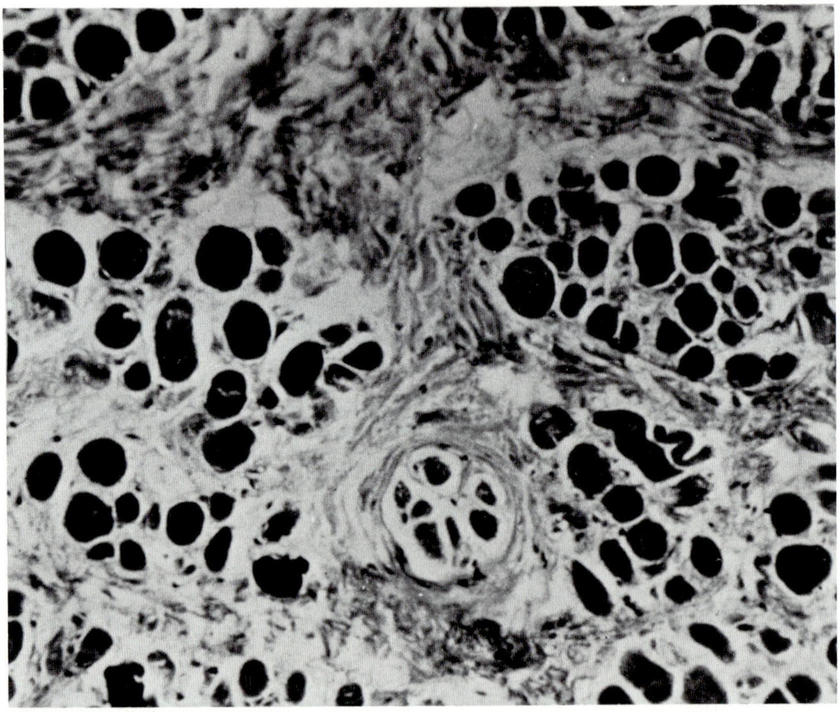

FIG. 18. Transverse section of a quadriceps biopsy from a child aged 3 years with congenital muscular dystrophy. Paraffin section; picro-Mallory stain; × 360. Scattered muscle fibers of varying size and shape are embedded in a dense background of connective tissue. A relatively well preserved muscle spindle is seen.

or 8 years, in none of which have we been able to identify any of the specific histologic characteristics associated with well recognized morphologic entities such as myotubular myopathy, central core disease, and nemaline myopathy. Histologic investigations of muscle biopsy specimens in such cases often demonstrate no more than an over-all smallness of the muscle fibers, both in total fiber area and in diameter, associated with a degree of variation in fiber size and shape which is outside of the normal range and with occasional central nucleation and a modest increase in interstitial connective tissue and fat. Active necrosis and regeneration of muscle is rarely if ever seen. For the moment we can do little more than to record and to recognize cases which fall into this obscure group and to hope that further investigations will clarify the nature of the disease process and will enable us to rationalize the classification of these cases on a more firmly established etiological basis.

FACIO-SCAPULO-HUMERAL MUSCULAR DYSTROPHY

This, the most benign variety of muscular dystrophy, which involves the facial and scapular muscles first in the upper limbs and the anterior tibial muscles in the lower, is inherited by an autosomal dominant mechanism. Some cases are mild and show relatively little progression for many years (abortive cases), while others are rather more severe and result in moderate disability in middle life. Yet again, the histologic changes observed in muscle biopsy sections from such patients lack specificity. By the time the biopsy is taken, muscle fiber necrosis is scanty, if indeed it is found, and regeneration is virtually absent. The histologic appearances differ in no significant respect from those observed in moderately advanced limb-girdle dystrophy and in severe advanced Duchenne dystrophy. It is in cases of this type that the most striking degree of variation in fiber size, shape, and configuration is often seen (Fig. 19), with relatively little else to indicate active and progressive pathologic change.

### Distal muscular dystrophy

My own personal experience of this rare form of muscular dystrophy is small, but in the few cases I have examined, as in those reported by Welander,[54] the changes are again nonspecific. By the time one has an opportunity of examining muscle biopsy material from such cases one usually finds a nonspecific appearance of advanced dystrophy in which muscle fibers of varying shape and size are embedded in a dense background of connective tissue (Fig. 3) but with relatively little and scanty fatty infiltration.

### Ocular muscular dystrophy

In 1951 Kiloh and Nevin[26] suggested that most patients previously diagnosed as suffering from progressive nuclear ophthalmoplegia were in fact suffering from a progressive dystrophy of the external ocular muscles and based their

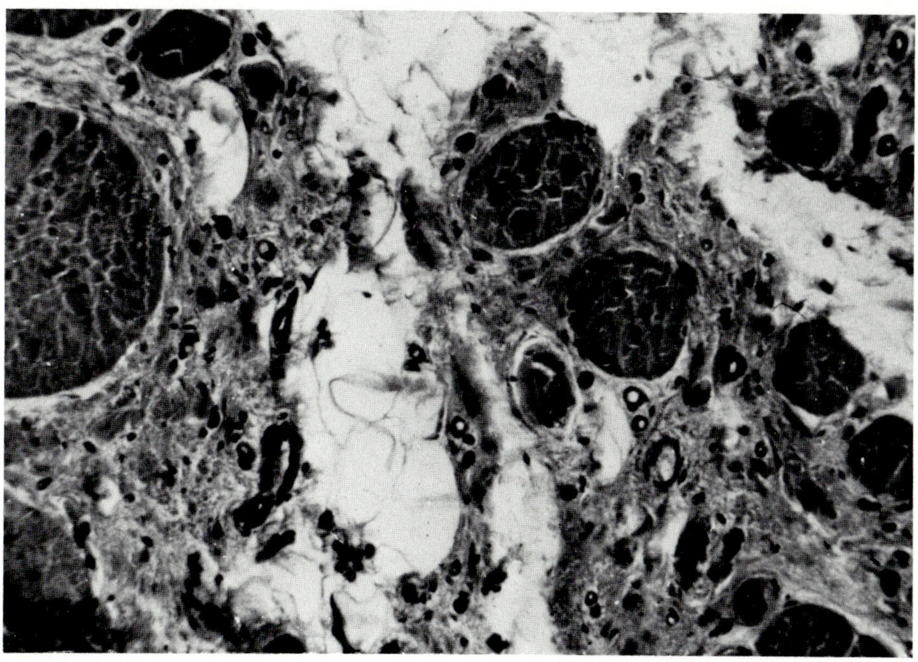

Fig. 19. Transverse section of a biceps brachii biopsy from an advanced case of facio-scapulohumeral dystrophy. Paraffin section; hemalum and eosin stain; × 508. There is marked random variation of muscle fiber size and shape and the nuclear remnants of grossly atrophic fibers are seen scattered amongst dense infiltrating connective tissue and fat.

conclusions upon histologic examination of samples of the levator palpebrae superioris obtained from such cases at operation carried out to elevate the upper lids. The histologic changes which they described lacked specificity but were similar to those of advanced muscular dystrophy as described above. Subsequently many authors have concluded that progressive external ophthalmoplegia occurring in any setting was almost invariably dystrophic in type, but recently, following experimental work in animals in which the ocular muscles were denervated, Drachman, Wetzel, Wasserman, and Naito[11] have shown that external ocular muscle responds to denervation by producing histologic changes which many pathologists would construe as being myopathic. Thus while one must accept that progressive dystrophy of the external ocular muscles does occur, the possibility that some such cases, particularly when associated with evidence of disease in the central nervous system, may be due to degeneration in the ocular motor nuclei cannot be denied.

The histologic changes observed in the external ocular muscles in long standing cases of neostigmine-resistant ocular myasthenia gravis are also indistinguishable from those observed in a primary ocular myopathy. Furthermore there is no means at present available to distinguish pathologically between the changes which occur in ocular myopathy of early onset on the one hand and in so-called oculopharyngeal muscular dystrophy,[49] a dominantly inherited disorder with onset usually in middle age, on the other. Clearly histologic study of the external ocular muscles in these various disorders has added little to our understanding of the disease processes which are responsible for their presenting clinical features.

## CONCLUSIONS

While the many subvarieties of human muscular dystrophy differ widely in their mode of inheritance, clinical presentation, and course and may well prove eventually to be different diseases whose etiology (in terms of fundamental molecular biology and biochemistry) differs from one type to another, the nature of the pathologic processes revealed within skeletal muscle specimens obtained by surgical biopsy or at autopsy in cases of all types in basically similar. Thus the fundamental abnormality seems to be one of necrosis of a part or the whole of a muscle cell followed by a regenerative response of varying extent and efficacy on the part of the damaged muscle. The interplay of these processes, combined with hypertrophy (probably compensatory) of some surviving fibers and a progressive infiltration with connective tissue and fat, leads to the familiar picture of advanced muscular dystrophy in which there are scattered muscle fibers of all shapes and sizes, some apparently showing splitting and others central nuclei; all are usually separated one from another by connective tissue and lipocytes. Occasionally (in the Becker type of muscular dystrophy) there appears to be an initial stage of widespread fiber hypertrophy. Otherwise, apart from the widespread necrosis and regeneration which is seen only in the preclinical or early cases of the Duchenne type, there are no distinguishing features by means of which the histologic picture of any subvariety of muscular dystrophy can be distinguished from that of the others. It is the tempo rather than the nature of the process which varies from one variety to another.

Despite the lack of specificity of the changes described and despite the fact that the findings in subacute polymyositis and the secondary myopathic change seen in cases of long standing neurogenic atrophy may sometimes be indistinguishable from dystrophic features when only a small single biopsy specimen is examined, the muscle biopsy technique and studies of conventional histologic type with the light microscope can be of very great diagnostic value. It is essential first that the biopsy technique, the fixation, blocking, and embedding of the tissue be immaculate, that full information is available concerning the results of clinical examination, serum enzyme studies, and electromyography for correlation with the histologic findings, and that the pathologist should be aware of the stage of the disease at which the biopsy sections were obtained. While the "blind" interpretation of muscle biopsies is an exercise which is invaluable to us all, the conclusions adduced after such a procedure can often be at best vague or speculative and may lead to the suggestion that several different disease processes could give such a picture. Armed with full information about the patient, however, as outlined above, the interpretation of the findings can be much more precise and at times specific. The diagnosis of muscle disease is a process which necessitates the careful correlation of information from clinical, genetic, biochemical, electrophysiologic, and pathologic sources; only rarely is an accurate diagnosis possible from one or two of these sources alone.

## ACKNOWLEDGEMENTS

I am grateful to Dr. P. Hudgson for allowing me to study many of the muscle sections referred to in this chapter and for his advice. The histological preparations and microphotographs were prepared by Mr. J. J. Fulthorpe and Mr. G. Henderson. I am also deeply indebted to Miss R. Allan for secretarial assistance.

### REFERENCES

1. Adams, R. D.: Pathological reactions of the skeletal muscle fibre in man. In *Disorders of Voluntary Muscle*, ed. by Walton, J. N., 1st ed., pp. 114–162. London, Churchill, 1964.
2. Adams, R. D.: Pathological reactions of the skeletal muscle fibre in man. In *Disorders of Voluntary Muscle*, ed. by Walton, J. N., 2nd ed., pp. 143–202. London, Churchill, 1969.
3. Adams, R. D., Denny-Brown, D., and Pearson, C. M.: *Diseases of Muscle: A Study in Pathology*, 1st ed. New York, Hoeber, 1953.
4. Adams, R. D., Denny-Brown, D., and Pearson, C. M.: *Diseases of Muscle: A Study in Pathology*, 2nd ed. New York, Hoeber, 1962.
5. Bell, C. D., and Conen, P. E.: Histopathological changes in Duchenne muscular dystrophy. *J. Neurol. Sci. 7:* 526–544, 1968.
6. Bethlem, J., and Van Wijngaarden, G. K.: The incidence of ringed fibres and sarcoplasmic masses in normal and diseased muscle. *J. Neurol. Neurosurg. Psychiatry 26:* 326–332, 1963.
7. Bonsett, C. A.: Pseudohypertrophic muscular dystrophy. Distribution of degenerative features as revealed by an anatomical study. *Neurology (Minneap.) 13:* 728–738, 1963.

7a. Brain, W. R.: *Diseases of the Nervous System*, revised by the late Lord W. R. Brain (1895–1966) and J. N. Walton, 7th ed., Series Oxford Medical Publications. London Oxford University Press, 1969.

8. Cazzato, G., and Walton, J. N.: The pathology of the muscle spindle. A study of biopsy material in various muscular and neuromuscular diseases. *J. Neurol. Sci. 7:* 15–70, 1968.
9. Coërs, C., and Woolfe, A. L.: *The Innervation of Muscle.* Oxford, Blackwell, 1959.
10. Drachman, D. A., Murphy, S. R., Nigam, M. P., and Hills, J. R.: Myopathic changes in chronically denervated muscle. *Arch. Neurol. 16:* 14–24, 1967.
11. Drachman, D. A., Wetzel, N., Wasserman, M., and Naito, H.: Experimental denervation of ocular muscles. A critique of the concept of "ocular myopathy." *Arch. Neurol. 21:* 170–183, 1969.
12. Dubowitz, V.: Myopathic changes in muscular dystrophy carriers. *Proc. Roy. Soc. Med. 56:* 810–812, 1963.
13. Duchenne, G.: Recherches sur le paralysie musculaire pseudohypertrophique ou paralysie myosclerosique. *Arch. Gén. Méd. 11:* 532–573, 1968.
14. Emery, A. E. H.: Clinical manifestations in two carriers of Duchenne muscular dystrophy. *Lancet 1:* 1126–1128, 1963.
15. Emery, A. E. H.: The electrophoretic pattern of lactic dehydrogenase in carriers and patients with Duchenne muscular dystropy. *Nature (Lond.) 201:* 1044–1045, 1964.
16. Emery, A. E. H.: Muscle histology in carriers of Duchenne muscular dystrophy. *J. Med. Genet. 2:* 1–7, 1965.
17. Emery, A. E. H.: Genetic counselling in X-linked muscular dystrophy. *J. Neurol. Sci. 8:* 579–587, 1969.
18. Engel, W. K.: Muscle biopsies in neuromuscular diseases. *Pediatr. Clin. North Am. 14:* 963–995, 1967.
19. Erb, W. H.: Ueber die "juvenile Form" der progressiven Muskelatrophie ihre Beziehungen zur sogenannten Pseudohypertrophie der Muskeln. *Dtsch. Arch. Klin. Med. 34:* 467–480, 1883.
20. Gardner-Medwin, D.: Studies of the carrier state in the Duchenne type of muscular dystrophy. II. Quantitative electromyography as a method of carrier detection. *J. Neurol. Neurosurg. Psychiatry 31:* 124–134, 1968.
21. Grundmann, E., and Beckmann, R.: Zur pathologischen Anatomie der Dystrophia musculorum progressiva Erb. *Beitr. Pathol. Anat. 127:* 335–350, 1962.
22. Gubbay, S. S., Walton, J. N., and Pearce, G. W.: Clinical and pathological study of a case of congenital muscular dystrophy. *J. Neurol. Neurosurg. Psychiatry 29:* 500–508, 1966.
23. Hudgson, P., Pearce, G. W., and Walton, J. N.: Pre-clinical muscular dystrophy. Histopathological changes observed on muscle biopsy. *Brain 90:* 565–576, 1967.
24. Johnston, H. A., Wilkinson, J. H., Withycombe, W. A., and Raymond, S.: Alpha-hydroxybutyrate dehydrogenase activity in sex-linked muscular dystrophy. *J. Clin. Pathol. 19:* 250–256, 1966.
25. Kakulas, B. A., Knight, J. O., Gubbay, S. S., and Mastaglia, F. L.: The detection of female carriers of pseudohypertrophic muscular dystrophy. *Proc. Aust. Assoc. Neurol. 5:* 545–551, 1968.
26. Kiloh, L. G., and Nevin, S.: Progressive dystrophy of the external ocular muscles (ocular myopathy). *Brain 74:* 115–143, 1951.
27. Kowalewski, S., Rotthauwe, H. W., Mölbert, E., and Mumenthaler, M.: Female carriers of muscular dystrophy. *Lancet 1:* 1216–1217, 1966.
28. Lapresle, J., and Milhaud, M.: Pathologie du fuseau neuromusculaire. *Rev. Neurol. (Paris) 110:* 97–122, 1964.
29. Lyon, M. F.: Gene action in the X-chromosome of the mouse. *Nature (Lond.) 190:* 372–373, 1961.
30. Macciotta, A., Costa, V., Cao, A., Sforza, F., and Scano, V.: Studio istologico sul tessuto muscolare dei genitori di un bambino con miodistrofia pelvifemorale tipo Duchenne. *Ann. Ital. Pediatr. 17:* 1–11, 1964.
31. Mannucci, P. M., Ideo, G., Cao, A., and Macciotta, A.: Gli isoenzimi della latticodeidrogenasi (LDH) e della transaminasi glutammico-ossalacetica (TGO) nel musculo fetale, adulto, e di soggetti affetti da distrofia muscolare progressiva tipo Duchenne. *Rass. Med. Sarda 68:* 287–300, 1965.
32. Mastaglia, F. L., and Kakulas, B. A.: The natural history of human muscle diseases studied by means of serial biopsy. *Proc. Aust. Assoc. Neurol. 5:* 537–543, 1968.

33. Mastaglia, F. L., and Kakulas, B. A.: Regeneration in Duchenne muscular dystrophy. A histological and histochemical study. *Brain 92:* 809–818, 1969.
34. Milhorat, A. T., Shafiq, S. A., and Goldstone, L.: Changes in muscle structure in dystrophic patients, carriers and normal siblings seen by electron microscopy; correlation with levels of serum creatinephospho-kinase (CPK). *Ann. N. Y. Acad. Sci. 138:* 246–292, 1966.
35. Morris, W. R.: Striated annulets (ringbinden). Their experimental production in mammalian muscle. *A. M. A. Arch. Pathol. 68:* 438–444, 1959.
36. Mumenthaler, M.: L'histopathologie des myopathies a propos de 139 observations personnelles. *Paediat. Fortbild. Prax. 18:* 47–62, 1966.
37. Patel, A. N., Lalitha, V. S., and Dastur, D. K.: The spindle in normal and pathological muscle. An assessment of the histological changes *Brain 91:* 737–750, 1968.
38. Pearce, G. W.: Histopathology of voluntary muscle. *Postgrad. Med. J. 41:* 294–312, 1965.
39. Pearce, G. W., Pearce, J. M. S., and Walton, J. N.: The Duchenne type muscular dystrophy. Histopathological studies of the carrier state. *Brain 89:* 109–120, 1966.
40. Pearce, G. W., and Walton, J. N.: Progressive muscular dystrophy. The histopathological changes in skeletal muscle obtained by biopsy. *J. Pathol. Bacteriol. 83:* 535–550, 1962.
41. Pearson, C. M.: Histopathological features of muscle in the preclinical stages of muscular dystrophy. *Brain 85:* 109–120, 1962.
42. Pearson, C. M., Fowler, W. M., and Wright, S. W.: X-chromosome mosaicism in females with muscular dystrophy. *Proc. Nat. Acad. Sci. U. S. A.* 50: 24–31, 1963.
43. Pearson, C. M., and Kar, N. C.: Isoenzymes. General considerations and alterations in human and animal myopathies. *Ann. N. Y. Acad. Sci. 138:* 293–303, 1966.
44. Price, H. M., Howes, E. L., Sheldon, D. B., Hutson, O. D., Fitzgerald, R. T., Blumberg, J. M., and Pearson, C. M.: An improved biopsy technique for light and electron microscopic studies of human skeletal muscle. *Lab. Invest. 14:* 194–199, 1965.
45. Radu, H., Migea, S., Torok, Z., Bordeianu, L., and Radu, A.: Carrier detection in X-linked Duchenne type muscular dystrophy. *J. Neurol. Sci. 6:* 289–300, 1968.
46. Samaha, F. J., Xchroeder, J. M., Rebeiz, J., and Adams, R. D.: Studies on myotonia. Biochemical and electron microscopic studies on myotonia congenita and myotonia dystrophia. *Arch. Neurol. 17:* 22–33, 1967.
47. Smith, H. L., Amick, L. D., and Johnson, W. W.: Detection of subclinical and carrier states in Duchenne muscular dystrophy. *J. Pediatr. 69:* 67–79, 1966.
48. Stephens, J., and Lewin, E.: Serum enzyme variations and histological abnormalities in the carrier state in Duchenne dystrophy. *J. Neurol. Neurosurg. Psychiatry 28*, 104–108, 1965.
49. Victor, M., Hayes, R., and Adams, R. D.: Oculopharyngeal muscular dystrophy. A familial disease of late life characterized by dysphagia and progressive ptosis of the eyelids. *New Engl. J. Med. 267:* 1267–1272, 1962.
50. Walker, B. E., and Drager, G. A.: Evidence of regeneration in repeat biopsies of dystrophic human muscle. *Neurology (Minneap.) 12:* 381–384, 1962.
51. Walton, J. N., and Ayyar, D. R.: Aspects anatomopathologiques de quelques maladies musculaires chez l'homme. In *Les Atteintes des Muscles des Ceintures chez l'Adulte*, ed. by Serratrice, G., pp. 9–42. Paris, L'Expansion Scientifique Francaise, 1969.
52. Walton, J. N., and Gardner-Medwin, D.: Second thoughts on classification of the muscular dystrophies. In *Research in Muscular Dystrophy*, ed. by members of the Research Committee of the Muscular Dystrophy Group of Great Britain, pp. 45–71. London, Pitman, 1968.
53. Walton, J. N., and Gardner-Medwin, D.: Progressive muscular dystrophy and the myotonic disorders. In *Disorders of Voluntary Muscle*, ed. by Walton, J. N., 2nd ed., pp. 455–499. London, Churchill, 1969.
54. Welander, L.: Myopathia distalis tarda hereditaria; 249 examined cases in 72 pedigrees. *Acta Med. Scand. 141 (Suppl. 265):* 1–124, 1951.
55. Zellweger, H., Afifi, A. K., and McCormick, W. F.: Congenital muscular dystrophy. In *Proceedings of the Second International Congress of Neuro-genetics and Neuro-ophthalmology*, ed. by Barbeau, A., and Brunette, J. R., pp. 235–237. Amsterdam, Excerpta Medica, 1969.

## Chapter 15

# The Pathologic Substratum of Polymyositis

RAYMOND D. ADAMS

The fact that skeletal muscles may be the seat of inflammation excited by a number of infective agents, including viruses, bacteria, and parasites, is a matter of common knowledge. In contrast the existence of another type of inflammation, this one notably subacute, paralytic, often associated with dermatitis (*viz.*, polymyositis and dermatomyositis) but of undetermined cause, has remained controversial. While well documented in the medical literature of the late nineteenth and early twentieth centuries and made the subject of a monograph by Walton and Adams[5] in 1958, Greenfield, Shy, Alvord and Berg[2] and others have denied the occurrence of a proximal myositic disease of this type, declaring that such cases are indistinguishable from other myopathies and especially from the progressive muscular dystrophies. Opinion on this matter is still divided; the majority of pathologists and clinicians tend now to accept it as an entity but with the reservation that there is a considerable imprecision in diagnosis, as might be expected of any disease in which cause is unknown and reliable biologic tests are lacking.

The clinical and pathologic material of the Massachusetts General Hospital has offered many examples of idiopathic polymyositis and dermatomyositis in the past 20 years, and our observations permit the description of a disease entity as follows.

Clinically, the condition develops relatively rapidly over a period of weeks. Weakness of muscles of shoulder and pelvic girdles in a person of adolescent or adult years stands as the clinical prototype. Often the laryngeal and pharyngeal muscles are involved as well, but many variations in the clinical picture have been observed. In some of the most severe forms all muscles of the limbs and trunk have been involved with no predilection for proximal or distal groups, and in a few myoglobinuria and even cardiac disorder (electrocardiogram changes and arrhythmia) have been added. Muscle atrophy is seldom prominent. In other patients the disease has been restricted to only the cervical or pelvic girdle muscles or, apparently, even to single muscles such as the quadriceps. Another combination is weakness of pelvic and thigh muscles with feebleness of hand grips and only slight fatigability of shoulders. The clinical course has also varied. Instead of being subacute in evolution, in a few the onset has been insidious, the course chronic and progressive; the clinical picture is then one of atrophic paralysis, first of all leg and thigh muscles and later those of trunk and upper extremities. Twice we have observed a prostigmine-responsive myasthenia gravis of ocular and other cranial muscles with polymyositis, and several patients have had the inverse myasthenia of the Eaton-Lambert syndrome.

Skin changes have tended to be mild and are also variable in degree and type. In more than half of the cases they are absent; in a few they are severe, diffuse, and either maculo-papular or exfoliative. Periungual erythema and butterfly pattern of erythema of face, suggestive of lupus erythematosus, have been most characteristic.

Some form of connective tissue disease such as rheumatoid arthritis, rheumatic fever, scleroderma, or lupus erythematosus has provided a background for the polymyositis in 35 to 40% of our patients and a small number, estimated not greater than 5%, have had or developed an occult tumor, the latter appearing as late as 2 years after the appearance of the myositic weakness. Pain is seldom associated with muscle weakness and, when present, usually signifies polyarthritis or some other disease such as polymyalgia rheumatica; the most characteristic and severe cases have neither pain nor tenderness. The majority of our patients have been adults of middle or advanced age, and the disease has been rare in children. The combination, in children, of weakness of limbs with necrosis of bowel and kidneys has more the aspect of a polyarteritis than polymyositis.

Serum enzymes, especially aldolase and creatine phosphokinase, have been consistently elevated in all of the more severe forms during the active phase, due presumably to leakage of cytoplasmic contents of injured fibers. Exceptionally, however, a moderately severe restricted polymyositis causes no rise in serum enzymes. The electromyogram has disclosed brief myopathic action potentials and fibrillations.

Responsivity to corticosteroid therapy has been another gratifying feature of the majority of our patients and relatively few have died (not more than 5% of more than 300). Therapeutic effects are admittedly difficult to judge, however, for natural remissions may be a source of confusion. In a small number of patients clinical progression occurs while the patient receives full doses of corticosteroids. Relapse and intermittent progression of the disease has featured the clinical course of many. Whereas a few patients have eventually made a complete recovery, the majority have been left with some weakness. Fortunately it is not usually disabling, but a few patients have had to live out their lives as permanent cripples.

Some of our most important ideas about this class of disease have been obtained from the study of its pathology. Tissue sections and cultures have not revealed a causative agent, such as have been observed in the infective inflammations of muscle. Following a recent demonstration of virus-like structures in the muscle fibers of one case (Chou[1] and Mastaglia and Walton[3]), we have subjected the tissue of four others to examination under the electron microscope, all except one with negative results. Pearson[4] has failed to see the virus in about 20 other cases. Whether this means that these 3 reported positive cases were atypical or that the virus is elusive and visualizable only at a certain stage of the disease cannot be decided at this time. We are compelled, therefore, to retain the term *idiopathic*, but we are quite prepared to abandon it at a moment's notice.

The essential lesions (Figs. 1 to 6) consist of perivenous and interstitial infiltrations of mononuclear leucocytes, lymphocytes, and plasma cells in combination with segmental necrosis of muscle fibers (Figs. 1 and 2). The latter undergoes repair by so-called regeneration. Later loss of muscle fibers occurs and they are eventually replaced by connective tissue and fat cells (Figs. 3 and 6). But, in a study of chronic cases that have come to autopsy years after the onset, in several we have still found evidence of fiber necrosis and inflammation. Thus it would seem that the inflammatory necrosis must recur over a period of years and is the presumed basis of the familiar clinical relapse.

In order to obtain information concerning the relation of inflammation to muscle fiber necrosis a number of our pathologic specimens have been serially sectioned. Whereas in single sections this relationship may be obscure, in serial sections inflammation and segmental necrosis are invariably contiguous. Perivenous inflammation was found to occur alone but necrosis of muscle fibers was seen only in relation to migration of mononuclear cells through the vein wall. The capillary or venular endothelium is altered; the basement membrane is thickened and endothelial cells contain multivesicular bodies and antophagic vacuoles (Shafiq *et al.*[6]). From this type of stereologic survey, in which were checked more than 100 lesions in six specimens, it was concluded that cell migration is early and may even play a role in initiating necrosis of muscle fibers. It would seem that the inflammation may bear the same relation to the paren-

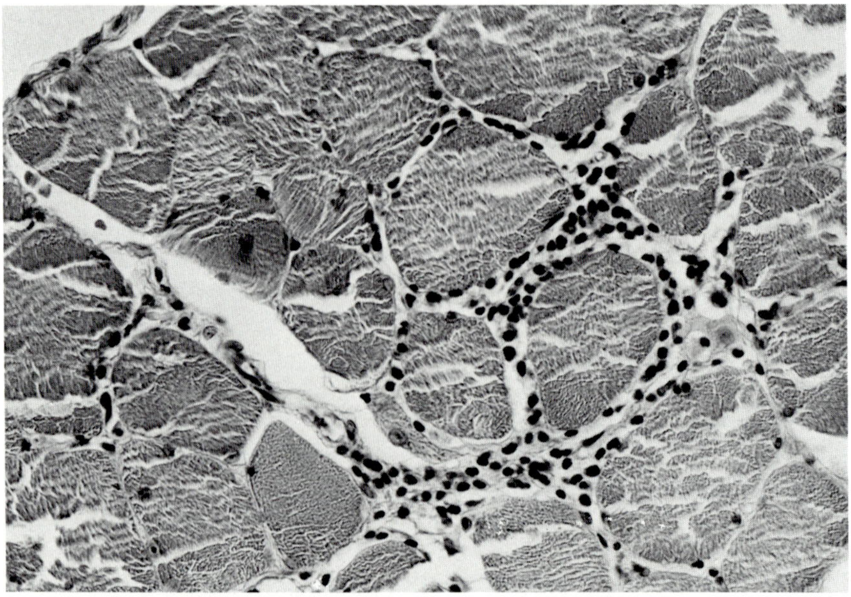

FIG. 1. Transverse section of quadriceps muscle biopsy to show infiltration of inflammatory cells (lymphocytes and mononuclear leucocytes) among well preserved muscle fibers. In the left half of the field, one damaged fiber with large sarcolemmal nucleus is seen. Hematoxylin and eosin stain.

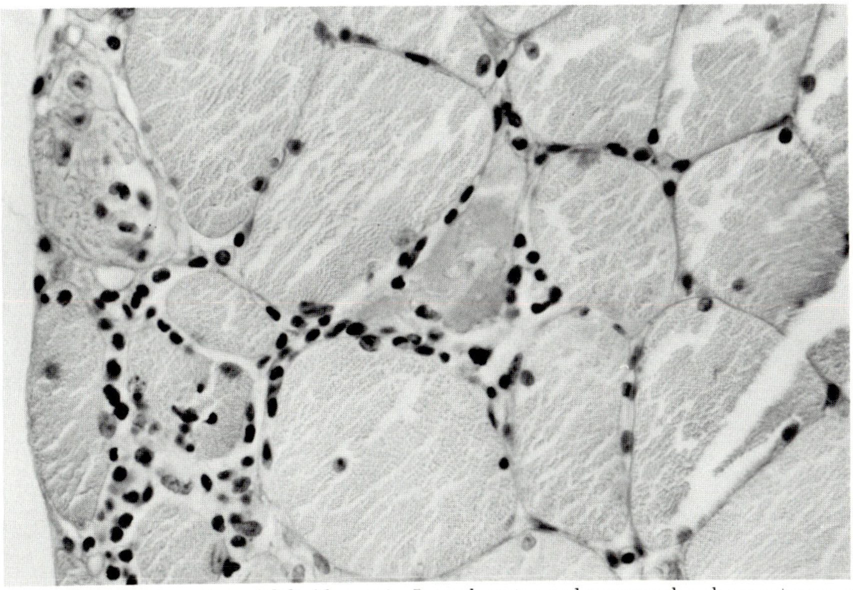

FIG. 2. Transverse section of deltoid muscle. Lymphocytes and mononuclear leucocytes separate muscle fibers. One fiber in the center shows vitreous change (necrosis) and two on the right are being invaded by histiocytes. Sarcolemmal nuclei (pale nuclei with large nucleoli) are visible in several places. Hematoxylin and eosin stain.

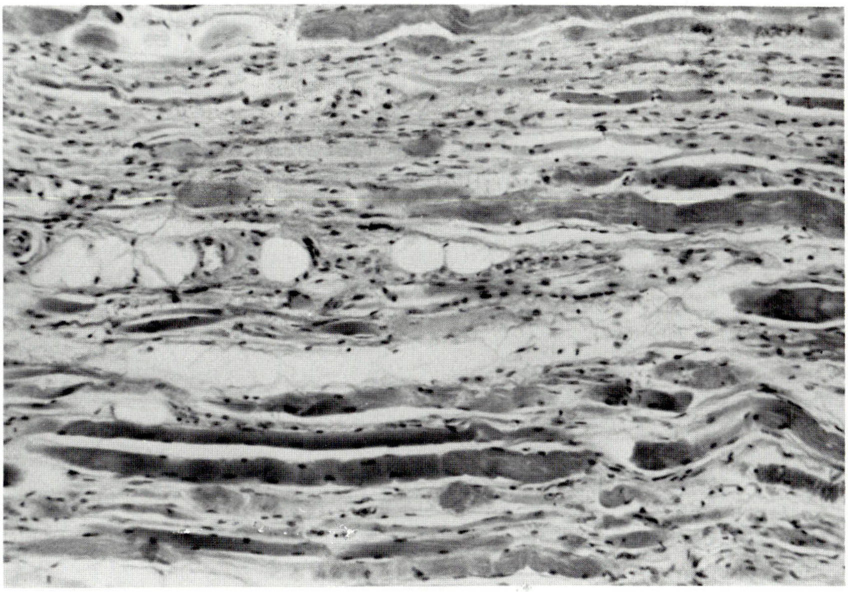

FIG. 3. Longitudinal section of biopsy from sternomastoid in a chronic polymyositis. Many muscle fibers have disappeared; others are extremely thin. Fat cells are abundant. Segments of larger fibers are well preserved. No inflammatory cells. Hematoxylin and eosin stain.

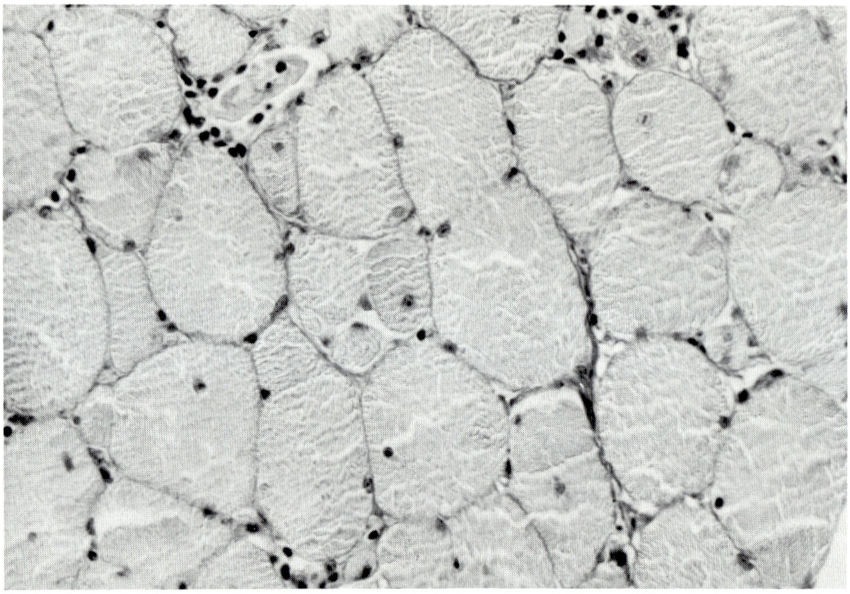

FIG. 4. Transverse section of biopsy from sternomastoid. Vessel near left upper center is surrounded by a few lymphocytes. Note single small fibers or groups of small fibers and central nuclei. Hematoxylin and eosin stain.

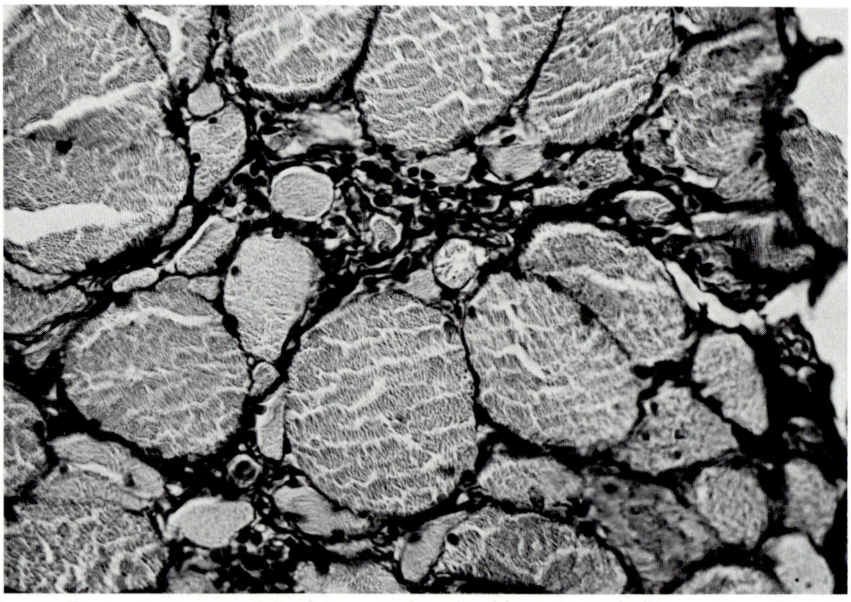

FIG. 5. Transverse section of quadriceps to show focal loss of muscle fibers and condensation of connective tissue. The larger fibers exceed 90 $\mu$ in diameter; the smallest ones are about 15 $\mu$. Mallory trichrome stain.

chymatous lesion as it does in experimental allergic polyneuritis. One cannot, however, dismiss the possibility of the inflammation being due to a primary viral infection of muscle.

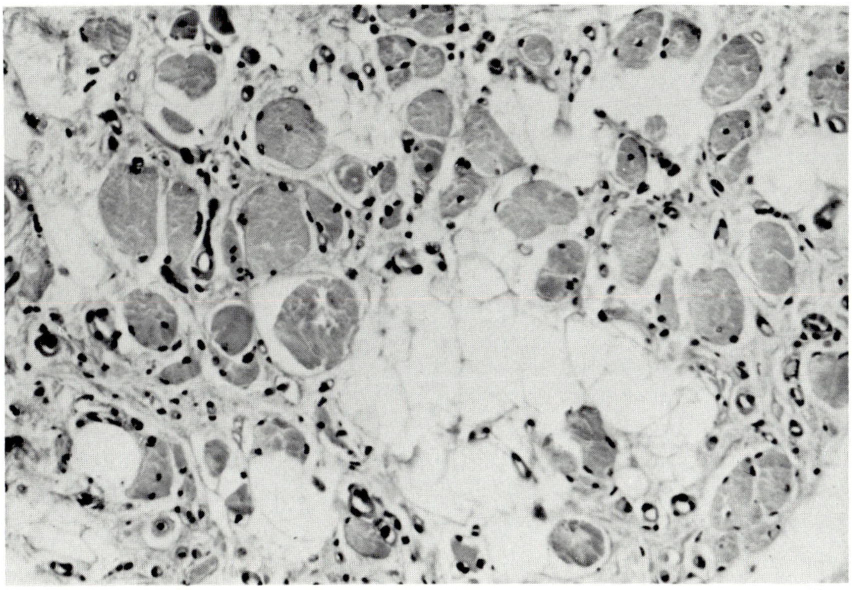

FIG. 6. Transverse section of quadriceps from case of chronic polymyositis—inactive. Note loss of muscle fibers, fat cells, extreme variation in fiber size, and lack of inflammatory cells. Hematoxylin and eosin stain.

We have come to believe that if the inflammatory infiltrates are lacking altogether or are insignificant during the clinically active phase of the disease, the segmental fiber necrosis signifies some other polymyopathy or dystrophy. As is well known, necrosis may appear as a finding in a number of such diseases; but, if it is extensive, most frequently it is a manifestation of polymyositis in which, because of inadequacy of the biopsy specimen, all aspects of the lesion are not revealed. And, of course, in small biopsies none of the characteristic lesions may be seen. It is necessary always to keep in mind the extreme limitations of the biopsy procedure as a means of study the pathology of this or any other muscle disease.

The necrosis of fibers, which we have termed necrobiosis in past writings (others speak of rhabdomyolysis), usually involves many sarcomeres and extends the entire width of the fiber. All sacroplasmic components including some of the sarcolemmal nuclei are destroyed. Under the electron microscope myofibrils are shredded, Z bands are fragmented, and endoplasmic reticulum vacuolated (Rose *et al.*[5]). The necrotic material swells, acquires a bright eosinophilia, and appears vitreous; and fracture lines may separate it from parts of the fiber that remain normal. The lesion may be restricted to only a part of the transverse diameter of a fiber involving just the center and leaving the peripheral sarcolemmal nuclei, myofibrils, endoplasmic reticulum, and mitochondria unscathed. The necrotic sarcoplasm invariably excites a histiocytic reaction, visible within 2 to 3 days, and soon thereafter macrophages crowd into the whole of the necrotic segment. These cells, easily identified by their pleomorphic nuclei, lack of prominent nucleolus, and phagocytic activity, are helpful in establishing the vitreous change as pathologic and not just a crush artifact due to the biopsy procedure. Particles of sarcoplasm may be visible in the cytoplasm of macrophages. Large parts of fibers persist, some still connected with their neuron.

Restoration of destroyed segments begins shortly after the necrosis occurs. Surviving sarcolemmal nuclei, presumably of satellite cells, multiply and rapidly join in chains and soon form sarcoplasm that is basophilic because of richness in RNA; myofilaments then reappear. This type of regeneration has been termed *embryonal* because of its resemblance to the manner in which the muscle fibers form in embryonic life, and it restores the damaged segments, thereby connecting the ends of intact parts of the fiber. Regeneration by the formation of buds may also be seen at the ends of healthy parts of the fiber if destruction has been complete and has involved all of the sarcolemmal nuclei and sarcolemma for a considerable distance. This process, called *regeneration by budding* or *muscle giant cell formation*, is much less frequent in polymyositis than is the embryonal type. From the end of an intact fiber four or five new small fibers then begin to form and grow longitudinally. Forking can be seen in fortuitous planes of section. In cross-section the branches look like a small group (subunit) of denervated fibers. Some of the branches wither, others presumably make contact with tendon or connective tissue sheath, restoring effective contractibility to the affected fiber. The necrobiotic process may also affect previously injured or regenerating fibers.

If the inflammatory process is arrested, the muscle regains normal structure and function. The thin fibers, containing myofilaments in disarray, gradually reconstitute organized myofibrils. The latter are at first sparse and conspicuous because of size and separateness (so-called protofibrils). Eventually the myofibrillar arrangement becomes normal with eosinophilic sarcoplasm and small, peripheral muscle nuclei.

A continuation of the destructive process appears eventually to exhaust the recuperative powers of muscle and many of the fibers die. Since most of the muscle cells are postmitotic, if completely destroyed they are never replaced. The more chronic relapsing form of polymyositis thus comes to exhibit a complex pathologic picture consisting of acute inflammation and necrosis together with fields of many thin regenerating fibers, connective tissue, and fat cells (Fig. 3). The connective tissue increase is more or less relative; at no point is there active nuclear multiplication such as is seen in grossly destructive lesions of muscle fibers and their endomysial sheaths. Yet the interstitial tissue changes do have a reactive aspect with some fibroblastic nuclei being larger and slightly more numerous and the collagen more abundant than would result from simple outfall of muscle fibers and condensation of their sheaths (Fig. 6).

Late phases of polymyositis are characterized pathologically by severe depopulation of fibers and those which survive are more often small than large. But, there are often a few scattered fibers in the fields of small ones that measure from 75 to 125 $\mu$. The abundance of small fibers relates to injury and forking as well as to segmental denervation. At this stage the inflammatory exudates may have disappeared. The pathologic picture is then indistinguishable from that of a late stage of progressive muscular dystrophy and differs from denervation atrophy with secondary dystrophic change only with respect to less evident motor unit grouping, *viz*., group atrophy. In reality this constellation of pathologic changes represents *end-stage disease* and the point that merits emphasis is that all forms of chronic muscle disease, even denervation and disuse atrophy, tend in the end to lose their distinguishing features. In fact, most of the attributes considered typical of the pathology of progressive muscle dystrophy, *i.e.*, loss of fibers, haphazard arrangements of small and large fibers, augmentation of connective tissue, and fat cell replacement, relate more to the chronicity of muscle disease than to any one feature of the dystrophic process. The pathologic picture of dystrophy may be regarded as the consequence of repeated noninflammatory segmental necrosis extending over a period of years while that of polymyositis results from repeated inflammatory necrosis more often taking place over a period of months. Even in chronic denervation or disuse, wherein a dystrophic process ultimately overtakes pure atrophy, the end-stage pathology is relatively nondescript. The late dystrophic changes after denervation have led many research workers to the erroneous conclusion that a purely peripheral nerve disease has a primary myopathic aspect.

The topography of the muscle affection in polymyositis deserves also to be remarked upon for it is predominantly proximal, not unlike that of many dystrophic diseases. However, there are no clues as to why some of the fibers in the pelvic and shoulder girdle, posterior cervical, laryngeal, and pharyngeal muscles should be so vulnerable to this disease. This susceptibility seems not to be understandable in terms of fiber size, arrangement, vascular supply, nerve-muscle motor unit ratios, or histochemical type. Although in their normal state the affected muscles look in every respect like nonaffected ones, the topography of these diseases informs us that these affected muscles must differ in some structural quality. If this disposition could be explained it would surely provide some further insight into the pathogenesis of this disease as well as the progressive muscular dystrophies.

No differences have been found in the muscle lesions in the pure cases of polymyositis and dermatomyositis and those associated with neoplasm and connective tissue disease. The only qualification which might be applied to this statement concerns the muscle lesion of lupus erythematosus, where involvement of the periphery of fascicles is more prominent than in other forms of polymyositis and the affected fibers are curiously vacuolated. Evidence of denervation atrophy in conjunction with polymyositis is to be found in occult neoplasia, especially of lung.

As to the relation between the muscle lesions and the clinical manifestations, only a few generalizations are possible. *Paresis* and *paralysis* correlate in the early stages of disease with muscle fiber necrosis, and power of contraction returns as segmental regeneration takes place. The speed and degree of reversibility of paresis can be explained best by the vigor of this regenerative process. In the later stages paresis (or paralysis) is due to loss of muscle fibers, the power being more or less proportional to the number of intact, surviving fibers. However, thinness of fibers and isolation of long stretches of fibers from their nerve supply or tendinous insertions must at all times be a factor in the production of paresis, and probably the only good evidence of it is the fibrillation potential in the electromyogram. Shortening of muscle, here pseudocontracture, is usually ascribed to fibrosis, although it is well known that the persistent inactivity of even healthy muscle may lead to shortening, as can be seen in limbs kept for weeks in a plaster cast. Active stretching can reverse much of the pseudocontracture of polymyositis. Strangely the inflammatory reaction no matter how intense is painless except when it extends to joint surfaces, tendon sheaths, etc., and seldom are there systemic signs of it. Fever does not occur, nor anemia, and even the sedimentation rate, white cell count, and differential are normal in many cases.

The most likely possibilities as to the causation of this peculiar malady of muscle are: an obscure infection, perhaps viral, of a type yet to be identified, or an autoallergic reaction in which the patient becomes sensitized to some component of his own skeletal muscles. In other communications at this symposium the results of attempts to affirm or refute one or other of these two hypotheses will be presented.

In summary, our studies continue to corroborate the existence of a large group of inflammatory diseases of muscle in which an infective agent cannot be identified. The clinical picture is often reminiscent of muscular dystrophy except that the disease evolves more rapidly, remits and relapses sometimes in response to corticosteroid therapy, may affect skin, and tends to occur at an age period when genetic muscle diseases are rare. The essential lesion, an inflammatory segmental necrosis of muscle cells, is at first highly distinctive; but later when it leads to fiber atrophy and loss (the end-stage pathology), except for persistence of the inflammatory reaction, it becomes virtually indistinguishable from other myopathies, dystrophies, and atrophies. The failure of pathologists to appreciate these points concerning the end-stage of muscle diseases has lead to considerable confusion about both polymyositis and progressive muscular dystrophy. Two theories of causation, viral and allergic, have emerged but both remain unproved. They are not mutually exclusive, for one or more viral infections may trigger an autoimmune reaction.

## REFERENCES

1. Chou, S-M.: Myxovirus-like structures in a case of human chronic polymyositis. *Science 158:* 1453, 1967.
2. Greenfield, J. G., Shy, C. M., Alvord, F. C., and Berg, L.: *An Atlas of Muscle Pathology.* Edinburgh and London, E. & S. Livingstone, 1957.
3. Mastaglia, F., and Walton, J.: Coxsackie virus-like particles in skeletal muscle from a case of polymyositis. *J. Neurol. Sci. 11:* 593, 1970.
4. Pearson, C. M.: Personal communication.
5. Rose, A. L. Walton, J. M., and Pearce, G. W.: Polymyositis: an ultramicroscopic study of muscle biopsy material. *J. Nuerol. Sci. 5:* 457, 1967.
6. Shafiq, S. A., Mulhorat, A. T., and Gorycki, M. A.: An electron microscopic study of muscle degeneration and changes in blood vessels in polymyositis. *J. Pathol. Bacteriol. 94:* 139, 1967.
7. Walton, J., and Adams, R. D.: *Polymyositis.* Edinburgh and London, E. & S. Livingstone, 1958.

Chapter 16

# Vacuolar Myopathies: Multiple Etiologies and Sequential Structural Studies*

ANDREW G. ENGEL

## INTRODUCTION

*Definitions.* A vacuolar myopathy can be defined as one in which fiber vacuolation is the typical or predominant pathologic change. A vacuole is usually defined as a space, or cavity, formed within the cell. Because recognition of vacuoles in muscle fibers depends, in part, on methods of tissue processing and the optics of examination, for this discussion a vacuole is defined as either an abnormal membrane-limited space with varying contents or a non-membrane-limited space wherein an abnormal excess of some cellular component or material is accumulating but which appears empty, or partially empty, by at least one optical method of examination.

*Classification.* A clinical classification of vacuolar myopathies is shown in Table 1. Vacuolated muscle fibers also occur in other myopathies (muscular dystrophy, polymyositis, paroxysmal myoglobinuria, steroid-induced myopathy, or in any other myopathy); trauma, freezing, and ischemia also can result in fiber vacuolation.[2] In this presentation, experimental chloroquine myopathy, type 2 glycogenosis, lipid storage myopathy, and the permanent vacuolar myopathy of primary hypokalemic periodic paralysis will be considered in detail.

From a clinical classification of vacuolar myopathies one can proceed to a morphologic classification of vacuoles (Table 2). Here two points require emphasis. First, although no one vacuole is specific for any one disease, certain types of vacuoles tend to be associated with certain types of vacuolar myopathies. Second, during evolution of the pathologic process, vacuolar morphology (dimensions, boundaries, and contents) can change—one type of vacuole can become converted into a different type.

* This work was supported in part by Research Grant NS-6277 from the National Institutes of Health, Public Health Service, and by a grant from the Louis W. and Maud Hill Family Foundation.

TABLE 1. CLINICAL CLASSIFICATION OF VACUOLAR MYOPATHIES

| |
|---|
| Chloroquine myopathy |
| Glycogenoses, types 2, 3, 4, 5, and 7 |
| Lipid storage myopathy |
| Periodic paralyses |
| Vacuolar myopathy of systemic lupus erythematosus |

TABLE 2. MORPHOLOGIC CLASSIFICATION OF VACUOLES

| |
|---|
| Size, shape, number, and position in fiber |
| Boundary: |
| Membranous—continuous or discontinuous |
| Nonmembranous |
| Contents: |
| Nondegraded normal or abnormal organelles |
| Cytoplasmic degradation products |
| Glycogen |
| Acidic mucosubstance |
| Lipid material |
| Fluid—communicating or noncommunicating with extracellular fluid |
| Indeterminate or unidentifiable |
| Combinations of above |

*Methods of study.* An adequate study of vacuolar myopathies requires ordinary bright-field, phase-contrast, and electron microscopy and a cytochemical study of vacuolar contents. Electron cytochemical markers are used as aids in determining the origin and character of vacuolar limiting membranes and contents. Finally, biochemical studies are utilized in quantitating abnormal accumulations of cellular material (such as lipid or glycogen) or in establishing enzyme deficiency (as in glycogen storage diseases) or excess (as acid phosphatase excess in type 2 glycogenosis). Table 3 lists the light microscopic cytochemical procedures used in this laboratory in studying vacuolar contents. The methods are relatively simple and can be performed quickly in the pathology laboratory.[3, 24, 31, 39]

The horseradish peroxidase technique of Graham and Karnovsky,[24, 31] as adapted to a study of experimental[32] and human myopathies,[12] was used for labeling the extracellular space, transverse tubular system (T system), membrane networks derived from the T system, and spaces or vacuoles communicating with the above. The endogenous peroxidatic activity of hemoglobin, myoglobin, lipofuscin granules, and peroxisomes does not interfere with procedure provided that suitable controls are used to display endogenous activity.

The biochemical properties of a magnesium-dependent ATPase associated with the isolated sarcoplasmic reticulum (SR) were utilized for *in situ* localization of the enzyme.[17, 53] The enzyme serves as an electron cytochemical marker for the SR in human and experimental myopathies. Under the experimental conditions used, the nuclear envelope, the surface membrane, and occasionally the mitochondria also react, but T system membranes do not.

TABLE 3. CYTOCHEMICAL STUDY OF VACUOLAR CONTENTS (LIGHT MICROSCOPY)

| Vacuole Content | Stain or Procedure | Preferred Method for Sectioning and Fixation[a] |
|---|---|---|
| Glycogen | Periodic acid-Schiff (PAS) and diastase + PAS | Cryostat s; Newcomer's fluid pf |
| | PAS | Glutaraldehyde and osmium f, Epon s |
| Acidic mucosubstances | Alcian blue, pH 2.6 | |
| | Acridine orange | Cryostat s; Newcomer's fluid pf |
| | Cresyl violet acetate, pH 2.9 (metachromasia) | |
| Lipid material | | |
| Neutral fats | Oil red O | Cryostat s; formalin pf |
| Neutral fats and phospholipids | Sudan black B | Cryostat s; formalin pf |
| Acidic lipids | Nile blue sulfate | Cryostat s; formalin pf |
| Acetal phosphatides | Plasmal reaction | Formalin f, post-chroming, fffs |
| Phospholipids | Acid hematin | Cryostat s; formalin pf |
| Cholesterol esters | Perchloric acid-naphthoquinone | Formalin f; fffs |
| Free cholesterol | Digitonin method | Formalin f; fffs |
| Fatty acids | Copper-rubeanic acid | Formalin f; fffs |
| Glycolipids | Hydroxylamine-PAS | Cryostat s; formalin pf |
| Calcific minerals | von Kossa | Cryostat s; alcohol pf |
| Lysosomal acid phosphatase activity | Barka-Anderson simultaneous-coupling azo-dye technique | Fresh cryostat s |
| Communication with extracellular space | Ingress of peroxidase-labeled extracellular fluid into fresh tissue; incubation of fixed tissue in diaminobenzidine medium | Glutaraldehyde f of loaded tissue; osmium f after diaminobenzidine; Epon |

[a] S, section; f, fixation; pf, postfixation; fffs, free-floating frozen section.

Barka's modification of Gomori's method for acid phosphatase[4] was used as marker for primary lysosomes and for carriers, storage sites, or targets of lysosomal enzymes. Membrane-bound vacuoles in muscle fibers containing cytoplasmic degradation products and displaying acid phosphatase activity within their confines can be identified as secondary lysosomes or autophagic vacuoles.[8]

## EXPERIMENTAL CHLOROQUINE MYOPATHY

In collaboration with R. D. Macdonald,[32] experimental chloroquine myopathy was used to investigate autophagic mechanisms in striated muscle. Previous studies indicated that vacuoles in this myopathy contain cytoplasmic degradation products, including myeloid bodies,[23, 45] while vacuoles considered to be autophagic in nature had been produced experimentally by chloroquine in different human and animal cells.[1, 20-22] In the present study, chloroquine hydrochloride (25 mg per kg) was administered intraperitoneally to 200- to 250-g female Sprague-Dawley rats for 1 to 100 days. Biopsy specimens were taken from soleus, an essentially red muscle, and superficial medial gastrocnemius, a predominantly white muscle, at suitable intervals.

*Light and phase-contrast microscopy*. Three distinct abnormalities were noted, each more severe in soleus than in gastrocnemius.

1. Multiple, small, central or peripheral vacuoles appeared in isolated fibers after 5 days of treatment; after 20 days, 70% of soleus fibers in a given plane were vacuolated. With continued treatment both small and larger (up to 20 $\mu$ wide and 50 $\mu$ long) vacuoles were found (Fig. 1).

2. Fiber splitting appeared in soleus after 20 days of treatment, when vacuolation was well established; by day 70 more than half of soleus fibers in a given plane were so affected. Fiber splitting resulted in a variation of fiber size initially and in the appearance of clusters of small fibers later. The presence of molding of adjacent small fibers to one another and study of serial sections aided in identifying fiber splitting (Figs. 2 and 3).

3. Phase-optically dense granules, initially small but eventually up to 10 $\mu$ in size, appeared after day 5 in increasing numbers in abnormal fibers (Fig. 4).

*Histochemistry*. Intense acid phosphatase activity was observed in vacuoles of affected fibers during treatment (Fig. 5). Periodic acid-Schiff (PAS)-positive material, mostly diastase-fast, occurred in numerous vacuoles. The phase-optically dense granules had staining properties of phospholipid material (positive with Baker's hematin and in the plasmal reaction and faintly positive with Sudan black B).

*Electron microscopy*. After 5 days of treatment, proliferation of SR and T system membranes occurred in both red and white muscle fibers (Figs. 6, 7, and 8). At the same time, small vacuoles limited by double membranes were seen (Fig. 8). The source of the double membrane could not be determined until electron cytochemical studies (described below) were done. The early vacuoles contained recognizable cytoplasmic organelles, many still of normal appearance (Figs. 8 and 9). The small double-membrane-bound spaces gradually enlarged and fused to form larger, lobulated vacuoles with scalloped borders limited mostly by a single, but in places still by a double, membrane (Fig. 10). Fraying, thickening, or discontinuities in the inner membrane of smaller double-membrane-lined vacuoles suggested that degradation of the inner membrane might result in conversion to a single limiting membrane. A wreathlike array of tubular networks, now recognized to be of T system origin,[14, 30, 42, 48, 49] appeared around some of the larger vacuoles (Fig. 10) and in places merged with the vacuolar limiting membrane. Vacuolar contents now consisted of heterogeneous degraded cytoplasmic material.

The phase-optically dense granules proved to be myeloid structures. These were found in vacuoles with heterogeneous contents (Figs. 10 and 11) or they filled some membrane-bound spaces completely (Fig. 11). Golgi complexes, SR membranes, or T system networks often abutted on the myeloid formations, suggesting that the formations were derived from more than one type of membranous organelle.

With continued chloroquine treatment, those vacuoles with heterogeneous contents and those containing only myeloid bodies both enlarged. Transitions between the two types of vacuoles developed, and small vacuoles and myeloid

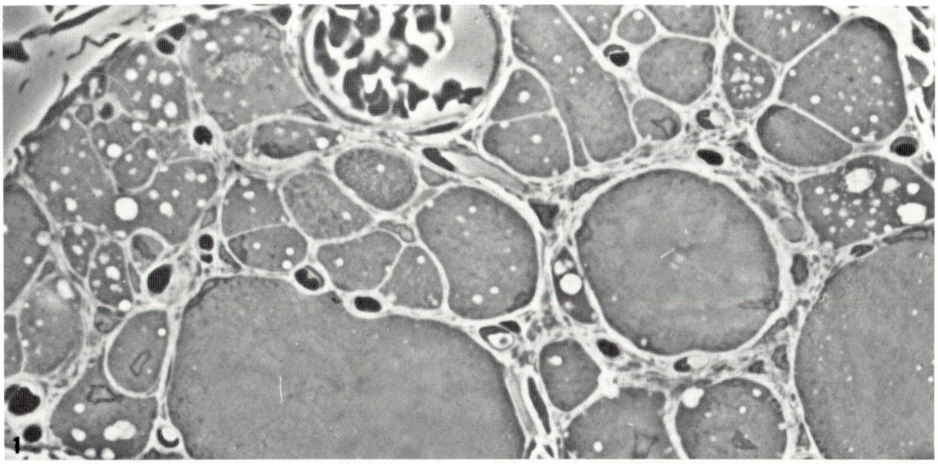

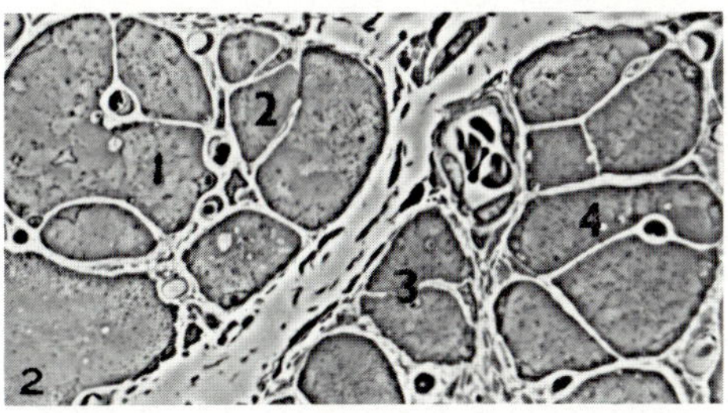

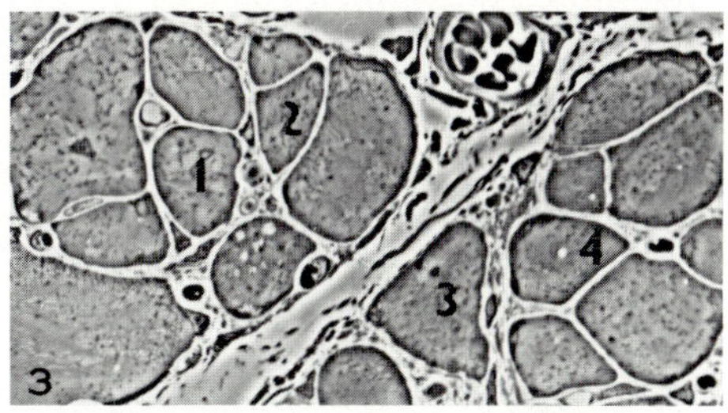

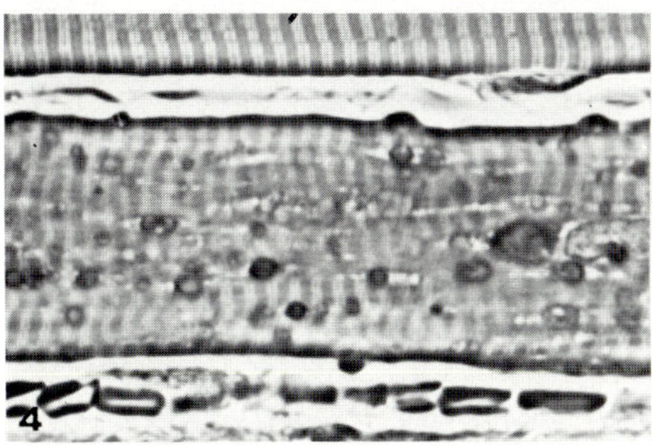

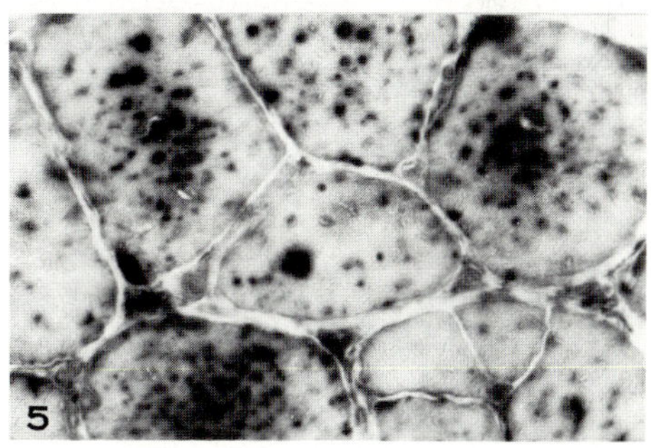

FIG. 1. Soleus muscle from chloroquine-treated rat. Majority of fibers contain vacuoles of various sizes. Fiber diameter varies greatly and smaller fibers are molded together. × 650. (Reproduced from Macdonald and Engel[32] by permission of the Association of the Journal of Neuropathology and Experimental Neurology, Inc.)

FIGS. 2 and 3. Soleus muscle of chloroquine-treated rat. Nonconsecutive serial sections separated by 60-μ interval. Fibers 1, 2, and 4 can be seen to split off from parent fiber. Incomplete subdivision of fiber 3 is noted in Figure 2. × 520. (Reproduced from Macdonald and Engel[32] by permission of the Association of the Journal of Neuropathology and Experimental Neurology, Inc.)

FIG. 4. Gastrocnemius muscle from rat treated with chloroquine for 100 days. Optically dense structures of different dimensions are present in fiber. Some of these are in or adjacent to small clear spaces. × 820. (Reproduced from Macdonald and Engel[32] by permission of the Association of the Journal of Neuropathology and Experimental Pathology, Inc.)

FIG. 5. Soleus muscle from rat treated with chloroquine for 45 days; acid phosphatase reaction. Intense enzyme activity is present in circumscribed areas of all fibers. × 420. (Reproduced from Macdonald and Engel[32] by permission of the Association of the Journal of Neuropathology and Experimental Neurology, Inc.)

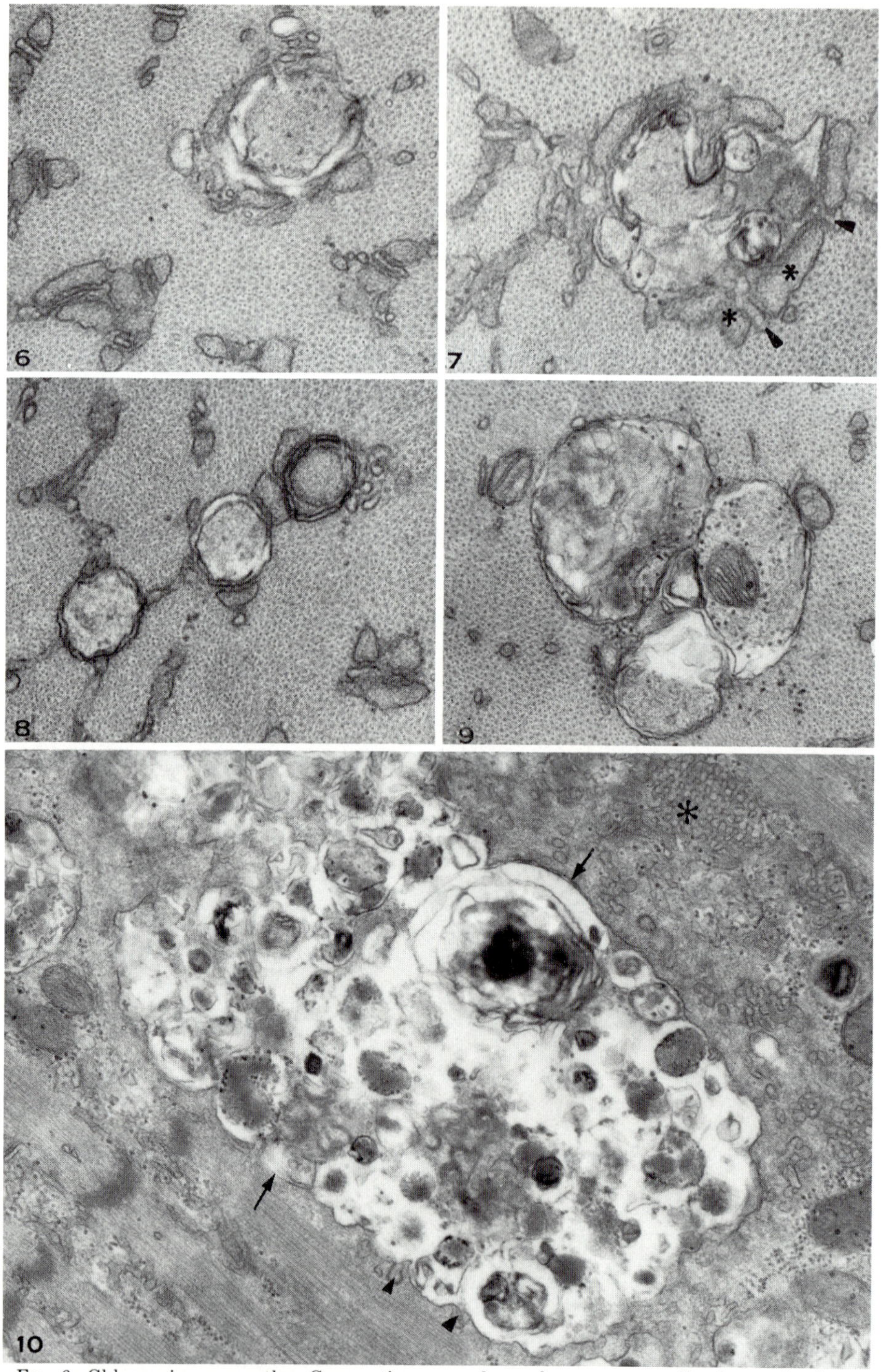


FIG. 6. Chloroquine myopathy. Crescentic array of membranous organelles incompletely surrounds small area of cytoplasm. × 49,000. (Reproduced from Macdonald and Engel[32] by permission of the Association of the Journal of Neuropathology and Experimental Neurology, Inc.)

FIG. 7. Chloroquine myopathy. Alternating profiles of T tubules (*arrowheads*) and lateral vesicles of SR (*asterisks*) as well as other membranes of undetermined origin encircle small fiber region. × 49,000. (Reproduced from Macdonald and Engel[32] by permission of the Association of the Journal of Neuropathology and Experimental Neurology, Inc.)

structures continued to increase in number (Fig. 11). After 3 weeks of treatment, vacuoles abutted on the fiber surface where they were covered by basement membrane only, allowing for extrusion of their contents into the extracellular space (exocytosis). When fiber splitting was noted, the separating fiber regions usually contained vacuoles and myeloid structures, indicating that fiber splitting, like exocytosis, might be a mechanism whereby the fiber rids itself of abnormal components.

*Electron cytochemistry.* The observations proved that the vacuoles were truly digestive, or autophagic, in character by virtue of their high acid phosphatase content (Fig. 12). The myeloid structures also were reactive for acid phosphatase and thus could be regarded as secondary lysosomes. Whether chloroquine produced myeloid bodies by direct chemical effect on membranes which were then further degraded by digestive enzymes or the drug and digestive enzymes acted in concert to produce myeloid bodies was not determined. The manner in which digestive enzymes were formed by the muscle fiber and delivered to the autophagic vacuoles was not elucidated.

When the T system was marked by peroxidase, proliferating tubules were seen partly or completely encircling small areas of cytoplasm (Figs. 13 and 14) and contributing to those membranes bounding larger vacuoles and surrounding myeloid bodies (Fig. 15). By contrast, labeled SR membranes, although often abutting on vacuoles, were not continuous with vacuolar membranes. Direct encirclement of cytoplasmic regions by Golgi cisternae was not observed.

## ACID MALTASE DEFICIENCY (TYPE 2 GLYCOGENOSIS)

Another myopathy in which an autophagic mechanism becomes activated is that due to acid maltase deficiency (AMD). While marked cardiomegaly, generalized glycogen storage, and death in infancy were thought to be hallmarks of the disease as recently as 1959,[47] since 1963, when Hers[26] discovered AMD in affected infants, the concept of type 2 glycogenosis has undergone a revision. Between 1963 and 1968 children with AMD but with little cardiac damage were found to survive into the second decade of life.[27] The first case of AMD with onset of weakness in adulthood was reported in 1968 by A. G. Engel and Dale.[14] Another case of AMD presenting with weakness in adult life was reported later in the same year by Hudgson, Gardner-Medwin, Worsfold, Pennington, and

FIG. 8. Chloroquine myopathy. Three small abnormal spaces surrounded by double membrane. Several tubular profiles abut on outer limiting membranes of spaces. × 49,000. (Reproduced from Macdonald and Engel[32] by permission of the Association of the Journal of Neuropathology and Experimental Neurology, Inc.)

FIG. 9. Chloroquine myopathy. Three small spaces limited by frayed double membrane are coalescing. Spaces contain mitochondrion, glycogen granules, and amorphous or degraded cytoplasmic material. × 34,600. (Reproduced from Macdonald and Engel[32] by permission of the Association of the Journal of Neuropathology and Experimental Neurology, Inc.)

FIG. 10. Soleus muscle of rat treated with chloroquine for 73 days. Typical autophagic vacuole with scalloped border and heterogeneous contents is limited by single or, in some places, double (*arrows*) membrane. Small vesicles (*arrowheads*) are fusing with limiting membrane. Well developed tubular networks (*asterisks*) occupy field adjacent to upper right vacuolar border. × 19,800, (Reproduced from Macdonald and Engel[32] by permission of the Association of the Journal of Neuropathology and Experimental Neurology, Inc.)

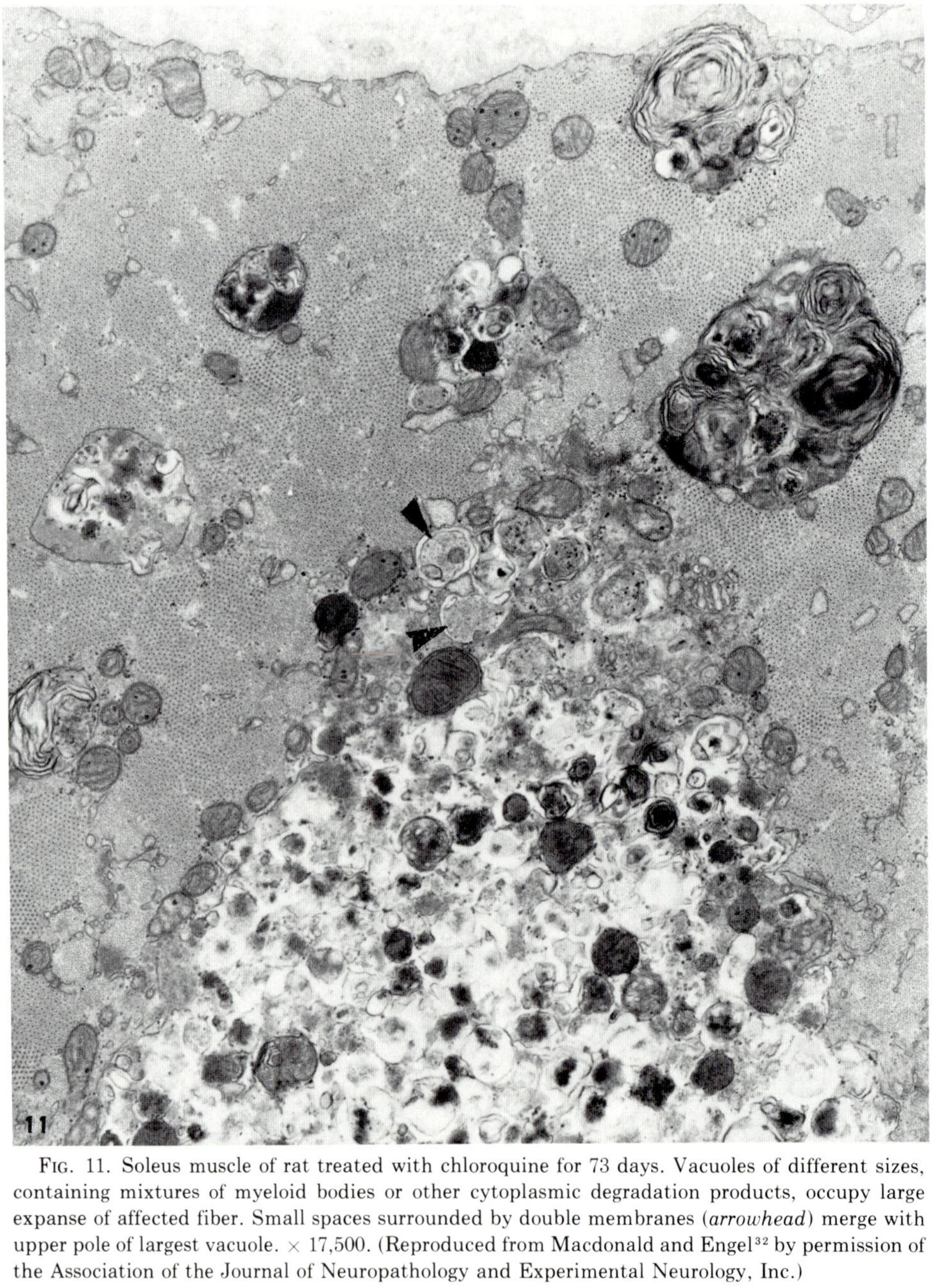


FIG. 11. Soleus muscle of rat treated with chloroquine for 73 days. Vacuoles of different sizes, containing mixtures of myeloid bodies or other cytoplasmic degradation products, occupy large expanse of affected fiber. Small spaces surrounded by double membranes (*arrowhead*) merge with upper pole of largest vacuole. × 17,500. (Reproduced from Macdonald and Engel[32] by permission of the Association of the Journal of Neuropathology and Experimental Neurology, Inc.)

Walton.[29] A total of 10 cases of AMD (2 in infants, 1 in a child, and 7 in adults) have been investigated in this laboratogy to date and the clinical, biochemical, and morphologic features of the adult syndrome have been reported.[10, 11, 14, 14a]

*Light and phase-contrast microscopy.* All striated muscle fibers were vacuolated in infants; in children and adults only some of the fibers in a given plane

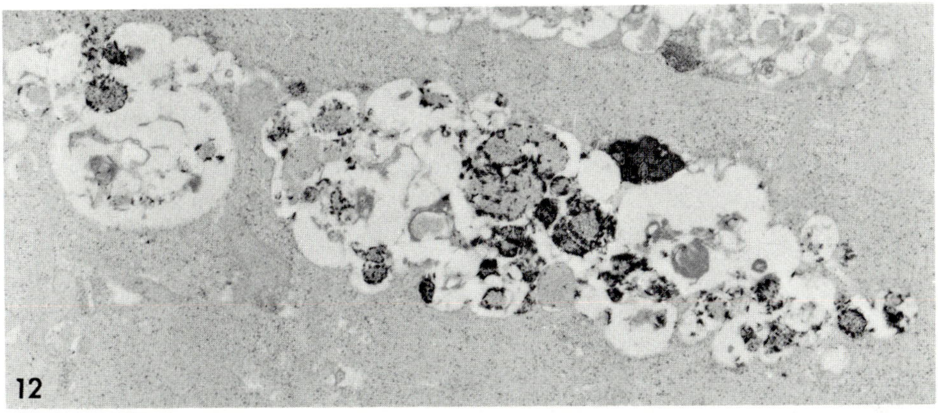

FIG. 12. Soleus muscle of rat treated with chloroquine for 45 days. Acid phosphatase activity appears in confines of autophagic vacuole and is associated with vacuolar contents. In this unstained section traces of extraneous precipitate are seen over other fiber regions. × 17,500.

displayed vacuoles (Fig. 16). The abnormal spaces varied from less than 2 to more than 60 μ in width and from 3 to more than 100 μ in length, and the larger spaces were subdivided by incomplete septa. In general, abnormal fibers contained either multiple, relatively large or small vacuoles; the total vacuolar volume and the proportion of vacuolated fibers were greater in the weaker than in the stronger muscles. The most severely affected muscles in adults also were extensively replaced by fibrous and fatty connective tissue, and here weakness was due to fiber loss as well as to the pathologic state of the surviving fibers. In some clinically unaffected muscles, vacuoles were so rare or small that they could be overlooked or confused with artifacts. Fiber splitting occurred in affected muscles and was more frequent in vacuolated than in nonvacuolated fibers. Regenerating fibers, necrosis, and phagocytosis were relatively uncommon.

*Histochemistry*. In PAS-stained Epon sections of tissue fixed in glutaraldehyde plus osmium, most but not all vacuoles contained PAS-reactive material (Fig. 17). In such sections, small glycogen-containing vacuoles were best displayed. More abundant but less well localized PAS-reactive material was found in fresh-frozen sections of tissue fixed in Newcomer's fluid. Regardless of the fixative used, in paraffin sections preservation of such material was inconsistent. Diastase digestion removed most of the PAS-positive material from frozen and paraffin sections, proving that it was glycogen.

Abundant acid phosphatase activity was localized in many vacuoles indicating that these were autophagic in character, or secondary lysosomes (Fig. 18). In some but not all cases in adults, in less severely affected muscles the red (type 1) fibers were more highly vacuolated than the white (type 2) fibers, and the former fibers also were overly reactive (even for red fibers) for mitochondrial oxidative enzymes (Fig. 19). Acidic mucosubstances (alcian blue-positive and metachromatic) were found in numerous vacuolated fibers in infantile and childhood (Fig. 20) cases but only sparse amounts were found in occasional vacuoles in 4 of 7 adult cases.

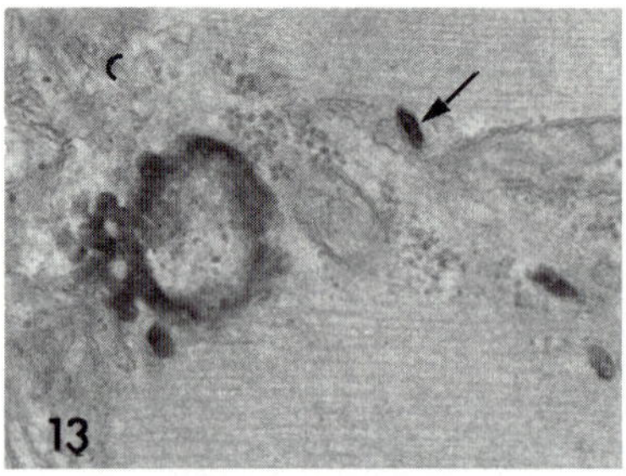

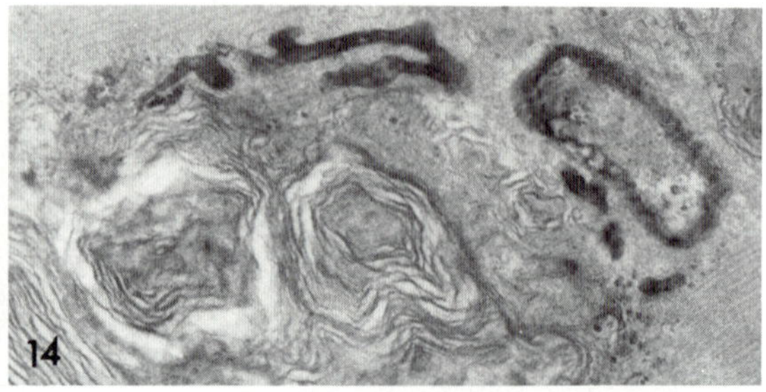

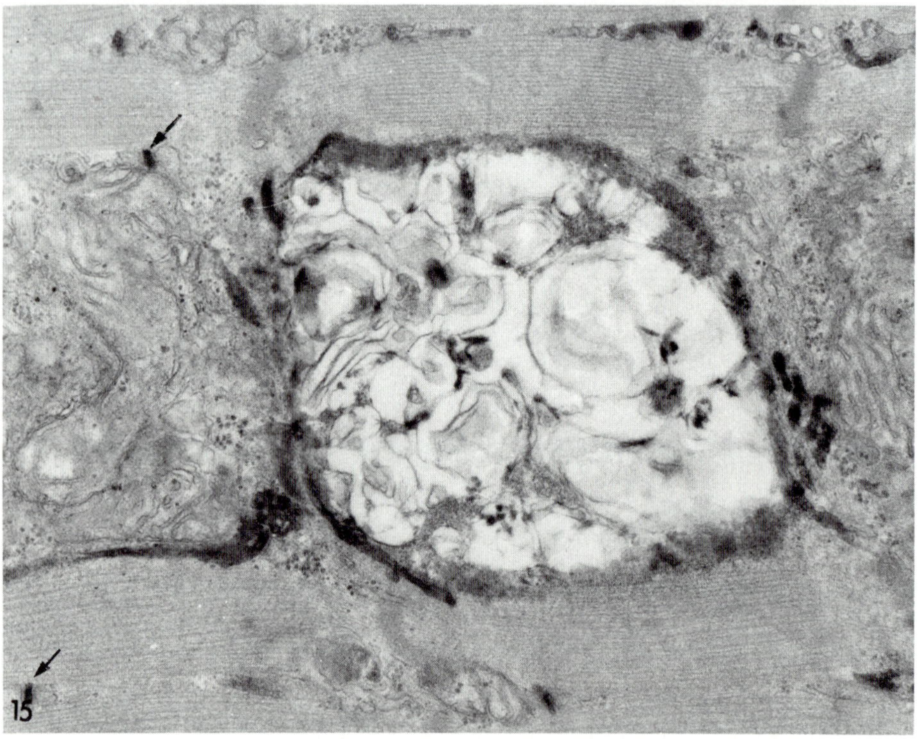

FIG. 13. Chloroquine myopathy. Peroxidase reaction, unstained section. Proliferating T system is continuous with tubule enclosing small portion of cytoplasm (*arrow* marks peroxidase-filled T tubule in triad). × 31,000. (Reproduced from Macdonald and Engel[32] by permission of the Association of the Journal of Neuropathology and Experimental Neurology, Inc.)

FIG. 14. Chloroquine myopathy. Peroxidase-containing tubular profiles partially or completely surround small areas of cytoplasm adjacent to myeloid bodies. × 31,000. (Reproduced from Macdonald and Engel[32] by permission of the Association of the Journal of Neuropathology and Experimental Neurology, Inc.)

FIG. 15. Chloroquine myopathy. Peroxidase-filled T system-derived tubules surround and limit autophagic vacuole. Small T networks appear on either side of central vacuole. *Arrows* mark T tubules in triads. × 31,000. (Reproduced from Engel and Macdonald[16a] by permission of the Excerpta Medica Foundation.)

*Electron microscopy*. Glycogen accumulated in four types of spaces.

Type 1 space was unlimited by membrane except where it abutted on the fiber surface; it usually contained glycogen or a mixture of glycogen and other cytoplasmic organelles or degradation products (Fig. 21). The glycogen dis-

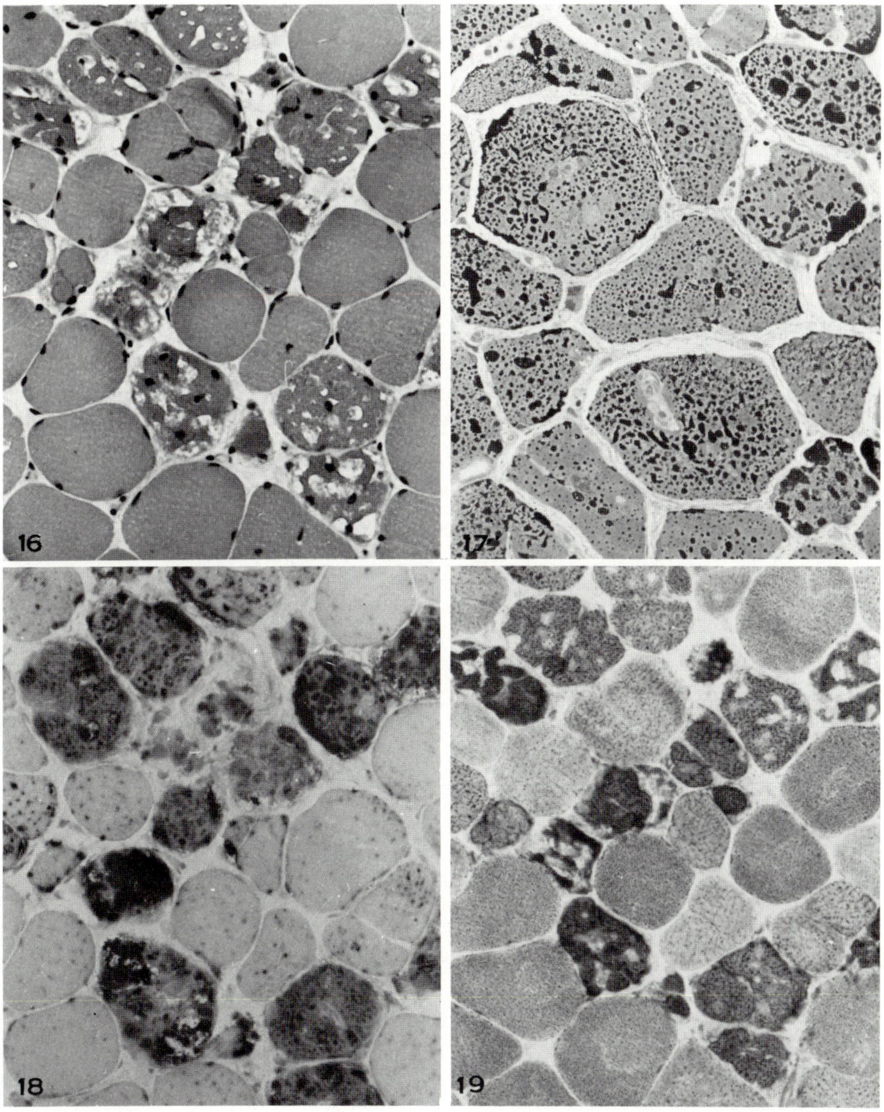

FIG. 16. Acid maltase deficiency, adult type. Some fibers contain multiple deep and superficial vacuoles of varying size. Several fibers are splitting. Hematoxylin and eosin stain, frozen section; × 175. (Reproduced from Engel[10] by permission of the Excerpta Medica Foundation.)

FIG. 17. Acid maltase deficiency, adult type. Intravacuolar glycogen is clearly localized. Periodic acid-Schiff stain, Epon section; × 270.

FIG. 18. Acid maltase deficiency, adult type; frozen section serial to that in Figure 16; acid phosphatase reaction. The vacuolated fibers are highly reactive. × 175. (Reproduced from Engel[10] by permission of the Excerpta Medica Foundation.)

FIG. 19. Acid maltase deficiency, adult type; frozen section serial to that in Figures 16 and 17; NADH dehydrogenase reaction. The vacuolated fibers are overly reactive. × 175. (Reproduced from Engel[10] by permission of the Excerpta Medica Foundation.)

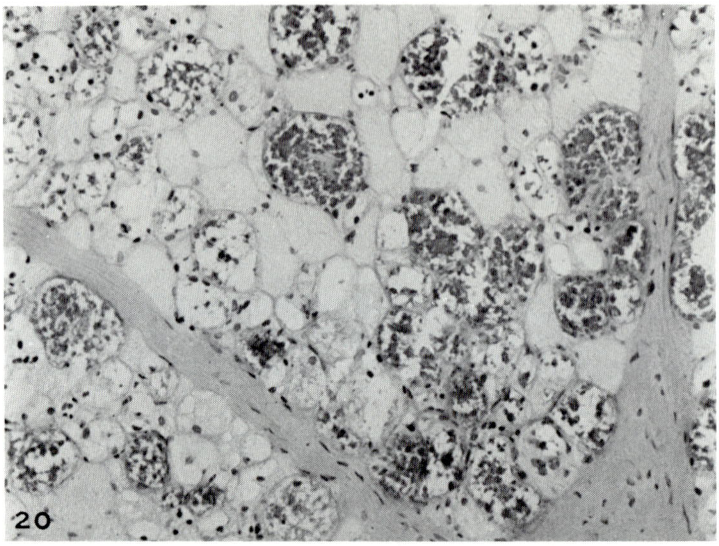

FIG. 20. Acid maltase deficiency, childhood type. Acidic mucosubstance is present in numerous vacuolated fibers. Alcian blue stain, frozen section; × 180.

placed, replaced, and compressed preexisting structures. Irregularly curving and disrupted linear membrane profiles often were seen scattered in such spaces, possibly indicating that this glycogen was previously membrane-bound (Fig. 21).

Type 2 space had smooth contours and was limited by a continuous or discontinuous single or double membrane (Figs. 21 through 26). The smallest type 2 space was approximately 0.5 $\mu$ wide, usually abutted on I bands, and represented the only departure from normality in least-affected fibers (Fig. 22). Satellite cells also contained small type 2 spaces (Fig. 21). Larger type 2 spaces, some as wide as 7 $\mu$, either appeared in clusters or formed a loculated network (Fig. 23).

Two sources for the membrane that limited type 2 spaces were encountered. Networks of T system origin were found to be continuous with tubules surrounding small type 2 spaces and to fuse with boundaries of larger type 2 spaces (Fig. 24). Other small type 2 spaces limited by double membrane arose by encirclement of glycogen by an inner Golgi cisterna (Fig. 25). Golgi cisternae also were encountered near or between smaller type 2 spaces, and here inner vesicles of the Golgi complex appeared to fuse with membranous boundaries of type 2 spaces (Fig. 26).

Type 3 spaces resembled ordinary autophagic vacuoles and had heterogeneous contents. Their border was smooth or scalloped and in some sectors the scalloped border consisted of double membrane (Fig. 27). The vacuoles contained small dense bodies, some studded with glycogen, myeloid figures, membranous fragments, amorphous material, and varying amounts of glycogen not attached to dense granules. Tubular networks of T system origin again were seen to fuse with boundaries of this space (Fig. 27). Some of the type 3 spaces

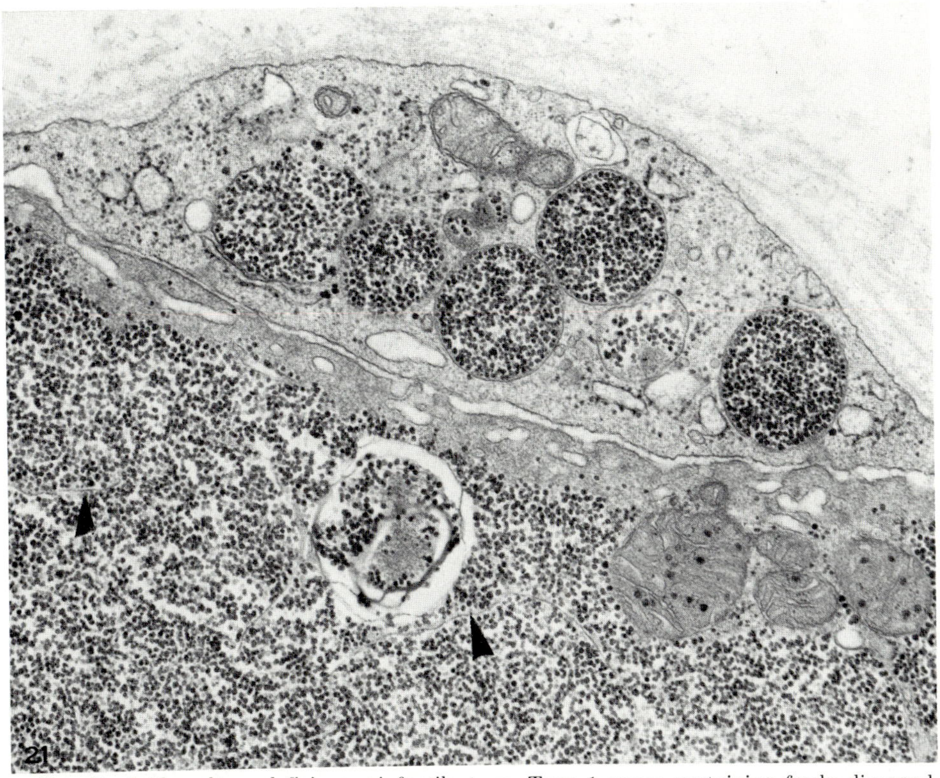


FIG. 21. Acid maltase deficiency, infantile type. Type 1 space containing freely dispersed glycogen is seen in fiber (low in field). Membrane profiles (*arrowheads*) occur in glycogen lake. Satellite cell (top of field) contains six small type 2 spaces. These are membrane-limited and contain only glycogen. × 21,000.

abutted on the cell surface, discharging their contents outside of the fiber (exocytosis) (Fig. 28).

Type 4 spaces represented transitions between the above types of spaces; membrane-bound and freely dispersed glycogen and a variety of normal and abnormal organelles intermingled.

Other features were as follows. In those fibers heavily laden with glycogen, large mitochondria with electron-dense granules appeared within lakes of freely dispersed glycogen. Giant, crystalloid-containing mitochondria were found in 1 adult case only.

*Electron cytochemistry.* Acid phosphatase activity was demonstrated within type 3 spaces (autophagic vacuoles with heterogeneous contents) (Fig. 29) but not in type 2 spaces (which contained only glycogen). Reaction product also was seen in the lumina of small vesicles adjacent to and fusing with type 3 spaces, and some of these reactive vesicles were seen to derive from nearby proliferating T tubules (Fig. 29). The small vesicles were thus carriers of the hydrolytic enzyme to the digestive vacuoles.

Horseradish peroxidase-labeled T tubules and T networks partially encircled small lakes of glycogen and contributed to boundaries of larger type 2 and 3

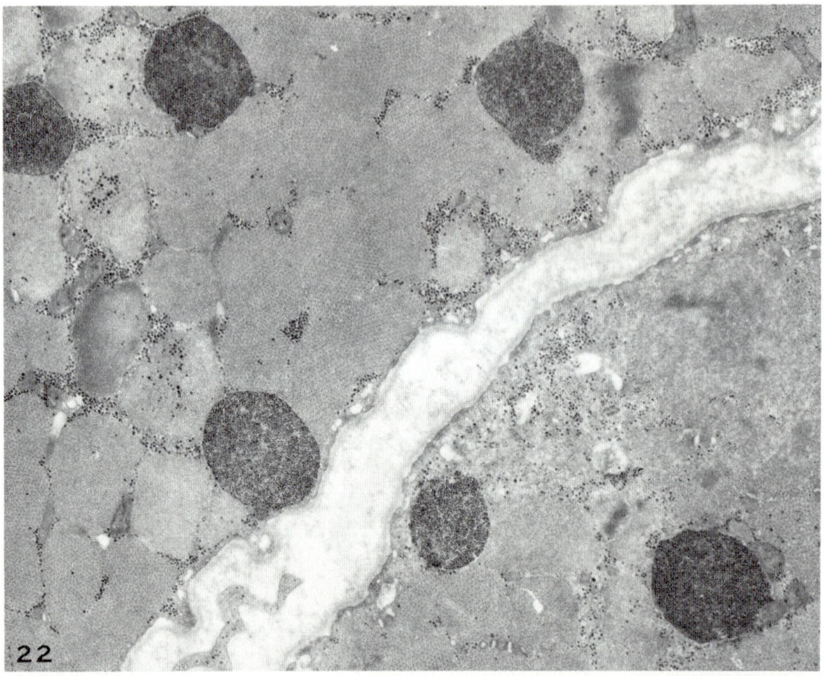

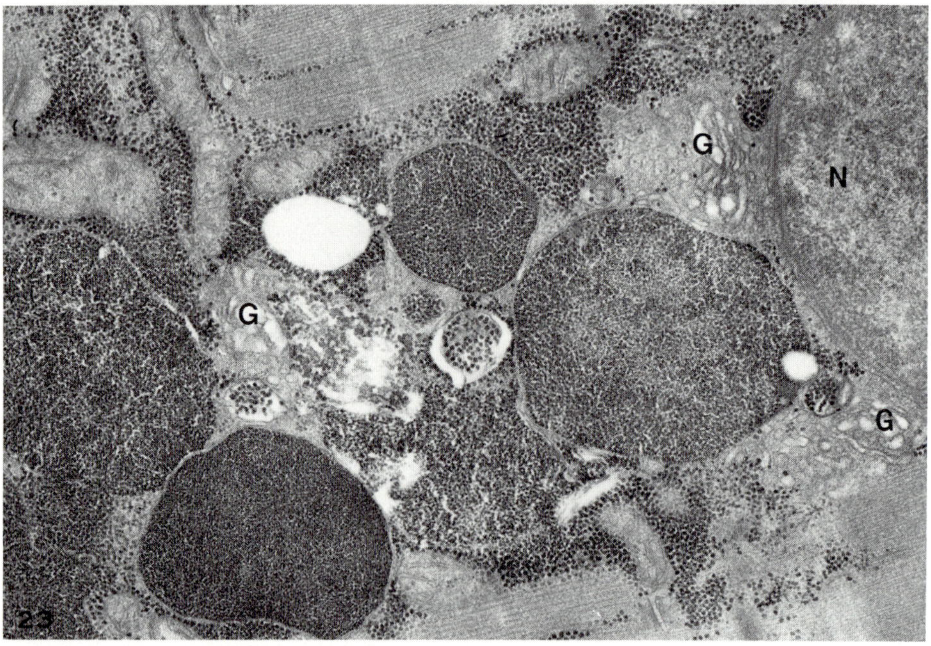


FIG. 22. Acid maltase deficiency, childhood type. Multiple small type 2 spaces containing only glycogen represent only abnormality in two adjacent fibers. × 13,600.

FIG. 23. Acid maltase deficiency, adult type. Larger type 2 spaces in hof region. Freely dispersed glycogen appears at lower left. *N*, nucleus; *G*, Golgi complex. × 24,400. (Reproduced from Engel[10] by permission of the Excerpta Medica Foundation.)

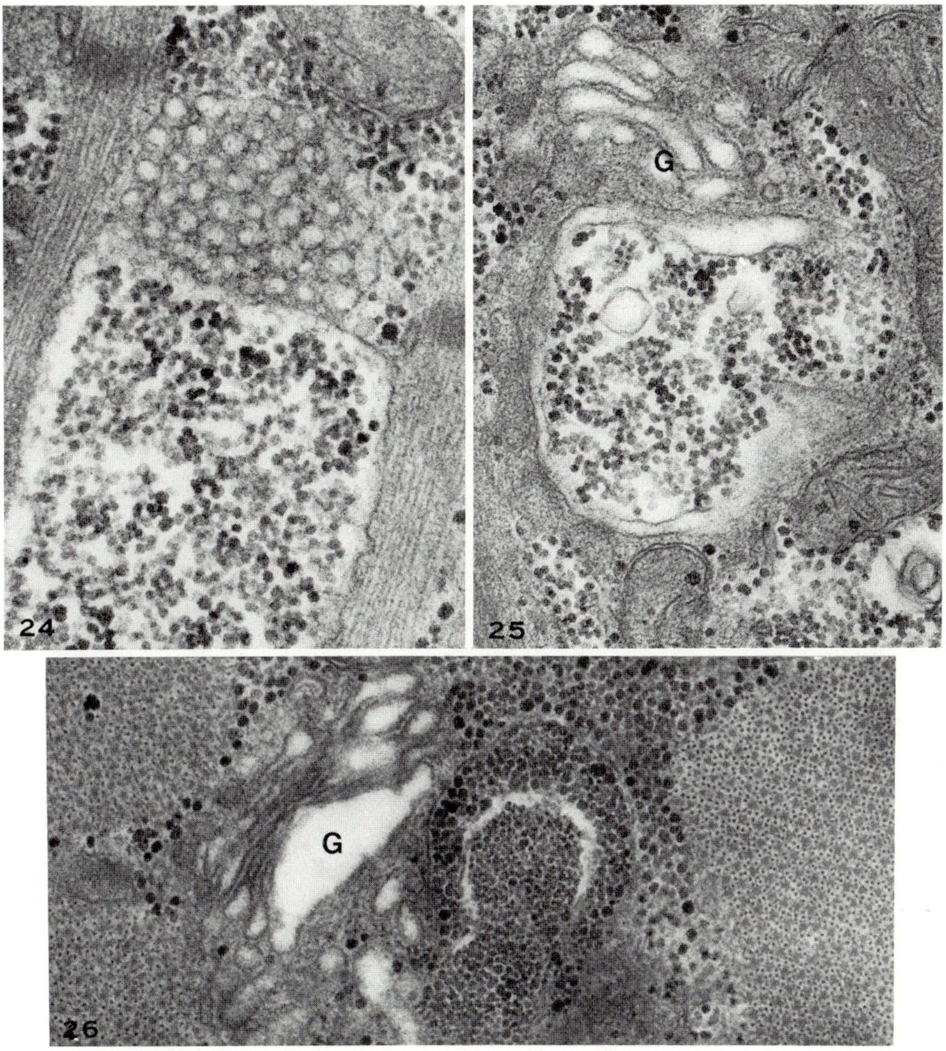


FIG. 24. T system network merges with membranous boundary of type 2 space. × 59,000. (Reproduced from Engel[11] by permission of Macmillan and Co.)

FIG. 25. Small glycogen deposit near Golgi complex (*G*) is incompletely enveloped by double membrane which appears to have arisen from inner Golgi cisterna. × 44,700. (Reproduced from Engel[11] by permission of Macmillan and Co.)

FIG. 26. Dilated inner cisterna of altered Golgi complex (*G*) abuts on small incompletely formed type 2 space. Intrasaccular glycogen granules are decreased in size. × 56,000. (Reproduced from Engel[10] by permission of the Excerpta Medica Foundation.)

spaces, confirming that the T system acted as a membrane source to these spaces (Figs. 30 through 33). In addition, a significant amount of peroxidase entered ordinary autophagic vacuoles, indicating that this type of space was in functional communication with the extracellular fluid via T system tubules and networks (Fig. 33). Labeled membranes of the SR did not contribute to boundaries of any of the abnormal spaces (Fig. 34).

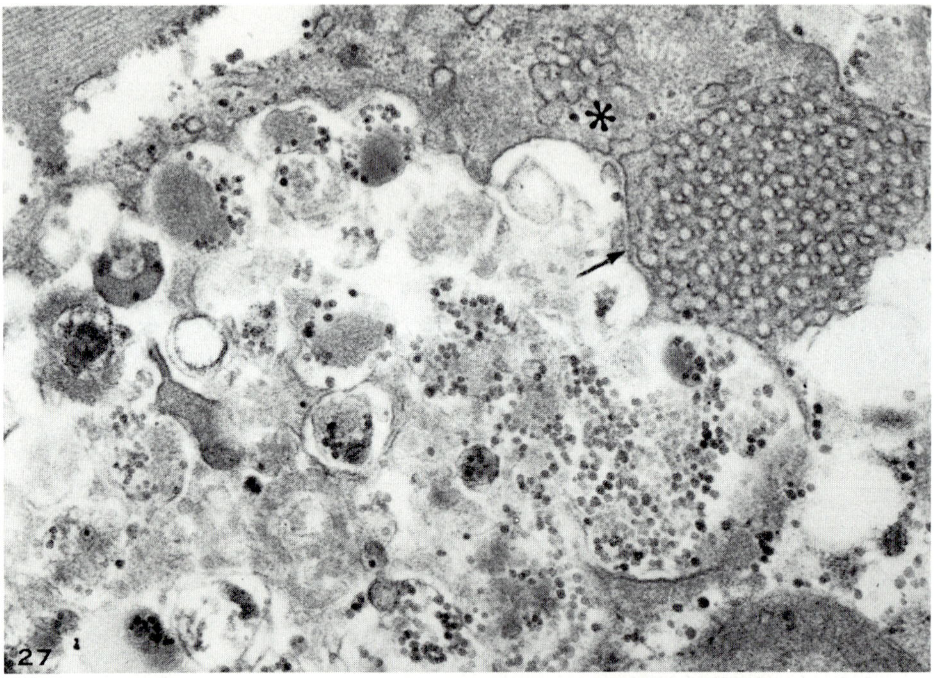


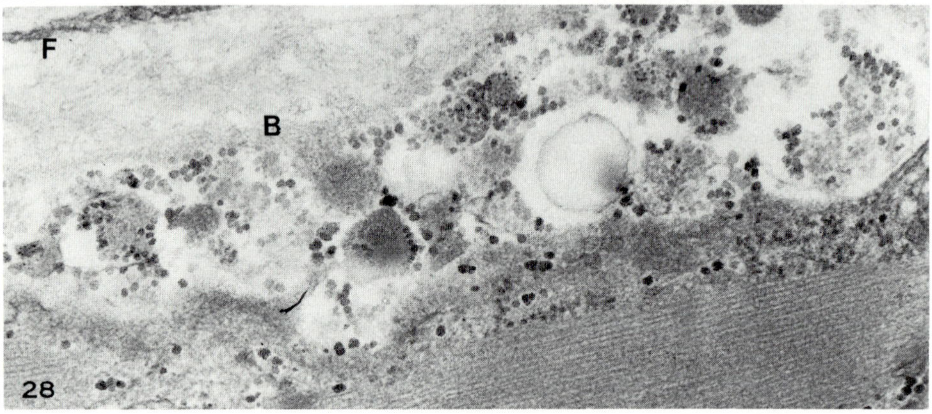


FIG. 27. Type 3 space, autophagic vacuole with heterogeneous contents. T system network merges with vacuole border. Tubule apposed to border (*arrow*) results in double limiting membrane. To left of T system network, tubules and vesicles derived from network (*asterisk*) also are fusing with vacuole. × 46,000. (Reproduced from Engel[11] by permission of Macmillan and Co.)

FIG. 28. Acid maltase deficiency, adult type. Type 3 space is separated from extracellular space only by basement membrane (*B*). Fibroblast process (*F*) is seen at upper left. × 39,400.

## LIPID STORAGE MYOPATHY

Human myopathies characterized by an excessive accumulation of lipid material, especially in red (type 1) fibers, have been described by Shy, Gonatas, and Perez,[50] by Price, Gordon, Munsat, and Pearson,[44] and by Bradley, Hudgson, Gardner-Medwin, and Walton.[7] Intramitochondrial crystalloid inclusions in at times abnormally large mitochondria also were found in affected fibers. Circular, moderately electron-dense inclusions and prominent dense granules in mitochondria also were noted by Shy, Gonatas, and Perez while Bradley, Hudgson, Gardner-Medwin, and Walton noticed poor definition of mitochondrial cristae. The morphologic nonspecificity of the mitochondrial alterations has been reviewed.[14, 43]

In this laboratory, striking muscle lipid excess, especially in red fibers, was observed in a 20-year-old woman.[16b] Although she had always had difficulty in sitting up from a recumbent position, progressive weakness eventually affecting cranial and respiratory muscles developed only 8 months before her examination at the Mayo Clinic. Electromyographic abnormalities were mild relative to the weakness, but serum enzyme levels were moderately increased.

*Light and phase-contrast microscopy.* Myriads of abnormal empty spaces, from less than 1 to 7 $\mu$ in diameter, were present in numerous fibers. In longitudinal sections the clear spaces appeared circular and were aligned along the length of the fiber in parallel rows. In some regions, especially under the surface membrane, adjacent small spaces coalesced to form larger, irregularly lobulated vacuoles. The connective tissue elements and blood vessels were normal. Inflammatory changes or necrosis and phagocytosis of individual fibers was not observed.

*Histochemistry.* The abnormal spaces occurred especially, but not exclusively, in red (type 1) fibers. Their contents stained positively with Sudan black B, oil red O (Fig. 35), and Nile blue sulfate, indicating the presence of neutral fat and presumably fatty acids. Stains for cholesterol, phospholipids, and glycolipids were negative.

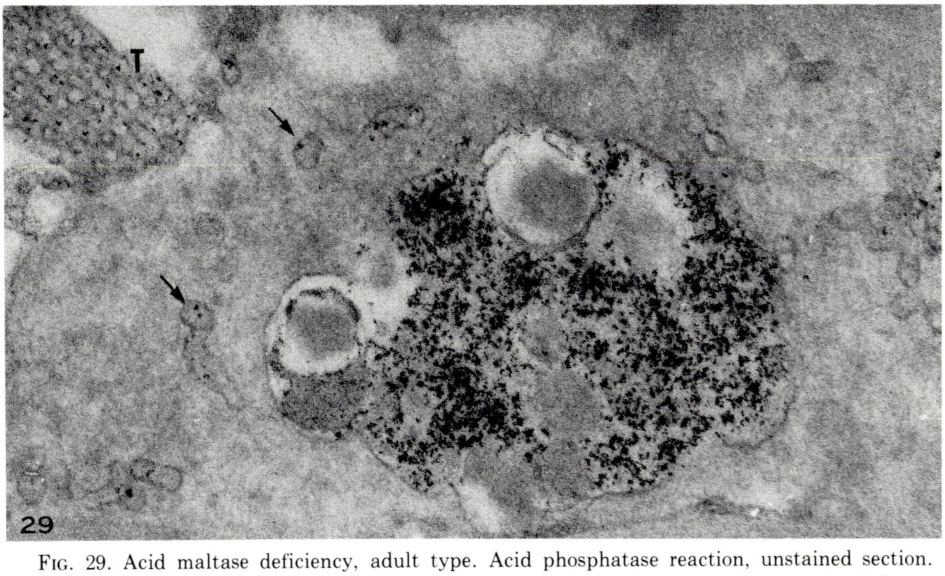


FIG. 29. Acid maltase deficiency, adult type. Acid phosphatase reaction, unstained section. Reaction product appears in autophagic vacuole, in small vesicles (*arrows*) some of which are coalescing with vacuole, and in T system-derived network (*T*). Vesicles are delivering digestive enzyme to vacuole. T network could be either source or carrier of enzyme. × 52,500.

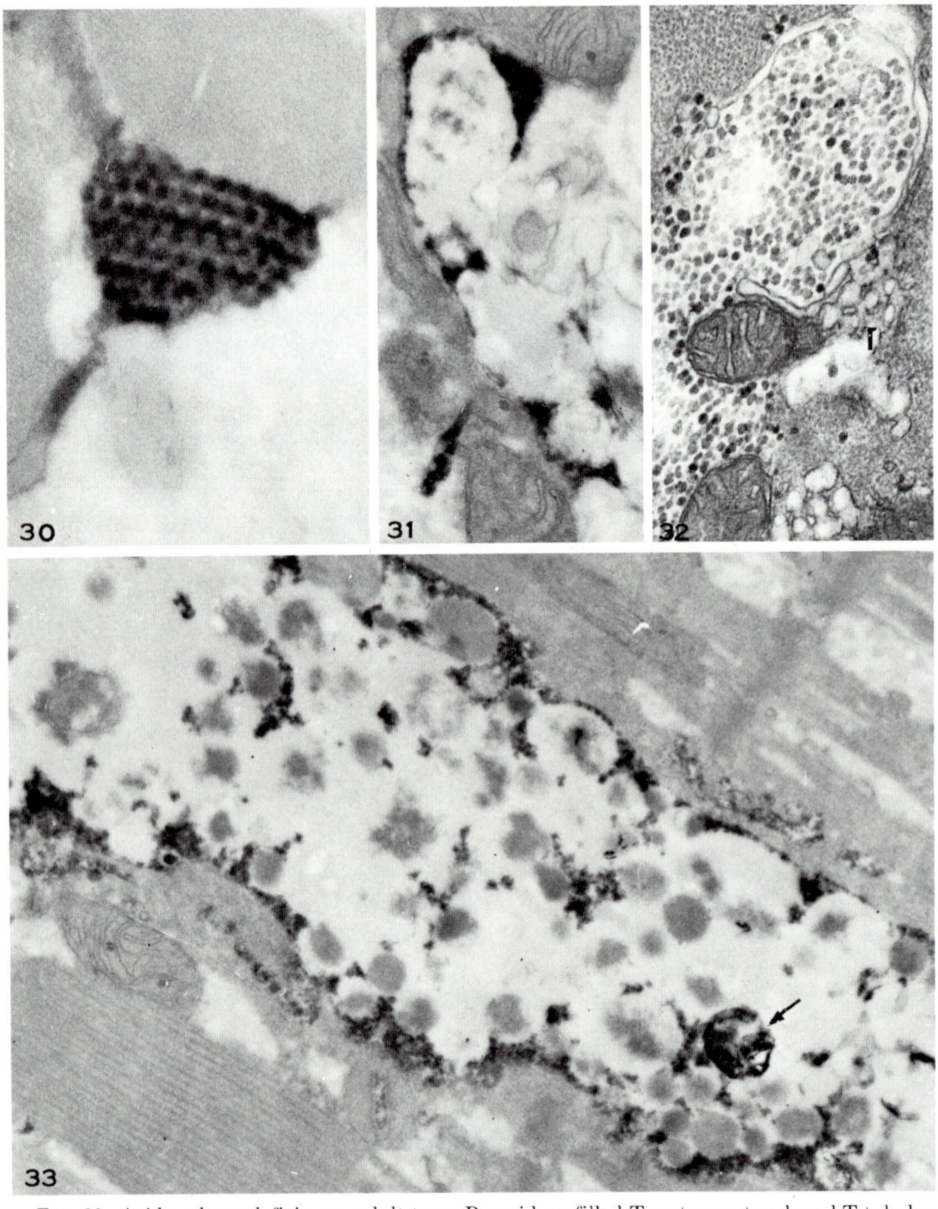

FIG. 30. Acid maltase deficiency, adult type. Peroxidase-filled T system network and T tubules continuous with network embrace vacuolar border. Unstained section; × 54,300.

FIG. 31. Acid maltase deficiency, adult type. Glycogen sequestration by peroxidase-filled T tubule which is continuous with small T network. Unstained section; × 43,200.

FIG. 32. Acid maltase deficiency, infantile type. Identical configuration to that seen in Figure 31 but in stained section not loaded with peroxidase. × 43,200. (Reproduced from Engel and Macdonald[16a] by permission of the Excerpta Medica Foundation.)

FIG. 33. Acid maltase deficiency, adult type. Peroxidase-filled tubules surround and limit autophagic vacuole (type 3 space). Tubules also open into vacuole and some peroxidase has entered vacuolar internum. Circular dense deposit (*arrow*) may indicate endogenous peroxidase activity. Unstained section; × 32,000.

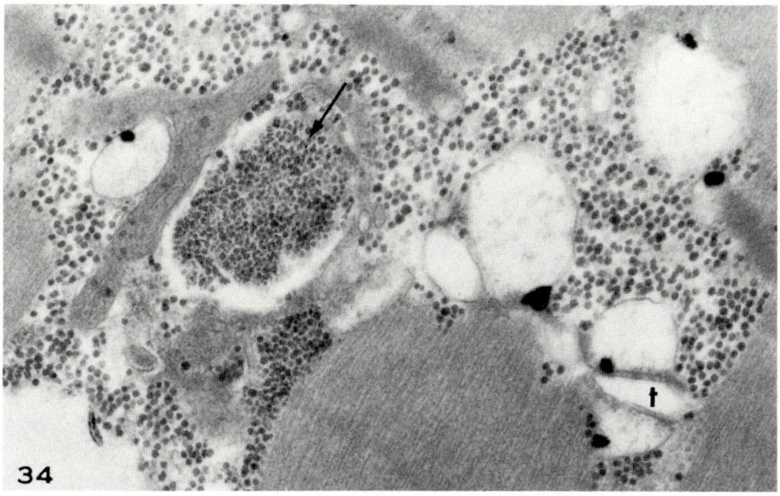

FIG. 34. Acid maltase deficiency, adult type; sarcoplasmic reticulum ATPase reaction. Electron-dense reaction product is present in dilated lateral vesicles. Transverse tubule (*t*) is not reacting. Glycogen-containing sac (*arrow*) displays no reaction product. × 21,800.

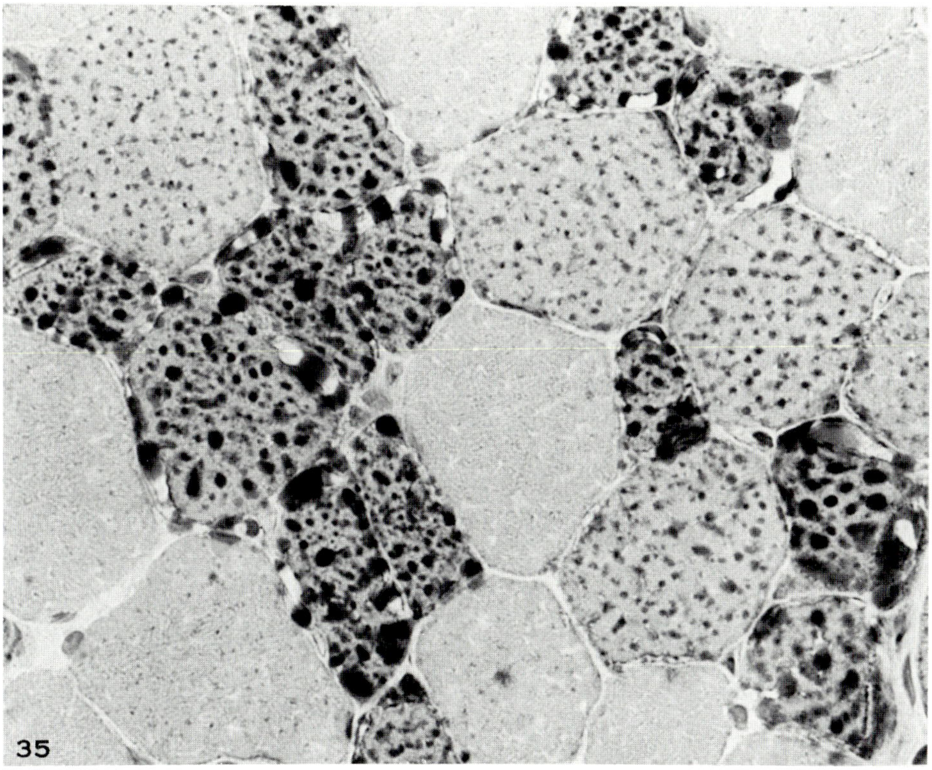

FIG. 35. Lipid storage myopathy. Depending on degree of storage, fibers fall into three types. Oil red O stain; × 480.

*Electron microscopy*. The abnormal spaces were either empty or contained traces of low density material and lacked clearly discernible limiting membranes (Fig. 36). The deposits often were adjacent to or surrounded by mitochondria with indistinct cristae, some also containing circular, moderately

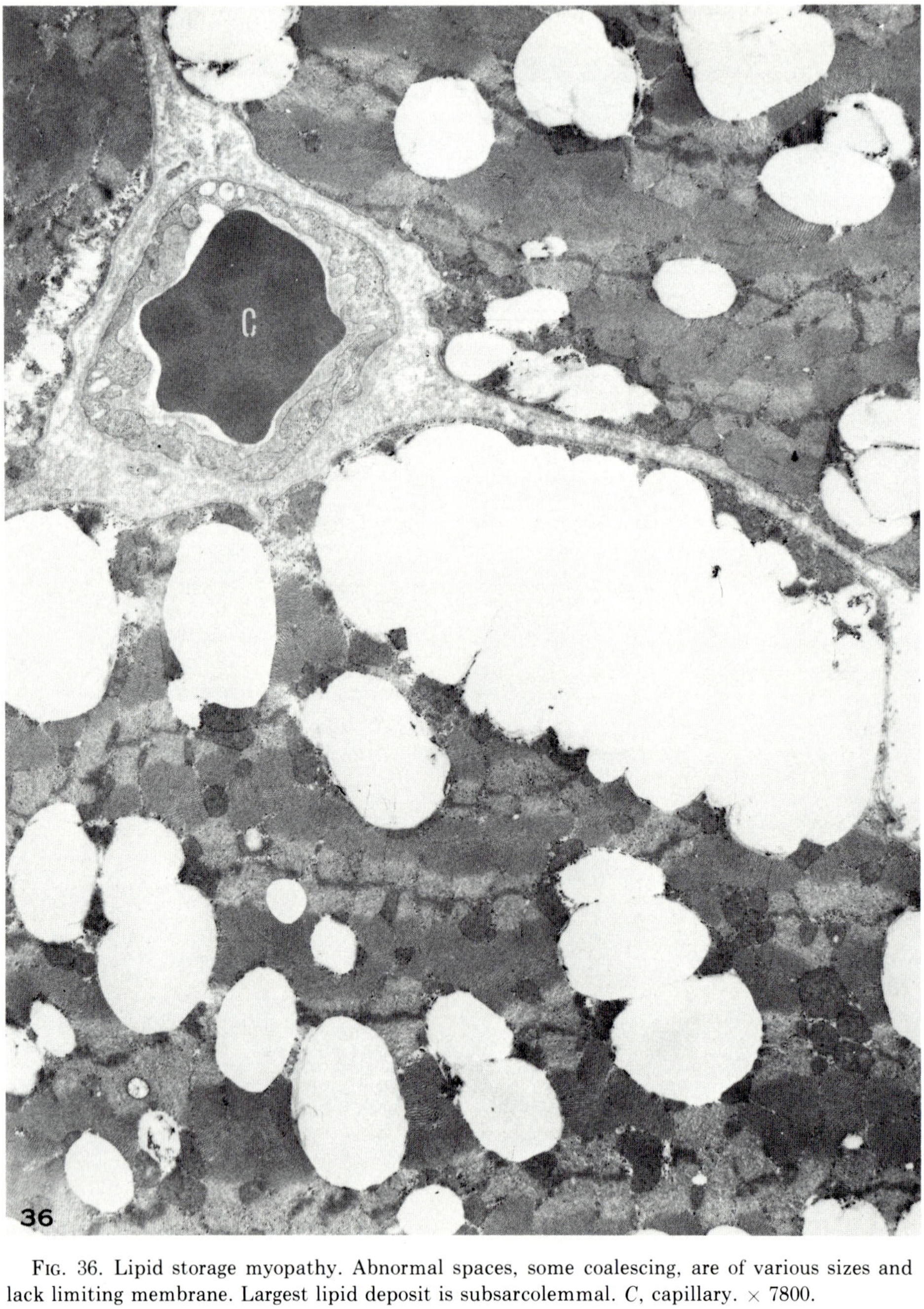


FIG. 36. Lipid storage myopathy. Abnormal spaces, some coalescing, are of various sizes and lack limiting membrane. Largest lipid deposit is subsarcolemmal. *C*, capillary. × 7800.

dense inclusions. Intramitochondrial crystalloids were not seen. The T system and SR appeared unaffected.

## PRIMARY HYPOKALEMIC PERIODIC PARALYSIS

A vacuolar myopathy occurs in either primary or secondary types of periodic paralysis, more commonly in the former than in the latter. Vacuolation is more consistently associated with the permanent myopathy that develops after repeated attacks than with the acute paralysis.[13, 16, 37, 41] Furthermore, because the origin in time of a given morphologic change in any one fiber is not known and because excitability of muscle fibers could fluctuate even without clinical attacks, one cannot be certain of the relationship of a pathologic change to a preceding clinical attack.

The factor which initiates the process that eventually leads to vacuolation is undoubtedly related to the heretofore unidentified basic metabolic error of the disease. During attacks, muscle fibers become electrically inexcitable but, in primary hypokalemic periodic paralysis (-KP(Pri)), delivery of calcium to the myofilament space can still activate the contractile mechanism.[15] From this it can be inferred that failure of excitation must reside in the surface membrane, T system, or SR, or in all three, and it might be logical to search for an early pathologic change in one or more of these organelles.

Three modes of vacuolar origin had been proposed by previous investigators: (1) coalescence of dilated SR components,[9, 28, 35, 36, 38, 51, 54] (2) fusion of proliferating T system elements,[5] and (3) focal fiber destruction.[33, 34, 48] Schutta and Armitage[48] proposed that limiting membranes of well developed vacuoles in thyrotoxic periodic paralysis were related to the T system.[48] Different investigators observed proliferating SR tubules,[19, 25, 36] dilated SR vesicles,[6, 9, 28, 33, 34, 36, 38, 48, 51] and abundant networks of T system[5, 13, 33, 34, 48] origin in abnormal fibers. Finally, the presence of glycogen granules in some vacuoles led to the speculation that such granules conditioned an influx of water and electrolytes into vacuoles during paralytic attacks.[51]

In a recent study of -KP(Pri), the evolution of vacuoles was reinvestigated and the contributions by the SR and the T system to vacuolar boundaries and contents were reassessed.[12] The steps identified were (1) evolving vacuole, (2) intermediate-stage vacuole, (3) mature vacuole, and (4) remodeling. The findings are summarized below.

### Light and phase-contrast microscopy

In paraffin sections a varying proportion of fibers (more than 50% in severely affected muscle) displayed vacuoles ranging from less than 10 $\mu$ in diameter to a width nearly encompassing the entire fiber (Fig. 37). In a given plane the vacuoles usually were central and solitary, but up to five vacuoles could occur in one fiber and some vacuoles abutted on the fiber surface. Most vacuoles appeared empty or contained varying amounts of granular material. On higher magnification, some nonvacuolated fibers displayed rarefied regions while some well formed vacuoles contained few to many dome-shaped or round ($<1$ to 40 $\mu$ wide) compartments, many of which abutted on the vacuolar rim. Regenerating

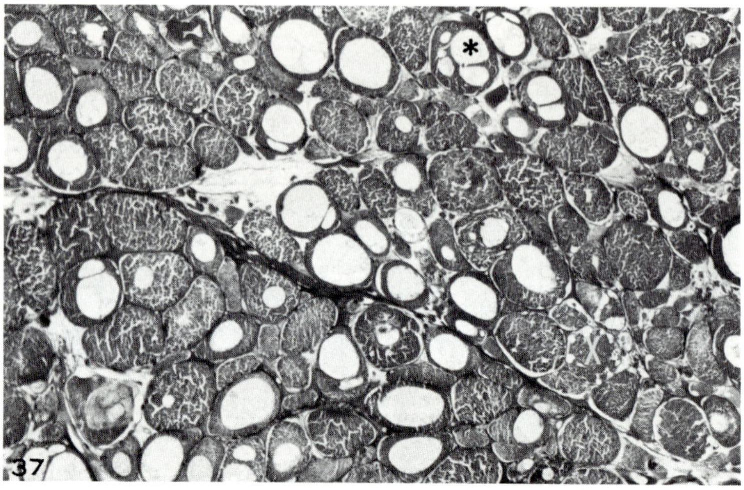

FIG. 37. Primary hypokalemic periodic paralysis. Most vacuoles are large, single, and central. Five vacuoles appear in one fiber (*asterisk*). Small superficial vacuoles occur in one fiber. Trichrome stain; × 130. (Reproduced from Engel[12] by permission of the Mayo Foundation.)

fibers, necrotic fibers, and intravacuolar macrophages were rare. Fiber splitting, in a given plane, occurred in 4 to 14% of fibers.

In Epon sections viewed by phase-contrast microscopy there were small, abnormal fiber regions containing irregularly curving lamellae, dense bodies of various shapes, myriads of vesicles, or a combination of these. Similar formations also were observed in larger, abnormal fiber regions (Fig. 38). In other fibers, myriads of vesicles, light or dark or concentrically light and dark, occupied the entire abnormal zone (Fig. 39). Some vacuoles with distinct borders had loculated interiors and heterogeneous contents (Fig. 40). The most common vacuole was one with a simple interior filled with homogeneous matrix and with few to many round or dome-shaped compartments abutting on its rim (Fig. 41).

HISTOCHEMISTRY

Large acid phosphatase-reactive regions were found in fibers of either histochemical type adjacent to a well formed vacuole, at one pole of a larger vacuole, or near the center of a vacuole (Fig. 42). These regions were rare relative to the total number of vacuoles. In addition, many fibers (more than half of fibers in one case and all fibers in another case) contained small ($<5$ μ wide) multiple, deep or superficial reactive regions (Fig. 42). Finally, there were fibers less than 20 μ wide diffusely reactive for acid phsophatase, which arose by splitting from larger parent fibers (Fig. 42).

Intense NADH dehydrogenase reaction was observed in circumscribed regions of some white (type 2, as adjudged by the myofibrillar ATPase reaction) fibers in central, eccentric, or pheripheral locations (Figs. 43 through 47). Parts or all of some of the less-than-20-μ-wide fibers also showed high activity of this enzyme. In deliberately underreacted sections, myriads of vesicles, $<0.3$ to 2 μ, were resolved in the reactive regions (Figs. 43 and 44). In serial sections, the vesicle aggregates also were reactive for NADPH dehydrogenase (although less intensely than for NADH) and for cytochrome oxidase but were unreactive for succinic dehydrogenase or in any of the above systems without substrate. The same regions were often, but not invariably, reactive for acid phosphatase. Serial sections again showed that in larger fibers such regions were either adjacent to or within a vacuole (Figs. 45 and 46) and that some of the highly reactive less-than-20-$\mu$-wide fibers arose by splitting from eccentric regions of parent fibers (Fig. 47). Regions with similar oxidative enzyme activities had been previously described in periodic paralysis, myotonia congenita, porphyria, in patients exposed to drugs, and in a case of transient muscle weakness and swelling.[18, 19, 40] They were first thought to contain abnormal mitochondrial aggregates[18] but more recently the aggregates were found to contain SR-derived tubules.[19, 40] A relationship of such regions to evolving or other vacuoles was not noted by the above authors. In the present study, as described below, vesicle (not tubule) aggregates were found in such regions in -KP(Pri). However, in hyperkalemic periodic paralysis, tubular aggregates were observed by us.[46]

Clusters of calcific mineral deposits (von Kossa method) were found in a few vacuoles, filling them partly or almost completely (Figs. 48 and 49).

In Epon sections, PAS-reactive material was seen (1) in small subsarcolemmal or deeper intermyofibrillar spaces (Fig. 50), (2) within those round or dome-shaped spaces which often abutted on the vacuolar rim (Fig. 50), or (3) rarely, filling the entire vacuole.

Lipid stains showed numerous, small sudanophilic droplets in some fibers and some lipid material in rare vacuoles. Stains for acidic mucosubstances and for metachromatic material were negative.

### Electron microscopy

*Nonvacuolated fiber regions*. Some fibers showed no abnormality. Common early changes were dilation (up to 0.6 $\mu$ in width and 2 $\mu$ in length) of SR components (Figs. 51 and 52) and proliferation and dilation of t-tubules (Fig. 53 and 54). Glycogen granules often surrounded the dilated SR vesicles (Fig. 51). Complexes of glycogen, abnormal T- and SR-derivatives, and scattered mitochondria were seen in occasional fibers.

*Evolving vacuole*. This was a large, central or eccentric fiber region without well demarcated borders. Different types of evolving vacuoles were encountered.

Type 1 evolving vacuole was composed predominantly of innumerable vesicles, some coalescing or rupturing into one another (Figs. 55 through 60). Some vesicles contained amorphous material (Figs. 59 and 60) as seen in the SR, but from other vesicles such material had partially or completely disappeared (Figs. 55 through 58). At the periphery of the abnormal region, triadic arrangements of the vesicles with T tubules attested to the SR origin of the vesicles (Figs. 56, 57, and 58). However, some vesicles without amorphous content and not in triadic relation to T tubules might have been derived from mitochondria. Intermingled with the vesicles there were one or more abnormal nuclei, as well as mitochondria, glycogen granules, irregularly curving osmiophilic lamellae, and debris (Fig. 55). Study of parallel thick and thin sections

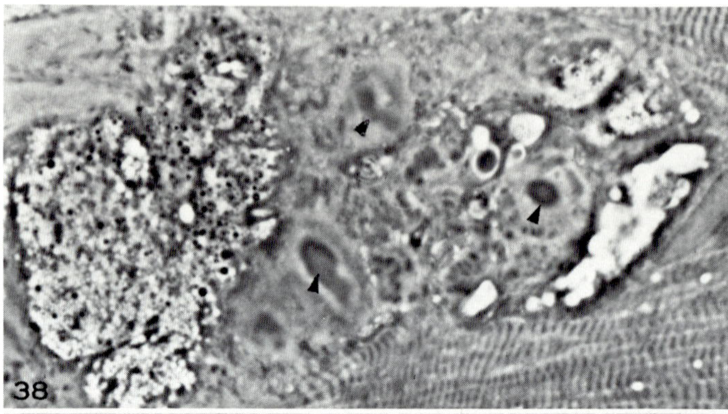


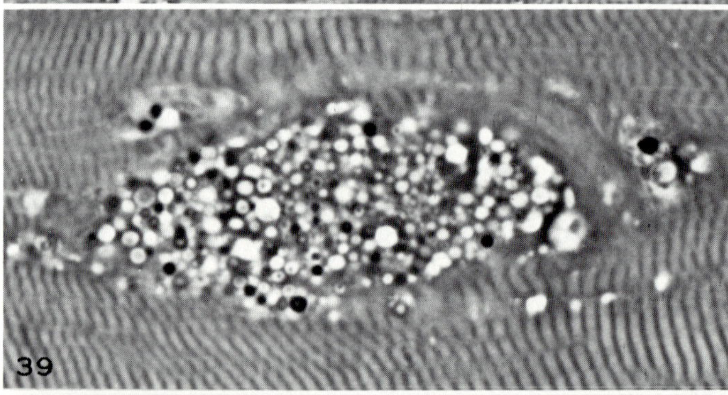


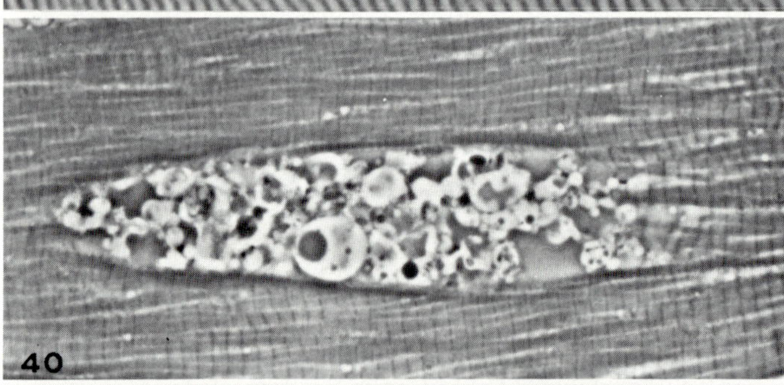


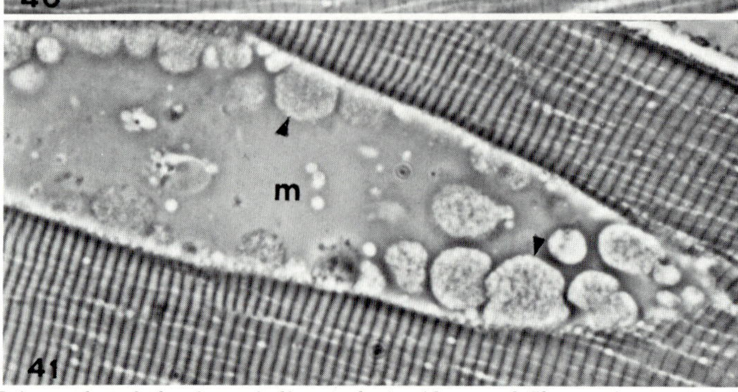


FIG. 38. Primary hypokalemic periodic paralysis. Abnormal fiber region containing myriads of black or light vesicles (left side) and other abnormal material (center and right side). *Arrows* mark cytoplasmic bodies (evolving vacuole). Phase micrograph; × 1350. (Reproduced from Engel[12] by permission of the Mayo Foundation.)

proved that the vesicle aggregates seen by electron microscopy corresponded to the vesicle aggregates seen in the phase-contrast microscope, which, in turn, corresponded to the NADH dehydrogenase-reactive vesicle aggregates seen in frozen sections.

Some of the vesicles contained highly electron-dense deposits in their interior with many deposits appearing in multiple, concentric, dense to highly dense rings (Figs. 59 and 60). Transitions from dilated SR vesicles containing amorphous material to others containing concentric deposits were seen (Figs. 59 and 60). The latter resembled the Liesegang rings which form when an inorganic salt precipitates in a supersaturated colloidal gel.[52] That the precipitates contained calcium, possibly calcium phosphate, was strongly suggested by results of the von Kossa stain on frozen sections plus the fact that the SR serves as a calcium sink in muscle.

Type 2 evolving vacuole was composed of irregularly curving, bizarre tubules and osmiophilic lamellae plus a variable admixture of vesicles (Figs. 61 and 62). At the periphery of the affected zone the bizarre tubules were continuous with recognizable T system-derived tubules while the osmiophilic lamellae arose by degradation of the bizarre tubular membranes (Fig. 62). Some of the vesicles in the abnormal zone were of SR origin by virtue of their content or relationship to the bizarre tubules (Fig. 62). Other vesicles and larger membrane-bound spaces, some also apposed to tubules, were filled with homogeneous material of intermediate density and unknown character (Fig. 62).

Type 3 evolving vacuole contained small dense bodies, amorphous material, cytoplasmic degradation products of undetermined origin, sparse glycogen granules, and occasional T system networks.

Finally, there were evolving vacuoles in which transitions between the above types of spaces were seen.

*Intermediate-stage vacuoles.* This was a well demarcated space, from few to more than 30 $\mu$ wide and from few to more than 100 $\mu$ long, with smoothly curving or irregularly undulating borders (Figs. 63 through 67). It was characterized by (1) a clearly defined limiting membrane; (2) a flocculent, low electron density matrix in its interior; and (3) embedded in the matrix, relics or derivatives of structures encountered in evolving vacuoles. The limiting membrane merged with T system tubules and networks that formed a wreathlike array around the vacuole and communicated with the vacuolar lumen (Fig. 63) and was studded with pinocytotic vesicles (Fig. 64), or was

---

FIG. 39. Primary hypokalemic periodic paralysis. Myriads of black or light vesicles, some containing central dark dots or concentric dark and light bands, occupy region with indistinct boundary (evolving vacuole). Phase micrograph; × 1840. (Reproduced from Engel[12] by permission of the Mayo Foundation.)

FIG. 40. Primary hypokalemic periodic paralysis. Loculated vacuole with distinct boundaries and heterogeneous contents (intermediate-stage vacuole). Phase micrograph; × 1150. (Reproduced from Engel[12] by permission of the Mayo Foundation.)

FIG. 41. Primary hypokalemic periodic paralysis. Most common type of vacuole contains matrix (*m*) which surrounds round or dome-shaped spaces that contain finely granular material. Several of these abut on vacuolar rim (*arrows*) (mature vacuole with sarcoplasmic invaginations). Phase micrograph; × 1150. (Reproduced from Engel[12] by permission of the Mayo Foundation.)

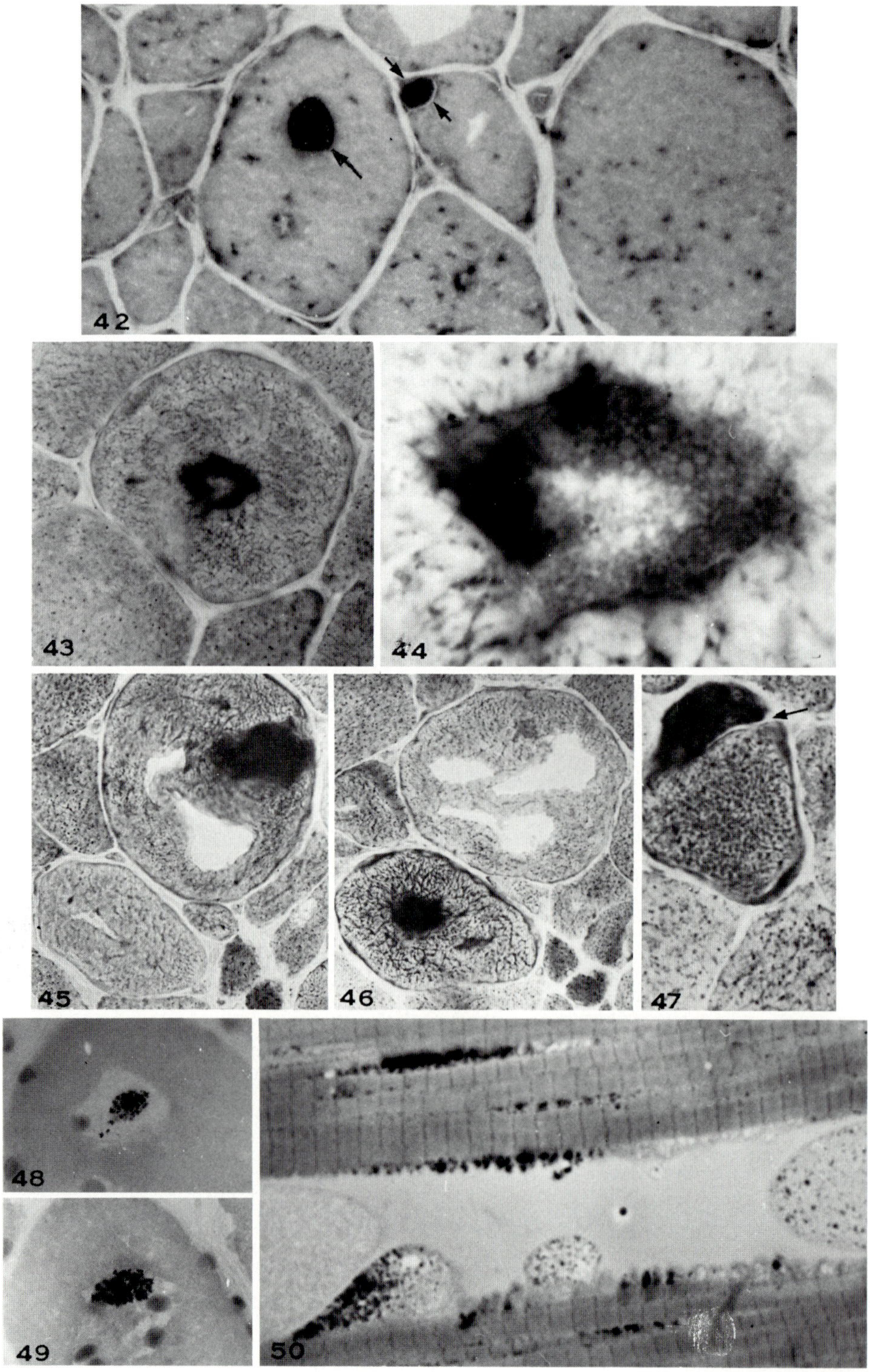
42
43
44
45
46
47
48
49
50

smooth (Fig. 65). Thus, as in chloroquine myopathy and in acid maltase deficiency, the T system again acted as a source of vacuolar limiting membrane. The intravacuolar residues consisted of myeloid structures, circular dense bodies of varying size (Fig. 65), recognizable dilated SR vesicles (Fig. 66), and mineralized SR vesicles displaying Liesegang rings (Fig. 67). In other vacuoles there were irregularly curving and branching bilaminar membranes which, at the edge of the vacuole, could be traced to apposed membranes binding pairs of adjacent T tubules or T tubule-derived spaces (Fig. 63).

*Mature vacuole.* This differed from the intermediate-stage vacuole by containing only matrix in its interior. T tubules and networks often abutted on the rim or communicated with the interior of this space (Figs. 68, 69, and 70).

*Remodeling.* Sarcoplasmic invaginations into intermediate-stage and mature vacuoles were often observed. The limiting membrane of these two types of vacuoles was displaced toward the lumen, producing a dome-shaped space (convex toward the center of the vacuole) (Figs. 71 and 72). The convexity was separated from the matrix compartment by the inwardly displaced vacuolar membrane while the base of the dome rested on the vacuolar rim and was continuous with the myofilament space. The invaginations were small (base $<1$ $\mu$) and festooned the vacuolar rim or were larger (base up to 40 $\mu$) and occupied much of the vacuolar internum (Fig. 72). Glycogen granules, irregularly curving membranous profiles, and sometimes mitochondria were present in the invaginated space. The invaginations corresponded to the dome-shaped or spherical spaces with PAS-positive content seen in Epon sections by light microscopy (Fig. 50).

Rupture of the inwardly displaced vacuolar membrane allowed sarcoplasmic

---

FIG. 42. Primary hypokalemic periodic paralysis. Acid phosphatase-reactive regions in central vacuole (*arrow*), in small fiber separating from parent fiber by splitting (*two arrows*), and scattered in all fibers. Fresh-frozen section; × 340. (Reproduced from Engel[12] by permission of the Mayo Foundation.)

FIG. 43. Primary hypokalemic periodic paralysis. Intense NADH dehydrogenase reaction occurs in central fiber region. Fresh-frozen section; × 400. (Reproduced from Engel[12] by permission of the Mayo Foundation.)

FIG. 44. Primary hypokalemic periodic paralysis. Higher magnification of Figure 43 shows myriads of vesicles filling abnormal region. Fresh-frozen section; × 1800. (Reproduced from Engel[12] by permission of the Mayo Foundation.)

FIGS. 45 and 46. Primary hypokalemic periodic paralysis. NADH dehydrogenase reaction in fresh-frozen 10-$\mu$ sections in series, separated by 30-$\mu$ interval. Intense enzyme activity is present in abnormal fiber regions adjacent to well formed vacuoles. × 160. (Reproduced from Engel[12] by permission of the Mayo Foundation.)

FIG. 47. Primary hypokalemic periodic paralysis. Small fiber intensely reactive for NADH dehydrogenase is arising by splitting from abnormal peripheral zone of larger parent fiber. *Arrow* indicates cleft between separating fibers. Fresh-frozen section; × 520. (Reproduced from Engel[12] by permission of the Mayo Foundation.)

FIGS. 48 and 49. Primary hypokalemic periodic paralysis. Calcific intravacuolar deposits. von Kossa stain, frozen sections; × 390. (Reproduced from Engel[12] by permission of the Mayo Foundation.)

FIG. 50. Primary hypokalemic periodic paralysis. Sarcoplasmic invaginations into vacuoles contain periodic acid-Schiff-positive material. There are small intermyofibrillar PAS-positive deposits. PAS-azure II-methylene blue stain, Epon section; × 1700. (Reproduced from Engel[12] by permission of the Mayo Foundation.)

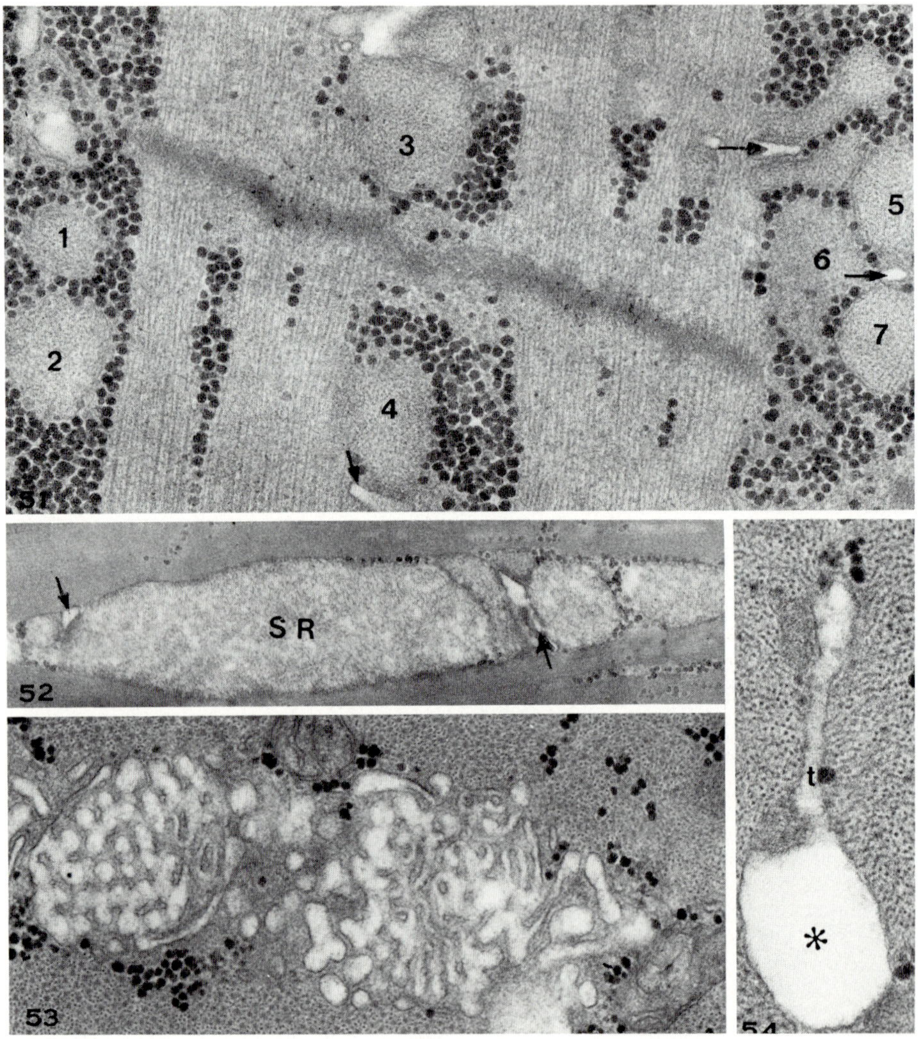


FIG. 51. Primary hypokalemic periodic paralysis. Seven dilated sarcoplasmic reticulum vesicles (*1*, *2*, *3*, *4*, *5*, *6*, and *7*) are shown. Glycogen granules surround vesicles. *Arrows* indicate T tubules in triadic relationship with SR vesicles. × 45,000. (Reproduced from Engel[12] by permission of the Mayo Foundation.)

FIG. 52. Primary hypokalemic periodic paralysis. Dilated round or markedly elongated sarcoplasmic reticulum components (*SR*) appear adjacent to myofibrils. *Arrows* indicate T tubules in triads. × 29,400. (Reproduced from Engel[12] by permission of the Mayo Foundation.)

FIG. 53. Primary hypokalemic periodic paralysis. Large T system networks occur between myofibrils. × 42,500. (Reproduced from Engel[12] by permission of the Mayo Foundation.)

FIG. 54. Primary hypokalemic periodic paralysis. Vesicular dilation (*asterisk*) of T tubule (*t*). × 53,000. (Reproduced from Engel[12] by permission of the Mayo Foundation.)

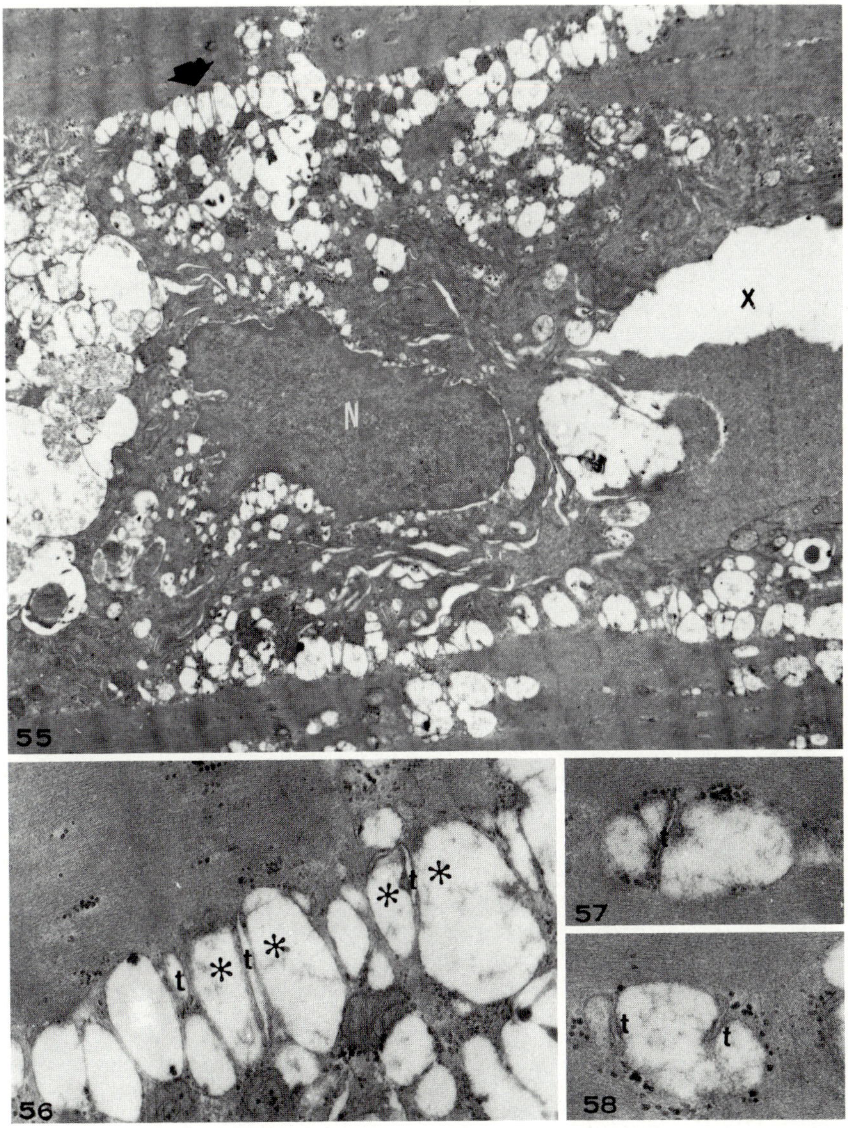


FIG. 55. Primary hypokalemic periodic paralysis. Type 1 evolving vacuole. At low magnification appears to be composed predominantly of innumerable vesicles. Larger, irregularly loculated aces appear at left. In center, nucleus (*N*) with scalloped border and dilated envelope is surɔnded by irregularly curving osmiophilic lamellae. Origin of wedge-shaped space at right (*X*) is known. × 7100. (Reproduced from Engel[12] by permission of the Mayo Foundation.)

FIG. 56. Primary hypokalemic periodic paralysis. Higher magnification of portion of Figure 55 dicated by *arrow*. Abnormal vesicles (*asterisks*) are in triadic relationship to T tubules (*t*). 28,400. (Reproduced from Engel[12] by permission of the Mayo Foundation.)

FIGS. 57 and 58. Primary hypokalemic periodic paralysis. In nearby fiber regions, dilated sarcoasmic reticulum vesicles have partially lost their amorphous content and abut on T tubules (*t*). 32,000. (Reproduced from Engel[12] by permission of the Mayo Foundation.)

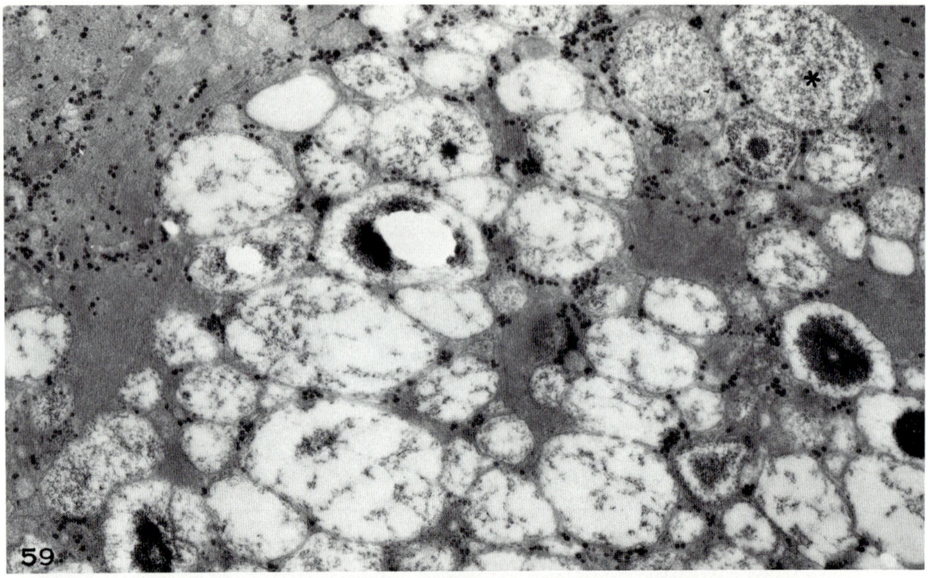

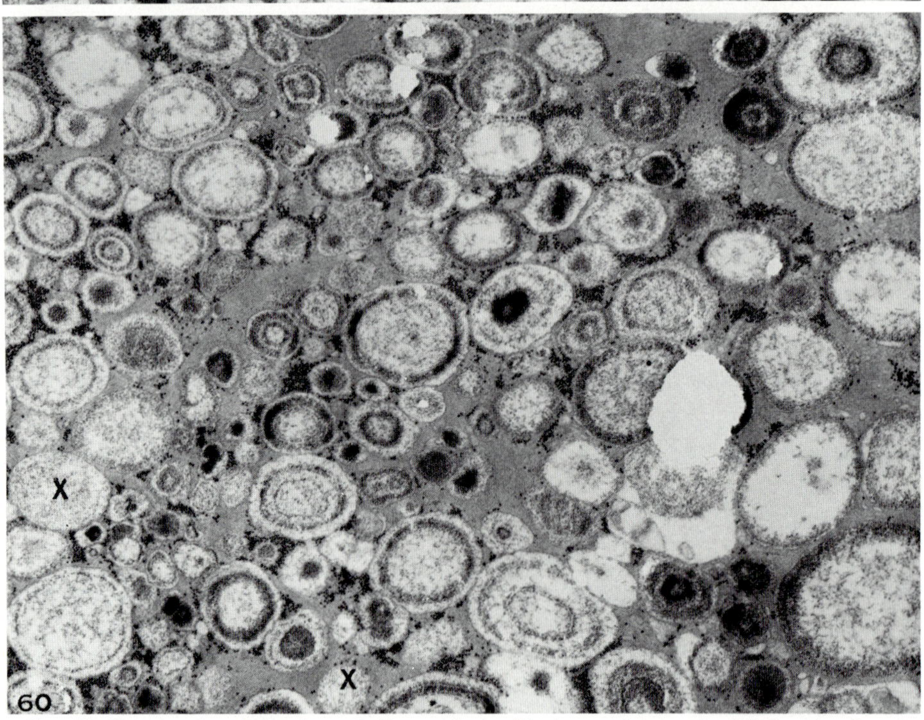

FIG. 59. Primary hypokalemic periodic paralysis. Electron micrograph of left side of region shown in Figure 38. Type 1 evolving vacuole composed predominantly of vesicles in various stages of mineralization. Some vesicles retain small amounts of amorphous material in their interior. Early mineralization occurs in vesicle marked by *asterisk*. Empty spaces are due to loss of part of section during exposure to electron beam. × 21,400.

FIG. 60. Primary hypokalemic periodic paralysis. Electron micrograph of abnormal fiber region shown in Figure 39. Type 1 evolving vacuole filled with myriads of vesicles in various stages of dilation and mineralization. Amorphous content of some vesicles is preserved (*X*) and indicates their sarcoplasmic reticulum origin. One to four concentric bands of dense precipitate (Liesegang rings) appear in majority of vesicles. × 10,900. (Reproduced from Engel[12] by permission of the Mayo Foundation.)

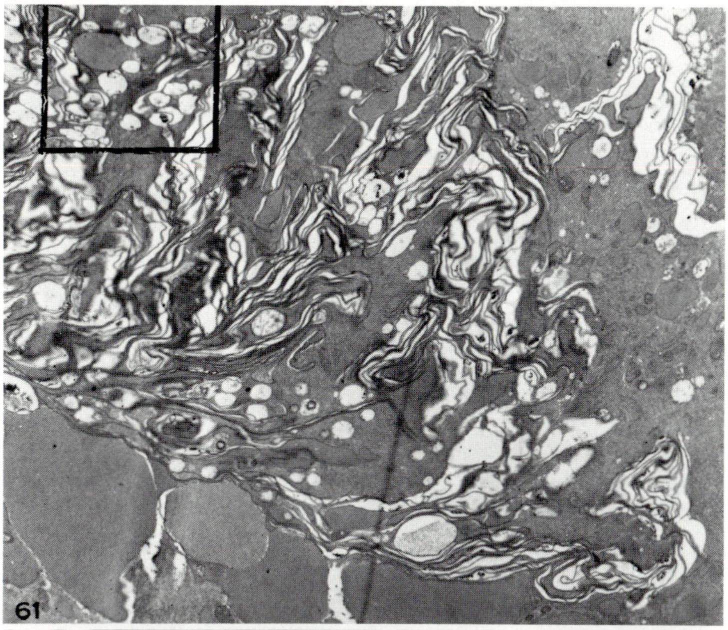


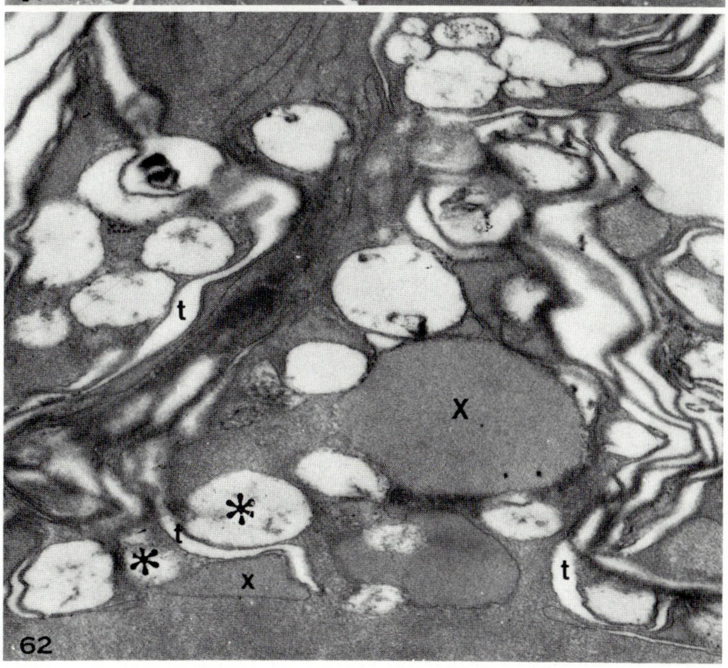


FIG. 61. Primary hypokalemic periodic paralysis. Type 2 evolving vacuole, composed of bizarre tubules, irregularly curving osmiophilic lamellae, and vesicles. × 5000. (Reproduced from Engel[12] by permission of the Mayo Foundation.)

FIG. 62. Primary hypokalemic periodic paralysis. Higher magnification of bracketed region in Figure 61 shows transitions from tubular membranes (*t*) to irregularly curving lamellae. Two vesicles (*asterisks*) are in triadic relation to tubules. Large, membrane-bound structure (*X*) contains amorphous material of intermediate density and undetermined character. Smaller space of same type (*x*) abuts on tubule. × 22,900. (Reproduced from Engel[12] by permission of the Mayo Foundation.)

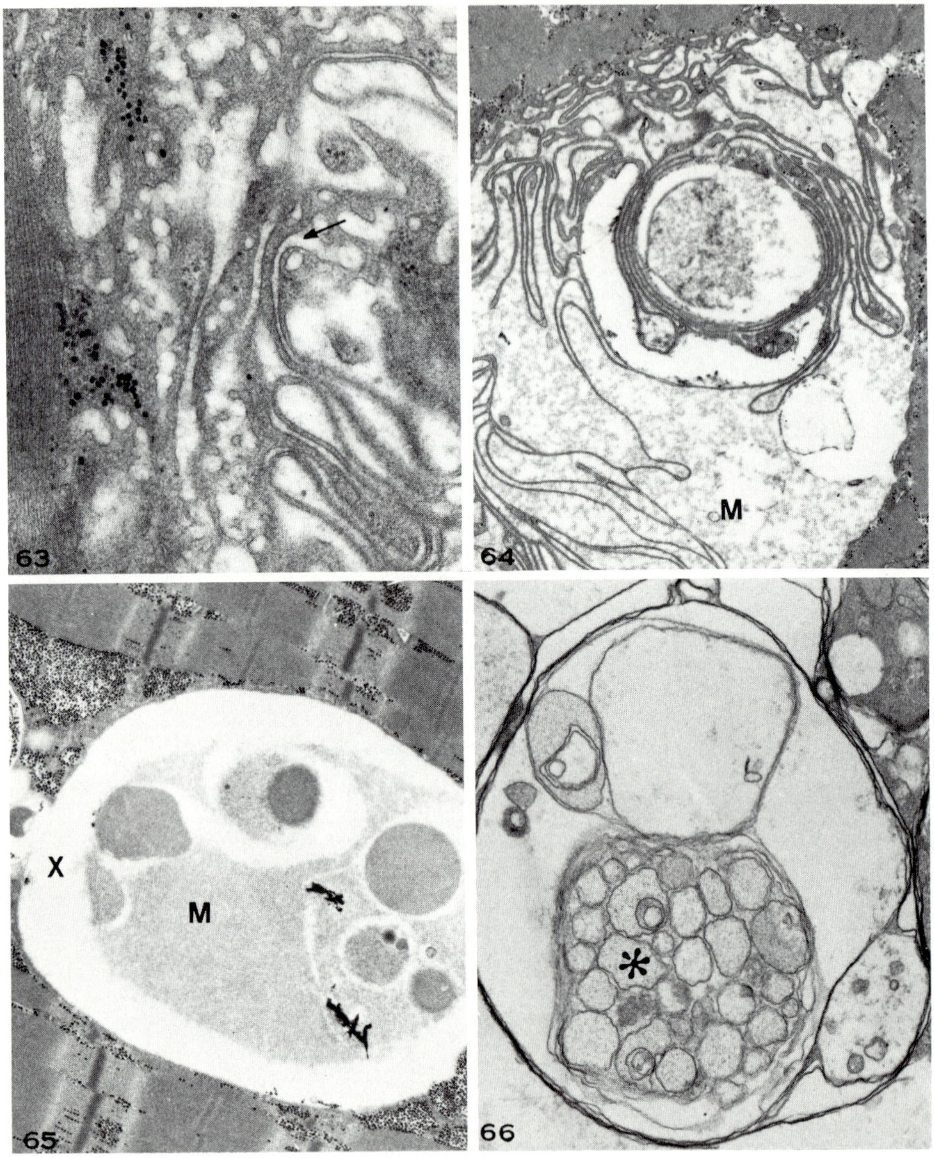


FIG. 63. Primary hypokalemic periodic paralysis. Intermediate-stage vacuole. Curving, bilaminar membranes in vacuole lumen can be traced to membranes limiting adjacent tubules or tubule-derived spaces at edge of vacuole (*arrow*). × 37,000. (Reproduced from Engel[12] by permission of the Mayo Foundation.)

FIG. 64. Primary hypokalemic periodic paralysis. Lumen of intermediate-stage vacuole contains irregularly curving and anastomosing bilaminar membranes. *M*, matrix. × 10,000. (Reproduced from Engel[12] by permission of the Mayo Foundation.)

FIG. 65. Primary hypokalemic periodic paralysis. Intermediate-stage vacuole contains dense bodies of unknown origin. Increase in density of matrix (*M*) could be due to condensation during fixation, resulting in empty space (*X*) between matrix and vacuolar rim. × 9600. (Reproduced from Engel[12] by permission of the Mayo Foundation.)

FIG. 66. Primary hypokalemic periodic paralysis. Cluster of vesicles, resembling dilated sarcoplasmic reticulum vesicles (*asterisk*), surrounded by frayed multilamellar membranes of increased density and other degraded structures appears in lumen of intermediate-stage vacuole. Matrix material is sparse or absent. × 13,500. (Reproduced from Engel[12] by permission of the Mayo Foundation.)

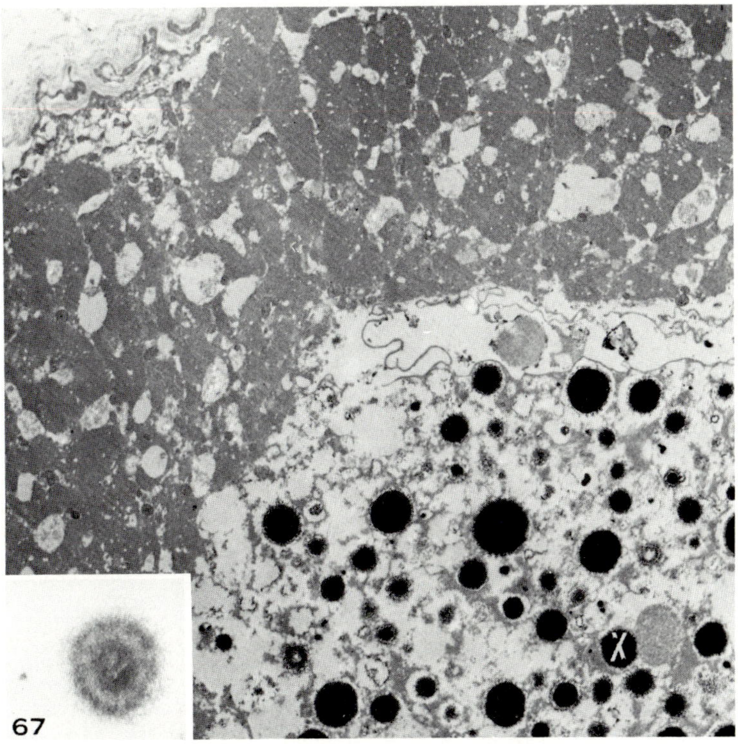


FIG. 67. Primary hypokalemic periodic paralysis. Intermediate-stage vacuole (unstained section) containing numerous round, electron-opaque mineral deposits of varying size, two small dense bodies, and poorly preserved matrix. Fiber around vacuole contains numerous small abnormal spaces. Deposit marked *X* is reprinted in *inset* with lower contrast, shorter exposure, and higher magnification (× 24,000). Liesegang rings are seen in deposit. × 7300. (Reproduced from Engel[12] by permission of the Mayo Foundation.)

contents and vacuolar matrix to intermingle. Thus, when the invaginated membrane was discontinuous, glycogen granules were scattered in the matrix and matrix material appeared in the invaginations. Myofibrils at the edge of some of these vacuoles were highly contracted. The sarcoplasmic bulging and eventual rupture into the matrix compartment of membrane-limited vacuoles must indicate a transient or permanent pressure gradient between the two adjacent spaces, with sarcoplasmic pressure higher. The findings also indicate that intravacuolar glycogen granules are not the cause of fluid and electrolyte shifts into vacuoles.

In each biopsy specimen examined, mature vacuoles, with or without sarcoplasmic invaginations, were more frequent than intermediate-stage vacuoles. Evolving vacuoles were rare in two specimens and could not be detected in another specimen. This observation could be explained if the mature vacuole represented a persistent (or relatively persistent) end-stage of the morphologic sequence. Conversely, if the evolving vacuoles had a brief existence, it would be infrequent at any given point in time.

*Abnormal small fibers*. Those less-than-20-$\mu$-wide fibers, which on phase-con-

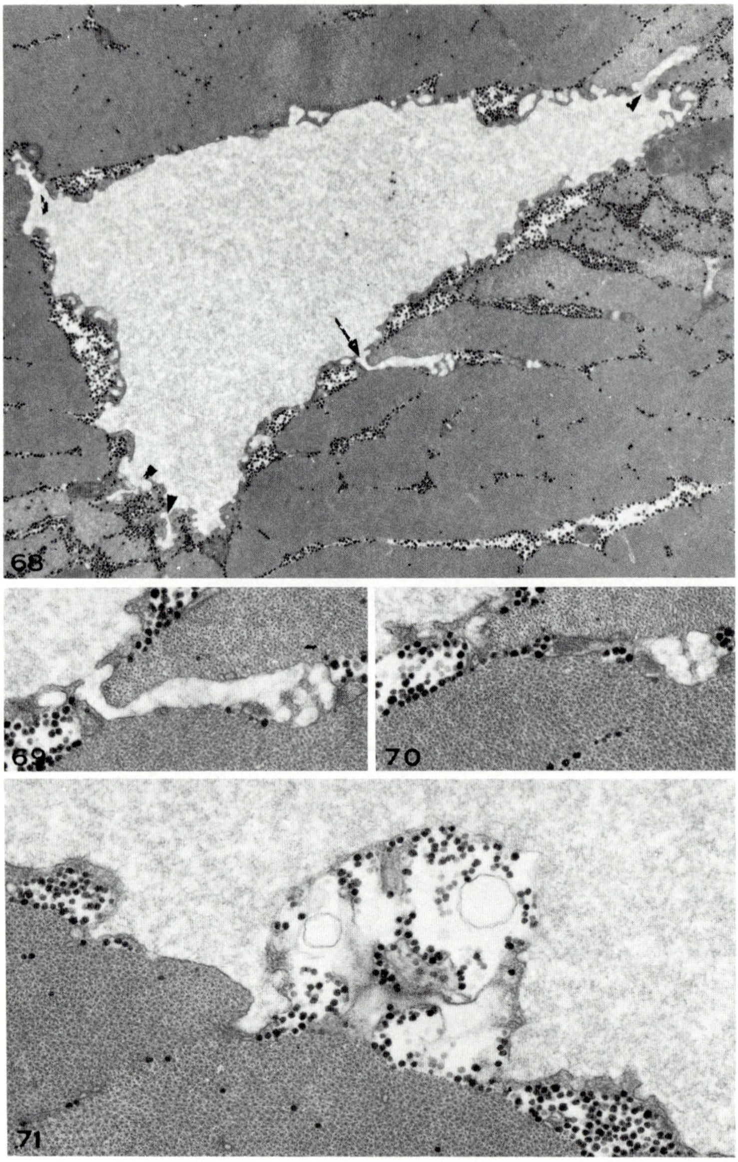


FIG. 68. Primary hypokalemic periodic paralysis. Mature vacuole contains only matrix and is membrane limited. Several T tubules or their derivatives communicate with vacuole (*arrowheads* and *arrow*). × 12,000. (Reproduced from Engel[12] by permission of the Mayo Foundation.)

FIG. 69. Primary hypokalemic periodic paralysis. Higher magnification of region marked by *arrow* in Figure 68. T tubule leads from T system network to vacuole. Note matrix material in T tubule lumen. × 32,000. (Reproduced from Engel[12] by permission of the Mayo Foundation.)

FIG. 70. Primary hypokalemic periodic paralysis. Serial section following that in Figure 69. T network now appears isolated from vacuole. Findings indicate that T networks around mature vacuoles can communicate with vacuolar lumen. × 32,000. (Reproduced from Engel[12] by permission of the Mayo Foundation.)

FIG. 71. Primary hypokalemic periodic paralysis. Small, dome-shaped invagination of sarcoplasm into mature vacuole. Dome contains glycogen granules and few circular or curving membranes. Convexity of dome is separated from vacuolar matrix by inwardly displaced vacuolar limiting membrane. Base of dome rests on vacuolar rim, and dome interior is continuous with myofilament space. × 31,200. (Reproduced from Engel[12] by permission of the Mayo Foundation.)

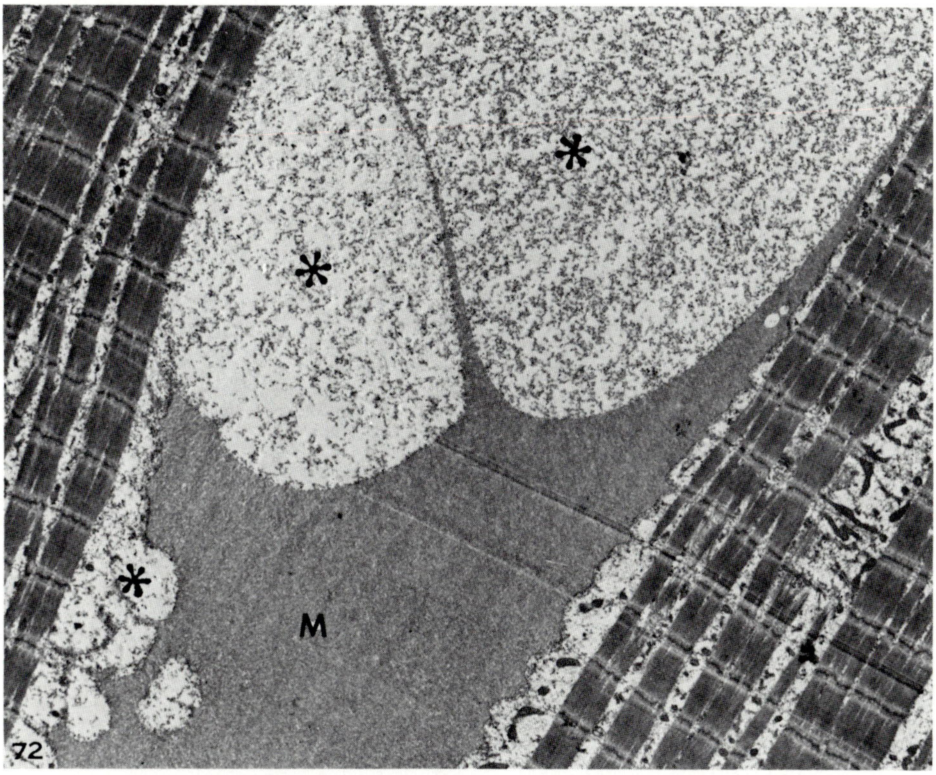


FIG. 72. Primary hypokalemic periodic paralysis. Large and small sarcoplasmic invaginations (*asterisks*) into matrix compartment (*M*) of mature vacuole contain glycogen granules and few mitochondria. This is an electron micrograph of region shown in periodic acid-Schiff-stained thick section in Figure 50. × 4100. (Reproduced from Engel[12] by permission of the Mayo Foundation.)

trast microscopy were filled with myriads of vesicles, contained the same type of vesicle aggregates noted in evolving vacuoles, with some vesicles showing mineralization and Liesegang ring formation (Figs. 73, 74, and 75). The fact that these fibers arose by splitting from larger fibers again suggested that fiber splitting, like sequestration in membrane-bound vacuoles and like exocytosis, might represent a mechanism whereby the fiber attempts to eliminate abnormal components. Portions of some small fibers showed regenerative activity. However, the vesicle aggregates in these fibers were unlike the SR vesicles seen in regenerating fibers on account of their size, abundance, and tendency to mineralize.

ELECTRON CYTOCHEMISTRY

Light microscopic examination of peroxidase-loaded blocks of tissue incubated in diaminobenzidine medium indicated that peroxidase had entered parts or all of most vacuoles in those fibers that were also surrounded by reaction product (Fig. 76). In the electron microscope, in such fibers the reaction product was observed in T tubules, T system networks, vesicular dilations of T tubules, T tubules and T networks communicating with intermediate-stage and mature vacuoles, and matrix of these two types of vacuoles (Figs. 77 and 78).

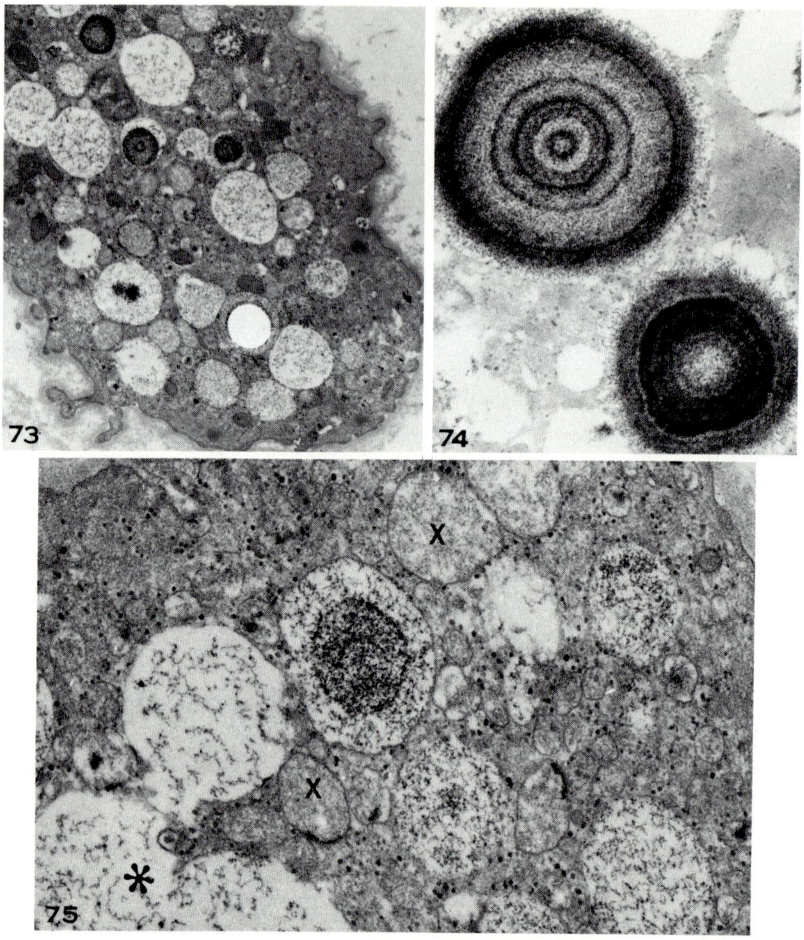


FIG. 73. Primary hypokalemic periodic paralysis. Sarcoplasmic reticulum vesicle aggregate in small fiber. Vesicles are in various stages of dilation and mineralization and some display Liesegang rings. × 8000. (Reproduced from Engel[12] by permission of the Mayo Foundation.)

FIG. 74. Primary hypokalemic periodic paralysis. Unstained section displays multiple concentric Liesegang rings in two mineralized vesicles. × 16,700. (Reproduced from Engel[12] by permission of the Mayo Foundation.)

FIG. 75. Primary hypokalemic periodic paralysis. Serial section to region shown in Figure 73. Note various stages of vesicle mineralization. Some vesicles (*X*) retain amorphous content of normal sarcoplasmic reticulum. Coalescing vesicles are indicated by *asterisk*. Glycogen granules appear between vesicles. × 27,600. (Reproduced from Engel[12] by permission of the Mayo Foundation.)

Reaction product was excluded from those vacuolar areas occupied by organelle remnants or by sarcoplasmic invaginations (Fig. 78). Little reaction product was found in those vacuoles where sarcoplasmic invaginations were limited by perforated membranes, possibly due to escape of peroxidase from the vacuoles into the myofilament space. The prompt entry of peroxidase-labeled extracellular fluid into the matrix compartment of intermediate-stage and mature vacuoles, which were the prevalent vacuoles, clearly establishes that most of the vacuolar volume in -KP (Pri) is filled with extracellular fluid. The intermediate-

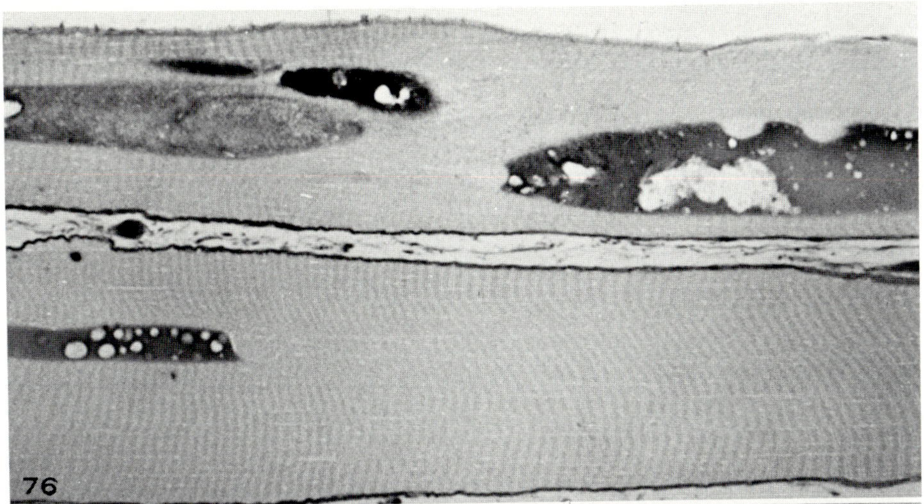


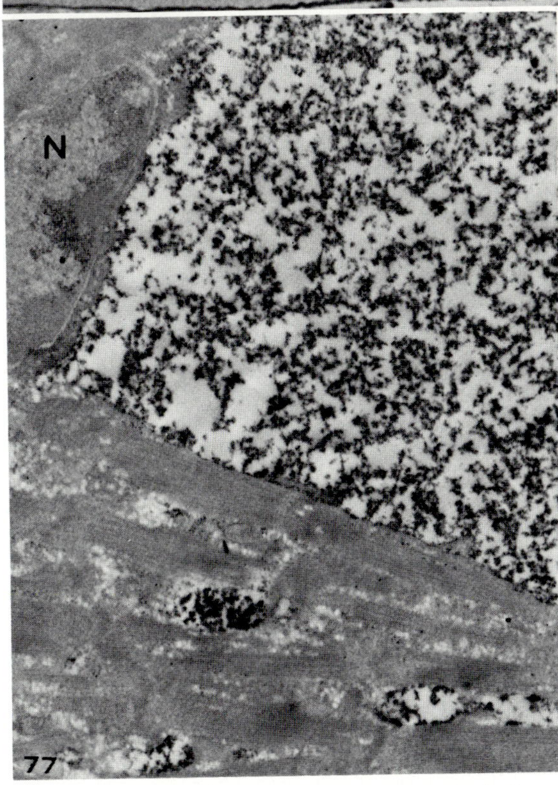


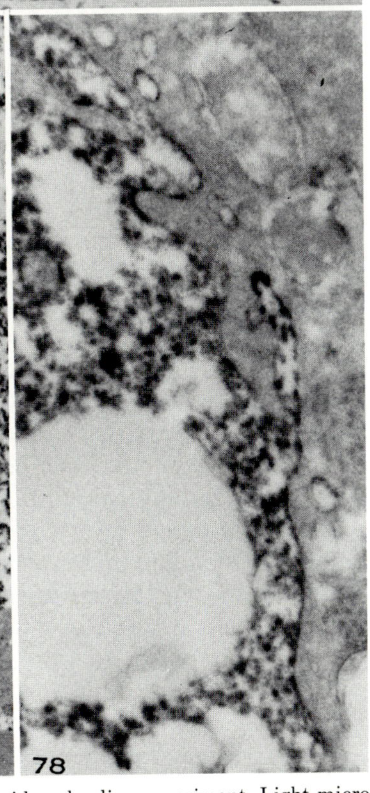


FIG. 76. Primary hypokalemic periodic paralysis. Peroxidase-loading experiment. Light micrograph, unstained Epon section. All vacuoles in two fibers contain peroxidase, but reaction product is excluded from round or irregularly lobulated regions of vacuoles. × 430. (Reproduced from Engel[12] by permission of the Mayo Foundation.)

FIG. 77. Primary hypokalemic periodic paralysis. Peroxidase-loaded fiber (unstained section). Vacuole and dilated T elements around it are filled with peroxidase. *N*, nucleus. × 15,700. (Reproduced from Engel[12] by permission of the Mayo Foundation.)

FIG. 78. Primary hypokalemic periodic paralysis. Peroxidase-loaded fibers (unstained sections). Peroxidase-containing tubular profiles communicate with peroxidase-containing vacuolar compartment. × 43,000. (Reproduced from Engel[12] by permission of the Mayo Foundation.)

stage and mature vacuoles could thus act as large conduits for fluid and electrolyte movements in affected fibers, and the reason for their persistence, once they are formed, might be that they subserve such a function. The findings also explain the previously observed and heretofore puzzling abnormally high muscle chloride space in -KP(Pri). The nature of the vacuolar matrix material is not known, but it may simply represent a proteinaceous precipitate.

## SUMMARY

The myopathies described fall into two major groups. The first group includes chloroquine myopathy, acid maltase deficiency (AMD), and primary hypokalemic periodic paralysis (-KP(Pri)). Membrane-bound vacuoles arise, an autophagic mechanism becomes excited, multiple organelles are affected, and vacuolar morphology changes during evolution of the pathologic process. The second group is represented by lipid storage myopathy. In this case, the accumulating material is non-membrane-bound, autophagy does not occur, and only mitochondria react pathologically; the vacuoles are stereotyped and, except for coalescing, do not change.

The initiating factors are clearly divergent in the three autophagic myopathies. Following chloroquine treatment, autophagy is excited artificially. In AMD, an inborn lysosomal enzyme defect initiates the process, but the strength of the initiating factor or of the factors which compensate for it differs in the infantile, childhood, and adult types of AMD. In -KP(Pri) the initiating factor is thought to be related to an abnormality of electrically excitable membranes, but the character of the abnormality is not yet clearly defined.

The content of early vacuoles is also different in the three autophagic myopathies. Soon after the start of chloroquine therapy, normal organelles or small myeloid bodies appear in membrane-bound spaces. In AMD, the smallest membrane-bound sacs contain glycogen. In -KP(Pri), relatively large fiber regions unlimited by membrane and containing an abundance of altered organelles derived from the SR or T system or other organelles arise before trapping into a membrane-bound vacuole takes place.

In each of the three autophagic myopathies, the T system acts as a source of the membrane which limits vacuoles. In addition, the Golgi apparatus also provides membrane for trapping glycogen in AMD. A similar role for the Golgi system has not been found in -KP(Pri) or in chloroquine myopathy.

The site of synthesis and storage of digestive enzymes in muscle is not yet known. However, it is of considerable interest that, in AMD, delivery of acid phosphatase and hence presumably of other lysosomal enzymes to the autophagic vacuole occurs via T system-derived vesicles. Further studies are required to elucidate the source and delivery of lysosomal enzymes to vacuoles in chloroquine myopathy and in -KP(Pri).

The degradation of vacuolar contents is partial in chloroquine myopathy and, as far as glycogen is concerned, unsuccessful in AMD. On the other hand, complete digestion of vacuolar contents can occur in -KP(Pri). Perhaps it is for this reason that exocytosis is seen in AMD and chloroquine myopathy but is not notable in -KP(Pri). Fiber splitting is common in each of the three myopathies and, like sequestration in a vacuole, may represent an attempt to eliminate abnormal fiber components.

Degraded, and presumably liquefied, vacuolar contents can exit via T system networks and tubules that connect the vacuole to the extracellular space while extracellular fluid can enter vacuoles through the same channels. Thus, it is not surprising that a large fraction of the vacuolar volume in -KP(Pri) fills with peroxidase-loaded extracellular fluid, while only small amounts of such fluid enter vacuoles in AMD and in chloroquine myopathy, in which degradation of vacuolar contents is less complete.

Since the pathologic process that leads to fiber vacuolation is either continuous or recurs repeatedly and the vacuoles, once formed, may continue to evolve, a variety of vacuoles in different stages of evolution can be encountered in a given biopsy. It also follows that, at any given time, the prevalence of a stage in vacuole evolution is proportional to the length of time that that stage persists. Consequently, repeated sampling and protracted examination of biopsy specimens are required to observe steps in vacuole evolution that exist only briefly.

## REFERENCES

1. Abraham, R., Hendy, R., and Grasso, P.: Formation of myeloid bodies in rat liver lysosomes after chloroquine administration. *Exp. Mol. Pathol. 9:* 212–229, 1968.
2. Adams, R. D., Denny-Brown, D., and Pearson, C. M.: *Diseases of Muscle: A Study in Pathology*, 2nd ed. New York, Harper & Row, 1962.
3. Bancroft, J. D.: *An Introduction to Histochemical Technique*. London, Butterworth & Co., Ltd., 1967.
4. Barka, T.: Electron histochemical localization of acid phosphatase activity in the small intestine of mouse. *J. Histochem. Cytochem. 12:* 229–238, 1964.
5. Biczyskowa, W., Fidzianska, A., and Jedrzejowska, H.: Light and electron microscopic study of the muscles in hypokalemic periodic paralysis. *Acta Neuropathol. (Berl.) 12:* 329–338, 1969.
6. Bradley, W. G. Ultrastructural changes in adynamia episodica hereditaria and normokalaemic familial periodic paralysis. *Brain 92:* 379–390, 1969.
7. Bradley, W. G., Hudgson, P., Gardner-Medwin, D., and Walton, J. N.: Myopathy associated with abnormal lipid metabolism in skeletal muscle. *Lancet 1:* 495–498, 1969.
8. De Duve, C., and Wattiaux, R.: Functions of lysosomes. *Annu. Rev. Physiol. 28:* 435–492, 1966.
9. Engel, A. G.: Electron microscopic observations in primary hypokalemic and thyrotoxic periodic paralyses. *Mayo Clin. Proc. 41:* 797–808, 1966.
10. Engel, A. G.: Acid maltase deficiency in adult life. Morphologic and biochemical data in 3 cases of a syndrome simulating other myopathies. *Excerpta Med. Int. Congr. Ser. 199:* 236–245, 1969.
11. Engel, A. G.: Acid maltase deficiency in adults. Studies in four cases of a syndrome which may mimic muscular dystrophy or other myopathies. *Brain 93:* 599–616, 1970.
12. Engel, A. G.: Evolution and content of vacuoles in primary hypokalemic periodic paralysis. *Mayo Clin. Proc. 45:* 774–814, 1970.
13. Engel, A. G.: Hypokalemic and hyperkalemic periodic paralyses. In *The Cellular and Molecular Basis of Neurologic Disease*, ed. by Shy, G. M., Goldensohn, E. S., and Appel, S. H. Philadelphia, Lea & Febiger, in press.
14. Engel, A. G., and Dale, A. J. D.: Autophagic glycogenosis of late onset with mitochondrial abnormalities. Light and electron microscopic observations. *Mayo Clin. Proc. 43:* 233–279, 1968.

14a. Engel, A. G., Gomez, M. R., Seybold, M. E., and Lambert, E. H.: The spectrum and diagnosis of acid maltase deficiency. *Neurology (Minneap.) 23:* 95–106, 1973.

15. Engel, A. G., and Lambert, E. H.: Calcium activation of electrically inexcitable muscle fibers in

primary hypokalemic periodic paralysis. *Neurology (Minneap.) 19:* 851–858, 1969.

16. Engel, A. G., Lambert, E. H., Rosevear, J. W., and Tauxe, W. N.: Clinical and electromyographic studies in a patient with primary hypokalemic periodic paralysis. *Am. J. Med. 38:* 626–640, 1965.

16a. Engel, A. G., and Macdonald, R. D.: Ultrastructural reactions in muscle diseases and their light microscopic correlates. *Excerpta Med. Int. Congr. Ser. 199:* 71–89, 1969.

16b. Engel, A. G. and Siekert, R. G.: Lipid storage myopathy responsive to prednisone. *Arch. Neurol. 27:* 174–181, 1972.

17. Engel, A. G., and Tice, L. W.: Cytochemistry of phosphatases of the sarcoplasmic reticulum. I. Biochemical studies. *J. Cell Biol. 31:* 473–487, 1966.

18. Engel, W. K.: Mitochondrial aggregates in muscle disease. *J. Histochem. Cytochem. 12:* 46–48, 1964.

19. Engel, W. K., Bishop, D. W., and Cunningham, G. G.: Tubular aggregates in type II muscle fibers. Ultrastructural and histochemical correlation. *J. Ultrastruct. Res. 31:* 507–525, 1970.

20. Fedorko, M. E.: Effect of chloroquine on morphology of leukocytes and pancreatic exocrine cells from the rat. *Lab. Invest. 18:* 27–37, 1968.

21. Fedorko, M. E., Hirsch, J. G., and Cohn, Z. A.: Autophagic vacuoles produced *in vitro*. I. Studies on cultured macrophages exposed to chloroquine. *J. Cell Biol. 38:* 377–391, 1968.

22. Fedorko, M. E., Hirsch, J. G., and Cohn, Z. A.: Autophagic vacuoles produced *in vitro*. II. Studies on the mechanism of formation of autophagic vacuoles produced by chloroquine. *J. Cell Biol. 38:* 392–402, 1968.

23. Garcin, R., Rondot, P., and Fardeau, M.: Sur les accidents neuro-musculaires et en particulier sur une "myopathie vacuolaire" observés au cours d'un traitement prolongé par la chloroquine. Amélioration rapide après arrêt du médicament. *Rev. Neurol. (Paris) 111:* 177–195, 1964.

24. Graham, R. C., Jr., and Karnovsky, M. J.: The early stages of absorption of injected horseradish peroxidase in the proximal tubules of mouse kidney. Ultrastructural cytochemistry by a new technique. *J. Histochem. Cytochem. 14:* 291–302, 1966.

25. Gruner, J.-E.: Anomalies du réticulum sarcoplasmique et prolifération de tubules dans le muscle d'une paralysie périodique familiale. *C. R. Soc. Biol. (Paris) 160:* 193–196, 1966.

26. Hers, H. G.: α-Glucosidase deficiency in generalized glycogen-storage disease (Pompe's disease). *Biochem. J. 86:* 11–16, 1963.

27. Hers, H. G., and van Hoof, F.: Glycogen storage diseases. Type II and Type VI glycogenosis. In *Carbohydrate Metabolism and Its Disorders*, ed. by Dickens, F., Randle, P. J., and Whelan, W. J., vol. 2, pp. 151–168. New York, Academic Press Inc., 1968.

28. Howes, E. L., Jr., Price, H. M., Pearson, C. M., and Blumberg, J. M.: Hypokalemic periodic paralysis. Electromicroscopic changes in the sarcoplasm. *Neurology (Minneap.) 16:* 242–256, 1966.

29. Hudgson, P., Gardner-Medwin, D., Worsfold, M., Pennington, R. J. T., and Walton, J. N.: Adult myopathy from glycogen storage disease due to acid maltase deficiency. *Brain 91:* 435–462, 1968.

30. Ishikawa, H.: Formation of elaborate networks of T system tubules in cultured skeletal muscle with special reference to the T system formation. *J. Cell Biol. 38:* 51–66, 1968.

31. Karnovsky, M. J.: The ultrastructural basis of capillary permeability studies with peroxidase as a tracer. *J. Cell Biol. 35:* 213–236, 1967.

32. Macdonald, R. D., and Engel, A. G.: Experimental chloroquine myopathy. *J. Neuropathol. Exp. Neurol. 29:* 479–499, 1970.

33. Macdonald, R. D., Rewcastle, N. B., and Humphrey, J. G.: The myopathy of hyperkalemic periodic paralysis. An electron microscopic study. *Arch. Neurol. 19:* 274–283, 1968.

34. Macdonald, R. D., Rewcastle, N. B., and Humphrey, J. G.: Myopathy of hypokalemic periodic paralysis. An electron microscopic study. *Arch. Neurol. 20:* 565–585, 1969.

35. Norris, F. H., Jr., Panner, B. J., and Stormont, J. M.: Thyrotoxic periodic paralysis. Metabolic and ultrastructural studies. *Arch. Neurol. 19:* 88–98, 1968.

36. Odor, D. L., Patel, A. N., and Pearce, L. A.: Familial hypokalemic periodic paralysis with permanent myopathy. A clinical and ultrastructural study. *J. Neuropathol. Exp. Neurol. 26:* 98–114, 1967.

37. Olivarius, B. F., and Christensen, E.: Histopathological muscular changes in familial, periodic paralysis. *Acta Neurol. Scand. 41:* 1–18, 1965.
38. Pearce, G. W.: Electron microscopy in the study of muscular dystrophy. In *Muscular Dystrophy in Man and Animals*, ed. by Bourne, G. H., and Golarz, M. N., pp. 160–176. New York, Hafner Publishing Company, Inc., 1963.
39. Pearse, A. G. E.: *Histochemistry: Theoretical and Applied*, 2nd ed. Boston, Little, Brown & Company, 1961.
40. Pearse, A. G. E., and Johnson, M.: Histochemistry in the study of normal and diseased muscle, with special reference to myopathy with tubular aggregates. *Excerpta Med. Int. Congr. Ser. 199:* 25–32, 1970.
41. Pearson, C. M.: The periodic paralyses. Differential features and pathological observations in permanent myopathic weakness. *Brain 87:* 341–354, 1964.
42. Pellegrino, C., and Franzini, C.: An electron microscope study of denervation atrophy in red and white skeletal muscle fibers. *J. Cell Biol. 17:* 327–349, 1963.
43. Price, H. M.: Mitochondrial myopathies in man? A review of the evidence. *Excerpta Med. Int. Congr. Ser. 147:* 341–350, 1966.
44. Price, H. M., Gordon, G. R., Munsat, T. L., and Pearson, C. M.: Myopathy with atypical mitochondria in type I skeletal muscle fibers. A histochemical and ultrastructural study. *J. Neuropathol. Exp. Neurol. 26:* 475–497, 1967.
45. Rewcastle, N. B., and Humphrey, J. G.: Vacuolar myopathy. Clinical, histochemical, and microscopic study. *Arch. Neurol. 12:* 570–582, 1965.
46. Santa, T., and Engel, A. G.: Unpublished data.
47. Sant'Agnese, P. A. di: Diseases of glycogen storage with special reference to the cardiac type of generalized glycogenosis. *Ann. N. Y. Acad. Sci. 72:* 439–450, 1959.
48. Schutta, H. S., and Armitage, J. L.: Thyrotoxic hypokalemic periodic paralysis. A fine structure study. *J. Neuropathol. Exp. Neurol. 28:* 321–326, 1969.
49. Schutta, H. S., Kelly, A. M., and Zacks, S. I.: Necrosis and regeneration of muscle in paroxysmal idiopathic myoglobinuria. Electron microscopic observations. *Brain 92:* 191–200, 1969.
50. Shy, G. M., Gonatas, N. K., and Perez, M.: Two childhood myopathies with abnormal mitochondria. I. Megaconial myopathy. II. Pleoconial myopathy. *Brain 89:* 133–158, 1966.
51. Shy, G. M., Wanko, T., Rowley, P. T., and Engel, A. G.: Studies in familial periodic paralysis. *Exp. Neurol. 3:* 53–121, 1961.
52. Stern, K. H.: Liesegang rings. In *Encyclopaedic Dictionary of Physics*, ed. by Thewlis, J., vol. 4, pp. 276–277. New York, Pergamon Press, Inc., 1961.
53. Tice, L. W., and Engel, A. G.: Cytochemistry of phosphatases of the sarcoplasmic reticulum. II. *In situ* localization of the Mg-dependent enzyme. *J. Cell Biol. 31:* 489–499, 1966.
54. Van Bogaert, L., Kissel, P., Schmitt, Y., Martin, G. G., Claes, C., Due, M., Royer, R., Barrucana, D., Grignon, J., and Floquet, Y.: Myopathic syndrome in the course of Gamstorp's disease. *Acta Neurol. Belg. 67:* 133–142, 1967.

## Chapter 17

# Normal Ultrastructural Aspect of Human Motor End-plate and Its Pathologic Modifications

MICHEL FARDEAU

The term of "plaque terminale" (end-plate) is more than a century old; it was introduced by Rouget[58] to describe the flattened heap of granular nucleated substance located at the neuromuscular junction. The term "motorische endplatte" (motor end-plate) appeared the next year, in 1863, with Krause's[41] description of the terminal nerve branching in cat ocular muscle. However, it took nearly a century before a clear picture developed of the relationship between the terminal nerves and the muscle fiber. It is well known that long and ardent controversies arose among histologists after introduction of the neurofibrillar staining methods. Crucial progress was achieved with the description by Couteaux,[17] using postvital staining, of a special lamellated structure located at the boundary between nerve and muscle, the subneural apparatus ("appareil sous-neural"). Based on biochemical and experimental data, Couteaux proposed that acetylcholinesterase is located at the level of the subneural apparatus. The histochemical demonstration of this hypothesis was advanced by Koelle and Friedenwald[39a] using acetylthiocholine as the substrate.

Early electron microscopic studies by Palade and Palay,[53] Robertson,[57] and Reger[55] brought a brilliant confirmation of the structural discontinuity between the terminal nerve and the muscle fiber. They showed the absence of any interpositioned teloglia between these two cells and confirmed the presence of a subneural apparatus formed by a succession of narrow infoldings of the sarcolemma. The first pictures of the ultrastructure of normal human motor end-plates were performed by de Harven and Coërs[35].

While basic knowledge on end-plate structure was growing, pathologic changes in terminal innervation were studied in light microscopy by Coërs and Woolf[13] with methylene blue and Koëlle's technique. Similarly, combined use of silver impregnation and histochemical technique can be useful for these studies.[22, 26, 37, 43] These studies have emphasized the diagnostic value of these techniques in human neuromuscular pathology, especially in two conditions: (1) the early detection of neural changes in a nonatrophic tissue, and (2) the differentiation between neurogenic and myogenic processes in severe, destructive atrophies. Many other stimulating points were brought out by these studies, such as changes in the terminal innervation pattern in the muscular dystrophies and the abnormal elongation of end-plates in myasthenia gravis.

Electron microscopic studies of pathologic end-plate alterations are to date very few. The rare papers including end-plate pictures do not always take into account normal, physiologic, or artifactual modifications of these structures. Nevertheless, it seems difficult to ignore the study of these structures in neuromuscular pathology.

## METHODS

Ultrastructural study of motor end-plates in human neuromuscular biopsies implies three successive stages:

1. Locating the innervation zone during the biopsy either by electrical stimulation[13, 15] or, more easily, by a mechanical detection. We have used the following technique. During dissociation of the muscular fascicles, contact of the scalpel with terminal nerves causes a brief twitch contraction. A 7- to 8-mm long muscle strip is cut off in this zone, after *in situ* fixation to prevent any contraction or ischemic artifact during excision (the subjacent fascicles are protected by a metallic gutter).

2. Locating the terminal nerve bundles in the muscle strip. After fixation, or postfixation, by the osmium tetroxide, terminal nerves can be easily located under a stereomicroscope, owing to the osmic impregnation of their myelin sheaths. Small blocks for electron microscopic study are trimmed in the branching zone of the terminal nerves, each block containing a small nerve in its middle.

3. After Araldite embedding, nerve trunks and end-plates are detected on 0.5- to 1-$\mu$ thick sections, with phase contrast or after short staining by toluidine blue; serial or semiserial thin sections are then cut from the blocks.

This technique allows a regular electron microscopic study of end-plates on pathologic muscle. It is time consuming and certainly cannot be used as a routine technique. The main problem occurs in advanced denervation atrophies, in which it may be extremely difficult to localize the tiny, degenerated intramuscular nerve trunks.

This paper includes results of 30 biopsies; two of these biopsies were considered normal on clinical, light and electron microscopic basis. Many of these biopsies were examined in collaboration with J. Godet-Guillain.

## NORMAL STRUCTURE OF HUMAN END-PLATE

To appreciate the details seen in electron micrographs, it is necessary to give a short summary of the general lay-out of the end-plate as shown by light microscopy. The myelinated motor nerve approaches the muscle fiber and ramifies at a short distance from it, giving off a bunch of tiny nonmyelinated terminal branches, the extremities of which make contact with the muscle surface (Fig. 1). Facing the nerve endings, the muscle fiber presents a rounded heap of granular sarcoplasm, containing several large nuclei ("fundamental" nuclei of Ranvier) and numerous mitochondria; its surface forms small cups or gutters in which the nerve endings lie. At the axoplasm-sarcoplasm interface is

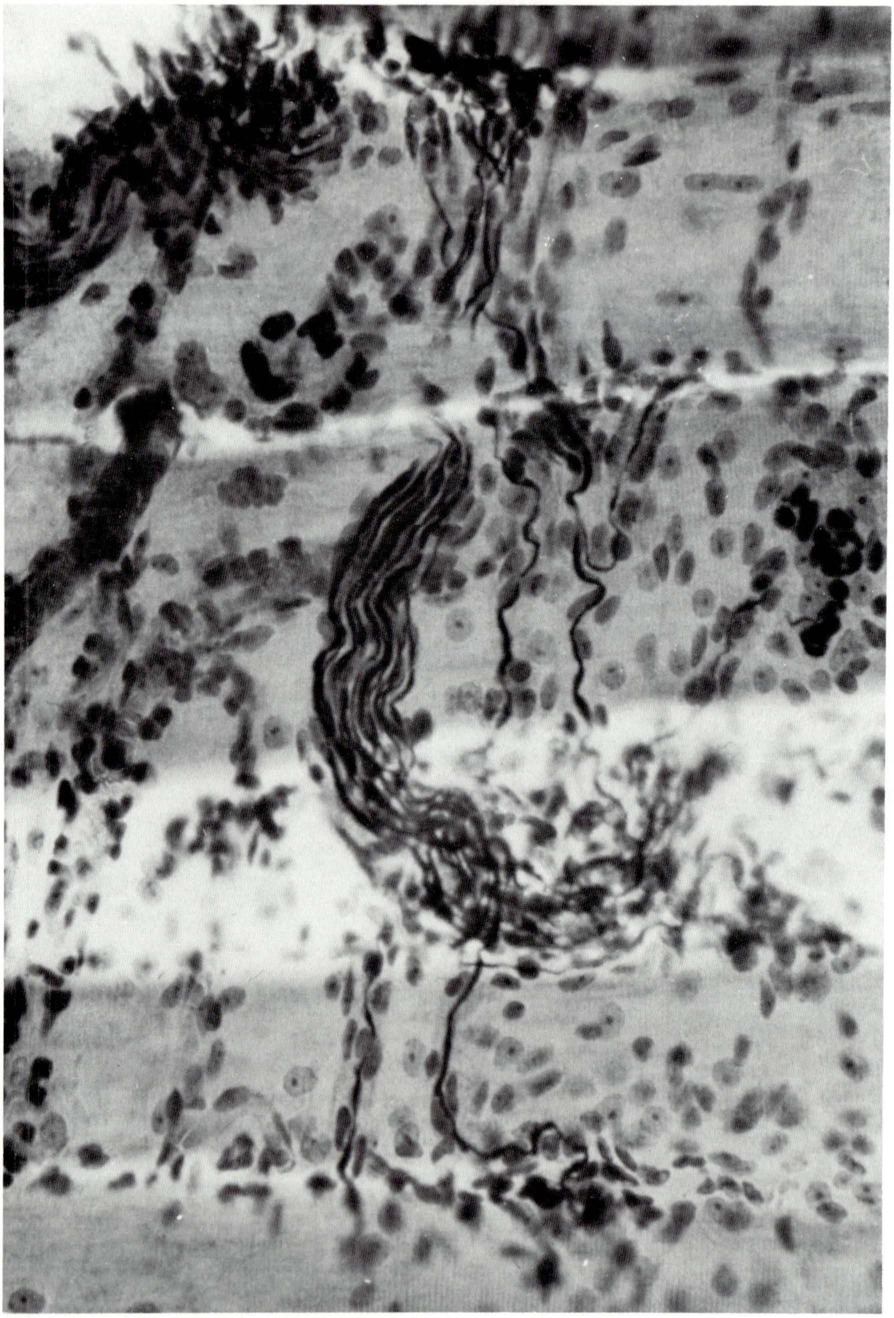

FIG. 1. Normal terminal innervation pattern in human muscle. Silver impregnation Bodian combined with Koelle technique. Four end-plates are visible in the picture, with their tiny terminal nerves. × 400.

developed the subneural apparatus, whose ribbon-shaped lamellae are perpendicular to the interface. This apparatus shows up with specific cholinesterase-staining methods (Fig. 2).[21, 39a]

In electron microscopy, each digitation of the end-plate has the classical three-cell arrangement (Fig. 3): (1) the nerve ending lies in a depression of the muscle surface, the synaptic trough; (2) a teloglial Schwann process forms a coat about the nerve ending, closing the synaptic trough. (3) facing the nerve

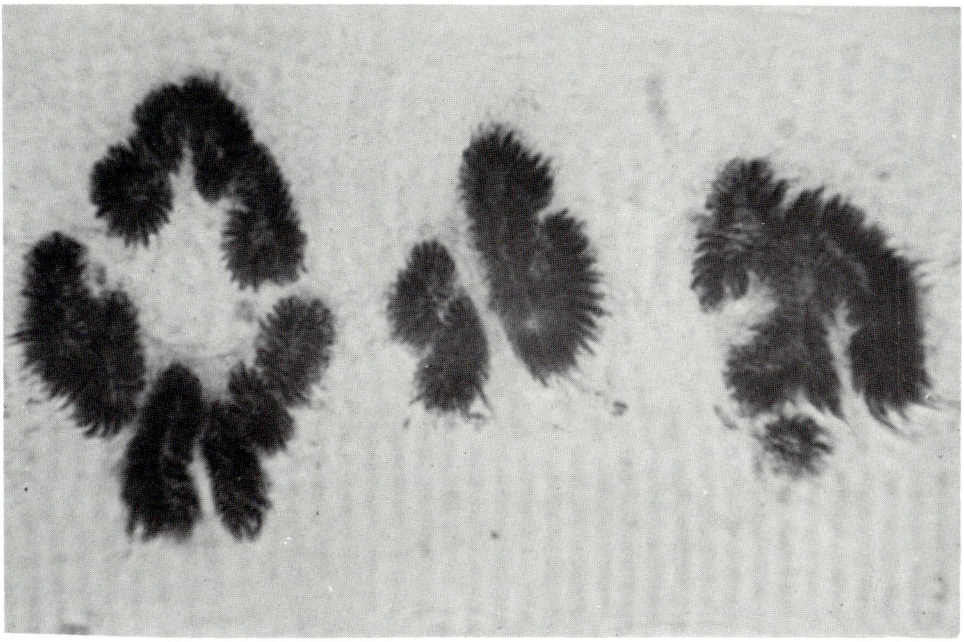

FIG. 2. Normal subneural apparatus (human muscle) showing the regular disposal of its lamellae. Koelle's technique. × 2000.

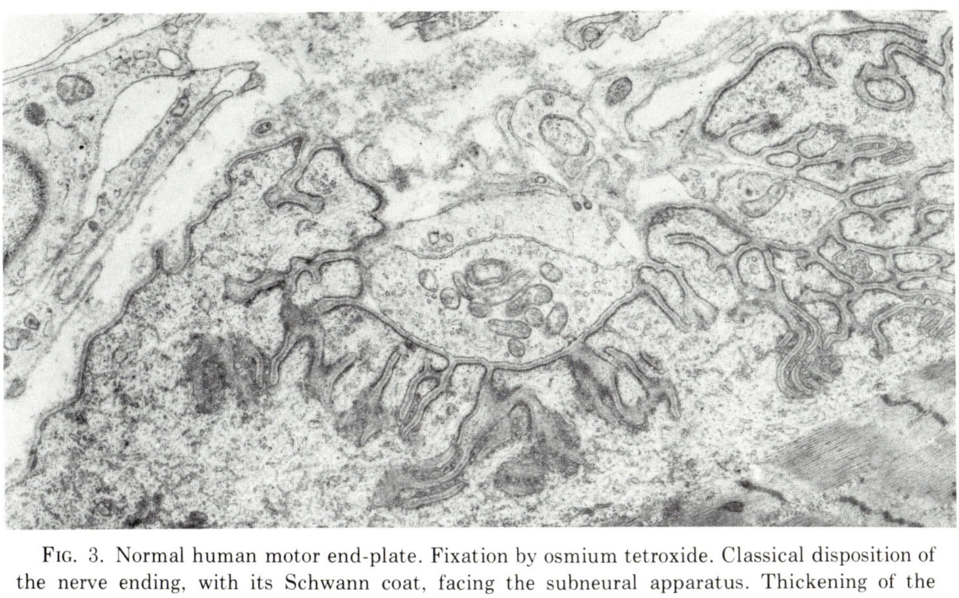

FIG. 3. Normal human motor end-plate. Fixation by osmium tetroxide. Classical disposition of the nerve ending, with its Schwann coat, facing the subneural apparatus. Thickening of the postsynaptic membrane. × 16,000.

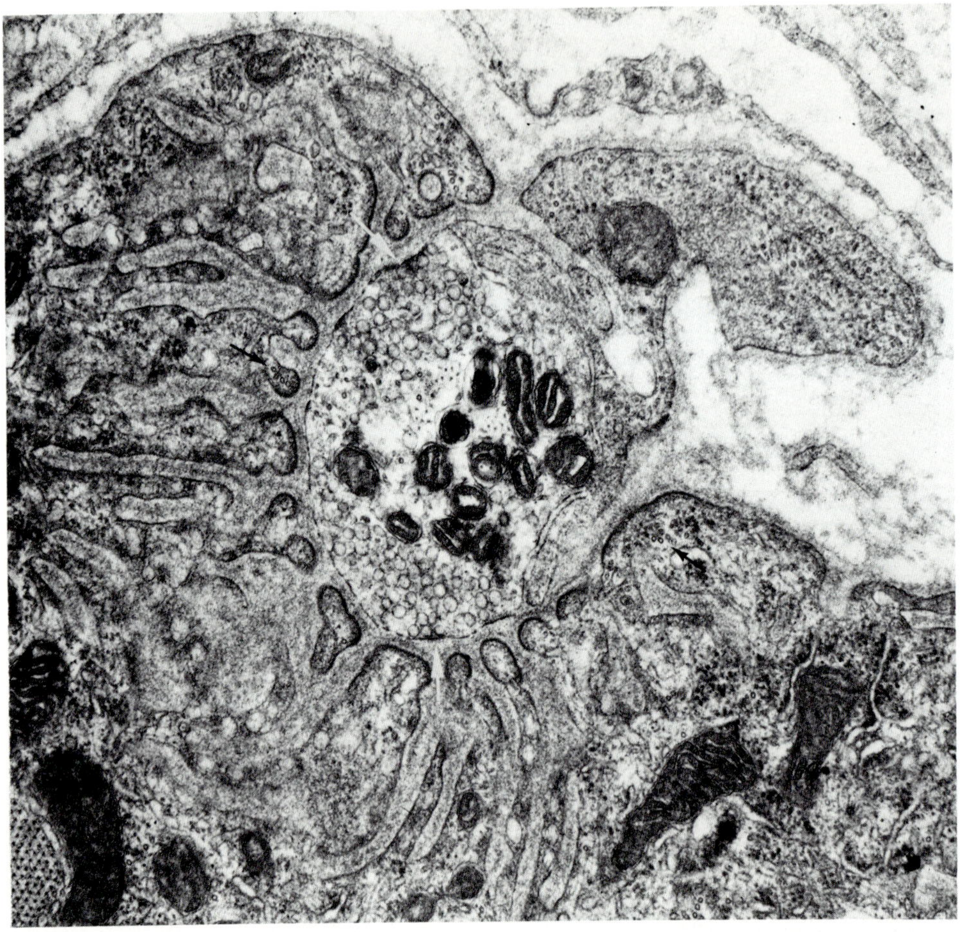

FIG. 4. Guinea pig musculus peroneus. Fixation by glutaraldehyde perfusion and postfixation by osmium tetroxide. Peripheral disposition of the synaptic vesicles, with agglutination against dense spots of the presynaptic membrane (*white arrows*) facing the aperture of junctional folds. Thickening of the postsynaptic membrane in the proximal part of the subneural apparatus. Numerous vesicular profiles in the sarcoplasm along the junctional folds, amidst a lot of microtubules (*black arrows*), filaments, and granules. × 31,000.

ending, the muscle fiber membrane is thrown into multiple folds. (4) the interval between axolemma and sarcolemma forms the synaptic cleft, which is divided into primary and secondary synaptic clefts; (5) the secondary clefts, 1 to 2 $\mu$ long, plunge into the sole-plate sarcoplasm.[18, 28]

*The nerve ending* contains mitochondria, smooth reticulum profiles, microtubules, filaments, and synaptic vesicles. Schematically, microtubules and filaments occupy the core, elongated mitochondria surround this core, and vesicles are particularly numerous at the periphery of the nerve ending. Such a distribution clearly appears on guinea pig perfused muscles (Fig. 4).

The axolemmal membrane often looks crenated. From place to place, it presents a dense thickened spot, and vesicles are concentrated in small stacks focused on these dense spots.[47] This aspect is clear after aldehyde perfusion on guinea pig motor end-plates (Fig. 4). Similar aspects have been described by Birks, Huxley, and Katz[7] in frog muscle, and more recently emphasized by Akert, Moor, Pfenninger, and Sandri[3] in the rat diaphragm. These "zones actives" (Couteaux[17]), reminiscent of other types of synapses, regularly face a secondary synaptic fold. Their number varies greatly from one digitation to another.

In the early electron microscopic descriptions of mammalian end-plates, filaments and tubules were not mentioned. With aldehyde fixation, they are regularly demonstrated. They probably form the terminal loops, rings, etc. visible after metallic impregnations. The relationship between neurotubules and synaptic vesicles, suggested by Pellegrino, de Iraldi, and de Robertis,[54] cannot be established from pictures of end-plates. Rare dense-core vesicles may be present in the nerve endings. Their diameter (500 to 1000 Å) is larger than those of synaptic vesicles (400 Å).

*Schwann coat.* Above the nerve ending, a thin Schwann process closes the synaptic gutter as an operculum. In normal conditions, it never penetrates into the synaptic cleft between axoplasm and sarcoplasm. The Schwann cytoplasm contains filaments and some microtubules; frequently, coated vesicles indent its surface. As a rule, one teloglial nucleus ("noyau de l'arborisation," according to Ranvier) looks down upon the terminal nerve branching. A double membrane separates the Schwann cytoplasm from the underlying axoplasm; it may display one or two desmosomes along its way. On the other side, facing the connective tissue space, the cell surface is bordered by the Henlé sheath.

*The Henlé sheath* prolongs the perineural lamellar connective sheath; this sheath takes the form of an inverted funnel, whose edge is attached to sarcolemma and covers the motor end-plate. Along its edge, in normal conditions, satellite cells are frequently seen, separated from the muscle fiber by a double membrane. Possibly they play the role of "stretchers" for this peripheral sheath. Some collagen fibril and fibroblastic processes are visible in the space between the Henlé sheath and the Schwann cell.

*The primary synaptic cleft* is about 600 Å wide (Fig. 5). It is composed of three layers of varying electron density. A basement membrane occupies the middle layer, in continuity with the basement membrane which covers the whole sarcolemma and the Schwann coat. The secondary synaptic clefts have the same internal aspect. Their deep part is sometimes slightly wider, and the basement membrane is split in two parts. As de Harven and Coërs[35] observed, we have occasionally seen some dense bodies 250 to 300 Å diameter in the lumen of the secondary clefts.

*The subneural apparatus* is formed by thin lamellae of sarcoplasm interdigitating with the secondary clefts. It is a very elaborate and complex structure in human muscle. It can be schematically divided into three different successive zones. In the first zone (Fig. 5), in close vicinity to the nerve ending, subneural folds have a precise orientation perpendicular to the edges of the synaptic trough. Tangential sections grazing the nerve ending show the reticular disposal of the apertures of these folds in the primary synaptic clefts (Fig. 6*a*). At a short distance (about 3000 Å) from the primary synaptic clefts, the folds branch and anastomose. Sections parallel to the preceding one and passing through this

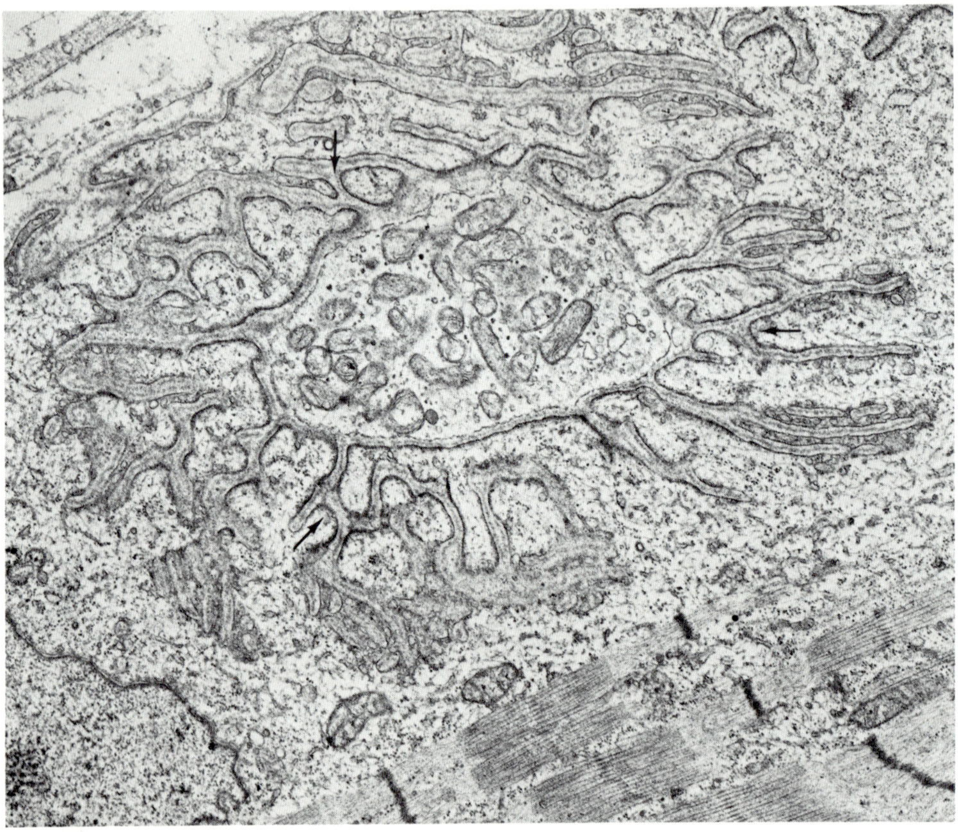

FIG. 5. "Frontal" section of a normal human motor end-plate. Fixation by osmium tetroxide. Elaborate subneural apparatus. Numerous anastomoses between its folds (*arrows*). × 20,000.

anastomotic (second) zone exhibit a rather regular reticular network. A preferential orientation of the folds stands out in this zone (Fig. 6*b*). In the third, deeper part, of the subneural apparatus, synaptic folds lie parallel to each other, frequently grouped in small clusters of three or four, in the junctional sarcoplasm.

Electron density and thickness of the membrane are different in the proximal and the distal zones of the folds. In the proximal zone the plasmic membrane is thickened and underlined by a sarcoplasmic densification, heavily stained after osmium tetroxide, as previously pointed out by Birks, Huxley, and Katz[7] and by Couteaux and Pecot-Dechavassine[20] in frog muscle. After osmic fixation (2%, in cacodylate buffer, 0°, 2 hours), it is frequent to observe (Fig. 6*c*) a regular periodicity in this subplasmic densification (200-Å interval). It must be noticed that this membrane thickening is the first visible sign of the neuromuscular junction in chick embryos[36] and appears when the nerve approaches the muscle fiber at a distance of 2000 to 3000 Å. This thickened part of the sarcolemma extends slightly laterally beyond the nerve ending contact. In the synaptic folds, it stops after the anastomotic zone. In the deep part, the plasmic membrane has its normal aspect.

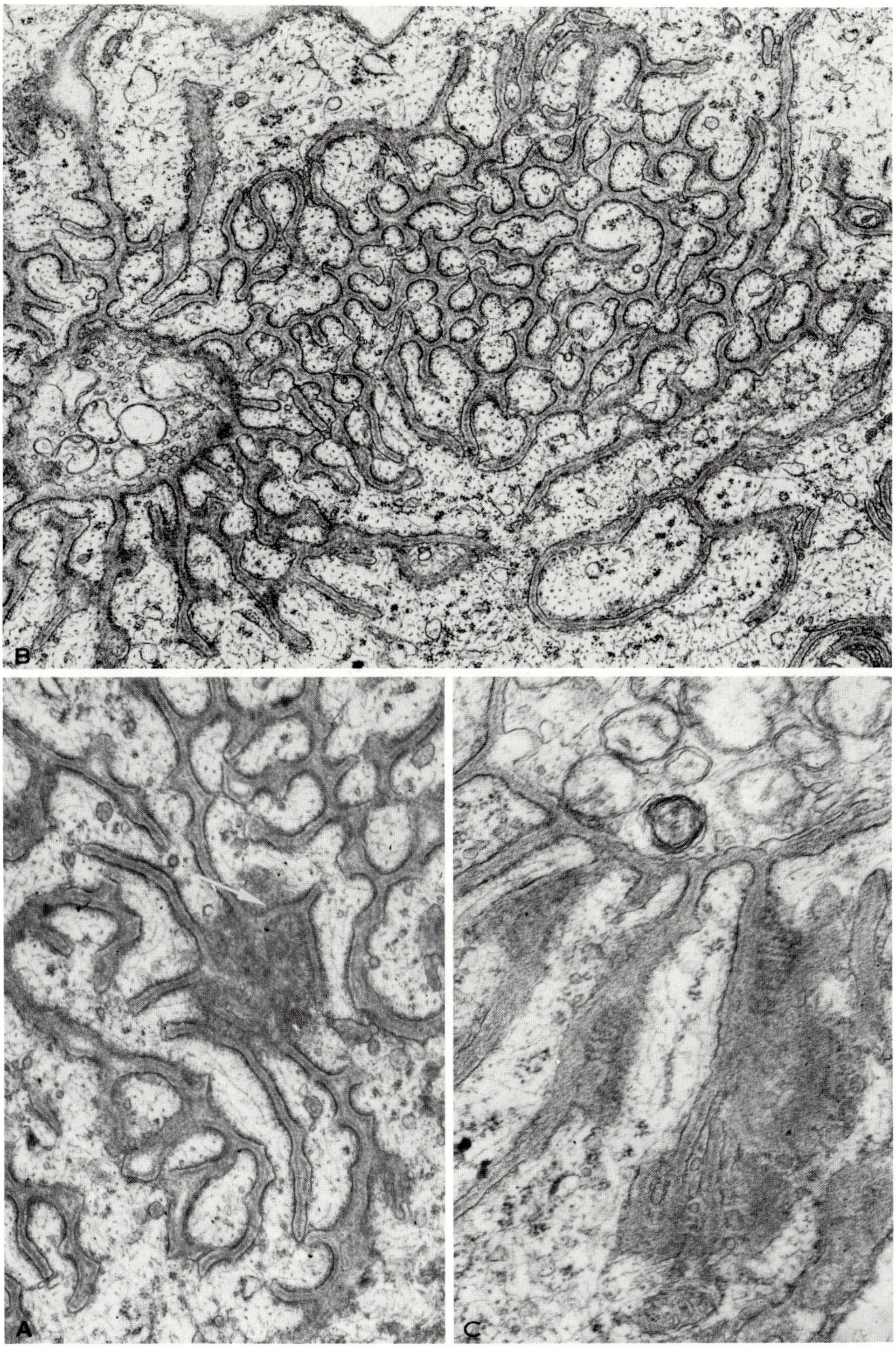



Fig. 6. Normal motor human end-plate. Fixation by osmium tetroxide. *a*, tangential (to the nerve ending) section. Reticular disposal of the junctional folds. Density is due to the section through the subplasmic densification of the postsynaptic membrane. Periodicity visible (*arrows*). × 20,000. *b*, subneural network, through anastomotic zone of the subneural apparatus. × 16,000. *c*, periodicity observed in grazing section through the subplasmic densification. × 37,000.

Likewise, connections of the subneural apparatus differ in the three zones. The proximal zone is only connected with the primary synaptic cleft; no vesicles are visible along the thickened membrane. The anastomotic zone forms a network between the folds below each nerve ending. As the more superficial of these folds normally outcrop at the surface of the sole-plate around the nerve ending, this network communicates with the external space on either side of the nerve terminal, as suggested by Couteaux.[19]

In their deep part, beyond anastomoses and branching, the folds are in tight connections with vesicular profiles, 500 to 800 Å diameter, often arranged in rows along them (Fig. 4). Some are seen opening into the lumen of the folds. Reger[55, 56] called them "subsynaptic" vesicles. They are limited by a smooth membrane, and are possibly connected with the sarcoplasmic reticulum of the muscle fiber, as previously suggested by de Harven and Coërs.[13, 15] Pictures by Couteaux and Pecot-Dechavassine on frog neuromuscular junction, and some pictures (Fig. 11a) of pathologic human end-plates favor this assumption. It might be suggested that this "sole reticular network" was equivalent to the networks demonstrated with metallic impregnations, "réseau intraplaculaire" described by Cajal[12] and "periterminal network" by Boëke[10] connecting, according to the latter author's thesis, nerve fibrils and myofibrils.

*The junctional sarcoplasm* or "sole-plate" contains several rounded voluminous nuclei, with one or two large nucleoli. Mitochondria are often located around the poles of the nuclei, with a variable amount of lipid droplets and lipopigments. Filaments of 50 Å and microtubules of 200 Å in diameter are accumulated in this area (the microtubules are visible only after aldehyde fixation). They are disposed parallel to the junctional folds. Similar bundles of filaments have been described by Couteaux and Pecot-Dechavassine[20] in frog sybsynaptic sarcoplasm. A dense tubuloreticular network winds between the secondary clefts, connected to the "subsynaptic" vesicles and possibly linked to the intermyofibrillar sarcoplasmic reticulum. Some coated vesicles and multivesicular bodies are observed communicating with the secondary synaptic clefts. Numerous dense granules are present in the junctional sarcoplasm. Many are glycogen granules, arranged in small clusters. Some of these "aposynaptic" granules are ribosomes, disposed along nuclear membranes or grouped in small chains.

## NORMAL VARIATIONS OF END-PLATE ULTRASTRUCTURE

Human end-plates do not always possess a uniform appearance. They vary by their shape and size and in number of subneural digitations in light microscopy. In electron micrographs, the density of mitochondria and vesicles in the nerve ending, the depth of the synaptic trough, the branching, and the development of the junctional folds vary greatly.

Two main factors must be kept in mind about these morphologic variations.

1. *Role of the fixation*. Results are, for instance, clearly different with osmic or aldehyde primary fixation, for the same specimen. In the nerve endings the density of the vesicles is much higher after aldehyde fixation and their presynaptic accumulation is preserved. After osmium fixation, the axonal cytoplasm is clearer and the mitochondria are often swollen. In the junctional

sarcoplasm, microtubules appear only after aldehyde fixation. A periodic striation in the postsynaptic membrane thickening is visible only after osmic fixation.

2. *Variations of end-plate development with "type" of fiber.* Histochemical studies have revealed[23] that mammalian muscle consists of different types of fibers, according to the relative activity of their oxidative, glycolytic, and other enzymes. Several electron microscopic studies have brought out the differences between the different types of fibers in mammalian muscle.[27, 48, 52] Morphologically, three types of fibers, "red," "white," and "intermediate," can be distinguished, according to their number of mitochondria, lipid droplets, development of sarcoplasmic reticulum, width of Z lines, and amount of glycogen granules. In human muscle, the same division has been proposed by differences in mitochondrial amount are not so clearly demonstrated as in other mammalian muscles.

As for the end-plate structure, Ogata, Hondo, and Seito,[49] Ogata and Murata,[50] and Padykula and Gauthier[52] have reported distinct differences between end-plates of the three types of fibers. We have studied by serial sections end-plates in guinea pig muscles, and we found clear differences (Fig. 7) in density, depth, and anastomoses of junctional folds between "red" (Fig. 7*a*) and "white" (Fig. 7*b*) fibers. It is more difficult to assume that significant differences exist between "intermediate" and "white" fiber end-plates.

In human muscle, differences in extent and complexity of the synaptic folds from one fiber to another are common. Murata and Ogata[46] have shown that these differences may be correlated with the fiber type as in other mammalian muscles. Even if quantitative study is necessary before a definite conclusion is made, it must be kept in mind that form and volume of the junctional infoldings may vary normally.

## PATHOLOGIC MODIFICATIONS

### End-plate modifications in neurogenic atrophies

Light microscopic studies have shown that degeneration of the distal part of the axon occurs within a few days after a nerve section.[15] Subneural apparatus modifications are visible only after a longer time, and their acetylcholinesterase activity may be detected for as long as 6 months after the denervation. In human pathology, denervated end-plates usually coexist with reinnervated ones, so that the terminal motor innervation exhibits important modifications, even if the muscle tissue itself has retained a normal or nearly normal appearance. Distal axons may show rarefaction, fragmentation, and spherical swellings. Remaining axons give extensive preterminal or ultraterminal ramifications. Small dense end-plates coexist with abnormally large and dispersed subneural apparatus.

Electron microscopic pictures of human denervated end-plates are infrequent, as their detection in biopsies is somewhat difficult. However, some have been published[16, 39] and we have studied, with J. Godet-Guillain, five cases of neurogenic atrophies. Three different kinds of alterations were observed in these biopsies.

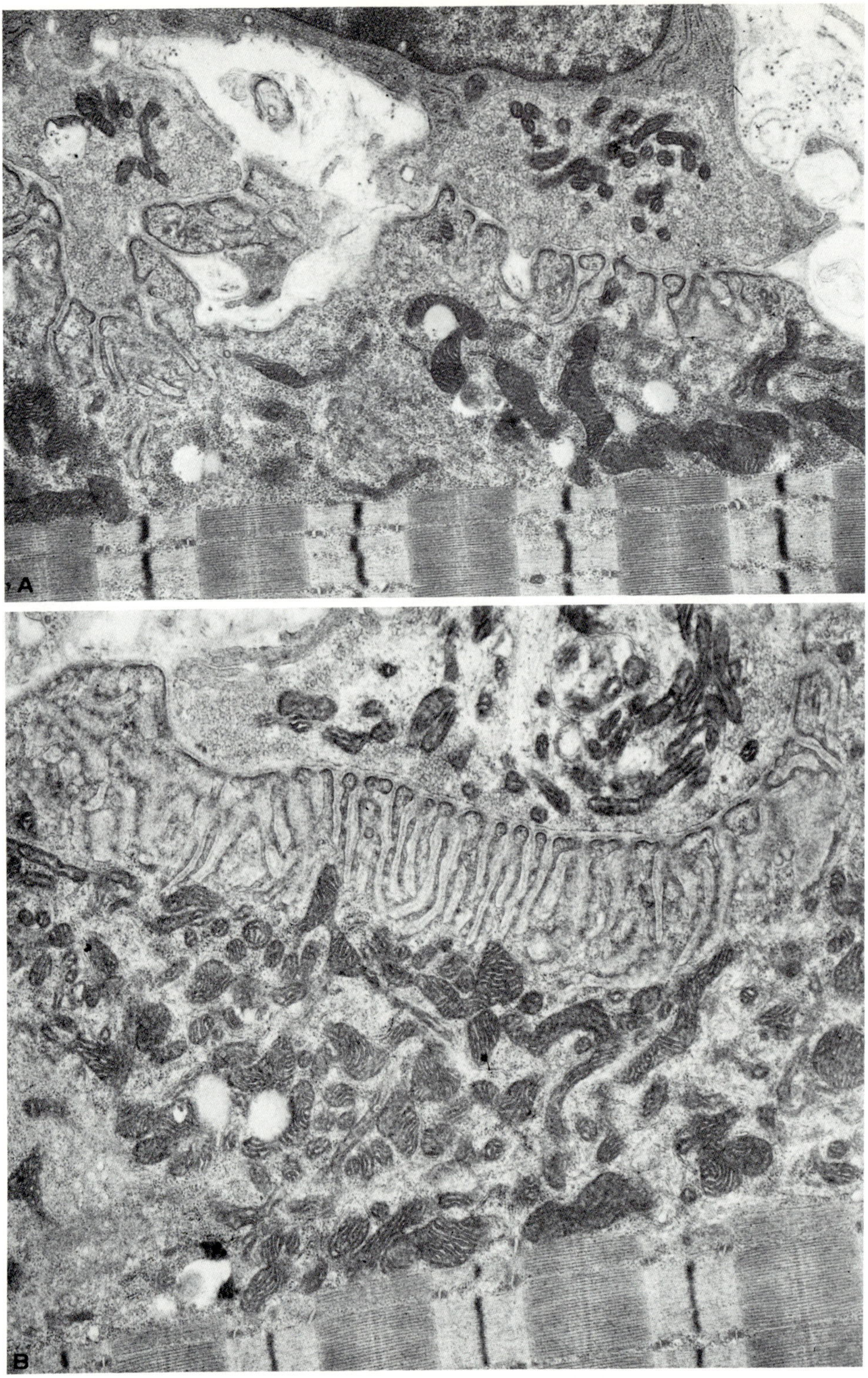


FIG. 7. Guinea pig musculus peroneus, fixed by aldehydic perfusion. *a*, "red" fiber, with thick Z lines. The junctional folds are scarce and rather simple, with rare anastomoses. Frequent widening of the deep part of the folds. × 13,000. *b*, "white" fiber, with thin Z lines. Well developed subneural apparatus, with numerous, deep, folds, × 13,000.

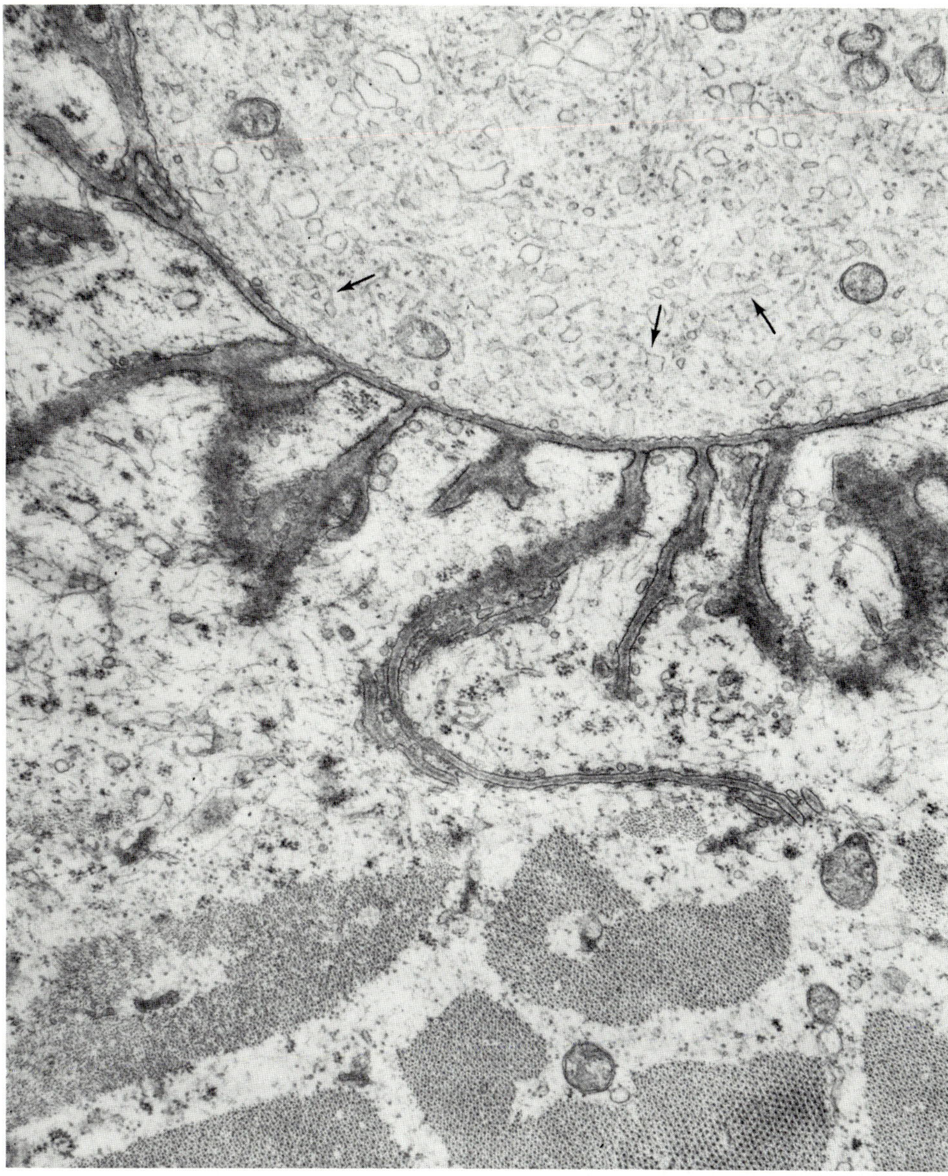

FIG. 8. Human muscle; neurogenic atrophy. Fixation by osmium tetroxide. Voluminous nerve endings, with reticular and tubular proliferation. Vesicular budding off of some tubules (*arrows*). The subneural folds, the sole reticular network, are reduced. × 24,000.

*Modification of nerve ending structure.* Most convincing of these modifications is a remarkable enlargement of the nerve ending, more than 10 $\mu$ in diameter, with a reticular and tubular proliferation (Fig. 8). Some aspects suggest a vesicular budding off of the tubules; mitochondria are nearly absent. Facing this nerve change, junctional folds are farther apart than normal. The numbers of subsynaptic vesicles, tubular network, and granules are reduced.

The subjacent myofibrils exhibit reduction in their diameters and some fragmentation.

*Schwannian interposition between axon and junctional folds.* An unusual plane of section may occasionally show a normally situated Schwann cell in contact with the muscle fiber.[7] However, in denervation, it is possible to demonstrate, by semiserial sections, that Schwann processes intrude into the synaptic cleft (Fig. 9*a*). The axonal ending is always modified and contains a few scattered vesicles. Axonal and Schwann processes are sometimes difficult to distinguish.

*End-plate emptiness.* As subneural folds normally extend beyond the nerve ending, this type of lesion can be identified only by semiserial sections. When it can be settled that the synaptic folds are "naked," a retraction space is sometimes apparent, as in Reger's illustrations.[56] A Schwann cell always overlays the junctional folds. It must be pointed out that usually this cell does not come into close contact with the folds, since collagen fibrils are usually visible between them. The sarcolemmal infolding persists, but greater modifications occur in the subjunctional sarcoplasmic area. Mitochondria and tubular reticulum have completely disappeared, and filament density is increased (Fig. 9*b*).

Some of these pathologic findings are very similar to the experimental data shown by Reger,[56] Birks, Katz, and Miledi,[8] Miledi and Slater,[45] and Waser and Nickel,[60] namely fragmentation of the axon terminal, retraction from the synaptic zone, replacement of the axon by the Schwann cell, and persistence of the postsynaptic folds. Subsarcolemmal accumulation of thin filaments is mentioned in denervated end-plates by Couteaux and Pecot-Dechavassine.[20] Reticulotubular proliferation in enlarged axonal endings may be compared to the accumulation visible in the proximal stump shortly after a nerve ligature.[54] However no experimental model has yet been reported for this end-plate modification, nor for the reduction and disorganization of the subneural folds facing an altered nerve ending.

Another unsolved problem in this electron microscopic study is the identification of reinnervated or newly formed end-plates; the vesicular content of the new axon terminals is similar to normal nerve endings.[7, 40]

### End-plate modifications in "primary" muscle diseases

In dystrophic diseases, studies with methylene blue or Koëlle's technique have demonstrated frequent abnormalities in the terminal innervation pattern. The proximal part of the intermuscular nerve bundles is usually normal, but an abnormal distal nerve sprouting is present. Large, complex, double, often multiple, subneural apparatuses are frequent on hypertrophic fibers, while small end-plates are visible on atrophic fibers. Finally, nerve fibers wander in the connective tissue, parallel to the degenerated muscle fascicles, making a normal contact when meeting a surviving fiber.

The number of dystrophic muscle end-plates examined by electron microscopy is still very limited. Jedrzejowska, Johnson, and Woolf[38] have reported 3 cases in progressive muscle dystrophies and Allen, Johnson, and Woolf[2] 7 cases

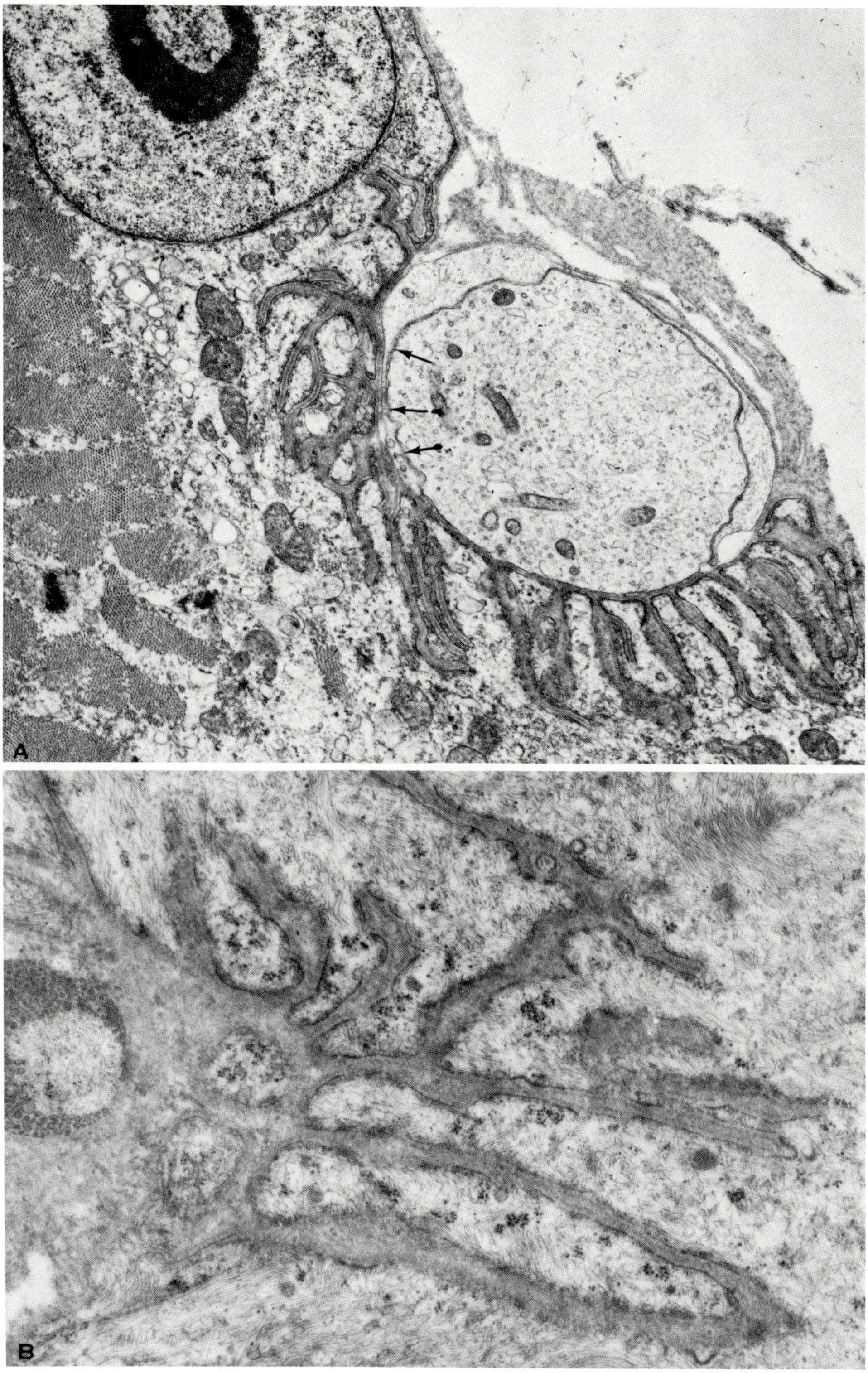


FIG. 9. Human muscle; neurogenic atrophy. Fixation by osmium tetroxide. *a*, Schwannian interposition (*arrows*) between a modified nerve ending and the junctional folds. × 11,000. *b*, "empty" end-plate. Numerous dense filaments between the remaining folds. Disparition of the "postsynaptic" vesicles. A cellular (? Schwann) process faces the sarcolemmal infolding. × 25,000.

in myotonic dystrophy. In collaboration with J. Godet-Guillain, we have examined 5 cases of myotonic dystrophy and 2 cases of Duchenne's dystrophy. Occasionally end-plate structure is quoted as normal in some cases of congenital myopathies[1, 27] and in Thomsen's disease.[30] In all of these cases, the general configuration of most of the end-plates is normal. The axonal ending, Schwann coat, and synaptic folds are in normal situation, even on an abnormal fiber. In some places, junctional folds lack their normal complexity. In 1 case of Allen, Johnson, and Woolf[2] and in 2 personal cases of myotonic dystrophy, several spherical dense bodies and cytolysomes were seen within the nerve ending (Fig. 10). In this latter disease, the secondary synaptic clefts were thought by Allen, Johnson, and Woolf[2] to be overdeveloped, *i.e.*, hypertrophic.

Whether these changes are primary or secondary to the muscle alterations cannot be decided from these morphologic data. With dystrophic mice, Wechsler[61] thought that peripheral nerves and end-plates "do not constitute a part of the primary process"; only in later stages of atrophy the secondary synaptic clefts are reduced. Bois, Roberge, and Cote,[11] in the same material, described an increase in the number of cytoplasmic microtubules in the junctional sarcoplasm and suggested an abnormal condition for the neuromuscular junction.

In nondystrophic muscle disorders, few observations have been reported. In one case of hypothyroid myopathy,[34] the general arrangement of the end-plate was normal, with a rather poorly developed subneural apparatus. In one case of paramyotonia congenita[32] end-plate structure was normal. On the other hand, end-plates exhibited marked changes in one case of chloroquine myopathy (Fig. 11) with numerous myelin figures and voluminous cytolysomes inside the nerve ending, facing very atrophic subneural folds.[33] The finding confirms involvement of the peripheral nerve in chloroquine intoxication, in addition to its predominant muscular toxicity. After vincristine intoxication, Anderson, Slotwiner, and Song[4] observed major ultrastructural changes inside the muscle fiber, with formation of numerous spheroid membranous bodies, while the terminal neural expansion was relatively unaffected and subneural folds only slightly widened.

### End-plate modifications in myasthenia gravis

Neuromuscular junctions were thought to be morphologically normal in myasthenia gravis, until the description of Coërs and Woolf[14] of elongated end-plates in vitally stained specimens as well as in cholinesterase preparations. Coërs and Desmedt[13] found that from 3 to 20% of end-plates were elongated independently of any muscle or interstitial abnormalities in 4 other cases. They concluded that these "dysplastic" end-plates were specific for myasthenia. Other end-plate modifications (expanded, multiple ramifications) were considered to be nonspecific and similar to those observed in muscle dystrophies. In later reports, Bickerstaff and Woolf[5] in 5 cases, MacDermot[42] in 8 cases, Fardeau[25] in 8 cases, Manolov and Haralanov[44] found similar modifications in most of their patients in variable proportion.

Furthermore, in some cases of long duration with associated muscle atrophy[25, 31] the nerve fibers form complex, extremely elongated end-plates (as large as 200 $\mu$) with fine ultraterminal sprouts, the subneural apparatus being

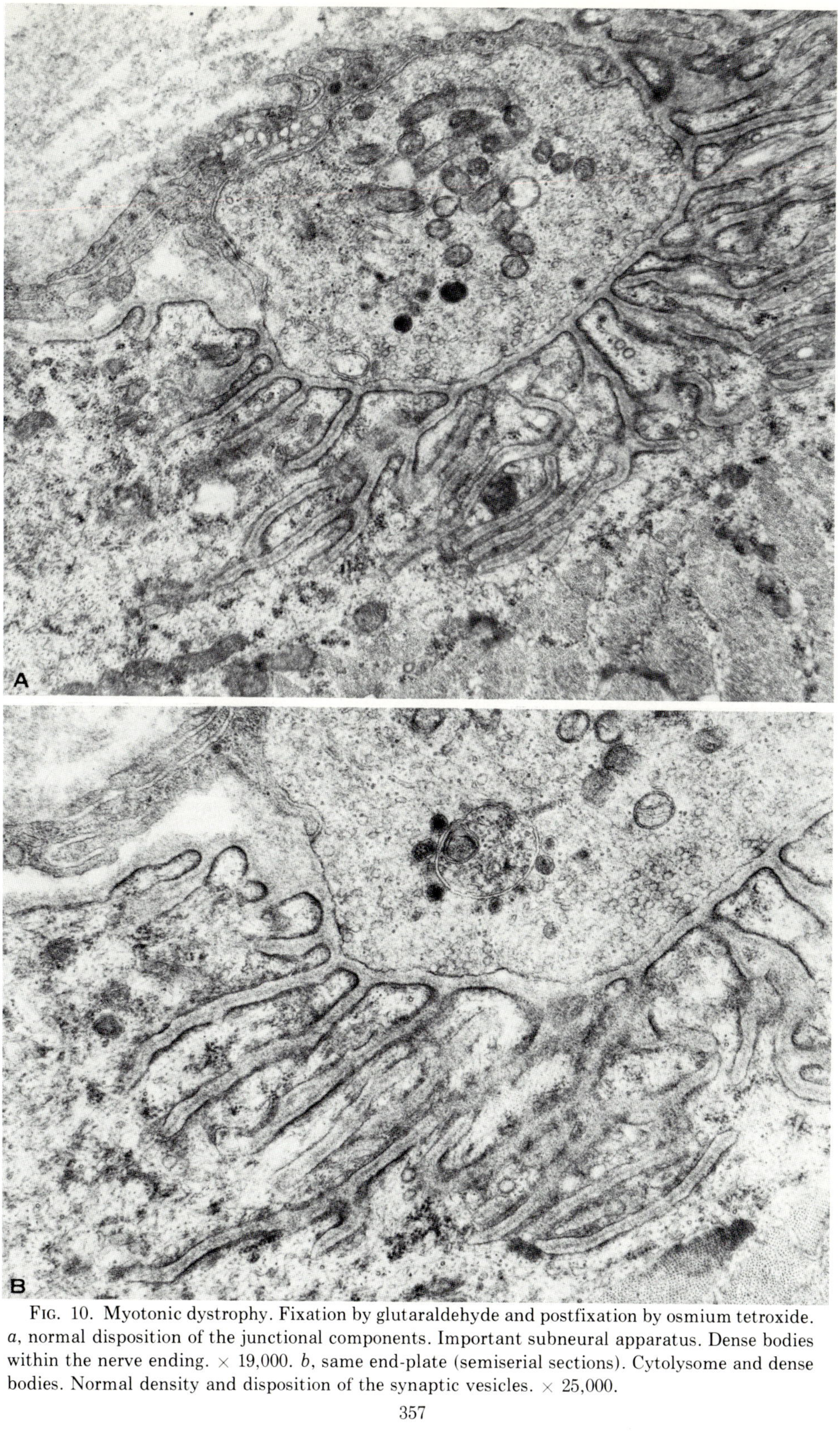


FIG. 10. Myotonic dystrophy. Fixation by glutaraldehyde and postfixation by osmium tetroxide. *a*, normal disposition of the junctional components. Important subneural apparatus. Dense bodies within the nerve ending. × 19,000. *b*, same end-plate (semiserial sections). Cytolysome and dense bodies. Normal density and disposition of the synaptic vesicles. × 25,000.

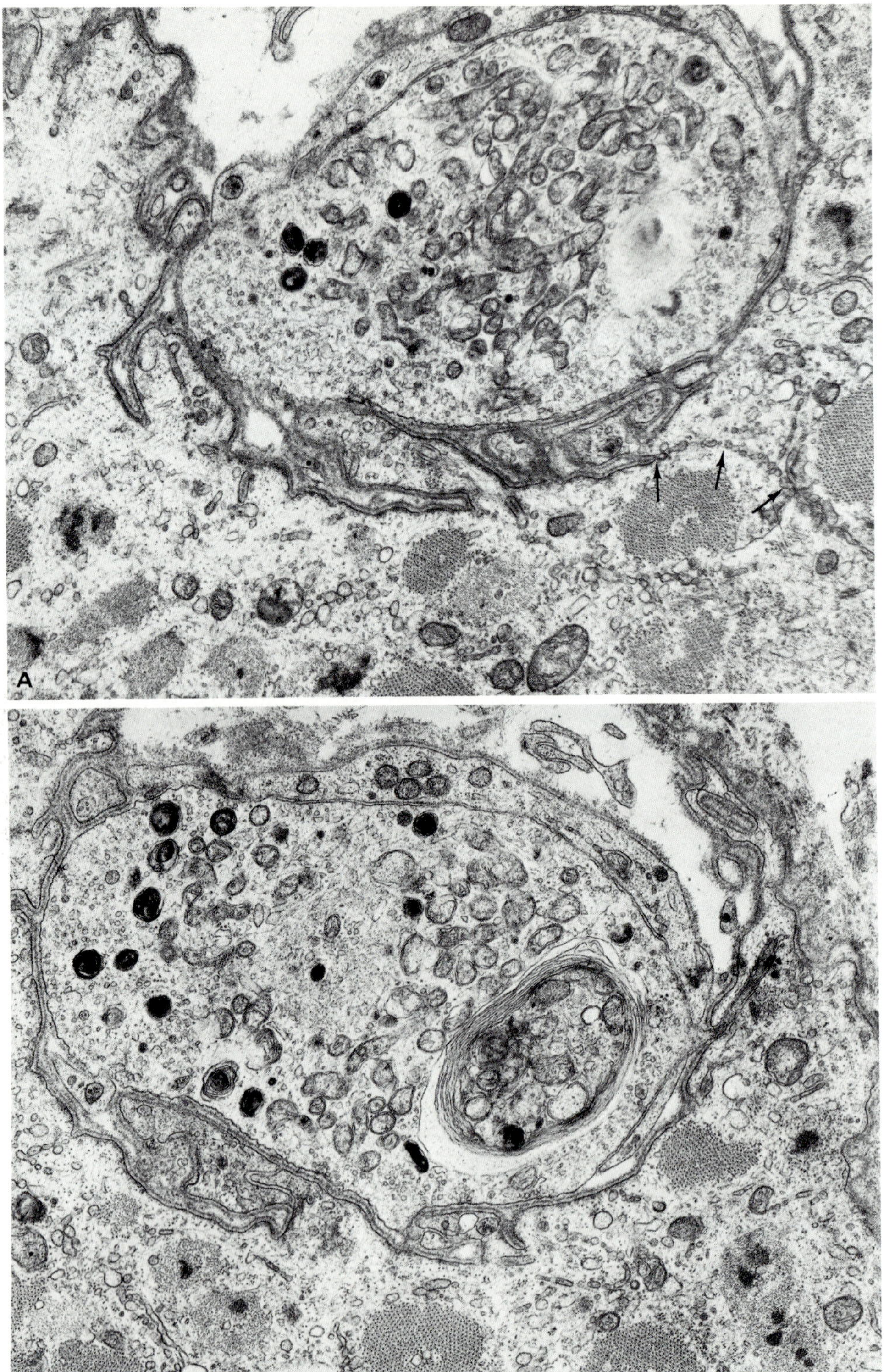


FIG. 11. Chloroquine neuromyopathy. Fixation by osmium tetroxide. Successive sections. *a*, large nerve ending, with a lot of mitochondria, vesicles, and filaments, facing a very atrophic subneural apparatus. Connection between postsynaptic vesicles and sarcoplasmic reticulum (*arrows*). × 16,000. *b*, same end-plate, with numerous dense bodies, and a voluminous cytolysome within the nerve ending. × 16,000.

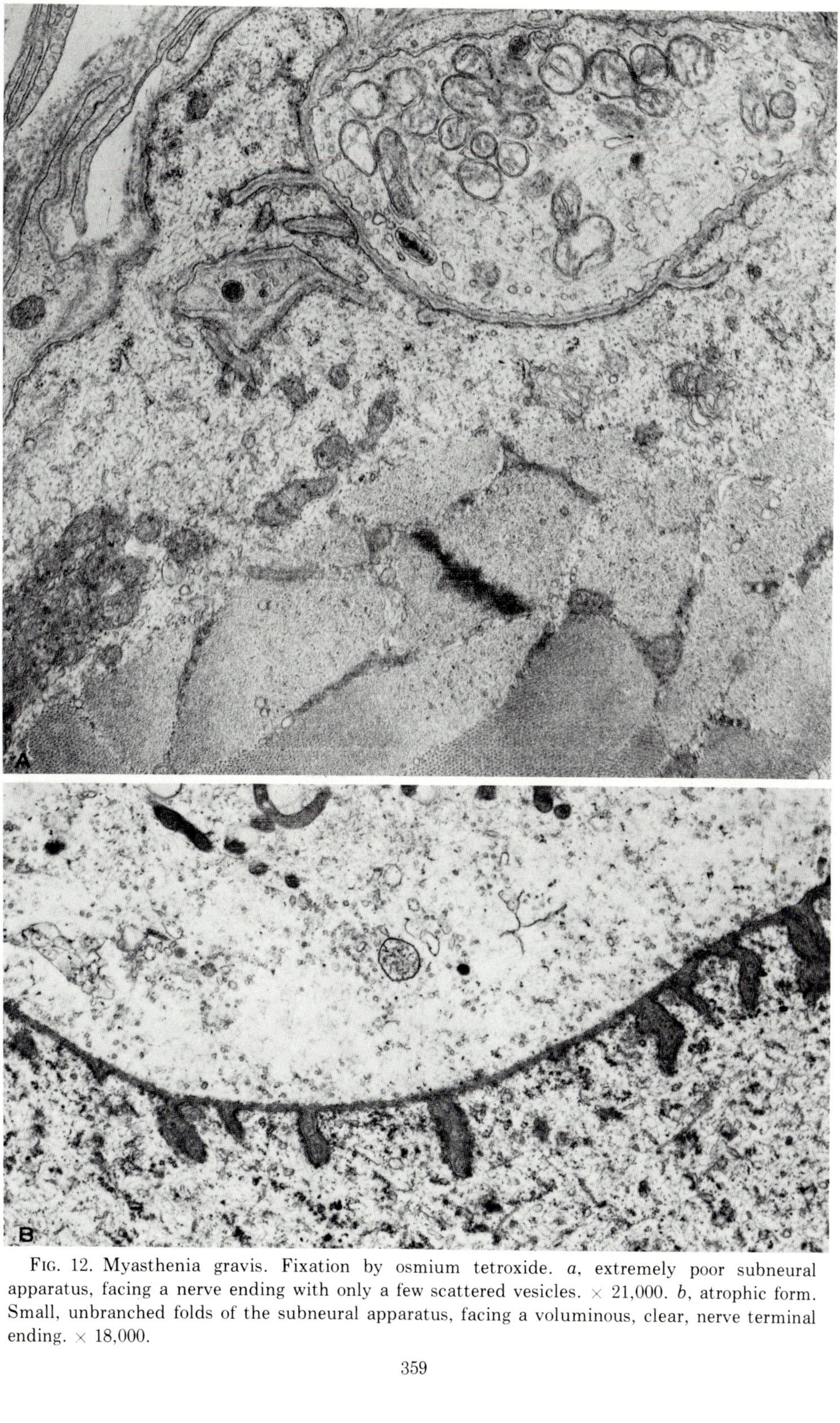


FIG. 12. Myasthenia gravis. Fixation by osmium tetroxide. *a*, extremely poor subneural apparatus, facing a nerve ending with only a few scattered vesicles. × 21,000. *b*, atrophic form. Small, unbranched folds of the subneural apparatus, facing a voluminous, clear, nerve terminal ending. × 18,000.

distributed in small digitations along the muscle fiber. Lastly, in rare cases[6, 24a, 29, 42, 62] the end-plates are shrunken and the subneural apparatuses are very small.

The first views of myasthenic end-plates by electron microscopy were achieved by Bickerstaff and Woolf[5] in 2 cases. Although their techniques were not as good as are possible now, they did note well preserved vesicles and a rather poor development of junctional folds. They concluded that fine structure of the motor end-plates is essentially normal. Zacks, Bauer, and Blumberg[64] and Zacks and Blumberg[65] thought they demonstrated a peculiar patchy decrease in the plasmic membrane within the secondary synaptic clefts. Their study was also limited by the imperfect techniques available. Today, these authors[9, 62] regard their earlier findings as not significant. However they did point out a decrease in the number of junctional folds and a marked widening of both primary and secondary synaptic clefts. In case 6 of Zacks,[63] a striking feature was the nearly complete absence of secondary clefts. In a Prostigmin-resistant case with small shrunken end-plates, Johnson and Woolf[39] observed the replacement of the terminal axonal expansion by a Schwann cell process, and again a paucity of synaptic folds.

In collaboration with J. Godet-Guillain, we have examined 5 cases of short duration myasthenia gravis without amyotrophy, and 2 cases of long duration with proximal amyotrophy. In each nonatrophic case we found a normal disposition of the three main components of the end-plates. Synaptic vesicle density in the nerve ending was sometimes poor, but we know it varies greatly with the fixation. The synaptic cleft was of normal width. The major change, illustrated in Figure 12*a*, was the reduction or paucity of subneural folds in many of the end-plates.

In atrophic cases, the lesions were extremely severe. The nerve endings were either clear or shrunken, the Schwann cell often intruded in the synaptic cleft, the subneural apparatus was reduced to very short folds, and the junctional sarcoplasm was without any mitochondria (Fig. 12*b*).

All of these observations clearly demonstrate that the end-plates show degenerative changes in myasthenia gravis. They may be compared to the severe abnormalities noted by Zacks, Shields, and Steinberg[66] in a myasthenic-like dog. As another sign suggestive of denervation, small groups of atrophic, angulated fibers have been observed by Fenichel[39] and by Engel and McFarlin[24] in human cases. Their significance remains obscure. All authors think that they probably do not constitute a primary lesion, but rather a secondary one. Woolf[62] suggested that elongation of the terminal nerve arborization might be induced by the defect of transmission, as a "compensatory" effect. Blumberg[9] thought that changes seen in advanced cases might be related to a decrease of available postsynaptic areas. By comparison with the other modifications observed in neuromuscular junctions, these changes tend to draw attention to a presynaptic defect in myasthenia gravis.

## REFERENCES

1. Afifi, A., Zellweger, H., Mc Cormick, W. F., and Mergner, W.: Congenital muscular dystrophy. Light and electron microscopic observations. *J. Neurol. Neurosurg. Psychiatry 32:* 273–280, 1969.

2. Allen, D. E., Johnson, A. G., and Woolf, A. L.: The intra-muscular nerve endings in dystrophia myotonica. A biopsy study by vital staining and electron microscopy. *J. Anat. 105:* 1–26, 1969.
3. Akert, K., Moor, H., Pfenninger, K., and Sandri, C.: Contributions of new impregnation methods and freeze-etching to the problems of synaptic fine structure. In *Progress in Brain Research, Mechanisms of Synaptic Transmission*, ed. by Akert, K. and Waser, P. G., vol. 31, pp. 223–240. Amsterdam, Elsevier, 1969.
4. Anderson, P. J., Slotwiner, P., and Song, S. K.: Structural parameters of neuromuscular diseases. Isolation and experimental reproduction. In *Exploratory Concepts in Muscular Dystrophy and Related Disorders*, ed. by Milhorat A. T., pp. 245–254. New York, Excerpta Medica Foundation, 1967.
5. Bickerstaff, E. R., and Woolf, A. L.: The intramuscular nerve endings in myasthenia gravis. *Brain 83:* 10–23, 1960.
6. Bickerstaff, E. R., Evans, J. V., and Woolf, A. L.: The ultrastructure of human myasthenic and non-myasthenic motor end-plates. *Brain 83:* 638–647, 1960.
7. Birks, R., Huxley, H. E., and Katz, B.: The fine structure of the neuromuscular junction of the frog. *J. Physiol. (Lond.) 150:* 134–144, 1960.
8. Birks, R., Katz, B., and Miledi, R.: Physiological and structural changes at the amphibian myoneural junctions in the course of nerve degeneration. *J. Physiol. (Lond.) 150:* 145–168, 1960.
9. Blumberg, J. M.: Discussion of the communication of A. L. Woolf. *Ann. N. Y. Acad. Sci. 135:* 57–58, 1966.
10. Boëke, J.: Nerve endings, motor and sensory. In *Cytology and Cellular Pathology of the Nervous System*, ed. by Penfield, W., vol. 1, pp. 241–315. New York, Hoeber, 1932.
11. Bois, P., Roberge, J. J., and Coté, M.: Distribution and ultrastructure of motor end-plates in the diaphragms of normal and dystrophic mice. *Anat. Rec. 154:* 453, 1966.
12. Cajal, S., and Ramon, Y.: *Histologie du Système Nerveux de l'Homme et des Vertébrés*, Tome I, pp. 371–375. Madrid, Instituto Ramon y Cajal, 1952.
13. Coers, C., and Desmedt, J. E.: Mise en évidence d'une malformation caractéristique de la jonction neuro-musculaire dans le myasthénie. *Acta Neurol. Belg. 59:* 539–561, 1959.
14. Coers, C., and Woolf, A. L.: Etude histologique et histochimique de la jonction neuro-musculaire dans deux cas de myasthénie. In *Comptes Rendus du Congrès des Médecins Aliénistes et Neurologistes*, pp. 500–511. Paris, Masson 2d., 1954.
15. Coers, C., and Woolf, A. L.: *The Innervation of Muscle. A Biopsy Study*. pp. 1–149. Oxford, Blackwell, 1959.
16. Cooke, W. T., Johnson, A. G., and Woolf, A. L.: Vital staining and electron microscopy of the intramuscular nerve endings in the neuropathy of adult coeliac disease. *Brain 89:* 663–682, 1966.
17. Couteaux, R.: Contribution à l'étude de la synapse myo-neurale. *Rev. Can. Biol. 6:* 563–711, 1947.
18. Couteaux, R.: Morphological and cytochemical observations on the postsynaptic membrane at motor end-plates and ganglionic synapses. *Exp. Cell Res. (Suppl.) 5:* 294–322, 1958.
19. Couteaux, R.: Motor end-plate structure. In *Structure and Function of Muscle*, ed. by Bourne G. H., vol. I, pp. 337–380. New York, Academic Press, 1960.
20. Couteaux, R., and Pecot-Dechavassine, M.: Particularités structurales du sarcoplasme sous-neural. *C. R. Acad. Sci. (Paris) 266:* 8–10, 1968.
21. Couteaux, R., and Taxi, J.: Recherches histochimiques sur la distribution des activités cholinestérasiques au niveau de la synapse myo-neurale. *Arch. Anat. Microsc. Morphol. Exp. 41:* 352–392, 1952.
22. Csillik, B.: *Functional Structure of the Post-synaptic Membrane in the Myoneural Junction*, pp. 1–154. Budapest, Hungarian Academy of Sciences, 1965.
23. Engel, W. K.: The essentiality of histo- and cytochemical studies of skeletal muscle in the investigation of neuro-muscular disease. *Neurology (Minneap.) 12:* 778–794, 1962.
24. Engel, W. K., and McFarlin, D. E.: Discussion of the communication of G. M. Fenichel. *Ann. N.Y. Acad. Sci. 135:* 68–77, 1966.
24a. Fardeau, M.: Personal observation.
25. Fardeau, M.: Données tirées de l'examen anatomo-pathologique du muscle squelettique dans la myasthénie. *Sem. Hôp. Paris 9:* 1141–1147, 1961.

26. Fardeau, M.: Coloration simultanée de l'innervation motrice terminale et des appareils sous-neuraux (technique de S. Manolov). Intérêt diagnostique en pathologie neuro-musculaire. *Rev. Neurol. (Paris) 111:* 501–506, 1964.
27. Fardeau, M.: Etude d'une nouvelle observation de "nemaline myopathy." II. Donnèes ultrastructurales. *Acta Neuropathol. (Berl.) 13:* 250–266, 1969.
28. Fardeau, M.: Ultrastructure des fibres musculaires squelettiques. *Presse Med. 77:* 1341–1344, 1563–1566, 1969.
29. Fenichel, G. M.: Muscle lesions in myasthenia gravis. *Ann. N. Y. Acad. Sci. 135:* 60–67, 1966.
30. Garcin, R., Dubois-Poulsen, A., and Fardeau, M.: Maladie de Thomsen de révélation néonatale avec atteinte particulière des muscles respiratoires. Documents ultrastructuraux concernant la biopsie musculaire. *Wien. Z. Nervenheilk. 25:* 2–4, 172–181, 1967.
31. Garcin, R., Fardeau, M., and Godet-Guillain, J.: A clinical and pathological study of a case of alternating and recurrent ophthalmoplegia, with amyotrophy of the limbs, observed for forty five years. Discussion of the relationship of this condition with myasthenia gravis. *Brain 88:* 739–752, 1965.
32. Garcin, R., Legrain, M., Rondot, P., and Fardeau, M.: Etude clinique et métabolique d'une observation de Paramyotonie congénitale d'Eulenburg. Documents ultrastructuraux concernant la biopsie musculaire. *Rev. Neurol. (Paris) 115:* 295–311, 1966.
33. Garcin, R., Rondot, P., and Fardeau, M.: Sur les accidents neuro-musculaires et en particulier sur une "myopathie vacuolaire" observée au cours d'un traitement prolongé par la chloroquine. *Rev. Neurol. (Paris) 111:* 177–195, 1964.
34. Godet-Guillain, J., and Fardeau, M.: Hypothyroid myopathy. Histological and ultrastructural study of an atrophic form. *Excerpta Medica International Congress Series No. 199, Muscle Diseases*, ed. by Walton, J. N., Canal, N., Scarlato, G., and Gleave, J. R. W., pp. 512–515. Amsterdam, Excerpta Medica Foundation, 1970.
35. Harven, E. De, and Coers, C.: Electron microscope study of the human neuro-muscular junction. *J. Biophys. Biochem. Cytol. 6:* 7–11, 1959.
36. Hirano, H.: Ultrastructural study on the morphogenesis of the neuro-muscular junction in the skeletal muscle of the chick. *Z. Zellforsch. Mikrosk. Anat. 79:* 198–208, 1967.
37. Ip, M. C.: A combined method for demonstrating the cholinesterase activity and the nervous structure of mammalian peripheral motor endings in teased preparations. *J. Physiol. (Lond.) 192:* 801–803, 1967.
38. Jedrzejowska, A., Johnson, A. G., and Woolf, A. L.: The intra-muscular nerve endings in muscular dystrophy. A biopsy study. *Acta Neuropathol. (Berl.) 5:* 225–242, 1965.
39. Johnson, A. G., and Woolf, A. L.: Replacement at the neuro-muscular synapse of the terminal axonic expansion by the Schwann cell. *Acta Neuropathol. (Berl.) 4:* 436–441, 1965.
39a. Koelle, G. B., and Friedenwald, J. S.: Histochemical localization of cholinesterase in ocular tissue. *Am. J. Ophthalmol. 33:* 253–256, 1950.
40. Koenig, J.: Contribution à l'étude de la néoformation expérimentale des plaques motrices du rat. *Arch. Anat. Microsc. Morphol. Exp. 60:* 1–26, 1970.
41. Krause, W.: Über die Endigung der Muskelnerven. *Z. Rechtsmedizin 18:* 136–160, 1863.
42. MacDermot, V.: The changes in the motor end-plate in myasthenia gravis. *Brain 83:* 24–36, 1960.
43. Manolov, S.: Détection des activités cholinestérasiques et imprégnation ultérieure des fibres nerveuses dans la musculature striée. *C. R. Acad. Bulg. Sci. 15:* 779–781, 1962.
44. Manolov, S., and Haralanov, H.: Changes in the myoneural synapses in myasthenia (clinical-anatomical parallels). *Neurol. Psikhiat. 6:* 49–55, 1967.
45. Miledi, R., and Slater, C. R.: Electrophysiology and electron-microscopy of rat neuro-muscular junctions after nerve degeneration. *Proc. Roy. Soc. Lond. (Biol.) 169:* 289–306, 1968.
46. Murata, F., and Ogata, T.: The ultrastructure of neuro-muscular junctions of human red, white and intermediate striated muscle fibers. *Tohoku J. Exp. Med. 99:* 289–301, 1969.
47. Nickel, E., Vogel, A., and Waser, P. G.: Coated vesicles in der umgebung der neuro-muskulären synapsen. *Z. Zellforsch. Mikrosk. Anat. 78:* 261–266, 1967.
48. Ogata, T.: An electron microscopic study on the red, white and intermediate muscle fibers of mouse. *Acta Med. Okayama 18:* 261–280, 1964.
49. Ogata, T., Hondo, T., and Seito, T.: An electron microscopic study on differences in the fine structure of motor end-plate in red, white and intermediate muscle fibers of rat intercostal muscle. A preliminary study. *Acta Med. Okayama 21:* 327–338, 1967.

50. Ogata, T., and Murata, F.: Fine structure of motor end-plate in red, white and intermediate fibers of mammalian fast muscle. *Tohoku J. Exp. Med. 98:* 107–115, 1969.
51. Ogata, T., and Murata, F.: Cytological features of three fiber types in human striated muscle. *Tohoku J. Exp. Med. 99:* 225–245, 1969.
52. Padykula, H. A., and Gauthier, G. F.: The ultrastructure of neuro-muscular junctions of mammalian red and white skeletal muscle fibers. *J. Cell Biol. 35:* 155A, 1967.
53. Palade, G. E., and Palay, S. L.: Electron microscope observations of interneuronal and neuro-muscular synapses. *Anat. Rec. 118:* 335–336, 1954.
54. Pellegrino De Iraldi, A., and De Robertis, E.: The neurotubular system of the axon and the origin of granulated and non-granulated vesicles in regenerating nerves. *Z. Zellforsch. Mikrosk. Anat. 87:* 330–344, 1968.
55. Reger, J. F.: Electron microscopy of the motor end-plate in intercostal muscle of the rat. *Anat. Rec. 118:* 344, 1954.
56. Reger, J. F.: Fine structure of normal and denervated neuro-muscular junctions from mouse gastrocnemius. *J. Ultrastruc. Res. 2:* 269–283, 1959.
57. Robertson, J. D.: The ultrastructure of reptilian myoneural junction. *J. Biophys. Biochem. Cytol. 2:* 381–394, 1956.
58. Rouget, C.: Note sur la terminaison des nerfs moteurs dans les muscles chez les reptiles, les oiseaux et les mammifères. *C. R. Acad. Sci. (Paris) 55:* 548–551, 1862.
59. Shafiq, S. A., Gorycki, M., Goldstone, L., and Milhorat, A. T.: Fine structure of fiber types in normal human muscle. *Anat. Rec. 156:* 283–301, 1966.
60. Waser, P. G., and Nickel, E.: Electronmicroscopic and autoradiographic studies of normal and denervated end-plates. In *Progress in Brain Research Vol. 31, Mechanisms of Synaptic Transmission*, ed. by Akert, K., and Waser, P. G., pp. 157–169. Amsterdam, Elsevier, 1969.
61. Wechsler, W.: Comparative electron microscopic studies on various forms of muscle atrophies and dystrophies in animals and man. *Ann. N. Y. Acad. Sci. 138:* 113–137, 1966.
62. Woolf, A. L.: Morphology of the myasthenic neuro-muscular junction. *Ann. N. Y. Acad. Sci. 135:* 35–56, 1966.
63. Zacks, S. I.: *The Motor End-plate*, pp. 1–321. Philadelphia, W. B. Saunders, 1964.
64. Zacks, S. I., Bauer, W. C., and Blumberg, J. M.: The fine structure of the myasthenic neuro-muscular junction. *J. Neuropathol. Exp. Neurol. 21:* 335–347, 1962.
65. Zacks, S. I., and Blumberg, J. M.: Observations on the fine structure of mouse and human neuro-muscular junctions. *J. Biophys. Biochem. Cytol. 10:* 517–528, 1961.
66. Zacks, S. I., Shields, D. R., and Steinberg, S. A.: A myasthenic syndrome in the dog. A case report with electron microscopic observations on motor end-plates and comparisons with the fine structure of end-plates in myasthenia gravis. *Ann. N. Y. Acad. Sci. 135:* 79–97, 1966.

Chapter 18

# Myopathies of Animals

W. J. HADLOW

---

Because of the relevance they may have to human muscular disease, the myopathic disorders of animals should be included in any comprehensive consideration of the myopathies. This belief, stated by Walton[116] 10 years ago, is sustained by the inclusion of this chapter in a monograph devoted to the myopathies of man. Accordingly, my intent here is to acquaint the medical reader with the varied naturally occurring myopathies of animals (mammals and birds).

In contrast to those of man, diseases of skeletal muscle of animals have received limited attention. Nevertheless, they are neither rare nor unimportant. Most major categories of muscular disease are represented by examples in animals. Certain of these seem analogous to those of man,[1] but some human muscular diseases, such as periodic paralysis and myopathies associated with endocrine disorders, apparently are without clear-cut counterparts in animals. On the other hand, some diseases of muscle, such as the nutritional myopathies and eosinophilic myositis, seem peculiar to animals.

In this chapter, I have used the term "myopathy" in a restricted sense to describe diseases that appear primary to the muscle fiber and not those of neural origin or those affecting mainly the interstitial tissue. Therefore, inflammatory diseases, whether infectious, parasitic, or of unknown cause, are excluded. Information on these and other diseases of skeletal muscle of animals can be found elsewhere.[42]

Myopathies have been described in many of the animal species that ordinarily come under human observation. Several affect farm animals. Those that occur enzootically in livestock have been studied intensively because of their great economic importance. Extremely few examples have been recognized in the dog and cat even though in veterinary practice they customarily receive more detailed clinical evaluation and individual attention than other species. Among small laboratory animals, several myopathic disorders have been studied either because of their resemblance to those of man or because they offer some insight into general problems of myopathology. Some that occur in poultry likewise have been studied as models of human disease. Lastly, free-living and captive wild animals, especially the larger mammals, are not immune to myopathic disorders. In the sections that follow, the varied myopathies affecting these groups of animals are considered under several broad categories.

## CONGENITAL MYOPATHIES

Various defects occur in skeletal muscle of newborn animals. Some represent imperfections in myogenesis, whereas others are a reflection of degenerative changes that affect fully differentiated muscle in the prenatal period. Many are hereditary. The more important ones are described here.

### Myopathy associated with arthrogryposis

Congenital deformities characterized mainly by rigid fixation of the limbs in abnormal postures frequently occur among domestic animals.[69] Often accompanied by other anatomic abnormalities, such deformities not only may be lethal for the fetus or newborn animal, but also often cause dystocia. Most examples of this syndrome, arthrogryposis, have not been adequately defined pathologically. In some, however, changes in skeletal muscle apparently account for the immobility of the limbs and axial deformities, such as torticollis and scoliosis, that may be associated with it. As in the similarly affected human infant, this syndrome in newborn animals may result either from primary disease of muscle[81, 82] or from denervation atrophy of muscle secondary to abnormalities in the spinal cord.[74, 120]

*Lambs.* A genetically determined primary myopathy causing muscular contractures in newborn lambs was reported in the Welsh Mountain breed[102, 103] and in Australian Merinos.[82] In each instance, the disease was inherited as an autosomal recessive trait. Affected lambs were stillborn or died shortly after birth at full term. Typically, one pair of limbs, or both, was rigidly fixed, usually in flexion. Muscles of the limbs were small, pale, and sometimes fibrous. Usually the joints became freely movable after the muscles and tendons were severed. Microscopically, affected muscles were composed of fibers of varying size, some of which lacked striations. Considerable fibrous tissue was present in some; others were largely replaced by fat.[81] The central nervous system and peripheral nerves were normal. This myopathy of newborn lambs appears analogous to the rare form of congenital muscular dystrophy associated with arthrogryposis in the human infant. I have not found recent observations on this disease of lambs, but presumably it still occurs sporadically and, like many other congenital abnormalities in domestic animals, is not reported or brought to the attention of interested persons.

*Calves.* Congenital deformities characterized largely by arthrogryposis have been frequently reported in calves of several breeds. Indeed, arthrogryposis is perhaps one of the more common congenital defects observed in cattle. Most reported examples have been hereditary ones, usually simple autosomal recessive traits.[37, 61, 65, 85, 106] But in none have the pathologic changes been adequately studied. Nevertheless, some most likely represent primary disease of muscle, like that in lambs. These hereditary deformities of cattle comprise a vast, virtually unexplored, field in which persons interested in comparative myopathology could find much material of potential value.

In contrast, nonhereditary and nonfamilial examples of arthrogryposis in calves have been studied more thoroughly. These have represented contractures caused by denervation atrophy rather than primary myopathy. Whittem[120] observed the syndrome in 30 newborn calves of several dairy breeds. In almost half of the calves, arthrogryposis was associated with hydranencephaly. Degeneration or complete absence of ventral horn neurons was found in segments of spinal cord that corresponded to affected muscles. Herzog and Adam[54] reported arthrogryposis in 20 newborn calves in which symmetrical muscle contractures of the hind limbs were associated with spina bifida and other abnormalities of the vertebral column and spinal cord. In 12 grade Charolais calves studied by Leipold, Cates, Radostits, and Howell,[74] the syndrome was associated with spinal dysraphism affecting mainly the cervical segments of the spinal cord.

*Other animals.* A lethal muscle contracture, inherited as an autosomal recessive trait, has been observed in a Swedish strain of the Large White breed of English swine.[45] The morphologic basis of the disease was not reported. Arthrogryposis in poultry is considered later in this chapter.

### Myopathy associated with hydrocephalus in calves

Congenital hydrocephalus has been known for many years in Hereford cattle in which it is inherited as an autosomal recessive trait.[11] Its recently recognized association with pronounced changes in the skeletal musculature is of particular interest. Present evidence suggests that these changes in muscle are primary and represent a generalized progressive dystrophy-like myopathy.

In 1957 Muth observed the myopathic changes in three hydrocephalic newborn calves from a closely bred herd of Hereford cattle in Oregon (see Hadlow[42]). Affected muscles, which were pale, soft, and somewhat spongy, were composed of fragments and segments of extremely small fibers irregularly scattered in a loose tangled network of fibrous tissue (Fig. 1). Striations were present in most small fibers, but a few larger ones were hyalinized. Whole fasciculi were virtually devoid of fibers. Fat cells were not seen.

A detailed account of this disease was reported by Urman and Grace[113] who examined 12 hydrocephalic Hereford calves in Nebraska. Affected calves were born alive and blind. They were unable to stand, though their limbs were mobile and joints normal. The hydrocephalus was associated with other neuropathologic changes, such as hypoplasia of the cerebellum, disseminated foci of decreased myelin particularly in the cerebellum and upper cervical segments of the spinal cord, atrophy of the optic nerves, and microphthalmia. Changes were widespread in the musculature, including the tongue and external ocular muscles. Various stages of the myopathic process could be found in any given animal. Hypertrophy of individual muscle fibers occurred early. Later, alternate areas of hypertrophy, hyaline degeneration, and atrophy with fibrosis were observed (Fig. 1). Neither fat cells nor evidence of regeneration of muscle fibers were present. Peripheral nerves in sections of affected muscle were unchanged. Despite distinct changes in the brain and white matter of the upper cervical segments of the spinal cord, ventral horns of the spinal cord were intact.

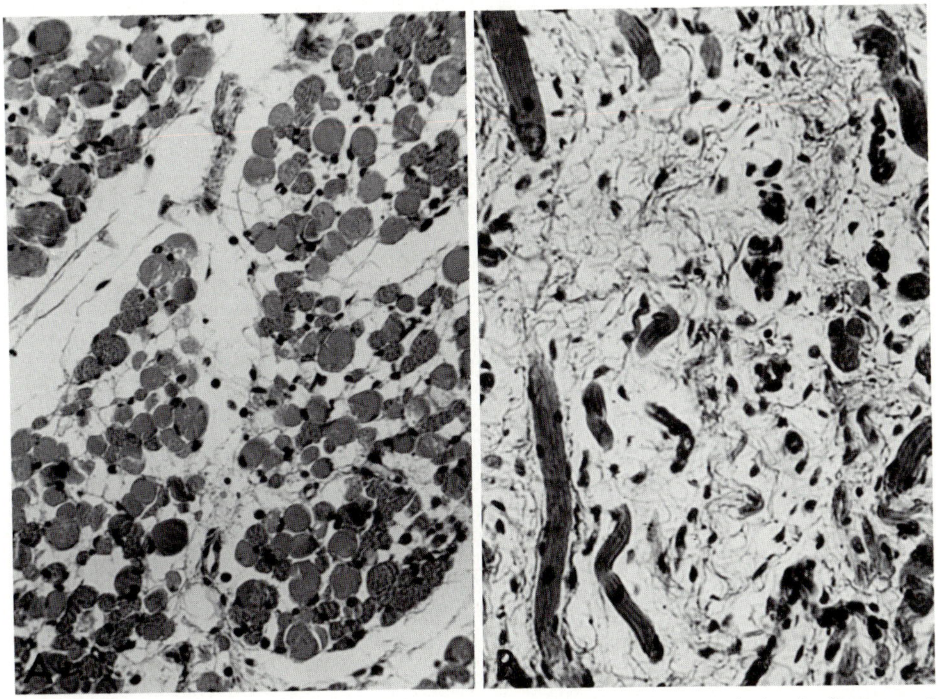

FIG. 1. Myopathy associated with hydrocephalus in calves. *A*, great variation in diameter of muscle fibers. Some enlarged ones are hyalinized. Nebraska calf. Hematoxylin and eosin stain; × 200. (Tissue, courtesy of Dr. Oliver D. Grace.) *B*, remnants of muscle fibers scattered in a tangled network of fibrous tissue as seen in late stage of the dystrophic process. Oregon calf. Hematoxylin and eosin stain; × 238. (Tissue, courtesy Dr. O. H. Muth.)

## MYOPATHY ASSOCIATED WITH SPLAYLEG IN NEWBORN PIGS

Inability of newborn pigs to stand or walk, ostensibly because of muscular weakness that allows the legs to become abducted or extended forward, is being recognized with increasing frequency.[70] Called spraddled-leg or splayleg, this syndrome occurs in otherwise normal baby pigs within a few hours after birth. Although the condition may completely disappear by the time the pigs are a week old, their early inability to move about makes them vulnerable to trauma from the sow and interferes with their chances to suckle. Presumably, the syndrome does not have a single cause. In one study, for example, poor footing on a slippery floor appeared to be the immediate cause that resulted in secondary trauma to joints and soft tissues of the legs.[70] In another study, the splayleg condition was associated with incompletely developed muscle fibers.

Under the term myofibrillar hypoplasia, Thurley, Gilbert, and Done[111] described the occurrence of this syndrome in many litters of two herds. Usually only one or two pigs in a litter were affected. An apparent deficiency in the number of myofibrils was found in muscle fibers of these pigs. This was rapidly resolved, however, for when the pigs were 4 days old the deficiency could no longer be detected.

The morphologic characteristics of this transient myopathy were described in detail by Thurley and Done.[110] Microscopically, muscle fibers of affected pigs contain less than normal numbers of myofibrils. Moreover, those present are abnormally small and are concentrated in one part of the fiber (Fig. 2). Parts free of myofibrils appear as clear spaces that do not contain fat or any other demonstrable substance. As judged by the cross-sectional area enclosed by the sarcolemma and endomysium, affected fibers are as large as normal ones. Within a single muscle, the number of myofibrils in individual fibers varies considerably. Also, the severity of the abnormality varies from one muscle to another. It is most pronounced in muscles of the hind legs, lumbar region, and fore legs, usually in that order of decreasing intensity. Other changes in muscle, either histologic or histochemical, have not been detected. The central nervous system and other organs are normal.

So far, the deficiency of myofibrils has not been associated with any really distinctive biochemical abnormality. Nevertheless, Patterson, Sweasey, Allen, Berrett, and Thurley[93] found evidence of altered distribution of tissue potassium. For example, the mean ratio of intracellular to extracellular potassium was about 49 : 1 in severely affected pigs, whereas it was 72 : 1 or more in normal pigs. Whether the hypoplasia of myofibrils is related to changes in membrane permeability, as suggested by this finding, is not known.

Actually, considerable variation occurs in the myofibrillar content of muscle fibers in newborn pigs. Thus, although fibers of the most severely disabled pigs contained the fewest myofibrils, a continuous gradation in number of myofibrils was observed between the least and the most severely affected pigs. Moreover, not all pigs whose muscle fibers contained few myofibrils were affected with the splayleg syndrome. Apparently then, splayleg is an inconstant clinical manifestation of the myofibrillar deficiency.

General growth of the pig seems to proceed normally before birth, but growth of myofibrils is retarded. Thus, the quantity of myofibrillar material present in fibers of severely affected pigs is comparable to that in muscle fibers of a 90-day fetus. The occurrence of the myopathy apparently is not affected by gestational age of the pig at birth or by the sow's diet during pregnancy. Studies so far have neither confirmed nor refuted the possibility of a hereditary predisposition to the condition suggested by some herd breeding data.[31, 110]

DEFECTIVE MYOGENESIS

In contrast to congenital myopathies affecting fully differentiated muscle fibers, two examples of muscular disease have been described in which degeneration occurs in incompletely developed muscle. Both are lethal mutations inherited as autosomal recessive traits.

*Muscular dysgenesis in the mouse.* A severe generalized disease of skeletal musculature in newborn mice, designated muscular dysgenesis, was described by Pai[89, 90] in 1965. The disease arose in an inbred stock of tailless mice ($T/t^9$). Mutant mice have loose skin and are shorter and broader than normal newborn sibs. Their limbs are short, edematous, and held in a clasping position. Micrognathia and skeletal abnormalities are present in all and cleft palate in most. The most striking characteristic, however, is a generalized and pronounced deficiency of skeletal musculature. Bones of the limbs and thorax are

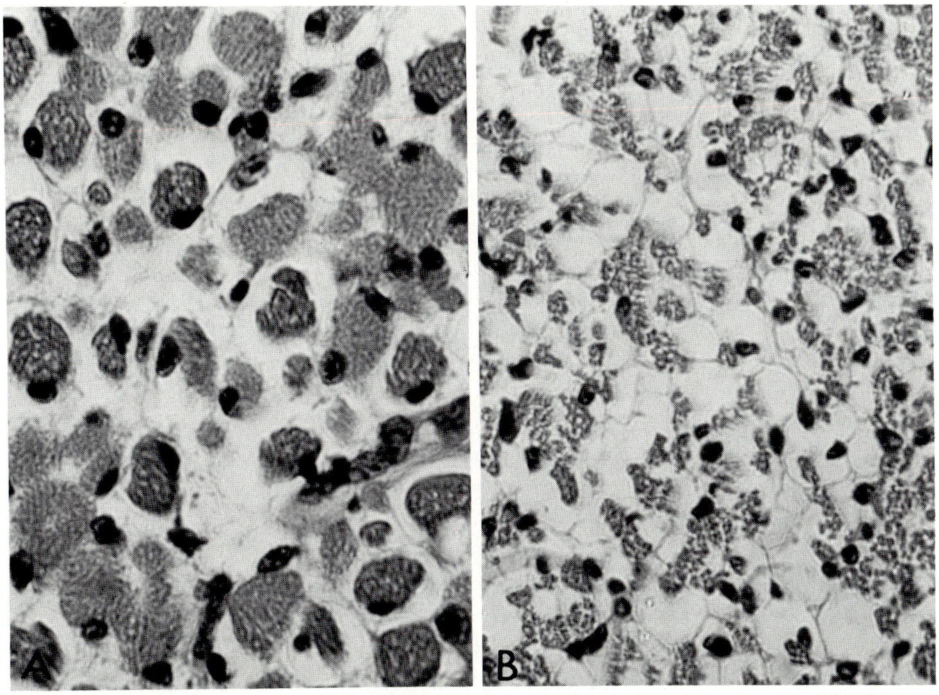


FIG. 2. Myopathy associated with splayleg in the newborn pig. *A*, appearance of muscle fibers in a normal pig. *B*, appearance of muscle fibers in a pig affected with splayleg. The fibers contain far fewer myofibrils than those of the normal pig. Hematoxylin and eosin stain; × 630. (Sections, courtesy of Mr. J. T. Done.)

almost bare of muscle; that present is soft and easily lifted from the skeleton. The tongue is small and soft, and the diaphragm is so thin as to be completely transparent. Affected mice die in the perinatal period, ostensibly because the poorly developed musculature of the thorax and diaphragm precludes normal breathing.

Because they are extremely edematous when about 13½ days old, mutant embryos can be readily distinguished from unaffected siblings. At this time, skeletal muscle masses appear less well organized microscopically than in normal mice. In the ensuing days until the end of gestation, the skeletal muscles become increasingly abnormal. Fewer myotubes are seen and they have less striations than normal. Some become atrophic, whereas others become swollen. Floccular degeneration, fragmentation, and rows and clumps of nuclei further distinguish the abnormal muscle cells. At birth, only the large muscles are still recognizable. They are composed of degenerating, fragmented pieces of muscle cells widely separated by connective tissue elements. Rarely, fully differentiated muscle fibers are seen, and these have centrally placed nuclei. Motor end-plates are not demonstrable, but the central and peripheral nervous systems are normal.

The basic process in this interesting myopathy appears to consist of abnormal differentiation and subsequent degeneration of embryonic muscle cells beginning at a stage in histogenesis when myoblasts differentiate into striated myotubes. This mutation should be extremely useful for studying the mechanisms of normal myogenesis and their genetic control.

*Crooked neck dwarfism in the chicken and turkey.* Although it might be interpreted differently in view of the scanty data available, crooked neck dwarfism in the chicken nevertheless seems best discussed here as another examples of defective myogenesis. This lethal mutation was reported in 1945 by Asmundson[4] in a strain of the New Hampshire breed. Affected embryos can be identified as early as the 9th day of incubation by their abnormally short necks, spindly legs, and edematous appearance. Growth becomes retarded and is reflected in a 30 to 40% reduction in muscle mass at this stage of development.[53] Further retardation in the growth of muscle begins on the 12th day. By the 20th day the embryos weigh half as much as their normal siblings. Pectoral muscles and those of the wings and legs are extremely small. The neck becomes bent and rigidly fixed, as do the hock (tibiotarsometatarsal) joints. Other features of the mutation include hydramnios, curly toes, and skeletal abnormalities. Affected embryos are unable to move normally and, therefore, fail to hatch; most die on the 20th or 21st day of incubation.

Pronounced microscopic changes are present in the skeletal musculature of 19-day embryos, although the sequence of events leading to these changes has not been studied. Rosenberg[104] found a few hyalinized muscle fibers, but prominent reduction in size of fibers through loss of myofibrils and sarcoplasm was the principal change. Severely affected muscles were composed of thin, stringy remnants of barely recognizable muscle fibers. Many muscle nuclei were pyknotic, but others contained skeinlike masses of chromatin. Cystlike structures of uncertain nature were present in some cells. The interstitial connective tissue was relatively more abundant. The central nervous system was not examined, but intramuscular nerves and blood vessels were normal. Rosenberg felt that the rapidly progressive degenerative changes were primary in muscle that had developed normally during the early stages of incubation. Although less well studied than muscular dysgenesis in the mouse, this myopathy affecting chicken embryos would seem to have much in common with it.

A crooked neck dwarf lethal mutation, similarly inherited as a simple autosomal recessive trait, has been observed also in the Broad Breasted Bronze turkey and in several strains descended in part from this breed.[7] It is essentially identical anatomically with that in the chicken, but I have not found a description of the microscopic appearance of muscle.

## HEREDITARY DYSTROPHY-LIKE MYOPATHIES (MUSCULAR DYSTROPHY)

In animal pathology, the term "muscular dystrophy" has been used to describe several spontaneous and experimentally induced myopathies whose natural histories and identifying clinicopathologic features are clearly unlike those of true muscular dystrophy in man.[42, 116] Most often the term has been applied to changes in skeletal muscle caused by nutritional deficiencies, particularly of vitamin E. Actually, of all examples of spontaneous primary myopathic conditions reported in the common domestic animals, only those few in the dog most closely resemble human progressive muscular dystrophy. Even though these canine examples fulfill some clinicopathologic criteria that distinguish the human disease, they have not been associated with a heredofamilial incidence. Thus, the 5 or 6 reported cases in the dog (see Hadlow[42] for details) remain isolated examples of an obscure, and presumably rare, disorder that is at least broadly comparable to human muscular dystrophy.

Nevertheless, several examples of genetically determined primary degenerative disease of skeletal muscle do exist among lower animals, such as inbred lines of laboratory rodents and domestic avian species. Because they more closely resemble forms of human muscular dystrophy than other myopathies of animals, they merit consideration in the context of true muscular dystrophy. In discussing them here, however, I shall use the term "hereditary dystrophy-like myopathy" as suggested by Homburger,[58] mainly to indicate that they probably are not identical with the human disease. Several authors have indicated clinical, histologic, and physiologic similarities of these myopathies with some of man. Yet, to attempt to equate these animal myopathies with any type of muscular dystrophy in man is perhaps premature. Such direct comparisons often are not warranted. Nevertheless, these animal myopathies undoubtedly represent true dystrophic processes with their own identifying features. Recognition of this seems more important than forcing identification with a specific myopathy of man.

### Mouse

In 1955 Michelson, Russell, and Harman[80] described a hereditary primary myopathy in the house mouse of the Bar Harbor inbred strain 129. Inheritance of the trait follows the pattern of an autosomal recessive gene, designated *dy*. Because this mutation was the first example of a clearly inherited animal myopathy resembling human muscular dystrophy, it has been studied intensively in many laboratories. Much of the information from these studies was recently summarized by Harman, Tassoni, Curtis, and Hollinshead[46] and West, Meier, and Hoag,[11] whose papers should be consulted for details.

This myopathy is characterized by early onset of rapidly progressive muscular weakness affecting mainly the hind legs. It is apparent 2 weeks after birth by the general inactivity and irregular, waddling gait of the dystrophic mouse. As the weakness becomes more pronounced, the mouse drags its hind legs and is easily exhausted when hanging from the sides of the cage by its hind feet. Extensor spasms of the hind legs also occur. Eventually, paresis of the hind legs becomes pronounced; locomotion then is accomplished entirely with the atrophied fore legs. A peculiar nodding of the head occurs in some mice. Kyphosis and severe contractures supervene as late manifestations.

Although dystrophic mice may be of normal size at birth, they never attain the weight of unaffected littermates. Because they become severely disabled physically, dystrophic mice, particularly males, are seldom capable of breeding naturally. Hence, artifical insemination and ovarian transplantation are used to produce adequate numbers of mutant mice for study. Characteristically, the lifespan of affected mice is shortened, although this varies somewhat with their genetic background and diet. Usually, however, they die when 3 to 5 months old, presumably from general debility and intercurrent infection.

Muscular atrophy progresses from the hind legs to the axial and fore leg musculature. The most conspicuous microscopic feature of affected muscles is wide variation in the diameter of muscle fibers (Fig. 3). This is evident shortly after birth. Extremely small and abnormally large rounded fibers are haphazardly distributed throughout the muscle. Increased numbers of enlarged muscle nuclei form prominent rows, often centrally in fibers (Fig. 3). Segments of single fibers undergo coagulation necrosis. The extent of this change and the age at which it is most conspicuous are partly influenced by diet. It has been found as early as a day or so before birth.[78] Regenerative activity occurs in association with such necrosis of fibers, but it is abortive. Condensation and proliferation of connective tissue, chiefly as thickening of the endomysium around small fibers, regularly occur but are never pronounced. Late in the course of the disease, a slight increase in interstitial fat is seen, notably between large fasciculi rather than between individual muscle fibers. Significantly, mild dystrophic changes also occur in muscles of some heterozygous or carrier mice.[78] Ultrastructural features of the myopathic process were considered in detail by Banker.[12]

Considerable biochemical and physiologic data have been obtained in an effort to characterize the dystrophic process. So far, however, they have not re-

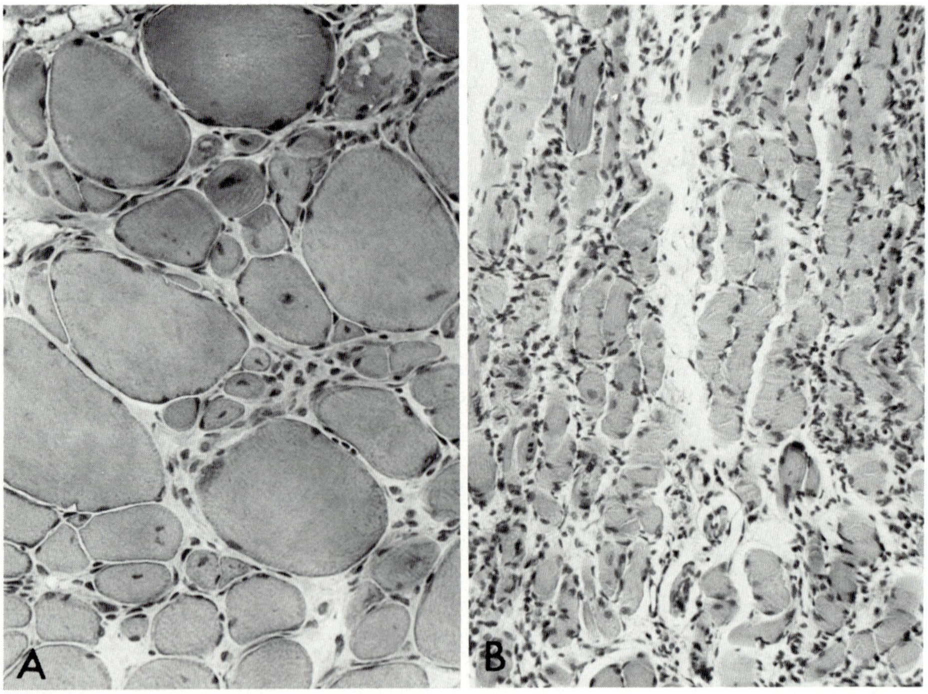


FIG. 3. Hereditary dystrophy-like myopathy of the mouse. *A*, extremely large and small muscle fibers in a 3-month-old mouse. Some fibers have internally located nuclei. Frozen section; Gomori trichrome stain; × 200. *B*, increased number of muscle nuclei and some atrophic fibers. Hematoxylin and eosin stain; × 160. (Sections, courtesy of Dr. Betty Q. Banker.)

vealed the primary defect or underlying mechanism of the disease. As might be expected, biochemically the metabolism of dystrophic mouse muscle is altered in many ways. Probably most of the changes are results of the degenerative process rather than indicative of the basic defect. Three biochemical changes that occur early in the myopathy (at 2 to 3 weeks of age) may more closely reflect the basic defect than those appearing later. These are: (1) elevated muscle adenosine monophosphate, (2) increased renal glycine transamidinase activity, and (3) accumulation in muscle of triose phosphates and decreased production of pyruvate. Quantitative functional differences between dystrophic and normal muscle have been difficult to assess because of the greater amount of connective tissue in dystrophic muscle. Qualitative differences include increased excitability of the dystrophic muscle fiber and a tendency to fibrillation. Although the contraction time is normal, the relaxation time is about 3 times greater than normal.

A milder, but otherwise essentially identical, myopathy was recently described by Meier and Southard.[77a] As an allelic variant of the original disease, it is caused by another recessive gene, designated $dy^{2J}$, that occupies the same locus as the *dy* gene. Onset of this myopathy occurs later, and its progression clinically and pathologically is slower. Thus, clinical signs appear when the mouse is 20 to 24 days old instead of at 14 days. Probably because of the slower course of the disease, affected mice of both sexes are capable of breeding naturally. Consequently, colonies of these dystrophic mice are much more easily maintained than are those of the original mutation. This occurrence in mice of alleles that give rise to myopathies differing mainly in their clinical expression provides a possible explanation for similar differences observed in forms of muscular dystrophy in man.

### Syrian hamster

In 1962 Homburger, Baker, Nixon, and Wilgram[59] described a primary hereditary myopathy in an inbred line (BIO 1.50) of the Syrian hamster. Since then, several other myopathic lines have been derived from the original one. Some of them have been studied intensively.[60] The disease is inherited as an autosomal recessive trait. Characteristically, hamsters of the dystrophic lines are smaller and lighter in weight than normal ones, especially during the first half of their life. Although reproduction is not impaired, female breeders produce relatively few litters. Muscular weakness, the main clinical sign, appears when the hamsters are 60 to 200 days old. Even in weanlings, however, such weakness can be demonstrated by forcing them to swim for an hour or more. After such vigorous exercise, the hind legs flap laterally, because of weakness of the adductor muscles, instead of remaining in a normal parallel position treading water.

As the disease progresses, weakness affecting mainly the hind legs becomes more pronounced, but only during advanced stages is locomotion impaired. The gait then is wobbly, and the hamsters are unable to raise themselves on their hind legs or to climb on sides of the cage. At about the same time, signs of cardiac decompensation, such as subcutaneous edema and ascites, supervene.

Affected hamsters generally live until they are about 150 to 200 days old, which is roughly one-third of the normal life span for this species. They usually succumb to congestive heart failure before changes in the skeletal musculature have reached the advanced degree seen in human muscular dystrophy and in the dystrophic mouse.

In biopsy specimens of the cheek pouch retractor muscle, microscopic changes can be detected in about 40% of hamsters up to 30 days old. From 70 to 100 days of age, 94% have lesions in muscle. The extent of the myopathic process and its rate of progression vary considerably from muscle to muscle, even in the same animal. The process is steadily progressive, however, and eventually all groups of skeletal muscles become affected.

Microscopically, the fully developed lesion is characterized by great variation in muscle fiber diameter, coagulation necrosis accompanied by accumulations of macrophages, sarcoplasmic basophilia, and formation of long chains of internally placed muscle nuclei bearing prominent nucleoli (Fig. 4). Proliferation of sarcoblasts accompanies focal necrosis of muscle fibers, but the regenerative process is incomplete. In late stages, mild degrees of fatty infiltration and proliferation of fibrous connective tissue are seen. As observed with the electron microscope, the degenerative changes in muscle are nonspecific and not unlike those seen in a variety of myopathies.[27]

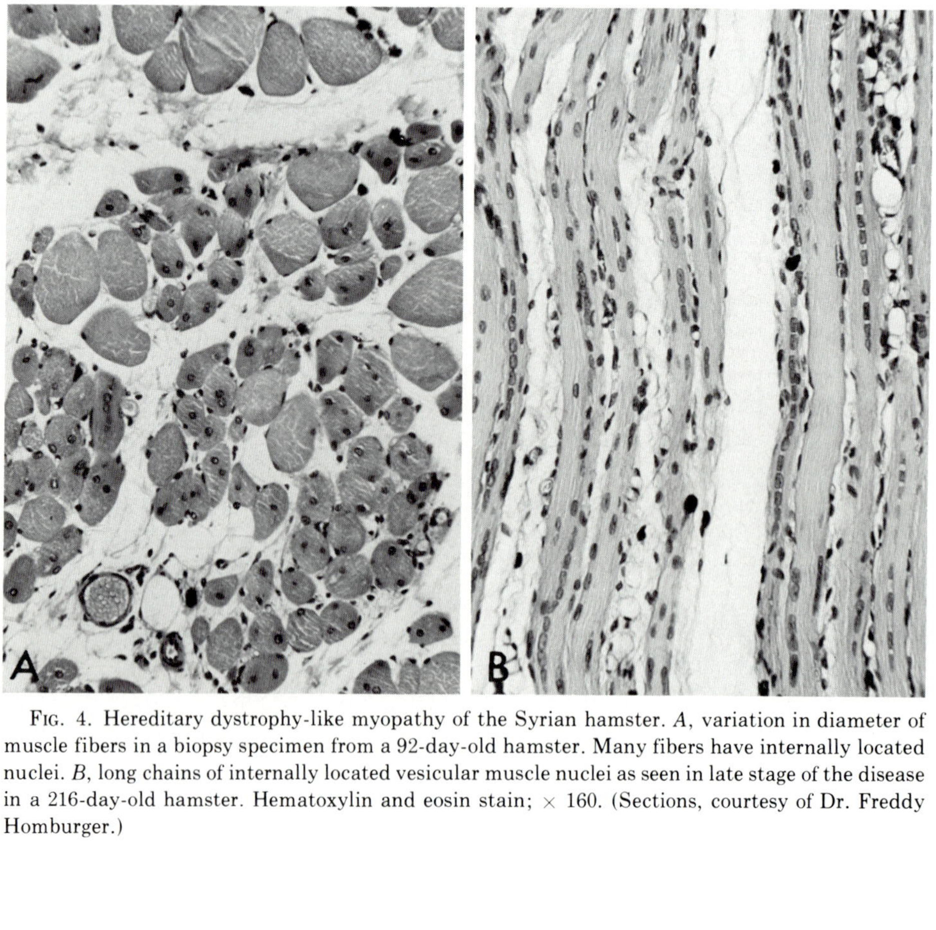


FIG. 4. Hereditary dystrophy-like myopathy of the Syrian hamster. *A*, variation in diameter of muscle fibers in a biopsy specimen from a 92-day-old hamster. Many fibers have internally located nuclei. *B*, long chains of internally located vesicular muscle nuclei as seen in late stage of the disease in a 216-day-old hamster. Hematoxylin and eosin stain; × 160. (Sections, courtesy of Dr. Freddy Homburger.)

Muscle levels of lactic dehydrogenase, malic dehydrogenase, aldolase, and creatine phosphokinase are normal, but serum levels of these enzymes are greatly increased. For example, serum creatine phosphokinase activity is significantly increased in hamsters 20 to 25 days old. Between 26 and 120 days of age extremely high levels are reached. These then begin to decline but always remain above normal up to 210 days. The serum levels are not closely correlated with the extent and severity of the changes in skeletal muscles. This pattern of creatine phosphokinase metabolism in the dystrophic hamster has been likened to that observed in patients with the Duchenne type of muscular dystrophy. Unlike the findings in man, however, the serum levels of this enzyme are not abnormal in hamsters heterozygous for the dystrophy gene (carriers).

Cardiac lesions, described in detail by Bajusz, Homburger, and Baker,[9] are detectable as early as 30 days of age in females and about 10 days later in males. When 70 to 80 days old, animals of both sexes have widespread lesions in the myocardium. These consist of myolytic foci unaccompanied by an appreciable cellular infiltration. Vacuolar degeneration of cardiac muscle fibers proceeds to complete disintegration of the fibers with only remnants of sarcolemma remaining. In advanced stages, fibrosis and calcification are seen. Because the cardiomyopathy occurs regularly in the dystrophic hamster, it has been proposed as a model for studying the pathogenesis and therapy of congestive heart failure.[8]

### Chicken

An inherited progressive myopathy of early onset in the domestic chicken was first reported by Asmundson and Julian[5] in 1956. It appeared in inbred strain 3 of the New Hampshire breed developed at the California Agricultural Experiment Station for heavy pectoral musculature. Ensuing genetic studies indicated that affected chickens are homozygous for a simple autosomal recessive gene designated *am*. Information from these and other studies was recently summarized by Julian and Asmundson[66] and by Asmundson, Kratzer, and Julian,[6] whose papers contain all pertinent references.

Affected chickens have a broad shallow body and short thick limb bones. They are unable to elevate their wings beyond a horizontal plane and to right themselves when supine on a flat surface. In the original mutant population, the myopathy was detectable clinically when the chickens were about 8 weeks old. Selective breeding, however, has produced other lines in which onset occurs as early as 1 week and as late as 12 weeks of age. The disease is progressive, but this varies among chickens and in individual muscles of a single chicken. Unlike the comparable myopathic disorder of the mouse, this avian myopathy does not necessarily result in a shortened lifespan.

In early stages of the disease, white muscles, primarily those of the pectoral girdle and wings, are much more severely affected than are the dark muscles of the legs. Eventually, however, most of the skeletal musculature becomes affected by some phase of the process. Striking enlargement of the superficial pectoral muscles (which form the bulk of "breast" musculature) occurs early in the disease before onset of overt clinical signs. It becomes progressively more pronounced with age. For instance, by the time chicks are 2 weeks old, the pectoral muscles are frequently 40% larger than normal size. At a variable period after the chickens are 12 weeks old, the enlarged pectoral muscles begin to decrease in size. This coincides with deposition of excessive fat in these muscles, which is more pronounced in females than in males.

Apart from obvious changes in size, affected muscles have an opacity that contrasts with the soft translucency of normal chicken muscle. In early stages, they appear grossly striated, apparently as a result of slight bulging of fasciculi bearing enlarged muscle fibers. Characteristically, affected muscles become tough, firm, and rubbery.

Extremely wide variation in the diameter of muscle fibers is the most striking microscopic feature (Fig. 5). Some fibers are much larger than normal, whereas others are distinctly smaller. The cross-sectional appearance of clusters of small fibers suggests that they arose by longitudinal splitting of a large fiber. Beginning early in the disease, the number of muscle nuclei is increased in both light and dark muscles. In the superficial pectoral muscle, for example, the number of muscle nuclei may reach twice the normal. Many of them are randomly distributed in the sarcoplasm. This feature does not, however, distinguish dystrophic muscle from normal muscle as it does in man, for this position of nuclei is normal in white muscles (pectoralis) of birds. Single and

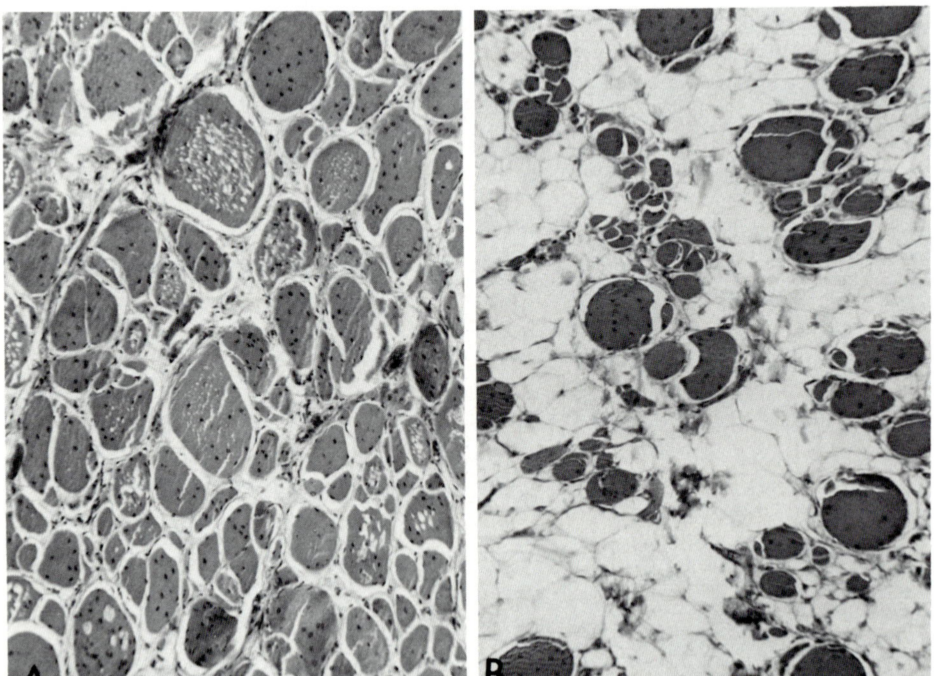


FIG. 5. Hereditary dystrophy-like myopathy of the New Hampshire chicken. *A*, great variation in diameter of muscle fibers, increased number of internally located muscle nuclei, and vacuolar degeneration in superficial pectoral muscle of a 28-week-old male. *B*, extensive replacement of muscle fibers by fat cells in superficial pectoral muscle of a 12-week-old female. Hematoxylin and eosin stain; × 135. (Sections, courtesy of Dr. Logan Julian.)

multiple vacuoles form in the dystrophic muscle fibers. At almost any stage of the myopathy, active destruction of muscle fibers is a prominent feature, but only segments are affected. They undergo necrosis and fragmentation. Late in the disease, ringed fibers ("ringbinden") have been observed. Accumulation of fat that replaces damaged muscle fibers is a variable but often pronounced feature of the dystrophic process after the chickens are 15 weeks old. In some muscles, such fat may account for 60% of their volume. Only a slight increase in connective tissue occurs, and this seems to be largely a condensation of interstitial tissue.

Levels of serum aldolase and glutamic oxalacetic transaminase are elevated in affected chickens. Such elevations occur before clinical signs are recognizable. Unlike the findings in mice heterozygous for the dystrophic gene, levels of these enzymes are increased as well in heterozygous chickens. Apparently, the dominance of the normal allele is incomplete in heterozygous chickens. Nevertheless, the heterozygotes remain normal clinically even though microscopic changes, less intense than those in clinically affected chickens, are present in some muscles, particularly the deep pectoral. Serum creatine phosphokinase activity is consistently higher in the dystrophic than in either the homozygous or heterozygous clinically normal chickens.

Although this myopathy of chickens is inherited as an autosomal recessive trait, other genes may modify its severity and the age at which it becomes apparent clinically. Moreover, the amount of fat deposited in the pectoralis muscle is also a highly heritable polygenic trait. The dystrophic gene, *am*, apparently also influences the size of muscle masses in both the homozygous recessive (*am:am*) and heterozygous (*Am:am*) combinations; heterozygous chickens have pectoral muscles intermediate in size between the normal (*Am:Am*) and dystrophic (*am:am*) birds of the New Hampshire breed.

Recently, Wagner and Peterson[115] briefly described a similar hereditary dystrophy-like myopathy in a flock of Cornish chickens at the West Virginia Agricultural Experiment Station. The main clinical manifestation is stiffness of the wings which becomes most apparent when the chickens are 10 to 12 weeks old. In severely affected birds, the wings cannot be manually elevated much above a horizontal plane. As judged macroscopically, the pectoralis and supracoracoideus are the most severely affected muscles. Some muscles are extremely hypertrophied, whereas others are extremely atrophied. Great variation in muscle fiber diameter, vacuolar degeneration and necrosis of fibers, and replacement of muscle fibers by fat cells are the main microscopic changes. Although the mode of inheritance of this myopathy of the Cornish chicken has not been established, preliminary breeding data indicate that, unlike the myopathy in the New Hampshire breed, it is not inherited as a simple autosomal recessive trait.

### Turkey

A third avian example of a hereditary dystrophy-like myopathy was reported in the domestic turkey by Harper and Parker[47] in 1964. It was first noted in 1960 in two 32-week-old Broad Breasted Bronze turkey females of a strain maintained at the Oregon Agricultural Experiment Station. Subsequently, various types of matings indicated that this mutation is caused by a single autosomal recessive gene.[48] So far, this myopathy has received only preliminary study.

The disease is characterized by atrophy of the pectoral muscles that first becomes evident between 8 and 16 weeks of age. Muscles of the wings are affected to a lesser degree; those of the legs are unaffected. When atrophy becomes most pronounced, the weight of the pectoral muscles is reduced 40 to 65%. In contrast to what is observed in the dystrophic chicken, hypertrophy of muscle does not precede the atrophy, as determined by width of the breast. The dystrophic process appears to become arrested once physical and sexual maturity is reached at 32 to 36 weeks of age. Despite atrophy of the pectoral and alar muscles, the wings are carried normally. Like the dystrophic chicken, however, affected turkeys are unable to right themselves from a supine position. Dystrophic turkeys have been kept for 3 years without abnormally high mortality. Fertility, hatchability, livability, and motility appear to be little affected by the disease. Bone growth is normal.

Affected muscles are paler than normal, have a stringy appearance, and lack the tonal quality to unfold that is characteristic of normal turkey muscle when folded. In material I examined, kindly provided by Professor Harper, great variation in the diameter of muscle fibers was the most conspicuous microscopic abnormality (Fig. 6). Some enlarged fibers were vacuolated, whereas others were hyalinized. Although fat almost completely replaced some fasciculi, it was not as abundant as that in muscles of the dystrophic chicken. Condensation of the endomysium was observed in the more severely affected portions, but frank fibrosis was absent. Much remains to be done to define more thoroughly this myopathy of the turkey.

## NUTRITIONAL MYOPATHIES

Of all myopathies affecting animals, those of nutritional origin are by far the most common and, certainly economically, the most important. Although often referred to as muscular dystrophy, these myopathies chiefly of cattle, sheep, swine, and horses are distinguished from true muscular dystrophy by their entirely different natural history and pathologic features.[42] All are acute, nonprogressive diseases without known genetic aspects. Typically, the nutritional myopathies affect young animals, particularly those of suckling age. They have been reported from many parts of the world but occur enzootically mostly in countries of temperate latitudes. All seem to occur under certain systems of management or husbandry practices that are peculiar to these areas. Regardless of species affected, the nutritional myopathies, at least as they occur in farm animals, have certain common clinicopathologic characteristics.

Depending on the distribution and severity of the muscular damage, clinical manifestations vary widely. Generally, however, they include rapidly developing symmetrical weakness of girdle muscles that results in a stiff or otherwise abnormal gait, postural changes such as difficulty in rising or inability to stand, and prostration and death within a few days to a week or so after onset. Inanition resulting from inability to nurse and complicating bronchopneumonia are frequently responsible for death. Acute cardiac failure is often the predominant clinical manifestation is some animals, particularly calves; they die suddenly after a brief period of respiratory distress or are simply found dead. Except when complications supervene, fever and other constitutional signs are not observed.

Lesions in the skeletal musculature vary from diffuse areas of pallor to white or gray-yellow streaks and large well demarcated firm patches that have a characteristic dry, parboiled appearance. Sometimes whole muscles are affected. In the more acute disease, some muscles may be moderately enlarged, apparently as a consequence of widespread swelling of muscle fibers. Although most muscles may be involved, macroscopic lesions are found mainly in the large muscles of the shoulder girdle, back, and thighs. Almost invariably they are bilaterally symmetrical. Often the diaphragm and tongue are affected too. In general, the distribution of the lesions correlates well with the main clinical signs.

Microscopically, waxy or hyaline degeneration of whole muscle fibers or of multiple segments is the characteristic change. The swollen, brightly acidophilic fibers undergo lumpy cleavage and granular disintegration, lose their affinity for stains, and disappear (Fig. 7). The process may affect a few fibers

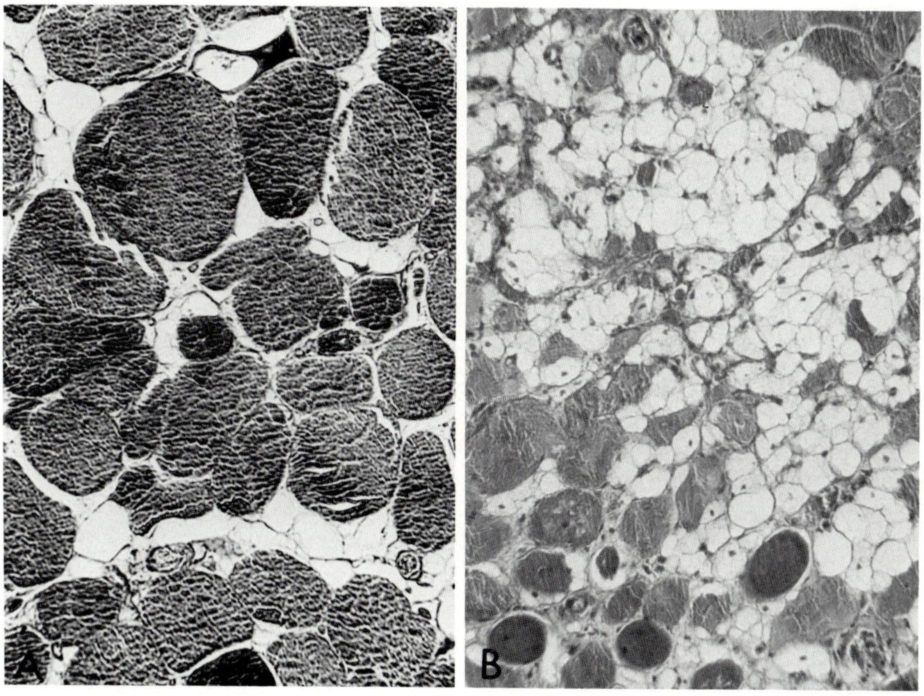


FIG. 6. Hereditary dystrophy-like myopathy of the Broad Breasted Bronze turkey. *A*, extremely large and small muscle fibers in the deep pectoral muscle of a 37-week-old female. Liisberg acid-fuchsin-aniline blue stain; × 200. *B*, replacement of muscle fibers by fat cells in a fasciculus of same muscle as *A*. Hematoxylin and eosin stain; × 160. (Tissues, courtesy of Professor J. A. Harper and Oregon Agricultural Experiment Station.)

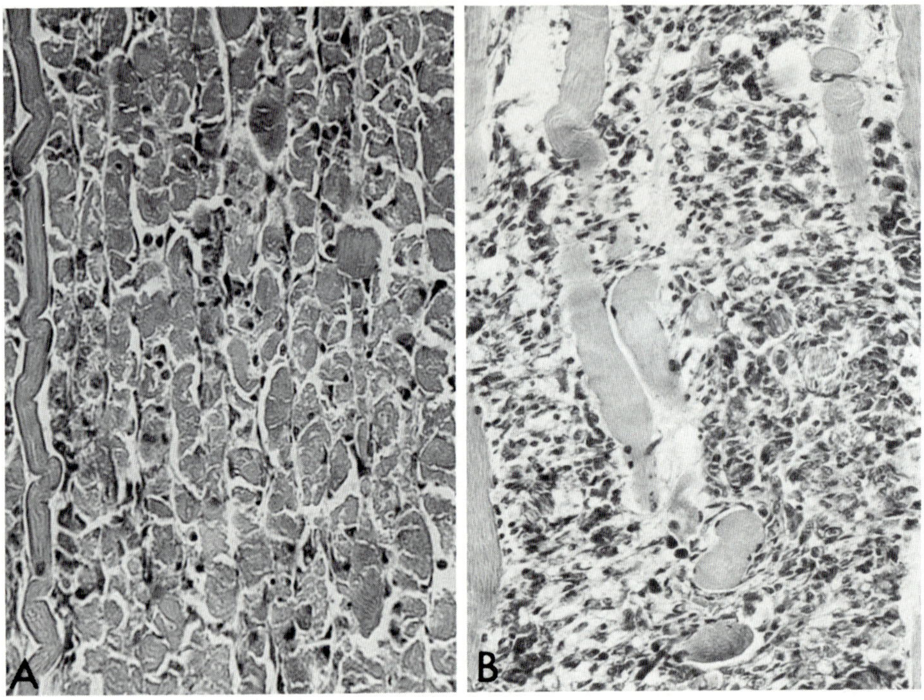


FIG. 7. Nutritional myopathy of calves. *A*, typical appearance of necrosis and lumpy fragmentation of skeletal muscle fibers in a 6-week-old calf. Azure-eosinate stain; × 200. *B*, pronounced proliferation of muscle nuclei in a 4-week-old calf. Azure-eosinate stain; × 150.

among many preserved ones, or may account for large areas of necrosis. Calcification in the form of fine granules arranged along myofibrils frequently occurs. Many histiocytes accumulate in the endomysium and perimysium, invade fragmented fibers, and actively remove necrotic muscle substance. In the more severely damaged areas, necrotic muscle undergoes complete lysis. Prominent proliferation of muscle nuclei is often seen. This and other elements of regeneration are particularly evident in calves and lambs in which they form a distinctive feature of the lesion (Fig. 7). Such regenerative activity often results in densely cellular masses that seemingly provide enough bulk to maintain the normal volume of the damaged muscle. Regeneration may be well advanced in animals that die suddenly. All stages of the myopathic process may be seen in different muscles of the same animal and in adjacent areas of a single macroscopic lesion.

Myocardial changes, which often accompany those in skeletal muscle, sometimes dominate the pathologic picture. Their macroscopic form and distribution vary with the species affected. In general, microscopic changes resemble those in skeletal muscle, but cardiac muscle fibers tend to be more coarsely granular rather than appreciably swollen and brightly acidophilic. Calcification often occurs in the form of beadlike granules along myofibrils or in transverse rows. The granules may completely fill segments of fibers which then appear as plaques. As in skeletal muscle, histiocytes and proliferating muscle nuclei sometimes completely replace large areas of necrotic myocardium.

Like most naturally occurring nutritional diseases, these myopathies are complex disorders involving several dietary and nondietary factors. Recent observations have made it increasingly clear that not all examples of nutritional myopathy, whether in different species or in a single species, have precisely the same cause. Factors implicated so far include: (1) deficiency of vitamin E, (2) deficiency of selenium, (3) presence of polyunsaturated fatty acids in the diet, (4) presence of legumes in the diet, and (5) unaccustomed muscular activity and probably other stresses. None is common to all of the several nutritional myopathies affecting animals. The exact role of each and their interrelationships are not certain. Because most examples of nutritional myopathy seem to be initiated or in some way associated with lipid peroxidation, the protective effect of vitamin E ($\alpha$-tocopherol), and possibly of selenium, is thought to result from their antioxidant properties. More complete discussions of the factors affecting the occurrence of nutritional myopathies can be found in several recent papers.[2, 35, 55, 73]

## Cattle

Nutritional myopathy of calves, usually termed "white muscle disease," has been recognized as a distinct clinicopathologic entity for at least 70 years (see Hadlow[42]). It occurs mainly in beef cattle because of the particular conditions of feeding and management under which they usually are maintained. In many countries where it is enzootic, this myopathy has been observed in offspring of cows over-wintered on restricted rations, usually but not necessarily of poor quality roughages. On the other hand, in such countries as the United States and New Zealand, the disease frequently occurs in calves whose dams were maintained on hay or pasture composed largely of legumes. It has been seen also in dairy calves fed liberal amounts of cod liver oil. This myopathy may affect calves from birth to 3 months old but occurs most often when they are 4 to 6 weeks old and still nursing. Two clinical forms of the disease are recognized. Of these, the cardiac form often predominates, particularly in younger calves. Sudden onset of labored breathing, blood-tinged nasal discharge, general weakness, and death (often within a few hours) constitute the usual clinical picture. The skeletal muscle form develops more slowly. Its clinical expression varies according to the muscles affected. Mildly affected calves often recover.

Depending on the clinical manifestations, the pathologic findings vary. In calves affected with the cardiac form, yellowish or gray-white streaks or irregularly outlined diffuse patches are conspicuous in the myocardium, especially in the subepicardial and subendocardial portions. Almost invariably, these lesions are restricted to the wall of the left ventricle and interventricular septum (Fig. 8). Extremely variable, the microscopic changes in the heart often appear to have existed long before onset of clinical signs. Pyknosis of cardiac muscle nuclei and granularity of the sarcoplasm sometimes accompanied by

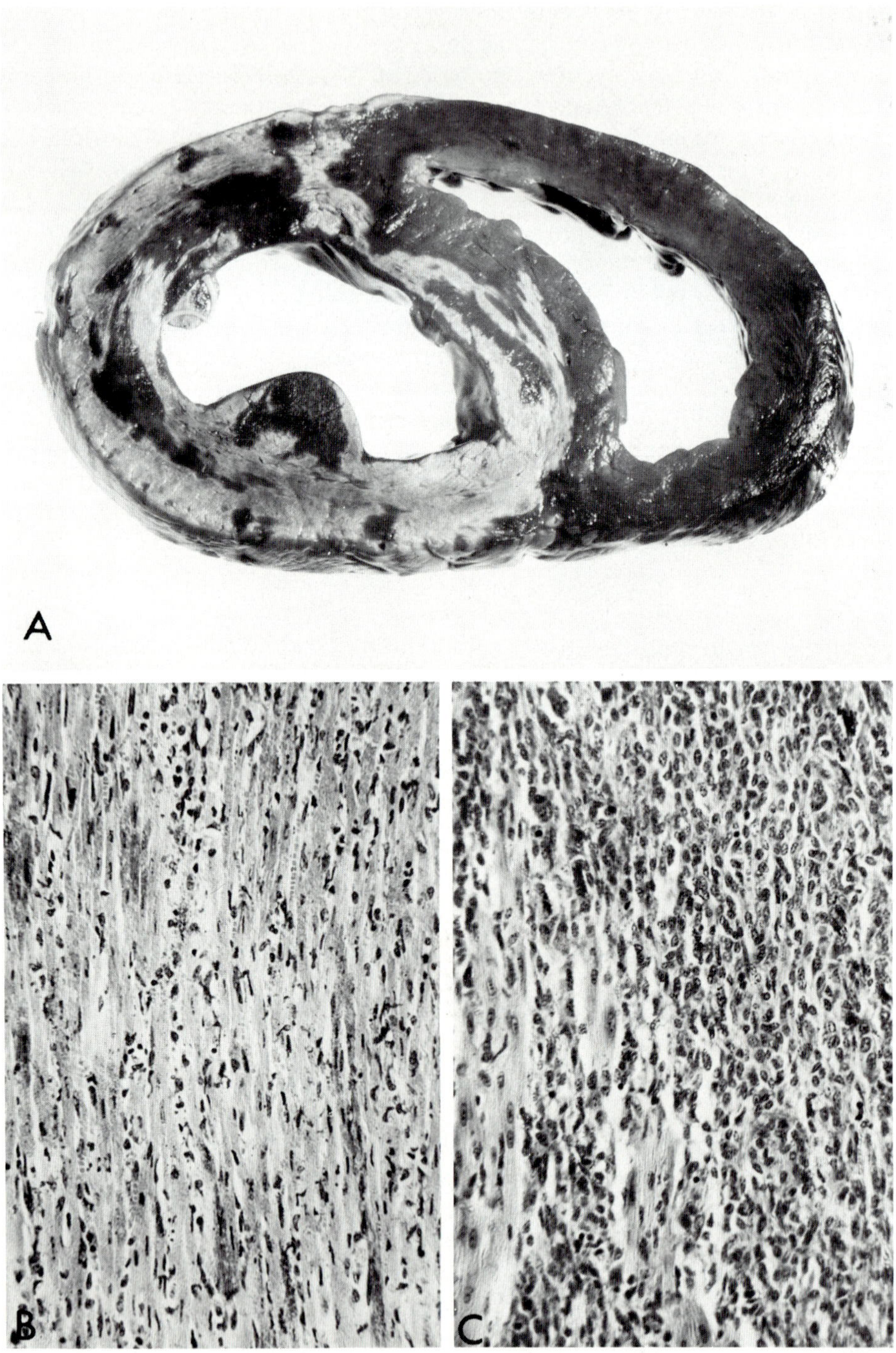


Fig. 8. Nutritional myopathy of calves. *A*, typical distribution of myocardial lesion in the wall of left ventricle as seen in transverse slice of heart. *B*, degeneration of cardiac muscle fibers, pyknosis of muscle nuclei, and beginning calcification of fibers in heart of a 3-week-old calf. *C*, myocardium of 6-week-old calf. Densely cellular area composed of histiocytes and many proliferating cardiac muscle nuclei. Azure-eosinate stain; × 212.

deposits of calcium may be the only apparent changes. In some calves, large areas of myocardium are completely devoid of muscle fibers; only the partially collapsed connective tissue stroma remains. Densely cellular areas composed of a few histiocytes and many regenerative elements of cardiac muscle predominate in other calves. Although frequently not obvious or extensive, lesions in the skeletal musculature usually can be found. In the skeletal muscle form, by contrast, myocardial lesions are either inconspicuous or absent. Prominent proliferation of muscle nuclei perhaps is the most significant microscopic feature of the lesion in skeletal muscle. This and other evidence of regeneration distinguish the acute myodegeneration in calves (and lambs) from otherwise comparable lesions occurring in older animals.

Although the nutritional origin of this myopathy was established early in its history, only recently has its specific cause become more clearly defined. For several reasons, vitamin E deficiency has been considered to have a causal relation to this myopathy. Feeding $\alpha$-tocopherol to preparturient cows has been effective in preventing the disease in their offspring. Similarly, oral doses of $\alpha$-tocopherol have reduced the incidence and severity of the myopathy in calves whose dams were fed known myopathy-producing legume hay during pregnancy. Also, rations fed to cows in some areas where the disease was enzootic were found to contain less vitamin E than rations fed to cows in adjacent areas where the myopathy did not occur. Nevertheless, in other areas the myopathy occurs under conditions of feeding and management that ensure an adequate dietary intake of vitamin E. In these areas, such as the United States, Canada, Scotland, and New Zealand, this myopathy is now considered to be caused by a deficiency of selenium, related either to inadequate dietary intake or to interference with the metabolism or biologic availability of the element. As little as 0.1 ppm of selenium, as sodium selenite, added to the ration of cows will prevent the disease in their calves under conditions requiring extremely large amounts of $\alpha$-tocopherol to produce the same effect. On the other hand, selenium does not prevent the myopathy associated with rations containing large amounts of unsaturated fat. More information on the effectiveness of selenium and its relation to vitamin E in the genesis of this myopathy can be found in the paper by Andrews, Hartley, and Grant.[2]

### Sheep

The nutritional myopathy of lambs to which the appellation "stiff lamb disease" is now applied in English-speaking countries was first observed in Germany 45 years ago (see Hadlow[42]). Essentially identical to the myopathy of calves, it occurs mainly in areas where pregnant ewes in small farm flocks are customarily kept in pens during the winter and fed hay, often legumes, and some grain. Nevertheless, this myopathy has been observed also in lambs born to ewes maintained on irrigated legume pasture. The disease typically affects lambs 2 to 4 weeks old, but it may be apparent at birth (so-called congenital myopathy) or occur in lambs up to 3 months old. As in calves, two clinical forms are seen. In lambs, the skeletal muscle form is the usual or more common one. The affected lamb has a stiff or stilted gait and is reluctant to move. Sometimes it can not elevate its head. Relaxation of the shoulder girdle muscles causes relative abduction and elevation of the suprascapular margins that results in a characteristic standing posture. Some lambs are unable to nurse or have difficulty in swallowing. Levels of serum glutamic oxalacetic transaminase are greatly elevated in affected lambs and may be detected before clinical signs appear.[109] Spontaneous recovery frequently occurs, though a flock mortality rate of 40 to 50% is not uncommon in the absence of effective treatment.

Apart from a tendency to be more widespread, lesions in the skeletal muscles are like those in calves (Fig. 9). They commonly occur in the semitendinosus, semimembranosus, supraspinatus, infraspinatus, longissimus dorsi, pectoral, intercostal, and cervical muscles as well as the diaphragm and tongue. Myocardial lesions, unlike those in calves, often occur mainly in the wall of the right ventricle, usually as partially calcified white subendocardial plaques.

The causal relation of deficiencies of vitamin E and of selenium to the occurrence of this myopathy in lambs parallels that in similarly affected calves. Extremely small amounts of selenium fed to ewes prevents the disease in their lambs. Likewise, a single dose of sodium selenite given subcutaneously to lambs at birth will prevent the myopathy (except for the congenital form). Despite these observations on the prophylactic effectiveness of selenium, a deficiency of selenium by itself may not be the only factor responsible for the occurrence of the disease.[56, 118, 121]

### Swine

Acute degeneration and necrosis of skeletal muscle occur as part of the pathologic pattern of several diseases of swine.[73] Because some have a nutritional basis, the definition of nutritional myopathy in this species has until recently been uncertain. Moreover, in the European veterinary literature frequent use of the term "muscular degeneration" to designate what is actually a postmortem phenomenon in swine muscle, which is unrelated to any specific nutritional disturbance, has contributed to this uncertainty (see "Transport Myopathy of Cattle, Sheep, and Swine"). As a result of studies in Sweden during the past decade, however, nutritional myopathy of swine is now recognized as a distinct entity. Its identifying features, clearly defined by Orstadius[88] and by Lannek,[73] resemble those of nutritional myopathy of cattle and sheep.

The disease affects principally rapidly growing pigs weighing 65 to 130 pounds. It may occur at any age, however, for it has been observed in suckling pigs and in sows at the time of parturition. As a disease problem, the myopathy seems to be limited to Scandinavia, where it is particularly widespread and economically important. Because it most likely has the same etiologic basis as nutritional myopathy of cattle and sheep, the lack of reports from other countries in which this disease is prevalent in young ruminants is not readily understood. Typically, the disease occurs in the autumn after the pigs have been fed newly harvested grain, oats and barley, for 2 to 3 weeks. Excessive rainfall during the summer growing season is generally thought to enhance the

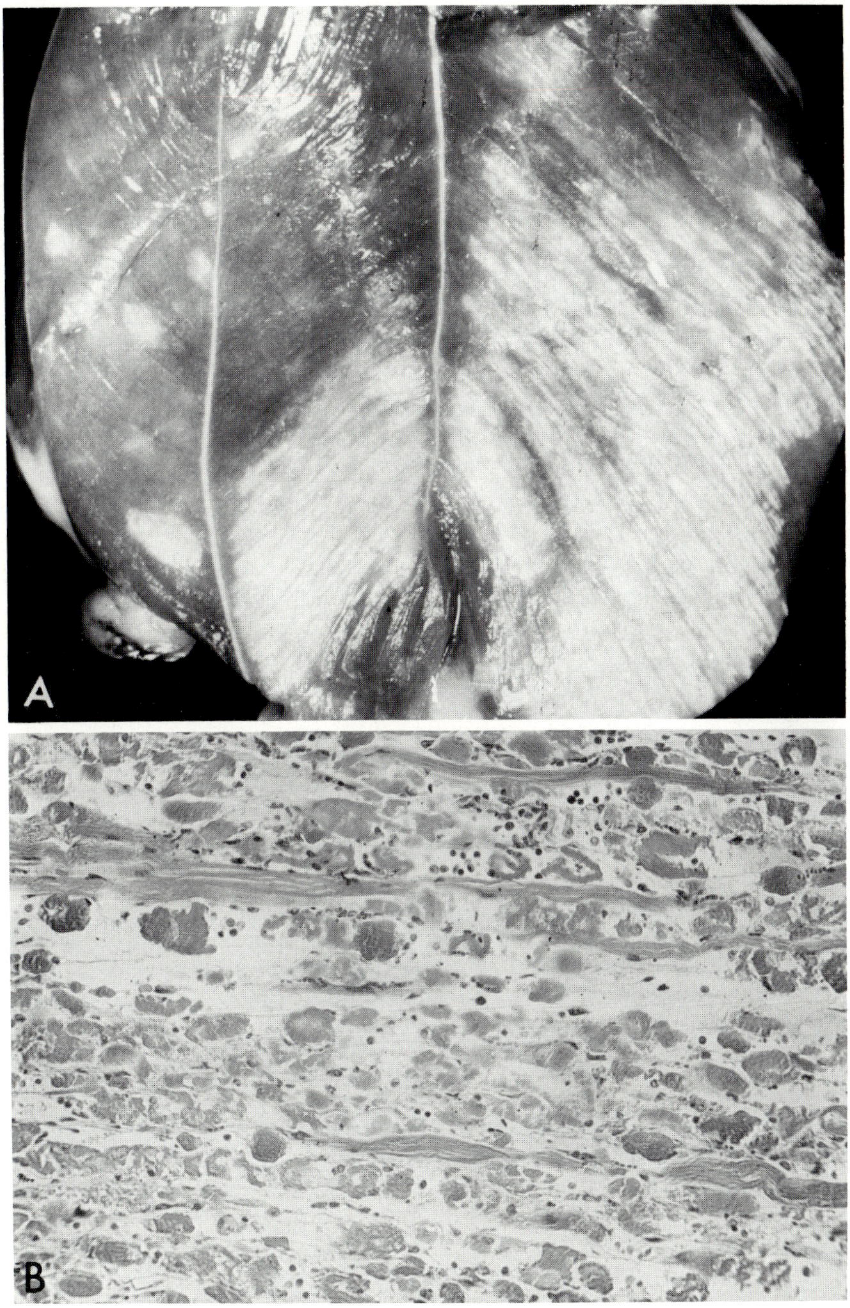


FIG. 9. Nutritional myopathy of lambs. *A*, typical macroscopic appearance of the lesion in quadriceps muscle. (Photo, courtesy of Dr. Rue Jensen.) *B*, necrosis and fragmentation of skeletal muscle fibers in a 2-week-old lamb. Hematoxylin and eosin stain; × 200.

myopathy-producing properties of the grain. During the winter and spring, the incidence of the disease decreases. This may be related to a supposed reduction in the myopathy-producing properties of grain during periods of storage.

The clincopathologic characteristics parallel those seen in nutritional myopathy of other farm animals. In a given herd, some pigs have a staggering gait affecting mainly the hind legs or are unable to rise from recumbency. Others, although clinically normal, are usually affected also, as is made evident by elevated levels of serum glutamic oxalacetic transaminase. Usually, overt disease is not associated with physical stress, such as driving, but the stress of parturition is thought to precipitate clinical myopathy in sows. Slightly affected pigs recover within a few days. More severely affected ones may die suddenly or slowly recover over a period of 2 to 3 weeks. Macroscopically, lesions in skeletal muscles may not be apparent, although pallor, edema, hemorrhage, and necrosis are sometimes seen. Microscopic changes are like those observed in other species affected with nutritional myopathy.

Occurrence of this disease in swine has been clearly related to grain, which is the main component of their diet. At one time, fats in the grain were considered important etiologically. In Sweden, however, the total content and composition of fatty acids of grains are remarkably constant in different geographic areas, whereas the incidence of the disease varies considerably from one area to another. Similarly, the $\alpha$-tocopherol content of the grain was found to have no consistent relationship to its myopathy-producing capacity. On the other hand, grain from areas in which the disease is common is consistently low in selenium, as are tissues from affected pigs. Moreover, selenium added to the feed will prevent the disease; when given parenterally, it provides effective treatment. Thus, in Sweden at least, nutritional myopathy of swine is considered to be caused primarily by dietary deficiency of this element.[72, 73]

### Horses

A myopathy usually accompanied by steatitis but otherwise resembling that of calves and lambs occurs principally in unweaned foals running with their dams on pasture. Dodd, Blakely, Thornbury, and Dewes[32] described its occurrence in New Zealand mainly in Thoroughbred foals 3 days to 5 months old. In the Netherlands, Kroneman and Wensvoort[71] observed it in Shetland ponies 2 to 6 months old. It is well known in Sweden[73] and has been recognized in Japan[122] and Great Britain.[10] Apart from a description of one case, its occurrence in the United States has not been reported. In western Montana where the myopathy of calves and lambs is enzootic, I observed the disease in two quarter horse foals that died within 36 hours after birth.

The clinical pattern of the myopathy varies according to its severity. The acute disease is characterized by a rapidly appearing state of dejection, accelerated heart and respiratory rates, stiff gait, prostration, and death within 1 to 7 days after onset. In many affected foals, swellings, painful on palpation, occur in the subcutaneous fat, particularly that along the nuchal crest and over the gluteal muscles and abdominal wall. Usually, myoglobinuria is present.

Levels of serum creatine phosphokinase are greatly elevated. Onset is more gradual and the clinical course more protracted in other foals. In these, the subcutaneous masses become extremely hard and knobby a few weeks after onset of disturbances in gait. Some foals are unable to suckle because they can not elevate their heads or use their tongues. Many of the less severely affected foals recover spontaneously, but their growth is retarded.

At necropsy, individual muscles or entire groups of them have the usual pale or chalky white appearance associated with hyaline degeneration and necrosis of myofibers (Fig. 10). Most muscles of the limbs, trunk, and neck are affected to varying degrees. In some foals, however, the changes occur mainly in muscles of mastication, external ocular muscles, and tongue.[71] Sometimes the myocardium is also affected. The most striking macroscopic changes are seen in the subcutaneous and visceral fat. The layers of fat are thickened, firm or hard, and yellowish brown. The nuchal crest is especially thickened. Microscopically, the altered fat is characterized by edema, massive infiltration of neutrophils, and necrosis and calcification of fat cells (Fig. 10). Kroneman and Wensvoort[71] mentioned finding in addition small amounts of acid-fast pigment.

As in cattle and sheep, dietary deficiency of either vitamin E or selenium or of both is thought to be related to the occurrence of this myopathy in foals. So far, however, the prophylactic effectiveness of these substances has not been adequately studied. Treatment of affected foals with $\alpha$-tocopherol has been disappointing.

Other animals

Although apparently infrequent, the occurrence of nutritional myopathy in other animals merits brief comment. Most examples are related to a deficiency of vitamin E. Acute myodegeneration is a constant lesion in the rabbit, guinea pig, Syrian hamster, rat, and mouse experimentally depleted of vitamin E, but the natural occurrence of a comparable myopathy in these species has seldom been reported.[42] Similarly, naturally occurring nutritional myopathy seems uncommon in the chicken, probably because its relatively well established nutrient requirements are met by commercial feeds commonly used as rations for poultry.[73] The disease is occasionally reported in ducklings in which myodegeneration characteristically is widespread. Myopathies considered to be caused by a deficiency of vitamin E have rarely been reported in the dog.[68, 77] In one instance, the disease affected four of six weanling Beagle pups whose dam had been fed a semisynthetic food before breeding and during gestation and lactation.[68] A congenital myopathy observed by Manktelow[75] in two litters of sheep dogs in New Zealand was characterized mainly by myocardial necrosis. Selenium deficiency was tentatively considered to be the cause.

Among captive wild animals, nutritional myopathy was reported in four tree kangaroos fed large amounts of fish oil (see Hadlow[42]). Its occurrence in the Rottnest quokka (*Setonix brachyurus*), a small nocturnal wallaby, has been studied intensively by Kakulas in western Australia.[67] The myopathic process can be reversed by treatment with $\alpha$-tocopherol but not with selenium. Dietary supplements of vitamin E prevent the disease. In a breeding herd of nyala

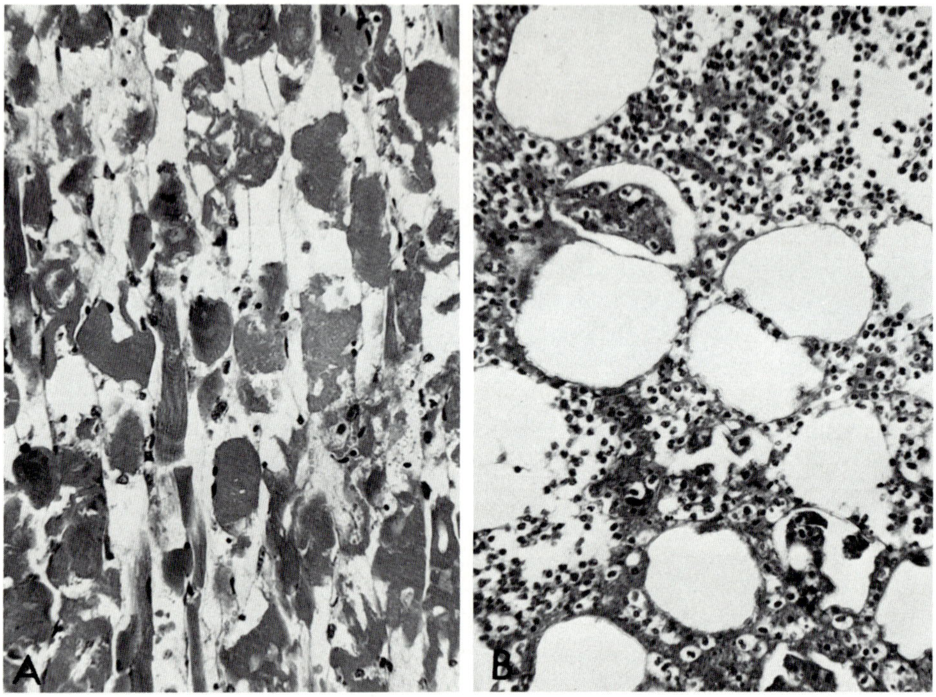


FIG. 10. Nutritional myopathy of foals. *A*, severe hyaline necrosis and fragmentation of muscle fibers in semitendinosus muscle. *B*, acute inflammation and necrosis of subcutaneous fat. Hematoxylin and eosin stain; × 200. (Sections, courtesy of Mr. W. J. Hartley.)

antelope (*Tragelaphus angasi*), Gandal[38] observed a myopathy resembling that of calves and lambs. It affected young antelope, usually those less than 5 months old. Parenteral use of a preparation containing vitamin E and selenium when the antelope were a week old seemed to have prophylactic value.

## TOXIC MYOPATHIES

Although several chemicals are known to have pronounced myotoxic properties when used experimentally,[64, 95] little is known about the causal relation of chemical poisons to naturally occurring myopathies in animals. The few examples of toxic myopathy that have been recognized in animals are considered here.

### Plant poisoning in ruminants

In many parts of the world poisonous plants are an extremely important cause of disease among farm animals. A number of these plants causes intoxications characterized clinically by signs of neuromuscular dysfunction, such as muscular spasm, muscular weakness, incoordination of gait, and inability to stand. Some of them contain neurotoxins that cause morphologic changes in the central or peripheral nervous system. Others are assumed to contain neurotoxins even though neuropathologic changes have not been evident. With the notable exceptions described here, the presence of myotoxic substances in plants has not been demonstrated, though often suspected.[33]

*Sheep and goats.* A syndrome characterized by progressive leg weakness, incoordination of gait, inability to stand, and death occurs in sheep and goats in southwest Texas. Called "limberleg," it has long been known to be caused by the ingestion of mature fruits of the coyotillo plant (*Karwinskia humboldtiana*), a spineless shrub indigenous to the area. Several early writers mention that children also were poisoned by eating the fruits, whose oily seeds were stated to contain a substance causing paralysis of motor nerves. Only recently, however, has the fruit of the coyotillo plant been shown to contain a myotoxic substance(s).[30]

Goats fed ground coyotillo fruit daily at levels of 0.025, 0.075, and 0.15% of their body weight became ataxic, usually in 7 days. Those that received the larger amounts became recumbent and died 1 to 2 days later. Goats fed the smallest amount survived 5 to 11 days after onset of clinical signs. In all, the levels of serum glutamic oxalacetic transaminase were greatly elevated. Lesions in muscle, usually not evident macroscopically, were widespread; all main groups of muscles, including diaphragm and external ocular muscles, were affected. Microscopically, the changes resembled those seen in other acute myodegenerations of animals. Similar, but less pronounced, changes were present also in the myocardium. Apart from minor alterations in axons of the spinal cord and peripheral nerves, the nervous system was intact. Thus, the syndrome caused by the ingestion of coyotillo fruit is now defined as a toxic myopathy.

*Cattle.* An economically important acute myopathy has been observed in yearling and older cattle in much of the eastern half of Texas at least since 1952. A similar disease has been observed in Florida for a number of years. Often considered in the past under the appellation "white muscle disease," the myopathy was recently shown to be caused by the ingestion of coffee senna or senna (*Cassia occidentalis*), a plant indigenous to southeastern United States and present in many parts of the world.[52, 79, 86] Of all parts of the plant, the beans are the most toxic. Although intoxication was produced experimentally in cattle, rabbits, sheep, and a goat, myodegeneration was found only in cattle. The myotoxic substance(s) has not been identified.

Typically, most cattle become affected during November and early December shortly after the first killing frost.[94] Apparently, at this time cattle are more apt to eat the plant. Often affecting 10 to 60% of animals in a herd, the disease is characterized by an afebrile course with gastrointestinal disturbances, including diarrhea, in the early stages. Muscular weakness becomes evident 5 to 7 days after the plant is first consumed. Affected cattle have a stumbling or swaying gait, have difficulty rising from recombency, and fall easily when forced to move rapidly. Myoglobinuria occurs, and serum levels of creatine phosphokinase, glutamic oxalacetic transaminase, and aldolase are elevated. The severity and rate of progression of the myopathy depend upon the amount of plant eaten—a feature confirmed experimentally.[51, 79] Animals may die suddenly 2 to 3 days after eating a large amount. Others that consume smaller amounts have some muscular weakness but remain standing. Most cattle that become recumbent die after several days.

Affected muscles are diffusely pale or have a stippled appearance as a result of many pale 1- to 2-mm foci regularly spaced along muscle fibers. Most large muscle masses are involved but especially those of the hind limbs. Severe widespread acute myodegeneration like that in goats poisoned by coyotillo fruit is seen microscopically (Fig. 11). Similar, but less severe and less extensive, changes also occur in the myocardium of some cattle, and account for the fulminant form of the toxicosis.

### Iron poisoning in baby pigs

Intramuscular injection of iron preparations is now widely used to prevent the hypochromic microcytic anemia that affects baby pigs when they are about 3 weeks old. Although usually without adverse reactions, this practice is attended by some risk. Acute poisoning, characterized mainly by widespread hyaline degeneration or coagulation necrosis of muscle fibers, has been reported from several countries, following such parenteral administration of iron.[3, 112] Poisoning also has resulted from the oral use of iron compounds. Iron-carbohydrate complexes (iron dextran) are less toxic than ionic iron. Moreover, considerably less risk is incurred when iron firmly bound to polysaccharide of high molecular weight is used, because of its slow rate of absorption. Although not adequately explained, iron is especially toxic to baby pigs deficient in vitamin E. This risk can be reduced by the use of natural and synthetic antioxidants and sodium selenite.

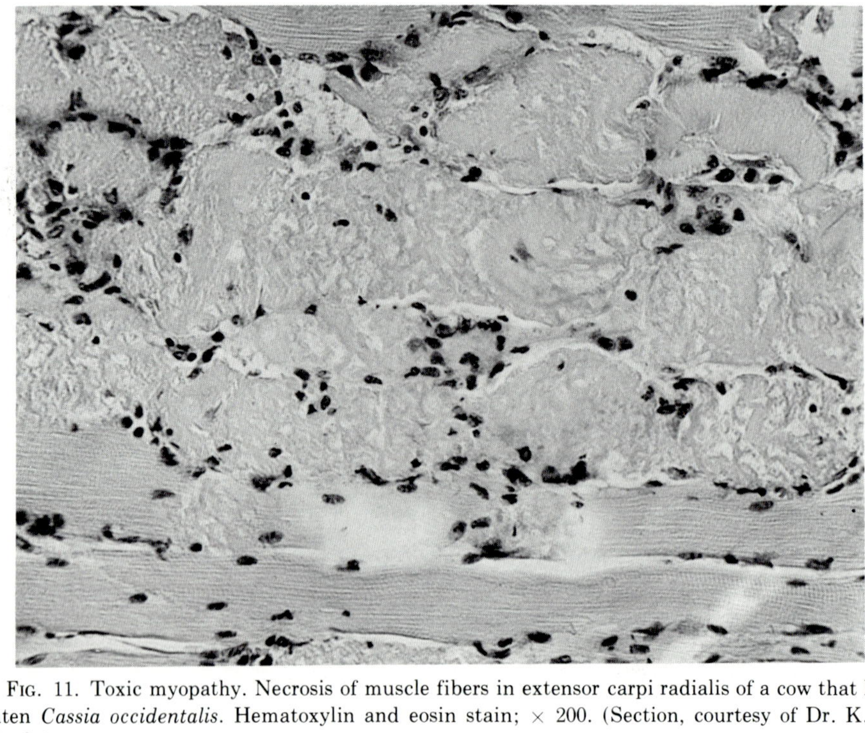

Fig. 11. Toxic myopathy. Necrosis of muscle fibers in extensor carpi radialis of a cow that had eaten *Cassia occidentalis*. Hematoxylin and eosin stain; × 200. (Section, courtesy of Dr. K. R. Pierce.)

Clinically, the toxicosis is characterized by muscular weakness and spasm that may proceed to prostration and death within a few hours to several days after the pigs receive the iron preparation. Macroscopic changes in the musculature are often absent, but small yellow-white streaks may be evident in muscles of pigs that live for several days. Severe hyaline degeneration or coagulation necrosis with fragmentation of muscle fibers is seen microscopically. Sometimes dystrophic calcification of damaged muscle fibers occurs. Beginning regeneration may be seen in pigs that live 4 to 5 days. In one experimental study,[92] damaged fibers were distinguished from normal fibers by a complete lack of glucosan phosphorylase staining, and succinic dehydrogenase activity was concentrated around the periphery of swollen fibers—an observation interpreted as suggesting clumping of mitochondria. In some reported cases, degenerative changes like those in skeletal muscle were found in the myocardium and resulted in pericardial and pleural effusions.

The mechanism of iron myotoxicity has been considered by Patterson *et al.*,[92] who suggested that iron apparently initiates myodegeneration by causing peroxidation damage; the concentration of "peroxides" in muscle of poisoned pigs was found to rise to more than twice the preinjection level. Their observation of release of acid phosphatase and $\beta$-glucuronidase into the plasma suggested damage to lysosomes. An increased affinity of skeletal muscle for iron may account for its enchanced toxicity in vitamin E-deficient pigs.

## ACUTE AND SUBACUTE MYOPATHIES OF UNCERTAIN CAUSE

Hyaline or waxy degeneration and segmental necrosis of muscle fibers with preservation of supporting connective tissues, like that seen in the nutritional and toxic myopathies, also characterizes other, largely acute, myopathies of animals.[42] In general, these occur sporadically, under varying circumstances, and affect mainly adults. Because of their morphologic resemblance to undoubted nutritional myopathies, they often have been wrongly considered as reflecting similar nutritional deficiencies. Actually, the cause of each is unknown or, at least, uncertain. Nevertheless, some may eventually be related to specific nutritional, or toxic, factors for which present evidence is only presumptive. Because of this uncertainty about their causes, they are considered here, admittedly somewhat arbitrarily, as a separate group. Several myopathies essentially similar, but of longer clinical duration, are also included.

### Paralytic myoglobinuria of the horse

First described over 100 years ago, paralytic myoglobinuria is a frequently fatal acute myopathy affecting mainly well nourished, regularly worked draft horses, especially during the spring (see Hadlow[42]). It is characterized by sudden onset of profound weakness of the hindquarters accompanied by profuse sweating, muscular spasm, and pronounced myoglobinuria. Typically, onset occurs within an hour after the horse is put to work following 2 to 5 days of rest while receiving a high caloric diet. Attacks vary widely in severity. They may disappear within a few hours or, more often, persist until the animal collapses. Horses that survive usually recover within 2 to 4 days, either completely or with some muscular atrophy. More severely affected horses die, usually after several days to a week, often from renal failure. In some respects, this equine myopathy resembles idiopathic paroxysmal myoglobinuria of man.

Lesions in the skeletal musculature may be widespread, but the most evident ones macroscopically are found in the lumbar, gluteal, and quadriceps groups. These vary from general pallor to grayish yellow friable patches that are sometimes mottled with small hemorrhages. Microscopically, the usual lesion consists of small and large groups of swollen, hyalinized, and fragmented muscle fibers (Fig. 12). In severely affected muscle, liquefaction of damaged fibers supervenes, and then only sarcolemma and endomysium remain. Infiltration of inflammatory cells is variable. Similar fiber degeneration sometimes accompanied by inflammation may also occur in the myocardium and result in death from cardiac failure. Complicating renal changes accompanied by oliguria or anuria consist of tubular degeneration progressing to complete necrosis and the formation of pigmented casts (Fig. 12).

Despite various explanations, the precise cause and pathogenesis remain obscure. The most widely held view is still that proposed by Carlström 40 years ago.[26] He believed the disease to be associated with the deposition of large amounts of glycogen in muscle during periods of rest. At the onset of exertion, this resulted in greatly increased concentrations of lactic acid in the blood—a feature not observed in the counterpart disease of man. The abnormally high concentrations of lactic acid, accompanied by local tissue hypoxia, were assumed to cause the muscular damage and release of myoglobin.

Because paralytic myoglobinuria is primarily a disease of draft horses, it is now less commonly seen than in the days before mechanical power. Nevertheless, a similar, but milder, syndrome is being increasingly recognized as an important entity among racing Thoroughbreds and other light horses under strenuous exercise or training.[18] Known in horsemen's parlance as "tying-up," the syndrome usually appears toward the end of a period of exercise or immediately afterward. Signs of the disease include rapid breathing, profuse sweating, stiff gait of the hind legs, reluctance to move, swelling of the lumbar and gluteal muscles, and sometimes myoglobinuria. Unlike paralytic myoglobinuria, however, this myopathy is seldom complicated by renal damage; most affected horses recover completely in a short time. Whether the "tying-up" syndrome and classic paralytic myoglobinuria represent variations of a single myopathic process or are fundamentally different is not known.[24]

For some years a syndrome resembling paralytic myoglobinuria has been observed also in Greyhounds used for racing.[40] Referred to as "pinched back" by Grehound fanciers, the disease occurs mainly during periods of training, particularly in warm weather. Typically, the syndrome appears within a day after the dog is strenuously exercised following a rest period of several days or weeks. Affected dogs have a stiff or ataxic gait of the hind legs; some are unable to stand. Lumbar muscles, and sometimes the gluteal muscles, are extremely firm and painful. Myoglobinuria occurs, and severely affected dogs may die from acute renal failure. Atrophy of the lumbar muscles ensues in some dogs

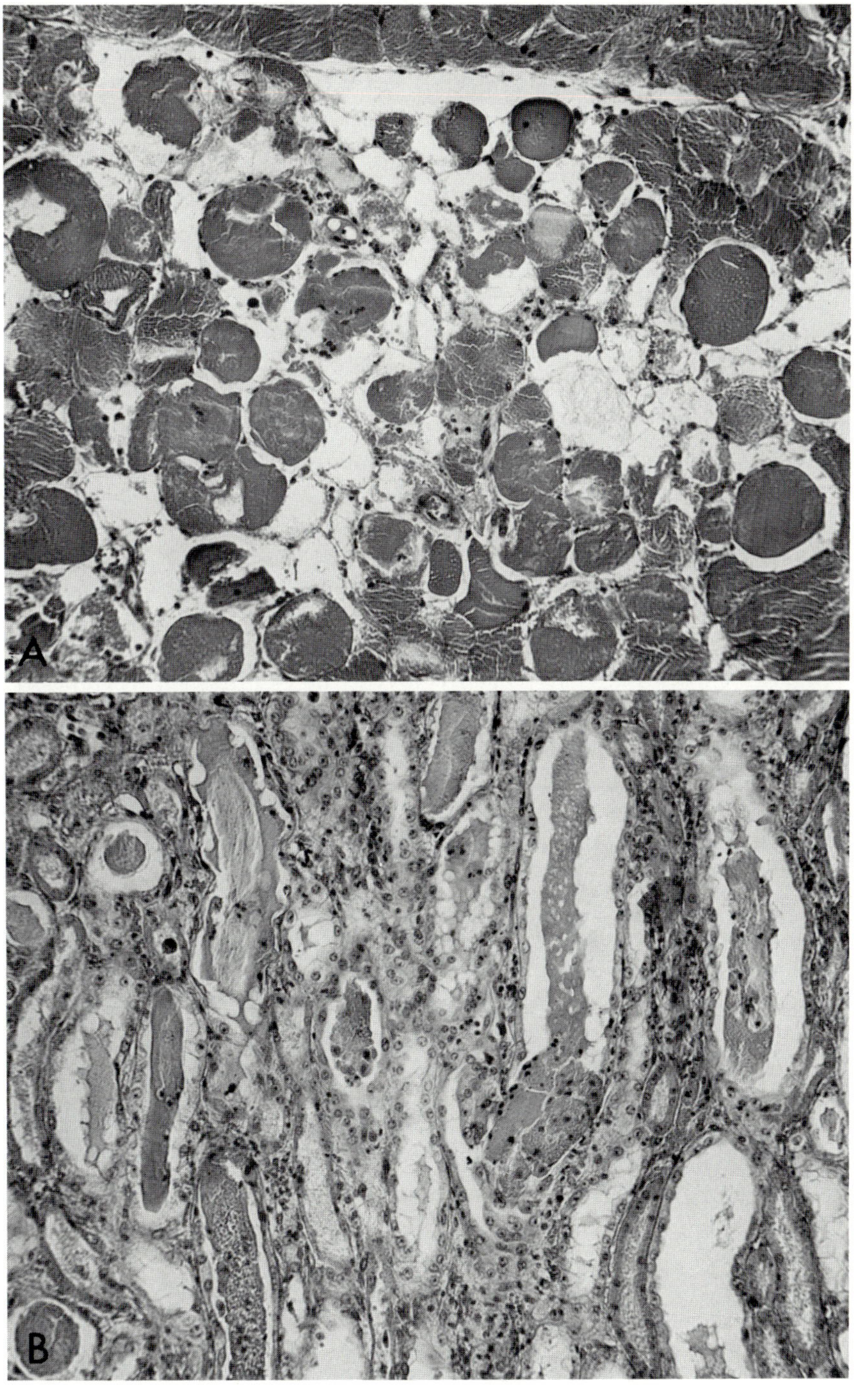


FIG. 12. Paralytic myoglobinuria in the horse. *A*, hyaline degeneration and lysis of skeletal muscle fibers. Hematoxylin and eosin stain; × 175. *B*, casts of myoglobin and degeneration of tubular epithelium in kidney. Hematoxylin and eosin stain; × 150.

that recover from an attack. Microscopic changes occurring in the skeletal musculature have not been described.

TRANSPORT MYOPATHY OF CATTLE, SHEEP, AND SWINE

In addition to the examples already cited, other acute myopathies have been recognized that are precipitated by muscular exertion ostensibly in the absence of some associated disease. Although the circumstances of their individual occurrence vary, they are grouped under this heading for convenience of description. In each, little is known about the underlying cause.

A myopathic disorder of sudden onset may affect young and adult cattle subjected to considerable muscular strain during transportation over long distances, especially under crowded conditions in railway cars.[29] A comparable condition may occur after prolonged driving or chasing or from violent struggling. The acute myopathic syndrome that has been observed in draft oxen under circumstances similar to those provoking paralytic myoglobinuria in the horse probably belongs to this general category too. Clinical signs vary from transient disturbances in gait to sudden collapse of the animal. Sometimes myoglobinuria occurs. Patchy areas of granular and hyaline degeneration of muscle fibers accompanied by fragmentation and lysis are found in the musculature of the shoulders, back, and thighs. Sometimes the myocardium is similarly affected.

An analogous syndrome has been observed in yearling sheep subjected to unaccustomed muscular activity after 3 to 5 months of winter feeding mainly on turnips.[49] In the South Island of New Zealand, where the syndrome has been recognized for many years, up to 50% of a flock may be affected. Onset is precipitated by driving, especially during cold and wet seasons, and by stresses associated with long distance transportation by truck or railroad. The main clinical features are listlessness, stiff gait, reluctance to move, and inability to stand. Myoglobinuria occurs, and serum glutamic oxalacetic transaminase levels are elevated. Some sheep become prostrate and often die within 24 hours. Those that are able to walk usually recover within 6 to 10 days. Affected portions of the musculature, often most evident in the thighs, are paler than normal and tend to be dry and granular. Severe acute degeneration and coagulation necrosis of muscle fibers are seen microscopically. Because a ration composed chiefly of turnips is low in vitamin E, a relative lack of this nutrient may play a pathogenetic role in this myopathy.

Swine transported from farm to abattoir may suffer from an acute, often fatal, disease while en route or shortly afterward. Occurring particularly during the summer months, it is characterized by sudden onset of labored breathing and muscular weakness. As observed in several European countries, however, so-called "transport muscular degeneration" of swine is in reality a problem of alterations in the skeletal musculature that become evident within a few hours after the animal is slaughtered, rather than a recognizable clinicopathologic entity. Muscles become pale and soft, and excessive fluid is liberated from their cut surfaces. Thus, although clearly related to the stresses of transportation and handling immediately before slaughter, the condition is to be distinguished from a true myopathic disorder.[72, 73, 114] The various factors responsible for this economically important postmortem change in swine muscle were discussed by Rückler.[105]

POLYMYOPATHY OF ADULT SHEEP

In 1956, Bosanquet, Daniel, and Parry[19] described a slowly progressive, almost always fatal, polymyopathy affecting adult sheep of either sex. Insidious onset, progressive muscular weakness leading to ataxia, pronounced muscular wasting, signs of cutaneous irritation, and death after 2 to 3 months were the clinical features most often observed. Macroscopic changes in the skeletal musculature, which tended to be bilaterally symmetrical, were widely scattered and varied greatly in severity and extent. Some muscles were enlarged and pale, whereas others were wasted, flabby, and translucent. Microscopically, the principal changes, which were not unlike those in clinically more acute degenerative polymyopathies, included granular and hyaline degeneration and fragmentation of muscle fibers, proliferation of muscle nuclei, and other signs of regenerative activity. Perivascular and interstitial accumulations of lymphocytes were infrequent; frank inflammatory changes were absent. Increased amounts of fibrous connective tissue and fat cells were not conspicuous.

Preliminary observations suggested that this myopathy was not of nutritional origin. Its strong familial incidence, however, led the authors to consider that genetic factors most likely were involved. Moreover, they regarded certain clinicopathologic features as indicating a general similarity of this ovine myopathy to polymyositis and to muscular dystrophy of man.

Significantly, most sheep in which the muscular degeneration was found were also affected with scrapie,[91] a disease now regarded as a slow viral infection of the central nervous system.[43] As a consequence, the clinical and epidemiologic characteristics described for the myopathy are in reality largely those long associated with scrapie, the neurologic disease. Thus, the cause and significance of this myopathy are far from clear. Actually, as shown by Stamp,[107] essentially identical degenerative changes may be widespread in muscles of adult sheep in the absence of the clinical signs described by Bosanquet, Daniel, and Parry[19] and of any epidemiologic findings indicative of a genetic origin.

POLYMYOPATHY OF THE DOG AND THE CAT

The frequent occurrence of degenerative polymyopathy in the large domestic animals stands in sharp contrast to its extreme rarity in the dog and the cat. In fact, as judged by the few reports, any kind of primary disease of muscle is rare in these species. Although the several examples of degenerative polymyopathy reported to date may have had different causes, they are considered together here because of their general morphologic similarity.

Among some 3000 dogs examined postmortem, Meier[77] found three with primary myopathies. One, a puppy affected with nutritional myodegeneration, was cited earlier. In the other two dogs, microscopic changes in the skeletal musculature were considered to resemble those of polymyositis. The cause was not apparent in either. One dog, an emaciated 8-month-old male German

Shepherd, had a stiff, stilted gait when first examined. This became progressively worse over a 2-month period until the dog was unable to stand or to elevate its head. The main microscopic changes in muscle included granular degeneration, fragmentation, and loss of muscle fibers with their replacement by densely cellular connective tissue. Fibers that remained were thin and contained many centrally placed nuclei. Little evidence of regeneration was seen, and inflammatory changes were virtually absent. The other dog, a 3-year-old female Boxer, had difficulty in climbing stairs, was unable to retain food in its mouth or to swallow, and over a 3-month period became emaciated. Extensive hyaline degeneration accompanied by prominent proliferation of muscle nuclei was observed in the wasted muscles. Many lymphocytes and neutrophils were present in some areas, but a widespread inflammatory response was not a conspicuous feature of the lesion.

Manktelow[75] reported an example of severe acute polymyopathy in a 2-year-old working Collie sheep dog that suddenly became unable to use its hind legs. Bilaterally symmetrical pale patches resembling those in calves and lambs affected with nutritional myopathy were observed in all main groups of muscles. Extensive acute myonecrosis unaccompanied by any appreciable interstitial cellular response was seen microscopically. In view of these pathologic changes, the dog's diet of mutton only, and origin of the dog from an area in which nutritional myopathy occurred in lambs, Manktelow considered dietary deficiency of selenium the probable cause. Whether an acute myopathy of this type in the dog is selenium-responsive is not known.

A strikingly similar acute myopathy in a 3-year-old cat, and the only reported example of a primary muscular disease I found in this species, was described by Gardner.[39] The cat, whose diet consisted mainly of raw mutton and milk, became affected with a febrile disorder lasting 10 days that was characterized by labored breathing, "twitching convulsions," and epistaxis. At necropsy, large, bilaterally symmetrical pale areas were present in several muscle groups. Swelling, coagulation necrosis, fragmentation, and calcification of muscle fibers were the essential microscopic changes. The cause of the myopathy was not determined, but like Manktelow, Gardner wondered about its relation to selenium deficiency.

### Hind leg myopathy of old rats

During studies on longevity in Sprague-Dawley rats, Berg[14] observed, among other diseases, a myopathy that was an important cause of death of senescent animals. It seldom appeared before rats were 2 years old. Although it occurred earlier and was more severe in males than in females, its incidence in the colony increased rapidly to 100% as the rats aged. Initially, rats were unsteady on their hind legs and were reluctant to move. As the weakness of the hindquarters became more pronounced, the rats acquired a waddling gait and tended to drag their hind legs. The forelegs were unaffected. Within 6 to 8 weeks after onset of weakness, paralysis of the hind legs became complete, and muscular control of the tail was lost. Muscles of the hindquarters became wasted, body weight declined abruptly, and the rats soon became moribund.

The myopathy affected mainly the gastrocnemius and adductor muscles of the hind legs, which were atrophic, brownish, and flabby. Early microscopic changes included loss of cross-striations from muscle fibers and proliferation of muscle nuclei. Later, muscle fibers became necrotic and fragmented, and then only collapsed sarcolemmal sheaths remained. Muscle nuclei, which were usually preserved, were often arranged in long rows; some formed clusters resembling giant cells. Various stages of regeneration of muscle fibers were evident. Sometimes fat cells replaced necrotic muscle fibers. Distribution of the myodegeneration was irregular or patchy, and its severity and extent varied considerably. In general, however, its severity paralleled the severity of the hind leg weakness.

Clinically and histologically, this myopathy resembles the chronic form of vitamin E deficiency in adult rats. Nevertheless, it was not prevented by large supplements of $\alpha$-tocopherol.[15] Moreover, the myopathy is believed to be unrelated to degeneration of myelin affecting the cauda equina, other spinal roots, and some peripheral nerves in the same animals. For example, dietary restriction delayed the onset of the myopathy but did not alter the incidence of the radiculoneuropathy.[16] Apparently, the cause of this condition affecting aging rats is still uncertain.

MYOPATHIES OF POULTRY

*Focal myopathy of the chicken, turkey, and duck.* Rigdon and associates[96, 98, 99, 100, 101] have described essentially identical myopathies in the chicken, turkey, and duck. In each, the most conspicuous morphologic change consists of focal swelling, vacuolar degeneration, necrosis, and liquefaction of individual muscle fibers. The process is seldom accompanied by an inflammatory reaction. Regeneration of necrotic fibers may occur, but apparently most of them are replaced by fat cells. The number of necrotic fibers varies widely not only from one bird to another, but also from one muscle to another in the same bird. Although in each species the myopathy may occur in association with contractures, torticollis, and arthrogryposis, the extent of the muscle fiber necrosis often bears no consistent relationship to their severity. Indeed, the causal relation of the focal necrosis in muscle to some of these conditions is uncertain. In all species, the myopathy usually is not accompanied by clinical manifestations. Because of these and other observations, the true nature and clinical significance of these avian myopathies remain obscure.

Their cause is not known. All seem unrelated to vitamin E deficiency, but a nutritional basis still appears worth considering. Rigdon suggested a genetic mechanism characterized by a local metabolic defect. Although he considers these myopathies somewhat akin to muscular dystrophy, their natural history and pathologic features readily distinguish them from the hereditary dystrophy-like myopathies of the chicken and turkey.

In the chicken, the myopathy has been seen mainly in several strains, some inbred, of the White Leghorn breed.[100] Occurring in birds of various ages, it sometimes is associated with arthrogryposis and various malformations of the feet but rarely with torticollis or a pendulous crop. Supposedly, another expression of this focal myopathy occurs in embryos and newly hatched chicks in which the musculus complexus, or "pipping" muscle, is the most severely affected muscle.[97]

Apparently common in the Broad Breasted Bronze turkey, the focal myopathy has been found in adult birds from several widely scattered areas in the United States.[99] It has occurred in turkeys that were lame or unable to stand. Other birds had enlarged hock (tibiotarsometatarsal) joints or a pendulous crop. Necrosis of the thin striated muscle that supports the crop apparently allows it to dilate and become pendulous. Maronpot, Bucci, and Stedham[76] found the myopathy in 10% of a large flock of young Broad Breasted White male turkeys that were lame and in generally poor physical condition. Some degreee of torticollis was also present in many of them. The disease was much milder and affected only 2% of the birds in an equally large flock of Broad Breasted Bronze turkeys on the same farm.

In the White Pekin duck, the focal areas of necrosis in muscle first become demonstrable when the ducks are 2 to 3 weeks old.[96] Thereafter, they can be seen in ducks of all ages, but are more numerous in the older ones. Although the myopathy occurs in 90% of the ducks, disabling contractures of one or both legs are found in only 10%. Unlike the turkey, the White Pekin duck is seldom affected by torticollis or pendulous crop. Similar focal lesions, although less extensive than those in the White Pekin, have been found also in high incidence in four other breeds or species of domestic ducks. In these, the myopathy was unaccompanied by leg weakness, contractures, or torticollis.

*Acute myopathy in the male turkey*. Bergeland and Barnes[17] described an acute myopathy that affects only male turkeys used for semen production during November through January. Of sudden onset, it usually occurs 5 to 10 days after the males are exposed to increasing hours of light to induce sexual activity outside the normal breeding season. During these artificially lengthened days, the male turkey begins to strut and fight as its testes enlarge. The incidence of the condition varies, but it commonly affects 5 to 10% of a flock, regardless of breed. Lameness and difficulty in maintaining normal posture of the head and neck are the main clinical signs. The turkey sits with its neck stretched out on the floor or rests on its hocks with its head pointed upward. When less severely affected, it tends to walk in circles, and its neck is deviated laterally. If removed from the flock and given good nursing care, most turkeys that are not severely affected recover, although sometimes lateral deviation of the neck persists. Acute degeneration and necrosis of muscle are widespread, but the dorsal cervical muscles and medial thigh muscles are most severely damaged (Fig. 13). Sometimes the leg muscles rupture. Although the cause of this myopathy is uncertain, it undoubtedly is precipitated by the sudden unaccustomed activity of neck and leg muscles, as in strutting and fighting, in well fed, heavily muscled birds.

### Myopathies of wild animals

Several reports clearly indicate that free-living wild animals are also affected by myopathies. These are characterized mainly by widespread segmental

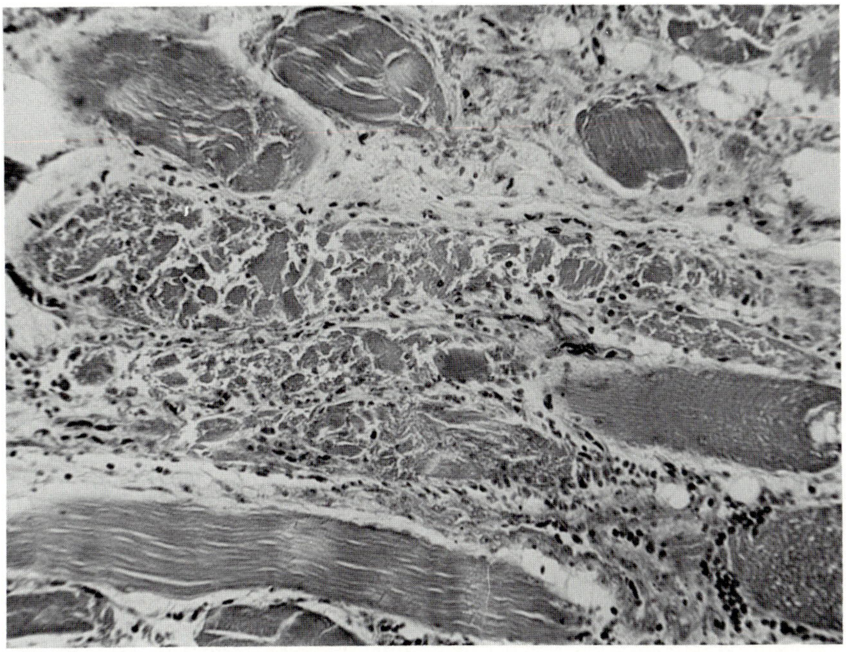

FIG. 13. Acute myopathy in a male turkey that was unable to elevate its head. Extensive necrosis in a cervical muscle. Hematoxylin and eosin stain; × 150. (Section, courtesy of Dr. M. E. Bergeland.)

degeneration and necrosis of skeletal, and sometimes cardiac, muscle fibers like that seen in acute myopathic disorders of domestic animals. Such nonspecific lesions have been observed in wild animals found dead or killed in their natural habitat. They also have been seen in animals that because of impaired locomotion were easily caught. In other instances, myopathic changes have been present in animals that became weak or suddenly collapsed and died within a few days to a week or so after their capture and transport to holding compounds. The clinical significance of the lesions in skeletal muscle is not altogether clear. Undoubtedly, they may be clinically inapparent. In other instances they may cause death. Mildly affected animals must be at a serious disadvantage in eluding predators.[84]

Presumably varied, the causes are as obscure as they are for some acute myopathies of domestic animals. Nutritional deficiencies and plant toxins are possibilities. Some examples of myopathy most likely are related to coexisting diseases often demonstrated at necropsy. In animals that ostensibly become affected only after they are captured and transported, excessive muscular exertion and stresses of handling undoubtedly are important causal factors.

Most reports have dealt with the occurrence of acute myopathy in wild ruminants. It seems to be especially common in several species of antelope and gazelle in Africa.[62, 63, 83, 84, 123] Muscular lesions have been reported also in a young roe deer in Scotland[36] and in an adult white-tailed deer in the United States.[41] Recently, I found widespread acute myonecrosis in a thin 2-year-old

Rocky Mountain bighorn ram that was unable to stand when caught in western Montana during late spring (Fig. 14).

## MISCELLANEOUS MYOPATHIES

Several myopathic disorders not readily classified elsewhere in this chapter are grouped under this heading. Functional disturbances presented by most of them are not associated with overt histologic changes in the muscle fibers but reflect more subtle abnormalities of muscle metabolism or neuromuscular transmission.

### Congenital myotonia in the goat

In 1904, White and Plaskett[119] reported a congenital myotonic disorder in the goat that has since been studied intensively as the counterpart of myotonia congenita (Thomsen's disease) of man (see Bryant, Lipicky, and Herzog[22] for references). Origin of the trait is obscure; supposedly, myotonic goats in existence today are descendants of four affected animals brought to Marshall County, Tennessee about 1880.

When startled into sudden movement by loud noise or threatening gesture, the goats become rigid and are unable to move. This may be momentary or last for as long as 30 seconds. All limbs become stiffly extended, and tonic contractions occur in muscles of the neck and trunk. If off balance, the goat usually topples over. Such episodes of rigidity, during which the goat remains conscious, recur only after a rest of 15 to 30 minutes. Percussing the surface of a

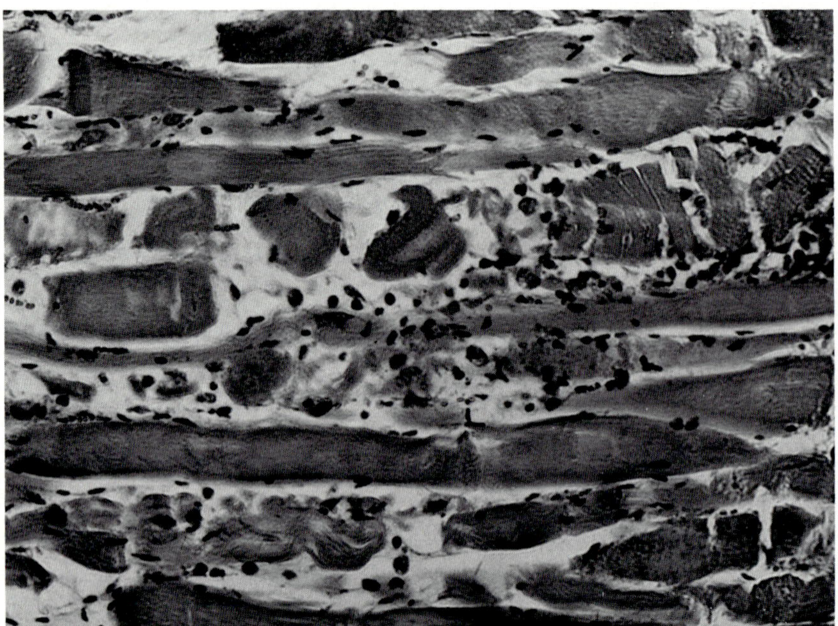

Fig. 14. Acute myopathy of unknown cause in a 2-year-old Rocky Mountain bighorn ram. Segmental necrosis of fibers in a cervical muscle. Azure-eosinate stain; × 200.

muscle elicits a typical myotonic response (localized contraction). Considerable day-to-day variation occurs in the intensity of the myotonic response that is not related to meteorologic changes, feed, or exercise.[22, 50] Dehydration, however, is known to lessen the intensity of the myotonic signs. By using several clinical and electrophysiologic criteria to assess the intensity of the myotonia, Bryant, Lipicky, and Herzog[22] were able to distinguish two classes of goats: mildly affected and severely affected. Overt myotonic signs accompanying movement were seldom seen when the percussion response persisted for less than 2 seconds.

Usually, the disorder becomes evident when kids are several weeks old and, according to some authors, it tends to become more severe with age. Both sexes are affected, but some observers mention a higher incidence of myotonia in males. Attacks do not occur, or are much less severe, during pregnancy and for a short time after parturition. Pathologic changes have not been found in the musculature,[23] but in two studies muscle fibers of myotonic goats were slightly larger than those of normal goats.

That the disorder is hereditary seems certain, but its mode of inheritance has not been established. Probably the trait is transmitted not as a simple character but by multiple factors. Such a genetic pattern was suggested by Clark, Luton, and Cutler,[28] who noted that the disorder tends to be transmitted in a "diluted" or less severe form when myotonic and normal goats are interbred. Burns, Dale, and Langley[23] did not see myotonia in progeny of a myotonic male and a normal female. Bryant, Lipicky, and Herzog[22] observed four normal male kids born to a myotonic female, and suggested that a single autosomal dominant gene with incomplete penetrance could account for the two intensities of myotonia observed in their herd of goats.

Various physiologic and pharmacologic studies have been made to define the mechanism of the myotonia. It is lessened or abolished by the administration of quinine but not by atropine; it is accentuated by prostigmine, pilocarpine, and diazacholesterol. The myotonia is not associated with lowered cholinesterase activity. Anesthetized goats respond with tetanic contractions when direct current is applied to their muscles, regardless of whether neuromuscular transmission is blocked by curare. Brown and Harvey[20] concluded that the disorder represents a dysfunction of the muscle fiber; the myotonic signs could be accounted for by some defect in the excitability and impulse-generating mechanisms of the muscle fiber membrane that leads to repetitive discharge. Added evidence supporting this view was recently discussed by Bryant and coworkers.[21, 22]

Little is known about the occurrence of congenital myotonia in other animals. It was described by Steinberg and Botelho[108] in a 3-week-old thoroughbred filly in which lameness was the principal clinical manifestation. The lameness was most pronounced after a period of rest and was less severe after periods of activity. Sustained contraction was observed in various muscles, notably those of the hindquarters, when stimulated mechanically.

### Myasthenic syndrome in the dog

Myasthenia gravis has been long considered a disease solely of man. Yet, among animals it does have a clinical counterpart in the strikingly similar myasthenic syndrome recently observed in the dog. Apparently rare, it has been reported only four times since 1961. In each instance, general muscular weakness exacerbated by exercise and improved by rest and prompt improvement after administration of neostigmine were the distinguishing clinical features. Ormrod[87] observed the disorder in a 14-month-old male Cocker Spaniel that had great difficulty eating and drinking. Hall and Walker[44] recognized it in a 2-year-old female Chow that responded abnormally to a depolarizing muscle relaxant. Hill[57] described the disorder in a 4-year-old male pointer in which onset was characterized by inability to stand or walk. None of these reports included observations on electromyographic or histologic examination of muscle, nor did they mention the status of the thymus.

The 8-month-old male of mixed breeding observed by Zacks, Shields, and Steinberg[125] was studied more thoroughly. Vomiting and general difficulty in swallowing solid food, associated with megaesophagus, were first noted when the dog was 5 weeks old. During the next few months its gait became progressively unsteady. When 7 months old, the dog was unable to stand or walk. Prolonged remission occurred after several months of medication with neostigmine. Histologic changes were not found in biopsy specimens of intercostal, esophageal, and anterior tibial muscles. By electron microscopy, however, structural abnormalities of uncertain significance were observed in the motor end-plates of the intercostal muscle. These involved both presynaptic and postsynaptic elements of the junctional region. Enlargement of the thymus was not detected. Decreasing amplitude of invoked muscle action potentials was found during preliminary electromyographic studies done while the dog was clinically weak. A slight decrease was found later during clinical remission.

### Motor end-plate disease myopathy of the mouse

Duchen, Searle, and Strich[34] recently reported a new hereditary neuromuscular disorder of the mouse. Inherited as a simple autosomal recessive trait, it is characterized by progressive muscular weakness beginning about 9 days after birth and ending in death 10 to 15 days later. Histologic changes were not found in peripheral nerves or the central nervous system. Nor was evidence of a primary myopathy seen, although scattered atrophic muscle fibers were present. Because abnormalities were demonstrated in motor end-plates, the disorder was designated MED for motor end-plate disease. In a more detailed study of this disease, Zacks, Sheff, Rhodes, and Saito[124] confirmed the absence of lesions in the central nervous system and peripheral nerves, but they did not find evidence of primary disease of the motor end-plate. Rather, they concluded that MED has features of a distinctive myopathy unlike any hitherto described in man or animals.

In their study, motor end-plates were found to have enlarged terminal axons but not atrophic or dystrophic abnormalities. The acetylcholinesterase staining and fine structure of the subneural apparatus of the end-plates were normal. Moreover, electromyographic observations indicated normal neuromuscular transmission. Muscle fibers in the intercostal and gastrocnemius muscles were generally smaller than those in normal littermate mice, but histologic changes indicative of neural atrophy were not seen. Likewise, necrosis, inflammation, and fibrosis of the musculature were absent. Electron microscopically, focal myofibrillar disorganization and accumulations of abnormal mitochondria and sarcoplasm in autophagic vacuoles were seen in the gastrocnemius muscle late in the course of the disease. Because the structural changes in myofibrils occurred only in the terminal stage, they most likely are not the primary cause of the pronounced muscular weakness that characterizes the disease. Some metabolic abnormality of muscle fibers is suspected.

TYPE II MUSCLE FIBER HYPOTROPHY IN THE DOG

Spasm or shortening of the pectineus muscle, an adductor, in the dog has been suggested as possibly playing a role in the genesis of hip dysplasia, a hereditary disease that commonly affects the German shepherd and other large breeds.[13] This suggestion prompted a detailed histologic and histochemical examination of the pectineus muscles of German shepherd and mongrel dogs to determine whether abnormalities were present.[25]

Pathologic changes were found in 21 of 23 (91%) German shepherds and in 3 of 16 (19%) mongrel dogs 37 to 118 days old but not in those younger or older. Pronounced variation in the size of component fibers was the most conspicuous histologic change. As defined by the adenosine triphosphatase reaction, the smallest fibers were mainly type II fibers, whereas the largest were mainly type I. This is in contrast to the pattern in normal dogs of comparable age in which type II fibers are larger than type I fibers. Actually, the size of type I fibers in affected muscles varied widely; some were even larger than normal. Most type II fibers, on the other hand, were much smaller and relatively more numerous than in the normal muscle.

The lesion varied from focal, *i.e.*, involving less than 25% of the cross-sectional area of the muscle belly, to diffuse, which involved more than 70% of the muscle. Other changes such as necrosis of muscle fibers, splitting of fibers, internally placed muscle nuclei, and inflammatory cell infiltration were seldom present. In some muscles, the amount of endomysial connective tissue was moderately increased. Usually both pectineus muscles were affected. Some were obviously atrophic. Changes were not found in other muscles innervated by the obturator nerve, or in the lumbar spinal cord and intramuscular nerves.

The size of type II fibers in affected pectineus muscles was about that of type II fibers in dogs 10 days old. Because of this, and the larger percentage of type II fibers in affected muscles, Cardinet, Wallace, Fredde, Guffy, and Bardens[25] suggested that failure or delay in growth and differentiation of neonatal type II fibers occurred. Absence of the lesion in dogs older than 118 days indicated that it was transient and became corrected with age. For these reasons, the lesion was classified as a type II muscle fiber hypotrophy. Such selective or preferential involvement of type II fibers is discussed elsewhere in this monograph.

The causal relation of the lesion in the pectineus muscle to hip dysplasia in German shepherd dogs is still uncertain. Nevertheless, its presence during the first 16 to 17 weeks after birth when bone is still susceptible to modeling forces suggests that dysfunction of this muscle might result in abnormalities of the coxofemoral joint. Apart from its immediate implications in the dog, this study may also provide clues about the cause of congenital hip dysplasia in man. Moreover, it illustrates the value of histochemical methods in examining skeletal muscles of animals.

## CONCLUDING COMMENTS

As is evident from the descriptions given in this chapter, a wide variety of myopathies has been recognized among domestic, laboratory, and wild animals. Although much is known about a few of them, most are still extremely poorly understood. Consequently, their immediate relevance to the myopathies of man is seldom clear. Yet, even now some are obviously more relevant than others. Once the myopathies of animals are better defined, especially in terms of cause and pathogenesis, their similarities and differences when compared with those of man will become more evident and misleading analogies less likely. Furthermore, continued evaluation of skeletal muscle of animals for additional examples of myopathy not only should contribute to myopathology generally, but also should provide some insight into myopathic processes in man. The more refined techniques of biochemistry, histochemistry, electrophysiology, and electron microscopy will play a significant role here, as they have in human medicine during the past 10 years. This, coupled with an increased awareness of the merits of studying naturally occurring diseases of animals quite apart from their immediate veterinary or economic considerations should yield much information of comparative value.

## ACKNOWLEDGMENTS

I am grateful to all of the persons (named elsewhere) who kindly provided me with tissue specimens used for personal study and for illustrating this chapter. The photomicrographs were prepared by Mr. Charles Taylor, Chief of Graphic Arts, Rocky Mountain Laboratory.

### REFERENCES

1. Adams, R. D., Denny-Brown, D., and Pearson, C. M.: *Diseases of Muscle: A Study in Pathology*, 2nd ed. New York, Harper & Brothers, 1962.
2. Andrews, E. D., Hartley, W. J., and Grant, A. B.: Selenium-responsive diseases of animals in New Zealand. *New Zeal. Vet. J. 16:* 3–17, 1968.
3. Arpi, T., and Tollerz, G.: Iron poisoning in piglets. Autopsy findings in experimental and spontaneous cases. *Acta Vet. Scand. 6:* 360–373, 1965.
4. Asmundson, V. S.: Crooked neck dwarf in the domestic fowl. *J. Hered. 36:* 173–176, 1945.
5. Asmundson, V. S., and Julian, L. M.: Inherited muscle abnormality in the domestic fowl. *J. Hered. 47:* 248–252, 1956.
6. Asmundson, V. S., Kratzer, F. H., and Julian, L. M.: Inherited myopathy in the chicken. *Ann. N. Y. Acad. Sci. 138:* 49–58, 1966.
7. Asmundson, V. S., and Pun, C. F.: Crooked neck dwarf in the turkey. *J. Exp. Zool. 131:* 225–238, 1956.
8. Bajusz, E.: Hereditary cardiomyopathy. A new disease model. *Am. Heart J. 77:* 686–696, 1969.
9. Bajusz, E., Homburger, F., and Baker, J. R.: The heart muscle in muscular dystrophy with special reference to involvement of the cardiovascular system in the hereditary myopathy of the hamster. *Ann. N. Y. Acad. Sci. 138:*213–231, 1966.

10. Baker, J. R.: Muscular dystrophy in the horse. *Vet. Rec. 84:* 488–490, 1969.
11. Baker, M. L., Payne, L. C., and Baker, G. N.: The inheritance of hyrocephalus in cattle. *J. Hered. 52:* 135–138, 1961.
12. Banker, B. Q.: A pathological study of muscular dystrophy in the Bar Harbor 129 house mouse with particular reference to the ultrastructural features. In *Modern Neurology*, ed. by Locke, S., pp. 241–259. Boston, Little, Brown & Co., 1969.
13. Bardens, J. W., and Hardwick, H.: New observations on the diagnosis and cause of hip dysplasia. *Vet. Med. 63:* 238–245, 1968.
14. Berg, B. N.: Muscular dystrophy in aging rats. *J. Gerontol. 11:* 134–139, 1956.
15. Berg, B. N.: Study of vitamin E supplements in relation to muscular dystrophy and other diseases in aging rats. *J. Gerontol. 14:* 174–180, 1959.
16. Berg, B. N., Wolf, A., and Simms, H. S.: Nutrition and longevity in the rat. IV. Food restriction and the radiculoneuropathy in aging rats. *J. Nutr. 77:* 439–442, 1962.
17. Bergeland, M., and Barnes, D.: Another problem appears—degenerative polymyopathy of breeder tom turkeys. *Gobbles 19:* 12, 33, 1965.
18. Bone, J. F. The "tying up" syndrome. In *Equine Medicine and Surgery*, ed. by Bone, J. F., Catcott, E. J., Gabel, A. A., Johnson, L. E., and Riley, W. F. Jr., pp. 475–476. Santa Barbara, Calif., American Veterinary Publications Inc., 1963.
19. Bosanquet, F. D., Daniel, P. M., and Parry, H. B.: Myopathy in sheep. Its relationship to scrapie and to dermatomyositis and muscular dystrophy. *Lancet 271:* 737–746, 1956.
20. Brown, G. L., and Harvey, A. M.: Congenital myotonia in the goat. *Brain 62:* 341–346, 1939.
21. Bryant, S. H.: Cable properties of external intercostal muscle fibers from myotonic and nonmyotonic goats. *J. Physiol. (Lond.) 204:* 539–550, 1969.
22. Bryant, S. H., Lipicky, R. J., and Herzog, W. H.: Variability of myotonic signs in myotonic goats. *Am. J. Vet. Res. 29:* 2371–2381, 1968.
23. Burns, T. W., Dale, H. E., and Langley, P. L.: Normal and myotonic goats receiving diazacholesterol. *Am. J. Physiol. 209:* 1227–1232, 1965.
24. Cardinet, G. H., Litterell, J. F., and Freedland, R. A.: Comparative investigations of serum creatine phosphokinase and glutamic-oxaloacetic transaminase activities in equine paralytic myoglobinuria. *Res. Vet. Sci. 8:* 219–226, 1967.
25. Cardinet, G. H., III, Wallace, L. J., Fredde, M. R., Guffy, M. M., and Bardens, J. W.: Developmental myopathy in the canine with type II muscle fiber hypotrophy. *Arch. Neurol. 21:* 620–630, 1969.
26. Carlström, B.: *Uber die Atiologie und Pathogenese der Kreuzlähme des Pferdes (Haemoglobinaemic paralytica)*. Berlin, De Gruyter, 1930. Also in: *Scand. Arch. Physiol. 61:* 161–224, 1931; *62:* 1–69, 1931; *63:*164–212, 1932.
27. Caulfield, J. B.: Electron microscopic observations on the dystrophic hamster muscle. *Ann. N. Y. Acad. Sci. 138:* 151–159, 1966.
28. Clark, S. L., Luton, F. H., and Cutler, J. T.: A form of congenital myotonia in goats. *J. Nerv. Ment. Dis. 90:* 297–309, 1939.
29. Colombo, S., and Gervasini, C.: Alterazioni delle masse muscolari nel bovino da transporto in ferrovia. *Atti Soc. Ital. Sci. Vet. 8:* 580–583, 1954.
30. Dewan, M. L., Henson, J. B., Dollahite, J. W., and Bridges, C. H.: Toxic myodegeneration in goats produced by feeding mature fruits from the coyotillo plant (*Karwinskia humboldtiana*). *Am. J. Pathol. 46:* 215–226, 1965.
31. Dobson, K. J.: Congenital splayleg of piglets. *Aust. Vet. J. 44:* 26–28, 1968.
32. Dodd, D. C., Blakely, A. A., Thornbury, R. S., and Dewes, H. F.: Muscle degeneration and yellow fat disease in foals. *New Zeal. Vet. J. 8:* 45–50, 1960.
33. Dollahite, J. W., and Henson, J. B.: Toxic plants as the etiologic agent of myopathies in animals. *Amer. J. Vet. Res. 26:* 749–752, 1965.
34. Duchen, L. W., Searle, A. G., and Strich, S. J.: An heriditary motor end-plate disease in the mouse. *J. Physiol. (Lond.) 189:* 4–6P, 1967.
35. Ewan, R. C., Baumann, C. A., and Pope, A. L.: Effects of selenium and vitamin E on nutritional muscular dystrophy in lambs. *J. Anim. Sci. 27:* 751–756, 1968.
36. Fairlie, G.: Myopathy in a roebuck. *Vet. Rec. 76:* 1147–1148, 1964.
37. Freeman, A. E.: Curved limbs—a lethal in dairy cattle. *J. Hered. 49:*229–232, 1958.

38. Gandal, C. P.: White muscle disease in a breeding herd of nyala antelope *Tragelaphus angasi* at New York Zoo. *Int. Zoo Yearbook 6:* 277–278, 1966.
39. Gardner, D. E.: Skeletal myonecrosis in a cat. *New Zeal. Vet. J. 15:* 211–213, 1967.
40. Gowing, G. M.: "Pinched back," an azoturia-like condition in coursing Greyhounds. *Southwest. Vet. 17:* 183–185, 1964.
41. Hadlow, W. J.: Degenerative myopathy in a white-tailed deer, *Odocoileus virginianus. Cornell Vet. 45:* 538–547, 1955.
42. Hadlow, W. J.: Diseases of skeletal muscle. In *Comparative Neuropathology*, ed. by Innes, J. R. M., and Saunders, L. Z., pp. 147–243. New York, Academic Press, 1962.
43. Hadlow, W. J., and Eklund, C. M.: Scrapie—a virus induced chronic encephalopathy of sheep. *Res. Publ. Assoc. Nerv. Ment. Dis. 44:* 281–304, 1968.
44. Hall, L. W., and Walker, R. G.: Suspected myasthenia gravis in a dog. *Vet. Rec. 74:* 501–503, 1962.
45. Hallqvist, C.: Ein Fall von Letalfaktoren beim Schwein. *Hereditas 18:*215–224, 1933.
46. Harman, P. J., Tassoni, J. P., Curtis, R. L., and Hollinshead, M. B.: Muscular dystrophy in the mouse. In *Muscular Dystrophy in Man and Animals*, ed. by Bourne, G. H., and Golarz, M. N., pp. 407–456. New York, Hafner Publishing Co., 1963.
47. Harper, J. A., and Parker, J. E.: Hereditary muscular dystrophy in the domestic turkey, *Meleagris gallopavo. Poult. Sci. 43:* 1326–1327, 1964.
48. Harper, J. A., and Parker, J. E.: Hereditary muscular dystrophy in the domestic turkey. *J. Hered. 58:* 189–193, 1967.
49. Hartley, W. J., and Dodd, D. C.: Muscular dystrophy in New Zealand livestock. *New Zeal. Vet. J. 5:* 61–66, 1957.
50. Hegyeli, A., and Szent-Györgyi, A.: Water and myotonia in goats. *Science 133:* 1011, 1961.
51. Henson, J. B., and Dollahite, J. W.: Toxic myodegeneration in calves produced by experimental *Cassia occidentalis* intoxication. *Am. J. Vet Res. 27:* 947–949, 1966.
52. Henson, J. B., Dollahite, J. W., Bridges, C. H., and Rao, R. R.: Myodegeneration in cattle grazing *Cassia* species. *J. Am. Vet. Med. Assoc. 147:* 142–145, 1965.
53. Herrmann, H., Clark, E. M., and Landauer, W.: Muscle development in the crooked neck dwarf mutant and in the acetylpyridine-treated chick embryo. *Acta Embryol. Morphol. Exp. 6:* 169–174, 1963.
54. Herzog, A., and Adam, R.: Zur Neuromyodysplasia congenita (kongenitale Arthrogrypose) der Hintergliedmassen beim Kalb. I. Pathologische Anatomie und Histologie der Muskulatur arthrogrypotischer Gliedmassen. *Dtsch. Tieraerztl. Wochernschr. 75:* 237–243, 1968.
55. Hidiroglou, M., and Jenkins, K. J.: Factors affecting the development of nutritional muscular dystrophy in northern Ontario. *Can. J. Anim. Sci. 48:* 7–14, 1968.
56. Hidiroglou, M., Jenkins, K. J., Carson, R. B., and Mackay, R. R.: Some aspects of selenium metabolism in normal and dystrophic sheep. *Can. J. Anim. Sci. 48:* 335–346, 1968.
57. Hill, W. J.: Myasthenia gravis in the canine. *Southwest. Vet. 20:* 319–320, 1967.
58. Homburger, F.: Concluding remarks. *Ann. N. Y. Acad. Sci. 138:* 356–359, 1966.
59. Homburger, F., Baker, J. R., Nixon, C. W., and Wilgram, G.: New hereditary disease of Syrian hamsters. Primary, generalized polymyopathy and cardiac necrosis. *Arch. Intern. Med. 110:* 660–662, 1962.
60. Homburger, F., Nixon, C. W., Eppenberger, M., and Baker, J. R.: Hereditary myopathy in the Syrian hamster. Studies on pathogenesis. *Ann. N. Y. Acad. Sci. 138:* 14–27, 1966.
61. Hutt, F. B.: A hereditary lethal muscle contracture in cattle. *J. Hered. 25:* 41–46, 1934.
62. Jarrett, W. F. H., Jennings, F. W., Murray, M., and Harthoorn, A. M.: Muscular dystrophy in wild Hunter's antelope. *East Aft. Wildl. J.* ×*:* 158–159, 1964.
63. Jarrett, W. F. H., and Murray, M.: Muscular dystrophy in antelope and gazelle in Kenya. *Vet. Rec. 80:* 483, 1967.
64. Jasmin, G., and Gareau, R.: Histopathological study of muscle lesions produced by paraphenylenediamine in rats. *Br. J. Exp. Pathol. 42:* 592–596, 1961.
65. Johnston, W. G., and Young, G. B.: A congenital muscle contracture and chondrodysplasia syndrome in cattle. *Vet. Rec. 70:* 1219–1220, 1958.
66. Julian, L. M., and Asmundson, V. S.: Muscular dystrophy of the chicken. In *Muscular*

*Dystrophy in Man and Animals*, ed. by Bourne, G. H., and Golarz, M. N., pp. 457–498. New York, Hafner Publishing Co., 1963.

67. Kakulas, B. A., and Adams, R. A.: Principles of myopathology as illustrated in the nutritional myopathy of the Rottnest quokka (*Setonix brachyurus*). *Ann. N. Y. Acad. Sci. 138:* 90–101, 1966.
68. Kaspar, L. V., and Lombard, L. S.: Nutritional myodegeneration in a litter of beagles. *J. Am. Vet. Med. Assoc. 143:* 284–288, 1963.
69. Koch, P., Fischer, H., and Schumann, H.: *Erbpathologie der landwirtschaftlichen Haustiere.* Berlin and Hamburg, Paul Parey, 1957.
70. Kohler, E. M., Cross, R. F., and Ferguson, L. C.: Experimental induction of spraddled-legs in newborn pigs. *J. Am. Vet. Med. Assoc. 155:* 139–142, 1969.
71. Kroneman, J., and Wensvoort, P.: Muscular dystrophy and yellow fat disease in Shetland pony foals. *Neth. J. Vet. Sci. 1:* 42–48, 1968.
72. Lannek, N.: Zwei Formen von Muskeldegenerationen bei Schweinen. *Dtsch, Tieraerztl. Wochenschr. 74:* 321–324, 1967.
73. Lannek, N.: Vitamin E deficiency in farm animals. *Wiss. Veroff. Deut. Ges. Ernahrung 16:* 145–160, 1967.
74. Leipold, H. W., Cates, W. F., Radostits, O. M., and Howell, W. E.: Spinal dysraphism, arthrogryposis and cleft palate in newborn Charolais calves. *Can. Vet. J. 10:* 268–273, 1969.
75. Manktelow, B. W.: Myopathy in dogs resembling white muscle disease in sheep. *New Zeal. Vet. J. 11:* 52–55, 1963.
76. Maronpot, R. R., Bucci, T. J., and Stedham, M. A.: Focal degenerative myopathy in turkeys. *Avian Dis. 12:* 96–103, 1968.
77. Meier, H.: Myopathies in the dog. *Cornell Vet. 48:* 313–330, 1958.
77a. Meier, H., and Southard, J. L.: Muscular dystrophy in the mouse caused by an allele at the *dy*-locus. *Life Sci.* (*II*) *9:* 137–144, 1970.
78. Meier, H., West, W. T., and Hoag, W. G.: Preclinical histopathology of mouse muscular dystrophy. *Arch. Pathol. 80:* 165–170, 1965.
79. Mercer, H. D., Neal, F. C., Himes, J. A., and Edds, G. T.: *Cassia occidentalis* toxicosis in cattle. *J. Am. Vet. Med. Assoc. 151:* 735–741, 1967.
80. Michelson, A. M., Russell, E. S., and Harman, P. J.: *Dystorphia muscularis.* A hereditary primary myopathy in the house mouse. *Proc. Nat. Acad. Sci. U.S.A. 41:* 1079–1084, 1955.
81. Middleton, D. S.: Studies on prenatal lesions of striated muscle as a cause of congenital deformity. *Edinburgh Med. J. 41:* 401–442, 1934.
82. Morley, F. H. W.: A new lethal factor in Australian Merino sheep. *Aust. Vet. J. 30:* 237–240, 1954.
83. Mugera, G. M., and Wandera, J. G.: Degenerative polymyopathies in East African domestic and wild animals. *Vet. Rec. 80:* 410–413, 1967.
84. Murray, M.: The pathology of some diseases found in wild animals in East Africa. *East Afr. Wildl. J. 5:* 37–45, 1967.
85. Nes, N. Spina bifida ledsaget av muskelkontraktur og andre defekter hos kalv. *Nord. Vet. Med. 11:* 33–54, 1959.
86. O'Hara, P. J., Pierce, K. R., and Read, W. K.: Degenerative myopathy associated with ingestion of *Cassia occidentalis* L. Clinical and pathologic features of the experimentally induced disease. *Am. J. Vet. Res. 30:* 2173–2180, 1969.
87. Ormrod, A. N.: Myasthenia gravis in a cocker spaniel. *Vet. Rec. 73:* 489–490, 1961.
88. Orstadius, K.: *Nutritional Muscular Dystrophy in Pigs. Studies on the Aetiology, Diagnosis, and Therapy.* Uppsala, Almqvist and Wiksells, 1961.
89. Pai, A. C.: Developmental genetics of a lethal mutation, muscular dysgenesis (mdg), in the mouse. I. Genetic analysis and gross morphology. *Dev. Biol. 11:* 82–92, 1965.
90. Pai, A. C.: Developmental genetics of a lethal mutation, muscular dysgenesis (mdg), in the mouse. II. Developmental analysis. *Dev. Biol. 11:* 93–109, 1965.
91. Parry, H. B.: Scrapie and related myopathies in sheep. Preliminary observations on their investigation and attempted control by a voluntary health scheme. *Vet. Rec. 69:* 43–55, 1957.
92. Patterson, D. S. P., Allen, W. M., Berrett, S., Sweasey, D., Thurley, D. C., and Done, J. T.: A

biochemical study of the pathogenesis of iron-induced myodegeneration of piglets. *Zentralbl. Veterinaermed.* (*A*) *16:* 199–214, 1969.

93. Patterson, D. S. P., Sweasey, D., Allen, W. M., Berrett, S., and Thurley, D. C.: The chemical composition of neonatal piglet muscle and some observations on the biochemistry of myofibrillar hypoplasia occurring in otherwise normal litters. *Zentralbl. Veterinaermed.* (*A*) *16:* 741–753, 1969.
94. Pierce, K. R., and O'Hara, P. J.: Toxic myopathy in Texas cattle. *Southwest. Vet. 20:* 179–184, 1967.
95. Rao, R. R., and Henson, J. B.: Studies on the pathogenesis of muscle necrosis in plasmocid intoxication. *Res. Vet. Sci. 8:* 369–374, 1967.
96. Rigdon, R. H.: Hereditary myopathy in the White Pekin duck. *Ann. N. Y. Acad. Sci. 138:* 28–48, 1966.
97. Rigdon, R. H.: Acute myopathy in embryos and newly hatched chicks. *Arch. Pathol. 84:* 663–637, 1967.
98. Rigdon, R. H., Ferguson, T. M., and Couch, J. R.: Pendulous crops in turkeys—an anatomic and pathologic study. *Am. J. Vet. Res. 21:* 979–986, 1960.
99. Rigdon, R. H., Ferguson, T. M., and Couch, J. R.: Spontaneous occurring muscular necroses and encephalomalacia in the turkey. *Poult. Sci. 40:* 766–771, 1961.
100. Rigdon, R. H., Ferguson, T. M., and Couch, J. R.: Spontaneous-occurring muscular necrosis in the chicken. *Poult. Sci. 41:* 398–409, 1962.
101. Rigdon, R. H., Schreiber, M. H., and Levy, A.: The arthrogryposis syndrome in chickens. *Arch. Pathol. 79:* 388–393, 1965.
102. Roberts, J. A. F.: A hereditary lethal deformity in new-born lambs. *J. Ministry Agr. 33:* 795–801, 1926.
103. Roberts, J. A. F.: The inheritance of a lethal muscle contracture in the sheep. *J. Genet. 21:* 57–69, 1929.
104. Rosenberg, L. E.: Histological studies of muscle from crooked neck dwarf fowl. *Anat. Rec. 97:* 277–282, 1947.
105. Rülcker, C.: The influence of physical training and short-time physical stress on colour, fluid loss, pH, adenosine triphosphate and glycogen of the gracilis muscle in pigs. An experimental study. *Acta Vet. Scand.* (*Suppl.*) *24:*1–44, 1968.
106. Schmahlstieg, R., and Meyer, H.: Angeborene Starre und Verkriimmung der Gelenke (Arthrogryposis congenita) beim Kalb. *Dtsch. Tieraerztl. Wochenschr. 67:* 41–44, 1960.
107. Stamp, J. T.: Muscular dystrophy in sheep and neonatal mortality. *J. Comp. Pathol. 70:* 296–304, 1960.
108. Steinberg, S., and Botelho, S.: Myotonia in a horse. *Science 137:* 979–980, 1962.
109. Swingle, K. F., Young, S., and Dang, H. G.: The relationship of serum glutamic oxalacetic transaminase to nutritional muscular dystrophy in lambs. *Am. J. Vet. Res. 20:* 75–77, 1959.
110. Thurley, D. C., and Done, J. T.: The histology of myofibrillar hypoplasia in newborn pigs. *Zentralbl. Veterinaermed.* (*A*) *16:* 732–740, 1969.
111. Thurley, D. C., Gilbert, F. R., and Done, J. T.: Congenital splayleg of piglets. Myofibrillar hypoplasia. *Vet. Rec. 80:* 302–304, 1967.
112. Tollerz, G.: *Studies on the Tolerance to Iron in Piglets and Mice with Special Reference to Vitamin E, Synthetic Antioxidants, and Sodium Selenite.* Uppsala, Almqvist and Wiksells, 1965.
113. Urman, H. K., and Grace, O. D.: Hereditary encephalomyopathy, a hydrocephalus syndrome in newborn calves. *Cornell Vet. 54:* 229–249, 1964.
114. van Logtestijn, J. G., and Sybesma, W.: Muskel-und Fleischdegeneration bei Schweinen. *Dtsch. Tieraerztl. Wochenschr. 74:* 377–385, 1967.
115. Wagner, W. D., and Peterson, R. A.: Muscular dystrophy syndrome in the Cornish chicken. *Am. J. Vet. Res. 31:* 331–338, 1970.
116. Walton, J. N.: Muscular dystrophy and its relation to other myopathies. In *Res. Publ. Assoc. Res. Nerv. Ment. Dis. 38:* 378–421, 1960.
117. West, W. T., Meier, H., and Hoag, W. G.: Hereditary mouse muscular dystrophy with particular emphasis on pathogenesis and attempts at therapy. *Ann. N. Y. Acad. Sci. 138:* 4–13, 1966.

118. Whanger, P. D., Muth, O. M., Oldfield, J. E., and Weswig, P. H.: Influence of sulfur on incidence of white muscle disease in lambs. *J. Nutr. 97:* 553–562, 1969.
119. White, G. R., and Plaskett, J.: "Nervous," "stiff-legged," or "fainting goats." *Am. Vet. Rev. 28:* 556–560, 1904.
120. Whittem, J. H.: Congenital abnormalities in calves. Arthrogryposis and hydranencephaly. *J. Pathol. Bacteriol. 73:* 375–387, 1957.
121. Wise, W. R., Weswig, P. H., Muth, O. M., and Oldfield, J. E.: Dietary interrelationship of cobalt and selenium in lambs. *J. Anim. Sci. 27:* 1462–1465, 1968.
122. Yamagiwa, S., Ono, T., and Kaneko, M.: Histopathological studies on peracute enzootic muscle dystrophy in foals. *Res. Bull. Obihiro Univ. 5:* 523–534, 1968. (In Japanese with English summary.)
123. Young, E.: Muscle necrosis in captive red hartebeeste (*Alcelaphus buselaphus*). *J. S. Afr. Vet. Med. Assoc. 37:* 101–103, 1966.
124. Zacks, S. I., Sheff, M. F., Rhodes, M., and Saito, A.: MED myopathy. A new hereditary myopathy. *Lab. Invest. 21:* 143–153, 1969.
125. Zacks, S. I., Shields, D. R., and Steinberg, S. A.: A myasthenic syndrome in the dog. A case report with electron microscopic observations on motor end-plates and comparisons with the fine structure of end plates in myasthenia gravis. *Ann. N. Y. Acad. Sci. 135:* 79–97, 1966.

Chapter 19

# Ultrastructure of Muscle in Glycogen Storage Diseases

DONALD L. SCHOTLAND

Glycogen storage diseases are inborn errors in which abnormal accumulations of glycogen result from a specific enzymatic defect. The principal glycogen storage diseases that affect muscle are types II, III, V, and VII all involving defects in glycogen degradation. The present discussion will review the morphologic abnormalities of muscle in these disorders.

Glycogen is a highly branched molecule containing glucose units joined by $\alpha$-1,4 linkages in straight chains[24] (Fig. 1). Branch points are formed when glucose units are joined by $\alpha$-1,6 linkages. Glycogen is synthesized from glucose-1-phosphate (Fig. 2). First a straight chain glucose polymer consisting of glucose units in $\alpha$-1, 4 linkages are formed with the aid of UDPG-glycogen transglucosylase. Then brancher enzyme joins glucose units at $\alpha$-1,6 linkages and a complicated branched structure is gradually formed.

The main path for the degradation of glycogen involves three enzymes: phosphorylase, oligo (1,4-1,4) transferase, and amylo-1,6-glucosidase, the latter two apparently inseparable and comprising the debrancher system. The $\alpha$-1,4 linkages of the outer chains (Fig. 1) are first broken by phosphorylase. When the outer chains are about 4 units long, 3 glucose units are transferred to other side chains by oligo (1,4-1,4) transferase and the residual glycogen molecule approaches the structure of a limited dextrin. Rupture of the $\alpha$-1,6 linkage of the remaining glucose unit in the outer chain is then carried out by debrancher enzyme and further breakdown continues by means of phosphorylase and transferase. In this pathway glycogen is mainly degraded to glucose-1-phosphate and is then metabolized to lactate. Another degradative pathway of importance is the lysosomal enzyme amylo-1,4-glucosidase or acid maltase that converts glycogen to glucose. This is the only lysosomal enzyme known to digest engulfed glycogen.

Acid maltase is absent in type II glycogenesis or Pompe's disease,[6] the debrancher system is absent in type III glycogenesis (also known as Forbe's disease, Cori's disease,[15, 22] and phosphofructokinase is absent in type VII disease.[27]

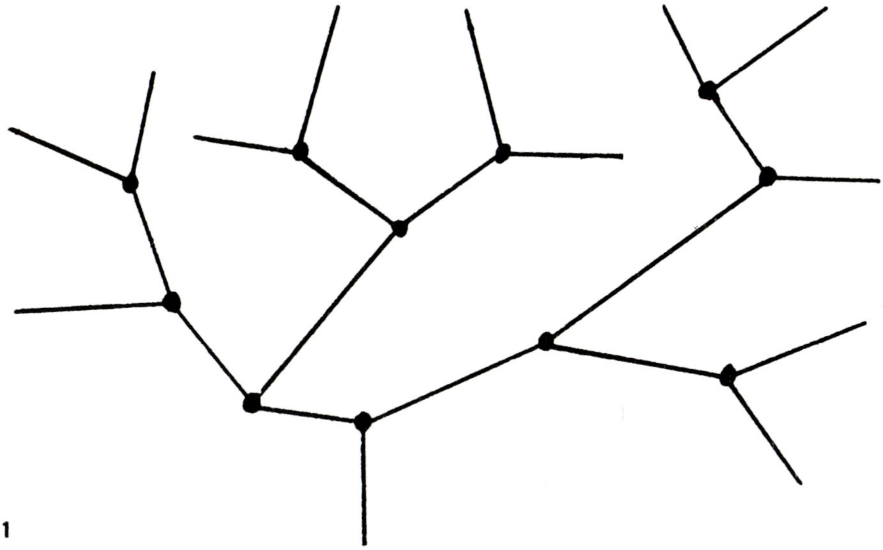

1

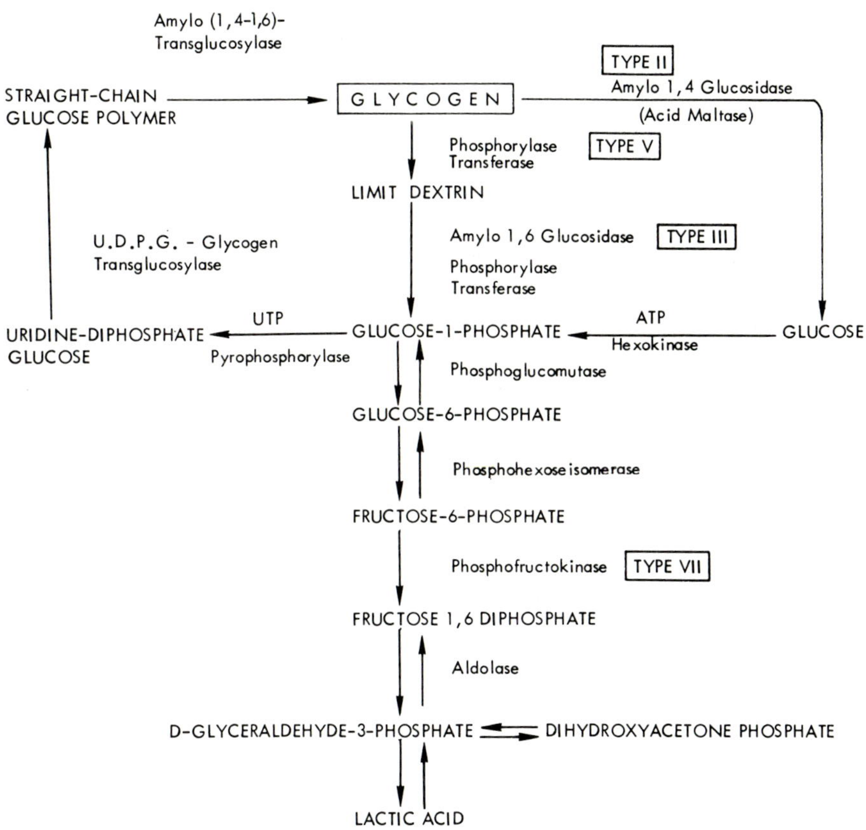


2

FIG. 1. Schematic model of glycogen molecule. ●, 1,6 bond.

FIG. 2. Diagram of glycogen metabolism. The glycogen storage diseases affecting muscle (types II, III, V, VII) are indicated.

## TYPE II (POMPE'S DISEASE)

Type II glycogenosis will be considered briefly since this is discussed elsewhere in this volume by Dr. Andrew Engel (Chapter 16). Pompe's disease is the most severe form of glycogen storage disease. All typical cases occur in infants. They are weak and hypotonic and death generally occurs in the first 2 years due to cardiac failure. At autopsy there is massive accumulation of glycogen of normal structure in many organs including skeletal, cardiac, and smooth muscle, central nervous system, liver, and kidney.[4, 8, 13, 25] In 1963 Hers[6] demonstrated the absence of the lysosomal enzyme acid maltase in the liver, heart, and skeletal muscle of five children with this disease. As discussed by Dr. Engel the same enzymatic defect may be found in a milder form of disease with muscular manifestations in older children or adults.

Light microscopic observation of skeletal muscle in the infantile form of the disease reveals massive accumulation of glycogen throughout the muscle fiber in the form of periodic acid-Schiff (PAS)-positive deposits removed in large part by diastase (Fig. 3). Acid mucopolysaccharide deposits may also be demonstrated by suitable staining procedures.[29]

Electron microscopic observations[4] reveal that skeletal muscle and other tissues contain numerous collections of glycogen enclosed in single membranes (Fig. 4) or autophagic vacuoles. Excessive free glycogen is also present in skeletal muscle as in other tissues (Fig. 5). The amount of free glycogen is much greater than the membrane-bound fraction.

Hers[7] has postulated that glycogen is normally engulfed by lysosomes and digested by acid maltase; in the absence of this enzyme glycogen accumulates, distends the lysosomes into large sacs, and causes disruption of adjacent muscle by mechanical pressure. The theory provides excellent correlation between the biochemical defect and a prominent ultrastructural abnormality. The large amount of free glycogen in muscle in this disease, however, is unexplained. Whether the free glycogen is due to an independent metabolic defect and is secondarily taken up by lysosomes or is derived from distended lysosomes that have ruptured remains to be determined.

## TYPE III (DEBRANCHER SYSTEM DEFICIENCY OR LIMIT DEXTRINOSIS)

Type III glycogenosis, described by Forbes in 1953,[3] chiefly affects the liver, heart, and skeletal muscle; in some instances the liver alone is involved.[28] In this disease the outer chains of glycogen are partially degraded by phosphorylase. Degradation then stops due to the deficiency of the debrancher system, and the residual glycogen molecule has many branch points and short outer chains. The patient presents in infancy with hepatomegaly, delayed growth, and some muscle hypotonia and weakness. There is also fasting hypoglycemia, hyperlipidemia, and acetonuria. Improvement usually occurs by adolescence.[18] The disease may also present in later life. Oliner[17] has described a patient seen at age 50 with chronic progressive myopathy. The deficiency of debrancher enzyme was established by Illingworth, Cori, and Cori in 1956.[10] More recently,[28] this disease has been classified into four types according to the

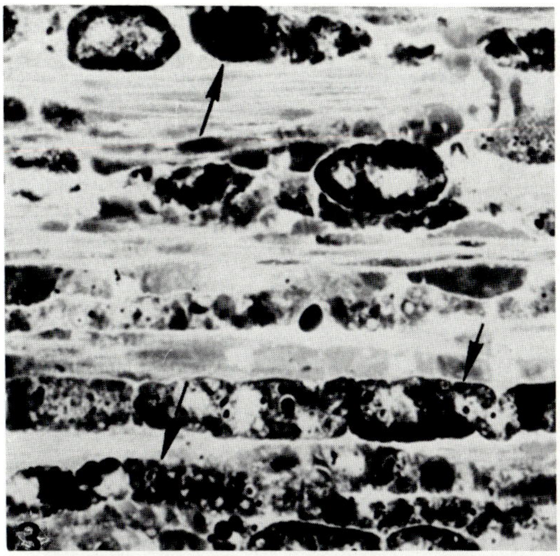


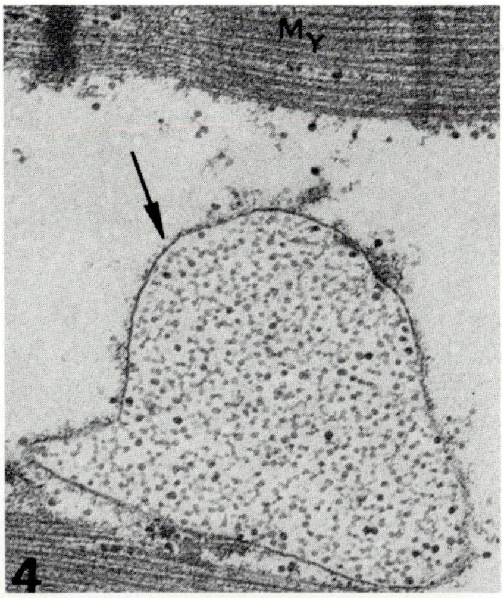


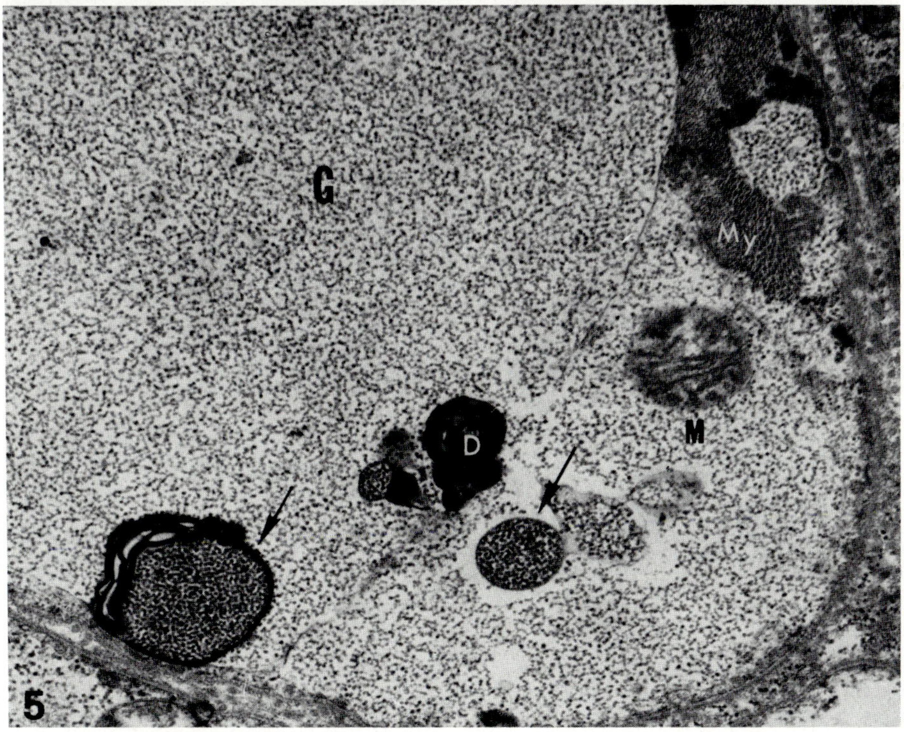


FIG. 3. Longitudinal section of Araldite-embedded skeletal muscle from a 3-month-old patient with acid maltase deficiency. Massive accumulation of glycogen occurs throughout the muscle fiber in the form of PAS-positive deposits (*single arrows*). The glycogen content of the muscle was 7.4 g%. Periodic acid-Schiff stain; × 290.

FIG. 4. Electron micrograph of glycogen enclosed in a single membrane (*single arrow*) in the intermyofibrillar space of skeletal muscle from a 3-month-old patient with acid maltase deficiency. *My*, Myofibril. × 32,500.

FIG. 5. Electron micrograph of free glycogen (*G*) within a skeletal muscle fiber from a 3-month-old patient with acid maltase deficiency. Membrane-bound sacs of glycogen (*single arrows*), myofibrillar remnants (*My*), a mitochondrion (*M*), and a dense body (*D*) of uncertain etiology are also present. × 12,000.

presence or absence of amylo-1,6-glucosidase and the closely associated transferase reaction in liver and muscle.

Skeletal muscle, as viewed in the light microscope, contains subsarcolemmal collections of glycogen (Fig. 6) and some increase in intermyofibrillar glycogen. Subsarcolemmal collections of glycogen shown here are also characteristic of type V and type VII disease and contrast with the more severe involvement of muscle in type II disease.

In electron microscope observations, skeletal muscle contains abundant glycogen in the intermyofibrillar space opposite the A band and I band (Fig. 7), between thin filaments in the I band (Fig. 8), and under the sarcolemma (Fig. 9). This distribution is similar to that in type V and type VII diseases.

Within the larger zones of free glycogen, scattered collections of membrane-bound glycogen are also present (Fig. 9). They contain single or multiple membranes and in some regions are numerous although not as abundant as in type II disease. These were also observed in a 9-month-old patient with type III disease and have been noted in previous reports by Hug, Garancis, Schubert, and Kaplan[9] and Neustein.[16] Since the muscle acid maltase activity in the presently reported patients was normal, occasional collections of membrane-bound glycogen in an electron micrograph do not establish a diagnosis of acid maltase deficiency.

The significance of membrane-bound glycogen in type III disease is uncertain. Perhaps this represents a nonspecific lysosomal response to glycogen accumulation as appears to occur in embryonic rat muscle between 18 and 22 days of fetal life. At this stage Kelly[11] has shown that the muscle fibers of the rat contain large accumulations of free glycogen (Fig. 10), and membrane-bound collections of glycogen are also present (Fig. 11). Other similarities between embryonic muscle and glycogen storage disease are being investigated.

## TYPE V DISEASE (PHOSPHORYLASE DEFICIENCY)

Type V glycogen storage disease or muscle phosphorylase deficiency was the first hereditary myopathy in which the metabolic defect was defined. The clinical picture was described in 1951 by McArdle[14] who encountered a patient with life-long intolerance to exercise. Physical and neurologic examination was otherwise normal. Ischemic exercise of the forearm resulted in shortening of the flexor muscles; they were electrically silent and thus in a state of physiologic contracture. In addition, ischemic exercise was not followed by the normal rise in venous lactic acid, and a block in the metabolic pathway from glycogen to lactic acid was postulated. Other features of this disorder may include myoglobinuria after exercise and occasionally permanent weakness after age 40.[19] In 1959 Schmid, Robbins, and Traut[22] and Mommaerts, Illingworth, Pearson, Guillory, and Seraydarian[15] demonstrated that skeletal muscle phosphorylase was absent in this disease. As a result glycogen cannot be broken down to glucose-1-phosphate and the muscle glycogen content is increased.

Skeletal muscle, viewed in the light microscope, contains PAS-positive subsarcolemmal blebs that disappear after treatment with diastase (Fig. 12). The histochemical reaction for phosphorylase[26] is absent (Fig. 13) as compared to control muscle incubated in the same solution (Fig. 14).

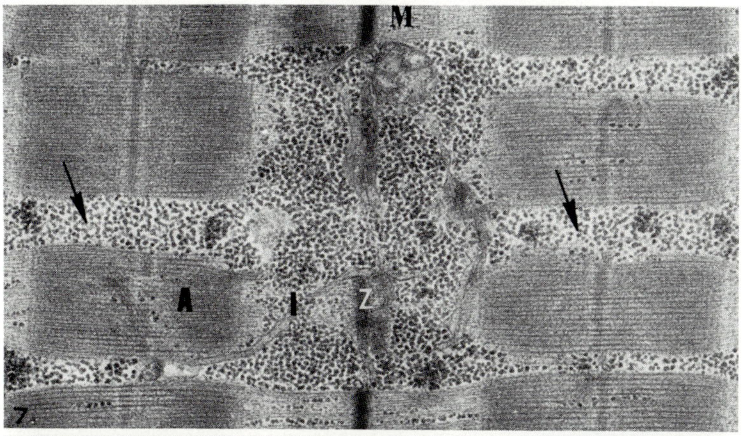


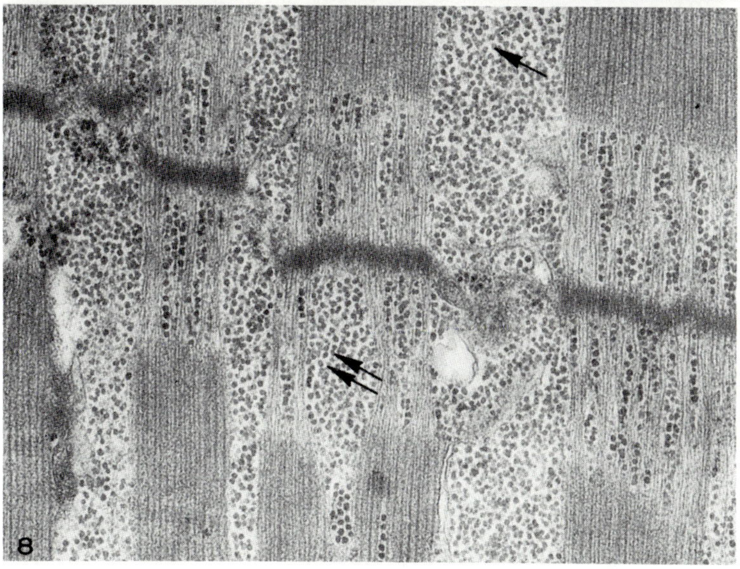


Fig. 6. Transverse section of Araldite-embedded quadriceps muscle from a 7-year-old patient with deficiency of liver and muscle debrancher enzyme. Numerous subsarcolemmal collections of PAS-positive material (*single arrows*) are present which can be removed with diastase. The intermyofibrillar glycogen is also increased. Periodic acid-Schiff stain; × 290.

Fig. 7. Electron micrograph of a longitudinal section of quadriceps muscle from a 7-year-old patient with type III glycogenesis. There is abundant glycogen in the intermyofibrillar space between the A band (*A*) (*single arrows*) and around the I band (*I*). *Z*, Z line. × 23,000.

Fig. 8. Electron micrograph of a longitudinal section of quadriceps muscle from a 7-year-old patient with type II glycogenosis. There is increased glycogen in the intermyofibrillar space (*single arrow*) and between thin filaments in the I band (*double arrow*). × 33,000.

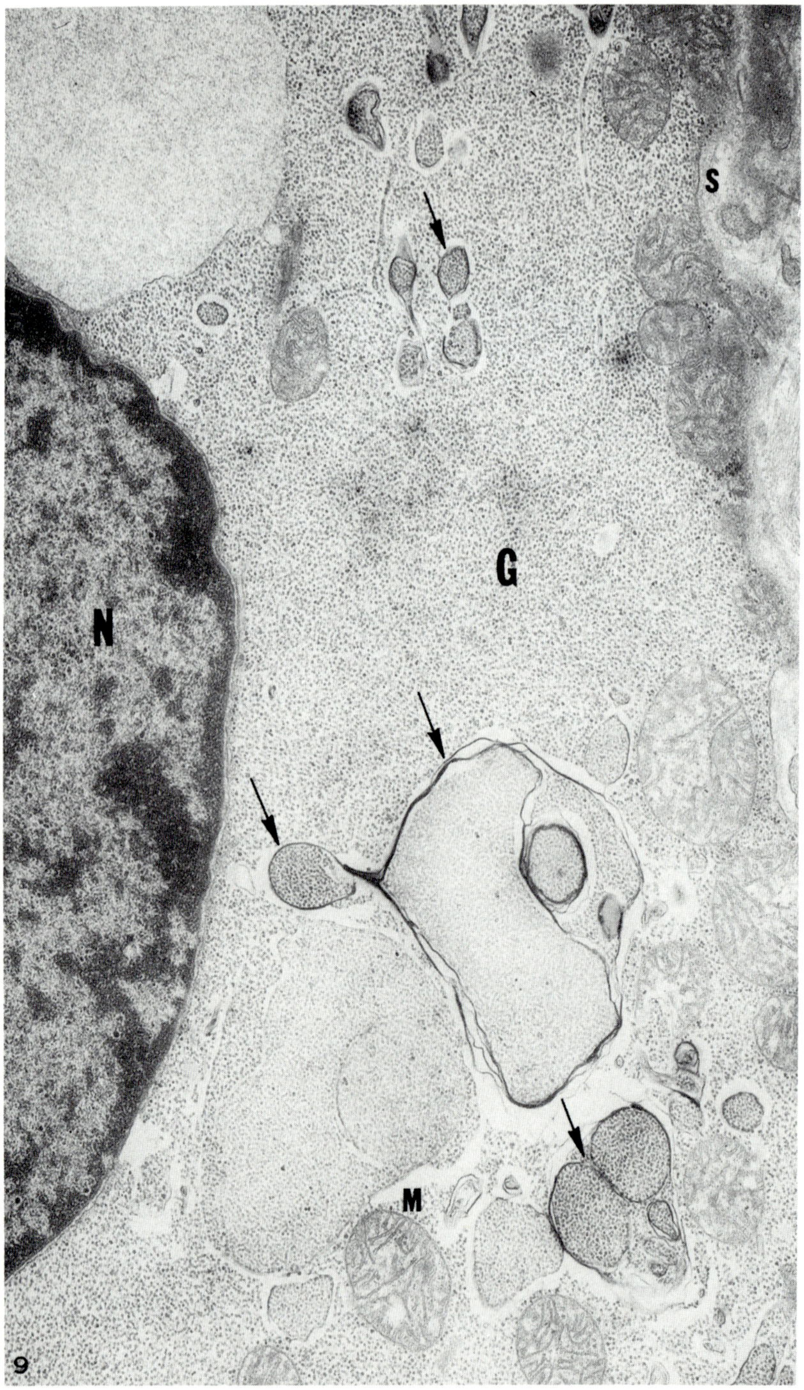


FIG. 9. Electron micrograph of quadriceps muscle from a 7-year-old patient with type III glycogenosis. There is increased glycogen (*G*) under the sarcolemma (*S*). Also present are single and multiple membrane-bound collections of glycogen (*single arrows*). *N*, nucleus; *M*, mitochondria. × 9000.

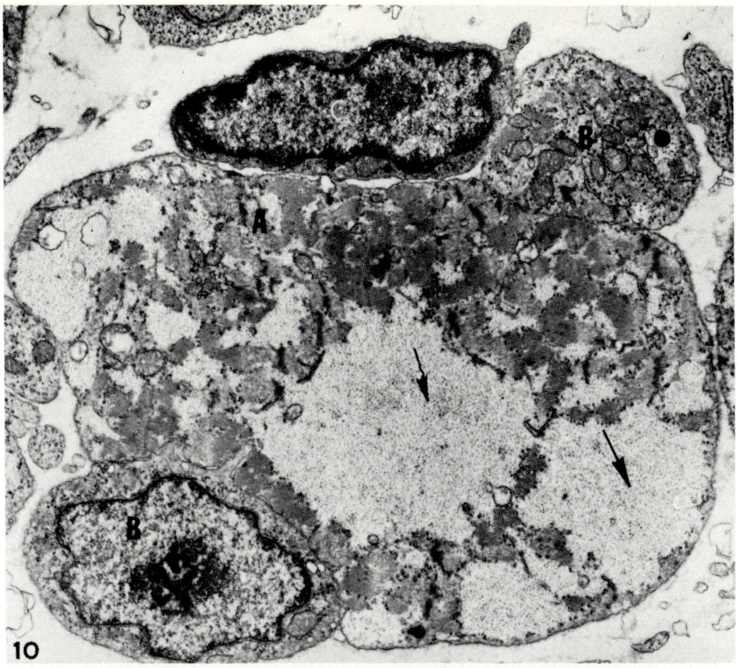

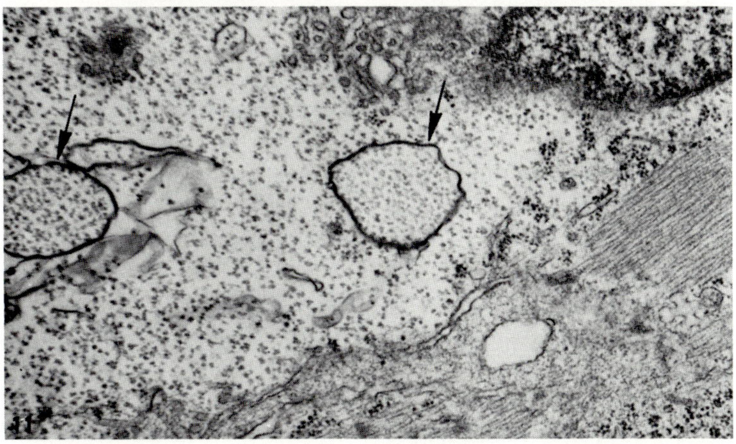

FIG. 10. Electron micrograph of a transverse section of 20-day fetal rat intercostal muscle. A large myotube (*A*) contains large collections of free glycogen (*single arrows*). Adjacent to the large myotube are two small myotubes (*B*) and an undifferentiated cell (*C*). × 9000. (Electron micrograph courtesy of Dr. Allan Kelly.)

FIG. 11. Electron micrograph of a longitudinal section of 20-day fetal rat intercostal muscle. Membrane-bound collections of glycogen (*single arrows*) are present within areas of free glycogen. × 27,000. (Electron micrograph courtesy of Dr. Allan Kelly.)

FIG. 12. Longitudinal paraffin section of the flexor carpi radialis muscle from a patient with McArdle's disease illustrating subsarcolemmal accumulation of glycogen (*arrows*). The subsarcolemmal masses disappeared after treatment with diastase. Periodic acid-Schiff stain; × 600. (Reproduced from Schotland, Spiro, and Rowland[23] by permission of the Association of the Journal of Neuropathology and Experimental Neurology, Inc.)

Ultrastructural investigation[2, 21, 23] reveals that skeletal muscle contains increased glycogen in the intermyofibrillar space (Fig. 15) between the thin filaments in the I band (Fig. 17), and under the sarcolemma (Fig. 16). Whether this distortion of thin filaments by glycogen has a mechanically disruptive effect on the muscle fiber as appears to be true in Pompe's disease is problematical. Membrane-bound collections of glycogen are also occasionally seen. Sarcolemmal defects, sufficient to account for myoglobinuria, have not been observed.

Mitochondrial abnormalities have been emphasized by Gruener, McArdle, Brenda, Ryman, and Weller[5] including dilatation and distortion of the cristae with accumulations of glycogen within the mitochondria in resting muscle and during contracture. These workers suggest that abnormal metabolic changes during exercise may cause mitochondrial degeneration which could eventually cause permanent weakness. We observed less severe mitochondrial changes during contracture in a previously studied case[23] but recently have confirmed the findings of Gruener and coworkers (Fig. 18).

The mechanisms responsible for the contracture in this disease are unknown. The absence of electrical activity in the shortened muscle indicates that it is a contracture, and an alteration in excitation contraction coupling should be considered. In an excellent paper Gruener and colleagues[5] have recently helped to clarify this defect.

Current theory suggests that the surface membrane potential spreads down the transverse tubular system (T system) and causes release of calcium in the sarcoplasmic reticulum (SR) which is directly adjacent to the T system in the triads. The released calcium then initiates contraction of the myofibrils. When the surface membrane repolarizes, calcium release from the SR ceases, calcium reaccumulates in the SR, and the muscle relaxes. The uptake of calcium is dependent on an abundant supply of ATP in the SR and an SR ATPase. Experimentally, caffeine causes release of calcium from the SR and inhibits reaccumulation.

Gruener and coworkers[5] studied the effect of calcium and caffeine on single

---

FIG. 13. Transverse section of phosphorylase-deficient flexor carpi radialis muscle incubated for phosphorylase reaction. No color is seen. × 75. (Reproduced from Rowland, Fahn, and Schotland[19] by permission of the American Neurological Association.)

FIG. 14. Transverse section of control flexor carpi radialis muscle incubated for phosphorylase reaction. The fibers show a positive reaction. × 75. (Reproduced from Rowland, Fahn, and Schotland[19] by permission of the American Medical Association.)

FIG. 15. Electron micrograph of a longitudinal section of flexor carpi radialis muscle from a patient with McArdle's disease. Increased amounts of glycogen are present in the intermyofibrillar at the level of the I band (*G*). In the upper portion of the picture excess glycogen is present in the intermyofibrillar space of the A band as well. *M*, mitochondria. × 28,710. (Reproduced from Schotland, Spiro, and Rowland[23] by permission of the Association of the Journal of Neuropathology and Experimental Neurology, Inc.)

FIG. 16. Electron micrograph of a transverse section of flexor carpi radialis muscle from a patient with McArdle's disease. The subsarcolemmal accumulation of glycogen (*G*) has displaced a sarcolemmal nucleus (*N*) from the sarcolemma toward the center of the fiber. A possible defect in the sarcolemma (*S*) is also present (*arrow*). × 14,320. (Reproduced from Schotland, Spiro, and Rowland[23] by permission of the Association of the Journal of Neuropathology and Experimental Neurology, Inc.)

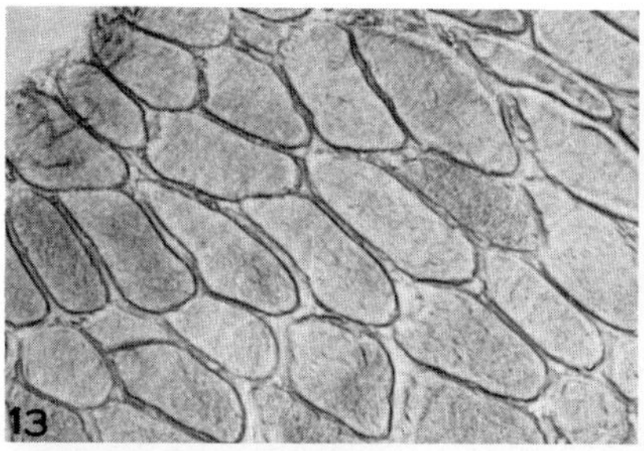
13

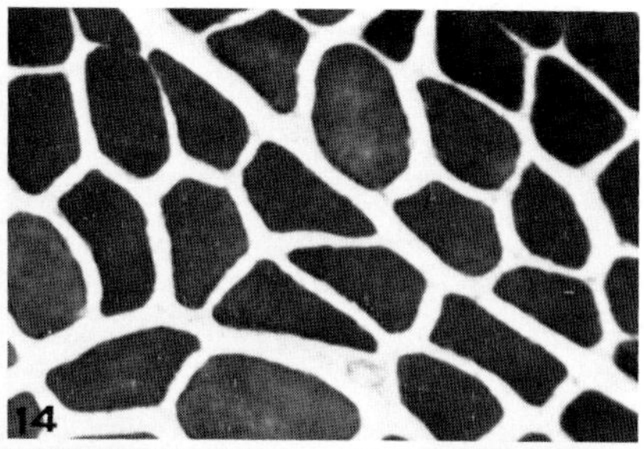
14

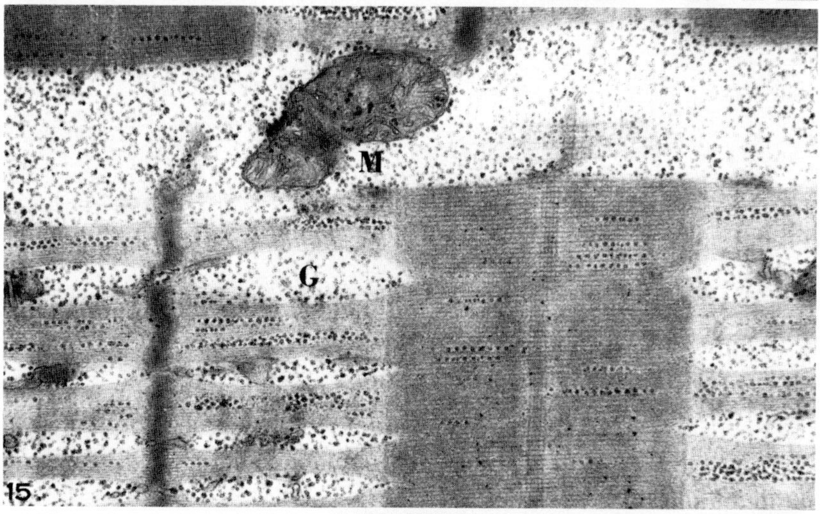
M
G
15

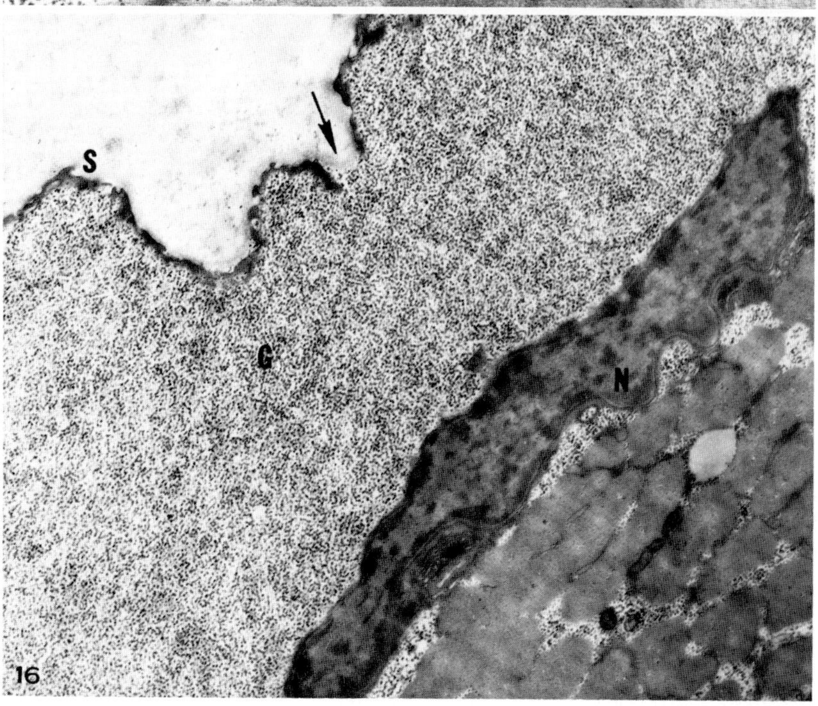
S
G
N
16

muscle fibers stripped of their surface membrane from a patient and from controls. In phosphorylase-deficient fibers, one to five applications of calcium or caffeine caused contractures that lasted as long as 1 minute; relaxation occurred after application of a medium containing ethylene glycol bis($\beta$-aminomethyl ether)-$N,N'$-tetraacetic acid and ATP. In control muscle, contractures lasted 1 to 3 seconds. These differences are best explained by a defect in the pump mechanism for reaccumulation of calcium in the SR of affected muscle.

Possible explanations for a pump defect include a local depletion of ATP in the SR secondary to the deficiency of phosphorylase and possibly secondary to the observed mitochondrial abnormalities as well. In this regard, ATP levels in extracts of whole muscle removed from 2 patients during contracture were unchanged, so local changes in SR were postulated.[20] A second explanation for failure of the calcium pump could be structural changes in the SR. We have some evidence that this may occur.[23] Although the sarcotubular system is usually normal in McArdle's disease, we observed dilatation of the SR in an isolated area of severely contracted muscle in 1 patient during contracture (Figs. 19 to 21); there was no evidence of shortening in the remainder of the specimen. To further study this possibility we are determining if ferritin, normally confined to the T system, can enter the SR during contracture.

## TYPE VII DISEASE (PHOSPHOFRUCTOKINASE DEFICIENCY)

The fourth type of glycogenosis that affects muscle is type VII disease or phosphofructokinase deficiency first described by Tarui *et al.* in 1965.[27] Phosphofructokinase catalyzes the conversion of fructose-6-phosphate to fructose-1,6 diphosphate; in the absence of this enzyme glycogen cannot be broken down to lactic acid. The clinical features are identical to phosphorylase deficiency.

We have had the opportunity to study a second family with this disease; the biochemical defect in an 18-year-old male was demonstrated by Layzer, Rowland, and Ranney.[12] Skeletal muscle viewed in the light microscope shows PAS-positive subsarcolemmal deposits (Fig. 22). A histochemical method for phosphofructokinase using fructose-6-phosphate as substrate and nitro blue tetrazolium as electron acceptor (Figs. 23 and 24) should facilitate future detection of this disease.[1]

---

FIG. 17. Electron micrograph of a transverse section of flexor carpi radialis muscle from a patient with McArdle's disease. There is extensive replacement of actin filaments by glycogen (*single arrows*) within two myofibrils at the I band level. × 30,100.

FIG. 18. Electron micrograph of a longitudinal section of gastrocnemius muscle from a patient with McArdle's disease. The mitochondria (*single arrows*) show loss of cristae, membrane distortion, and some glycogen accumulation. × 24,000.

FIG. 19. Electron micrograph of a longitudinal section of flexor carpi radialis muscle from a patient with McArdle's disease removed 30 minutes after the onset of ischemic exercise. In this area the sarcomeres are markedly shortened, the I bands are absent, and the Z lines are thickened due to the presence of contraction bands. × 28,600. (Reproduced from Schotland, Spiro, and Rowland[23] by permission of the Association of the Journal of Neuropathology and Experimental Neurology, Inc.)

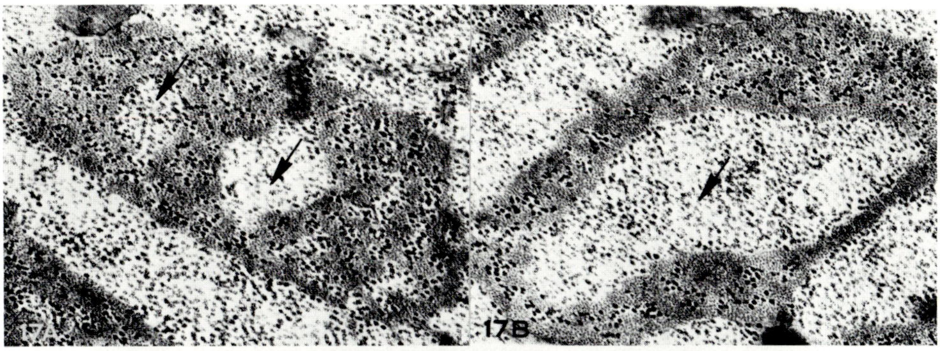
17B

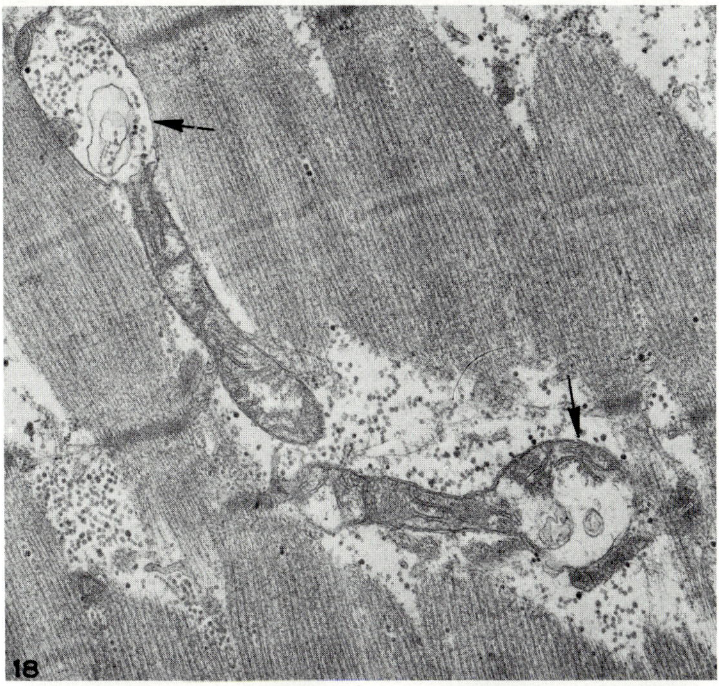
18

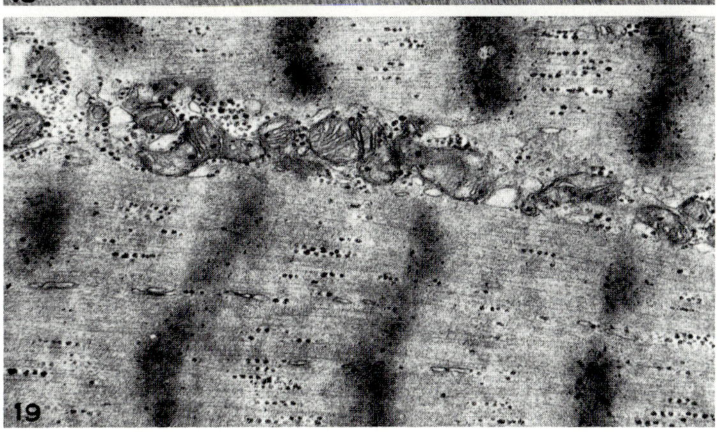
19

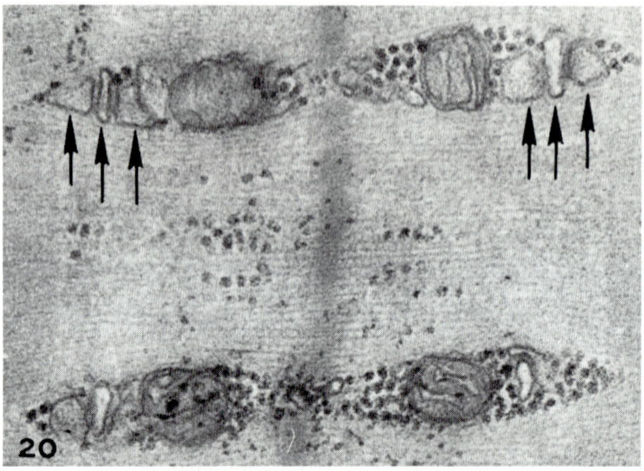
20

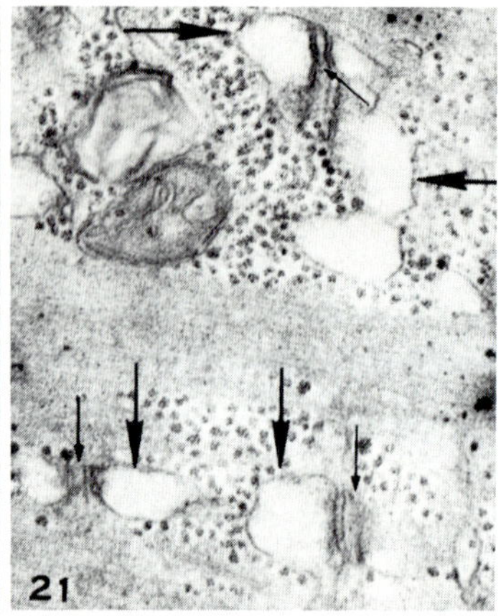
21

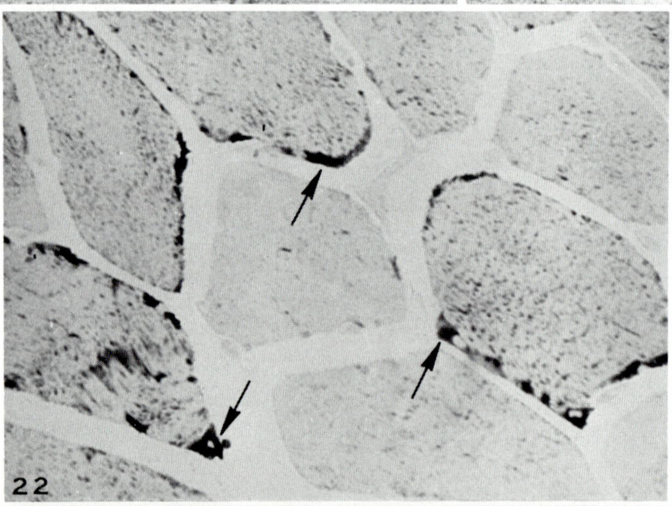
22

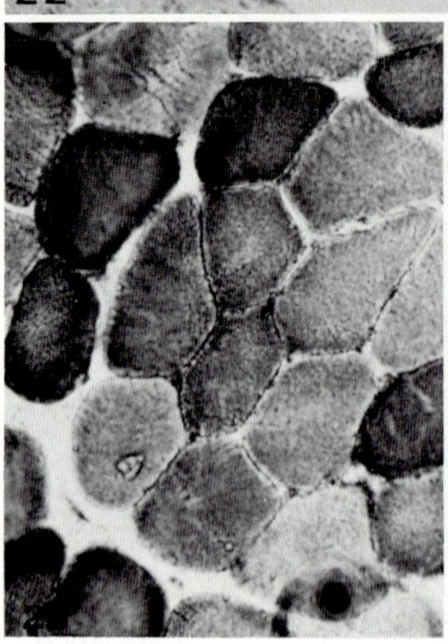

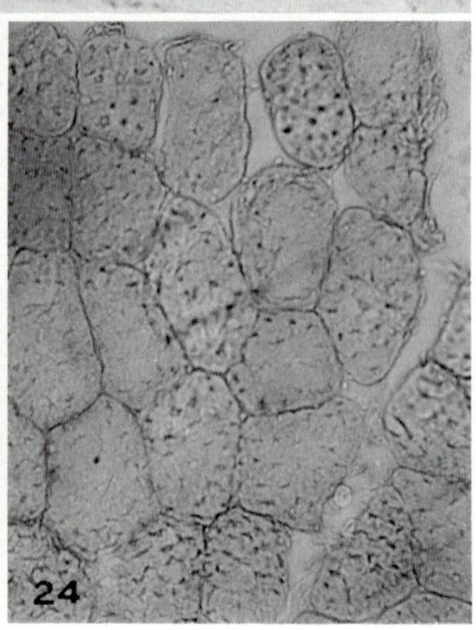
24

Electron microscopic study reveals a pattern of glycogen deposition similar to that observed in type III and type V disease. There is increased glycogen in the intermyofibrillar space opposite the A and I band (Fig. 25) between the thin filaments and under the sarcolemma (Fig. 26). Occasional collections of lysosome-like bodies and rare membrane-bound collections of glycogen are also seen.

## SUMMARY

Observations in the light microscope show that in type II glycogenesis PAS-positive vacuoles are distributed throughout the muscle fiber, presumably secondary to the greater accumulation of glycogen in this disease. In types III, V, and VII glycogen deposition is predominantly subsarcolemmal.

In the electron microscope type II disease is suspected if large areas of free glycogen and many single membrane-bound collections of glycogen are seen. The considerable amount of free glycogen is not explained by the lysosomal theory of this disease. In types III, V, and VII disease increased glycogen in the intermyofibrillar space, between thin filaments in the I band and under the sarcolemma is seen. Some membrane-bound collections of glycogen are also seen, especially in type III disease; these may be a secondary reaction to glycogen excess and cannot be considered a morphologic marker of acid maltase deficiency.

The morphologic characteristics of type II disease are distinctive and can be partially correlated with a biochemical defect, but the morphologic abnormalities in types III, V, and VII disease are fairly uniform. The ultimate understanding of these disorders will require further investigation.

---

FIG. 20. Electron micrograph of a longitudinal section of flexor carpi radialis muscle from a healthy adult male to illustrate elements of the sarcoplasmic reticulum and transverse tubule system which are present in the form of triads (*triple arrows*) adjacent to mitochondria. The central, lighter staining tubule is a portion of the T system, and the lateral sacs are part of the sarcoplasmic reticulum. × 57,750. (Reproduced from Schotland, Spiro, and Rowland[23] by permission of the Association of the Journal of Neuropathology and Experimental Neurology, Inc.)

FIG. 21. Electron micrograph of a longitudinal section of flexor carpi radialis muscle from a patient with McArdle's disease removed 30 minutes after the onset of ischemic exercise. The lateral sacs (*heavy arrows*) of the sarcoplasmic reticulum are dilated. The T system (*light arrows*) is unchanged. × 37,740. (Reproduced from Schotland, Spiro, and Rowland[23] by permission of the Association of Neuropathology and Experimental Neurology, Inc.)

FIG. 22. Transverse section of quadriceps muscle from a patient with phosphofructokinase deficiency. Subsarcolemmal deposits of PAS-positive material are present (*single arrows*) which can be removed with diastase. × 290.

FIG. 23. Transverse section of control quadriceps muscle incubated for phosphofructokinase reaction using fructose-6-phosphate as substrate. The fibers show a checkerboard pattern. × 240. (Reproduced from Bonilla and Schotland[1] by permission of the American Medical Association.)

FIG. 24. Transverse section of phosphofructokinase-deficient quadriceps muscle incubated for phosphofructokinase reaction. No color is seen. × 240. (Reproduced from Bonilla and Schotland[1] by permission of the American Medical Association.)

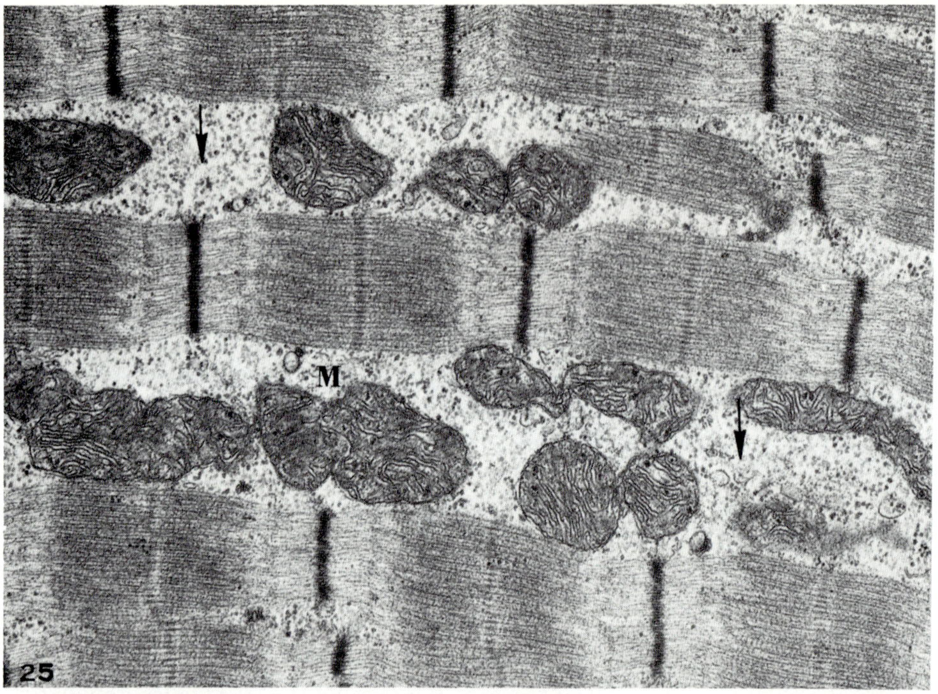


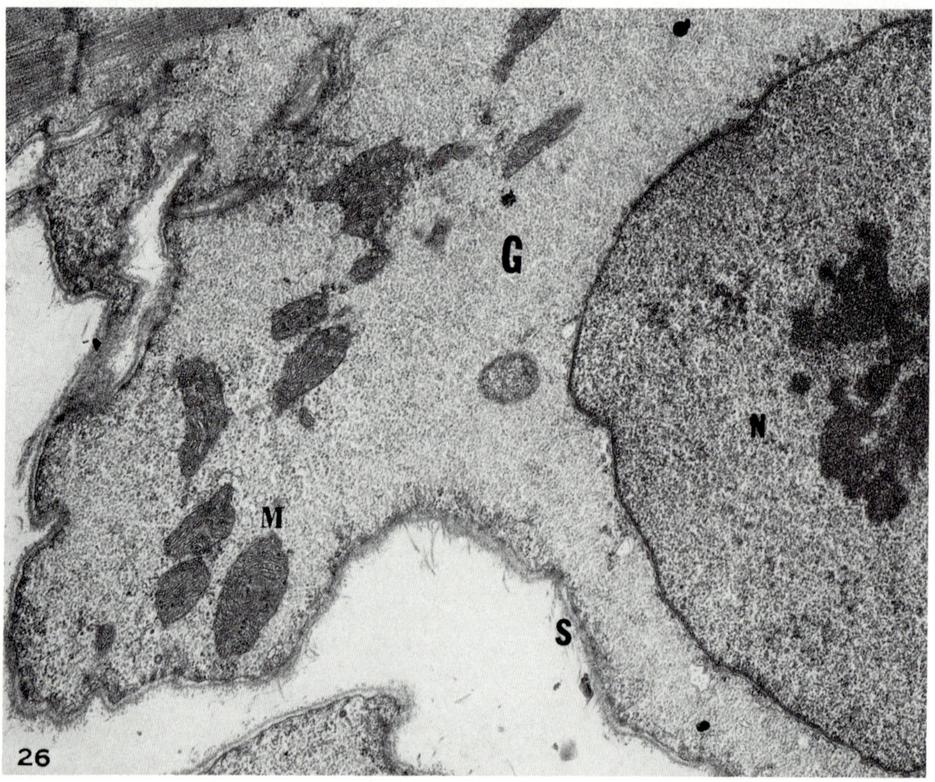


FIG. 25. Electron micrograph of a longitudinal section of quadriceps muscle from a patient with phosphofructokinase deficiency. There is increased glycogen in the intermyofibrillar space (*single arrows*). *M*, mitochondria. × 18,000.

FIG. 26. Electron micrograph of a longitudinal section of quadriceps muscle from a patient with phosphofructokinase deficiency. There is increased glycogen (*G*) under the sarcolemma (*S*). *M*, mitochondria; *N*, nucleus. × 15,000.

## ACKNOWLEDGMENTS

The biochemical studies were performed by Dr. Salvatore DiMauro. Miss Victoria Babaian and Miss Belen Aldover supplied technical assistance. Mr. Arthur Siegel aided in the photography.

REFERENCES

1. Bonilla, E., and Schotland, D. L.: Histochemical diagnosis of muscle phosphofructokinase deficiency. *Arch. Neurol. 22:* 8–12, 1970.
2. Brownell, B., Hughes, J. T., Goldby, F. S., and Woods, H. F.: McArdle's myopathy. A report of a case with observations on the muscle ultrastructure. *J. Neurol. Sci. 9:* 515–526, 1969.
3. Forbes, G. B.: Glycogen storage disease. Report of a case with abnormal glycogen structure in liver and skeletal muscle. *J. Pediatr. 42:* 645–653, 1953.
4. Garancis, J. C.: Type II glycogenosis. *Am. J. Med. 44:* 289–300, 1968.
5. Gruener, R., McArdle, B., Brenda, E., Ryman, E., and Weller, R. O.: Contracture of phosphorylase-deficient muscle. *J. Neurol. Neurosurg. Psychiatry 31:* 268–283, 1968.
6. Hers, H. G.: α-Glucosidase deficiency in generalized glycogen-storage disease. *Biochem. J. 86:* 11–16, 1963.
7. Hers, H. G.: Inborn lysosomal diseases. *Gastroenterology 48:* 625–633, 1965.
8. Hogan, G. R., Gutmann, L., Schmidt, R., and Gilbert, E.: Pompe's disease. *Neurology (Minneap.) 19:* 894–900, 1969.
9. Hug, G., Garancis, J. C., Schubert, W. K., and Kaplan, S.: Glycogen storage disease, types II, III, VIII and IX. *Am. J. Dis. Child. 3:* 457–474, 1966.
10. Illingworth, B., Cori, G. T., and Cori, C. F.: Amylo-1,6-glucosidase in muscle tissue in generalized glycogen storage disease. *J. Biol. Chem. 218:* 123–129, 1956.
11. Kelly, A. M.: Personal communication.
12. Layzer, R. B., Rowland, L. P., and Ranney, H. M.: Muscle phosphofructokinase deficiency. *Arch. Neurol. 17:* 512–523, 1967.
13. Mancall, E., Aponte, G., and Berry, R.: Pompe's disease (diffuse glycogenosis) with neuronal storage. *J. Neuropathol. Exp. Neurol. 24:* 85–96, 1965.
14. McArdle, B.: Myopathy due to a defect in muscle glycogen breakdown. *Clin. Sci. 10:* 13–35, 1951.
15. Mommaerts. W. F. H. M., Illingworth, B., Pearson, C. M., Guillory, R. J., and Seraydarian, K.: A functional disorder of muscle associated with the absence of phosphorylase. *Proc. Nat. Acad. Sci. U. S. A. 45:* 791–797, 1959.
16. Neustein, H. B.: Fine structure of skeletal muscle in type III glycogenosis. *Arch. Pathol. 88:* 130–136, 1969.
17. Oliner, L., Schulman, M., and Larner, J.: Myopathy associated with glycogen deposition resulting from generalized lack of amylo-1,6-glucosidase. *Clin. Res. 9:* 243, 1961.
18. Pearson, C. M.: Glycogen metabolism and storage diseases of types III, IV, and V. *Am. J. Clin. Pathol. 50:* 29–43, 1970.
19. Rowland, L. P., Fahn, S., and Schotland, D. L.: McArdle's disease. *Arch. Neurol. 9:* 325–342, 1963.
20. Rowland, L. P., Araki, S., and Carmel, P.: Contracture in McArdle's disease. *Arch. Neurol. 13:* 541–544, 1965.
21. Salter, R. H., Adamson, D. G., and Pearce, G. W.: McArdle's syndrome (myophosphorylase deficiency). *Q. J. Med. 36:* 565–578, 1967.
22. Schmid, R., Robbins, P. W., and Traut, R.: Glycogen synthesis in muscle lacking phosphorylase. *Proc. Nat. Acad. Sci. U.S.A. 45:* 1236–1240, 1959.
23. Schotland, D. L., Spiro, D., and Rowland, L. P.: Ultrastructural studies of muscle in McArdle's disease. *J. Neuropathol. Exp. Neurol. 24:* 629–644, 1965.
24. Smith, E. E., Taylor, P. M., and Whelan, W. J.: Enzymatic processes in glycogen metabolism. In Carbohydrate Metabolism and its Disorders, ed. by Dickens, F., Randle, P. J., and Whelan, W. J., pp. 89–138. London, Academic Press, 1968.

25. Steinitz, K., and Rutenberg, A.: Tissue $\alpha$-glucosidase activity and glycogen content in patients with generalized glycogenosis. *Israel J. Med. Sci. 3:* 411–420, 1967.

26. Takeuchi, T., and Kuriaki, H.: Histochemical detection of phosphorylase in animal tissues. *J. Histochem. Cytochem. 3:* 153–160, 1955.

27. Tarui, S., Okuno, G., Ikura, Y., Tanaka, T., Suda, M., and Nishikawa, M.: Phosphofructokinase deficiency in skeletal muscle. A new type of glycogenosis. *Biochem. Biophys. Res. Commun. 19:* 517–523, 1970.

28. Van Hoof, F., and Hers, H. G.: The subgroups of type III glycogenosis. *Eur. J. Biochem. 2:* 265–270, 1967.

29. Wolfe, H. J., and Cohen, R. B.: Nonglycogen polysaccharide storage in glycogenosis type II. *Arch. Pathol. 86:* 579–584, 1968.

Chapter 20

# The Pathologic Features of the Periodic Paralyses

CARL M. PEARSON

There are several forms of periodic paralysis (PP) which are characterized clinically by intermittent attacks of flaccid paralysis of skeletal muscles. Such attacks are due to a failure of excitation and contractility of muscle fibers. The pathogenesis of these disorders is unclear, but they appear to involve transmembrane shifts of potassium as well as sodium and probably other ions. Paralytic attacks characteristically involve the extremities symmetrically and only occasionally affect the bulbar or cranial musculature. Each of the primary forms of disease is transmitted through predominantly autosomal inheritance. Spontaneous remissions are common, and sensory and sphincter involvement are essentially unknown.

Since there are frequent alterations in serum potassium during attacks, three types of primary periodic paralysis have been recognized, namely, hypokalemic (Hypo PP), hyperkalemic (Hyper PP), and normokalemic (Normo PP) varieties. The syndrome of paramyotonia congenita, which is characterized by episodic attacks of massive paralysis upon exposure to cold, is also included as one subvariety. In addition, certain other forms of secondary periodic paralysis are associated with various disease states, including endocrine dysfunctions and alterations of potassium metabolism from various miscellaneous causes.

Histopathology of muscle in the periodic paralyses does not, normally, serve to separate these conditions. The following classification of the periodic paralyses has generally been accepted, based upon genetic, clinical and laboratory features[27]:

I. *Primary Periodic Paralyses*
   - A. Hypokalemic (familial) periodic paralysis, in which there is a fall in serum potassium during attacks.
   - B. Hyperkalemic periodic paralysis, with or without myotonia and characterized by a rise in serum potassium during attacks.
   - C. Normokalemic periodic paralysis, in which there is no significant alteration of serum potassium levels during attacks but in which there is a favorable clinical response to the administration of sodium.
   - D. Paramyotonia congenita with myotonia and massive weakness upon exposure to cold.

II. *Secondary Periodic Paralyses*
   A. Paralysis in endocrine dysfunctions, *e.g.* thyrotoxicosis and primary aldosteronism.
   B. Paralysis secondary to disorders affecting potassium metabolism, *e.g.* renal tubular acidosis, diabetic acidosis, and ureterosigmoidostomy.
   C. Licorice intoxication.

The clinical and laboratory features of many of these syndromes have been described in considerable detail elsewhere.[5, 6, 10, 12, 13, 20, 23, 26-28, 31, 33]

## PATHOGENESIS

The pathogenesis of several of the forms of periodic paralysis has been attributed to (1) defects in carbohydrate metabolism in skeletal muscle with consequent accumulation of metabolic cationic and anionic intermediates[31] or (2) to one or more defects in the sarcolemmal membrane or in the membranes of the sarcoplasmic reticulum.[5, 7, 10, 16] Such membrane defects would have, as a final expression, shifts in ionic concentrations of potassium, sodium, chloride, and maybe calcium. Such shifts could thus alter the polarization of the sarcolemmal or sarcoplasmic reticulum membranes which could consequently interfere with impulse propagation and ultimately interfere with muscle contraction.[16] Evidence for significant membrane depolarization in muscle fibers during an attack of paralysis has been amply recorded, at least in Hyper PP.[7]

## HISTORICAL AND CLINICAL FEATURES

Clinically, episodes of periodic paralysis were first described by Shakhnovitch in 1884,[30] Westphal in 1885,[34] Goldflam in 1890,[14] and Oppenheim in 1891.[25] These were at that time thought to be variants of the then delineated disease, spinal muscular atrophy. Probably all were observing cases of Hypo PP, but since the association with serum potassium decline was not known until almost a half a century later, one cannot be certain about the precise identity of these cases.

Hypo PP is the most common form to be clinically recognized. To date nearly 1,000 cases have been described in the literature. Hyper PP is less common, and probably 150 cases have been described so far. Normo PP is relatively uncommon, and only a few families with this syndrome have been recorded; the total number of cases is probably about 35.

## HISTOPATHOLOGY

In the early stages of the disease, usually after one or several paralytic episodes, there may be no or minimal pathologic alterations. On the other hand, it is recognized that after many paralytic attacks (varying from six to eight up to several hundred) permanent myopathic weakness may ensue. Permanent weakness does not develop in all cases, and spontaneous decline in clinical severity of attacks may ensue during later years of life.

Permanent weakness may appear at any interval after the onset of paroxysmal attacks, but it usually does not develop before at least 5 years after onset of periodic attacks. The youngest recorded incidence was Oppenheim's[25] case of a 15-year-old boy. There is often no close correlation between the frequency or the severity of paroxysmal attacks and the appearance of myopathic weakness.

The pathologic features of periodic paralysis have been studied intensively, especially within the past few years.[2, 8, 9, 18, 22, 24, 26, 31] Most of the literature on pathologic observation in these diseases revolves around studies of Hypo PP since that is the most frequent disease and since the pathology in that condition is the most dramatic.

As far back as 1895, soon after the clinical description of this disease syndrome, Goldflam[15] described the histopathology of muscle in periodic paralysis (probably Hypo PP). He noted at that time separation and rarefaction of myofibrils and vacuolization of muscle fibers. In his publications Goldflam described "rarefactions beginning between the myofibrils in the muscle fibers, splitting of Cohnheim's fields and vacuolation." His historic description of the basic pathology of advanced stages of periodic paralysis is reproduced in Figure 1 from Goldflam's publication in 1895. It is obvious that he recognized clearly the vacuolization that occurs in the muscle fibers in this disease state. Goldflam mentioned that the vacuoles were often round or oval, were of varying size, and at times appeared to be membrane-bound. In regular histologic sections, the vacuoles frequently distended and distored the adjacent myofibrils and other structures within the muscle fiber and appeared to be either empty or filled with a granular or a hyaline material. He thought that the vacuoles presented an advanced stage of the same process which caused the separation of the myofibrils.

Although autopsies or muscle biopsies were infrequently performed in patients with these diseases in the 50 years after Goldflam's description, a number of authors specifically commented upon the normalcy of the histology of their cases.[1, 17] Workers such as Biemond and Daniels[3] even considered that Goldflam's beautiful demonstrations and conclusions merely represented artifacts of tissue fixation or processing, even though they themselves observed mild muscle fiber vacuolization in their material, as did certain other authors during that time interval. Some authors did, however, describe the regular occurrence of muscle fiber vacuolization and attempted to convince their readers that these changes were significant.[32]

More recently, from about 1959 onward, muscle fiber vacuolization has been recognized as an impressive aspect of Hypo PP by many authors.[2, 5, 8, 9, 18, 24, 26, 31] This phase was culminated by an elegant and extensive review as well as a detailed pathologic, histochemical, and ultrastructural evaluation of the vacuolar myopathy by A. G. Engel in 1970,[9] in which a number of new conclusions were reached.

In all cases studied by various authors in which permanent myopathic weakness has occurred following episodes of Hypo PP the histologic alterations are impressive. The most striking change by light microscopy is the presence of a large number of vacuoles which distort and damage the majority of the skeletal muscle fibers (Figs. 2 to 5). In many instances about 40% of the fibers contain one or several small or large elongated or multiloculated vacuoles which

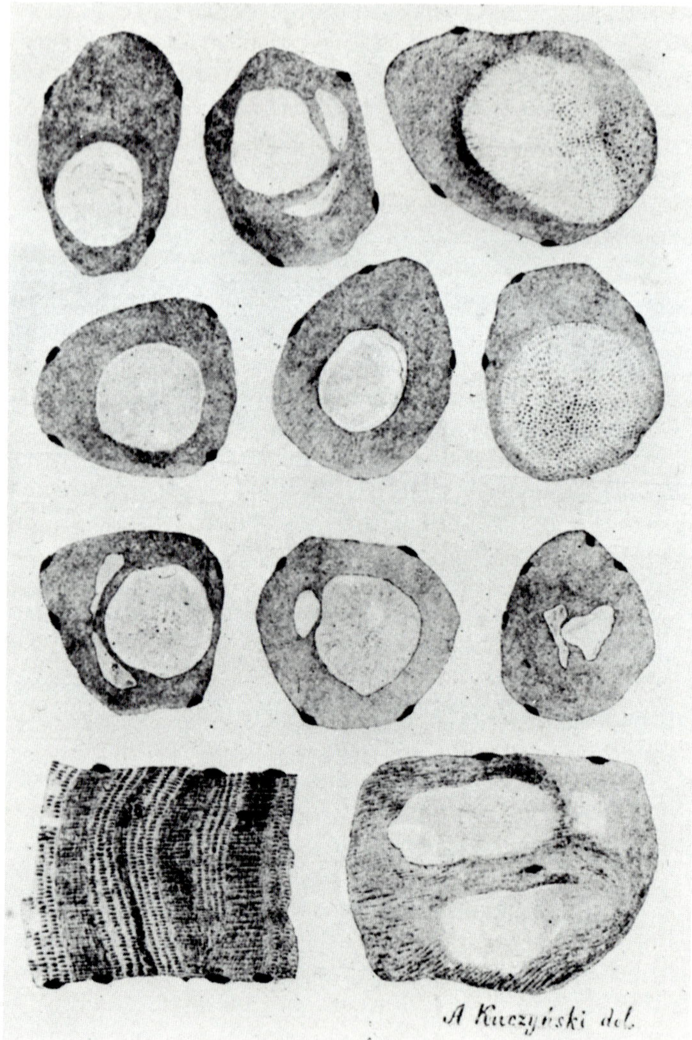


FIG. 1. Reproduction of the 1895 drawings of vacuoles in muscle fibers. (Reproduced from Goldflam.[15])

often distend and expand the fiber dimensions and compress myofibrillar bundles into condensed hyaline rims around the vacuole margins (Figs. 2 and 3). When small vacuoles are present, they are often fusiform in shape in the longitudinal axis and frequently, but not always, are perinuclear in location. Some fibers reveal a loss of cross-striations and evidence of hyaline degeneration. In transverse sections the myopathy is identical to that depicted 75 years ago by Goldflam[15] (Figs. 2 and 5). Many fibers exhibit a reduction in over-all diameter, and in conjunction with this, there appears to be a significant increase in endomysial connective tissue. In many fibers appropriately stained, there appears to be a finely dispersed periodic acid-Schiff (PAS)-positive and diastase-sensitive material (*i.e.*, glycogen) in the sarcoplasm, with a similar staining material in slightly larger aggregates in some but not all of the vacuoles.

Less striking alterations consist of (1) an increase in sarcolemmal nuclei, many of which were centrally positioned and pyknotic; (2) multiple contraction bands in some fibers; (3) occasional regenerating fibers; (4) very rare intravacuolar macrophages; and (5) clusters or myriads of vacuoles or vesicles rather than single large ones in some fibers, especially as seen in transverse section.

FIG. 2. Medium magnification of striated muscle from a patient with Hypo PP demonstrating extreme vacuolization, severe destruction of many muscle fibers, and some interstitial fibrosis.

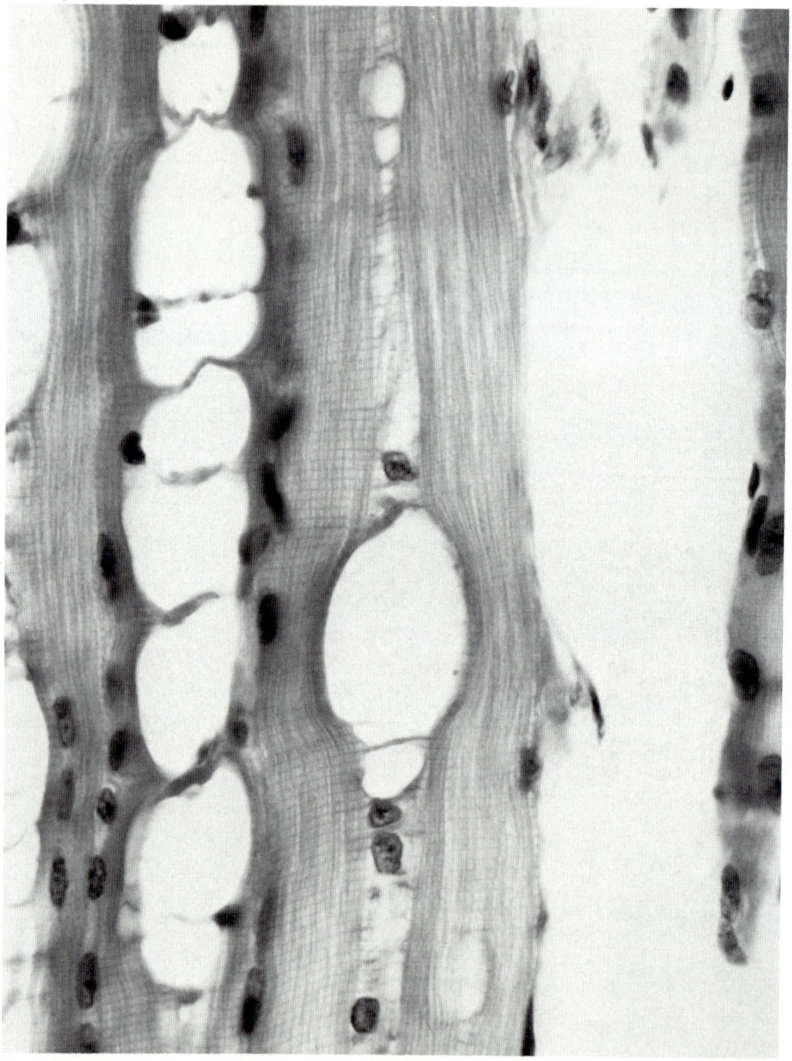

FIG. 3. Higher magnification of muscle fibers from the patient described in Figure 2. Illustrated here is the multilocular nature of the vacuoles and the evolution from small, centrally positioned vacuoles to large ones. Compression of adjacent myofibrils is obvious.

## HISTOCHEMISTRY

A. G. Engel has done the most extensive histochemical studies in Hypo PP[9] and has described acid phosphatase reactive regions, from a few microns to more than 20 $\mu$ wide, which are present in some fibers and usually adjacent to a well formed vacuole at one pole or near the center of a vacuole. In addition occasional fibers contain multiple, deep, or superficial reactive regions within the nonvacuolated fiber substance. Occasionally fibers of small diameter (less than 20 $\mu$ wide) are diffusely reactive for acid phosphatase, and some of these are seen to arise by splitting from larger parent fibers. Such reactive regions are often present near myelin-like structures (see below) which contained clusters of irregular curving and anastomosing bilaminar membranes.

When the NADH dehydrogenase reaction is utilized there is intense focal staining in some muscle fibers. Such regions are either central, eccentric, or peripheral in location and measure up to 30 $\mu$ in diameter. A positive staining for cytochrome oxidase also frequently corresponds with the NADH dehydrogen-

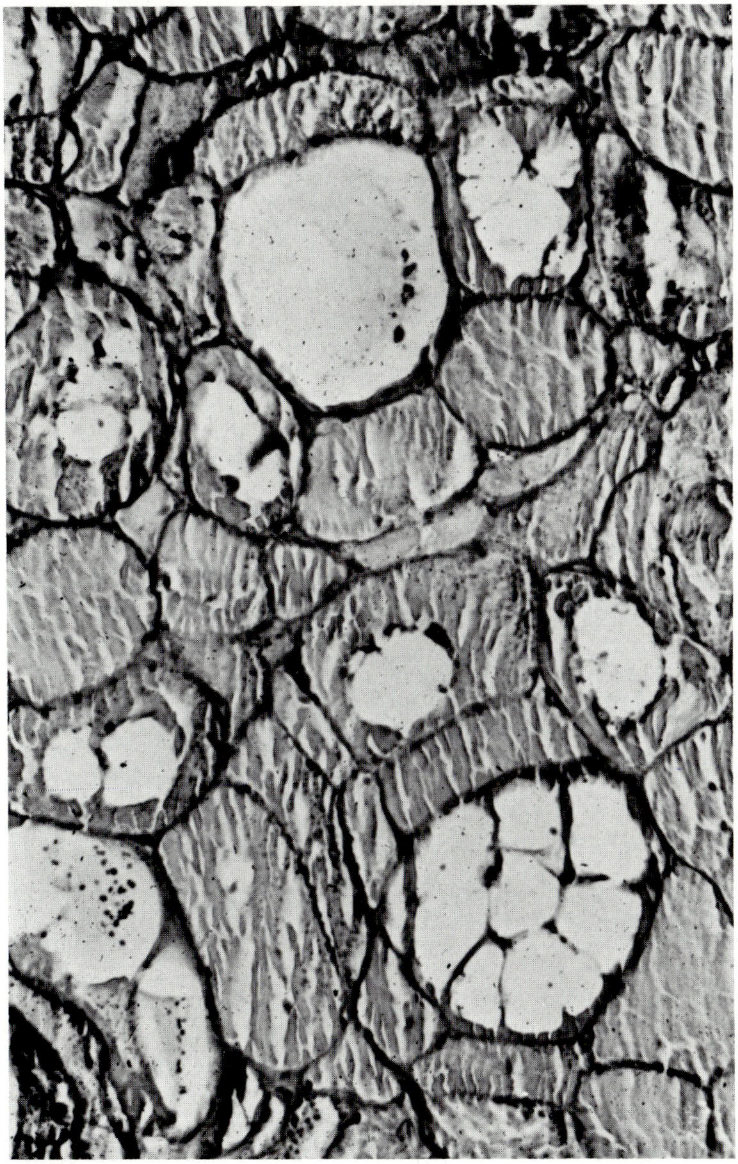

FIG. 4. Transverse section from a patient with Hypo PP demonstrating massive vacuolization and multiple loculation of some vacuoles.

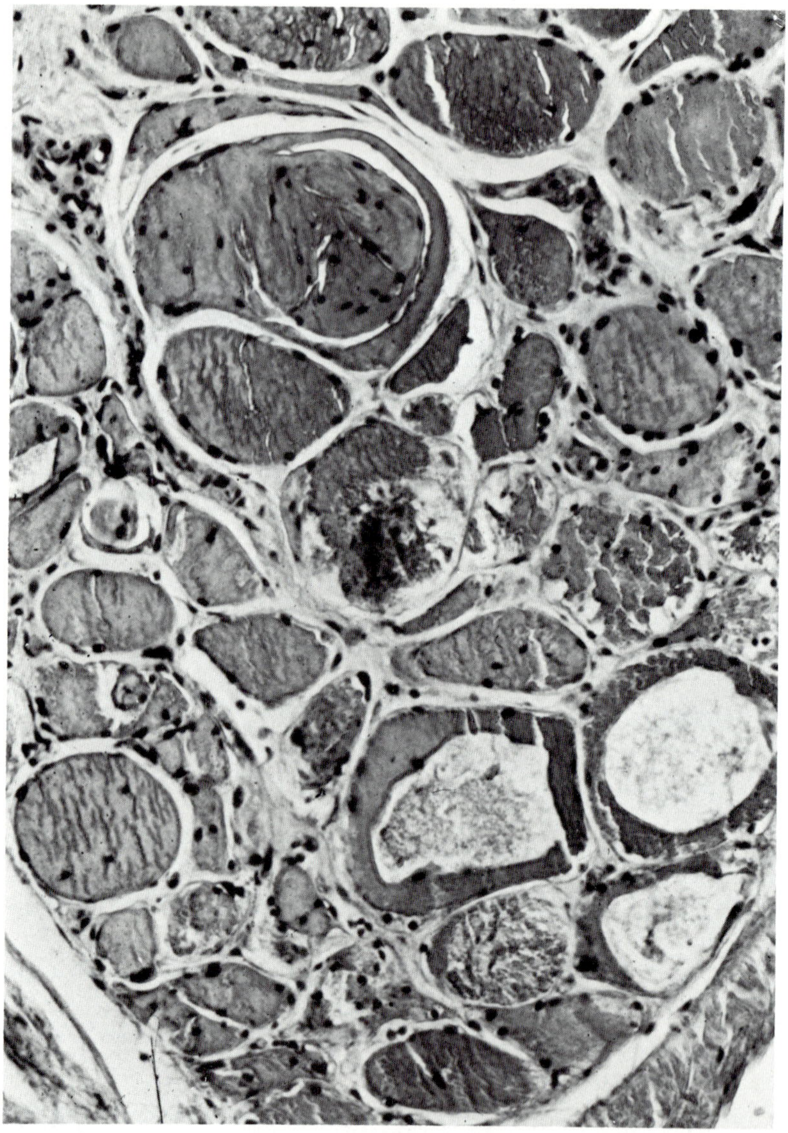

FIG. 5. An additional transverse section from the patient with Hypo PP demonstrating again some vacuolization but also overwhelming muscle fiber hyalinization, floccular degeneration, and marked central nucleation.

ase reaction, but similar areas are said to be always unreactive for succinic dehydrogenase. These regions are frequently, but not invariably, also reactive for acid phosphatase.

Utilizing staining procedures for calcific mineral deposits by the von Kossa method, a few vacuoles contained small deposits or massive concentrations of calcium.

PAS-reactive glycogen-like material is seen in all biopsy specimens in three locations: (1) in small subsarcolemmal or deeper intermyofibrillar spaces, (2) within round or dome-shaped intravacuolar compartments on the vacuolar rim, or (3) occasionally filling the entire vacuole (Fig. 6). Histochemically, also, small lipid droplets are found fairly frequently (by the oil red O method) in the nonvacuolated regions of some biopsy speciemns and only very occasionally within the vacuoles.

## ELECTRON MICROSCOPY

Ultrastructural studies of muscle in Hypo PP have disclosed much about the origin, evolution, and nature of the remarkable vacuolation of striated muscle

FIG. 6. Periodic acid-Schiff stain of muscle in Hypo PP demonstrating fine granularity of nearly all visible fibers and coarse clumps of PAS glycogen material in most vacuoles.

fibers that occurs in this condition. Some authors[29] felt that since the limiting membranes of the vacuoles are related to the T system they could not have originated from the sarcoplasmic reticulum (SR). Others, on the other hand,[18, 31] suggested that vesicles arising from the SR solely contribute to formation of the vacuoles.

A careful study by A. G. Engel has shed considerable light on this entire problem.[9] He has demonstrated that the following sequence apparently is followed in the evolution of vacuoles. (1) Abnormalities develop in certain regions of the muscle fiber which appear to contain clusters of dilated and at times mineralized SR vesicles (Fig. 7) and later masses of bizarre tubules and lamellae of T system origin. These are *the evolving vacuoles.* (2) The T system proliferates and acts as a membrane source which usually completely entraps the evolving vacuole. Thus, a membrane-bound space is formed whose matrix contains the remnants of the evolving vacuole (Fig. 8). This is *the intermediate-stage vacuole.* (3) Eventually the trapped contents of the intermediate-stage vacuole disappear through a variety of disposal mechanisms and the resultant vacuole which remains is a large dilated space which is often filled with amorphous precipitate (Fig. 9). This is *the mature vacuole.* It is the type that is seen most frequently in regular histologic sections as the large dilated vacuoles that are characteristic of the late stages of this disease syndrome (Figs. 1, 2, and 5).

Engel has clearly demonstrated that a communication exists between the T tubules, the T networks, the lumen of the mature vacuole, and the extracellular fluid. He demonstrated by ultrastructure cytochemical methods that when the large marker molecule, horseradish peroxidase, was placed in the extracellular fluid surrounding intact vacuolated muscle fibers this marker soon appeared within the lumens; thus indicating free communication of large molecules between the vacuole contents and the extracellular space and labeled the mature vacuole, at least, as an extracellular fluid reservoir.[9] Some remodeling of intermediate-stage and mature vacuoles apparently occurs by invaginations of the limiting membrane which may introduce glycogen and presumably other soluble components into the matrix lumen from the adjacent sarcoplasm. In turn, due to this remodeling process, extracellular fluid may enter the myofilament space surrounding vacuoles.

Engel has also noted that some of the small dilated SR vesicles appear to contain a mineral deposit, as demonstrated by an occasional finding of a crystalline pattern in the outer margins of the vesicle lumens. Frozen sections stained by the von Kossa method showed that these deposits contained calcium salts, possibly calcium phosphate. Since the SR serves as a calcium repository in muscle, and since there are also phosphatases associated with the SR,[11] a source for both calcium cations and phosphate anions exists in the SR membrane and likely contributes to these mineralizations that are found in some of the SR vesicles.

In summary, the factors which initiate morphologic changes that lead to vacuolization in muscle fibers in Hypo PP are so far unknown. Such factors are undoubtedly related to the basic metabolic error that occurs (probably in the

FIG. 7. Electron micrograph of the earliest vacuolated areas which are seen as dilatations of the sarcoplasmic reticulum principally at the A-I band junctions. × 11,500. (Reproduced from Howes, Price, Blumberg, and Pearson[18] by permission of the American Academy of Neurology.)

membrane) in this condition. Either these peculiar morphologic changes are a delayed consequence of the paralytic attacks (and this seems most likely), or less likely, they may not be related at all to clinically recognizable attacks of paralysis. At any rate, since there is ample time for structural remodeling and modification into this striking myopathy, in conjunction with intermittent variablity in electrical excitability of membranes on and in muscle fibers (the membranes being relatively unexcitable during attacks and regaining excitability again following attacks), one can predict with some degree of certainty

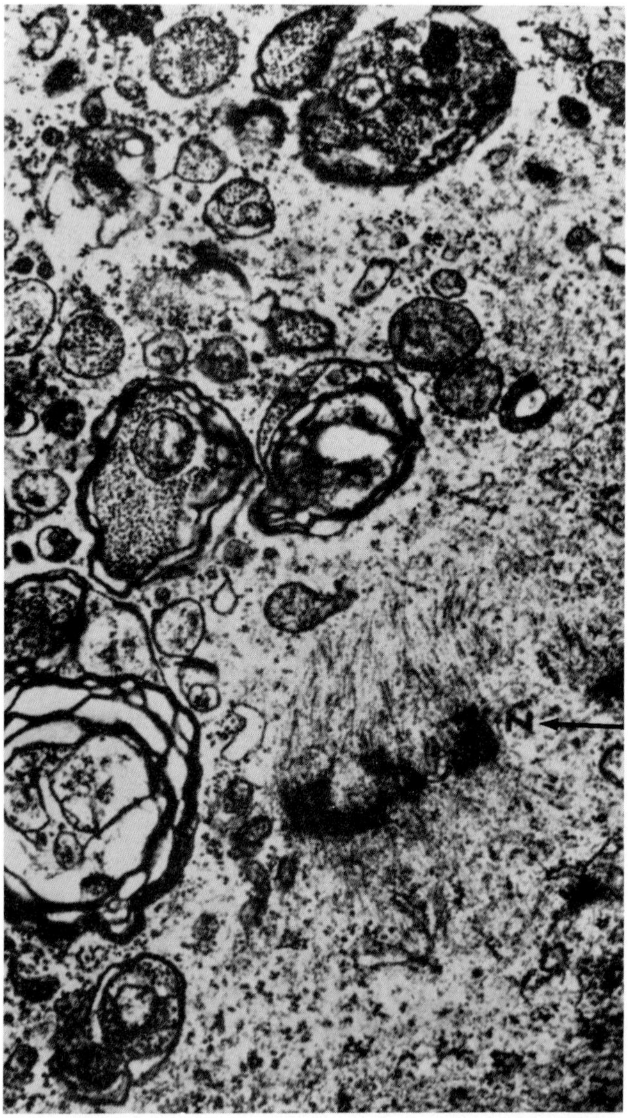


FIG. 8. A segment of a degenerating muscle fiber from a patient with Hypo PP. A degenerating Z band is labeled. Irregular clusters of tubular structures are shown in swirling forms in much of the remainder of the section. These "myelin-like" figures, some of which contain debris or other inclusion material, are what A. Engel would call intermediate-stage vacuoles. ×17,500. (Reproduced from Howes, Price, Blumberg, and Pearson[18] by permission of the American Academy of Neurology.)

that there exists a reasonably close relationship between these pathologic changes and prior periodic attacks of paralysis.

The pathologic vacuolization points to: (1) a disordered SR system with accumulation of some metabolic products in it, (2) an altered and proliferated T tubule system, (3) combinations of both of these systems within the vacuoles, and (4) a communication by the vacuoles with the extracellular spaces. The evolution of these changes that have been worked out so well by A. G. Engel clearly contributes a significant step to our understanding of the morphology and soon, it is hoped, to the metabolic defect in Hypo PP.

## PATHOLOGIC OBSERVATIONS IN HYPER PP AND NORMO PP

Fewer studies of the pathology are available in either of these two forms of periodic paralysis and none is as thoroughly analyzed as in Hypo PP. From among the several studies that are available[4, 19, 21, 26] the pathologic alterations, even in the late stages of either of these diseases, are in general less striking than in Hypo PP. Vacuolization occurs in a smaller number of muscle fibers, and it is not usually so severe, even after many years of recurrent episodes of paralysis.

Most authors have, however, observed some vacuolization of fibers, and in an occasional instance it appears to have been as extreme as in Hypo PP.[21] The general consensus is that the vacuoles originate in the longitudinal elements of the SR. In this respect these early changes are similar to those already described in Hypo PP. On the other hand fewer workers have presented descriptions of intermediate-stage or mature vacuoles, although Macdonald, Rewcastle, and Humphrey[21] have shown some convincing illustrations of the former in which complex, bizarre masses of swirling tubular structures occurred.

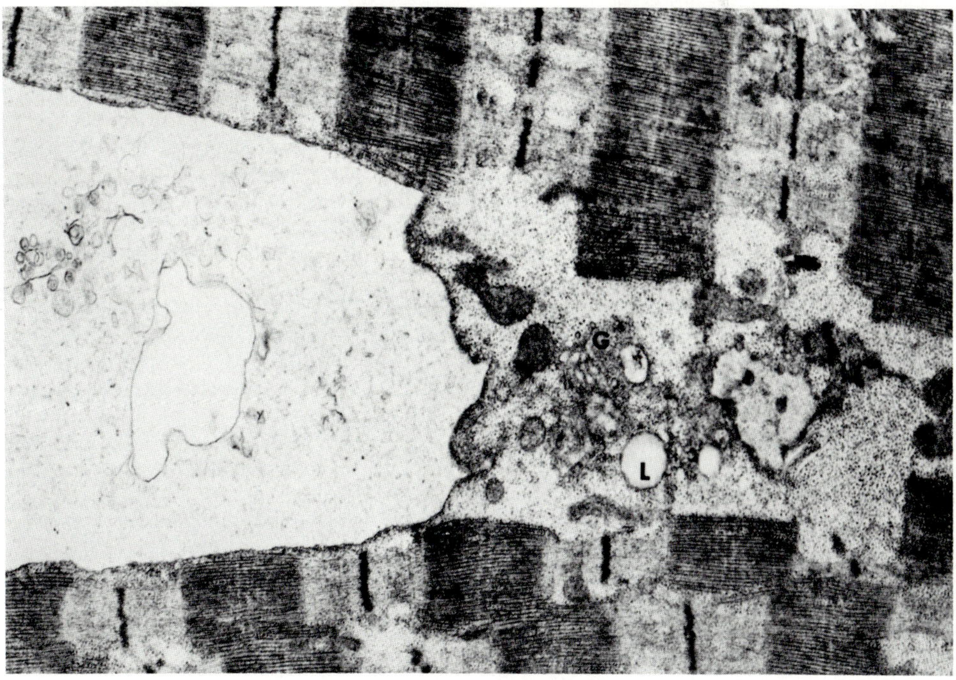


FIG. 9. A portion of a large vacuole which can be seen to contain finely granular as well as fibrillar material and a number of membrane forms. Glycogen granules (*G*) are present in varying numbers in the adjacent sarcoplasm and one lipid droplet (*L*) is also present. This vacuole can be considered to be a mature vacuole in A. Engel's terminology. (Reproduced from Howes, Price, Blumberg, and Pearson[18] by permission of the American Academy of Neurology.)

Thus, the histologic and ultrastructural findings in all three forms of periodic paralysis cannot in general be distinguished one from another, although usually the muscle fiber alterations in Hypo PP are more severe and extensive. The early changes in all seem to arise from dilatation of the longitudinal elements of the SR with, of course, consequent interference of excitation-contraction coupling and paralytic attacks.

As in Hypo PP, glycogen deposits, interference with surrounding contractile protein myofilaments, and other secondary changes involving mitochondria and lysosomes are found in the other two diseases, but they are definitely less severe. This, along with the less extensive vacuolization, seems to correspond well with the clinical observation that permanent myopathic weakness occurs much less commonly or extremely in these latter two diseases even after many years of recurrent paralytic attacks.

## REFERENCES


1. Benedek, L., and von Angyal, L.: Beiträge zur Pathogenese der paroxysmalen Lähmung (Myoplegia familaris). *Z. Gesamte Neurol. Psychiat. 174:* 213–228, 1942.
2. Biczyskowa, W., Fidzianska, A., and Jedrzejowska, H.: Light and electron microscopic study of the muscles in hypokalemic periodic paralysis. *Acta Neuropathol. (Berl.) 12:* 329–338, 1969.
3. Biemond, A., and Daniels, A. P.: Familial periodic paralysis and its transition into spinal muscular atrophy. *Brain 57:* 91–108, 1934.
4. Bradley, W. G.: Ultrastructural changes in adynamia episodica hereditaria and normokalemic familial periodic paralysis. *Brain 92:* 379–390, 1969.
5. Bradley, W. G.: Adynamia episodica hereditaria. Clinical, pathological and electrophysiological studies in an affected family. *Brain 92:* 345–378, 1969.
6. Carson, M. J., and Pearson, C. M.: Familial hyperkalemic periodic paralysis with myotonic features. *J. Pediatr. 64:* 853–865, 1964.
7. Creutzfeldt, O. D., Abbott, B. C., Fowler, W. M., and Pearson, C. M.: Muscle membrane potentials in episodic adynamia. *Electroencephalogr. Clin. Neurophysiol. 15:* 508–519, 1963.
8. Engel, A. G.: Electron microscopic observations in primary hypokalemic and thyrotoxic periodic paralyses. *Mayo Clin. Proc. 41:* 797–808, 1966.
9. Engel, A. G.: Evolution and content of vacuoles in primary hypokalemic periodic paralysis. *Mayo Clin. Proc. 45:* 774, 1970.
10. Engel, A. G., Lambert, E. H., Rosevear, J. W., and Newlon Tauxe, W.: Clinical and electromyographic studies in a patient with primary hypokalemic periodic paralysis. *Am. J. Med. 38:* 626–640, 1965.
11. Engel, A. G., and Tice, L. W.: Cytochemistry of phosphatases of the sarcoplasmic reticulum. I. Biochemical studies. *J. Cell Biol. 31:* 473–499, 1966.
12. Gamstorp, I.: Adynamia episodica hereditaria. Acta Paediatr. (*Suppl. 108*) *45:* 1–126, 1956.
13. Gass, H., Cherasky, M., and Savitsky, N.: Potassium and periodic paralysis. A metabolic study and physiological considerations. *Medicine 27:* 105–137, 1948.
14. Goldflam. S.: Ueber eine Eigenthiümliche Form von periodischer, familiärer, wahrscheinlich autointoxicatorischer Paralyse. *Wien. Med. Presse 31:* 1418, 1457, 1500, 1536, 1890.
15. Goldflam, S.: Weitere Mittheilung über die paroxysmale, familiäre Lähmung. *Dtsch. Z. Nervenheilkd. 7:* 1–31, 1895.
16. Hofmann, W. W., and Smith, R. A.: Hypokalemic periodic paralysis, studies *in vitro*. *Brain 93:* 445–474, 1970.
17. Holtzapple, G. E.: Periodic paralysis. *J.A.M.A. 45:* 1224, 1905.
18. Howes, E. L, Jr., Price, H. M., Blumberg, J. M., and Pearson, C. M.: Hypokalemic periodic paralysis. Electromicroscopic changes in the sarcoplasm. *Neurology* (*Minneap.*) *16:* 242–256, 1966.
19. Jaffurs, W. J., Herman, R. H., McDowell, M. K., and Blumberg, J. E.: Hyperkalemic paralysis (adynamia episodica hereditaria). Ultrastructural studies of muscle in two cases. *Metabolism 12:* 740–750, 1963.

20. Layzer, R. B., Lovelace, R. E., and Rowland, L. P.: Hyperkalemic periodic paralysis. *Arch. Neurol. 16:* 455–472, 1967.
21. Macdonald, R. D., Rewcastle, N. B., and Humphrey, J. G.: The myopathy of hyperkalemic periodic paralysis. *Arch. Neurol. 19:* 274–283, 1968.
22. Deleted in proof. See Reference 21.
23. McArdle, B.: Familial periodic paralysis. *Br. Med. Bull. 12:* 226–229, 1956.
24. McArdle, B.: Adynamia episodica hereditaria and its treatment. *Brain 85:* 121–148, 1962.
25. Oppenheim, H.: Neue Mitteilungen über den von Professor Westphal beschriebenen Fall von periodischer Lähmung aller vier Extremitaten. *Charite-Ann 16:* 350–372, 1891.
26. Pearson, C. M.: The periodic paralyses. Differential features and pathological observations in permanent myopathic weakness. *Brain 87:* 341–354, 1964.
27. Pearson, C. M., and Kalyanaraman, K.: *The periodic paralyses.* In *The Metabolic Basis of Inherited Disease*, ed. by Stanbury, J., Wyngaarden, J., and Frederickson, D., Chapter 47, pp. 1181-1203. New York, McGraw-Hill, 1972.
28. Poskanzer, D. C., and Kerr, D. N.: Periodic paralysis with response to spironolactone. *Lancet 2:* 511–513, 1961.
29. Schutta, H. S., and Armitage, J. L.: The sarcoplasmic reticulum in thyrotoxic hypokalemic periodic paralysis. *Metabolism 18:* 81–83, 1969.
30. Shakhnovitch: On a case of intermittent paraplegia. *Russk. Vrach.* 32, 537, 1882; abstracted in *London Med. Rec. 12:* 130, 1884.
31. Shy, G. M., Wanko, T., Rowley, P. T., and Engel, A. G.: Studies in familial periodic paralysis. *Exp. Neurol. 3:* 53–121, 1961.
32. Singer, H. D., and Goodbody, F. W.: A case of family periodic paralysis with a critical digest of the literature. *Brain 24:* 257–285, 1901.
33. Talbott, J. H.: Periodic paralysis. A clinical syndrome. *Medicine 20:* 85–143, 1941.
34. Westphal. C.: Ueber einen merk würdigen Fall von periodscher Lähmung aller vier Extremitaten mit gleichzeitigem Erläschen der elektrischen Erregbarkeit während der Lähmung. *Berlin. Klin. Wochenschr. 22:* 489–509, 1885.

Chapter 21

# Congenital Myopathies

THEODORE L. MUNSAT

During the past 10 years a considerable effort has been made to increase our knowledge of those patients who demonstrate disease of the motor unit early in life. Within the large number of publications on this subject, certain trends can be delineated.

1. Nosology has improved considerably, and in the process, the nonexistence of several established "diseases" (*i.e.*, amyotonia congenita, universal hypoplasia of muscle) has been demonstrated.
2. Several "experiments of nature" have been uncovered and have provided an additional stimulus for the study of normal structure and function (*i.e.*, McArdle's disease and "myotubular" myopathy).
3. Histochemical and cytochemical studies have been successfully applied as in no other area of medicine.
4. Structural and histochemical findings initially thought to be disease specific or unique have been demonstrated in an increasing variety of unrelated clinical and experimental settings.
5. The important role of the nervous system in controlling the muscle fiber through as yet unifentified "trophic" substances has been established.
6. The clinical, electromyographic, and structural differentiation between chronic anterior horn cell disease and primary myopathy has become increasingly difficult, if not at times impossible.
7. Although most of these disorders continue to be incurable, improved management has resulted.

Table 1 lists those disorders which have been loosely grouped under the descriptive clinical heading of Congenital Nonprogressive Myopathies. This list constitutes a wide spectrum of apparently unrelated disorders which present in a similar clinical manner. At this time a clinical classification is much more meaningful than a structural one. Hopefully, with time, a biochemical classification will become possible. Although the majority of patients present with symptoms at birth, weakness may begin or become apparent to the family later in life. Often it is extremely difficult to accurately date the onset of definite motor impairment. Although the weakness tends to be stationary or slowly progressive, more malignant clinical courses do occur. Lastly, although these disorders are termed "myopathies," increasing evidence suggests that disordered neural control may be a factor in the pathogenesis of several of these conditions.

The *chronic neurogenic muscular atrophies* ("benign" Werdnig-Hoffman, Kugelberg-Welander, certain cases of scapuloperonal muscular atrophy) may be indistinguishable in all aspects from a congenital myopathy[19] (Fig. 1) and are intimately involved with the clinical and pathologic differential diagnosis.

Because of the large numbers of disorders to be considered, I have elected to concentrate on only a few and to discuss the others superficially. The *myotonic*

TABLE 1. CONGENITAL MYOPATHIES

| |
|---|
| Nemaline (Z band) myopathy |
| Centronuclear (myotubular) myopathy |
| Central core disease |
| Mitochondrial myopathies |
| Myotonic disorders |
| Periodic paralyses |
| Glycogen storage diseases |
| Miscellaneous |

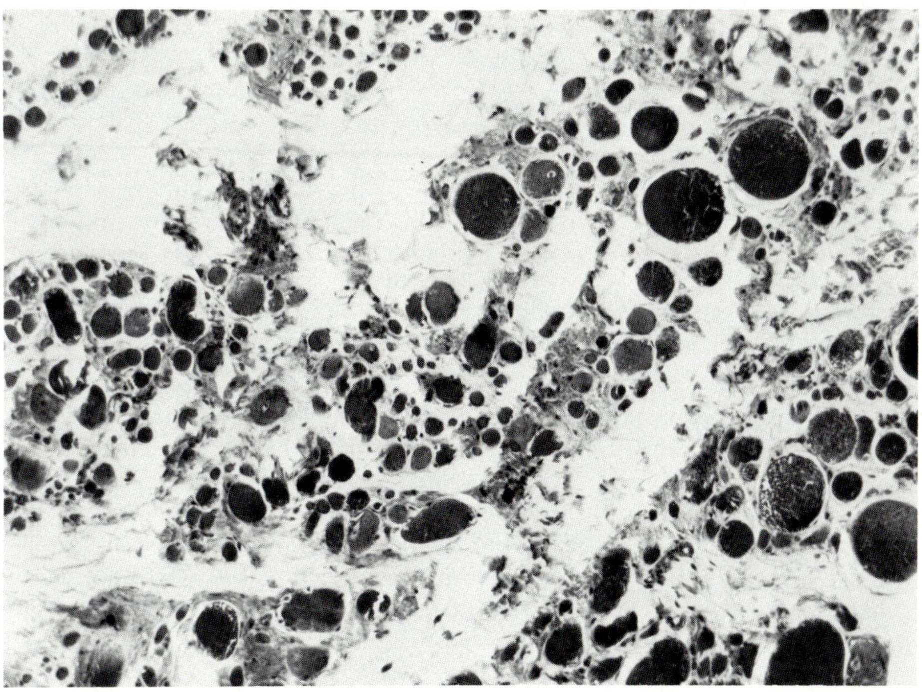

FIG. 1. Gastrocnemius biopsy from patient with chronic neurogenic muscular atrophy ("benign" Werdnig-Hoffmann disease). Alterations are indistinguishable from primary myopathy such as Duchenne dystrophy. Note variation of fiber size, increased endomysial connective tissue, central nuclei, necrosis, fiber splitting, and fat infiltration. Hematoxylin and eosin stain; × 136. (Reproduced from Munsat *et al.*[19] by permission of Macmillan and Company.)

*disorders* are included for completeness. In fact, these patients rarely present with generalized weakness early in life. It is important to realize that classical myotonic muscular dystrophy rarely presents with clinical dysfunction within the first few years of life. Neither does myotonia congenita (Thomsen's disease) which is primarily a physiologic disturbance of muscle with minimal if any pathologic alterations. Congenital muscle dysfunction, especially involving the cranial nerves, does occur in the variety of myopathy called "myotonic dysembryoplasia" by Pruzanski,[24] a syndrome seen in the newborn of parents with the classical disorder.

The *periodic paralyses* and *glycogen storage diseases* have been discussed by others in the symposium (Chapters 19 and 20).

I would suggest that the term "dystrophy" is presently only meaningful as applied to the clinical disorder Duchenne muscular dystrophy. Thus, I would limit the term "*congenital dystrophy*" to those patients with Duchenne dystrophy symptomatic at birth. Since there is extremely little evidence to suggest that this ever occurs, the term "congenital muscular dystrophy" in this context becomes confusing and meaningless and should be replaced by the less specific "congenital myopathy."

*Central core disease* originally described by Shy and Magee in 1956[31] is the grandfather of the congenital myopathy syndrome. This muscle fiber alteration is characterized by a central zone without oxidative or glycolytic enzyme activity containing disordered myofibrils. More than one "core" may be observed in a single fiber. The central core alteration is somewhat similar to the target fiber and can be seen in clinical and experimental denervation.[8] Thus, evidence suggests that the central core change may be a phenomenon of disordered neural control rather than a primary myopathy. It also occurs in association with clinical[1] and experimentally induced[27] nemaline bodies. We have not observed a patient with clear-cut central core "disease." We have seen, however, central core alterations in patients who by other clinical and histologic criteria had a chronic denervating process.

*Mitochondrial alterations* in the congenital nonprogressive myopathy syndrome have been reported with increasing frequency since Shy's original description of "megaconial" and "pleoconial" myopathies.[30] Alterations have included changes in number, size, and profile of mitochondria, changes in the configuration of mitochondrial cristae, and several varieties of inclusion bodies. Associated derangement of lipid and carbohydrate metabolism has been suggested in several cases.[5, 22] Histochemical reactions may reveal increased oxidative enzyme activity corresponding to mitochondrial aggregates in the subsarcolemmal position (Fig. 2). Entirely similar mitochondrial alterations have been reported in striated muscle in widely dissimilar clinical and experimental settings as well as in other tissues.[5, 22] The sorting process has begun but will require considerable effort to clarify the increasing number of "mitochondrial myopathies."

*Nemaline* (*Z band, rod-body*) *myopathy* has aroused the interest of several investigators because of the unusual nature of these sarcoplasmic bodies which cytochemically and structurally appear very similar if not identical to the Z band. The clinical picture[21] is one of diffuse weakness involving almost all striated muscle, usually present at birth, slowly progressive if at all, and associated with a variety of relatively nonspecific skeletal alterations such as pectus excavatum, high arched palate, and pes planus. Serum enzymes are usually normal and the electromyogram shows nonspecific myopathic changes. Both autosomal dominant and autosomal recessive inheritance occur in the familial cases. As one might expect, with additional reports, variation from this clinical pattern has occurred. Thus, onset of clinical weakness in nemaline myopathy has been reported as late as 61 years.[4] Rarely the clinical progression may be rapid and malignant with an early demise.[26] Nemaline changes have been reported in association with central core alterations (see above) and in a setting of electromyographic and histologic denervation.[11] Nemaline bodies occur with collagen-vascular diseases and following tenotomy, although much less prominently.[7]

The histopathologic diagnosis of nemaline myopathy should not be difficult.[20] With proper handling of the biopsy, routine fixation and paraffin embedding, the rods can easily be seen with both understained phosphotungstic acid-hematoxylin, where they appear dark blue-black, or with the trichrome (Gomori or Masson's) reaction, where they are usually deep red in color (Fig. 3). Rods cannot be easily identified with hematoxylin and eosin. In our screening program of early onset motor unit disease, we routinely process paraffin-embedded biopsies with hematoxylin and eosin, phosphotungstic acid-hema-

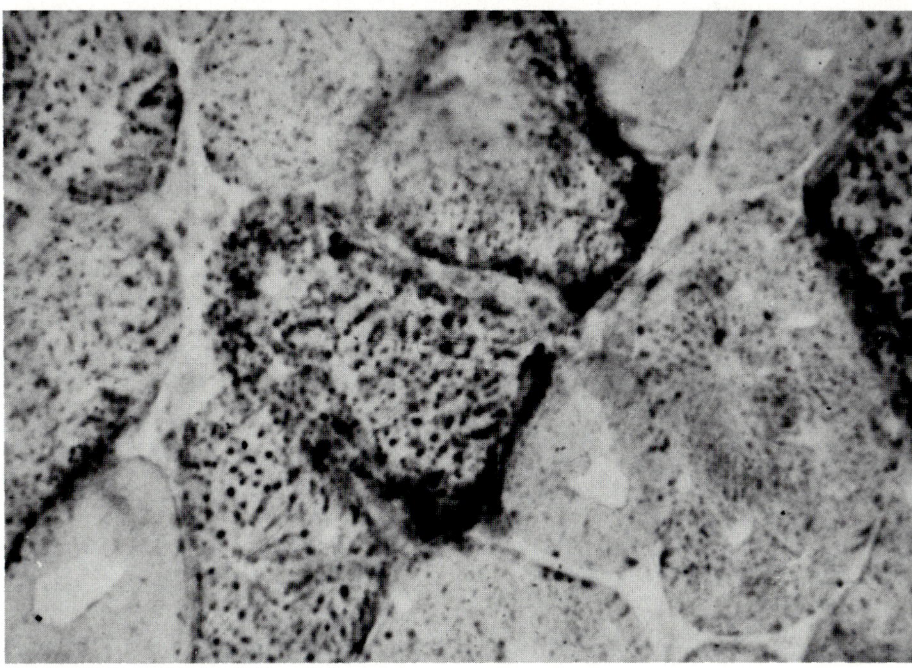

FIG. 2. Mitochondrial myopathy. Cytochrome oxidase reaction. Subsarcolemmal accumulations of oxidative enzyme activity may indicate mitochondrial abnormalities. In this patient, electron microscopy revealed increased numbers of abnormal mitochondria.[22]

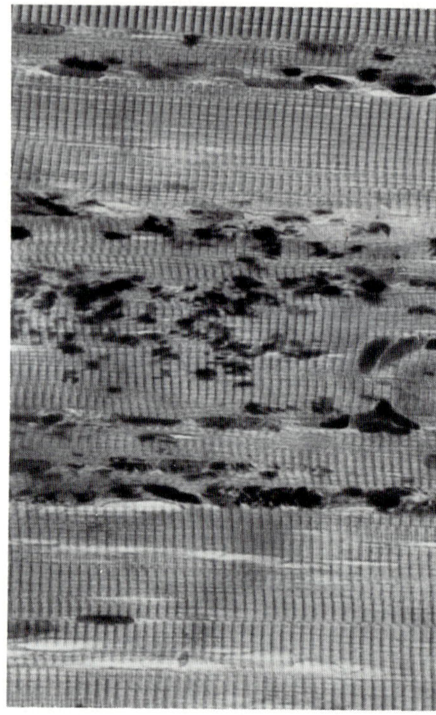

FIG. 3. Nemaline myopathy. Masson's trichrome stain; × 480. Fixed in 10% formalin and paraffin embedded. Dark staining (red) rods are easily seen in center field.

toxylin, and trichrome stains. With routine fixation and paraffin embedding, the rods measure approximately 1 by 5 $\mu$ and are usually cigar-shaped or oval in configuration. They are most frequently arranged in groups in the form of amorphous clumps, in a radial configuration, or in palisades. They may be located in any region of the muscle fiber. Frequently small stacks of rods will be arranged in a regular, repeating manner several sarcomeres apart. There is poor correlation between clinical weakness and the degree of rod formation.[20] A normal biopsy can be seen in a muscle that shows considerable clinical weakness. Classical "myopathic" features, *i.e.*, variation in fiber size, intracellular nuclei, degeneration and necrosis, increased connective tissue, etc., are minimal. There often is a tendency for grouped or scattered small fibers suggestive of a denervating element. It is important to emphasize that a normal hematoxylin and eosin preparation does not rule out nemaline myopathy.

Histochemical reactions have revealed a tendency for poor differentiation of fiber types with both intermediate and uniform responses occurring. Rods occur more frequently in type 1 fibers.[7] Histochemical studies have demonstrated that the rods have neither oxidative nor glycolytic enzyme activity.[20] They do not react with lipid or carbohydrate stains. They consistently react in a manner similar to the Z band. This is especially prominent with phosphotungstic acid-hematoxylin and trichrome stains. Both z band and rods show no reaction for protein-bound amino groups or arginine, but do react for tyrosine.[20, 29] Tryptophan reactivity varies with the technique used.[26] The rods do not show myosin ATPase activity.[20, 29] Thus, histochemical studies suggest that rods are composed of a protein, probably not actin or myosin, but a protein normally occurring in the Z band region of the sarcomere. With phase microscopy the rods are optically dense and may be refractile[6, 15] or not.[13] With polarized light the structures are not anisotropic[14, 29] as is myosin or are anisotropic in a different position from the myofibrils.[10]

Electron microscopic observations have furthered our understanding of the rods. In 1965 Price, Gordon, Pearson, Munsat, and Blumberg[23] clearly demonstrated that not only did the rods have ultrastructural similarity to the Z band, but they actually appeared to "arise" from it. Thus, the earliest ultrastructural changes were those of widening or enlargement of an apparently normal Z band (Fig. 4). In later stages, as the rod enlarged, it often maintained a thin

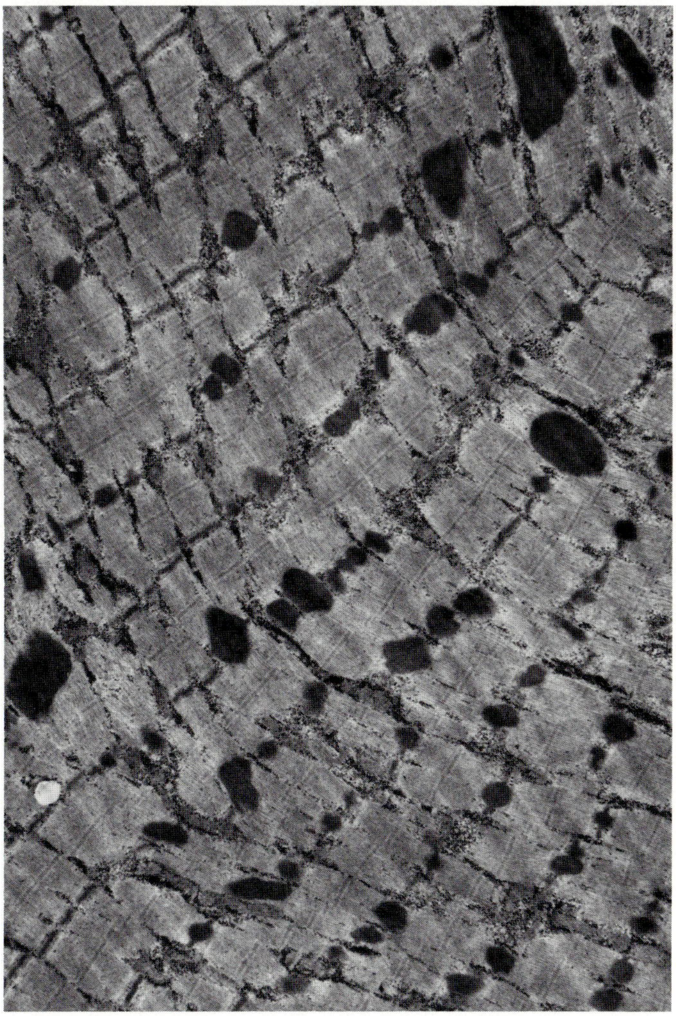

FIG. 4. Nemaline myopathy. Electron micrograph. × 14,500. Early stage of rod formation with multiple areas of focal Z band enlargement. This differs from less specific Z band "streaming."

connection with the "parent" Z band. Larger rods lay free in the sarcoplasm with little disturbance of the adjacent myofibrils and extended over two or more sarcomeres. Subsequent electron microscopic observations have consistently demonstrated the structural similarity and frequent continuity between Z band and rods.[4, 6, 13]

Although individual reports have varied somewhat, general agreement regarding the substructure of the rods is present. They are usually oriented in the same direction as adjacent normal appearing myofibrils and are oval or rectangular in configuration with a rather clear 120 to 180 Å periodicity perpendicular to the long axis of the rod (Fig. 5) and a less clear parallel periodicity of 100 to 120 Å. On cross-section, this protein crystal lattice configuration is maintained with a series of boxlike structures. The structural similarity of the rods to purified animal tropomyosin[23, 29] suggests that the rod protein could contain tropomyosin as has been suggested for the Z band.[16] However, flourescein-labeled clam antitropomyosin antibody does not attach to the rods.[29]

A recent observation[4] has demonstrated thin (presumably actin) myofilaments frequently attached to the ends of the rods. Since actin filaments are strongly attached to normal Z band, this provides additional evidence for the theory that the rods are Z band related.

Few direct biochemical studies have been carried out to date. It has been shown by Engel[6] that neither glycerination nor myosin extraction alters the rods significantly. Actin-extracting solutions, however, did result in a loss of integrity of both rods and Z bands. These studies would suggest that the rods are not composed of myosin but could be actin, tropomyosin, or one of the less well defined contractile or sarcoplasmic proteins.

*Centronuclear (myotubular) myopathy.* In 1966 Spiro, Shy, and Gonatas[32]

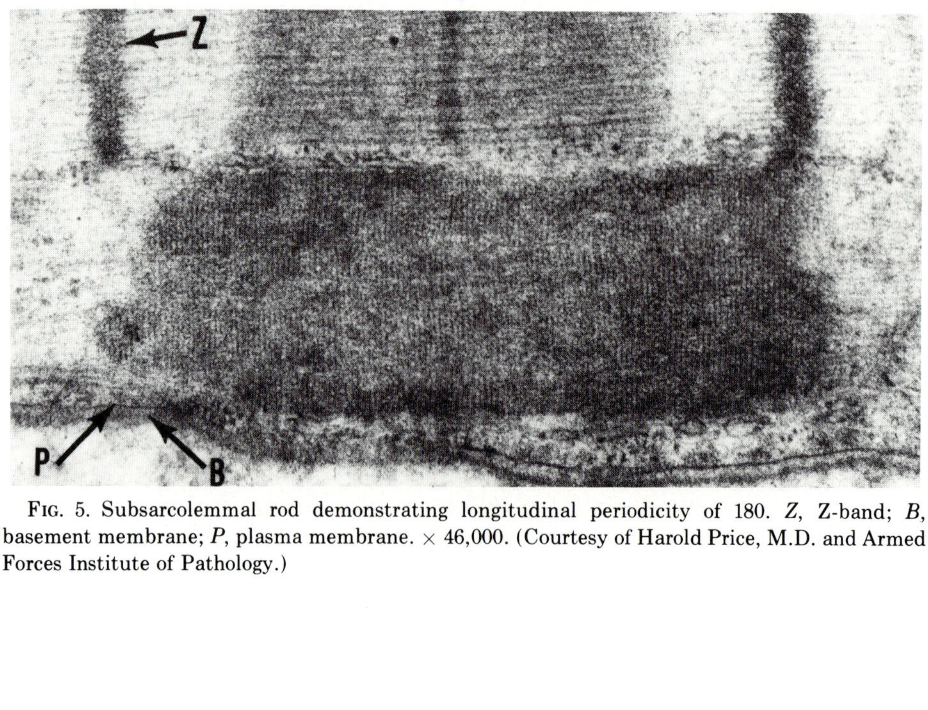


FIG. 5. Subsarcolemmal rod demonstrating longitudinal periodicity of 180. *Z*, Z-band; *B*, basement membrane; *P*, plasma membrane. × 46,000. (Courtesy of Harold Price, M.D. and Armed Forces Institute of Pathology.)

reported a slowly progressive, early onset myopathy in a 12-year-old boy characterized by muscle fibers with centrally placed nuclei (Fig. 6). Cytochemical and electron microscopic studies led the authors to postulate an arrest of muscle fiber maturation at the myotubular (10 to 12 week) stage of embryogenesis. Subsequent case reports have tended to cloud rather than strengthen this fascinating hypothesis.[3, 17, 18, 28, 33] Although the patients of Sher, Rimalovski, Athanassiades, and Aronson[28] demonstrated similar clinical and pathologic

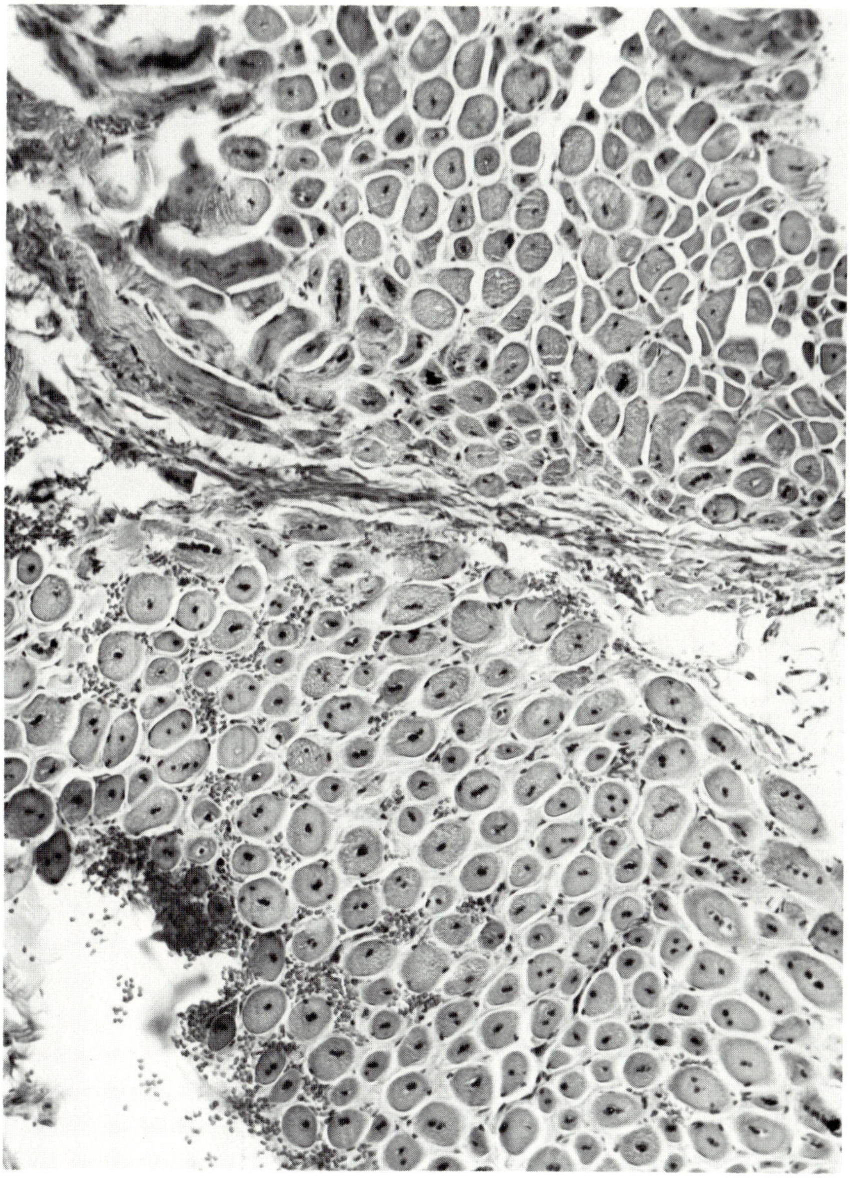

FIG. 6. Centronuclear myopathy. Almost every fiber contains one or more centrally positioned nuclei. Hematoxylin and eosin stain; × 165.

lesions, Sher correctly advised caution in assuming these structures to be myotubes and offered the less specific designation of "centronuclear myopathy," which has been accepted by most subsequent authors. Munsat, Thompson, and Coleman[18] pointed out several aspects by which the muscle fibers in their patient differed from true human myotubes. (1) A clear-cut "hole" could not be demonstrated in most fibers but rather the structural and histochemical derangement could be explained by the presence of central nuclei. (2) The presence of a "halo" in only paraffin-embedded tissue suggested shrinkage artifact. (3) Histochemical differentiation suggested a stage of maturity of at least 20 weeks, which is beyond the myotubular stage. (4) Fetal myoglobin could not be demonstrated. (5) Fiber size was nearly adult. (6) Central nuclei occur in a variety of unrelated disorders (Fig. 7). In addition, it was pointed out that the reported two-fiber population of myotubes with either high or low central oxidative enzyme activity was most likely related to the level of fiber section and whether a nucleus (no activity) or internuclear area (many mitochondria and high activity) was sectioned.

Campbell, Rebeiz, and Walton,[3] and to a lesser extent others, have demonstrated necrotic, destructive changes with electron microscopy which clearly show that a degenerative process is occurring within the central part of the fiber. This is further evidence to suggest that these fibers do not simply represent an arrest of normal development. Reports by Engel, Gold, and Karpati[9] and Bethlem, van Wijngaarden, Meÿer, and Hülsmann[2] have introduced yet another dimension to the concept of centronuclear myopathy. They have reported patients with preferential involvement of type I fibers which, in addition to being centrally nucleated, are abnormally small.

One cannot help but be impressed with the fact that several congenital motor unit syndromes demonstrate a peculiar propensity of the central part of the muscle fiber to behave differently from the perimeter which usually remains normal. This occurs in centronuclear myopathy, central core disease, and has

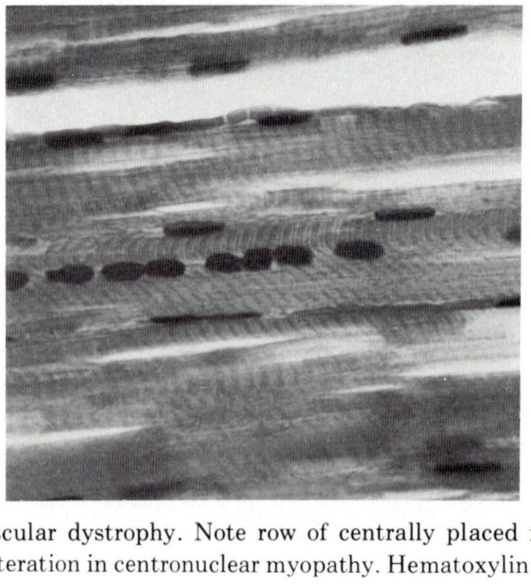

FIG. 7. Myotonic muscular dystrophy. Note row of centrally placed nuclei in center of field, indistinguishable from alteration in centronuclear myopathy. Hematoxylin and eosin stain; × 480.

been reported in a patient with congenital target fiber disease.[25] Of even greater interest is the fact that in all of these conditions, there is evidence to suggest that altered nervous system control is present. The first several cases of centronuclear myopathy all demonstrated either clinical seizures and/or electroencephalogram abnormalites, and in several patients frank electromyogram denervation has been observed.[18] The association of the central core change and target fibers with denervation is fairly clearly established. Arrest at the myotubular stage can be produced by neonatal denervation of guinea pig soleus[9] and myotube-like structures have been observed in patients with perinatal central nervous system damage.[12] Thus, information continues to accumulate suggesting that pathologic changes classically attributed to primary muscle disease may be caused by neural "trophic" disturbances, and that in certain cases, these changes may affect, preferentially, constituents of the central part of the muscle fiber.

## REFERENCES


1. Afifi, A. K., Smith, J. W., and Zellweger, H.: Congenital nonprogressive myopathy. Central core disease and nemaline myopathy in one family. *Neurolgy (Minneap.) 15:* 371–381, 1965.
2. Bethlem, J., van Wijngaarden, G. K., Meijer, A. E. F. H., and Hülsmann, W. C.: Neuromuscular disease with type I fiber atrophy, central nuclei, and myotube-like structures. *Neurology (Minneap.) 19:* 705–710, 1969.
3. Campbell, M. J., Rebeiz, J. J., and Walton, J. N.: Myotubular centronuclear or peri-centronuclear myopathy? *J. Neurol. Sci. 8:* 425–443, 1969.
4. Engel, A. G.: Late-onset rod myopathy (A new syndrome?). Light and electron microscopic observations in two cases. *Mayo Clin. Proc. 41:* 713–741, 1966.
5. Engel, A. G., and Dale, A. J. D.: Autophagic glycogenosis of late onset with mitochondrial abnormalities. Light and electron microscopic observations. *Mayo Clin. Proc. 43:* 233–279, 1968.
6. Engel, A. G., and Gomez, M. R.: Nemaline (Z disk) myopathy. Observations on the origin, structure, and solubility properties of the nemaline structures. *J. Neuropathol. Exp. Neurol. 26:* 601–619, 1967.
7. Engel, W. K.: A critique of congenital myopathies and other disorders. In *Exploratory Concepts in Muscular Dystrophy and Related Disorders*, ed. by Milhorat, A. T., ICS 147, p. 27, Amsterdam, Excerpta Medica, 1967.
8. Engel, W. K.: Selective and nonselective susceptibility of muscle fiber types. *Arch. Neurol. 22:* 97–117, 1970.
9. Engel, W. K., Gold, G. N., and Karpati, G.: Type I fiber hypotrophy and central nuclei. *Arch. Neurol. 18:* 435–444, 1968.
10. Engel, W. K., Wanko, T., and Fenichel, G. M.: Nemaline myopathy. A second case. *Arch. Neurol. 11:* 22–39, 1964.
11. Feigenbaum, J., and Munsat, T. L.: A neuromuscular syndrome with scapuloperoneal distribution. *Bull. Los Angeles Neurol. Soc. 35:* 47–57, 1970.
12. Fenichel, G. M.: Abnormalities of skeletal muscle maturation in brain damaged. children. A histochemical study. *Dev. Med. Child Neurol. 9:* 419–426, 1967.
13. Fulthorpe, J. J., Gardner-Medwin, D., Hudgson, P., and Walton, J. N.: Nemaline myopathy. A histological and ultrastructural study of skeletal muscle from a case presenting with infantile hypotonia. *Neurology (Minneap.) 19:* 735–748, 1969.
14. Gonatas, N. K., Shy, G. M., and Godfrey, E. H.: Nemaline myopathy. The origin of nemaline structures. *New ENgl. J. Med. 274:* 535–539, 1966.
15. Hudgson, P., Gardner-Medwin, D., Fulthorpe, J. J., and Walton, J. N.: Nemaline myopathy. *Nuerology (Minneap.) 17:* 1125–1142, 1967.
16. Huxley, H. E.: Electron microscope studies on the structure of natural and synthetic protein filaments from striated muscle. *J. Mol. Biol. 7:* 281–308, 1963.

17. Kinoshita, M., and Cadman, T. E.: Myotubular myopathy. *Arch. Neurol. 18:*265–271, 1968.
18. Munsat, T. L., Thompson, L. R., and Coleman, R. F.: Centronuclear ("myotubular") myopathy. *Arch. Neurol. 20:* 120–131, 1969.
19. Munsat, T. L., Woods, R., Fowler, W., and Pearson, C. M.: Neurogenic muscular atrophy of infancy with prolonged survival. *Brain 92:* 9–24, 1969.
20. Nienhuis, A. W., Coleman, R. F., Brown, W. J., Munsat, T. L., and Pearson, C. M.: Nemaline myopathy. A histopathologic and histochemical study. *Am. J. Clin. Pathol. 48:* 1–13, 1967.
21. Peterson, D. I., and Munsat, T. L.: The clinical presentation of nemaline myopathy. *Bull. Los Angeles Neurol. Soc. 43:* 39–45, 1969.
22. Price, H. M., Gordon, G. B., Munsat, T. L., and Pearson, C. M.: Myopathy with atypical mitochondria in type I skeletal muscle fibers. A histochemical and ultrastructural study. *J. Neuropathol. Exp. Neurol. 26:* 475–497, 1967.
23. Price, H. M., Gordon, G. B., Pearson, C. M., Munsat, T. L., and Blumberg, J. M.: New evidence for excessive accumulation of Z-band material in nemaline myopathy. *Proc. Natl. Acad. Sci. U.S.A. 54:* 1398–1406, 1965.
24. Pruzanski, W.: Variants of myotonic dystrophy in pre-adolescent life. (The syndrome of myotonic dysembryoplasia.) *Brain 89:* 563–568, 1966.
25. Schotland, D. L.: Congenital myopathy with target fibers. *Trans. Am. Neurol. Assoc. 92:* 107–111, 1967.
26. Shafiq, S. A., Dubowitz, V., Peterson, H. de C., and Milhorat, A. T.: Nemaline myopathy. Report of a fatal case with histochemical and electron microscopic studies. *Brain 90:* 817–828, 1967.
27. Shafiq, S. A., Gorycki, M. A., Asiedu, S. A., and Milhorat, A. T.: Tenotomy. Effect on the fine structure of the soleus of the rat. *Arch. Neurol. 20:*625–633, 1969.
28. Sher, J. H., Rimalovski, A. B., Athanassiades, T. J., and Aronson, S. M.: Familial centronuclear myopathy. A clinical and pathological study. *Neurology (Minneap.) 17:* 727–742, 1967.
29. Shy, G. M., Engel, W. K., Somers, J. E., and Wanko, T.: Nemaline myopathy. A new congenital myopathy. *Brain 86:* 793–821, 1963.
30. Shy, G. M., Gonatas, N. K., and Perez, M.: Two childhood myopathies with abnormal mitochondria. I. Megaconial myopathy. II. Pleoconial myopathy. *Brain 89:* 133–158, 1966.
31. Shy, G. M., and Magee, K. R.: A new congenital non-progressive myopathy. *Brain 79:* 610–621, 1956.
32. Spiro, A. J., Shy, G. M., and Gonatas, N. K.: Myotubular myopathy. *Arch. Neurol. 14:* 1–14, 1966.
33. van Wijingaarden, G. K., Fleury, P., Bethlem, J., and Meijer, A. E. F. H.: Familial "myotubular" myopathy. *Neurology (minneap.) 19:* 901–908, 1969.

Chapter 22

# Duchenne Muscular Dystrophy: A Histologically Based Ischemia Hypothesis and Comparison with Experimental Ischemia Myopathy

W. KING ENGEL

With the Technical Assistance of GUY G. CUNNINGHAM, DONALD W. BISHOP, and THELMA R. FLETCHER

A hypothesis contrary to the usually accepted notion of the cause of Duchenne pseudohypertrophic muscular dystrophy is proposed. It is based on evidence obtained from histochemical and electron microscopic studies of muscle biopsies from patients with typical Duchenne dystrophy and carriers of the disease who appear normal clinically. Since the hypothesis states that the muscle damage is caused by abnormalities of small blood vessels within the muscles, the muscle damage is considered to be secondary rather than due to a primary defect of muscle fiber metabolism as is generally thought.

## DUCHENNE DYSTROPHY: ISCHEMIA HYPOTHESIS

1. Small arterial vessel ischemia causes necrosis of muscle fibers in groups (or scattered).
2. Collateral blood supply allows regeneration of such fibers in groups (or scattered).
3. Repeated ischemic insults cause repeated necroses and regenerations.
4. Lack of adequate collateral blood supply eventually causes prominent endomysial connective tissue proliferation (scarring) during healing phases.
5. Blood vessels in some skeletal muscles (*e.g.*, proximal muscles) are more susceptible to the abnormality than in others.
6. Heart muscle blood vessels may be involved, and possibly some cerebral vessels in occasional cases.

The histologic evidence favoring small vessel ischemia in the patients is a grouped pattern of the degenerating (Figs. 1 to 4, 7, and 8) and regenerating (Figs. 11 to 13, 15, 17, 18, 20 to 22, and 52) muscle fibers as the earliest morphologic sign of disease[7] and a disproportionately large increase of connective tissue (scar tissue) between muscle fibers after the earliest stage of disease (Fig. 26). As support for the hypothesis, muscle lesions identical both to the

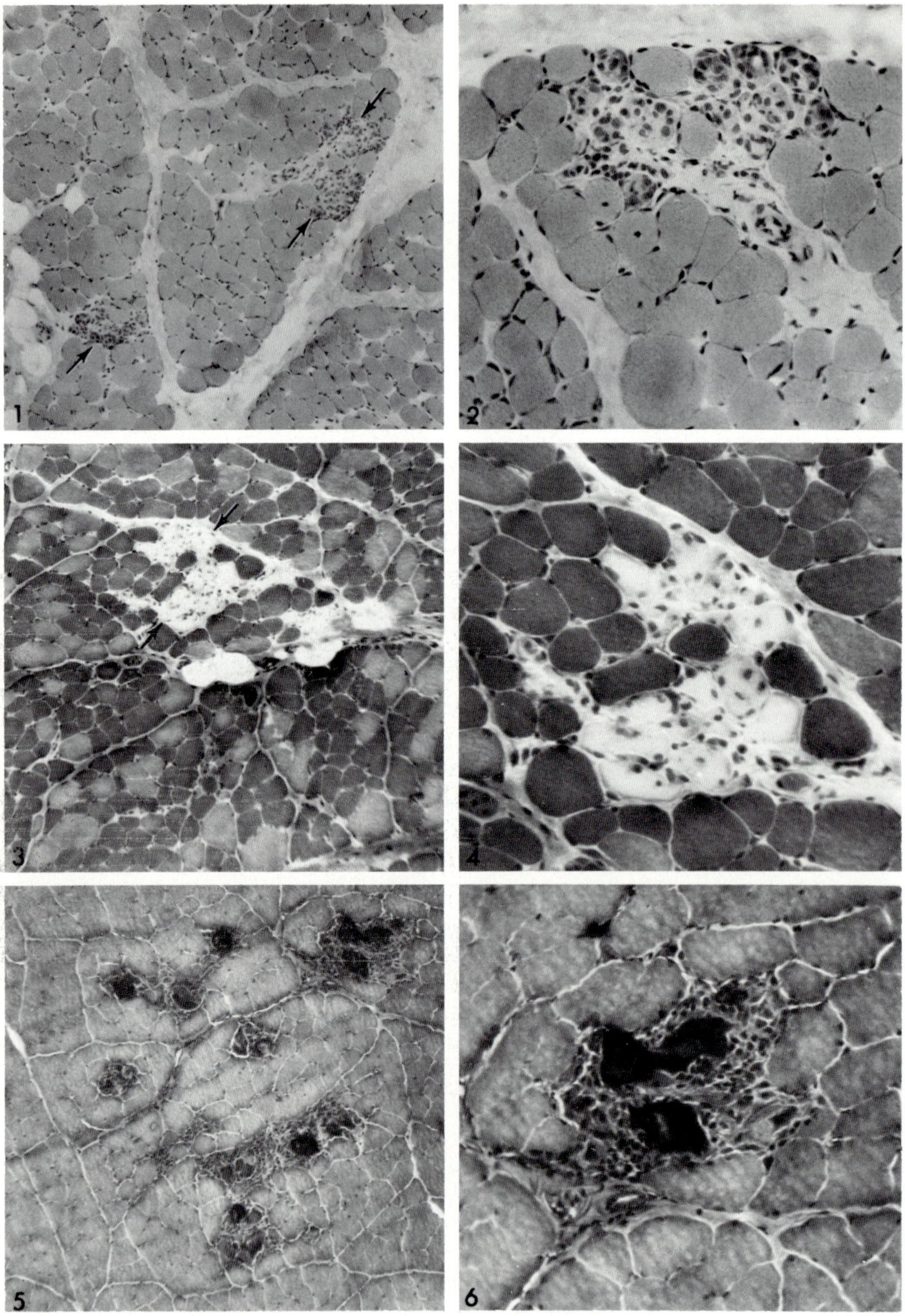

All figures are of fresh frozen sections unless noted otherwise.

FIGS. 1 to 4. Duchenne dystrophy. Grouping of muscle fibers (light) undergoing acute necrosis in a field of essentially normal fibers (*cf.*, Figs. 5, 6). 1, 2—hematoxylin-eosin; 3, 4—modified trichrome. 2, 4—higher powers of 1, 3. 1, 3—× 68; 2, 4—× 173.

FIGS. 5 and 6. Embolized rabbit. Grouping of muscle fibers (dark) undergoing acute necrosis and early regeneration in a field of essentially normal fibers. The dark necrotic fibers probably represent an earlier stage of degeneration than the light necrotic fibers of Figs. 3, 4. 6—higher power of 5. Modified trichrome. 5—× 68; 6—× 173.

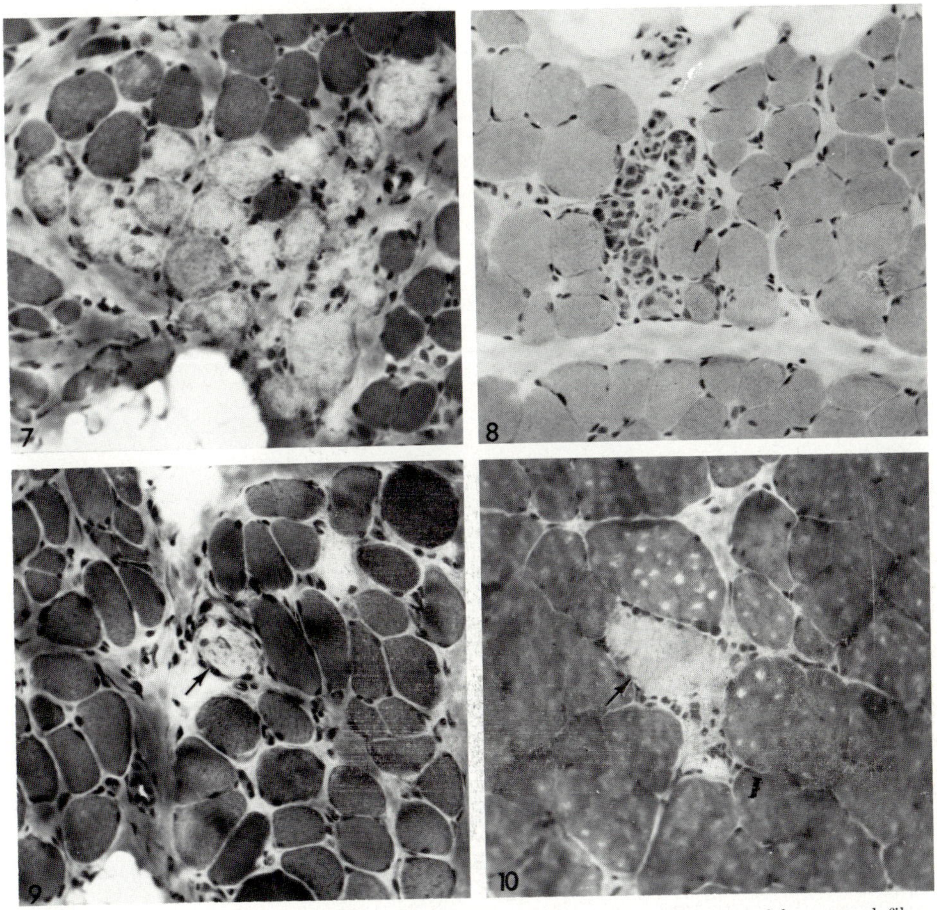


FIGS. 7 and 8. Duchenne dystrophy. Grouping of necrotic fibers surrounded by normal fibers. Fig. 8, a higher power of 1, shows signs of early regeneration in the necrotic fibers. 7—modified trichrome, × 173; 8—hematoxylin-eosin, × 173.

FIGS. 9 and 10. In the field, a single necrotic fiber (white) being phagocytosed, surrounded by intact fibers. 9—Duchenne dystrophy, × 173; 10—embolized rabbit. Modified trichrome, × 173.

early grouped degenerating fibers (Figs. 5 and 6) and grouped regenerating fibers (Figs. 14, 16, and 19) and to the mid stage of disproportionate endomysial fibrosis (Fig. 27) were produced by Hathaway, Engel, and Zellweger.[13] These changes resulted respectively early after one and later after repeated occlusions of muscle arterial vessels by injecting microscopic beads of dextran (Sephadex G-25 fine) 20 to 80 $\mu$ in diameter (Figs. 30 and 31) into rabbit femoral arteries.[13] Intermediate "healing" stages were also comparable (Figs. 23 to 25, 28, and 29). Occasionally in Duchenne patients, Duchenne carriers (see below), and the embolized rabbits a single fiber in the field is necrotic (Figs. 9, 10, and 64 to 66) and/or regenerating (Figs. 53 to 63, and 67 to 69), the latter especially being typical of the carriers (Figs. 53 to 58, 60 to 63, and 67 to 69). (Care was taken that none of the muscle from patients or animals had been needled prior to histologic or histochemical examination, because needling itself is known to produce focal myopathy.[6]

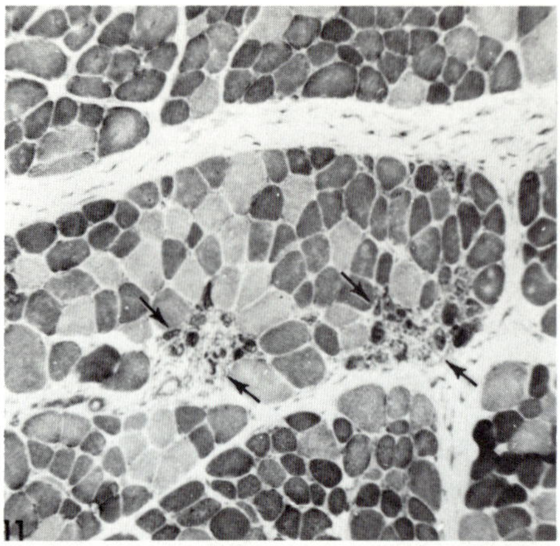 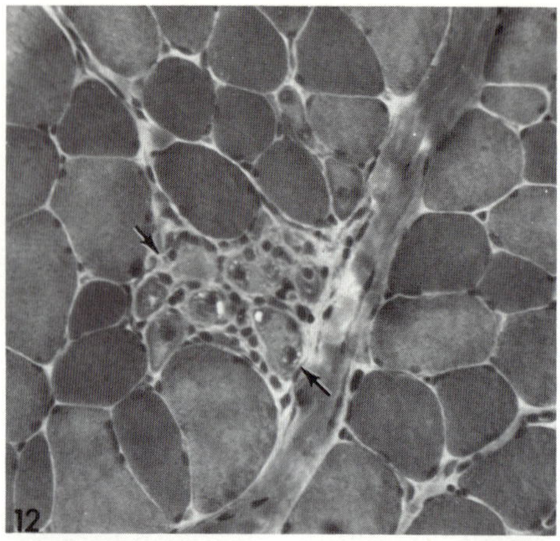

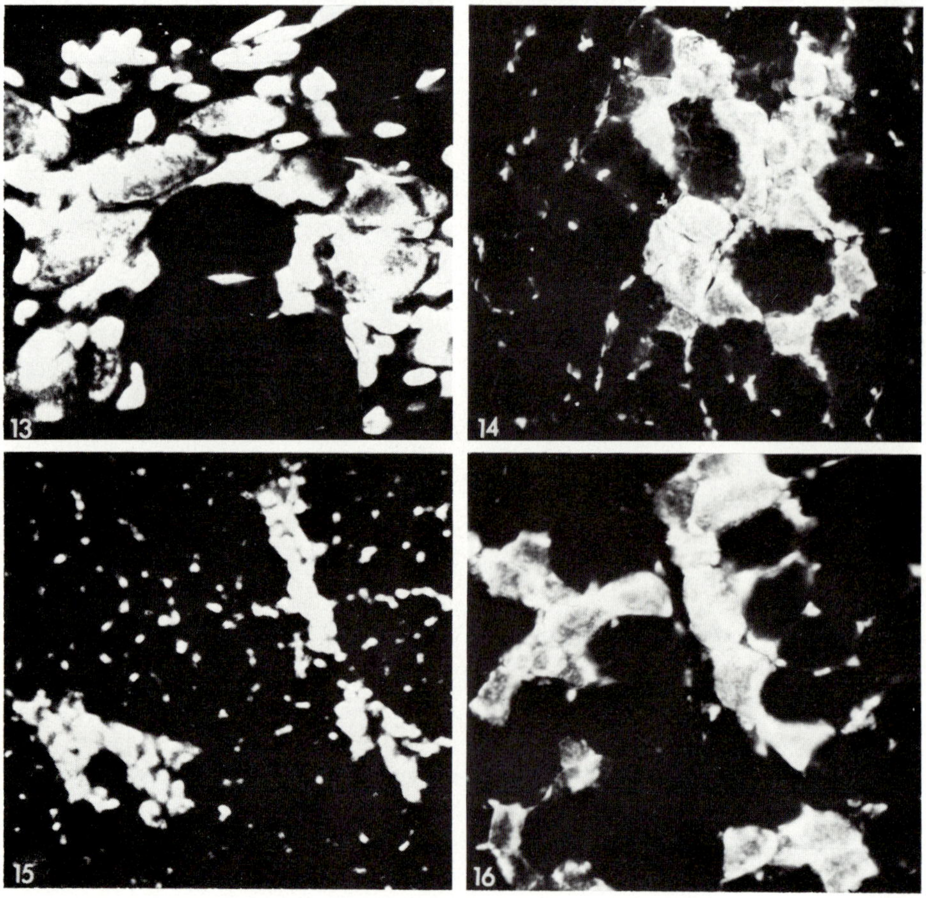

FIG. 11. Duchenne dystrophy. Two groups of regenerating muscle fibers in a field of intact fibers. Reduced diphosphonucleotide-tetrazolium reductase (DPNH-TR), × 68.

FIG. 12. Duchenne dystrophy. Group of "regen-degen" fibers in a field of intact fibers. Modified trichrome, × 173.

FIGS. 13 to 16. Grouping of regenerating muscle fibers (white patches, representing orange fluorescence of cytoplasmic RNA) in a field of normal fibers (black). Small white dots are nuclei (green fluorescence of DNA). Acridine orange stain. 13 (× about 432), 15 (× about 114)—Duchenne dystrophy. 14, 16—embolized rabbit, × 173.

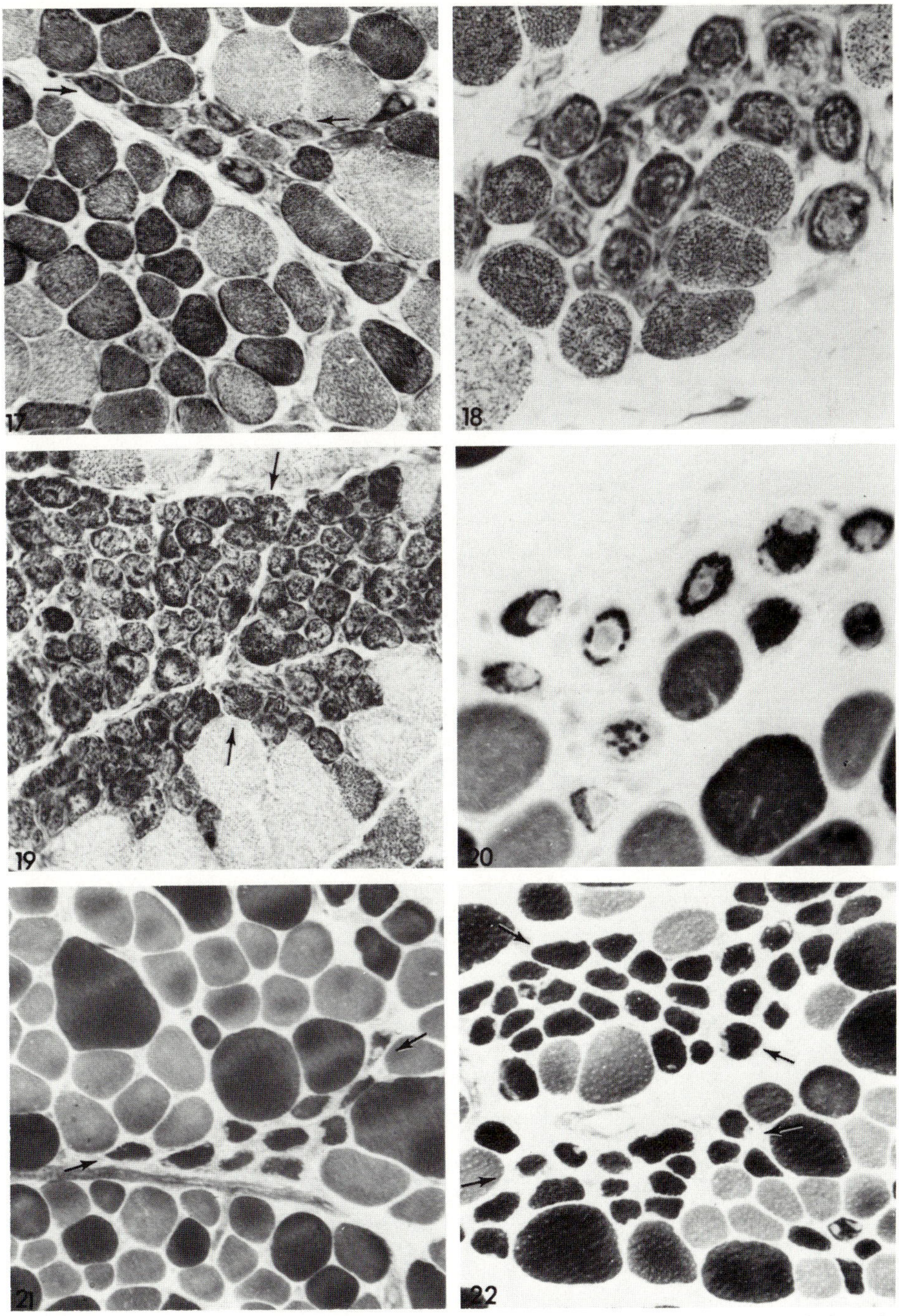


FIGS. 17 to 22. Groups of small fibers, presumably regenerating. They all appear as dark as the normally dark fibers with both the myofibrillar ATPase and DPNH-TR reactions. (Even later these fibers are not as easily separable into histochemical types as in normals.) 17, 18, 19—DPNH-TR. 20, 21, 22—myofibrillar ATPase (pH 9.4). 17, 18, 20, 21, 22—Duchenne dystrophy. 19—embolized rabbit. Some small fibers have central nuclei (unstained) or central collections of oxidative enzyme material (dark with DPNH-TR). Such fibers and slightly more mature ones (Fig. 18) have coarsening and irregularity of the intermyofibrillar network, *i.e.*, "granular fibers." All of these resemble regenerating fibers produced experimentally.[15]

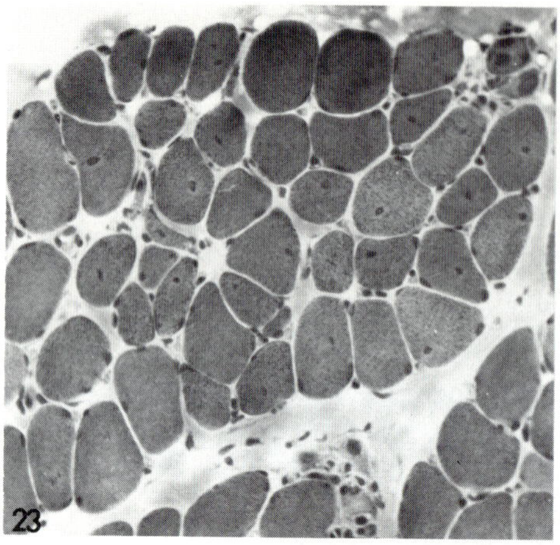

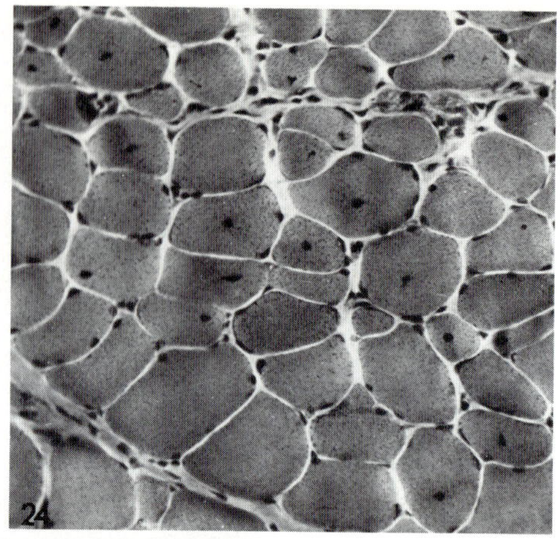

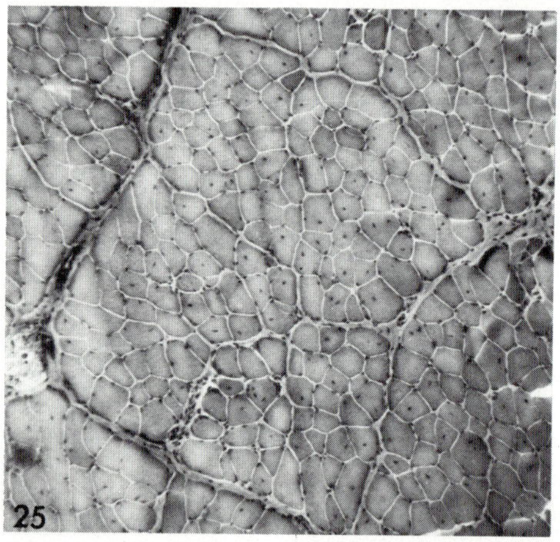

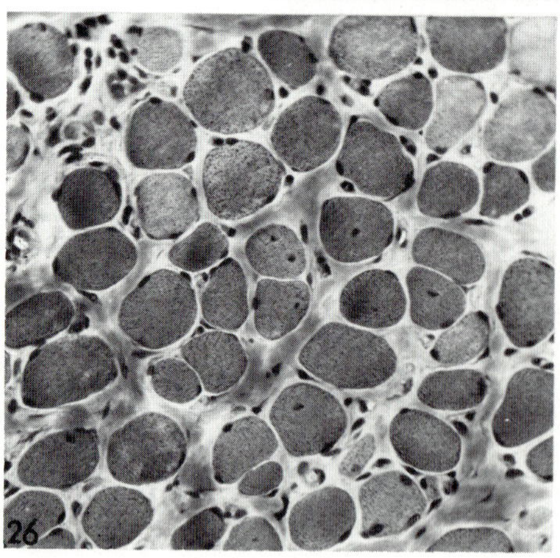

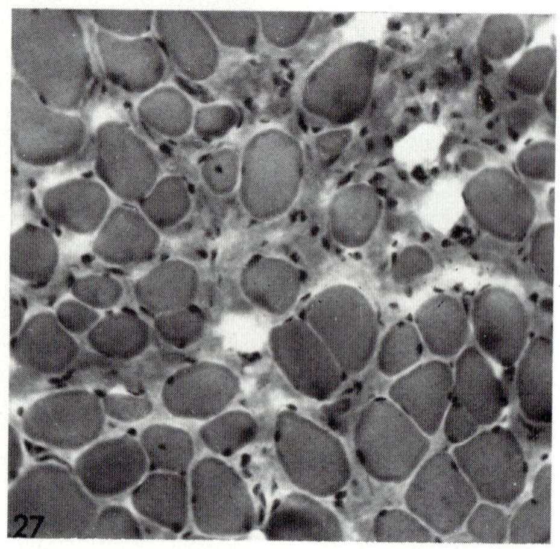

FIG. 23 and 24. Intact muscle fibers, some with single central nuclei, and no or minimal increase of endomysial connective tissue. 23—Duchenne dystrophy. 24—rabbit muscle during recovery after embolization. Modified trichrome, × 173.

FIG. 25. All fibers in the field are intact and most contain single central nuclei. Rabbit muscle during recovery after embolization. Modified trichrome, × 68.

FIGS. 26 and 27. All fibers in the field are intact. Some are abnormally small or large. Some contain internal nuclei. Prominent increase of endomysial connective tissue. 26—Duchenne dystrophy. 27—rabbit muscle 3 months after three monthly embolizations. Modified trichrome, × 173.

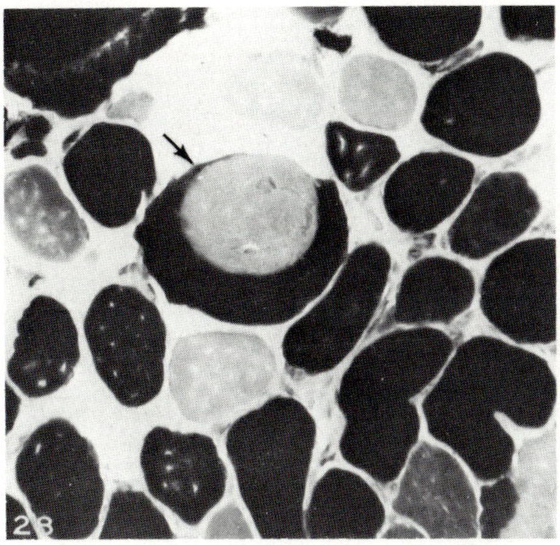

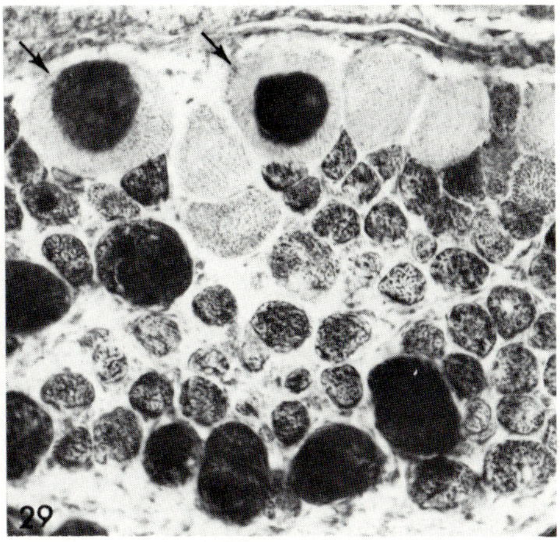


FIGS. 28 and 29. Snake-coil fibers; central and peripheral portions stain differently. 28—Duchenne dystrophy, EDTA low-pH activated ATPase, × 273. 29—embolized rabbit muscle, DPNH-TR, × 173.

On the basis of histologic evidence, I believe the best explanation for the pathogenesis of Duchenne muscular dystrophy is the hypothesis of ischemia at the level of small vessels of the arterial tree.[8, 13] While there were logical reasons earlier for proposing, *a priori*, that the disease might not be a primary disorder of the skeletal muscle fiber,[4, 5] a distinct histologically based hypothesis now allows logical positivity to replace negativity. If the hypothesis is correct, the postulated abnormality of muscle blood vessels could be functional alone or accompanied by structural change. A morphologic lesion, if present, could in turn be based on a pharmacologic-biochemical abnormality. Thus, morphologic lesions of blood vessels are not required to support the hypothesis, but the vascular findings observed in biopsies of Duchenne dystrophy patients *may* be related to the pathogenesis. Further study is required to determine what relationship that might be.

This hypothesis based on skeletal muscle histology has been formulated independent of other evidence. Hence, confirmation of suggestions that effective muscle blood flow through nutritional capillaries is decreased in Duchenne dystrophy and Duchenne carriers as compared to neurogenic atrophy[2, 3, 17] and that the heart muscle pathology resembles ischemic disease[14, 16] could be used as supportive evidence, but lack of such confirmation would not detract.

The observed arterial vascular findings in Duchenne dystrophy are of five types. All can be seen in early and mid stages of disease.

*Inflammatory cell reaction.* Perivascular inflammatory cell reaction, which also may be within the wall, is sometimes seen in Duchenne dystrophy (Figs. 32 to 34) and in carriers (Figs. 69 and 70). This occurs rarely around an arterial vessel and more commonly around a venous one, especially near a muscle fiber undergoing necrosis or regeneration, or both (Fig. 69). Increased cellularity is also seen in the embolized rabbit vessels (Figs. 30, 31 and 35). Muscle fibers surrounding a rabbit vessel containing an embolus are usually normal (Figs. 30 and 31), probably because the muscle fibers damaged by the ischemia are

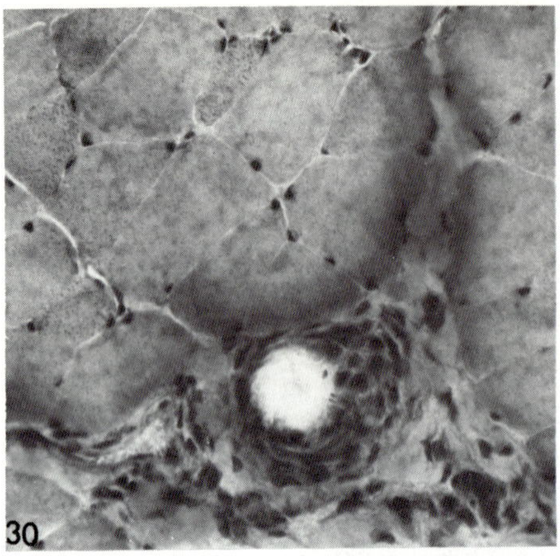

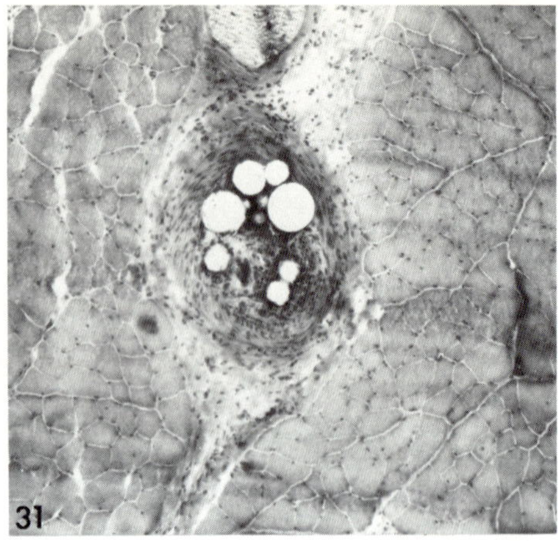

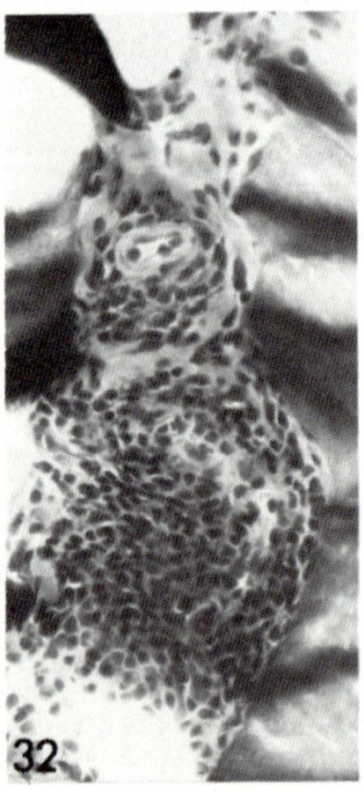

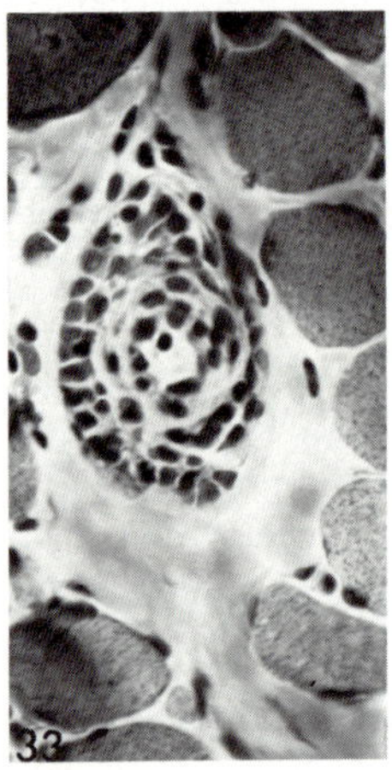

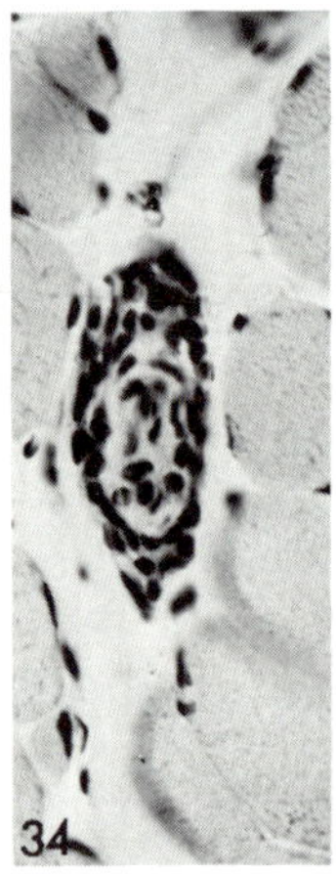

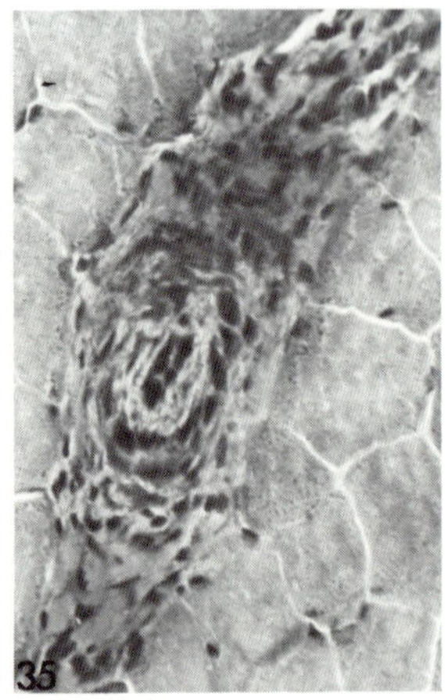

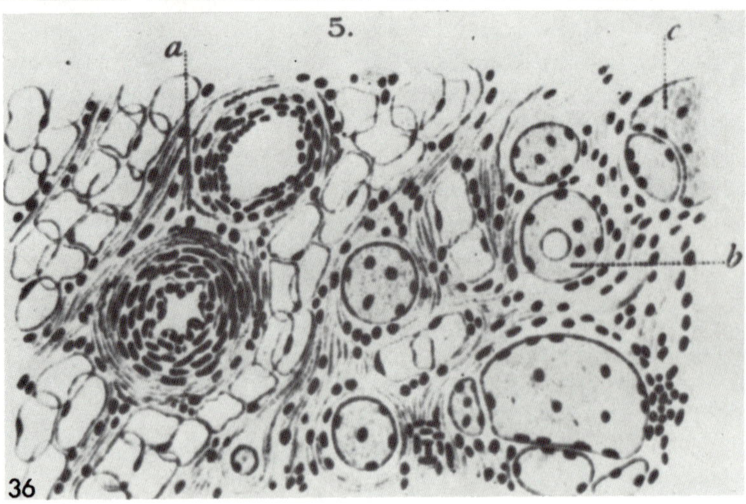


FIGS. 30 and 31. Embolized rabbit muscle. 30—one embolus in an arteriole. 31—several emboli in a small artery. Slight cellular reaction in the vessel walls. Adjacent muscle fibers are not affected, the damage from these occlusions presumably occurring farther distally. Modified trichrome. 30—× 273; 31—× 68.

FIGS. 32 to 34. Duchenne dystrophy. Perivascular mononuclear cell inflammatory reaction. 32, 33—modified trichrome; 34—hematoxylin-eosin. 32—× 173; 33, 34—× 273.

FIG. 35. Cellular reaction around arteriole of embolized rabbit. Modified trichrome, × 273.

FIG. 36. Wall of blood vessel showing cellular reaction and thickening, as described by Erb in 1891 in Duchenne dystrophy.[12]

located more distally. Increased cellularity and thickening of muscle blood vessels in Duchenne dystrophy, both early childhood and slower juvenile forms, has been described in 1891 by Erb[12] (Fig. 36).

*Thickened arterioles.* These are occasionally seen in both Duchenne dystrophy and the embolized rabbits (Figs. 37 to 40).

*Endothelial proliferation.* The observation by light microscopy and electron microscopy of numerous endothelial cells filling the original small artery lumen, with one or two slitlike channels remaining amongst the proliferated cells, has been interpreted as endothelial proliferation (Figs. 41 and 42). There is no inflammatory cell reaction.

*Possible occlusion by a plug.* The lumen of a small artery or arteriole is sometimes filled with cellular and granular material as seen in fresh-frozen sections stained with a modified trichrome reaction[10] (Figs. 43 to 45). It is also seen in the embolized rabbit (Fig. 46). Somewhat similar changes, with more suggestion of organization, are seen in paraffin sections of Duchenne dystrophy (Figs. 47 and 48). There is no inflammatory cell reaction. Although this appearance possibly indicates occlusion by a plug, further study is required to determine whether the finding represents an *in vivo* change. It has not yet been studied by electron microscopy. If the finding is an *in vivo* phenomenon, it must be determined whether or not it is a variation of endothelial proliferation.

*Obturation by organized cells.* This change in small and medium sized arteries has been seen by light microscopy in about 20% of Duchenne dystrophy biopsies. It consists of complete or partial filling of the lumen inside the normally placed elastica by a mass of cells resembling endothelial cells or smooth muscle (Figs. 49 to 51). (By electron microscopy, endothelial cells of human intramuscular arterioles and small arteries are virtually identical to smooth muscle cells of the media.[8a]) Sometimes the intralumenal mass is circumscribed by a layer of elastica and in other examples this layer is not seen. In some, a single or double layer of flat cells lies immediately central to the original elastica. Whether such cells are endothelial is not certain. The intralumenal mass may or may not appear to be contiguous in places with the normal wall. Inflammatory cell reaction does not occur. This finding has not been studied by electron microscopy. The interpretation of it is not certain. Whether some or all such changes represent (1) elastica duplication and endothelial proliferation, similar to "Endothelial Proliferation" above (pathologic change); (2) tucked-in end of a cut artery (artifact); (3) longitudinal slippage, *i.e.*, intussusception, of the inner portion of a cut artery (artifact); (4) bulge in the wall (artifact); or (5) an arterial bifurcation (normal variant), is being analyzed. Even if artifact, preliminary survey suggests it is more common in Duchenne dystrophy muscle biopsies than in other neuromuscular diseases.

Further study is required to determine more precisely the frequency of all of these arterial changes in Duchenne dystrophy and in comparison to other types of myopathies and neurogenic atrophies, except the collagen-vascular diseases. None is unique to Duchenne dystrophy, but it is our initial impression that they are more frequently found there. What relationship each may have to the pathogenesis of Duchene dystrophy remains to be determined, as well as

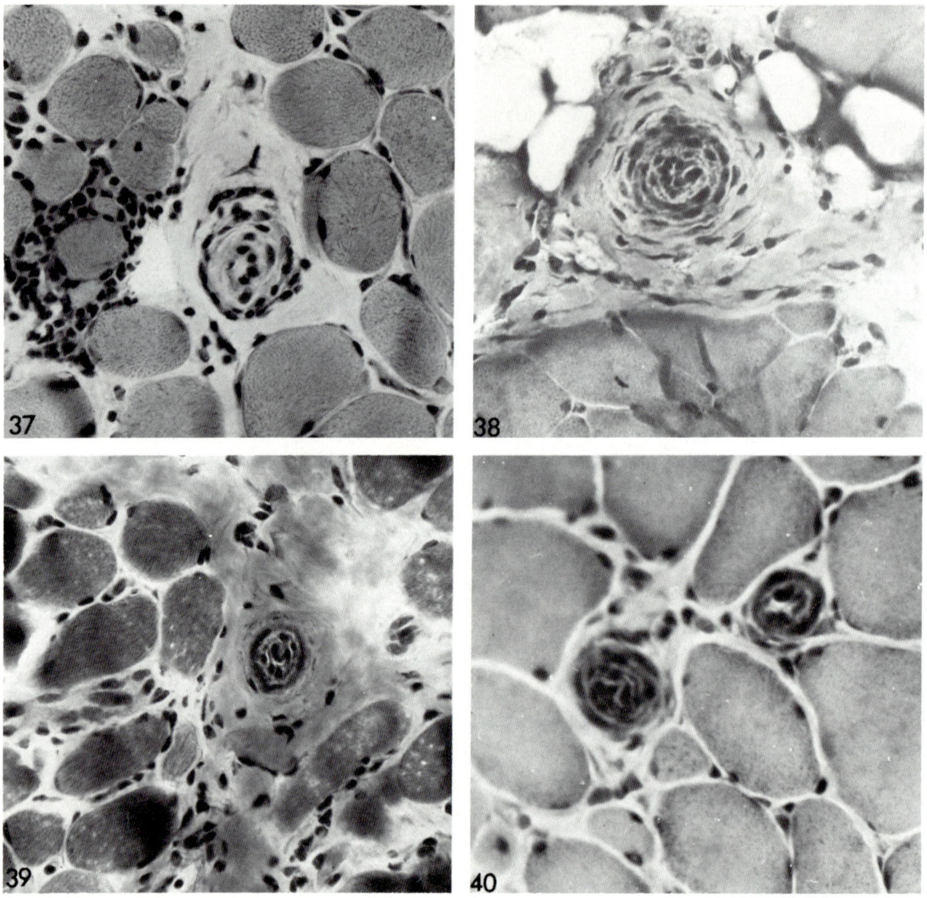

FIGS. 37 to 40. Thickened arterioles. 37, 39—Duchenne dystrophy; 38, 40—embolized rabbit. Modified trichrome. 37, 38, 39—× 273; 40—× about 364.

whether they precede or follow muscle fiber damage.

Quantitatively similar lesions of muscle fibers occur to a very minor degree in clinically normal carriers of Duchenne dystrophy (Figs. 53 to 55, 57 to 63, 65 and 67 to 70); these lesions, too, were reproduced in minimally abnormal embolized rabbit muscle (Figs. 59 and 66). In the carriers the lesions include single fibers undergoing phagocytosis (Fig. 65) and scattered fibers, one to three per group, showing less severe changes (Figs. 53 to 55, 57, 67, and 68). These latter fibers seem to have features of both degeneration and regeneration, as if they still have residual degeneration while beginning to show regenerative changes, namely, large nuclei (often internal) and nucleoli, coarsened intermyofibrillar network with the DPNH-tetrazolium reductase reaction, low over-all phosphorylase activity, high DPNH-tetrazolium reductase and myofibrillar ATPase (pH 9.4) activities, and cytoplasmic basophilia. To emphasize both aspects of change, these fibers are designated as "regen-degen fibers" or "degen-regen fibers," naming first the process which seems more prominent. These fibers are seen in

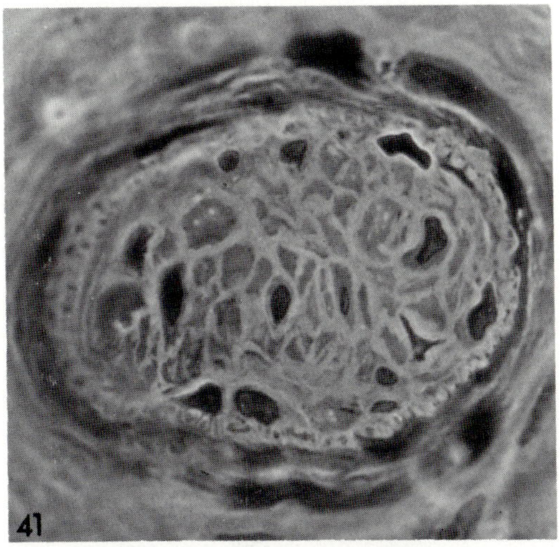

FIGS. 41 and 42. Duchenne dystrophy. Occlusion of small artery by proliferated endothelial cells. 41—plastic-embedded section, toluidine blue, phase contrast, × 100. 42—electron micrograph of same vessel shows two slitlike channels (*arrows*) are all that remain of the lumen, × 2278.

nearly all active myopathies and occasionally in neurogenic atrophy, but in Duchenne carriers a few such fibers are the most common and typical change (Figs. 53 to 55, 57, 58, 60 to 63, 67, and 69). In minimally affected Duchenne dystrophy muscle they have a similar appearance (Fig. 56). One way of highlighting some, but not all, of these regen-degen fibers is with their positive staining by the alkaline phosphatase reaction,[9] normal fibers being unstained, as are ones undergoing simple denervation atrophy (Figs. 52 to 55). Duchenne

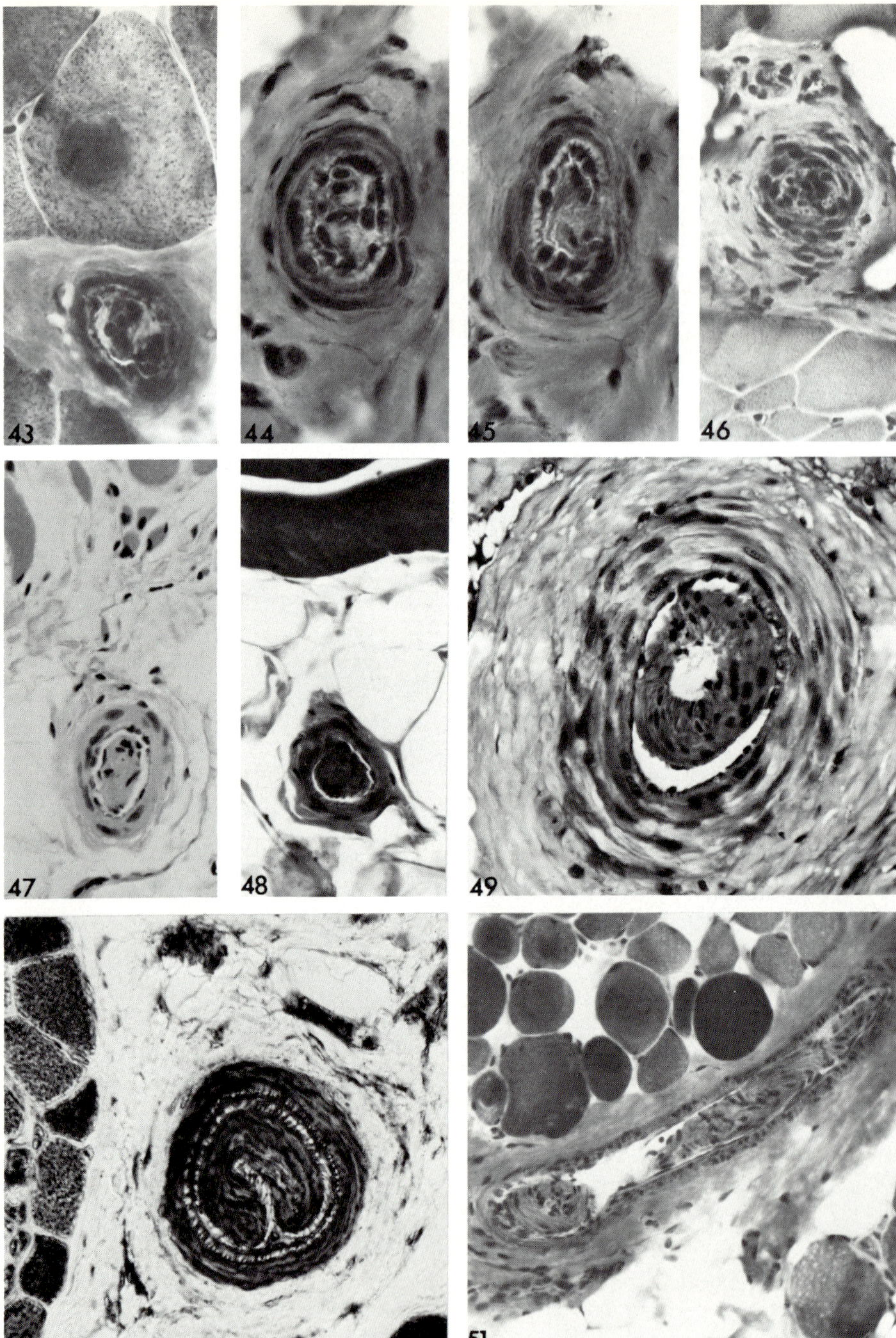


FIGS. 43 to 45. Duchenne dystrophy. Lumens of small arteries are filled with cellular and granular material. 43 also has early snake coil in muscle fiber (*cf.*, Figs. 28, 29). Modified trichrome, × 437.

FIG. 46. Embolized rabbit, lumen of small artery is occupied by cellular and granular material. Modified trichrome, × 273.

FIGS. 47 and 48. Duchenne dystrophy. Occupying the lumen of small arteries is more organized cellular and amorphorus material. 47—hematoxylin-eosin, × 273. 48—Gomori trichrome, × 273. Paraffin sections.

FIGS. 49 to 51. Duchenne dystrophy. Lumen of the artery is filled with well organized cells resembling endothelial or smooth muscle cells. The possibility that this finding is an artifact requires further study. 49—Gomori trichrome of paraffin section, × 273. 50—DPNH-TR, polarized light to emphasize anisotropic (white) elastica in two layers, × 273, 51—longitudinal section, modified trichrome, × 173.

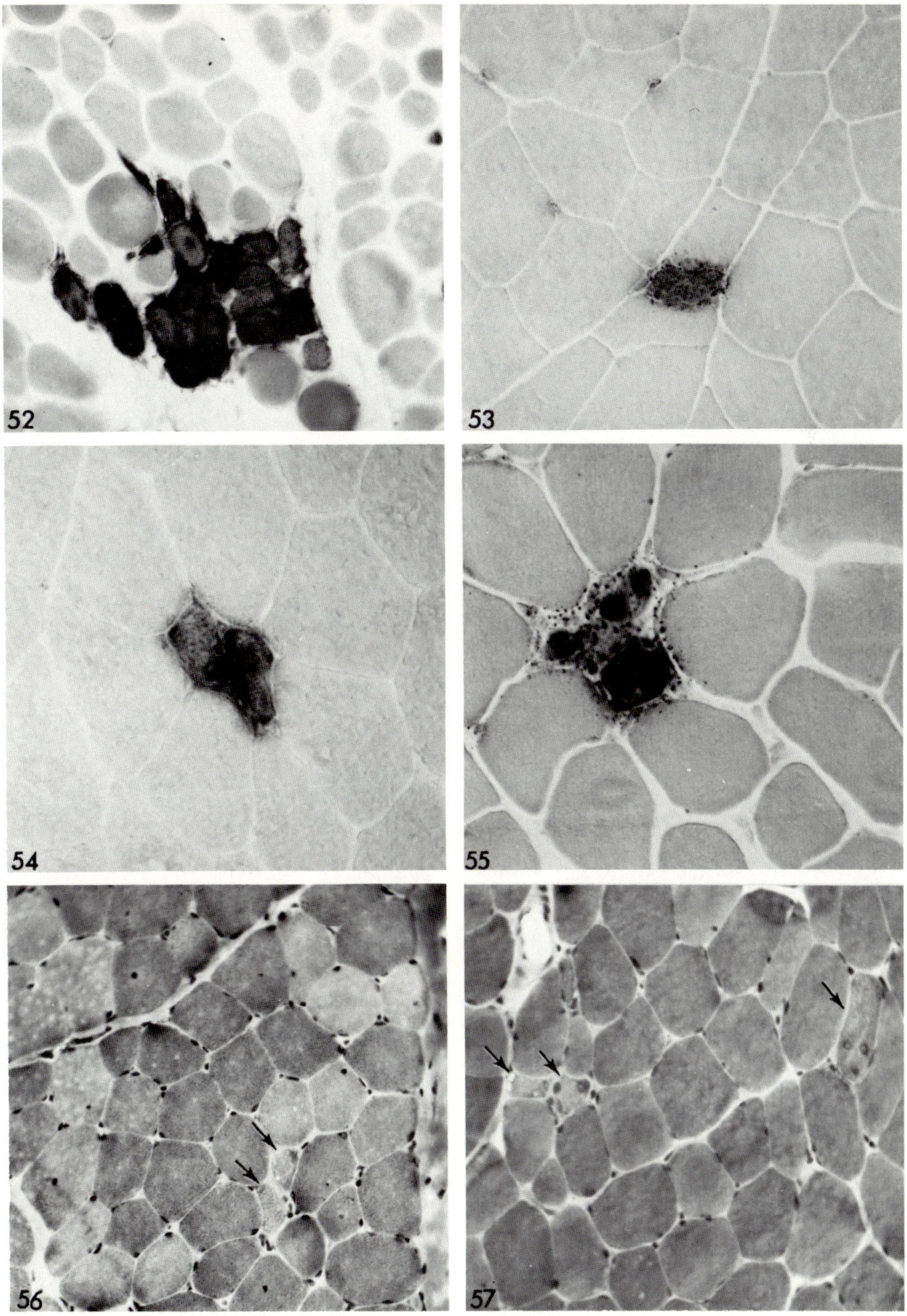


FIGS. 52 to 55. Alkaline phosphatase (AP)-positive fibers are abnormal and probably represent a certain phase of regeneration (possibly with simultaneous degenerative aspects), *i.e.*, regen-degen fibers.[9] The reaction is an excellent way to identify scattered abnormal fibers. In Duchenne dystrophy patients AP-positive fibers are abundant, in groups (Fig. 52) or scattered, while in Duchenne carriers they are rare and single (Fig. 53) or in two's or three's (Fig. 55). 52—× 173; 53, 54, 55—× 273.

FIGS. 56 and 57. In each field, paired abnormal regen-degen fibers. An additional single regen-degen fiber in Fig. 57 (*arrow*). 56—minimally affected Duchenne dystrophy patient, × 173. 57—Duchenne carrier, × 300. Modified trichrome.

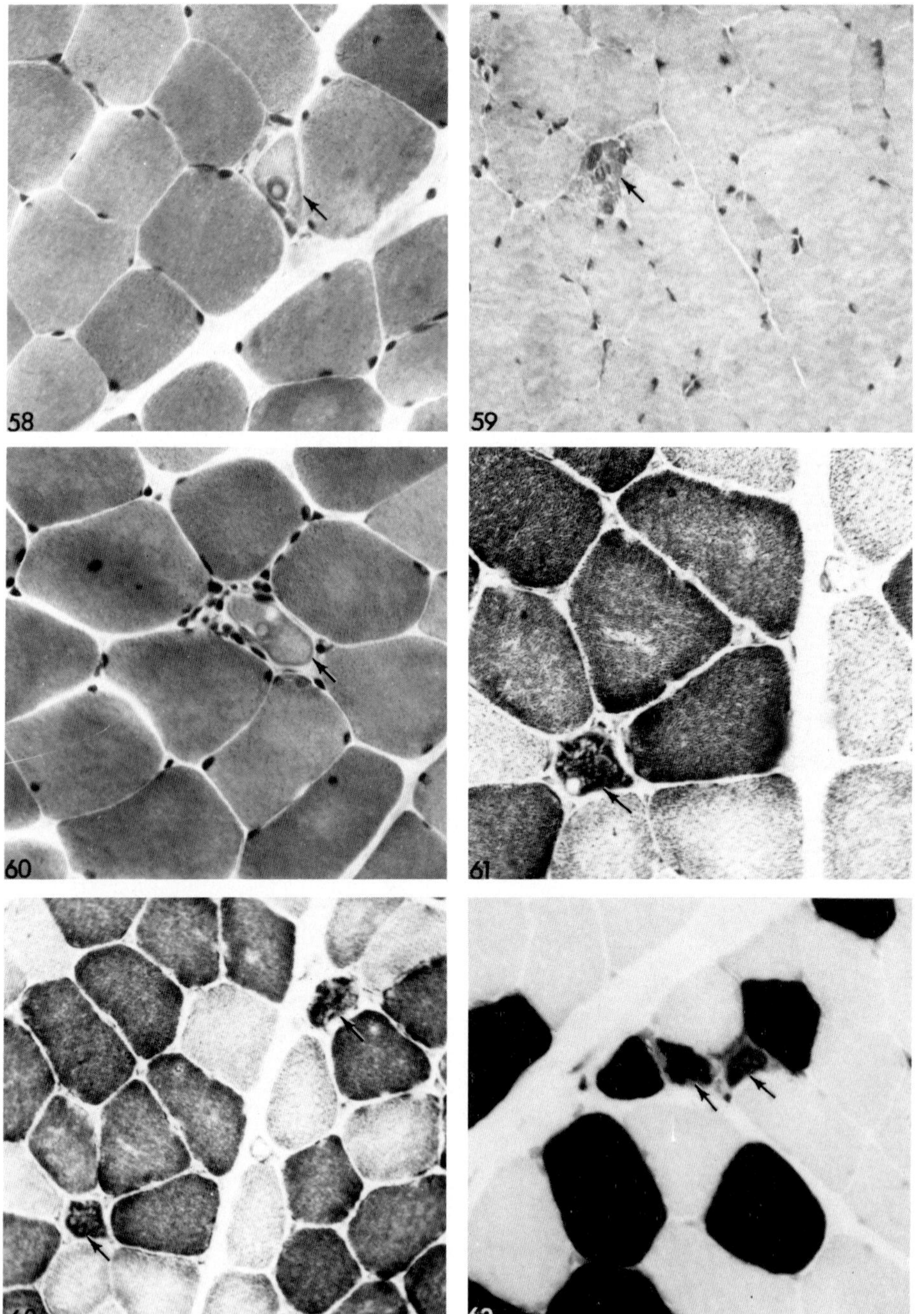

FIGS. 58 to 60. Single abnormal regen-degen fiber in each field, modified trichrome. 58, 60—Duchenne carrier, × 273, 59—embolized rabbit, × 273. 60 shows very early mononuclear cell reaction around the abnormal fiber.

FIGS. 61 and 62. Duchenne carrier. Isolated small abnormal regen-degen fibers, with coarsening and irregularity of the intermyofibrillar network. 61 is detailed view of one fiber in 62. (Compare with Figs. 53 to 58, 60, 63.) DPNH-TR. 61—× 273; 62—× 173.

FIG. 63. A pair of regen-degen fibers in Duchenne carrier. With the EDTA low-pH activated ATPase they are often of intermediate to dark color even though they are also dark with the regular myofibrillar ATPase. × 273.

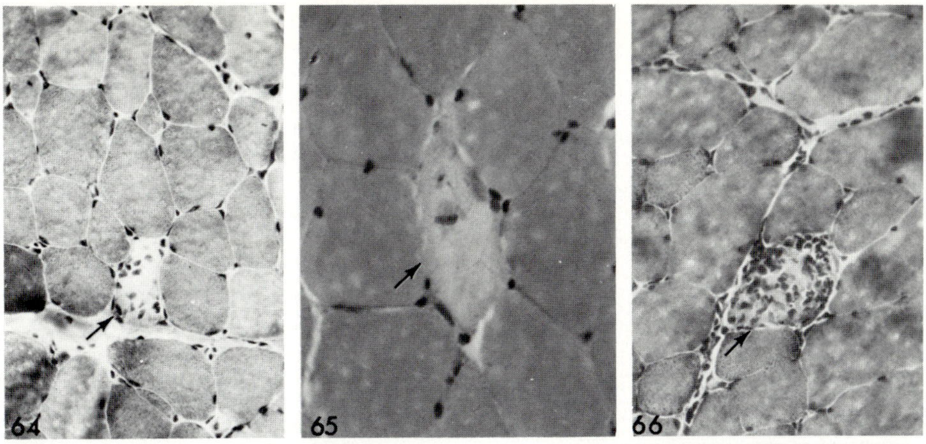

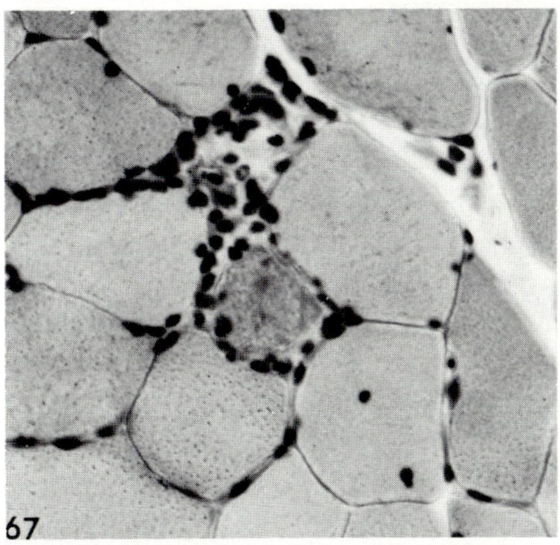

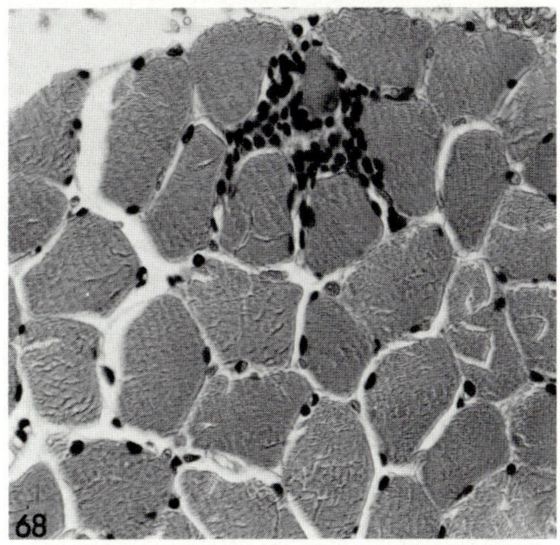

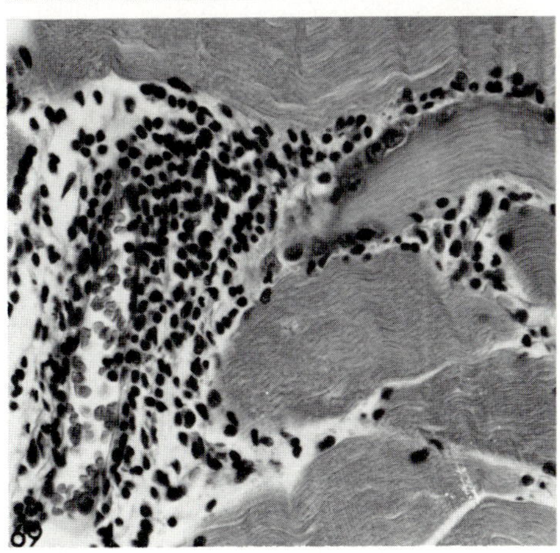

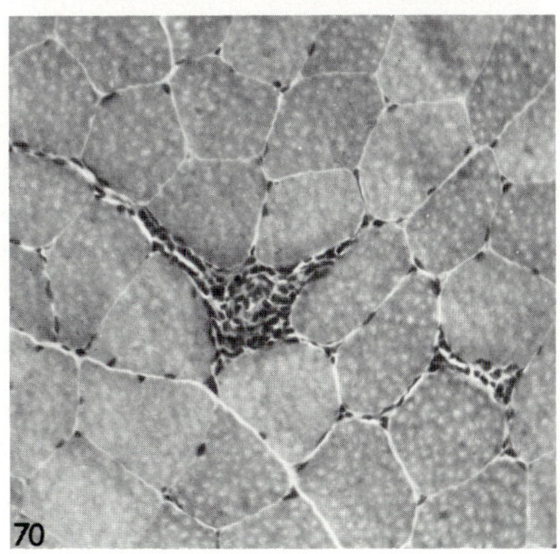

FIGS. 64 to 66. Single fiber in the field undergoing necrosis and phagocytosis, all others appearing intact. Modified trichrome. 64—Duchenne dystrophy, × 173. 65—Duchenne carrier, × 273. 66—embolized rabbit, × 173.

FIG. 67. Duchenne carrier. Two abnormal degen-regen fibers, one above surrounded by mononuclear inflammatory cells and perhaps showing early phagocytosis. Hematoxylin-eosin, × 437.

FIG. 68. Duchenne carrier. Regen-degen fiber with prominent mononuclear inflammatory cell reaction around it. Hematoxylin-eosin, paraffin section, × 273.

FIG. 69. Duchenne carrier. Loose perivascular mononuclear cellular reaction adjacent to an abnormal regen-degen fiber. Hematoxylin-eosin, paraffin section, × 273.

FIG. 70. Duchenne carrier. Compact perivascular mononuclear cellular reaction. Modified trichrome, × 173.

carriers infrequently showed a perivascular cellular reaction (Figs. 69 and 70).

If a vascular abnormality is an essential step in the pathogenesis of Duchene dystrophy, many types are theoretically possible. They can be categorized as (1) functional alone, (2) structural alone, and (3) both (either causing the other). Since the disease is inherited as an X-linked recessive trait, the genetically determined abnormality could be in cells of the blood vessels or in cells influencing blood vessels by neural or other pharmacologic-biochemical mechanisms. And two or more abnormal factors may be interacting, as in some "pluricausal experimental myopathies"[18] involving vascular mechanisms.

In the rabbit muscle 3 months after three embolizations there were both (1) focal areas of increased endomysial connective tissue between large and small muscle fibers of essentially normal structure except for internal nuclei and (2) normal areas (Fig. 71). In the normal areas (Fig. 72), capillaries and arterioles were darkly stained with the alkaline phosphatase reaction[9] as in normal rabbit muscle, but no staining was seen in the abnormal areas even though they contained essentially intact muscle fibers[13] (Fig. 73). The loss of staining from capillaries and arterioles is interpreted as a secondary effect distal to blockage of arterioles or small or large arteries. Thus the loss of alkaline phosphatase staining from capillaries and arterioles as a very early sign in chicken muscular dystrophy[1] also could be secondary to lesions more proximally in the arterial tree. A parallel cannot be drawn in human Duchenne dystrophy because human capillaries are not stained by the alkaline phosphatase reaction (even after 14 hours of incubation compared to the routine incubation of 1 hour) and the arterioles normally are widely and irregularly dispersed in skeletal muscle. No abnormality of arteriolar distribution or staining was apparent with the alkaline phosphatase method in Duchenne dystrophy biopsies.

The finding of many otherwise normal fibers with single central nuclei on a given cross-section in the rabbits some weeks after recovery from one episode of embolization (Figs. 24 and 25) resembles the pathologic change which has been referred to as "centronuclear myopathy." Such an appearance in the healing (? healed) phase of this and other experimental myopathies[15] indicates that the centronuclear change is not specific and should not be used to designate a disease entity. In some human cases possibly that change may be the nonspecific result of incomplete healing of muscle fibers in a slowly progressive or even self-limited myopathic process.

Of relevance to the hypothesis that functional or structural abnormality of intramuscular arterial blood vessels causes Duchenne dystrophy is the newly recognized possiblity that in rare instances amyloidosis of intramuscular blood vessels also may cause, by an intramuscular ischemia mechanism, a progressive myopathy indistinguishable from limb-girdle dystrophy. This explanation was prompted by the following patient. He was a 26-year-old male of above average muscle strength until age 21 when he gradually developed proximal muscle weakness and wasting of both girdle regions and proximal upper and lower limbs. He had no sensory symptoms or signs. Serum "muscle enzymes" were elevated, electromyogram was "myopathic," and muscle biopsy showed slightly active muscle fiber degeneration and regeneration with prominent increase of endomysial connective tissue. Amyloid was found by crystal violet

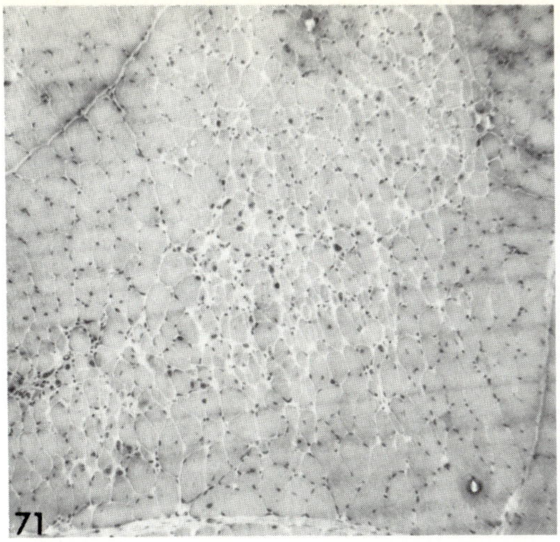

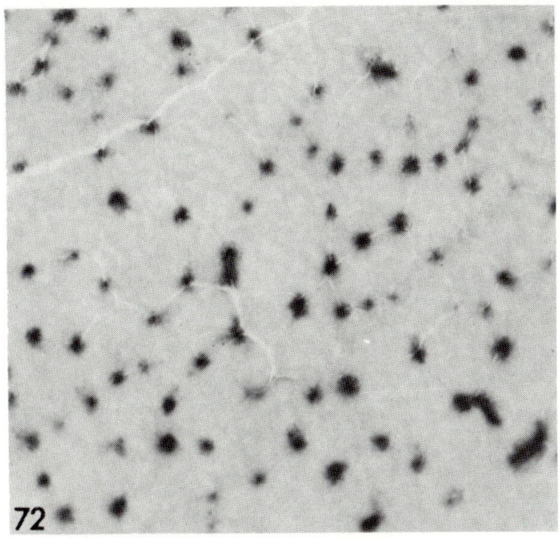

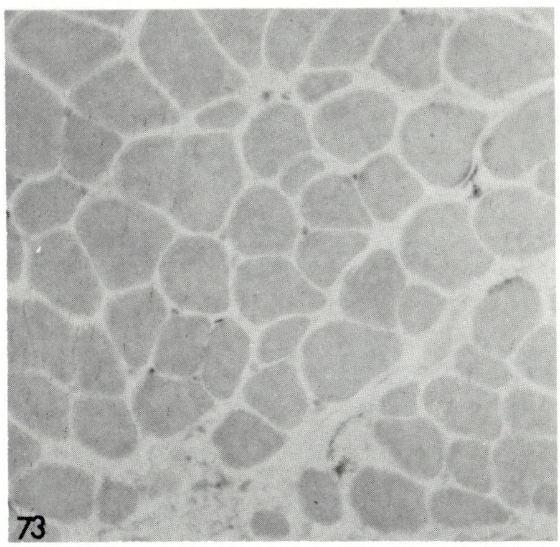

Fig. 71. Rabbit muscle 3 months after triple monthly embolization, demonstrating regional variation of involvement. Peripheral area of fascicle appears unaffected, while center shows abnormally large and small muscle fibers, some with internal nuclei, and increased endomysial connective tissue. Sometimes in other portions of the tissue the peripheral parts of the muscle fascicles were preferentially involved and the centers relatively intact. Modified trichrome, × 75.

Figs. 72 and 73. Rabbit muscle 3 months after triple monthly embolization, two areas from same muscle section and photographed under same conditions. 72—unaffected area contains abundant darkly stained capillaries and arterioles. 73—abnormal area, with intact muscles of abnormally large and small size and increased endomysial connective tissue (as in Fig. 27 and the center of Fig. 71), does not contain any stained capillaries and arterioles. Alkaline phosphatase reaction, × 190.

(fresh-frozen sections), congo red (anisotrophic dichroism in polarized light), and thioflavin T methods in walls of many small arteries, arterioles, veins, and venuoles of two skeletal muscle biopsies from different areas 1 year apart, but it was not found in gingival, rectal, skin, or sural nerve biopsies. None of the typical clinical features of primary amyloidosis was present and no predisposing cause for a secondary amyloidosis was found. His brother has recently developed a similar clinical disease at age 19, but it is not known whether he has amyloid in intramuscular vessels. Although every muscle biopsy from our last 700 patients with various neuromuscular disease (including at least 20 each with Duchenne dystrophy and limb-girdle dystrophy) has been routinely stained for amyloid (crystal violet on fresh-frozen sections), no amyloid has been seen in any other case of myopathy (dystrophy, inflammatory, uncertain type, or special types). If the possibility of an "amyloid ischemia myopathy" in this patient eventually proves to be correct, it would be a rare disease but an important mechanism in reference to the ischemia hypothesis of Duchenne dystrophy.

Recently in patients with acute psychoses, especially schizophrenia, a subclinical myopathy has been described to be histochemically similar to the subclinical myopathy in carriers of Duchenne dystrophy.[11] It seems to be paralleled by a rise of serum creatinine phosphokinase of the skeletal muscle type. In spite of the clinical dissimilarity between acute psychosis and Duchenne dystrophy, the similarity of muscle changes ought not be readily dismissed. Rather, consideration should be given to the possibility that some acute psychoses, *e.g.*, schizophrenic ones, are caused by a functional abnormality of small branches of the arterial tree in specific areas of the brain (*e.g.*, limbic system, hypothalamus, or midline brainstem nuclei) and in skeletal muscle, following the general principle postulated herein for Duchenne dystrophy and carriers thereof. Perhaps in some psychotic patients blood vessels of both organs are peculiarly susceptible to a circulating vasoconstrictor substance.

## CONCLUSION

If the ischemia hypothesis proves correct for the pathogenesis of Duchenne muscular dystrophy, then the disease cannot be considered a "primary disorder of the skeletal muscle cell." It would have to be classified as a "secondary myopathy"—secondary to vascular abnormality.

It must be emphasized that the ischemia hypothesis is currently being applied only to Duchenne dystrophy; whether ischemia will be considered important in some other muscular dystrophies remains to be determined.

Two implications of this hypothesis are that (1) additional studies should be devoted to confirming a role of small blood vessel abnormality in this disease, and (2) consideration should be given to possible therapeutic agents which might enhance the blood supply to muscle fibers in Duchenne dystrophy.

## SUMMARY

On the basis of histologic findings, an hypothesis of ischemia at the level of intramuscular small arteries and arterioles is proposed for the pathogenesis of

Duchenne muscular dystrophy. Identical histologic findings to those of Duchenne dystrophy, both early (*groups* of fibers undergoing necrosis or regeneration) and late (disproportionately greater increase of endomysial connective tissue) were reproduced, respectively, by arterial embolization of rabbit skeletal muscle once, or three times at monthly intervals. The hypothesis can also be applied to carriers of Duchenne dystrophy and is compatible with the abnormalities seen in their muscle biopsies. A preliminary description of findings in intramuscular arterial blood vessels of Duchenne dystrophy is given.

## NOTE ADDED IN PROOF

Since the time this article was submitted for publication, the histologically based ischemia hypothesis for the pathogenesis of Duchenne dystrophy has been further refined, in collaboration with Drs. Jerry Mendell and Dennis Murphy and Miss Carolyn Derrer.[13a-d] The structural lesions in blood vessels, as described above, have been abundant enough to suggest that they could be the major cause of ischemia. Consequently, a model in rats of functional ischemia of muscle was utilized. It combined a subthreshold vascular abnormality (aortic ligation) and a subthreshold amount of vasoactive amine (serotonin or norepinephrine) and resulted in grouped lesions (or in minimally affected regions scattered lesions) identical to those described above, but without any observable morphologic abnormality in intramuscular blood vessels.[13a] Serum "muscle enzymes" were markedly elevated in the model.[13b] Thus a comparable uni- or multifactorial mechanism of functional ischemia of intramuscular arterial blood vessels is possible in Duchenne dystrophy. Although 5-hydroxyindoacetic acid, epinephrine, and norepinephrine are not elevated in urine and 5-hydroxyindoacetic acid and homovanillic acid not increased in spinal fluid of Duchenne dystrophy patients,[13c] their platelets have a decreased rate of uptake and content of serotonin.[13d] Conceivably, abnormality of certain uptake mechanisms in platelets or other cells that are responsible for inactivating vasoactive amines affecting blood vessels of muscle could contribute to muscle damage in Duchenne muscular dystrophy. There could also be an inherited susceptibility of blood vessels in the muscles preferentially involved in the disease.

## REFERENCES

1. Ashmore, C. R., Doerr, L., and Somes, R. G., Jr.: Microcirculation. Loss of enzyme activity in chickens with hereditary muscular dystrophy. *Science 160:* 319–320, 1968.
2. Demos, J., and Ecoiffier, J.: Troubles circulatoires au cours de la myopathie, Etudies artériographiques. *Rév. Franc. Étud. Clin. Biol. 2:* 489–494, 1957.
3. Demos, J., Treumann, F., and Schroeder, W.: Anomalies de regulation de la micro-circulation musculaire chez les enfants atteints de dystrophie musculaire progressive par rapport a des enfants normaux du meme age. *Rév. Franc. Étud. Clin. Biol. 13:* 467–483, 1968.
4. Engel, W. K.: A clinical approach to the myopathies. *Clin. Orthop. 39:* 6–18, 1965.
5. Engel, W. K.: A critique of congenital myopathies and other disorders. In *Exploratory Concepts in Muscular Dystrophy and Related Disorders*, ed. by Milhorat, A. T., pp. 27–40. New York, Excerpta Medica Foundation, 1967.
6. Engel, W. K.: Local myopathic changes produced by electromyographic and hypodermic needles (needle myopathy). *Arch. Neurol. 16:* 509–511, 1967.
7. Engel, W. K.: Muscle biopsy. *Clin. Orthop. 39:* 80–105, 1965.

8. Engel, W. K.: Selective and nonselective susceptibility of muscle fiber types. A new approach to human neuromuscular diseases. *Arch. Neurol. 22:* 97–117, 1970.
8a. Engel, W. K., and Bishop, D. W.: Unpublished observations.
9. Engel, W. K., and Cunningham, G. G.: Alkaline phosphatase-positive abnormal muscle fibers of humans. *J. Histochem. Cytochem. 18:* 55–57, 1970.
10. Engel, W. K., and Cunningham, G. G.: Rapid examination of muscle tissue. An improved trichrome method for rapid diagnosis of muscle biopsy fresh-frozen sections. *Neurology (Minneap.) 13:* 919–923, 1963.
10 . Engel, W. K., and Hathaway, P. W.: Unpublished observations.
11. Engel, W. K., and Meltzer, H.: Histochemical abnormalities of skeletal muscle in patients with acute psychoses. *Science 168:* 273–276, 1970.
12. Erb, W. H.: Dystrophia muscularis progressiva. Klinische und pathologisch-anatomische studien. *Dtsch. Z. Nervenheilkd. 1:* 13–94, 173–261, 1891.
13. Hathaway, P. W., Engel, W. K., and Zellweger, H.: Experimental myopathy after microarterial embolization. *Arch. Neurol. 22:* 365–378, 1970.
13a. Mendell, J. R., Engel, W. K., and Derrer, E. C.: Duchenne muscular dystrophy: functional ischemia reproduces its characteristic lesions. *Science 172:* 1143–1145, 1971.
13b. Mendell, J. R., Engel, W. K., and Derrer, E. C.: Plasma enzyme elevations in rats with functional ischemia—a possible model of Duchenne muscular dystrophy. *Nature 239:* 522–524, 1972.
13c. Mendell, J. R., Murphy, D. L., Engel, W. K., Chase, T. N., and Gordon, E.: Catecholamines and indoleamines in patients with Duchenne muscular dystrophy. *Arch. Neurol. 27:* 518–520, 1972.
13d. Murphy, D. L., Mendell, J. R., and Engel, W. K.: Serotonin and platelet function in patients with Duchenne muscular dystrophy. *Arch. Neurol.*, 1972. In press.
14. Perloff, J. K., Roberts, W. C., DeLeon, A. C., and O'Doherty, D.: The distinctive electrocardiogram of Duchenne's progressive muscular dystrophy. *Am. J. Med. 42:* 179–188, 1967.
15. Resnik, M., and Engel, W. K.: Ultrastructural and histochemical correlation of experimental muscle regeneration. *J. Neurol. Sci. 11:* 167–185, 1970.
16. Rossi, L., and James, T. N.: Neurovascular pathology of the heart in progressive muscular dystrophy. *Panminerva Med. 6:* 357–360, 1964.
17. Schroeder, W., Treumann, F., and Demos, J.: Der wirkungsgrad der skelettmuskeldurchblutung bei der dystrophia musculorum progressiva typ Duchenne. *Arzneim. Forsch. 23:* 185–190, 1969.
18. Selye, H.: Pluricausal experimental myopathies. In *Progress in Neuro-Genetics*, ed. by Barbeau, A., and Brunette, J. R., pp. 79–92. Amsterdam, Excerpta Medica Foundation, 1969.

Chapter 23

# Some Observations on Normal and Diseased Human Muscle in Tissue Culture*

VICTOR DUBOWITZ

Fibroblast culture from skin has aroused much interest in recent years in the investigation of patients with disorders of the central nervous system associated with abnormalities of mucopolysaccharide or lipid metabolism.

Although as yet we have no clue as to the fundamental abnormality in muscular dystrophy and other neuromuscular disorders, it would seem reasonable to pursue the possibility of some structural or biochemical abnormality in the cultured muscle cells from these patients.

Muscle explants can readily be grown from the developing muscle of chick, mouse, or other embryos and also from some mature animals. The cells initially growing out from the explant are mononucleate myoblasts which frequently contain mitotic figures. These then fuse to form multinucleate myoblasts and subsequently ribbon or strap forms (myotubes). Cross-striations may appear in either the multinucleate myoblasts or the myotubes.

In recent years a number of workers have succeeded in growing mature human muscle in culture, and a few attempts have also been made to sustain dystrophic human muscle in culture, but without much success.

Geiger and Garvin[1] found that, although normal and dystrophic muscle initially appeared to grow in a similar way, the cells of the dystrophic muscle were more heavily granulated and failed to develop the cross-striations seen in the normal muscle. After 4 weeks, the cells from the dystrophic muscle hypertrophied, reaching lengths up to 3,000 $\mu$, and then degenerated. When subcultured the dystrophic muscle only survived one passage and then degenerated, whereas the normal muscle could be maintained for as long as a year through as many as 12 passages.

O'Steen[7] grew human muscle in diffusion chambers inserted into the peritoneal cavity of mice. He observed that in cultures from dystrophic muscle the size of mononucleate myoblasts was similar to that of normal cultures,

* This work was supported by grants from the Muscular Dystrophy Group of Great Britain, the Medical Research Council, and the Sheffield University Research Fund.

whereas the multinucleate myoblasts were larger and the myotubes shorter. In a study of cultures from four normal and three dystrophic muscles, Goyle, Kalra, and Singh[2] found that myoblasts from dystrophic muscle were smaller and more variable in size and shape than the normal, and that there were fewer multinucleate cells. More recently, Kakulas, Papadimitriou, Knight, and Mastaglia[6] studied the primary explants of normal and dystrophic human muscle by time-lapse cinemicrophotography and observed differences both in the morphology and in the pattern of growth of dystrophic cells.

There has also been some divergence in observations on the lag period between the initiation of the culture and the first cell growth phase. Thus Geiger and Garvin[1] observed a lag of 2 to 6 days in dystrophic cultures compared to 7 to 20 days in the normal, while Goyle, Kalra, and Singh[2] found a lag phase of 3 to 5 days in both normal and dystrophic cultures and O'Steen[7] a lag phase of 2 to 4 days in both. Herrmann[4] also found a similar lag phase in both normal and dystrophic muscle, and Herrmann, Konigsberg, and Robinson[5] observed no striking difference between normal and dystrophic muscle in either the number of cultures showing outgrowth after 1 month or the maximal duration of *in vitro* maintenance of the culture.

A number of previous workers have observed cross-striations in normal human muscle grown *in vitro*[1, 5, 7, 8] but not in dystrophic muscle.[1, 5, 7] Goyle, Kalra, and Singh[3] claimed to have seen cross-striations in stained preparations from both normal and dystrophic muscle under polarized light, but the appearances illustrated were irregular and somewhat bizarre. Kakulas, Papadimitriou, Knight, and Mastaglia[6] noted the presence of multinucleated myotubes with cross-striations in the primary explants of dystrophic muscle after several days growth.

In a recent study we[9] found that diseased human muscle was capable of differentiation in culture in the same way as normal muscle, and we were unable to detect any apparent differences in their morphologic appearance. Cross-striations were observed in a number of cultures from both normal and diseased muscle.

We have now completed a study of a much larger series of normal and diseased muscles and have attempted a quantitative analysis of a number of morphologic parameters.

## CLINICAL MATERIAL

Muscle was obtained at biopsy from 34 patients with a variety of muscle disorders (Table 1) and from 19 controls in the course of surgical procedures (Table 2).

## TECHNIQUE OF TISSUE CULTURE OF MUSCLE

The biopsy was placed in TC.199 in a Bijou bottle and subsequently cut into small pieces of approximately 1 $mm^3$ in a Petri dish. It was then transferred by Pasteur pipet to a Carrell flask, where it was clotted by overnight incubation with a drop of human plasma and 1% calcium chloride (5:1). The plasma was made by mixing 2 parts of blood with 1 part of Alsever's solution and separating the plasma by centrifugation. The cultures were initiated in a medium of 7 parts of TC.199 : 2 parts of human AB serum : 1 part of chick embryo extract. When growth had commenced it was maintained with 8 parts of TC.199 : 2 parts of AB serum, and the culture medium was changed twice weekly.

After variable periods *in vitro* (from 15 to 147 days in the present series), and after two to eight passages, the cells were trypsinized with 0.04% trypsin and 0.04% Versene (EDTA) for approximately 5 minutes with gentle agitation. A concentration of 250,000 cells in 6 ml of medium was transferred from the large Carrell flasks into a 95-mm Petri dish containing coverslips attached by plasma clot. The Petri dishes were kept in a humid atmosphere of 5% $CO_2$ in air.

TABLE 1. CLINICAL AND CULTURE DATA: DISEASED MUSCLE

| Diagnosis | Patient No. | Sex | Age | Muscle | Lag Phase | Total Duration *in Vitro* | No. of Passages | Ultimate Fate of Culture |
|---|---|---|---|---|---|---|---|---|
| | | | | | *days* | *days* | | |
| Duchenne muscular dystrophy | 2 | M | 10y | Gastrocnemius | 26 | 37 | 0 | Died |
| | 24 | M | 8y | Gastrocnemius | 21 | 19 | 0 | Died |
| | 54 | M | 8¾y | Gastrocnemius | 2 | 142 | 4 | Infected |
| | 55 | M | 5½y | Gastrocnemius | 4 | [a] 69, 144 | [a] 5, 7 | PTAH[b] |
| | 64 | M | 3y | Gastrocnemius | 21 | [a] 84, 60 | [a] 3, 4 | PTAH, R.C. |
| | 92 | M | 10y | Gastrocnemius | 6 | 25 | 0 | Died |
| | 93 | M | 7½y | Gastrocnemius | 6 | 58 | 2 | R.C. |
| | 94 | M | 11y | Gastrocnemius | 7 | 107 | 3 | R.C. |
| | 98 | M | 6y | Gastrocnemius | 5 | [a] 130, 130 | [a] 3, 4 | PTAH |
| | 111 | M | 5y | Gastrocnemius | 4 | [a] 74, 77, 140, 165 | [a] 3, 4, 7, 8 | PTAH |
| | 126 | M | 5y | Gastrocnemius | 9 | [a] 72, 76, 86, 82 | [a] 4, 4, 3, 8 | PTAH |
| | 132 | M | 13y | Gastrocnemius | 9 | 120 | 5 | PTAH |
| | 142 | M | 4y | Gastrocnemius | 12 | 79 | 4 | Infected |
| Muscular dystrophy carriers | 10 | F | 16y | Rectus femoris | 7 | 48 | 1 | Infected |
| | 21 | F | 40y | Gastrocnemius | 5 | 54 | 0 | Died |
| | 85 | F | 37y | Gastrocnemius | 11 | 57 | 3 | PTAH |
| | 87 | F | 25y | Gastrocnemius | 8 | [a] 98, 70, 76 | [a] 7, 3, 6 | PTAH, R.C. |
| Spinal muscular atrophy | 63 | F | 7m | Rectus femoris | 18 | [a] 89, 101, 40, 79, 108 | [a] 2, 5, 2, 5, 5 | PTAH, R.C. |
| | 74 | F | 3m | Rectus femoris | 10 | 38 | 2 | PTAH |
| | 115 | M | 1¾y | Rectus femoris | 5 | [a] 55, 69, 75, 94, 103, 132 | [a] 3, 4, 5, 6, 7, 11 | PTAH |
| | 119 | F | 8m | Rectus femoris | 9 | [a] 61, 68, 85 | [a] 4, 5, 7 | PTAH |
| | 131 | M | 5y | Rectus femoris | 19 | 48 | 2 | Excess fibroblasts |
| | 145 | F | 8m | Rectus femoris | 25 | 52 | 3 | Infected |
| | 148 | M | 3m | Rectus femoris | 6 | 40 | 3 | Infected |
| Central core disease | 73 | F | 29y | Rectus femoris | 10 | [a] 62, 84, 35, 49, 56 | [a] 2, 3, 4, 5, 2 | PTAH, R.C. |
| | 82 | M | 4y | Rectus femoris | 7 | [a] 103, 96, 2 | 2 | PTAH, R.C. |
| Congenital myopathy (nonspecific) | 9 | M | 9m | Rectus femoris | 5 | 48 | 2 | Infected |
| | 108 | M | 12y | Rectus femoris | 15 | [a] 115, 142 | [a] 5, 6 | PTAH, R.C. |
| | 122 | M | 10w | Rectus femoris | 5 | 52 | 2 | PTAH, R.C. |
| Peroneal muscular atrophy | 26 | F | 5y | Flexor digitorum longus | 7 | 50 | 3 | Died |
| | 114 | F | 17y | Soleus | 5 | 114 | 4 | PTAH |
| Muscular dystrophy (Becker type) | 150 | M | 5y | Gastrocnemius | 15 | 36 | 3 | Infected |
| Dystrophia myotonica | 121 | M | 41y | Deltoid | 9 | 113 | 4 | PTAH |
| Myotubular myopathy | 60 | M | 10y | Tibialis anterior | 14 | [a] 59, 102, 93, 94 | [a] 3, 4, 5, 4 | PTAH, R.C. |

[a] Multiple subcultures from the same explant.
[b] PTAH, preparations stained with phosphotungstic acid and hematoxylin; R.C., ring culture (living).

TABLE 2. CLINICAL AND CULTURE DATA: CONTROL MUSCLE

| Diagnosis | Patient No. | Sex | Age | Muscle | Lag Phase | Total Duration *in Vitro* | No. of Passages | Ultimate Fate of Culture |
|---|---|---|---|---|---|---|---|---|
| | | | | | *days* | *days* | | |
| Control | 5 | F | 11m | Rectus femoris | 5 | 46 | 2 | Infected |
| | 6 | F | 6y | Gastrocnemius | 4 | 89 | 3 | Infected |
| | 7 | M | 7y | Rectus femoris | 7 | 77 | 4 | Infected |
| | 8 | M | 5y | Rectus femoris | 14 | 20 | 0 | Infected |
| | 16 | M | 4½y | Tibialis posterior | 19 | 61 | 1 | Died |
| | 18 | M | 2y | Abductor hallucis | 19 | 61 | 2 | Died |
| | 19 | F | 8y | Rectus femoris | 18 | 111 | 0 | Died |
| | 67 | F | 5¼y | Vastus Intermedius | 24 | [a] 88, 139 | [a] 3, 5 | PTAH, R.C.[b] |
| | 81 | F | 2½y | Iliacus | 8 | [a] 78, 74, 40, 15, 18, 22, 58 | [a] 8, 8, 3, 2, 3, 2, 8 | PTAH, R.C. |
| | 83 | M | 3y | Gastrocnemius | 11 | [a] 57, 64, 71, 71 92, 96, 36 | [a] 3, 3, 4, 4, 8, 8, 3 | PTAH, R.C. |
| | 84 | F | 10m | Quadriceps | 17 | 36 | 0 | Died |
| | 101 | | | Gluteus medius | 4 | 51 | 0 | Infected |
| | 102 | M | 7m | Adductor brevis | 10 | [a] 124, 119 | [a] 6, 6 | PTAH |
| | 113 | M | 2⅓y | Abductor hallucis | 8 | [a] 77, 68, 74 | [a] 4, 5, 6 | PTAH |
| | 117 | | 1d | Spinal extensor | 7 | [a] 61, 86 | [a] 2, 3 | PTAH |
| | 118 | | 3½y | Flexor digitorum | 7 | 56 | 2 | PTAH |
| | 135 | M | 3y | Rectus femoris | 12 | [a] 86, 91 | [a] 6, 7 | PTAH |
| | 136 | M | | Vastus medialis | 27 | 90 | 4 | Infected |
| | 141 | F | | Vastus medialis | 10 | 15 | 3 | Infected |

[a] Multiple subcultures from the same explant.

[b] PTAH, preparations stained with phosphotungstic acid-hematoxylin; R.C., ring culture (living).

After an arbitrary 7, 14, and 21 days from initiation of subculture, coverslips were fixed for 10 minutes in methanol and stained with phosphotungstic acid-hematoxylin (PTAH) for morphologic and quantitative study.

In order to observe cross-striations in the living cell, ring cultures were set up for high power phase contrast microscopy. A sterile glass ring of 26-mm outside diameter was sealed onto a sterile slide with silicone grease, and the top surface of the ring was also greased. Twenty thousand cells from a subculture were placed in the cavity and it was filled with medium. A coverslip was then lowered onto the ring and pressed down firmly. The slide was inverted to allow the cells to grow onto the coverslip.

*Measurements.* In each of the preparations stained with PTAH, the length and average breadth of a random selection of 10 multinucleate cells were measured with an eyepiece micrometer. Multinucleate myoblasts were differentiated from myotubes by their cluster of rounded nuclei, usually near the center of the cell, while the myotube tended to be longer and narrower, with oval nuclei spread along its length. A count was also made of the number of nuclei in a random selection of 10 multinucleate myoblasts and 10 myotubes in each preparation.

*Storage.* After successful preparations of coverslips were obtained, healthy subcultures from the biopsies were cooled down to −30°C at 1° per minute by the standard techniques and then stored in liquid nitrogen. When reconstituted, these cultures regenerated and differentiated again as before.

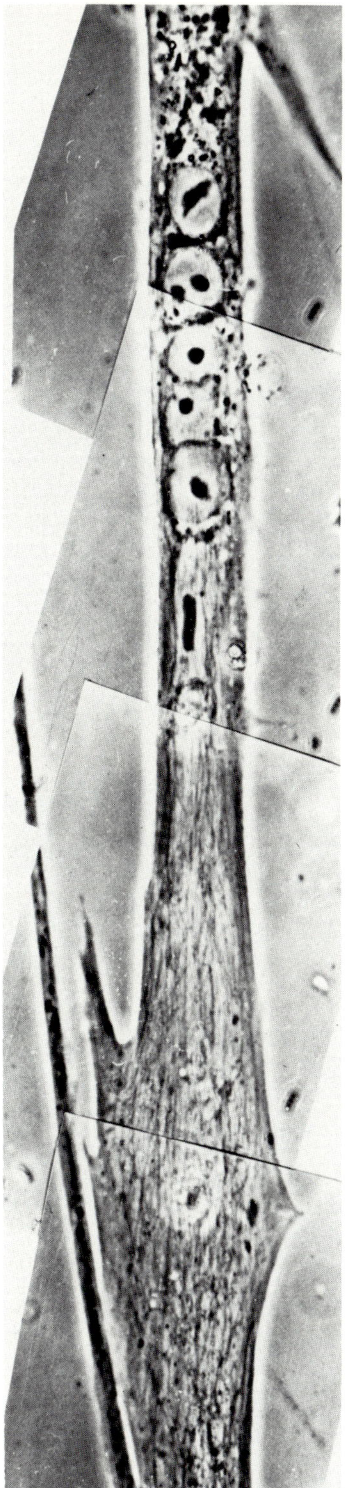

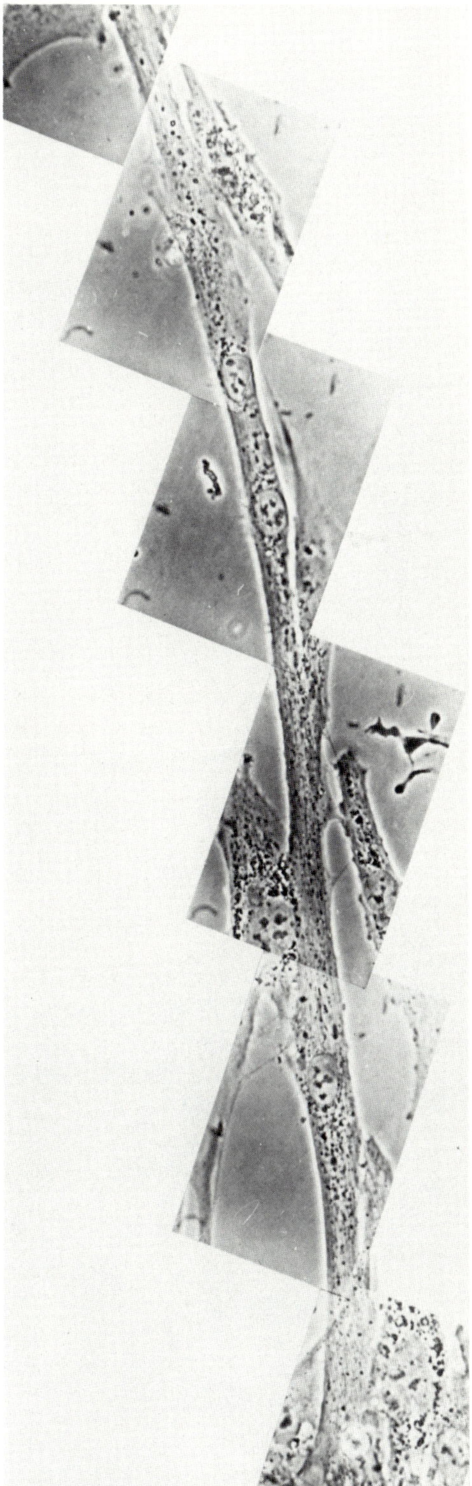

FIG. 1. Multinucleate myoblast; case number 64; Duchenne muscular dystrophy; age 3 years; gastrocnemius; 42 days *in vitro*; three passages in culture; 27 days in final subculture. Living culture, phase contrast; × 750. (Reproduced from Skeate, Bishop, and Dubowitz[9] by permission of Blackwell Scientific Publications.)

FIG. 2. Myotube; case number 64; Duchenne muscular dystrophy; age 3 years; gastrocnemius; 42 days *in vitro*; four passages in culture; 5 days in final subculture. Living culture; phase contrast; × 440. (Reproduction from Skeate, Bishop, and Dubowitz[9] by permission of Blackwell Scientific Publications.)

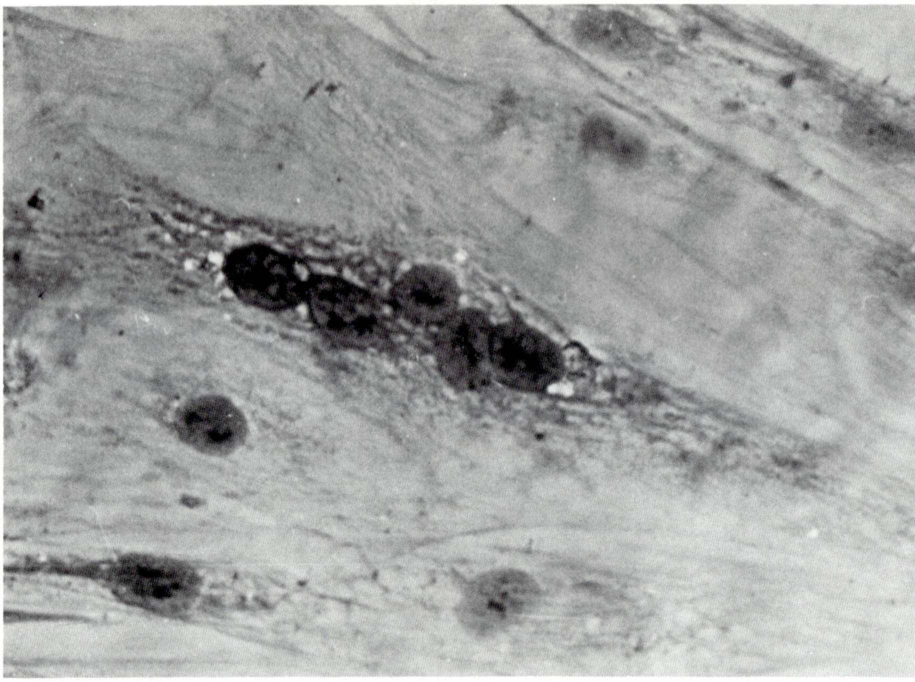

FIG. 3. Multinucleate myoblast; case number 55; Duchenne muscular dystrophy; age 5½ years; gastrocnemius; 144 days *in vitro*; seven passages in culture; 13 days in final subculture; note cluster of nuclei. Phosphotungstic acid-hematoxylin stain; phase contrast; × 400.

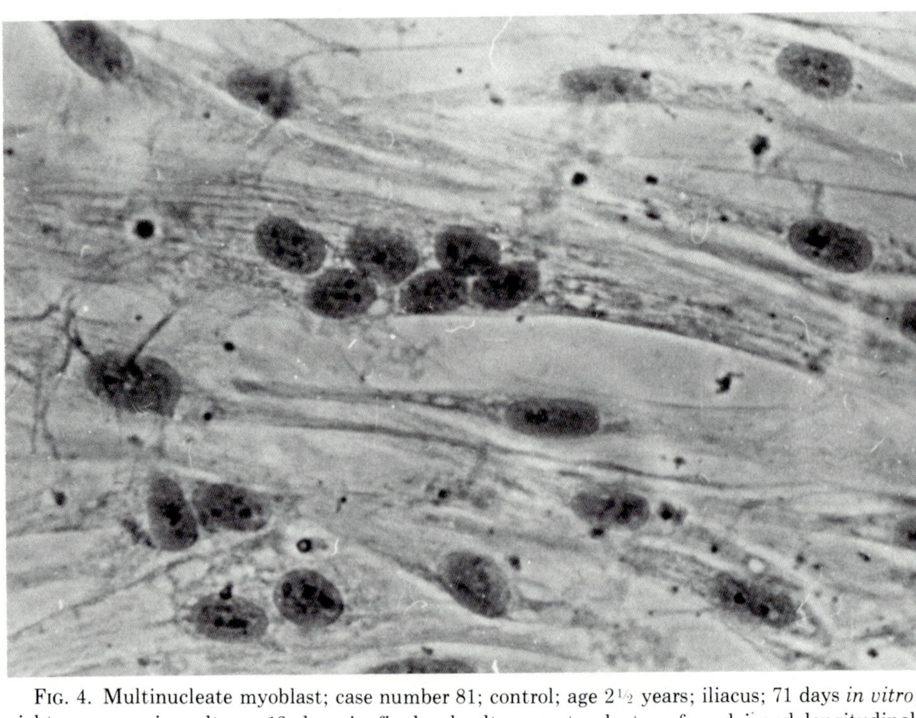

FIG. 4. Multinucleate myoblast; case number 81; control; age 2½ years; iliacus; 71 days *in vitro*; eight passages in culture; 13 days in final subculture; note cluster of nuclei and longitudinal striations. Phosphotungstic acid-hematoxylin stain; phase contrast; × 400.

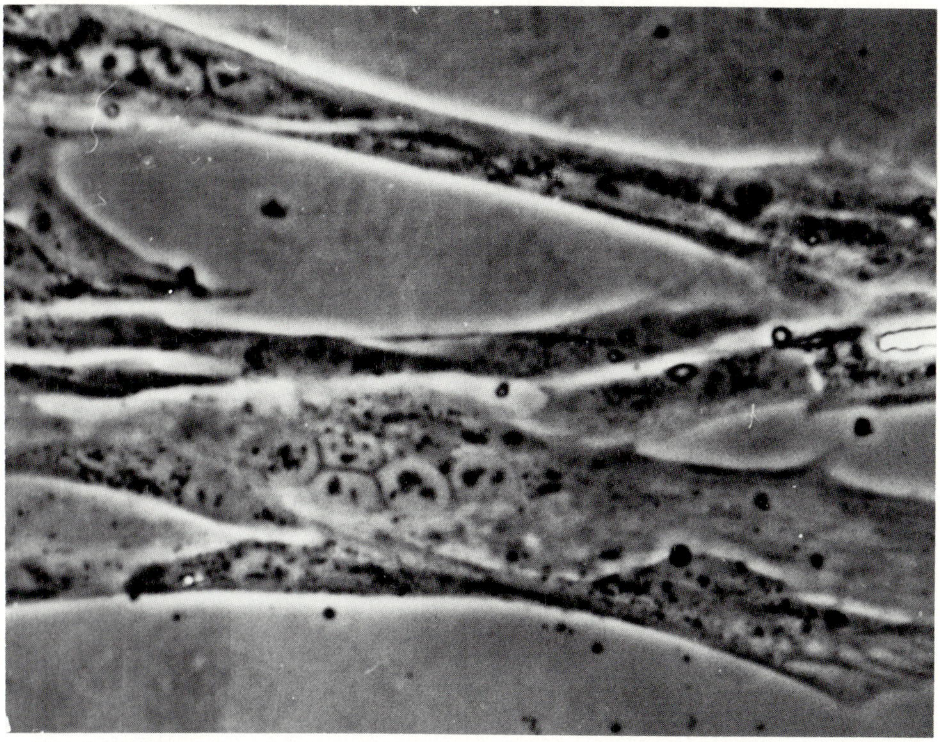

FIG. 5. Multinucleate myoblast; case number 7; control; age 7 years; rectus femoris; 53 days *in vitro*; two passages in culture; 21 days in final subculture. Living culture; phase contrast; × 600.

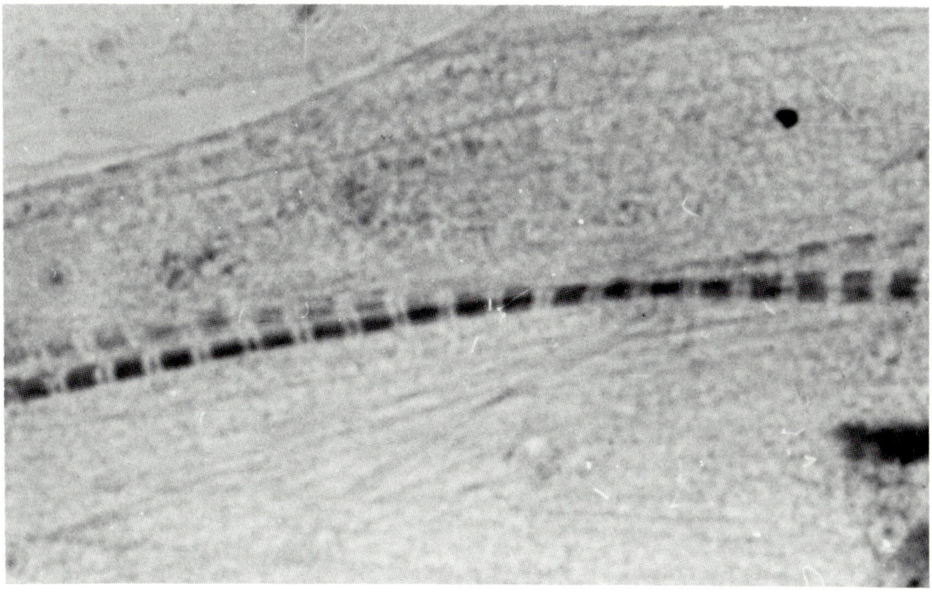

FIG. 6. Cross-striations, showing A band and Z line; case number 64; Duchenne muscular dystrophy; age 3 years; gastrocnemius; 79 days *in vitro*; three passages in culture; 36 days in final subculture. Phosphotungstic acid-hematoxylin stain; phase contrast; × 3000.

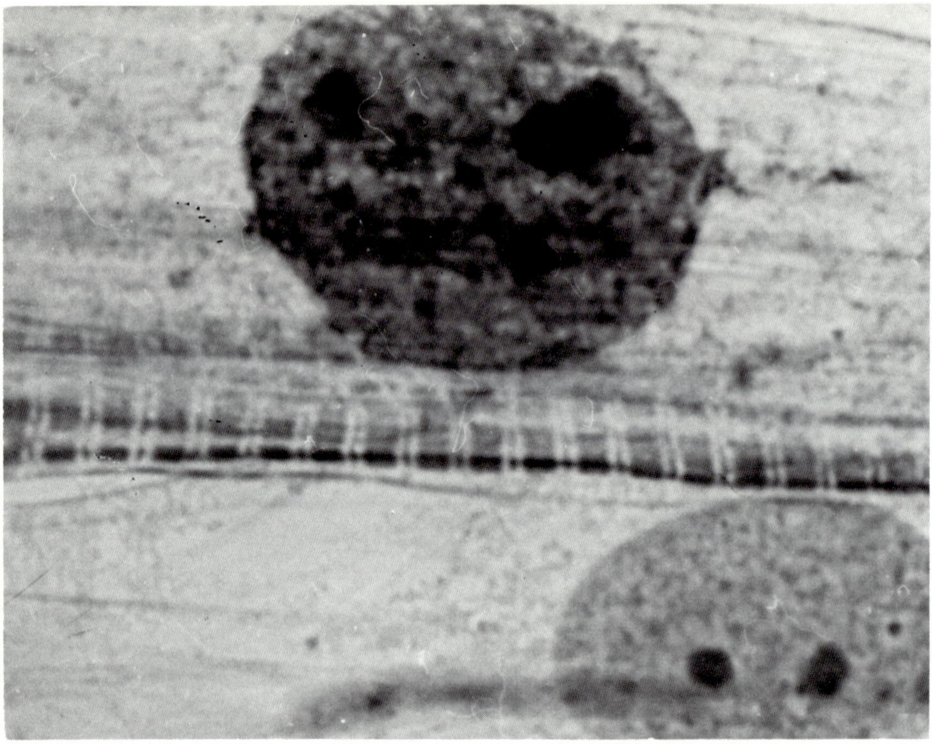

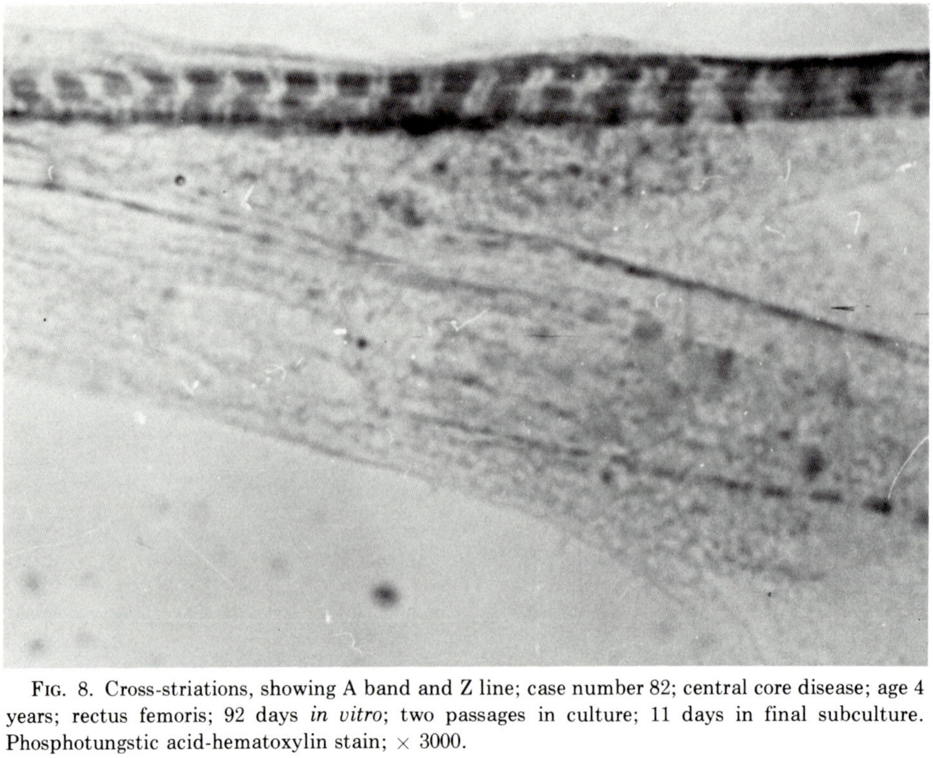

FIG. 8. Cross-striations, showing A band and Z line; case number 82; central core disease; age 4 years; rectus femoris; 92 days *in vitro*; two passages in culture; 11 days in final subculture. Phosphotungstic acid-hematoxylin stain; × 3000.

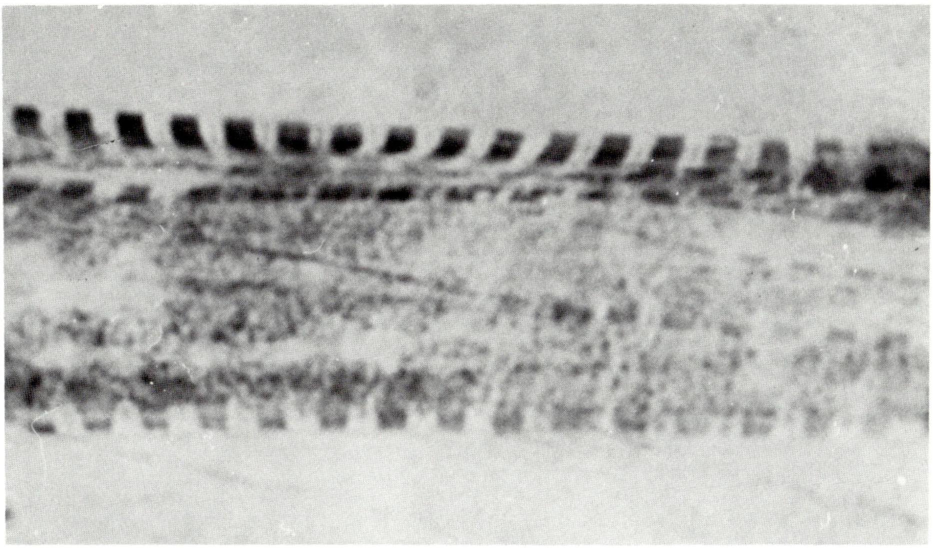

FIG. 9. Cross-striations, showing A band and Z line; case number 81; control; age 2½ years; iliacus; 84 days *in vitro*; eight passages in culture; 13 days in final subculture. Phosphotungstic acid-hematoxylin stain; × 3000.

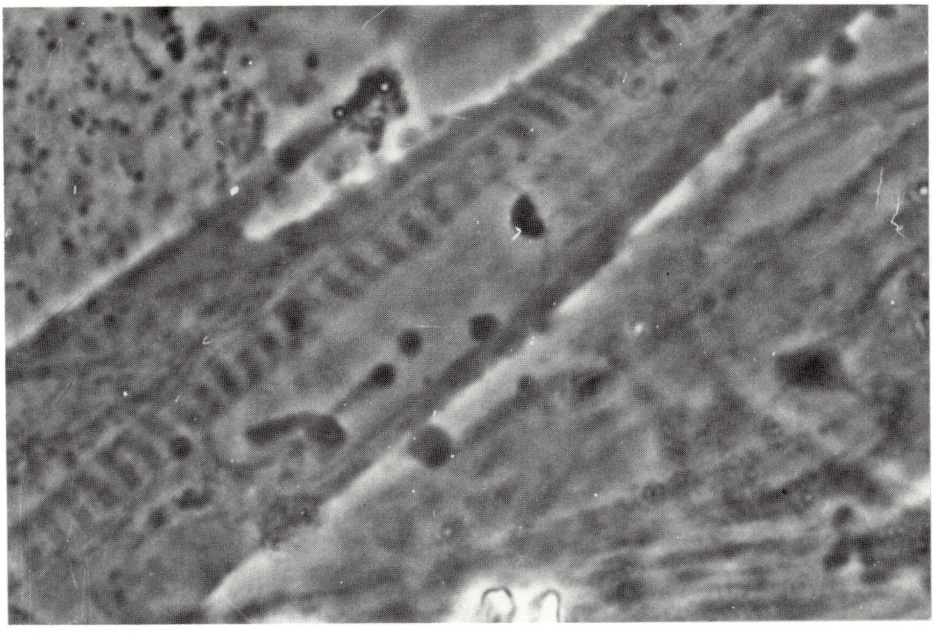

FIG. 10. Cross-striations, showing A band and Z line; case number 81; control; age 2½ years; iliacus; 104 days *in vitro*; eight passages in culture; 46 days in final subculture. Living culture; phase contrast; × 1500.

## RESULTS

Of a total of 81 explants, 53 grew successfully (61%). Subcultures from cases of muscular dystrophy, muscular dystrophy carriers, neurogenic atrophies, central

TABLE 3. LAG PHASE

| Diagnosis | No. of Cases | Lag Phase (Range) |
|---|---|---|
| | | *days* |
| Duchenne muscular dystrophy | 12 | 2 to 26 |
| Muscular dystrophy carriers | 4 | 5 to 11 |
| Spinal muscular atrophy | 7 | 5 to 25 |
| Controls | 19 | 4 to 27 |

TABLE 4. MEDIAN LENGTH OF MULTINUCLEATE CELLS

| Diagnosis | No. of Biopsies Studied | No. of Sub-cultures | Median Length | | |
|---|---|---|---|---|---|
| | | | 7 Days | 14 Days | 21 Days |
| | | | | $\mu$ | |
| Duchenne dystrophy | 4 | 8 | 275 | 350 | 250 |
| Duchenne dystrophy carriers | 2 | 2 | 200 | 250 | 250 |
| Spinal muscular atrophy | 3 | 5 | 225 | 250 | 225 |
| Controls | 7 | 15 | 275 | 350 | 250 |

core disease, and myotubular myopathy were observed for up to 5 to 7 weeks. In some cases multiple subcultures from the same biopsy were studied. The rate of growth, differentiation, and ability to develop cross-striations in the diseased muscle did not appear to differ from the controls (Figs. 1 to 10).

## ANALYSIS

*Lag Phase*. The range in time between the initiation of culture and the phase of initial growth did not show any significant difference between the various diseased muscles and the controls (Tables 1 to 3).

*Lengths*. There was no significant difference in the lengths of the multinucleate cells at 7, 14, and 21 days growth of any particular subculture (Table 4). The results for the 3 separate days were then pooled for further comparison between patients (Table 5). The myotubes showed a very wide range in length with marked overlap between the normal and diseased groups. The multinucleate myoblasts did not show such a wide range in length and could be subjected to statistical analysis. This revealed that (1) the difference in length between cultures within patients was significantly greater than the variance within individual cultures; (2) the difference between cultures from different patients was significantly greater than the difference between cultures from the same patient; and (3) there was no significant difference between the lengths of multinucleate myoblasts of cultures of normal and diseased muscle.

*Breadths*. The individual breadths of the differentiated cells were very variable in both normal and dystrophic cultures (Table 5). In spite of a wide range, the medians of the various groups were fairly close to each other.

*Nuclei*. There were no morphologic differences in the nuclei of cultures from the diseased and normal muscle, and counts of the nuclei in random selections of cells from individual culture showed no significant difference between the normal and diseased groups.

*Myotubes versus multinucleate myoblasts*. The proportion of myotubes to multinucleate myoblasts increased slightly with the duration of subculturing in all groups. There was no significant difference between individual groups.

*Cross-striations.* These were observed in 14 subcultures from 7 control patients and in 17 subcultures from 14 patients with neuromuscular disorder. They were clearly visible both in PTAH-stained preparations as well as in living cells under phase contrast microscopy. They appeared in mononucleate myoblasts, multinucleate myoblasts, and myotubes. Frequently they were localized along one or other side of the cell and rarely extended the full breadth of the cell. In the ring cultures observed serially under phase contrast microscopy, the striations seemed to disappear in some of the living cells and to reappear again later.

There was no apparent difference in the time of first appearance or the frequency of observation of cross-striations in various groups of diseased or normal muscle (Table 6).

## CONCLUSIONS

It is somewhat surprising that dystrophic muscle, with its striking pathologic changes *in vivo*, should be so remarkably normal when grown in culture. The same applies to the biopsies from the other neuromuscular disorders studied.

There are a number of possible explanations for this. One possibility is that

TABLE 5. RANGE AND MEDIAN LENGTH AND BREADTH IN MULTINUCLEATE CELLS[a]

| Diagnosis | No. of Biopsies | No. of Subcultures | Length | | Breadth | |
|---|---|---|---|---|---|---|
| | | | Multinucleate myoblasts | Myotubes | Multinucleate myoblasts | Myotubes |
| Duchenne dystrophy | 4 | 8 | 75–610 (225) | 100–1350 (375) | 5.0–95.0 (20.0) | 7.5–28.0 (12.5) |
| Duchenne dystrophy carriers | 2 | 2 | 100–375 (225) | 175–475 (275) | 7.5–55.0 (15.0) | 7.5–15.0 (10.0) |
| Spinal muscular atrophy | 3 | 5 | 50–900 (225) | 125–1095 (475) | 5.0–85.0 (15.0) | 2.5–94.0 (12.5) |
| Controls | 7 | 15 | 75–1350 (250) | 150–1025 (425) | 4.5–95.0 (20.0) | 5.0–95.0 (12.5) |

[a] Combined data for 7, 14, and 21 days growth of subculture.

TABLE 6. FIRST OBSERVATION OF CROSS-STRIATIONS AFTER SUBCULTURE

| Diagnosis | Days after Subculturing | |
|---|---|---|
| | PTAH[a]-stained coverslips | Ring cultures |
| Controls | 5 | 4 |
| Duchenne dystrophy | 11 | 16 |
| Duchenne carriers | 14 | |
| Spinal muscular atrophy | 5 | 16 |
| Peroneal muscular atrophy | 6 | |
| Central core disease | 8 | 16 |
| Myotubular myopathy | 14 | |

[a] PTAH, phosphotungstic acid-hematoxylin.

the myoblasts which grow and differentiate from explants of diseased muscle are derived from normal fibers within the muscle and that the abnormal fibers do not grow. This seems improbable since in Duchenne muscular dystrophy all fibers are potentially abnormal and will tend to degenerate in the later stages of the disease. Biopsies taken from older children gave similar results to those from early cases. Moreover, in one of the cases of central core disease studied, 99% of the muscle fibers had central cores in them on histologic study—yet the muscle explant grew quite normally in culture.

Another possibility is that the myoblasts are in fact abnormal but that the parameters we have used are not sensitive enough to detect the abnormality. A study of various biochemical aspects, such as enzyme levels and RNA synthesis, may yet show differences between the normal and diseased muscle.

A third possibility is that the period in subculture has not been long enough to allow the dystrophic changes to develop. Perhaps if one could maintain an individual subculture for a much more prolonged period, the myotubes might mature further and eventually manifest dystrophic features.

Finally, there is the possibility that the diseased muscle may in fact revert to a normal state when cultured *in vitro*. This would imply that some environmental factor *in vivo* may be responsible for perpetuating the dystrophic state of the muscle. Identification of such a factor might well be an important stepping-stone towards unravelling the mystery of muscular dystrophy and other neuromuscular disorders.

## ACKNOWLEDGMENTS

I am pleased to acknowledge the collaboration in this program of Mrs. A. Bishop, B.Sc., Miss B. Gallup, B.Sc., and Miss Y. Skeate, B.Sc.

## REFERENCES

1. Geiger, R. S., and Garvin, J. S.: Pattern of regeneration of muscle from progressive muscular dystrophy patients cultivated *in vitro* as compared to normal human skeletal muscle. *J. Neuropathol. exp. Neurol. 16:* 532–543, 1957.
2. Goyle, S., Kalra, S. L., and Singh, B.: The growth of normal and dystrophic human skeletal muscle in tissue culture. *Neurology India 15:* 149–151, 1967.
3. Goyle, S., Kalra, S. L., and Singh, B.: Further studies on normal and dystrophic human skeletal muscle in tissue culture. *Neurology India 16:* 87–88, 1968.
4. Herrmann, H.: An evaluation of tissue culture of muscle. *Res. Publ. Assoc. Res. Nerv. Ment. Dis. 38:* 697–703, 1958.
5. Herrmann, H., Konigsberg, U. R., and Robinson, G.: Observations on culture *in vitro* of normal and dystrophic muscle tissue. *Proc. Soc. Exp. Biol. Med. 105:* 217–221, 1960.
6. Kakulas, B. A., Papadimitriou, J. M., Knight, J. O., and Mastaglia, F. L.: Normal and abnormal human muscle in tissue culture. *Proc. Aust. Assoc. Neurol. 5:* 79–85, 1968.
7. O'Steen, W. K.: Growth activity of normal and dystrophic muscle implants in normal and dystrophic hosts. *Lab. Invest. 11:* 412–419, 1962.
8. Pogogeff, I. A., and Murray, M. R.: Form and behavior of adult mammalian skeletal muscle *in vitro*. *Anat. Rec. 95:* 321–329, 1946.
9. Skeate, Y., Bishop, A., and Dubowitz, V.: Differentiation of diseased human muscle in culture. *Cell Tissue Kinet. 2:* 307–310, 1969.

Chapter 24

# Observations on the Etiology of Polymyositis*

BYRON A. KAKULAS

Polymyositis has been known for more than 100 years[16, 39-41] as an important cause of nonhereditary progressive muscular weakness. Clinically about two-thirds of patients have cutaneous lesions,[4] in which case the disease is designated dermatomyositis. A remarkable feature is the high incidence of associated visceral malignancy.[3] Histologically, the muscle lesions consist of segmental muscle fiber necrosis with phagocytosis of necrotic sarcoplasm and regenerative changes. Cellular infiltration is variable and consists of lymphocytes, mononuclear cells, and plasmacytes with fewer neutrophils and eosinophils. Polymyositis may be acute or chronic and often occurs in association with connective tissue diseases such as rheumatoid arthritis, polyarteritis nodosa, and systemic lupus erythematosus. In this case the clinical manifestations of the associated connective tissue disease may be minor or dominant.[42]

In recent years, the suggestion has been advanced that polymyositis is due to a disorder of immune function. In support of this concept much clinical, laboratory, and experimental evidence has accumulated. Such evidence includes the presence of elevated serum globulins, the frequent dramatic response to treatment with steroid hormones, and the clinical association of the disease with other conditions of suspected immunologic origin. In addition, the high incidence of carcinoma occurring in relationship to polymyositis and the sensitivity to tumor extracts[8, 14] may reflect an immunologic disorder.

Early attempts to induce an experimental disease of muscle by immunologic methods in a laboratory animal were hampered by technical difficulties.[31, 38] In a later study, Dawkins[9] succeeded in producing an experimental autoallergic myositis which he showed was associated with hypersensitivity to muscle by skin testing. The muscle lesions in such guinea pigs treated with repeated injections of heterologous skeletal muscle and Freund's adjuvant resembled those which occurred in human polymyositis. In a later series of experiments it was demonstrated that lymph node cells of rats with the immunologic disease,

* This work was supported by the National Health and Medical Research Council of Australia and by the Muscular Dystrophy Research Association of Western Australia, Inc.

when added to tissue cultures of differentiated fetal rat skeletal muscle, caused destruction of the outgrowth in the neighborhood of the explant.[18] Pursuing this approach it was later found that cytopathic properties also existed for human peripheral (blood) lymphocytes derived from patients with polymyositis or systemic lupus erythematosus when added to cultures of human fetal muscle.[19] On this basis it is now possible to bring forward a reasonable formulation of a probable etiology and pathogenesis for the muscle changes which occur in these conditions.

## CLINICAL EVIDENCE

In 1954, Eaton[13] reviewed the problem of polymyositis and provided a clinical classification which was further developed by Walton and Adams.[4-] More recently, Pearson[32] carefully documented the clinical features of polymyositis and has surveyed the laboratory and therapeutic evidence in favor of an immune origin for the disease. In this regard, the elevated $\alpha$-2- and $\gamma$-globulins, the presence of circulating rheumatoid factor, and the general, although not universal, clinical improvement when steroid hormones are administered are the most noteworthy.

## PATHOLOGIC OBSERVATIONS

Considerable variability exists in the histomorphology of biopsy specimens of which the common denominator is muscle fiber necrosis. This manifests as architectural disorganization of the muscle fiber in both light and electron microscopic preparations. The ultrastructural changes appear first in the

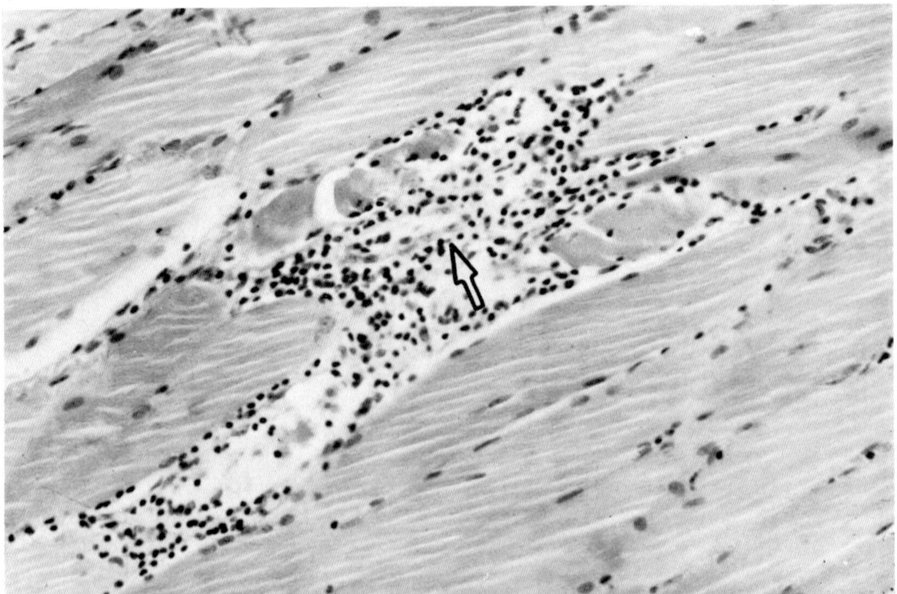

FIG. 1. Muscle biopsy of a patient with polymyositis. In the center of a field is a small venule (*arrow*) surrounded by a cellular exudate consisting predominantly of lymphoid cells. Focally necrotic muscle fibers are in close proximity to both the vessel and the cellular exudate. Hematoxylin and eosin stain; $\times$ 450.

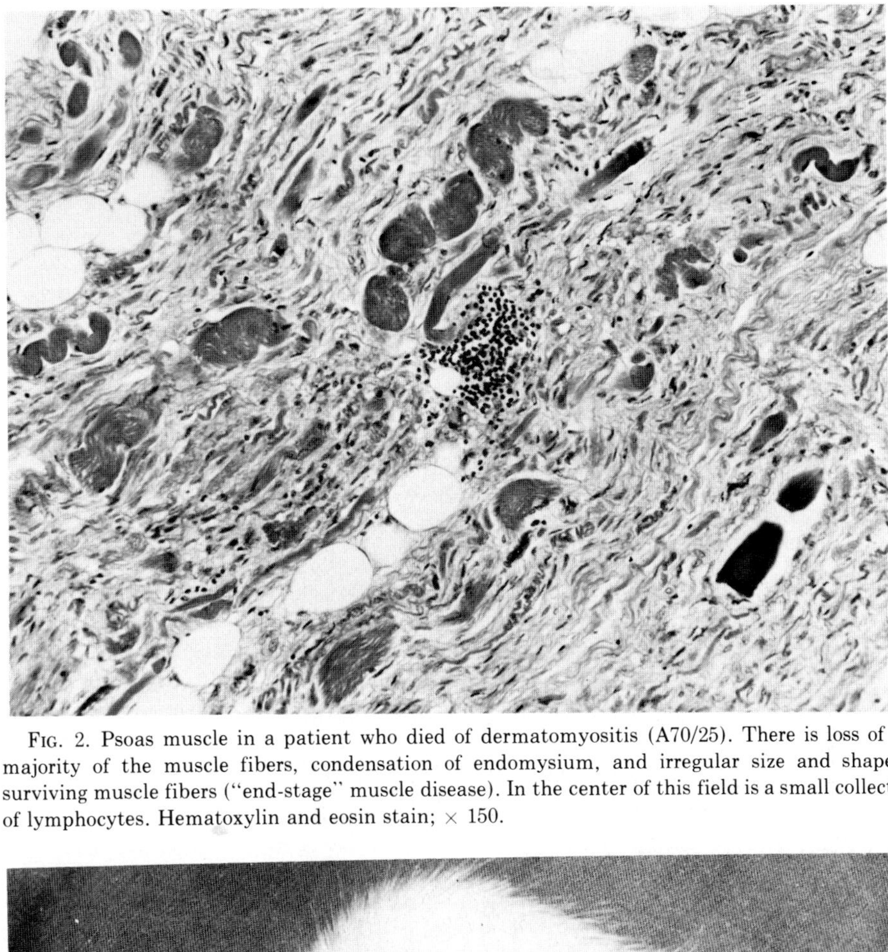

FIG. 2. Psoas muscle in a patient who died of dermatomyositis (A70/25). There is loss of the majority of the muscle fibers, condensation of endomysium, and irregular size and shape of surviving muscle fibers ("end-stage" muscle disease). In the center of this field is a small collection of lymphocytes. Hematoxylin and eosin stain; × 150.

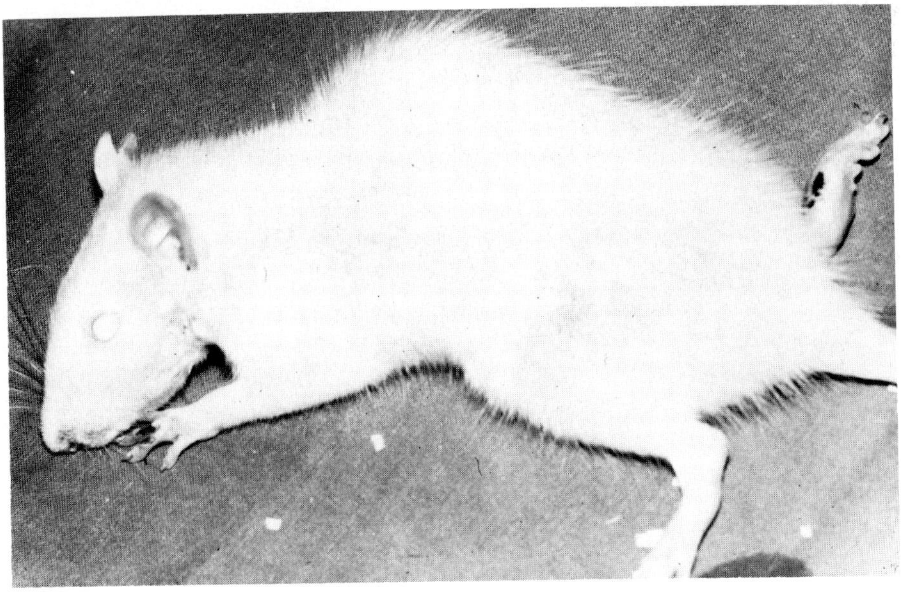

FIG. 3. Rat with profound muscle weakness after sensitization to rabbit muscle and Freund's adjuvant by repeated footpad injections.

myofibrils which are rarified due to loss of myofilaments, and Z bands are obscured. Other changes consist of mitochondrial alterations and dilation of the sarcoplasmic reticulum together with lipid vacuoles and myelin "bodies." It is noteworthy that the sarcolemma remains intact.[34] Macrophages are observed in the necrotic areas and regenerating muscle fibers are common. Regeneration which is related to the intact sarcolemmal membrane is an almost invariable element of the lesion.[27] Regeneration *ad integro* occurs over a period of time when treatment with prednisone is successful.[28] Lymphocytes, plasma cells, mononuclear cells, and less commonly neutrophils and eosinophils, complete the microscopic picture. The histologic changes are therefore those of a necrotizing myopathy and closely resemble the changes which occur in a wide variety of conditions including nutritional disorders in animals.[20]

Additional microscopic features sometimes encountered in biopsy specimens are vacuolar changes in the sarcoplasm[32] and the presence of islands of atrophied fibers. Grouped atrophy, when present, suggests complicating peripheral nerve disease. This combination is most frequently observed when carcinoma is associated.[11] A very important feature of the lesions which may only be adequately observed in serial sections of muscle biopsies is the close proximity of the focal necrotic muscle lesions to the perivascular cellular infiltrates (Fig. 1).[1]

Necropsy studies, although less numerous than biopsy examinations, reveal a wide spectrum of changes. The final morphology in fatal cases may be termed "end-stage" muscle disease, which is characterized by over-all loss of muscle fibers and replacement by connective tissue and fat. The few surviving fibers show excessive variation in size and shape, and nuclei are irregularly distributed. Leukocytic infiltrates are generally in the form of focal collections of round cells (Fig. 2). Changes in other orgnas are the result of intercurrent disease or are due to associated conditions, *e.g.*, Hashimoto's thyroiditis, rheumatoid arthritis, atrophic gastritis, and polyarteritis nodosa have been observed in 1 patient.[29] It is also relevant that in a necropsy study of 122 patients with neoplastic disease, 21 showed changes of necrosis and degeneration in the vastus lateralis or deltoid muscles. The findings suggested a high subclinical incidence of polymyositis.[21]

## EXPERIMENTAL OBSERVATIONS

Although attempts to produce myositis by immunologic methods were undertaken as early as 1929,[17] it was not until 1956 that further interest was shown when Pearson briefly reported "focal polymyositis" in rats that had received muscle mixed with Freund's adjuvant.[31] However, more attention was devoted to an associated arthritis which could be produced by the administration of adjuvant alone. A "dystrophy-like" picture in rabbits and guinea pigs receiving homologous and heterologous muscle mixed with Freund's adjuvant was reported in 1962. However, the changes described were primarily atrophic and did not resemble the histologic findings in cases of polymyositis.[38]

Dawkins,[9, 10] in a well controlled series of experiments undertaken in 1963, conclusively demonstrated that myositis could be induced on an autoallergic basis in the guinea pig. His work included a study of the "graft *versus* host" reaction where myopathic changes in immunologically "runted" mice were observed to be widespread. Muscle fiber necrosis was a prominent change in the lesions. Similar muscle lesions did not occur in the muscles of "runted" rats. The mouse "runts" were produced in neonates, less than 24 hours old, by the intraperitoneal injection by about $50 \times 10^6$ lymph node or spleen cells from adult donor mice of different strain to the recipient. Necrotic changes occurred in many parenchymal organs and as well as in skeletal muscle.

Lesions similar to those immunologically induced in guinea pigs were also produced in rats by repeated footpad injections of homogenized rabbit muscle and Freund's adjuvant.[18] Rarely, rats treated in this manner showed signs of muscle weakness (Fig. 3). Histologically, multifocal myositic changes with muscle fiber necrosis and a variable leukocytic infiltrate were consistent (Fig. 4). Careful light microscopic analysis of the development of the muscle changes showed that leukocytic infiltrates which were characteristically perivascular preceded signs of muscle fiber necrosis. The lesions, although generally similar to the changes which occurred in human polymyositis, tended to be more focal. Electron microscopy of the experimental disease in rats revealed widespread alterations in the organelle systems, with rarifaction of myofibrils due to loss of myofilaments (Fig. 5), membranous structures (Fig. 6), and dilated vesicles. The ultrastructural changes closely resembled those which occur in biopsy specimens of patients with polymyositis.[34]

*In vitro* studies demonstrated that lymph node cells of rat sensitized to rabbit muscle are cytopathic to tissue cultures of fetal rat skeletal muscle.[18] In this

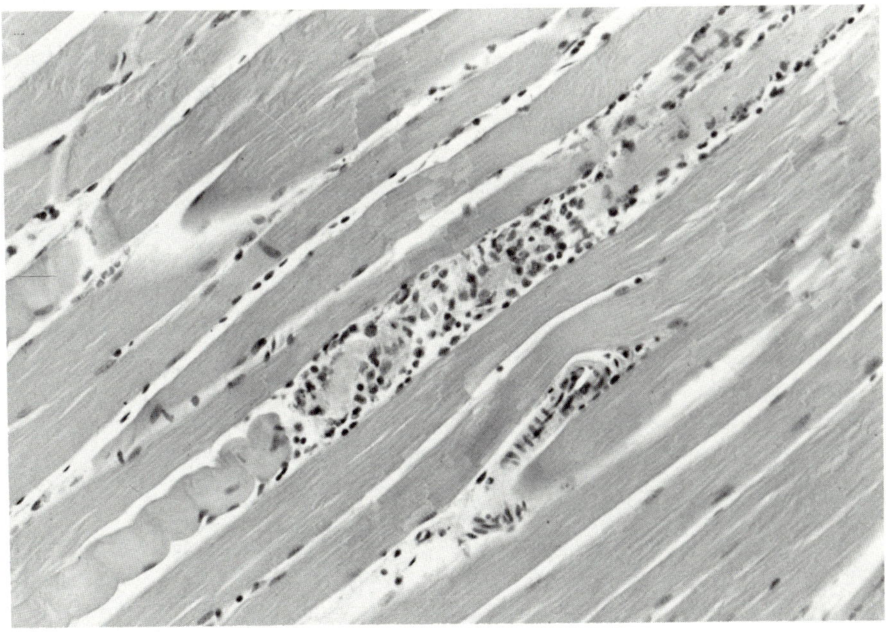

FIG. 4. Focal necrosis of a muscle fiber in the pectoralis major muscle of a rat sensitized to rabbit muscle and Freund's adjuvant by repeated footpad injections. The necrotic focus in the center of the field is associated with an infiltrate of round cells and macrophages. Hematoxylin and eosin stain; × 450.

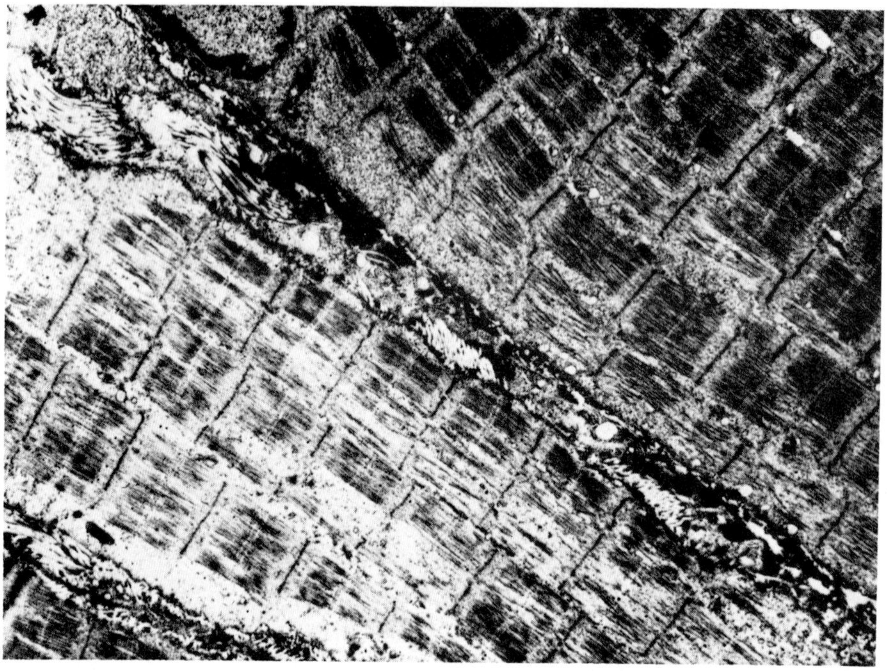

FIG. 5. Electron micrograph of vastus lateralis muscle of rat sensitized to rabbit muscle showing rarifaction of myofibrils due to loss of myofilaments and disappearance of mitochondria and sarcoplasmic reticulum. The changes represent early necrosis. Uranyl acetate-lead citrate stain; × 4000.

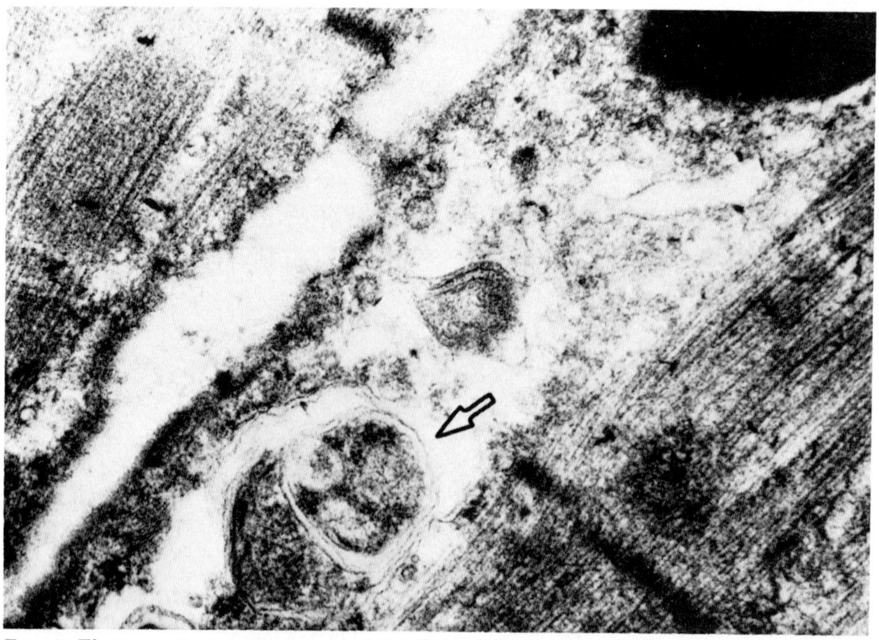

FIG. 6. Electron micrograph of vastus lateralis muscle of rat sensitized to rabbit muscle showing loss of mitochondria dilatations of the sarcoplasmic reticulum, disappearance of myofilaments, and the presence of membranous structures. Uranyl acetate-lead citrate stain; × 16,000.

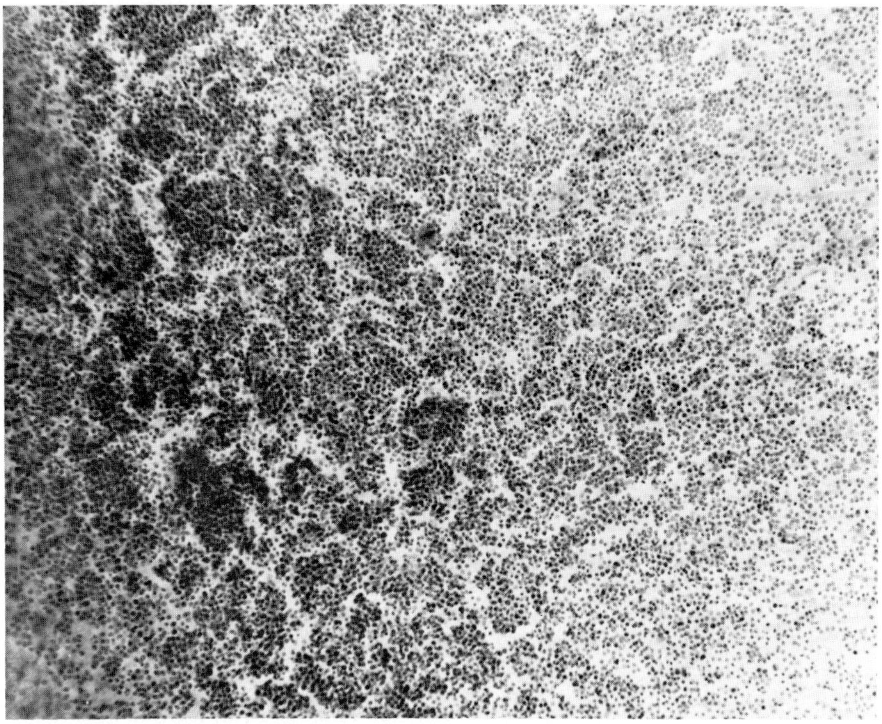

FIG. 7. Destruction of fetal rat skeletal muscle *in vitro* by the addition of lymph node cells of rats sensitized to rabbit muscle. Phosphotungstic acid-hematoxylin stain; × 150.

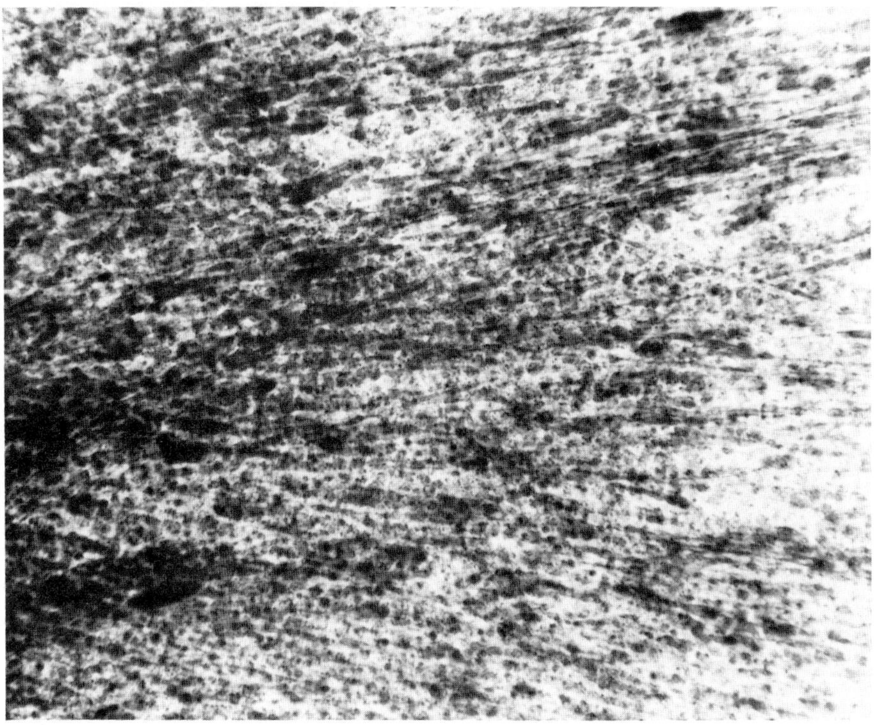

FIG. 8. Fetal rat skeletal muscle fibers *in vitro* remain intact after treatment with lymph node cells of a rat sensitized to rabbit kidney and adjuvant (compare with Fig. 7). Phosphotungstic acid-hematoxylin stain; × 150.

work, approximately $4 \times 10^6$ lymph node cells derived from rats given three injections of homogenized rabbit muscle and adjuvant were added to each tissue culture. Of 40 cultures treated with cells from rats sensitized to muscle, 38 were destroyed within 2 to 6 days (Fig. 7). In contrast, control studies in which tissue cultures were treated with lymph node cells of animals injected with adjuvant only, or which were sensitized to rabbit kidney as well as cells of untreated rats, showed destructive changes in a much smaller number (Fig. 8). The results indicated that suitably sensitized cells are able to destroy muscle *in vitro* and suggested that a similar pathogenic mechanism may occur in human polymyositis. The results of the animal experiments are summarized in Table 1.

A logical development of the passive "*in vitro*" transfer work in the rat was the exploration for similar properties of human lymphoid cells obtained from patients with polymyositis and related "autoimmune" diseases. In a large number of such experiments, in which the target tissue was human fetal skeletal muscle, similar destructive changes occurred. The lymphoid cells were concentrated from about 20 ml of blood obtained by venipuncture and were added in lots of approximately 0.5 to $1.0 \times 10^6$ to each culture. The muscle was grown on collagen-coated coverslips in Maximov assemblies and, depending on the supply of satisfactory cultures, approximately six were used for the cells of each patient. The cultures were observed microscopically daily and on the 6th day they were fixed and stained with phosphotungstic acid-hematoxylin. The cytodestructive effects were essentially similar to those observed in the experimental disorder (Fig. 9). The majority of cultures treated with lymphoid cells from normal subjects serving as controls remained intact (Fig. 10). The cells added to the cultures with cytopathic changes rapidly underwent enlarge-

TABLE 1. RAT EXPERIMENTS: EFFECT OF ADDING TO FETAL RAT SKELETAL MUSCLE TISSUE CULTURE $4 \times 10^6$ LYMPH NODE CELLS OR 0.02 ML OF SERUM FROM RATS SENSITIZED TO RABBIT MUSCLE OR FROM NORMAL RATS

| Donor rats (groups of 10) | Proportion of donor rats showing muscle lesions in : | | Donor material added to culture | Number of cultures | |
|---|---|---|---|---|---|
| | fore-limb | hind limb | | destroyed by 6th day | intact on 6th day |
| Received injections of rabbit muscle plus adjuvant | 9/10 | 10/10 | Cells<br>Serum | 38<br>4 | 2<br>16 |
| Received injections of saline only | 0/9 | 0/9 | Cells<br>Serum | 7<br>2* | 33<br>18 |
| Received injections of adjuvant only | 2/10 | 3/10 | Cells<br>Serum | 6<br>3 | 34<br>17 |
| Received injections of rabbit kidney plus adjuvant | 0/10 | 1/10 | Cells<br>Serum | 10**<br>3 | 30<br>17 |

* *One of these two cultures was destroyed by bacterial contamination*
** *Two of these ten cultures were destroyed by bacterial contamination*

TABLE 2. HUMAN CELL EXPERIMENTS (TARGET TISSUE: HUMAN FETAL SKELETAL MUSCLE

| Disorder | Number of patients | Number of cultures | |
|---|---|---|---|
| | | Destroyed by 6th day | Intact on 6th day |
| Polymyositis | 7 | 33 | 9 |
| Systemic Lupus | 6 | 27 | 9 |
| Non-specific Muscle Pain | 4 | 8 | 17 |
| NORMAL CONTROLS | 8 | 6 (5)* | 29 |

* Contaminated with fungi.

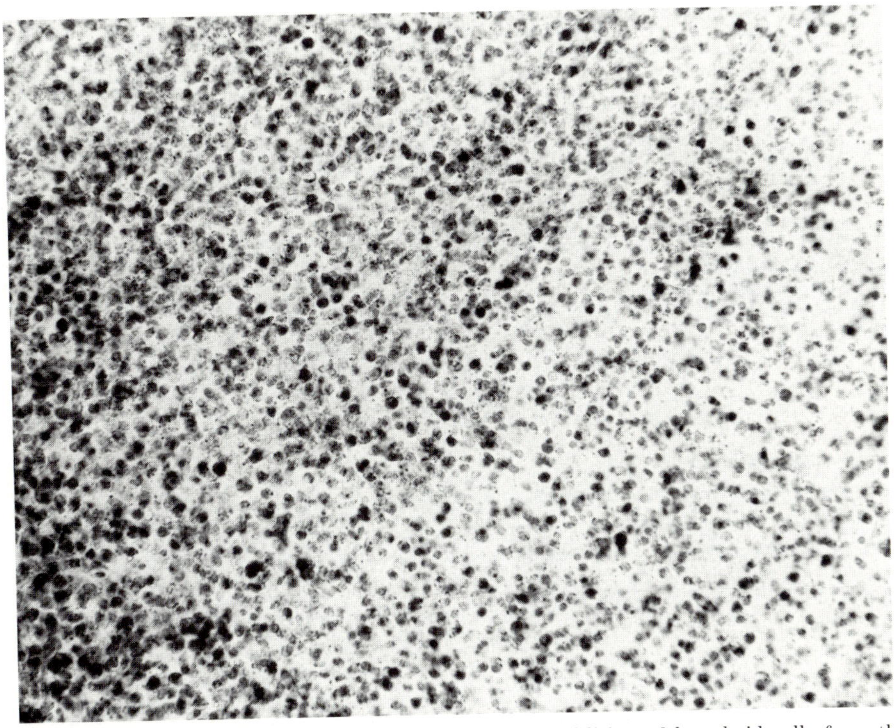

FIG. 9. Human fetal skeletal muscle destroyed by the addition of lymphoid cells from the peripheral blood of a patient with polymyositis. The cells which remain are enlarged and contain heavily staining material (compare with Fig. 10). Phosphotungstic acid-hematoxylin stain; × 150.

ment and transformation, whereas the lymphocytes of the normal controls added to the cultures which remained intact generally remained small and inactive. The results are summarized in Table 2.

## DISCUSSION

The muscle changes which occurred in the immunologically induced disease in the experimental animal strongly resembled those which arose "spontaneously" in human polymyositis. Differences were minor and included a more focal character to the changes. In some instances there was a more acute cellular response with polymorphs being more abundant than in the human disease. This laboratory model for human polymyositis was subsequently studied in a number of laboratories.[33, 37] Similarly, the cytopathic properties of peripheral blood lymphocytes of patients with polymyositis were recently corroborated by Currie and Walton.[7] Analogous results are reported for a number of other human diseases including multiple sclerosis[24] and idiopathic polyneuritis.[2] In this work, Arneson, Winkler, and Hadler[2] extended their studies of experimental allergic neuritis to include human buffy coat cells of patients with idiopathic polyneuritis. Such cells were myelinodestructive in 48 to 96 hours. It was evident, therefore, that cytopathic mechanisms similar to those demonstrated for polymyositis also existed for other human diseases.

Wilson[43] and Maclennan and Loewi[25] studied the mechanisms involved in cellular cytotoxicity while Dumonde *et al.*[12] used the term "lymphokine" for the low molecular weight soluble factors generated by antigen-activated lymphocytes which are believed to be the toxic agents and are different from classical antibodies.

The results of the human lymphocyte work reported here render it possible to formulate an etiology and pathogenesis for human polymyositis. Thus, in this disease, sensitized lymphoid cells accumulate in the perivascular and interstitial spaces of skeletal muscle and become activated on contact with muscle antigens leading to the secretion of albumen-like factors ("lymphokines"[12]) which cause the muscle fiber necrosis. Other histologic elements in the lesions, *e.g.*, regeneration, are an inherent response of muscle to injury.[17] In the more chronic stages of polymyositis the architectural alterations with loss of fibers, fat, and fibrous tissue replacement are due to a continuation of the destructive process, and the lesions simply represent "end-stage" muscle disease. As such, the "burnt-out" lesions have much in common with the muscle changes in progressive muscular dystrophy.

Difficulties in this hypothesis are concerned, firstly, with differences in the natural history between the experimental and spontaneously occurring diseases and, secondly, with the initiating mechanism in the human disorder. For instance, in the experimental condition the initiating factor is the repeated injection of foreign muscle combined with adjuvant. Such myositic changes which are induced are also self-limiting with recovery occurring on cessation of the injections. In contrast, the naturally occurring disease in man is seemingly self-perpetuating. However, it may be arrested by use of steroids which block the immune mechanism. It is possible that the "triggering" agent for the autoimmune process in man is an infection with "complete" or "incomplete" virus. In support of this suggestion is the observation of Chou[6] who found virus-like structures in a muscle biopsy of a patient with polymyositis. Virus-like structures are also reported in a patient with systemic lupus erythematosus and suggest that a similar role for a virus exists for this condition.[22] The experimental findings of Stanley and Walters[35] with reovirus 3 support the concept of virus induction of both autoimmune disease and neoplasia.

FIG. 10. Human fetal skeletal muscle remains intact after addition with lymphoid cells of a normal subject. Phosphotungstic acid-hematoxylin stain; × 150.

Another difficulty in the present theory is related to myasthenia gravis, which is clinically diverse from polymyositis but for which much evidence suggesting an immunologic origin has accumulated.[17] However, Dawkins,[10] in his experimental work, demonstrated increased curare sensitivity in his guinea pigs sensitized to muscle. Other evidence for a "bridge" between myasthenia gravis and polymyositis exists on clinical grounds, as patients with both diseases occurring in association with thymoma have been found.[23] Associated myasthenia gravis and systemic lupus erythematosus have also been documented.[15]

The obviously complex nature of the problem is underlined by the frequent finding of serum antibodies to muscle in patients with a wide variety of disorders and point to the nonspecificity of the findings.[5, 36] Such reports suggest that cellular immune factors, unassociated with classical antibodies, on the lines presented above are more important in the pathogenesis of these diseases.

Recent application of the immune theory is the use of immunosuppressive drugs which have been used clinically with good results in patients with polymyositis.[26, 30] The search for a possible virus underlying polymyositis is currently in progress. Similarly, the cytodestructive properties of peripheral blood lymphocytes in patients with a variety of diseases using the *in vitro* methods outlined above may shed further light on the pathogenesis of these conditions.

## SUMMARY

Clinical, morphologic, and experimental evidence is presented in support of a possible immune etiology for human polymyositis. For the first time direct and objective evidence exists in the human subject for this suggestion on the basis of observed cytodestructive effects *in vitro* of lymphoid cells derived from patients with the disease. Presumably, similar cytopathic effects occur *in vivo* as a result of the perivenular accumulation of the lymphoid cells and are the cause of the necrotic lesions in muscle. Unexplained, at present, is the initiating or "triggering" mechanism for such a process, but the recently observed virus particles in polymyositis may be important in this regard. This report, in addition to providing evidence of an immune etiology of human polymyositis, describes a method that may be applicable to a wide variety of other human diseases for which the cause is at present unknown.

## ACKNOWLEDGMENTS

Gratitude is expressed to Dr. R. D. Adams for discussions and to Dr. K-E. Astrom for assistance with the electron microscopy. Dr. Arneson provided facilities and Mr. H. Upenieks prepared the photomicrographs.

## REFERENCES

1. Adams, R. D.: Personal communication, 1969.
2. Arneson, B. G. W., Winkler, G. F., and Hadler, N. M.: Cell-mediated demyelination of peripheral nerve in tissue culture. *Lab. Invest. 21:* 1–10, 1969.
3. Arundell, F. D., Wilkinson, R. D., and Haserick, J. R.: Dermatomyositis and malignant neoplasms in adults. *Arch. Dermatol. 82:* 772–775, 1960.
4. Barwick, D. D., and Walton, J. N.: Polymyositis. *Am. J. Med. 35:* 646–660, 1963.
5. Caspary, E. A., Gubbay, S. S., and Stern, G. M.: Circulating antibodies in polymyositis and other muscle-wasting disorders. *Lancet 2:* 941, 1964.
6. Chou, S. M.: Myxovirus-like structures in a case of human chronic polymyositis. *Science 158:* 1453–1455, 1967.
7. Currie, S., and Walton, J. N.: Personal communication, 1969.
8. Curtis, A. C., Heckaman, H. J., and Wheeler, A. H.: A study in the autoimmune reaction in dermatomyositis. *J. A. M. A. 178:* 571–573, 1961.
9. Dawkins, R. L.: Experimental myositis associated with hypersensitivity to muscle. *J. Pathol. Bacteriol. 90:* 619–625, 1965.
10. Dawkins, R. L.: Immunofactive disease of muscle. Thesis, University of Western Australia, 1963.
11. Denny-Brown, D.: Primary sensory neuropathy with muscular changes associated with carcinoma. *J. Neurol. Neurosurg. Psychiatry. 11:* 73–87, 1948.
12. Dumonde, D. C., Wolstencroft, R. A., Panayi, G. S., Matthew, M., Morley, J., and Howson, W. T.: "Lymphokines." Non-antibody mediators of cellular immunity generated by lymphocyte activation. *Nature (Lond.) 224:* 38–42, 1969.
13. Eaton, L. M.: The perspective of neurology in regard to polymyositis; study of 41 cases. *Neurology (Minneap.) 4:* 245–263, 1954.
14. Grace, J. T., and Dao, T. L.: Dermatomyositis in a cancer; a possible aetiological mechanism. *Cancer 12:* 648–650, 1959.
15. Hausmanowa-Petrusewicz, I., Chorzelski, T., and Strugalska, H.: Three year observation of a myasthenic syndrome concurrent with other auto-immune syndromes in a patient with thymoma. *J. Neurol. Sci. 9:* 273–284, 1969.
16. Jackson, H.: Myositis universalis acuta infectiosa with a case. *Boston Med. Surg. J. 116:* 498–500, 1887.
17. Kakulas, B. A.: Experimental myopathies. In *Disorders of Voluntary Muscle*, ed. by Walton, J. N., 2nd ed., pp. 363–384. London, J. & A. Churchill Ltd., 1969.
18. Kakulas, B. A.: Destruction of differentiated muscle cultures by sensitized lymphoid cells. *J. Pathol. Bacteriol. 91:* 495–503, 1966.
19. Kakulas, B. A.: The pathogenesis of human muscle disease. In *Muscle Diseases* (Proceedings

of International Congress on Muscle Diseases, Milan 1969), ed. by Walton, J. N., Canal, N., Scarlato, G., and Gleave, J. R. W., pp. 377–382. Amsterdam, Exerpta Medica, 1970.

20. Kakulas, B. A.: Regeneration of skeletal muscle in the Rottnest quokka. *Aust. J. Biol. Med. Sci. 44:* 673–688, 1966.
21. Kakulas, B. A., and Mastaglia, F. L.: The type and incidence of lesions in a human necropsy survey of skeletal muscle. *Proc. Aust. Assoc. Neurol. 4:* 35–47, 1966.
22. Kawano, K., Miller, L., and Kimmelstiel, P.: Virus-like structures in lupus erythematosus. *New Engl. J. Med. 281:* 1228–1229, 1969.
23. Klein, J. J., Gottlieb, A. J., Mones, R. J., Appel, S. H., and Osserman, K. E.: Thymoma and polymyositis. Onset of myasthenia gravis after thymectomy. Report of two cases. *Arch. Intern. Med. 113:* 142–152, 1964.
24. Koprowski, H., and Fernandes, M. V.: Autosensitization reaction *in vitro*. Contactual agglutination of sensitized lymph node cells in brain tissue culture accompanied by destruction of glial elements. *J. Exp. Med. 116:* 467–476, 1962.
25. Maclennan, I. C., and Loewi, G.: The effect of peripheral lymphocytes from patients with inflammatory joint disease on human target cells *in vitro*. *Clin. Exp. Immunol. 3:* 385–391, 1968.
26. Malaviya, A. N., Many, A., and Schwartz, R. S.: Treatment of dermatomyositis with methotrexate. *Lancet 2:* 485–488, 1968.
27. Mastaglia, F. L., and Kakulas, B. A.: A histological and histochemical study of skeletal muscle regeneration in polymyositis. *J. Neurol. Sci. 10:* 471–487, 1970.
28. Mastaglia, F. L., and Kakulas, B. A.: The natural history of human muscle diseases studied by means of serial biopsy. *Proc. Aust. Assoc. Neurol. 5:* 537–543, 1968.
29. Mastaglia, F. L., and Kakulas, B. A.: Rheumatoid arthritis, polymyositis, polyarteritis, atrophic gastritis and Hashimoto's thyroiditis. *Med. J. Aust. 1:* 1135–1136, 1967.
30. McFarlin, D. E., and Griggs, R. C.: Treatment of inflammatory myopathies with azathioprine. *Trans. Am. Neurol. Assoc. 93:* 244–246, 1968.
31. Pearson, C. M.: Development of arthritis, periarthritis and periostites in rats given adjuvants. *Proc. Soc. Exp. Biol. Med. 91:* 95–101, 1956.
32. Pearson, C. M.: Polymyositis and related disorders. In *Disorders of Voluntary Muscle*, ed. by Walton, J. N., 2nd ed., pp. 501–539. London, J. & A. Churchill, Ltd., 1969.
33. Pearson, C. M., Peter, J. B., and Morgan, G.: Personal communication, 1969.
34. Rose, A. L., Walton, J. N., and Pearce, G. W.: Polymyositis. An ultramicroscopic study of muscle biopsy material. *J. Neurol. Sci. 5:* 457–472, 1967.
35. Stanley, N. F., and Walters, M. N-I.: Virus induction of auto-immune disease and neoplasia. *Lancet 1:* 962–963, 1966.
36. Stern, G. M., Rose, A. L., and Jacobs, K.: Circulating antibodies in polymyositis. *J. Neurol. Sci. 5:* 181–183, 1967.
37. Takayanagi, T.: Immunohistological studies of experimental myositis in relation to human polymyositis. *Folia Psychiatr. Neurol. Jap. 21:* 118–127, 1967.
38. Tal, C., and Liban, E.: Experimental production of muscular dystrophy-like lesions in rabbits and guinea pigs by an autoimmune process. *Br. J. Exp. Pathol. 43:* 525–529, 1962.
39. Unverricht, H.: Polymyositis acuta progressiva. *Z. Klin. Med. 12:* 533–549, 1887.
40. Wagner, E.: Ein fall von acuter polymyositis. *Dtsch. Arch. Klin. Med. 40:* 241–266, 1887.
41. Wagner, E.: Fall einer seltnen muskelkrankheit. *Arch. Heilkunde. 4:* 282–283, 1863.
42. Walton, J. N., and Adams, R. D.: *Polymyositis*. Edinburgh and London, E. & S. Livingstone, Ltd., 1958.
43. Wilson, D. B.: Quantitative studies on the behaviour of sensitized lymphocytes *in vitro*. 1. Relationship of the degree of destruction of homologous target cells to the number of lymphocytes and to the time of contact in culture and consideration of the effects of isoimmune serum. *J. Exp. Med. 122:* 143–166, 1965.

## Chapter 25

# Myasthenia Gravis: Evidence for Thymus Regulation of Neuromuscular Transmission

GIDEON GOLDSTEIN

---

Experimental studies related to the disease myasthenia gravis have suggested the following relationship between the thymus and neuromuscular transmission (Fig. 1).[23]

1. The thymus is an endocrine organ which secretes a hormone for which the name thymin has been proposed.
2. Thymin physiologically inhibits neuromuscular synaptic transmission and, in excess, causes both neuromuscular block and myopathy, as found in myasthenia gravis and experimental autoimmune thymitis.
3. An abnormal reaction of the immune system against the thymus, autoimmune thymitis, causes the release of excess thymin into the circulation in myasthenia gravis and experimental autoimmune thymitis.

This proposed relationship of the thymus to myasthenia gravis closely resembles that of the thyroid to thyrotoxicosis and perhaps myasthenia gravis could be renamed "thymotoxicosis" to emphasize that the pathogenesis of the neuromuscular lesions of myasthenia gravis involves hypersecretion of a thymic hormone.

### EXPERIMENTAL MYASTHENIA GRAVIS

The above interpretations are based largely on experimental studies[21] which were developed to test hypotheses derived from the study of human patients with myasthenia gravis.[15, 16] Figure 2 shows a plan of the two models of experimental myasthenia gravis that were developed.

In the first, experimental autoimmune thymitis, advantage was taken of the ill understood but well documented effect of Freund's complete adjuvant (an emulsion of water in oil containing killed mycobacterium) to produce a tissue-specific autoimmune disease. The immune system, stimulated by thymic antigens in Freund's complete adjuvant, reacted immunologically with the animal's own thymus to produce thymitis. Thymitis was associated with release of excess thymin from the thymus resulting in a partial neuromuscular block.[21]

In the second model, extracts of normal bovine thymus were injected parenterally to produce neuromuscular lesions directly,[17] the interpretation being that these extracts contained bovine thymin.

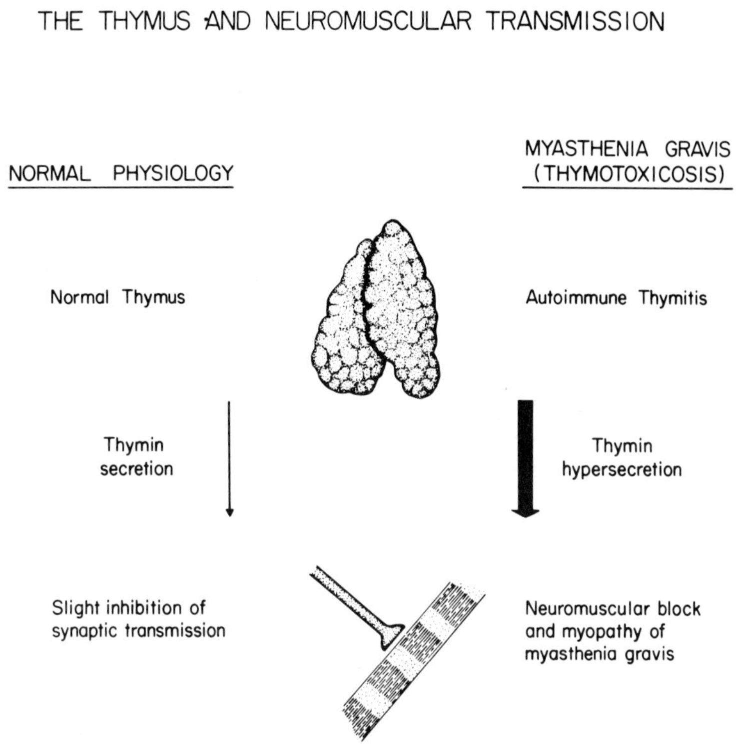


FIG. 1. Diagram showing proposed endocrine relation of thymus to neuromuscular transmission. Myasthenia gravis is considered to represent thymotoxicosis. (Reproduced from Goldstein and Hofmann[21] by permission of the Williams and Wilkins Co.)

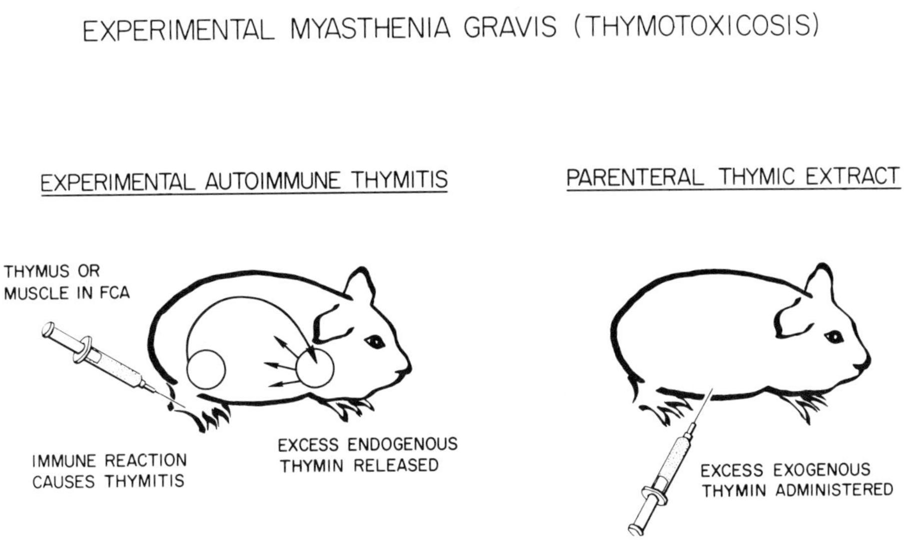


FIG. 2. Diagram showing interpretation of two models of experimental myasthenia gravis. Excess of a thymic hormone thymin, endogenous or exogenous, is considered to cause a neuromuscular lesion similar to that of myasthenia gravis. (Reproduced from Goldstein and Hofmann[21] by permission of the Williams and Wilkins Co.)

These models, interpreted in parallel with the findings in human myasthenia gravis, provide evidence that suggests an endocrine function for the thymus in regulating neuromuscular transmission.

The similarities between the experimental models and human myasthenia gravis will be considered in relation to findings in the thymus, autoimmune reactions with thymus, the neuromuscular lesions related to thymitis, and the evidence that a thymic humoral factor, thymin, affects neuromuscular transmission.

## THE THYMUS (TABLE 1)

The thymus in 10% of patients with myasthenia gravis contains a thymoma.[4] The thymus adjacent to this thymoma shares certain histopathologic features with the nontumorous thymus found in the remaining 90% of patients with maysthenia gravis.[1, 5, 16] In the absence of a thymoma, the thymic weight is normal (Fig. 3)[5, 44]; it is *not* hyperplastic by the criterion of weight and early classifications of the thymus in myasthenia gravis as hyperplastic derived from erroneous concepts concerning the normal weight of the thymus.[15, 23] The thymus undergoes rapid involution with the stress of terminal illness, so that the "small" thymus found in most hospital necropsies represents involution whereas the "large" thymus found with sudden death or at operation represents the "normal" weight of the thymus.[23] In experimental autoimmune thymitis, as in myasthenia gravis, the weight of the thymus is normal.[26, 34]

In myasthenia gravis the thymus shows the normal degree of involution with age, with a normal ratio of cortex to medulla and a normal density of Hassall's corpuscles.[22] The histologic abnormalities are the presence, in the medulla, of an increased number of lymphocytes[44] and plasma cells[14, 18] and the presence of numerous large germinal centers (Fig. 4).[5, 16, 44] These changes have also been termed hyperplasia, but it would seem more appropriate to interpret them as thymitis by analogy with the similar interpretation of lymphocyte and plasma cell infiltrations with lymphoid germinal center formation in thyroid or salivary gland as thyroiditis or sialadenitis. In experimental autoimmune thymitis, there is an increased density of lymphocytes in the thymic medulla detectable in randomized coded sections (Fig. 5).[25, 26, 34] The plasma cells and lymphoid germinal centers of the human disease are not found.

An additional finding suggests that the increased medullary lymphocytes in

TABLE 1. THE THYMUS IN MYASTHENIA GRAVIS

| | Human | Experimental |
|---|---|---|
| Weight | Normal[5, 44a] | Normal[26, 34] |
| Histology | Normal age involution[22b] | Normal age involution[26, 34] |
| | In medulla—aggregates of lymphocytes,[44] plasma cells,[14, 18] and numerous large lymphoid germinal centers[5, 16, 44b] | In medulla—lymphocyte aggregates[26, 34] |
| Thymocyte transformation with phytohemagglutinin | Increased[35] | Increased[39] |

[a] Increased only with thymoma.

[b] Also found in thymus adjacent to thymoma.

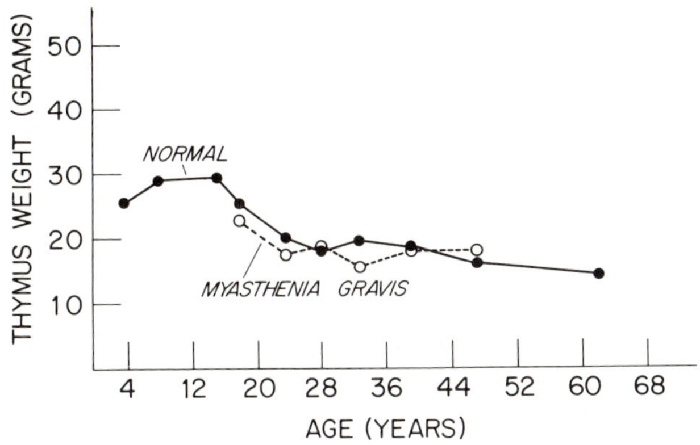


FIG. 3. Graph showing the mean thymic weight at various ages in normals (data of Hammar[28]) and patients with myasthenia gravis (data of Castleman and Norris[5]). The thymus is *not* enlarged in myasthenia gravis in the absence of a thymoma.

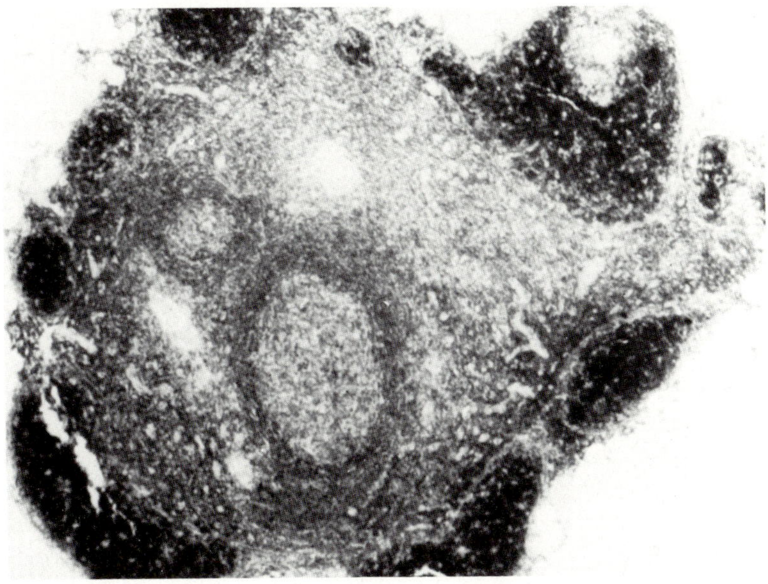

FIG. 4. Section of thymus adjacent to a thymoma in a male aged 44 with myasthenia gravis. There is normal age involution, the abnormality being an aggregation of lymphocytes (darker area) occupying half of the medulla and the presence of two lymphoid germinal centers. × 40.

the thymus in myasthenia gravis and experimental autoimmune thymitis represent an inflammatory infiltration. The mitogen phytohemagglutinin (PHA) causes peripheral or lymph node lymphocytes to transform *in vitro* to blast cells and to undergo mitotic division. Normal thymus lymphocytes are poorly responsive to PHA *in vitro*, but the thymus lymphocytes in myasthenia gravis[35] and in experimental autoimmune thymitis[39] were found to have an increased response *in vitro* to PHA, suggesting that a population of responsive lymphocytes was present within the thymus.

Thus the findings in the thymus in human myasthenia gravis suggest chronic inflammation or thymitis, both in the nontumorous thymus and in the thymus

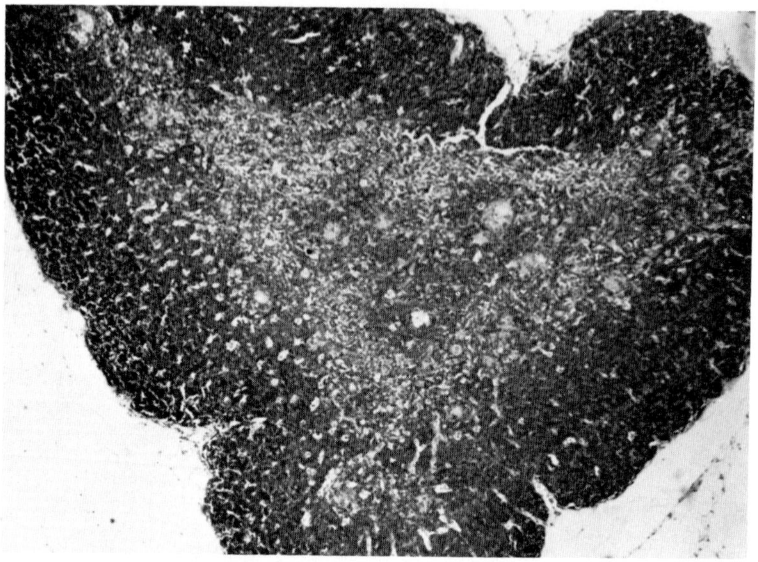

FIG. 5. Section of thymus from a guinea pig immunized with thymus in Freund's complete adjuvant. There are aggregates of lymphocytes forming darker areas in the central medulla. × 50.

adjacent to a thymoma. Furthermore, there is a close parallel between these thymic changes in myasthenia gravis and those found in experimentally induced autoimmune thymitis.

## IMMUNE REACTION TO THYMUS (TABLE 2)

The serum of at least 30% of patients with myasthenia gravis contains an autoantibody reactive with the striations of skeletal muscles.[47] This autoantibody also reacts with myoid cells of the thymus.[51, 52] These cells, normally present in the thymic medulla, have rounded heads and tails which may show the actual structural features of striated muscle[3, 12, 29, 46, 53]; they are antigenically identical with striated muscle (Fig. 6).[52] Thus, the myoid autoantibody of myasthenia gravis, which does not react with the neuromuscular synapse,[37] also represents a *thymic* autoantibody since it is reactive with thymic myoid cells which are normal components of the thymus (Fig. 6). A number of patients with myasthenia gravis also have serum autoantibodies reactive with antigens of nuclei,[2] thyroid cell cytoplasm and colloid,[50] and gastric parietal cells.[11]

In animals with experimental autoimmune thymitis, serum autoantibodies were detected in animals immunized with bovine tissues but not in guinea pigs immunized with homologous tissues.[24] Guinea pigs immunized with bovine muscle in Freund's complete adjuvant (FCA) developed thymitis (since muscle is equivalent to the thymic antigen of myoid cells) and also developed circulating autoantibodies reactive with muscle striations and thymic myoid cells.[25, 26, 34] Guinea pigs immunized with bovine thymus in FCA did *not* develop myoid antibodies but did develop antibodies reactive with the cytoplasm of epithelial cells and lymphocytes in the thymus.[25, 26, 34] Thus, animals with experimental autoimmune thymitis may have myoid antibody

TABLE 2. IMMUNE REACTION TO THYMUS IN MYASTHENIA GRAVIS

| | Human | Experimental |
|---|---|---|
| Serum antibodies to: | {Muscle striations[47] | {Muscle striations[26, 34] |
| | {Thymic myoid cells[48, 52] | {Thymic myoid cells[26, 34] |
| | Cell nuclei[2] | Thymic epithelial cells[26, 34] |
| | Thyroid cytoplasm[50] | Thymic lymphocytes[26, 34] or none[26, 34] |
| | Thyroid colloid[50] | |
| | Gastric parietal cells[11] or none[48] | |

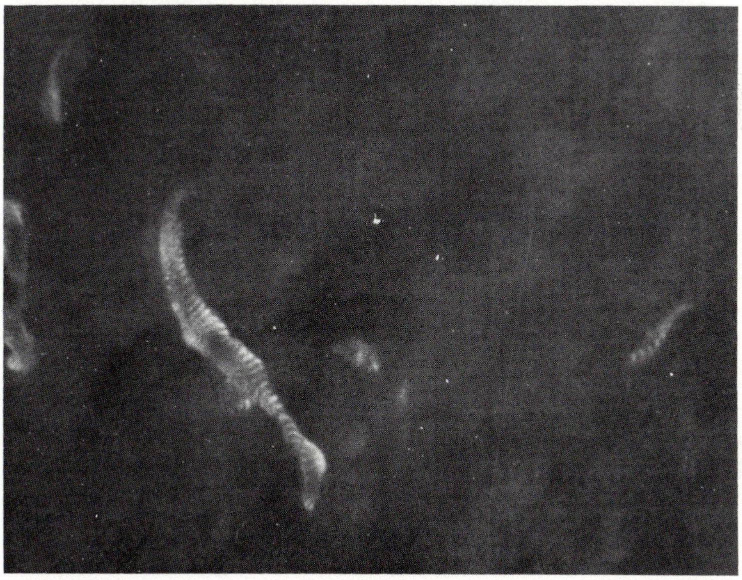

FIG. 6. Immunofluorescent section showing reactivity of myoid antibody from a patient with myasthenia gravis with the striations of myoid cells in the thymus of a 16-year-old boy dying of leukemia. × 900. (Reproduced courtesy of Dr. Senga Whittingham from Goldstein and MacKay[23] by permission of the Warren H. Green Co.)

similar to some patients with myasthenia gravis or, like other patients with myasthenia gravis, they may have other autoantibodies or no autoantibodies in the serum.

The serologic findings in myasthenia gravis are consistent with an autoimmune reaction to thymus and, in conjunction with the evidence for thymitis (*vide supra*), support the suggestion that autoimmune thymitis is the basic thymic lesion in masthenia gravis.[15, 16, 23] This conclusion is supported by the close parallel between the findings in human myasthenia gravis and those in animals when autoimmune thymitis is produced experimentally.

## INITIATION OF AUTOIMMUNE THYMITIS (TABLE 3)

In experimental autoimmune thymitis, the damaging autoimmune reaction is initiated by immunization with thymic antigens in FCA. Both thymus and muscle represent effective *thymic* antigens,[24] and heterologous,[25, 26, 34] isologous,[24] or homologous[19, 33] tissues can be used.

In human myasthenia gravis, thymitis may or may not be associated with a thymoma, which is present in 10% of cases.[4] Approximately one-third of all thymoma patients have myoid antibody in the serum[48] and, of these, one-half have myasthenia gravis, these being the patients in whom thymitis or germinal centers tend to be found in the adjacent thymus.[1, 5] In this group of patients, the tumor or thymoma appears to induce an autoimmune reaction to myoid antigen and thus initiates autoimmune thymitis.

In the remaining 90% of patients with myasthenia gravis in whom there is no thyoma, there is a clinical and serologic overlap with other autoimmune diseases,[23] but the initiating cause of the autoimmune thymitis remains unknown.

## NEUROMUSCULAR LESIONS (TABLE 4)

The main lesion producing the characteristic symptoms of myasthenia gravis is a defect in transmission at the neuromuscular synapse. There is controversy as to whether the lesion is primarily presynaptic,[7, 49] affecting acetylcholine production or release in the motor nerve terminal, or postsynaptic,[6, 27] affecting the postjunctional receptors of the motor end-plate. Despite this uncertainty as to the precise localization of the defect, there are a number of electrophysiologic

TABLE 3. INITIATION OF AUTOIMMUNE THYMITIS IN MYASTHENIA GRAVIS

| Human | Experimental |
|---|---|
| 10% thymoma[1, 4, 48]<br>90% unknown—clinical and serologic overlap with other autoimmune diseases[23] | Thymic antigens (thymus or muscle) in Freund's complete adjuvant[19, 24-26, 33, 34] |

TABLE 4. THE NEUROMUSCULAR LESION OF MYASTHENIA GRAVIS

| | Human | Experimental |
|---|---|---|
| Electromyography with supramaximal nerve stimulation: | | |
| Single stimulus, amplitude muscle action potential | Decreased[32] | Decreased[19, 24] |
| With anticholinesterase drugs | Increased[32] | Increased[25, 34] |
| 50/second nerve stimulus, amplitude successive muscle action potentials | Decremental[32] | Decremental[19, 25, 33, 34] |
| With anticholinesterase drugs | Sustained[32] | Sustained[25, 34] |
| Microelectrode studies of motor end-plates: | | |
| Resting membrane potential | Normal[8, 31] | Normal[19] |
| Miniature endplate potentials (mepps) | | |
| Amplitude | Decreased[8, 31] | Decreased[19] |
| Amplitude *in vitro* with washing | Decrease persists[8, 31] | Decrease persists[19] |
| Frequency | Normal[8, 31] | Normal[19] |
| *In vitro* neuromuscular preparation with nerve stimulation and muscle tension record: | | |
| Muscle tension developed | Reduced[30] | Reduced[17 a] |
| With washing | Reduction persists[30] | Reduction persists[17] |
| Effect of neostigmine | Marked increase[30] | Marked increase[17] |
| Effect of d-tubocurarine | Marked decrease[30] | Marked decrease[17] |
| Muscle histology | Myopathy[9, 13, 42] | Myopathy[17 a] |

[a] In model produced with bovine thymus extracts.

and pharmacologic findings in myasthenia gravis that clearly document the decreased safety margin of neuromuscular synaptic transmission in this disease.

By electromyography, with supramaximal nerve stimulation and recording of muscle action potentials, characteristic changes were found in human myasthenia gravis.[32] There was a small but statistically significant decrease in the muscle action potential following a single supramaximal nerve stimulus (Fig. 7), and, with repetitive stimulation, there was a decrement in the amplitude of successive muscle action potentials. Both of these abnormalities were improved by cholinesterase antagonists such as neostigmine or edrophonium chloride (Tensilon®). Furthermore, the muscle response to direct electrical stimulation of the muscle was unimpaired.

In animals with experimental autoimmune thymitis, similar electromyographic abnormalities were found. There was a reduced amplitude of the muscle action potential following a single nerve stimulus (Fig. 7),[19, 24] and this was significantly increased with neostigmine[25] or edrophonium chloride.[34] With repetitive nerve stimulation at 50 impulses per second there was a decrease in the amplitude of successive muscle action potentials (Fig. 8),[19, 24, 25, 33, 34] but with anticholinesterase drugs these were sustained.[25, 34] With direct stimulation of the muscle at 50 impulses per second, the amplitude of the muscle action potentials was sustained.[33] Thus, by all of these electromyographic and pharmacologic criteria, the neuromuscular block of human and experimental myasthenia gravis appeared similar.

Events at individual neuromuscular junctions have been studied in myas-

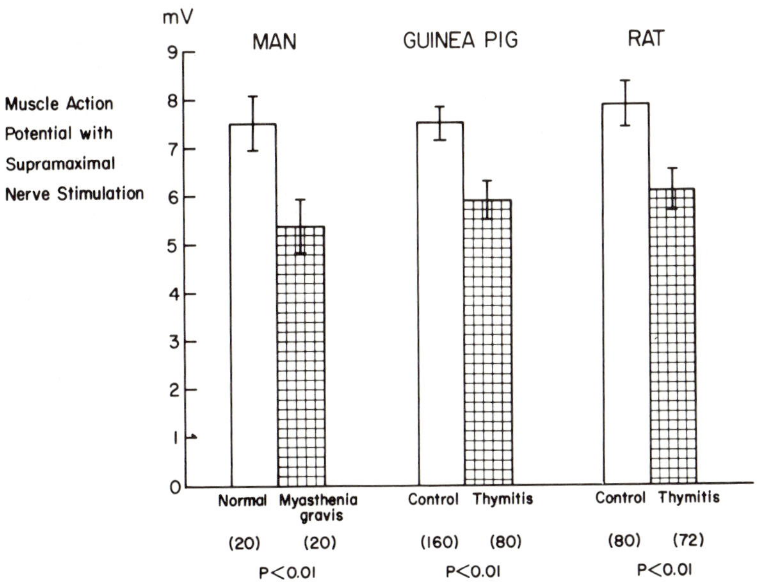


FIG. 7. Bar graph showing, in both human and experimental myasthenia gravis, a significant decrease in the amplitude of muscle action potentials following supramaximal nerve stimulation. Human data are from Johns, Grob, and Harvey,[32] guinea pig data from Goldstein, Strauss, and Pickeral,[24] and rat data from Goldstein and Hofmann.[19] Means and standard errors are shown together with the number of determinations in each group (in parentheses) and the *p* value of statistical significance of differences.

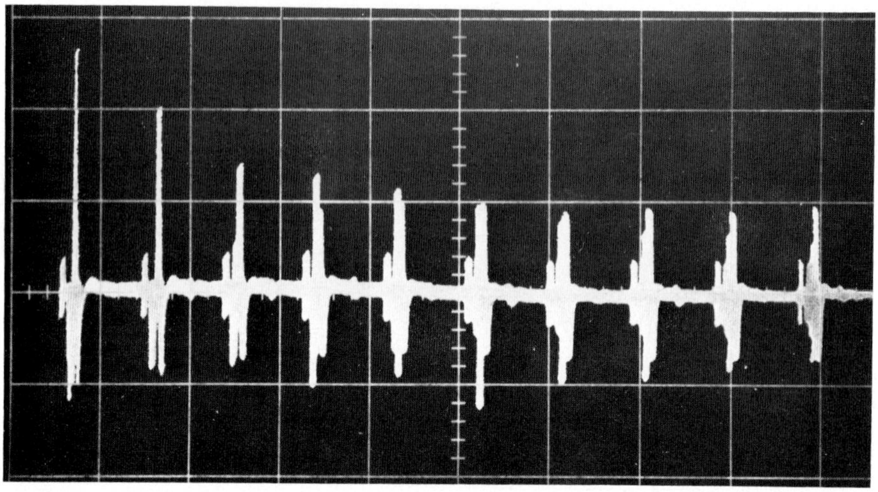

FIG. 8. Muscle action potentials, with 50 supramaximal nerve stimuli per second, of rat with experimental autoimmune thymitis. There is a decremental response similar to that of human myasthenia gravis, the amplitude of the tenth response being only 45% that of the first response. (One vertical division equals 2 millivolts; one horizontal division equals 20 milliseconds).

thenia gravis utilizing muscle biopsy specimens and capillary microelectrodes. Such microelectrodes record the resting membrane potential of the muscle cell and the spontaneous subthreshold depolarizations termed miniature end-plate potentials (mepps), which are believed to represent the depolarization caused by the smallest amount (quantum) of transmitter that can escape from a single release site.[10] In human myasthenia gravis, the findings were that the resting membrane potentials and mepp frequencies were normal, but there was a reduction in amplitude of mepps which persisted despite thorough washing *in vitro* of the muscle specimen.[8, 31] Similarly, in rats with experimental autoimmune thymitis, there were normal resting membrane potentials and normal mepp frequencies but decreased amplitude of mepps[19] (Fig. 9); these findings were present both *in vivo* and *in vitro*. The reduction in amplitude of mepps persisted *in vitro* despite thorough washing of the muscle specimen.[19]

The neuromuscular block of myasthenia gravis was further studied *in vitro* in an isolated nerve-muscle preparation obtained by biopsy from human patients.[30] The muscle tension was recorded following nerve stimulation, and the findings in human myasthenia gravis were (1) a decrease in muscle tension compared with controls, (2) a marked increase in muscle tension following the addition of neostigmine to the bath, and (3) a marked sensitivity of the preparation to the blocking action of d-tubocurarine in the bath.[30] These *in vitro* findings correspond with the known characteristics of the neuromuscular block in patients with myasthenia gravis, who show improvement with the anticholinesterase drug neostigmine and marked sensitivity to the neuromuscular blocking action of d-tubocurarine.

An *in vitro* nerve-muscle preparation was studied in the model of experimental myasthenia gravis produced by chronic injections of bovine thymus extract into guinea pigs.[17] The nerve of a phrenic nerve-hemidiaphragm preparation was stimulated, and the muscle tension was recorded. As with preparations from

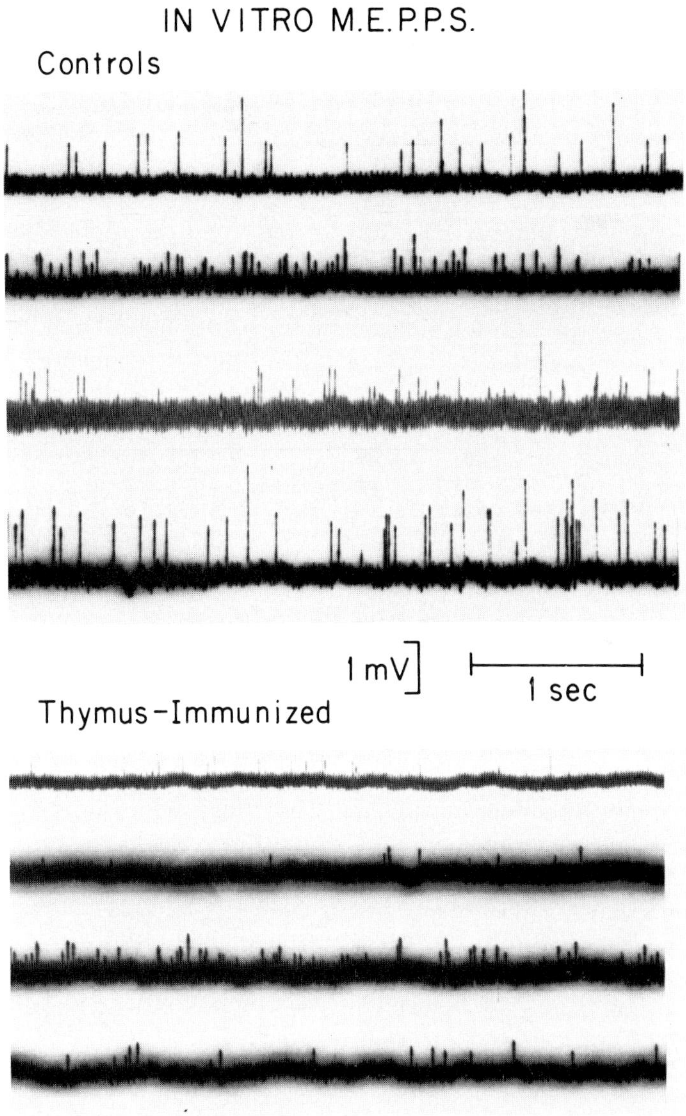


FIG. 9. Representative records of miniature end-plate potentials (mepps) *in vitro* showing normal frequency but decreased amplitude in animals with experimental autoimmune thymitis. This change is also found in human myasthenia gravis. (Reproduced from Goldstein and Hofmann[19] by permission of the British Medical Association.)

human myasthenia gravis, there was, with nerve stimulation, a decrease in the muscle tension developed, a marked improvement in tension with neostigmine in the bath, and a marked sensitivity to the blocking action of d-tubocurarine in the bath.[17]

Thus, by the various electromyographic, microphysiologic, and pharmacologic criteria studied there was a close parallel between the findings in human and experimental myasthenia gravis.

Finally, in addition to the classical functional neuromuscular lesion of myasthenia gravis, there are definite histophatologic changes in muscle in this disease.[42] Muscle lesions, which included muscle fiber atrophy or necrosis and infiltration with lymphocytes or macrophages, were detected by conventional light microscopy in 22 of 37 single biopsies from patients with myasthenia gravis.[13] Furthermore, by the use of histochemical techniques applied to fresh-frozen sections, abnormalities were found in all of 30 muscle biopsies from patients with myasthenia gravis.[9]

In animals with experimental autoimmune thymitis, there were either no muscle lesions by light microscopy[25, 26] or an occasional inflammatory focus in muscle.[34] In guinea pigs injected with extracts of bovine thymus, a severe myositis developed in addition to the neuromuscular block.[17]

These findings suggest that both in human and experimental myasthenia gravis there is a definite myopathic component in addition to the more striking neuromuscular block which produces the characteristic symptoms and response to drugs in myasthenia gravis.

## THYMIN—A THYMIC HORMONE DEPRESSING NEUROMUSCULAR TRANSMISSION

Thirty years ago the concept that the thymus secreted a substance affecting neuromuscular transmission was a popular one. McEachern in 1943[36] wrote "The thymus might conceivably produce some substance which acts as a chemical brake to muscular activity, especially during fetal development"; however, in a comprehensive study and review of the thymus in relation to myasthenia gravis, McEachern[36] was unable to find convincing experimental evidence for this concept. More recent findings in the fields of human and experimental myasthenia gravis (Table 5) provide support for the concept that the thymus is an endocrine gland and that a thymic hormone, termed thymin, depresses neuromuscular transmission.

The results of large series of thymectomies for myasthenia gravis are now available and these patients *do* show a striking benefit, with 70 to 80% having either complete remission or substantial improvement.[40, 43] This supports the concept of a thymic secretion causing neuromuscular block; an explanation for

TABLE 5. EVIDENCE IN MYASTHENIA GRAVIS FOR THYMIC SECRETION OF A CIRCULATING NEUROMUSCULAR BLOCKING SUBSTANCE—THYMIN

| | Human | Experimental |
|---|---|---|
| Thymectomy | 70% cured or markedly improved[40, 43a] | Before immunization—prevents NMB[25, 34b]<br>During NMB—speeds recovery from NMB[34] |
| Transfer of NMB | Transplacental—neonatal myasthenia gravis[38, 45c] | "Thymitis" thymus in Millipore chamber produces NMB[34]<br>Multiple histocompatible normal thymus grafts produce NMB[20] |
| Neonatal thymectomy | Not studied | Neuromuscular transmission facilitated[20] |
| Effect of thymus extracts | Not studied | Myasthenic NMB and myopathy[17] |

[a] Cases not responding may have ectopic thymus or irreversible neuromuscular lesions.
[b] NMB, neuromuscular block.
[c] Transplacental transfer of neonatal myasthenia gravis is *not* related to myoid antibody.

those cases not improving with thymectomy may lie in incomplete thymectomy, the frequent presence of ectopic thymus tissue in man,[41] or the presence of irreversible structural neuromuscular lesions with longstanding disease.[23]

In experimental autoimmune thymitis, thymectomy prior to immunization prevented the development of neuromuscular block, although these animals developed the same immune responses as thymus-intact animals and were capable of developing experimental autoimmune encephalomyelitis or thryoiditis if appropriately immunized.[25, 26, 34] When thymectomy was done at 2 weeks in animals with thymitis and neuromuscular block, the neuromuscular block cleared in 3 to 4 days, as contrasted with a usual persistence for 7 to 10 days in thymus-intact animals.[34] Thus, these experiments, in keeping with the thymectomy data in man, support the suggestion that a substance in thymus is released with thymitis to produce the neuromuscular lesions of myasthenia gravis. It should be noted, by way of contrast, that the model of experimental myasthenia gravis produced with repeated injections of extracts of bovine thymus (exogenous thymin) is *not* dependent on the recipient animal's own thymus as it can be produced in thymectomized guinea pigs.

Further evidence that a thymic secretion, thymin, is involved in myasthenic neuromuscular block comes from transfer "experiments." In man, neonatal myasthenia gravis[45] represents an "experiment of nature"; the newborn child of a mother with myasthenia gravis may develop transient myasthenic neuromuscular block, and it would seem that this is caused by a substance transmitted to the fetus through the placenta. This substance has been clearly shown *not* to be related to myoid antibody.[38, 45] In experimental studies, it was shown that transfer of a "thymitis" thymus within a Millipore chamber caused neuromuscular block in the recipient,[34] suggesting that the inflamed thymus was secreting a humoral substance which passed through the Millipore filter to cause neuromuscular block. Furthermore, animals bearing increased amounts of *normal* thymus in the form of histocompatible subcutaneous thymus grafts were shown to have depressed neuromuscular transmission,[20] suggesting that thymin was secreted by the normal thymus. This conclusion was supported by studies of neuromuscular transmission in neonatally thymectomized rats in which neuromuscular transmission was found to be facilitated.[20]

Finally, a substance causing myasthenic neuromuscular block was demonstrated in extracts of normal thymus.[17] Contrary to some previous reports, thymic extracts were *not* found to cause neuromuscular block directly *in vitro* but *were* found to produce myasthenic neuromuscular block if repeated injections were given *in vivo*.[17, 21] Thus, in these experiments, the substance termed thymin appeared to be unlike curare in that it did not rapidly produce neuromuscular block which could be reversed with washing. Rather the slow mode of action was more in keeping with the rate of action of most hormones in the body and the production of a myasthenic neuromuscular block which was not reversed *in vitro* with washing was also in keeping with the findings in the human disease (*vide supra*).

Experiments are in progress to purify and isolate thymin from extracts of thymus. Thymin appears to be a small protein, and it is hoped that methods could be developed to assay thymin in the serum. Hopefully, this should provide direct refutation or substantiation of the concept that the thymus secretes a hormone affecting neuromuscular transmission and that myasthenia gravis represents thymotoxicosis.

## SUMMARY

Human myasthenia gravis is reviewed in relation to the findings in two models of experimental myasthenia gravis.

The evidence from human and experimental myasthenia gravis suggests that the thymus has an endocrine function affecting neuromuscular transmission. The name thymin was proposed for a thymic hormone which at physiologic levels appears to depress slightly neuromuscular transmission slightly and in excess to cause the neuromuscular block and myopathy of myasthenia gravis.

In myasthenia gravis an autoimmune reaction in the thymus, autoimmune thymitis, is considered to cause the hypersecretion of thymin. By analogy with a similar pathogenesis in thyrotoxicosis, myasthenia gravis could be renamed "thymotoxicosis."

### REFERENCES

1. Bernatz, P. E., Harrison, E. G., and Clagett, O. T.: Thymoma. A clinicopathological study. *J. Thorac. Cardiovasc. Surg. 42:* 424–444, 1961.
2. Beutner, E. H., Witebsky, E., Ricken, D., and Adler, R. H.: Studies on autoantibodies in myasthenia gravis. *J. A. M. A. 182:* 46–58, 1962.
3. Bockman, D. E.: Myoid cells in adult human thymus. *Nature (Lond.) 218:* 286–287, 1968.
4. Castleman, B.: The pathology of the thymus gland in myasthenia gravis. *Ann. N. Y. Acad. Sci. 135:* 496–503, 1966.
5. Castleman, B., and Norris, E. H.: The pathology of the thymus in myasthenia gravis. A study of 35 cases. *Medicine (Balt.) 28:* 27–58, 1949.
6. Churchill-Davidson, H. C., and Richardson, A. T.: A study of neuromuscular transmission in one hundred cases of myasthenia gravis. In *Myasthenia gravis: The Second International Symposium Proceedings*, ed. by Viets, H. R., pp. 191–207. Springfield, Illinois, Charles C Thomas, 1961.
7. Desmedt, J. E.: Presynaptic mechanisms in myasthenia gravis. *Ann. N. Y. Acad. Sci. 135:* 209–246, 1966.
8. Elmqvist, D., Hofmann, W. W., Kugelberg, J., and Quastel, D. M. J.: An electrophysiological investigation of neuromuscular transmission in myasthenia gravis. *J. Physiol. (Lond.) 174:* 417–434, 1964.
9. Engel, W. K., and McFarlin, D. E.: In discussion of Muscle Lesions in Myasthenia Gravis by Fenichel, G. M. *Ann. N. Y. Acad. Sci. 135:* 68–78, 1966.
10. Fatt, P., and Katz, B.: Spontaneous subthreshold activity at motor nerve endings. *J. Physiol. (Lond.) 117:* 109–128, 1952.
11. Feltkamp, T. E. W.: *Idiopathic Autoimmune Diseases. A Study of Their Serological Relationship.* Amsterdam, Drukkerij, 1966.
12. Feltkamp-Vroom, T.: Myoid cells in human thymus. *Lancet 1:* 1320–1321, 1966.
13. Fenichel, G. M.: Muscle lesions in myasthenia gravis. *Ann. N. Y. Acad. Sci. 135:* 60–67, 1966.
14. Goldstein, G.: Plasma cells in the human thymus. *Aust. J. Exp. Biol. Med. Sci. 44:* 695–700, 1966.
15. Goldstein, G.: Thymitis and myasthenia gravis. *Lancet 2:* 1164–1167, 1966.
16. Goldstein, G.: Thymic germinal centres in myasthenia gravis. A correlative study. *Clin. Exp. Immunol. 2:* 103–107, 1967.
17. Goldstein, G.: The thymus and neuromuscular function. A substance in thymus which causes myositis and myasthenic neuromuscular block in guinea pigs. *Lancet 2:* 119–122, 1968.
18. Goldstein, G., Abbot, A., and Mackay, I. R.: An electron microscopic study of the human

thymus. Normal appearances and findings in myasthenia gravis and systemic lupus erythematosus. *J. Pathol. Bacteriol. 95:* 211–215, 1968.

19. Goldstein, G., and Hofmann, W. W.: Electrophysiological changes similar to those of myasthenia gravis in rats with experimental autoimmune thymitis. *J. Neurol. Neurosurg. Psychiatry 31:* 453–459, 1968.
20. Goldstein, G., and Hofmann, W. W.: Endocrine function of the thymus affecting neuromuscular transmission. *Clin. Exp. Immunol. 4:* 181–189, 1969.
21. Goldstein, G., and Hofmann, W. W.: Experimental myasthenia gravis. In *Immunological Disorders of the Nervous System*, ed. by Rowland, L. P. Baltimore, The Williams & Wilkins Company, 1971.
22. Goldstein, G., and Mackay, I. R.: Contrasting abnormalities in the thymus in systemic lupus erythematosus and myasthenia gravis. A quantitative histological study. *Aust. J. Exp. Biol. Med. Sci. 43:* 381–390, 1966.
23. Goldstein, G., and Mackay, I. R.: *The Human Thymus*. St. Louis, Warren H. Green, Inc., 1969.
24. Goldstein, G., Strauss, A. J. L., and Pickeral, S.: Antigens in thymus and muscle effective in inducing experimental autoimmune thymitis and the release of thymin. *Clin. Exp. Immunol. 4:* 3–16, 1969.
25. Goldstein, G., and Whittingham, S.: Experimental autoimmune thymitis. An animal model of human myasthenia gravis. *Lancet 2:* 315–318, 1966.
26. Goldstein, G., and Whittingham, S.: Histological and serological features of experimental autoimmune thymitis in guinea pigs. *Clin. Exp. Immunol. 2:* 257–268, 1967.
27. Grob, D., Namba, T., and Feldman, D. S.: Alterations in reactivity to acetylcholine in myasthenia gravis and carcinomatous myopathy. *Ann. N. Y. Acad. Sci. 135:* 247–274, 1966.
28. Hammar, J. W.: *Die Menschenthymus in Gesundheit und Krankheit*, p. 570. Leipsig, Akademische Verlagsgesellschlaft M.B.H., 1926.
29. Henry, K.: Mucin secretion and striated muscle in the human thymus. *Lancet 1:* 183–185, 1966.
30. Herrmann, C., Jr., Sabawala, P. B., Barker, W. F., and Dillon, J. B.: Reappraisal of *in vitro* muscle biopsy in myasthenia gravis. *Ann. N. Y. Acad. Sci. 135:* 302–311, 1966.
31. Hofmann, W. W., and Stemmer, E. A.: Subthreshold activity at normal and myasthenic end-plates. *Neurology (Minneap.) 13:* 227–236, 1963.
32. Johns, R. J., Grob, D., and Harvey, A. M.: Studies in neuromuscular function. II. Effects of nerve stimulation in normal subjects and in patients with myasthenia gravis. *Bull. Johns Hopkins Hosp. 99:* 125–135, 1956.
33. Kalden, J. R., and Irvine, W. J.: Experimental myasthenia gravis. *Lancet 2:* 638–639, 1969.
34. Kalden, J. R., Williamson, W. G., Johnston, R. J., and Irvine, W. J.: Studies on experimental autoimmune thymitis in guinea pigs. *Clin. Exp. Immunol. 5:* 319–340, 1969.
35. Knight, S., Bradley, J., Oppenheim, J. J., and Ling, N. R.: The *in vitro* transformation of thymocytes and lymphocytes from humans, rabbits, and guinea pigs and from thymomas. *Clin. Exp. Immunol. 3:* 323–341, 1968.
36. McEachern, D.: The thymus in relation to myasthenia gravis. *Medicine (Balt.) 22:* 1–25, 1943.
37. McFarlin, D. E., Engel, W. K., and Strauss, A. J. L.: Does myasthenia serum bind to the neuromuscular junction? *Ann. N. Y. Acad. Sci. 135:* 656–663, 1966.
38. Oosterhuis, H. J. G. H., Feltkamp, T. E. W., and Van Der Geld, H. R. W.: Muscle antibodies in myasthenic mothers and their babies. *Lancet 2:* 1226–1227, 1966.
39. Oppenheim, J. J., and Goldstein, G.: Enhanced thymic lymphocyte response to phytohaemagglutinin in experimental autoimmune thymitis. *Nature (Lond.) 222:* 192–193, 1969.
40. Perlo, V. P., Poskanzer, D. C., Schwab, R. S., Viets, H. R., Osserman, K. E., and Genkins, G.: Myasthenia gravis. Evaluation of treatment in 1,355 patients. *Neurology (Minneap.) 16:* 431–439, 1966.
41. Rieffel, H., and LeMee, J.: A propos de l'anatomie du thymus humain. *C.R. Acad. Sci. Paris 148:* 519, 1909.
42. Russell, D. S.: Histologic changes in the striped muscles in myasthenia gravis. *J. Pathol. Bacteriol. 65:* 279–289, 1966.
43. Simpson, J. A.: An evaluation of thymectomy in myasthenia gravis. *Brain 81:* 112–144, 1958.
44. Sloan, H. E., Jr.: The thymus in myasthenia gravis. With observations on the normal anatomy and histology of the thymus. *Surgery 13:* 154–174, 1943.

45. Stern, G. M., Hall, J. M., and Robinson, D. C.: Neonatal myasthenia gravis. *Br. Med. J. 2:* 284–286, 1964.
46. Strauss, A. J. L., Kemp, P. G., and Douglas, S. D.: Myasthenia gravis and thymic myoid cells. *Lancet 1:* 772–773, 1966.
47. Strauss, A. J. L., Seegal, B. C., Hsu, K. C., Burkolder, P. M., Nastuk, W. L., and Osserman, K. E.: Immunofluorescence demonstration of a muscle-binding complement-fixing serum globulin fraction in myasthenia gravis. *Proc. Soc. Exp. Biol. Med. 105:* 184–191, 1960.
48. Strauss, A. J. L., and Van Der Geld, H. W. R.: The thymus and human diseases with auto-immune concomitants, with special reference to myasthenia gravis. In *The Thymus: Experimental and Clinical Studies*, ed. by Wolstenholme, G. E. W., and Porter, R., pp. 416–439. London, J. & A. Churchill, Ltd., 1966.
49. Thesleff, S.: Acetylcholine utilization in myasthenia gravis. *Ann. N. Y. Acad. Sci. 135:* 195–206, 1966.
50. Van Der Geld, H., Feltkamp, T. E. W., Van Loghem, J. J., Oosterhuis, H. J. G. H., and Biemond, H.: Multiple antibody production in myasthenia gravis. *Lancet 2:* 373–375, 1963.
51. Van Der Geld, H., Feltkamp, T. E. W., Van Loghem, J. J., and Oosterhuis, H. J. G. H.: Reactivity of myasthenia gravis serum gamma globulin with skeletal muscle and thymus demonstrated by immunofluorescence. *Proc. Soc. Exp. Biol. Med. 115:* 782–785, 1964.
52. Van Der Geld, H. W. R., and Strauss, A. J. L.: Myasthenia gravis. Immunological relationship between striated muscle and thymus. *Lancet 1:* 57–60, 1966.
53. Van Der Velde, R. L., and Friedman, N. B.: The thymic "myoidzellen" and myasthenia gravis. *J. A. M. A. 198:* 287–288, 1966.

# Index

Page references in *italic* type refer to illustrations.


Acid maltase deficiency, *30*, *35*, 307–315, *311–319*, 412, *413*
Animal myopathies, 364–409
  acute and subacute of uncertain cause, 391–400
    hind leg myopathy of old rats, 396–397
    of poultry, 397–398, *399*
    of wild animals, 398–400, *400*
    paralytic myoglobinuria of horse, 391–394, *393*
    polymyopathy of adult sheep, 395
    polymyopathy of dog and cat, 395–396
    transport myopathy of cattle, sheep, and swine, 394–395
  congenital, 365–370, 400–401
    associated with arthrogryposis, 365–366
    associated with hydrocephalus in calves, 366, *367*
    associated with splayleg in pigs, 367–368, *369*
    defective myogenesis, 368–370
    myotonia in goat, 400–401
  hereditary dystrophy-like types, 370–378
    of chicken, 375–377, *376*
    of mouse, 371–373, *372*
    of Syrian hamster, 373–375, *374*
    of turkey, 377–378, *379*
  myasthenic syndrome in dog, 401–402
  miscellaneous, 400–404
  motor end-plate disease of mouse, 402–403
  nutritional, 378–388
    of cattle, *380*, 381–383, *382*
    of horses, 386–387, *388*
    of sheep, 383–384, *385*
    of swine, 384–386
  toxic, 388–391
    iron poisoning in baby pigs, 390–391
    plant poisoning in ruminants, 388–390, *390*
  type II fiber hypotrophy in dog, 403–404
Arthritis, rheumatoid, *100*, *113*
Arthrogryposis, animal myopathies and, 365–366
Atrophy
  disuse, 259, *260*
  factor, calculation of, *92*
  large group, *101*
  neurogenic (see Neurogenic atrophy)
  perifascicular, *101*, *112*
  spinal muscular, *480*

Becker dystrophy, 276–277, *279–280*
Biochemistry
  of skeletal muscle, 6–13
  relationship with histoenzymology, 31–32
Biopsy, 19–27
  artifacts, 22–24
  children, pathologic interpretation, 114–120, *116–117*
  fixation methods, 21–22
  histochemical priniciples and mechanisms of enzyme demonstration, *33*
  interpretation, problems in, 24–26
  sampling, 19–21
  staining methods, 22

Cardiomegaly, generalized hypotonia and, *35*
"Cell coat," comparison with external lamina, 138–141
Central core disease, *101*, 244, *245*, 444, *480*
Centronuclear myopathy, 448–451, *449–450*
Children, interpretation of biopsies, 114–120, *116–117*
Chloroquine myopathy, experimental, 303–307, *305–306*, *308–310*, *358*
Collagen vascular disease, myopathies associated with, 111–114, *112*
Congenital myopathies, 442–452
  of animals, 365–370, 400–401
Cori's disease, 412–414, *415–416*
Cytochemistry
  properties of external lamina of myofibers and neuromuscular junctions, 123–143
  study of vacuolar contents, 302–303

Debrancher system deficiency, 412–414, *415–416*
Degeneration, 98, 187–195
Deltoid muscle
  necrotic, *295*
  normal, *30*, *89*
Denervation (see also Innervation and Nervous system)
  comparison with disuse atrophy, 259, *260*
  features of muscle, 251–261
  stages of, 251–253, *252*, *254*
Dermatomyositis, *81*, *100–101*, *112*, *487*
Diagnosis, value of histoenzymology, 28–31
Distal muscular dystrophy, *269*, 287
Disuse atrophy, comparison with denervation, 259, *260*
Drugs, effect on muscle regeneration, 220–221
Duchenne dystrophy, *36*, *96–97*, 107–110, *108*, 150, *270–271*, 272–276, *273*, *275–276*, *477–479*
  ischemia hypothesis, 453–472, *454–460*, *462–467*, *469*
Dystrophy
  Becker, 276–277, *279–280*
  congenital, *278*, 284–286, *286*
  distal, *269*, 287

Dystrophy—*continued*
Duchenne (see Duchenne dystrophy)
facio-scapulo-humeral, *109*, 110, 286–287, *287*
genetically determined types, 106–111
limb girdle, *93*, *96–97*, *100*, 110, *277*, 279–283, *282–283*, *285*
muscular, progressive (see Muscular dystrophy, progressive)
myotonic, *82*, *109*, 111, *284*, *357*, *450*

Electron microscopy, of skeletal muscle, 4–6, 58–76
Embryogenesis, electron microscopic observations, 6
End-plate, motor, ultrastructure and pathologic modifications, 342–363
Enzyme deficiency (see also specific enzyme), fiber types, 78, *81*

Facio-scapulo-humeral dystrophy, *109*, 110, 286–287, *287*
Fibers
architecture, changes in, 98–99
atrophic
distribution of, 253–255
fascicles, *25*
type II, 102–103
biochemistry, 6–13
central core, 159–160
degeneration, 98
granular dissolution associated with excessive contraction, *23*
histology, *25*
intrafusal, 227–234, *228–234*
nuclear-bag, 227–229
nuclear-chain, 229–234
mutability, 15–16
regeneration, *20*, 98
size, changes in, 94–95, 253
subcellular components, 4–6
target, *97*, 158–159
terminology, 14
type grouping, 78–80, *81*, *105*, 255–256
types, 1–4
in clinical disorders of muscle, 77–85, *79*, *81*
relationship to contraction-relaxation kinetics of muscle, 14
ultrastructural pathologic characteristics, 144–184, *150*, *152–153*, *171*
central mofibrillar degeneration, 154–160, *155–158*
electron microscopy, value of, 145–146
fiber surface, 146–149, *148*, *152–153*
Golgi system, 165
I bands and Z lines, 151–154, *156–158*, 159
lysosomes and autophagic vacuoles, 167–170, *173*
mitochondria, *162–164*, 165–169
myofibrillar degeneration, 160–161
myofibrillar regeneration, 161
myofibrils, 149–161
nuclei, 174–175, *176–177*
sarcoplasmic particulate components, 172–174, *173*
satellite cells, 170–172
T system and sarcoplasmic reticulum, 161–165, *166*, *171*
with irregular contraction bands, *20*, *23*
with segmental expansion and loss of striations, *23*
Fibroblasts, ruthenium red staining, *127*
Flexor digitorum longus muscle, rabbit, *240–241*
Flexor hallucis longus muscle, cat, *238–239*
Forbes' disease, 412–414, *415–416*

Gastrocnemius muscle
injured, of mouse, *186*, *188*, *190–192*, *194*, *198*, *200*, *203–205*, *208*, *210*, *212*, *214*, *216–217*
normal, *35*
rat, *305*
α-1,4-Glucosidase deficiency, *30*, *35*, 307–315, *311–319*, 412, *413*
Glycogen, 172
Glycogenosis (see Glycogen storage diseases)
Glycogen storage diseases
acid maltase deficiency, *30*, *35*, 307–315, *311–319*, 412, *413*
debrancher system deficiency, 412–414, *415–416*
muscle ultrastructure, 410–426, *411*
myophosphorylase deficiency, *81*, 414–420, *417*, *419*, *421–422*
phosphofructokinase deficiency, 420–423, *422*, *424*
Golgi system, 165
Guillain-Barré syndrome, *255*

Henlé sheath, 347
Histochemistry
histoenzymology (see Histoenzymology)
of degenerating muscle, 194–195
pathologic interpretation, 86–122
abnormalities of distribution, 99–102
calculation of atrophy and hypertrophy factors, *92*
cellular reactions, 98
changes seen in disease, 102–114
children's biopsies, 114–120, *116–117*
degeneration and regeneration, 98
fiber architecture, changes in, 98–99
fiber size, changes in, 94–95
genetically determined dystrophies, 106–111
illustration of pathologic changes, 94–102
lower motor neuron diseases, 103–106
myopathies associated with collagen vascular disease and rheumatoid diseases, 111–114
normal human muscle, 86–94
nuclear changes, 98
type II fiber atrophy, 102–103
periodic paralyses, 432–435
of skeletal muscle, 1–4
principles and mechanisms of enzyme demonstration in muscle biopsies, *33*
Histoenzymology, 28–39
diagnostic value, 28–31

methodology, 33–34
mitochondrial activity in neuromuscular disease, 34–38
relationship with biochemistry, 31–32
Histopathology
denervated muscle, 251–261
neurogenic atrophy, 249–262
periodic paralyses, 428–431, *430*
progressive muscular dystrophy, 265–288, *266–267*, *269–271*
Hydrocephalus, animal myopathies and, 366, *367*
Hyperkalemic periodic paralysis, 439–440
Hypertrophy factor, calculation of, *92*
Hypokalemic periodic paralysis, *153*, *155*, *171*, 321–338, *322*, *324*, *326*, *328–337*, 429–439, *431–435*
Hypotonia, generalized, and cardiomegaly, *35*

I bands, 151–154, 156, 159
Innervation (see also Nervous system and Denervation), *252*, *254*
cross-innervation, 237–239, *237*, *239*
effect on muscle regeneration, 216–220, *219*
human motor end-plate, 342–363, *344*
mutability of skeletal muscle, 15–16
Intercostal muscle, mouse, *126*, *131–132*, *136–137*
Isometric muscle clamp, *20*

Kearn's and Sayre's ophthalmoplegia syndrome, *101*, *104*
Kugelberg-Welander syndrome, *267*

Lamina, external
of myofibers and neuromuscular junctions
"cell coat," 138–141
cytochemical properties, 123–143, *127*, *140*
denervation experiments, 124, 130
electron microscopy, 124
enzyme cytochemistry, 124–125, 130–134
optical microscopy, 123
sarcolemma, 125–128, 134, *140*
staining characteristics, 128–130
ultrastructural pathologic characteristics, 149
Limb girdle dystrophy, *93*, *96–97*, *100*, 110, *277*, 279–283, *282–283*, *285*
Limit dextrinosis, 412–414, *415–416*
Lipid particles, 172, *173*
Lipid storage myopathy, 316–321, *319–320*
Lipopigments, 172–174
Lysosomes
in normal muscle, 64–68, *67*
ultrastructural pathologic characteristics, 167–170

McArdle's disease, *81*, 414–420, *417*, *419*, *421–422*
Mitochondria
histoenzymatic activity in neuromuscular disease, 34–38
isolated, conformation changes, *8*
ultrastructural pathologic characteristics, *162–164*, 165–169
Mitochondrial myopathy, 444, *445*

Motor end-plate, human, 342–363
morphologic variations, 350–351
normal structure, 343–350, *344–345*, *348–349*
Henlé sheath, 347
junctional sarcoplasm, 350
nerve ending, 346–347
primary synaptic cleft, 347
Schwann coat, 347
subneural apparatus, 347–350
pathologic modifications, 351–360
in myasthenia gravis, 356–360, *359*
in neurogenic atrophies, 351–354, *353*, *355*
in primary muscle diseases, 354–356, *357–358*
Motor unit, normal, 249–250
Muscle (see specific name or type of muscle)
Muscle disorders
histochemical fiber types, 77–85, *79*
fiber predominance, 83, *83*, *105*
fiber type grouping, 78–80, *81*, *105*, 255–256
generalized alteration, 80, *81*
specific enzyme deficiency, 78, *81*
type I specific alteration, 80–82, *82*
type II specific alteration, 82–83, *83*
Muscle spindle, normal
ultrastructural aspects, 226–235, *227*
capsule, 226–227
intrafusal fibers, 227–234, *228–234*
nuclear-bag fibers, 227–229
nuclear-chain fibers, 229–234
Muscular dystrophy, progressive, 263–291
autosomal recessive types, 279–286
childhood (except Duchenne), 284
congenital, *278*, 284–286, *286*
limb girdle, *277*, 279–283, *282–283*, *285*
classification of disorders, 265
distal, *269*, 287
facio-scapulo-humeral, *109*, 110, 286–287, *287*
histopathology, 265–288, *266–267*, *269–271*
ocular, 287–288
X-linked types, 272–279
Becker, 276–277, *279–280*
carrier detection, 277–279, *281*
Duchenne, *270–271*, 272–276, *273*, *275–276*
Myasthenia gravis, *83*, 498–512
action of thymin, 508–510
experimental, 498–500, *499*
findings in the thymus, 500–502, *501–502*
immune reaction to thymus, 502–503, *503*
initiation of autoimmune thymitis, 503–504, *506–507*
neuromuscular lesions, 504–508, *505–507*
Myoblasts, 199–202, *200*, *204–205*, *208*, *210–211*
from chick embryo, *42*, *46*, *52*, *54*
Myofibers
cytochemical properties of external lamina, 123–143
general morphology, 60–62
in injured muscle, *190–194*, *197–198*, *203*
normal variations in structure, 62–72
lysosomal content of normal muscle, 64–68, *67*
satellite cells in normal muscle, 68–72, *69*, *71*
Z band spreading, 62–64

Myofibers—*continued*
regeneration, 207–211, *214–218*
Myofibrillogenesis, 207, *212*
Myofibrils
organization of myofilaments, 51–53
schematic illustration, *5*
ultrastructural pathologic characteristics, 149–161
central degeneration, 154–160, *155–158*
peripheral degeneration, 160–161
regeneration, 161
transverse amorphous degeneration, 161
Myofilaments
appearance in myogenesis, 41–43
organization into myofibrils, 51–53
Myogenesis
comparison of adult and embryonic myosin, 51
defective, animal myopathies and, 368–370
immunochemical and ultrastructural studies, 40–57
myofilaments, appearance of, 41–43
organization of myofilaments into myofibrils, 51–53
polyribosomes, large
appearance of, 43–45
correlation with myosin synthesis, 45–50
Myoglobinuria, paralytic, of horse, 391–394, *393*
Myopathy, *117*, 243–246
acid maltase deficiency, *30*, *35*, 307–315, *311–319*, 412, *412*
animal (see Animal myopathies)
associated with collagen vascular disease and rheumatoid diseases, 111–114
centronuclear, 448–451, *449–450*
chloroquine, experimental, 303–307, *305–306*, *308–310*, *358*
congenital, 442–452
familial, *36*, *176*
lipid storage, 316–321, *319–320*
mitochondrial, 444, *445*
myotubular, 448–451, *449–450*
nemaline, *82*, 444–448, *446–448*
nonfamilial degenerative, *36*
primary hypokalemic periodic paralysis, 321–338, *322*, *324*, *326*, *328–337*, 429–439, *431–435*
rod-body, *82*, 444–448, *446–448*
scapuloperoneal, *37*
vacuolar, 301–341
acid maltase deficiency, *30*, *35*, 307–315, *311–319*, 412, *413*,
classification, 301–302
experimental chloroquine, 303–307, *305–306*, *308–310*
lipid storage, 316–321, *319–320*
methods of study, 302–303
primary hypokalemic periodic paralysis, 321–338, *322*, *324*, *326*, *328–337*
Z band, *82*, 444–448, *446–448*
Myophosphorylase deficiency, *81*, 414–420, *417*, *419*, *421–422*
Myosin, comparison of adult and embryonic, 51
Myositis, acute, *30*
Myotomic muscle, from chick embryo, *42*
Myotonia, congenital, in goat, 400–401
Myotonic dystrophy, *82*, *109*, 111, *357*, *450*
Myotubular myopathy, 448–451, *449–450*

Nemaline myopathy, *82*, 444–448, *446–448*
Nervous system (see also Innervation and Denervation)
influence on muscle structure and function, 236–248
cross-innervation, 237–239, *237*, *239*
developing muscle, 239–240
histochemical studies, 236–240, *238*
myopathies, 243–246, *245*
neurogenic atrophies, 241–243, *242–244*
structural changes after cross-innervation, 240, *241*
motor end-plate, human, 342–363
Neurogenic atrophy, *37*, *83*, 241–243, *242–244*, 249–262, 351–354, *353*, *355*
histopathology, 249–262
denervated muscle, 251–261, *252*, *254–260*
normal motor unit, 249–250
pathologic reactions of motor neuron, 250–251
pathologic changes in motor end-plate, 351–354, *353*, *355*
Neuromuscular disease
coordinated investigation of, *29*
mitochondrial histoenzymatic activity in, 34–38
Neuromuscular junctions
cytochemical properties of external lamina, 123–143
mouse, *126–127*, *131–133*, *136–137*
Neuromuscular lesions, 504–508, *505–507*
Neuron, lower motor
diseases of, 103–106
pathologic reactions of, 250–251
Neuropathy, chronic, *96*, *105*
Neurovascular bundle, *25*
Normokalemic periodic paralysis, 439–440
Nucleus, 98, 174–175, *176–177*

Ophthalmoplegia syndrome, Kearn's and Sayre's *101*, *104*

Paralysis, periodic (see Periodic paralysis)
Parkinson's disease, *100*
Pectoralis major muscle, *489*
Periodic paralysis, 427–441
classification, 427–428
electron microscopy, 435–439, *437*, *439*
histochemistry, 432–435
histopathology, 428–431, *430*
hyperkalemic, 439–440
hypokalemic, *153*, *155*, *171*, 321–338, *322*, *324*, *326*, *328–337*, 429–439, *431–435*
normokalemic, 439–440
pathogenesis, 428
Peroneus muscle, guinea pig, *346*, *352*

Physiology, of skeletal muscle, 14
Polymyalgia rheumatica, *112–113*
Polymyopathy
  of adult sheep, 395
  of dog and cat, 395–396
Polymyositis, *37*, *97*, *112*, *177*
  etiology, 485–497, *486–487*, *489–491*, *493*, *495*
  pathologic substratum, 292–300, *294–297*
Polyribosomes, large, *48*
  appearance in myogenesis, 43–45
  correlation with myosin synthesis, 45–50
    morphology, 45–47
    precipitation experiments, 47–49, *50*
Pompe's glycogenosis, *30*, *35*, 307–315, *311–319*, 412, *413*
Proteins, structural, localization relative to actin and myosin, *6*
Psoas muscle, *487*

Quadriceps muscle, *294*, *296–297*

Regeneration, 98, 185–225, *186*
  continuous, 196, *197*
  discontinuous, 196–197, *197*
  effect of drugs, 220–221
  effect of innervation, 216–220, *219*
  inflammatory reaction, 195–196
  myoblasts, 199–202, *200*, *204–205*, *208*, *210–211*
  myofibers, 207–211, *214–218*
  myofibrillogenesis, 207, *212*
  necrosis and degenerative changes, 187–195
    electron microscopy, 187–194, *188*, *190–194*
    histochemical reactions, 194–195
    histologic alterations, 187
  principal cells in regenerating muscle, *201*
  sarcoblasts, 202–207, *208*, *212–213*
Rhabdomyoma, *157*
Rheumatoid arthritis, *100*, *113*
Rheumatoid diseases, myopathies associated with, 111–114
Ribosomes, *173*, 174

Sarcoblasts, 202–207, *208*, *212–213*
  myofibrillogenesis, 207
Sarcolemma
  interruptions in continuity, 147
  of external lamina of myofibers and neuromuscular junctions, 125–128, 134, *140*
  of human skeletal muscle, *13*
Sarcomere, *54*
Sarcoplasmic particulate components, 172–174
Sarcoplasmic reticulum
  fragmented, *10*
  ultrastructural pathologic characteristics, 161–165, *166*, *171*
Satellite cells
  in normal muscle, 68–72, *69*, *71*
  ultrastructural pathologic characteristics, 170–172
Schwann coat, 347
Sclerosis, amyotrophic lateral, *101*, *105*

Skeletal muscle, 1–18
  biochemistry, 6–13
  electron microscopy of, 4–6
  fibers, 144–184
  guinea pig, *3*
  histochemistry, 1–4
  human
    in tissue culture, 473–484
    normal, 86–94, *88–89*, *91*, *93*
    ultrastructure, 58–76, *61*
  mutability, 15–16
  necrosis and degeneration, 187–195
  physiology, 14
  rapidly frozen, *20*
  regeneration, 185–225
  structure and function, influence of nervous system on, 236–248
  terminology, 14
Soleus muscle
  cat, *238*
  rabbit, *241*
  rat, *305–306*, *308–309*
Somites, from chick embryo, *44–45*
Splayleg, animal myopathies and, 367–368, *369*
Staining
  characteristics of external lamina of myofibers and neuromuscular junctions, 128–130
  methods, 22
Sternomastoid muscle, *295–296*
Subsarcolemmal crystals, 174
Substratum, pathologic, of polymyositis, 292–300

T system, 161–165, *166*
Tibialis anterior muscle, rabbit, *193–194*, *213*, *215*, *218*
Tissue
  biopsy, 19–27
  fixation methods, 21–22
  sampling, 19–21
  staining methods, 22
Tissue culture
  human skeletal muscle, 473–484, *477–481*
    analysis, 482–483
    technique, 474–480

Ultrastructure
  human motor end-plate, 342–363
  human skeletal muscle, 58–76, *61*, *63*
    biopsy methods, 59–60
    general morphology of myofiber, 60–62
    in glycogen storage diseases, 410–426
    lysosomal content of normal muscle, 64–68, *67*
    normal variations in myofiber structure, 62–72
    satellite cells in normal muscle, 68–72, *69*, *71*
    subjects, 59
    Z band spreading, 62–64
  normal mammalian muscle spindle, 226–235
  skeletal muscle fiber, pathologic characteristics, 144–184

Vacuolar myopathies, 301–341
Vacuoles, autophagic, 167–170, *173*
Vastus lateralis muscle, *490*

Werdnig-Hoffmann's disease, *117*, 119, *443*

Z lines
  appearance in myogenesis, 53, *54*
  spreading, 62–64, *65–67*
  ultrastructural pathologic characteristics, 151–154, *156–158*, 159